Advanced Structured Materials

Volume 204

Common engineering materials are reaching their limits in many applications, and new developments are required to meet the increasing demands on engineering materials. The performance of materials can be improved by combining different materials to achieve better properties than with a single constituent, or by shaping the material or constituents into a specific structure. The interaction between material and structure can occur at different length scales, such as the micro, meso, or macro scale, and offers potential applications in very different fields.

This book series addresses the fundamental relationships between materials and their structure on overall properties (e.g., mechanical, thermal, chemical, electrical, or magnetic properties, etc.). Experimental data and procedures are presented, as well as methods for modeling structures and materials using numerical and analytical approaches. In addition, the series shows how these materials engineering and design processes are implemented and how new technologies can be used to optimize materials and processes.

Advanced Structured Materials is indexed in Google Scholar and Scopus.

Holm Altenbach · Viacheslav Bogdanov ·
Alexander Ya. Grigorenko · Roman M. Kushnir ·
Vladimir M. Nazarenko · Victor A. Eremeyev
Editors

Selected Problems of Solid Mechanics and Solving Methods

Springer

Editors
Holm Altenbach
Institute of Mechanics
University of Magdeburg
Magdeburg, Sachsen-Anhalt, Germany

Viacheslav Bogdanov
S. P. Timoshenko Institute of Mechanics
National Academy of Sciences of Ukraine
Kyiv, Ukraine

Alexander Ya. Grigorenko
S. P. Timoshenko Institute of Mechanics
National Academy of Sciences of Ukraine
Kyiv, Ukraine

Roman M. Kushnir
Ya. S. Pidstryhach Institute
National Academy of Sciences of Ukraine
Lviv, Ukraine

Vladimir M. Nazarenko
S. P. Timoshenko Institute of Mechanics
National Academy of Sciences of Ukraine
Kyiv, Ukraine

Victor A. Eremeyev
DICAAR
University of Cagliari
Cagliari, Italy

ISSN 1869-8433 ISSN 1869-8441 (electronic)
Advanced Structured Materials
ISBN 978-3-031-54062-2 ISBN 978-3-031-54063-9 (eBook)
https://doi.org/10.1007/978-3-031-54063-9

This Springer imprint is published by the registered company Springer Nature Switzerland AG
The registered company address is: Gewerbestrasse 11, 6330 Cham, Switzerland

In memory of Y. M. Grigorenko, our teacher, colleague, and friend, who began his life as an Ostarbeiter and became an outstanding scientist in computational mechanics.

Preface

Yaroslav Mykhaylovych Grigorenko, an outstanding Ukrainian scientist in Computational Mechanics, passed into his new world on January 18, 2022, at the age of 95. His childhood happened to overlap with the desperate times of World War 2. Despite being a child he was taken by force from occupied Kyiv, where he was born in 1927 and lived, for forced labor in Germany. Later on, after the liberation, he was subject to military service during the difficult and hungry post-war years. Despite the 7-year break in studies, Y. M. Grigorenko successfully completed school and then the mechanical and mathematical faculty of the Taras Shevcenko State University of Kyiv (since 1992—Taras Shevcenko National University of Kyiv). After graduating from the university, his entire scientific activity was inextricably linked with the Institute of Mechanics of the Academy of Sciences of the Ukrainian SSR in the city of Kyiv (since 1992, the Timoshenko Institute of Mechanics of the National Academy of Sciences of Ukraine). Based on the analytical and numerical studies of the stressed state of the conical shells with linearly varying thickness, he successfully defended his Ph.D. thesis in 1961.

At the same time, the Institute received one of the first large electronic computers BESM 2M (literally from the Russian—Highspeed Electronic Computational Machine), of which only a few were produced in the USSR. For that reason, a Department of Computational Methods was organized in the Institute and Y. M. Grigorenko was appointed the first head. He remained on this position until 2005. From that time on, he began to be one of the first in the world to develop numerical methods and successfully apply them to solve complex problems in the theory of shells, which were previously impossible to obtain using analytical approaches. In 1970, Y. M. Grigorenko defended his doctoral thesis. In 1973, he was awarded the academic title of professor. In 1978, he was elected a corresponding member, and in 1992, a full member (academician) of the National Academy of Sciences of Ukraine.

The scientific results of Y. M. Grigorenko are published in 34 monographs, textbooks, and technical dictionaries, and more than 500 scientific papers (including the abstracts of his scientific reports). He created a scientific school in computational mechanics, in which large-scale research continues work on the development of various aspects of the theory and improvements of computational methods with

Yaroslav Grigorenko with his students

regard to inhomogeneous anisotropic shells and elastic solids. He was an advisor to 10 doctors and 43 candidates of science (Ph.D.), collaborated fruitfully with many scientists from several academic and educational institutions in Ukraine, and assisted in their scientific growth. He contributed to numerous international conferences, symposiums, and seminars in Japan, the USA, Greece, Germany, Belgium, Great Britain, Austria, and Poland.

Scientific results of Y. M. Grigorenko are known to a wide range of specialists, and have been highly appreciated by the world scientific community. They are included in educational and reference manuals and other books. The contribution of Y. M. Grigorenko to the development of Ukrainian science was awarded a number of government awards and prizes.

The death of Y. M. Grigorenko is a painful loss for us—his colleagues and students. We found ourselves not only without a wise teacher and mentor in life, but also without a close, noble, sensitive, and sincere friend. This book presents articles by his many students and colleagues who recall with great warmth the time they communicated with this man, his openness, and goodwill. The good memory of Yaroslav Mykhaylovych Grigorenko, the outstanding scientist of our time, will remain for a long time in the hearts of those who knew him.

Magdeburg, Germany — Holm Altenbach
Kyiv, Ukraine — Viacheslav Bogdanov
Kyiv, Ukraine — Alexander Ya. Grigorenko
Kyiv, Ukraine — Vladimir M. Nazarenko
Lviv, Ukraine — Roman M. Kushnir
Cagliari, Italy — Victor A. Eremeyev

Acknowledgments The editors would like to take this opportunity to thank Dr. M. F. Selivanov (S. P. Timoshenko Institute of Mechanics, National Academy of Science of Ukraine, Kyiv, Ukraine) for his technical support in preparing the manuscript.

Contents

Chapter 1
Stages of Yaroslav M. Grigorenko's Life Path and the Main Results of His Scientific Research

Holm Altenbach, Viacheslav Bogdanov, Alexander Grigorenko, Olexandr Khimich, Roman Kushnir, and Vladimir Nazarenko

Abstract Yaroslav Mikhailovich Grigorenko was a Ukrainian scientist with many contributions to computational mechanics. This chapter provides a brief biographical overview of his most important scientific achievements and a bibliography.

Yaroslav Mikhailovich Grigorenko was born in the city of Kyiv on October 12, 1927. As a child, he and his parents experienced the Great Famine (Holodomor) of 1932 and 1933, forever in his memory, were left hungry, barely alive people who fled from the Ukrainian villages to Kyiv in the hope of escape from hunger. In the summer of 1941, he graduated from the sixth grade of the famous Kyiv school No. 20, founded in the center of Podol near the beginning of Andrew's descent. On June 22, the war with Germany began. Classes at the school lasted for two weeks in the new academic year; on September 19, German troops marched into Kyiv, and the terrible days of occupation began. A few days later, Khreshchatik—the main street of Kyiv—was bombed, and in late September, the tragedy of Babiy Yar was over, where thousands of Jews were killed. Y. M. Grigorenko witnessed how his friend and fellow Jewish boy with his mother and sister were going to Babyn Yar, not knowing what awaited them. There was a famine in Kyiv, and he walked with his relatives for 30 km from the city to neighboring villages to exchange things for food. So, it lasted all the freezing winter of 1941–1942 and the spring of 1942.

H. Altenbach (✉)
Otto-von-Guericke-Universität Magdeburg, Magdeburg, Germany
e-mail: holm.altenbach@ovgu.de

V. Bogdanov · A. Grigorenko · V. Nazarenko
S. P. Timoshenko Institute of Mechanics, National Academy of Science of Ukraine, Kyiv, Ukraine
e-mail: bogdanov@nas.gov.ua

R. Kushnir
Pidstryhach Institute for Applied Problems of Mechanics and Mathematics, National Academy of Sciences of Ukraine, Lviv, Ukraine
e-mail: dyrector@iapmm.lviv.ua

O. Khimich
V. M. Glushkov Institute of Cybernetics, National Academy of Science of Ukraine, Kyiv, Ukraine

H. Altenbach et al. (eds.), *Selected Problems of Solid Mechanics and Solving Methods*, Advanced Structured Materials 204, https://doi.org/10.1007/978-3-031-54063-9_1

In May 1942, he was arrested and, despite his fourteen-year-old age, was sent to Germany for forced labor. It is simply impossible to imagine what he endured. He finds himself working for a German farmer in the village of Pommerzig on the right bank of the river Oder (now Poland), 30 km from Grünberg (now Zielona Góra, Poland). In this place, Y. M. Grigorenko, who was almost a child, had to do hard physical work and was on the rights of a modern enslaved person. It should be especially noted that after that, he had no hatred or bitterness toward the German people. After that, he forged good, friendly relations with many of his German colleagues.

In January 1945, the Red Army entered the village; it was a real holiday. Y. M. Grigorenko wanted to stay in the army but was not yet old, so he had to go home. In February, he came to Kyiv, surrounded by his relatives, who no longer thought to see him alive. However, the laws of the USSR at the time equated citizens who had been sent to Germany to work as traitors, and they were not allowed to live in large cities. The militia ordered Y. M. Grigorenko to leave the city. Otherwise, he would be arrested. He did not wait for that. He went to the enlistment office and asked to be sent as a volunteer to the front, adding an extra year to his name. So, his military service began. However, the war ended on May 9, 1945. Then, there were another three and a half years of service in the difficult postwar period. He lived in a dugout with hungry soldiers' rations for over two years.

In the autumn of 1946, an important event occurred in his life. Night schools were established in the country as a result of the war years. To attend school, he had to travel 9 km on difficult roads. However, with his commanding officer's permission, he was able to attend. It was a challenging experience, not just physically, as he started studying in the first grade despite completing seven grades before the war. He dedicated all his free time to his studies and finished the first quarter with good and satisfactory grades, receiving a report card. Unfortunately, later on, his commanding officers forbade him from attending school. However, in the fall of 1948, he was discharged from the army for health reasons and could enroll in the ninth grade of Kyiv evening school No. 11, thanks to the report card he had received earlier.

It is necessary to mention Mikhail Savvich and Anna Ivanovna, the parents of Y. M. Grigorenko. Like many others in that problematic era, they had a modest income. However, they recognized their son's natural inclination toward learning and supported him by providing the minimum for his schooling. They insisted that he should not work to support himself but instead focus on his studies. Y. M. Grigorenko always cherished his parents' support and was grateful to them. After the war, he resumed his studies with a renewed zeal for knowledge and worked tirelessly. As a result of his hard work, he was awarded a gold medal for graduating from the secondary evening school.

Y. M. Grigorenko developed a fondness for mathematics even before the war, but it was not until his time at night school that it blossomed into his favorite subject. This passion led him to pursue a degree in the Faculty of Mechanics at Shevchenko Kyiv State University, and he successfully graduated in 1955. On September 14 of the same year, Y. M. Grigorenko joined the Institute of Structural Mechanics of AS Ukr. SSR, taking on the role of a senior engineer in the department led by Academician A.

D. Kovalenko. Employment after graduation was a formidable challenge for Y. M. Grigorenko due to his wartime experiences in Germany, which unfortunately closed doors to numerous scientific and educational opportunities. However, noteworthy is the commendable courage of A. D. Kovalenko, who fearlessly extended an invitation for Y. M. Grigorenko to join his team. Grigorenko, always grateful to his mentor, found his sole professional haven at the Institute. The unwavering support from Kovalenko marked a turning point, allowing Y. M. Grigorenko to establish a lifelong commitment to his work at the Institute.

Y. M. Grigorenko first learned about the theory of plates and shells at the Institute. He extensively studied all the available articles and monographs in this field. During his studies, he found out that there were significant computational difficulties in solving differential equations using series, which were used to describe various problems in plate and shell theory. Grigorenko then managed to simplify the process of determining the coefficients of these series and created recurrence relations for logarithmic solutions of the Fuchs class ordinary differential equations with polynomial coefficients. In 1956, he wrote his first article on this topic, which was published in the journal Soviet Applied Mechanics [1].

In 1958–1959, he was also engaged in solving a complicated problem of antisymmetric bending of a conic disk with consideration of tensile forces. He considered a circular plate of conic profile in antisymmetric bending in the field of forces acting in the median plane. Such a problem arises, for example, during the evolution of an airplane in a thin, fast-rotating disk of an axial turbocharger when the disk is loaded simultaneously by Coriolis and centrifugal forces, which cause its bending and stretching. In a strict formulation, such a problem is described by a nonlinear system of equations. Assuming that the forces acting in the median plane do not depend on transverse loads, the problem is reduced to a linear one. The problem is described by a complex differential equation and solved in series. The effect of tensile forces on disk bending is analyzed. In December 1958, the paper was submitted to the journal Soviet Applied Mechanics [2].

This was when the first electronic computing machines appeared in the USSR. By then, he had prepared a program for calculating conical shells of linearly variable thickness on the computer Strela. There were only a few machines of this type in the USSR. Therefore, half a year before the beginning of the computation work, Grigorenko had to reach an agreement with the Computing Center of the Academy of Sciences of the USSR in Moscow, which was founded in 1955 and had this machine. He had to wait half a year when it was possible to conduct the calculations. Grigorenko managed to obtain a handwritten program (that was not an easy task in those years) for the Strela computer and to prepare a program for solving the problem on the calculation of the stress–strain state of conical shells of linear-variable thickness under axisymmetrical loads. A general solution to this problem in complex hypergeometric Gauss functions had been obtained earlier by A. D. Kovalenko, but these functions had not been tabulated. In the summer of 1959, Grigorenko was one of the first in the USSR to conduct his first calculations using electronic computers in the theory of shells. Upon his return from Moscow, he wrote an article, "On solving the problem of the stress state of a conic shell on the computer Strela" [3].

Also, during this period, he conducted the solution to the problem of the vibration behavior of conic shells of linear-variable thickness under antisymmetric loads from the action of Coriolis forces. Such shells were widely used in gas turbine structures. This was a very complicated problem. In 1960, his three articles were published in the journal Soviet Applied Mechanics [2–4]. To the article, which was devoted to solving the conic shell stress state problem on the computer Strela, he added the articles on the solution of the problem on axisymmetric deformation of a conic shell with linearly variable thickness and on antisymmetric deformation of a conic shell.

At the end of 1960, the supervisor, A. D. Kovalenko, discussed the results of his five-year scientific work with Y. M. Grigorenko and offered to prepare the thesis for a doctoral dissertation. The thesis contained the results of solving complicated problems in the theory of circular plates and conic shells of linear-variable thickness under asymmetric loads based on precise analytical solutions in the form of series and hypergeometric functions. On April 6, 1961, the thesis was successfully defended for the Candidate of Physical and Mathematical Sciences degree.

In the summer of 1961, the Bureau of Academy of Sciences of the Ukrainian Socialist Soviet Republic (AS UkrSSR) decided to create a structural laboratory of computer engineering at the Institute of Mechanics. Y. M. Grigorenko, who had experience in conducting calculations on electronic computers, was appointed as the acting head of this laboratory on November 1, 1961. Later on, this laboratory was renamed the Department of Computational Methods.

The laboratory was responsible for operating the BESM-2 computer and was part of the scientific department. Therefore, the laboratory staff was assigned to carry out scientific research and implementation projects. The reason for creating the laboratory was that Prof. A. N. Golubentsev, the deputy director of the Institute, faced great difficulty in organizing the acquisition of the best three-address computer BESM-2M. This computer had 5,000 lamps and was the only one in Kyiv and one of the few in Ukraine then. According to the USSR Council of Ministers decree, the laboratory was given a plan to work for not less than 22 h per day. Employees from all institutions of Kyiv and other cities of Ukraine came to the laboratory to solve problems.

Gradually, all departments of the Institute of Mechanics began to use computers to solve their problems, and before that time, to solve mechanics problems, they used exact analytical, asymptotic, and other approximate methods; with the advent of computers, they began to use numerical methods, first of all, the finite-difference method. Considerable development of new technologies in rocket engineering and the creation of aircraft engines, aircraft, turbines, etc., occurred at this time in the USSR and abroad. Many problems were reduced to solving static and dynamic ones for thin and non-thin shell structures of various shapes. The Institute was also interested in these problems. In 1957, it began joint work on missile production with the Yuzhnoye Design Office. Even earlier, the Institute conducted works on calculating turbines for various purposes, where many elements are made in the form of thin and non-thin shells. These issues interested the laboratory. Naturally, the laboratory staff and their supervisor considered it necessary to pay special attention to numerical methods for calculating shell structures. At that time, many problems were solved using analytical methods, which allowed solving only simple problems for shells.

In this context, Y. M. Grigorenko, serving as the head of the laboratory, meticulously evaluated prevalent applied numerical methods for computing shells during that era. He systematically compared the results of applying these methods with solutions derived from analytical approaches. Consequently, around 1962, the laboratory commenced solving shell-related problems using diverse numerical techniques, such as Abramov and Godunov's method, finite differences, integral relations, and projective methods. The assessment of these methods hinged primarily on their implementation simplicity and result accuracy.

These comprehensive evaluations deduced that the discrete orthogonalization method, as outlined in S. K. Godunov's article [5], emerged as the most effective. Although lacking numerical examples, Y. M. Grigorenko and his students conducted practical assessments to substantiate its advantages in computations. This methodology, later coined the Godunov–Grigorenko method, found widespread application in calculating the mechanical behavior of a broad range of plate shells. Its efficacy garnered domestic and international recognition from researchers, solidifying its status as the preferred approach for shell structure computations from the late 1960s onwards.

Furthermore, a more refined approach to numerically solving non-axisymmetric deformation problems in shells of revolution was introduced in a 1968 article published in Soviet Applied Mechanics [6]. Remarkably, it was marked as the pioneering development of a solution for this specific class of shell theory problems. Subsequently, requests for assistance in calculating shell structures, predominantly related to rocket hardware elements, poured in from various organizations in the USSR, underscoring the practical significance and applicability of their groundbreaking contributions.

His first three monographs [7–9] were published in the Department of Mechanics, where he worked with his teacher and colleagues. Later on, his laboratory was renamed into the Department of Computational Methods. In [7], he calculated conic shells of linearly variable thickness under axisymmetric loads. In [8], he described the theoretical positions of conic shells and their applications in mechanical engineering. Additionally, he proposed approaches to solving problems of statics of conic shells under antisymmetric loads in [9]. It is worth noting that the last monograph was translated into English and published by the U.S. NASA Publishing House [10], which received high international recognition for the research conducted.

The department's staff collaborated with top organizations in the USSR to calculate gas aviation and stationary turbines. These collaborations included world-renowned rocket-space bureaus like Yuzhnoye Design Office and Research and Production Association Energiya. With the use of numerical methods, a significant leap was made in shell calculations. Before this, problems were mainly solved for shells made of isotropic materials of the simplest form with axisymmetric loads. However, numerical methods allowed finding solutions for anisotropic, layered, complex-form shells and various asymmetric and local loads. It is worth noting that, during that time, computing machinery had limited memory capacity and calculation speed capabilities. The Department of Computational Methods' research on shell calculations was reflected in one of the first monographs by Y. M. Grigorenko and his students, published in 1971. The book noted that the developed approach to solve problems of shell statics of rotation resulted from the rational joint use of advancements in the theory of thin isotropic and anisotropic shells, modern numerical methods and programming, and an understanding of the design bureaus' and computational departments' requirements. This book was utilized for shell structure calculations across many enterprises of the USSR.

By early 1970, Y. M. Grigorenko was preparing his doctoral thesis based on his 1962–1969 research. The thesis consisted of two parts. The first one contained results on the theory and analytical methods for solving problems of statics of conic shells of variable thickness under the action of asymmetric loads, which had been published in the three monographs mentioned above. The second part of the thesis described effective approaches to solving a broad class of problems concerning the stress state of multilayered shells of arbitrary shape with isotropic and anisotropic layers of variable stiffness under nonuniform force and temperature influences based on numerical analysis methods. The doctoral thesis defense was held on June 16, 1970, at the Institute of Mechanics of the AS of UkrSSR. The speakers and participants of the defense noted the effective application of the numerical methods for the solution of a new complex class of the problems of the shell theory.

In 1971, Y. M. Grigorenko began his teaching activity as a part-time teacher at Taras Shevchenko Kyiv State University at the Cybernetics Department of the Faculty of Computational Mathematics. He has read basic courses on methods of computation and various directions in programming. He also prepared and taught several special courses based on his research, later published as textbooks. His teaching activity was continued for more than thirty years. Some of his students were employed at the Department of Computational Methods at the Institute of Mechanics, where they defended their Ph.D. theses. In the spring of 1973, he was awarded the title of professor.

After successfully defending the doctoral thesis, Y. M. Grigorenko wrote and published a monograph called [12]. The monograph was focused on the theory and methods of solving statics problems related to thin layered shells of revolution with anisotropic layers of variable thickness under force and thermal loads. He developed a numerical solution method for boundary value problems based on the continuum scheme. This approach surpassed many other methods in the domestic and foreign literature at that time in their efficiency, stability, and accuracy.

It is worth noting that the application of numerical methods using computers made it possible to overcome the limitations of analytical methods, which could only be used to solve relatively simple problems by reducing them to tabulated functions. Up to that time, many structural elements in the form of rotating shells (such as turbines, rockets, etc.) were calculated as rods with specific integral properties. Such an approach gave a very rough approximation of the exact solution. However, calculating structural elements as shells of revolution made obtaining more accurate information about their tenseness and strength possible. This resulted in the creation of more solid and reliable structural elements. During this period, numerical methods gradually replaced analytical methods in calculating structural elements of modern technology.

In May 1973, Y. M. Grigorenko attended the Polish–Soviet Conference in Warsaw. It was his first trip abroad during peacetime. Interestingly, he was back in Warsaw, a city he had previously seen destroyed but now rebuilt from the ruins, which struck him profoundly.

At that time, he continued his research in the department on two-dimensional boundary value problems of statics of thin shells of more complicated shapes with variable parameters using numerical methods. Among others, he developed approaches to calculating the stress–strain state of noncircular cylindrical shells under certain boundary conditions at the ends, toroidal shells of arbitrary cross-section, and shells of variable thickness in two coordinate directions with a median surface as the rotation surface. The results of these studies are reflected in the monographs [13, 16]. In 1971, he and his colleagues began addressing issues related to calculating thick-walled shells based on the spatial theory of elasticity. Here, he had to use analytical approaches to reduce the boundary value problem to one-dimensional and solve the latter by a stable numerical method of discrete orthogonalization. Problems for isotropic and orthotropic cylindrical and spherical thick-walled shells with certain boundary conditions at the edges were considered. Some estimates of assumptions in the theory of homogeneous and layered shells were obtained. These results were published in monographs by Y. M. Grigorenko, A. T. Vasilenko, and N. D. Pankratova [14, 20].

Some new approaches were implemented to calculate thick-walled shells of any shape with different boundary conditions at the ends. Fourier series expansions were used for the circumferential coordinate, while a finite-difference approximation was used for the thickness. The one-dimensional formant problem was solved using a stable numerical method. The department also developed methods for calculating non-thin shells based on refined theories. For the entire package, the approach relied on the straight-line hypothesis, with an additional approach that considered the transverse shear deformations in the shell layers. This resulted in the discovery of some original results. A. T. Vasilenko, G. P. Urusova-Golub, and Y. M. Grigorenko all collaborated in the development of these approaches [18].

Under the guidance of Y. M. Grigorenko, various computational methods were developed and applied for the calculation of structural elements in many enterprises across the Soviet Union. Theoretical investigations and calculation methods produced concrete results in designing and creating modern equipment. Grigorenko regularly delivered lectures in several cities, including Moscow, Leningrad, Novosibirsk, Tbilisi, Yerevan, Tallinn, Riga, Lviv, and Tashkent. During this period, many of his students defended their Ph.D. theses under his supervision, and in 1978, his first student, A. T. Vasilenko, earned his doctorate.

In June 1977, there was a change in leadership at the Institute of Mechanics. A. N. Guz, an academician of the AS USSR, became the director, and Y. M. Grigorenko became the deputy director of the Institute for Scientific Work. Grigorenko continued to head the department while also working as the deputy director of the Institute for ten years. This was challenging as he had to balance scientific and organizational work.

In 1978, the Academy of Science of Ukraine held its next elections, and Y. M. Grigorenko was elected as a corresponding member. In 1979, he, together with A. N. Guz and N. A. Kilchevsky, was awarded the Ukrainian SSR State Prize in the field of science and technology for a series of works related to the general theory of shells and research on stress fields in shells of complex structures and shapes. In the same year, he published a textbook on methods for solving linear boundary value problems of shell theory using computers. The book is titled "Methods of Solving Linear Boundary Value Problems of Shell Theory on Computer" [15].

In 1981, the generalizing monograph [17] was published. The theory of shells of variable stiffness covers the results of many years of his research. It is devoted to the theory of isotropic and layered anisotropic shells and methods of solving some classes of problems by numerical analysis. A study of the stress–strain state of shells in the classical, refined, and spatial formulations under the action of force and heat is conducted. This monography was included in the five-volume edition and awarded the Ukr. SSR State Award on Science and Technology (1986).

Another monograph was published in 1983 as a textbook for universities and technical institutes of higher education [19]. This book considers methods of numerical solution of nonlinear boundary value problems for systems of ordinary differential equations and partial differential equations applied to study the deformation of flexible plates and shells. The basic information on the stress–strain state of shells in the geometrically nonlinear formulation is given. Some methods of solution of one- and two-dimensional problems are considered: reduction to Cauchy problems and the system of nonlinear algebraic equations, linearization, continuation of solution in parameter, asymptotic integration, variational, reduction to one-dimensional problems, finite differences, and finite elements. The issues of realization of these methods on the computer are covered, and examples of solutions to particular problems are given.

It should be noted that Y. M. Grigorenko frequently delivered lectures at various all-Union and international conferences held in the USSR between 1960 and 1988. Due to the nature of his research related to defense issues, he only participated in a few conferences outside the Soviet Union, namely in Poland and Bulgaria.

On April 26, 1986, a significant accident occurred at the nuclear power plant near Chornobyl, which is located around 100 km from Kyiv. Radiation started to spread in different directions. Initially, the wind was blowing toward the north, and the Swedes were the first to discover it. However, from May 3, the wind changed direction and started blowing toward the south. In Kyiv, the authorities ordered the measurement of radiation levels. Initially, on April 26, rumors about the accident started to spread. A few days later, the radio began to report it, but they advised the public very sparingly and asked them to wipe their feet and close the vents. On May 1, children were even allowed to March in a parade, almost naked, which caused panic among the people. Many people rushed to get their children out of the city and went on vacations at their own expense, although the authorities did not allow it. Permission was not granted until three months later. In mid-May, children of junior high schools and other educational institutions began to be taken south. At that time, Y. M. Grigorenko served as the director of the Institute for two weeks and had to break the law to let his staff go on leave at their own expense. He always thought he was doing the right thing, although various commissions noted this violation.

Y. M. Grigorenko received instructions from the Academy of Sciences in Moscow to embark on a two-week scientific trip to Poland. Despite several requests to postpone the trip until autumn due to the Chornobyl disaster, the response was that their plan was more important than any accident there. Regrettably, many individuals residing far from the epicenter of the disaster in the USSR failed to comprehend the tragedy.

Y. M. Grigorenko's scientific visit to Poland in June 1986 was highly productive. He visited major scientific centers in Poznan, Zielona Góra, Wroclaw, Krakow, and Warsaw, where he presented scientific reports and discussed topical problems in the field of shell theory with his Polish colleagues. During this visit, he also revisited the village near Zielona Góra, where he had worked as a teenager during World War II, and went to see the house where he had lived. It is hard to imagine the emotions he must have felt while doing so.

Grigorenko continues his vigorous scientific activity, which has resulted in the publication of several monographs in co-authorship with his students devoted to the numerical solution of a broad class of shell theory problems. Together with his students, he published some monographs during these years. In [20, 24], methods of solving problems of statics of inhomogeneous orthotropic cylindrical, spherical, and rectangular shells based on equations of elasticity theory under certain kinds of boundary conditions were presented. It studied hollow layered cylinders, spheres, cones, and rectangular planar shells and plates under certain boundary conditions and force and temperature loading. The effects in shells made of composite materials caused by non-uniformity and anisotropy of their elastic properties were studied. A comparison was made with the results of solutions of the tasks based on approximate models of shells. In [21], the investigation methodology for free vibrations of shell structures was considered. The basic principles and equations describing the free vibrations of the shells within the framework of classical and revised shell theories and on the basis of the spatial theory of elasticity were given. A numerical technique for solving problems is described. The application of this technique to the calculation of structural elements for various purposes was illustrated by specific examples. In [22], Grigorenko and co-authors discussed the challenges in constructing refined theories for multilayered anisotropic shells. They developed computational models for all proposed versions of the models in the non-classical formulation. Using these models, they obtained solutions for specific problems, investigating the stress–strain state of the shell elements that depend on variations in geometrical and mechanical parameters, applied forces, and different edge fixing methods.

During this period, the department published the monograph [23] that presented a numerical study of the boundary value problems of the statics of shells and plates in the geometrically nonlinear formulation. The stress–strain state of flexible layered orthotropic shells of arbitrary shape with stiffness that varies in two coordinate directions and noncircular cylindrical shells with variable parameters along the guide were investigated in the pre-critical region of deformation. The problems were solved based on the reduction of two-dimensional nonlinear boundary value problems to one-dimensional ones using the straight-line method, Vlasov–Kantorovich method, linearization method, and discrete orthogonalization method. The class of problems on axisymmetric deformation of flexible anisotropic shells of rotation was also considered. Additionally, approaches to constructing solutions for nonlinear problems of statics of shells in the noncritical region were proposed. The results of the solutions to particular problems of nonlinear deformation of shells from these classes were also presented.

In 1988, Y. M. Grigorenko, for the first time, had an opportunity to participate in the international conference in the USA on computational methods in engineering sciences and made a report there. He noted that over the past 25 years, computational mechanics had developed into a scientific discipline capable of producing significant changes in engineering practice and that the conference should become a forum for discussing scientific research in computational mechanics—a science that enriches engineering practice by applying mechanical and numerical methods. In November 1991, he attended and reported at the First International Symposium on Line, Surface, and Dimensionality Methods in Computational Mathematics and Mechanics in Athens, Greece.

In December 1991, the Soviet Union officially collapsed, which led to the formation of separate states, including Russia, Ukraine, Belarus, and other former socialist republics of the USSR. This caused a significant economic crisis, and many staff members of various institutes were laid off. However, despite the difficult situation, Y. M. Grigorenko and his remaining students continued their scientific activities. At that time, they published two important monographs. The first one, cited as [25], provides approaches to solve the problems of statics of shells made of anisotropic nonuniform materials in linear and geometrically nonlinear formulations under nonuniform force and temperature influences. Shells are considered both within the assumptions of the classical theory of non-deformable norms and based on refined models that consider transverse shear and compression. The developed approaches are implemented in software packages based on robust numerical methods. The results of studies of the stressed state of shells in a wide range of geometric and mechanical parameters and the calculations of specific structural elements are presented. The second monograph, cited as [26], focuses on methods of computer solution of nonlinear boundary value problems reduced to systems of ordinary and partial differential equations. It is applied to the study of the deformation of flexible plates and shells. The text includes basic information on the stress–strain state of shells in the geometrically nonlinear formulation and some methods for solving one- and two-dimensional problems. The text also covers Cauchy's problems, the system of nonlinear algebraic equations, linearization, continuation of solution over parameter, asymptotic integration, variational one-dimensional problems, finite differences, and finite elements. The authors also address the issues of implementing these methods on a computer and provide examples of solving specific problems.

In September 1992, Y. M. Grigorenko participated in two European conferences, "Modeling of shells under nonlinear behavior," organized by EUROMECH in Freising, Germany, and the First European Conference on Numerical Methods in Engineering in Brussels, Belgium. During the conferences, Grigorenko presented his reports on numerical modeling and methods for solving nonlinear problems of shell theory under static and dynamic actions. He also introduced his approach to solving linear and nonlinear problems of complex shape shell theory based on tensor representation of initial relations for the case of nonorthogonal parameterization and combined application of linearization methods, reduction to a one-dimensional problem, and its numerical integration. Grigorenko's scientific communications were well-received by E. Ramm, a prominent German professor in computational mechanics. They established a friendly relationship that lasted until Grigorenko's passing.

During the autumn of 1992, the National Academy of Science of Ukraine held regular elections. As a result of Grigorenko's outstanding scientific contributions in computational mechanics, he was elected as an academician of the National Academy of Sciences of Ukraine.

In 1994, Y. M. Grigorenko attended the EUROMECH 317 colloquium in Liverpool, England, on "Buckling resistance of shells sensitive to imperfections." During this event, he presented a report titled "Influence of shape, anisotropy, and heterogeneity on deformation of flexible shells." Later, in February 1995, he delivered a review report on "Some approaches to the numerical solution of linear and nonlinear problems of shell theory in classical and refined formulations" during a seminar on mechanics at the Berlin University of Technology.

It is worth noting that the scientist Y. M. Grigorenko has consistently applied his fundamental research to solve practical problems. Between 1998 and 2000, he supervised a sizeable scientific–technical project supported by a governmental agreement between the USA and Ukraine. The project was titled "Stress-strain state research, stability, and vibrations of shell systems; preparation of recommendations on the choice of rational parameters and creation of elements of shipbuilding constructions." Teams from the Institute of Mechanics NASU, Mykolaiv Shipyard, and the laboratory of Mykolaiv Pedagogical Institute carried out the project. The project was supported by the well-known American professor L. Librescu, with whom Y. M. Grigorenko had maintained close scientific ties since 1970. It was tragic news for Y. M. Grigorenko when he learned about the heroic death of L. Librescu, who had covered the departure of students from the shots of a maniac. In memory of L. Librescu, a three-volume monograph was dedicated in the future [31–33]. In 2001–2004, he supervised the research of contact problems for large-diameter tires for the famous French tire manufacturing firm Manufacture Francaise des Pneumatiques Michelin Corporation Technology Center, Clermont-Ferrand, France.

Grigorenko has always paid great attention to the training of scientific staff. Among his students, there are ten doctors and 43 candidates of sciences. Many of them became well-known professors and co-authors of his scientific works. His students recall with warmth his constant attention and desire to help, but, at the same time, the necessary exactingness.

During the early twenty-first century, the discrete–continuum numerical approach developed by Y. M. Grigorenko gained widespread use in solving static and dynamic problems for plates and shells of complex shapes and materials, which are often heterogeneous and anisotropic. This approach involves applying the spline collocation method and solving the boundary value problem for systems of ordinary differential equations. The method uses a stable and effective numerical technique of discrete orthogonalization with variable coefficients in the general case.

In 2005, Y. M. Grigorenko left the leadership of the Department of Computational Methods, which he had created, to his son, Professor A. Y. Grigorenko; he continued to work as a chief researcher of the department. Despite his age, Y. M. Grigorenko continued actively involved in scientific activity. The monographs published together with his son [27, 30] delve into the approaches for solving linear and geometrically nonlinear problems of shell mechanics (statics and dynamics) based on discrete–continuum methods in classical, refined, and spatial statements, as well as the main principles and equations of classical, refined, and spatial models for anisotropic nonuniform shells with variable geometrical and mechanical parameters. The developed approaches are implemented in software packages. The results of the investigations of the stress state and dynamic characteristics of shells of various shapes and structures depending on the variations of the basic parameters are presented. Calculations of particular structural elements are also fulfilled. The monograph "Problems of Mechanics for Anisotropic Inhomogeneous Shells on the Basis of Different Models" was awarded the State Prize of Ukraine in Science and Technology.

Grigorenko often acted as an opponent during the defenses of theses. He helped young scientists, many of whom later became well-known professors, improve the contents of their works. Academicians of the National Academy of Sciences of Ukraine R. M. Kushnir and A. N. Khimich and Academician of the Academy of Sciences of Armenia L. A. Agolovian always remember with gratitude his help and attention during the defense of their doctoral dissertations.

After Ukraine became an independent state, there was a need to develop scientific terminology in the Ukrainian language. During the Soviet era, Russian was the predominant language, and Ukrainian terminology was not a priority. Grigorenko has been passionate about the Ukrainian language since childhood. He spent a lot of time creating Russian–Ukrainian and Russian–Ukrainian–English dictionaries focused on mechanics. These dictionaries [28, 29] are widely used in many scientific and educational institutions throughout Ukraine.

Grigorenko was awarded the honorary title of Honored Worker of Science and Engineering of Ukraine for his many years of scientific and pedagogical activity by a decree of the President of Ukraine. In the last years of his life, due to his health, Grigorenko could not travel abroad. He often hosted the famous German professors H. Altenbach and W. H. Müller, who noted a warm, friendly atmosphere and friendliness in communication, knowing about his difficult childhood during the war years in forced labor in Germany.

Professors Y. M. Grigorenko, W. H. Müller, and A. Y. Grigorenko collaborated on a multi-physics monograph consisting of three books [31–33]. The books, edited by H. Altenbach, demonstrate effective numerical computations based on the discrete–continuum method for new problems in the theory of plates and shells. The examples use different modes and illustrate the potential of discrete–continuum numerical approaches to solve broad problems in the theory of shells and plates. While Y. M. Grigorenko acknowledged the importance of the finite element method in computational mechanics, he also identified some issues with its application, particularly in cases of an inhomogeneous structure of shells. For a large class of problems in the theory of shells, it is simpler and more efficient to use the discrete–continuum approach by reducing the original problem to a system of ordinary differential equations.

The words of the outstanding English scientist, Lord Kelvin, "I never get pleasure from formulas till I feel numerical values of magnitudes," can be applied to all the scientific work of Grigorenko. His contributions are reflected in his 35 books and 500 articles. All his monographs, articles, and theses by his students were devoted to finding numerical solutions for new, complex tasks in the theory of shells and plates—specifically, the spatial theory of elasticity. He is rightfully considered one of the pioneers of the new field of computational mechanics in the world of science.

Y. M. Grigorenko had a happy life for 67 years with his wife, Nina Feodorovna, who was a mathematics teacher. They raised their son, Alexander, who successfully carried on his father's work. Unfortunately, Y. M. Grigorenko passed away on January 18, 2022, after a severe illness. He was an outstanding scientist of Ukrainian and world science, a wise teacher and mentor, and a kind and sensitive friend. The memories of him will forever remain in the hearts of his family, students, and friends.

References

1. Grigorenko, Y.M.: Recurrence relations for logarithmic solutions in the problem of bending round plates. Sov. Appl. Mech. **3**(4), 400–407 (1957). [In Ukrainian]
2. Grigorenko, Y.M.: Antisesymmetric bending of conical profile disc taking into account tensile forces. Sov. Appl. Mech. **6**(1), 88–94 (1960). [In Ukranian]
3. Grigorenko, Y.M.: Solving the symmetric deformation of the conical shell of linearly variable thickness on the electronic machine"Strela." Sov. Appl. Mech. **6**(2), 224–228 (1960). [In Ukranian]
4. Grigorenko, Y.M.: Antisymmetric stress state of variable thickness conical shell. Sov. Appl. Mech. **6**(6), 385–392 (1960). [In Ukranian]
5. Godunov, S.K.: On numerical solution of boundary value problems for systems of linear differential equations. Adv. Math. Sci. **3**, 171–174 (1961). [In Russian]
6. Grigorenko, Y.M., Sudavtsova, G.K., Shinkar, A.I.: Digital-Computer solution for problems of the asymmetric stress-strain state of systems composed of shells of revolution. Sov. Appl. Mech. **13**(11), 11–16 (1968). [In Russian]
7. Kovalenko, A.D., Grigorenko, Y.M., Lobkova, N.A.: Calculation of Conic Shells of Linear Variable Thickness. Publishing house of the Ukrainian SSR Academy of Sciences, Kiev (1961). [In Russian]
8. Kovalenko, A.D., Grigorenko, Y.M., Ilyin, L.A.: Theory of Thin Tapered Shells and its Application in Mechanical Engineering. Publishing house of the Academy of Sciences of the USSR, Kiev (1963). [In Russian]
9. Kovalenko, A.D., Grigorenko, Y.M., Ilyin, L.A., Polishchuk, T.I.: Calculation of Tapered Shells under Antisymmetric Loads. Kiev. Nauk, Dumka (1966). [In Russian]
10. Kovalenko, A.D., Grigorenko, Y.M., Illin, L.A., Polischuck, T.I.: The design of conical shells subjected to antisymmetric loadings. National Aeronautics and Space Administration Technical Translation (1969)

11. Grigorenko, Y.M., Bespalova, E.I., Vasilenko, A.T., et al.: Numerical solution of boundary value problems of statics of orthotropic layered shells of rotation on computers of type M. Kiev. Nauk, Dumka (1971). [In Russian]
12. Grigorenko, Y.M.: Isotropic and Anisotropic Laminated Shells of Variable Stiffness. Naukova Dumka, Kiev (1973). [In Russian]
13. Grigorenko, Y.M., Vasilenko, A.T., Bespalova, E.I., et al.: Numerical Solution of Boundary Value Problems of Statics of Orthotropic Shells with Variable Parameters. Naukova Dumka, Kiev (1975). [In Russian]
14. Grigorenko, Y.M., Vasilenko, A.T., Pankratova, N.D.: Calculation of Noncircular Cylindrical Shells. Naukova Dumka, Kiev (1977). [In Russian]
15. Grigorenko, Y.M., Mukoed, A.D.: Problem Solution of the Theory of Shells on the Computer: Textbook. Kiev, Vyshcha shk (1979). [In Russian]
16. Grigorenko, Y.M., Kitaygorodsky, A.B., Semenova, V.V., et al.: Calculation of Orthotropic Layered Shells of Revolution with Variable Parameters on the ECM. Naukova Dumka, Kiev (1980). [In Russian]
17. Grigorenko, Y.M., Vasilenko, A.T.: Theory of Shells of Variable Stiffness. Kiev. Naukova Dumka. (Methods of Calculating Shells: In 5 vols.; Vol.4), (1981). [In Russian]
18. Grigorenko, Y.M., Vasilenko, A.T., Golub, G.P.: The Solution of Problems of Statics of Shells of Revolution in the Specifying Statement on the ECM. Naukova Dumka, Kiev (1982). [In Russian]
19. Grigorenko, Y.M., Mukoed, A.P.: The Solution of Nonlinear Problems of the Theory of Shells on the ECM: Textbook. Kiev, Vyshcha shk (1983). [In Russian]
20. Grigorenko, Y.M., Vasilenko, A.T., Pankratova, N.D.: Statics of Anisotropic Thick-Walled Shells. Kiev, Vischya Shk (1985). [In Russian]
21. Grigorenko, Y.M., Vasilenko, A.T., Pankratova, N.D.: Statics of Anisotropic Thick-Walled Shells. Kiev, Vischya Shk (1985). [In Russian]
22. Grigorenko, Y.M., Vasilenko, A.T., Golub, G.P.: Statics of Anisotropic Shells with Finite Shear Stiffness. Naukova Dumka, Kiev (1987). [In Russian]
23. Grigorenko, Y.M., Kryukov, N.N.: Numerical Solution of Problems of Static of Flexible Laminated Shells with Variable Parameters. Naukova Dumka, Kiev (1988). [In Russian]
24. Grigorenko, Y.M., Vasilenko, A.T., Pankratova, N.D.: Problems of Elasticity Theory of Non-Homogeneous Bodies. Naukova Dumka, Kiev (1991). [In Russian]
25. Grigorenko, Y.M., Vasilenko, A.T.: Problems of Statics of Anisotropic Non-Homogeneous Shells. Moscow. Nauka, Main edition of Physics and Mathematics (1992). [In Russian]
26. Grigorenko, Y.M., Mukoyed, A.P.: Solution of Linear and Nonlinear Problems of Envelope Theory on the EOM: Textbook. Kiev, Libid (1992). [In Ukrainian]
27. Grigorenko, Y.M., Vlaikov, G.G.: Grigorenko AYa,: Numerical Analytical Solution of Problems of Shell Mechanics on the Basis of Various Models. Kiev, Academperiodica (2006). [In Russian]
28. Grigorenko, Y.M., Bikhovets, N.M., Bikhovets, O.M.: Russian-Ukrainian Dictionary of Mechanics. Kiev, Libid (1995)
29. Grigorenko, Y.M., Bastun, V.M., Shirokov, V.A.: Russian-Ukrainian-English Glossary of Mechanics. Naukova Dumka, Kiev (2007)
30. Grigorenko, Y.M., Grigorenko, A.Y., Vlaikov, G.G.: Problems of Mechanics for Anisotropic Inhomogeneous Shells on the Basis of Different Models. Kyiv, Akademperiodika (2009)
31. Grigorenko, A.Y., Müller, W.H., Grigorenko, Y.M., Vlaikov, G.G.: Recent Developments in Anisotropic Heterogeneous Shell Theory. General Theory and Applications of Classical Theory. Volume I. Berlin, Springer (2016)
32. Grigorenko, A.Y., Müller, W.H., Grigorenko, Y.M., Vlaikov, G.G.: Recent Developments in Anisotropic Heterogeneous Shell Theory. Applications of Refined and Three-dimensional Theory. Volume II A. Berlin, Springer (2016)
33. Grigorenko, A.Y., Müller, W.H., Grigorenko, Y.M., Vlaikov, G.G.: Recent Developments in Anisotropic Heterogeneous Shell Theory. Applications of Refined and Three-dimensional Theory. Volume II B. Berlin, Springer (2016)

Chapter 2
Finite Element Modeling of Partially Conductive Interfacial Crack in Piezoelectric Bimaterial

Victor Adlucky, Natalia Guk, and Volodymyr Loboda

Abstract A plane strain state of an interface crack in a piezoelectric bimaterial under the action of electric loading parallel to the crack and mechanical loading orthogonal to it is analyzed using the finite element method (FEM). Both piezomaterials are polarized in the direction orthogonal to the crack faces. One part of the crack faces is electrically conductive, while the other is insulated. The fracture parameters of the crack are investigated and presented in graphical and tabular forms.

2.1 Introduction

Piezoelectric materials are widely used in modern microelectronic products such as piezoelectric resonators, transducers, and actuators. During maintenance, they are usually subjected to significant loads of both mechanical and electrical nature, and various violations of the material integrity, such as pores, cracks, and delaminations, may occur. It is known that piezoelectric materials are susceptible to brittle fracture [1, 2], so when analyzing the strength and reliability of piezoelectric structures, it is necessary to use fracture mechanics concepts. Available analytical solutions of the corresponding boundary value problems are most often obtained using methods of complex potentials for infinite domains with inhomogeneities of canonical form. Most of the works in which exact solutions for a wide range of both homogeneous and inhomogeneous piezomaterials with defects such as holes and cracks under given electromechanical loads are obtained belong to the class of two-dimensional problems [3–8]. The number of publications in this area is constantly increasing. Mainly, linear models of deformation and fracture of piezoelectric materials are considered.

V. Adlucky · N. Guk · V. Loboda (✉)
Oles Honchar Dnipro National University, Dnipro, Ukraine
e-mail: loboda@dnu.dp.ua

V. Adlucky
e-mail: adluckiivy_kkt@dnu.dp.ua

N. Guk
e-mail: natalyguk29@gmail.com

H. Altenbach et al. (eds.), *Selected Problems of Solid Mechanics and Solving Methods*, Advanced Structured Materials 204, https://doi.org/10.1007/978-3-031-54063-9_2

At the same time, the range of problems in fracture mechanics of piezomaterials is far from limited to the case of the infinite domain. The results for finite domains of complex shapes are of practical interest. Naturally, in such cases, there is a need to use numerical methods, in particular, the FEM [9]. Many studies consider various aspects of the implementation of the FEM to piezomaterials [10–16]. As a rule, problems for infinite domains with available analytical solutions are chosen to test the accuracy of finite element algorithms. The important problem in implementing FEM is the adequate modeling of electrical and mechanical conditions on the crack faces and interfaces between different materials. In addition, since the main interest in the problems related to the fracture is obtaining reliable fracture parameters, it is important to build effective algorithms for the calculation of such quantities as crack opening, stress intensity factors, J-integrals, and energy release rate (ERR) [17–21].

In some cases, a crack may appear due to the exfoliation of the electrode from piezoelectric material. If the electrode located at the interface is completely exfoliated along its entire length, the resulting crack can be considered electrically conductive. Electrically conductive interfacial cracks in the plane case were considered in [6, 8, 22] for open and contact crack models. In many cases, a multilayer electrode of finite length may partially exfoliate along its length with some part of the material interface, forming a partially conductive interfacial crack. This important problem has not yet been sufficiently studied. An interface crack with partially electrically conductive crack faces under antiplane mechanical and in-plane electric loadings was considered in [23].

The subject of the present article is finite element modeling and the study of fracture parameters of a partially conductive interfacial crack in piezoelectric bimaterial for the case of plane strain.

2.2 Problem Formulation

The plane deformation of a bimaterial composed of $x_3 > 0$ and $x_3 < 0$ half-spaces connected at the interface $x_3 = 0$ and filled with piezomaterials 1 and 2 is considered. Both materials are polarized along the x_3-axis. An interfacial crack is located on the interface segment AB: $x_1 \in L = (-l, l)$ (Fig. 2.1).

Governing differential equations of electromechanical coupling problem in the case of mass forces and free charge absence have the following form [2]:

$$\begin{gathered}\sigma_{ij} = c_{ijkl}\varepsilon_{kl} - e_{kij}E_k, \quad D_i = e_{ikl}\varepsilon_{kl} + \chi_{ik}E_k, \\ \sigma_{ij,i} = 0, \quad D_{i,i} = 0, \quad \varepsilon_{ij} = 0.5\left(u_{i,j} + u_{j,i}\right), \quad E_i = -\varphi_{,i}\end{gathered} \tag{2.1}$$

$(i, j, k, l = 1, 3)$, where u_i are components of displacement vector; σ_{ij} and ε_{ij} are components of stress and strain tensors, respectively; E_i and D_i are components of electric field vector and electric displacement vector, respectively; φ is electri-

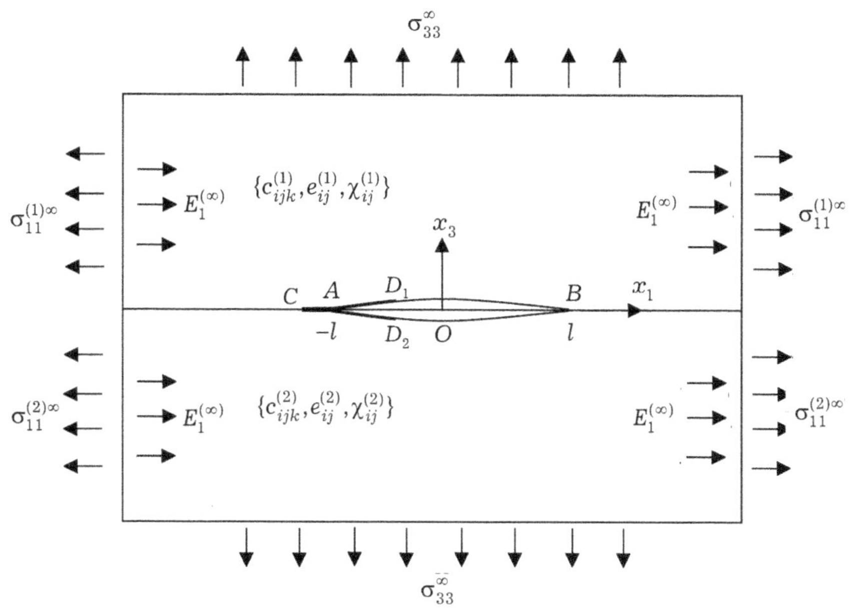

Fig. 2.1 Loading of piezoelectric bimaterial

cal potential; c_{ijkl}, e_{ijk}, and χ_{ij} are components of elastic stiffness, piezoelectric strain, and dielectric matrices, respectively. Commas denote derivatives with respect to corresponding coordinate variables. Under the conditions of plane deformation, $u_2 = 0$, $\sigma_{12} = 0$, $\sigma_{23} = 0$, $\varepsilon_{12} = 0$, $\varepsilon_{23} = 0$, $E_2 = 0$, $D_2 = 0$, and all other components are the functions of x_1 and x_3 variables only.

With the elimination of stresses and strains, Eq. (2.1) can be reduced to the following system of differential equations:

$$\left(c_{ijkl}u_k + e_{lij}\varphi\right)_{,li} = 0, \qquad \left(e_{ikl}u_k - \chi_{il}\varphi\right)_{,li} = 0. \tag{2.2}$$

The components of electric and mechanical fields $\sigma_{11}^{(1)\infty}$, $\sigma_{11}^{(2)\infty}$, σ_{33}^{∞}, and E_1^{∞} are specified at infinity. The boundary conditions on the interface for the case of an open crack are as follows:

$$\begin{aligned}
&\sigma_{i3}^{(1)}(x_1, 0) = \sigma_{i3}^{(2)}(x_1, 0), \\
&\varphi^{(1)}(x_1, 0) = \varphi^{(2)}(x_1, 0), \\
&D_3^{(1)}(x_1, 0) = D_3^{(2)}(x_1, 0), \\
&u_i^{\prime(1)}(x_1, 0) = u_i^{\prime(2)}(x_1, 0), \quad x_1 \in (-\infty, \infty) \setminus L, \\
&\sigma_{i3}^{(m)}(x_1, 0) = 0, \quad x_1 \in L \quad (m = 1, 2, \quad i = 1, 3).
\end{aligned} \tag{2.3}$$

The crack faces can be in frictionless contact for $x_1 \in (c, d)$ $(-l \leqslant c < d \leqslant l)$. Thus, the following mechanical conditions are added to the conditions at the interface:

$$
\begin{aligned}
&\sigma_{i3}^{(m)}(x_1,0)=0, \quad x_1\in[-l,c]\cup[d,l],\\
&u_3^{(1)}(x_1,0)=u_3^{(2)}(x_1,0),\\
&\sigma_{13}^{(m)}(x_1,0)=0,\\
&\sigma_{33}^{(1)}(x_1,0)=\sigma_{33}^{(1)}(x_1,0), \quad x_1\in(c,d) \quad (m=1,2,\ i=1,3).
\end{aligned} \tag{2.4}
$$

We also consider an electrode CD at the bimaterial interface completely or partially located in the crack zone. It is assumed that in the crack zone, the electrode is delaminated, forming electrically conductive parts of the crack faces (for example, CD_1 and CD_2 in Fig. 2.1). The remaining parts of the crack (D_1B and D_2B in Fig. 2.1) are considered to be insulated. We focus our attention on the following fracture parameters: crack opening $\Delta(x_1)=|u_3^{(2)}(x_1,0)-u_3^{(1)}(x_1,0)|$, ERR G^A and G^B at the crack tips A and B, respectively.

Piezomaterials with hexagonal crystal system and point group 6 mm are considered, for which with use of reduced index notation [2] $c_{11}=c_{1111}$, $c_{33}=c_{3333}$, $c_{13}=c_{1133}$, $c_{44}=c_{1313}$, $e_{31}=e_{311}$, $e_{33}=e_{333}$, $e_{15}=e_{113}$, and elastic, piezoelectric, and dielectric matrices take the form:

$$
\|c_{ij}\|=\begin{pmatrix}
c_{11} & c_{12} & c_{13} & 0 & 0 & 0\\
c_{12} & c_{11} & c_{13} & 0 & 0 & 0\\
c_{13} & c_{13} & c_{33} & 0 & 0 & 0\\
0 & 0 & 0 & c_{44} & 0 & 0\\
0 & 0 & 0 & 0 & c_{44} & 0\\
0 & 0 & 0 & 0 & 0 & \frac{1}{2}(c_{11}-c_{12})
\end{pmatrix},
$$

$$
\|e_{ij}\|=\begin{pmatrix}
0 & 0 & 0 & 0 & e_{15} & 0\\
0 & 0 & 0 & e_{15} & 0 & 0\\
e_{31} & e_{31} & e_{33} & 0 & 0 & 0
\end{pmatrix},
$$

$$
\|\chi_{ij}\|=\begin{pmatrix}
\chi_{11} & 0 & 0\\
0 & \chi_{11} & 0\\
0 & 0 & \chi_{33}
\end{pmatrix}.
$$

The e-type constitutive relations (which are usually called piezoelectric equations in the theory of piezoelectricity) take the form [2]

$$
\begin{aligned}
\sigma_{11}&=c_{11}\varepsilon_{11}+c_{13}\varepsilon_{33}-e_{31}E_3,\\
\sigma_{33}&=c_{13}\varepsilon_{11}+c_{33}\varepsilon_{33}-e_{33}E_3,\\
\sigma_{13}&=c_{44}\varepsilon_{13}-e_{15}E_1,\\
D_1&=e_{15}\varepsilon_{13}+\chi_{11}E_1,\\
D_3&=e_{31}\varepsilon_{11}+e_{33}\varepsilon_{33}+\chi_{33}E_3.
\end{aligned} \tag{2.5}
$$

Boundary conditions at infinity must satisfy the requirement that deformation along the bimaterial interface is continuous:

$$
\varepsilon_{11}^{(1)\infty}=\varepsilon_{11}^{(2)\infty}=\varepsilon_{11}^{\infty}. \tag{2.6}
$$

Table 2.1 The values of mechanical stresses that provide continuity of deformation ε_{11} at the interface of bimaterial PZT-4/PZT-5H

$\sigma_{11}^{(1)\infty}/\sigma_{33}^{\infty}$	$\sigma_{11}^{(2)\infty}/\sigma_{33}^{\infty}$
0.106	0
0	−0.100
−1.637	−1.637

Taking into account the continuity of stresses and electric displacements at the interface, we can obtain the relations for the components of electromechanical fields, which follow from Eqs. (2.5) and (2.6):

$$\begin{aligned}
\sigma_{11}^{(m)\infty} &= c_{11}^{(m)}\varepsilon_{11}^{\infty} + c_{13}^{(m)}\varepsilon_{33}^{(m)\infty} - e_{31}^{(m)}E_{3}^{(m)\infty},\\
\sigma_{33}^{(m)\infty} &= c_{13}^{(m)}\varepsilon_{11}^{\infty} + c_{33}^{(m)}\varepsilon_{33}^{(m)\infty} - e_{33}^{(m)}E_{3}^{(m)\infty},\\
D_{3}^{(m)\infty} &= e_{31}^{(m)}\varepsilon_{11}^{\infty} + e_{33}^{(m)}\varepsilon_{33}^{(m)\infty} + \chi_{33}^{(m)}E_{3}^{(m)\infty} \quad (m = 1, 2).
\end{aligned}$$

Solving these equations with respect to $\varepsilon_{11}^{\infty}$, $\varepsilon_{33}^{(1)\infty}$, $\varepsilon_{33}^{(2)\infty}$, $E_{3}^{(1)\infty}$, $E_{3}^{(2)\infty}$, and D_{3}^{∞}, we can obtain the following condition:

$$\begin{aligned}
&\Big\{\sigma_{11}^{(1)\infty}\Big[\Big(e_{33}^{(1)}\Big)^{2} + c_{33}^{(1)}\chi_{33}^{(1)}\Big] - \sigma_{33}^{\infty}\Big(e_{31}^{(1)}e_{33}^{(1)} + c_{13}^{(1)}\chi_{33}^{(1)}\Big)\\
&\qquad + D_{3}^{\infty}\Big(c_{33}^{(1)}e_{31}^{(1)} - c_{13}^{(1)}e_{33}^{(1)}\Big)\Big\}(\delta^{(1)})^{-1} =\\
&\Big\{\sigma_{11}^{(2)\infty}\Big[\Big(e_{33}^{(2)}\Big)^{2} + c_{33}^{(2)}\chi_{33}^{(2)}\Big] - \sigma_{33}^{\infty}\Big(e_{31}^{(2)}e_{33}^{(2)} + c_{13}^{(2)}\chi_{33}^{(2)}\Big)\\
&\qquad + D_{3}^{\infty}\Big(c_{33}^{(2)}e_{31}^{(2)} - c_{13}^{(2)}e_{33}^{(2)}\Big)\Big\}(\delta^{(2)})^{-1},\\
&\delta^{(m)} = c_{33}^{(m)}\Big(e_{31}^{(m)}\Big)^{2} - 2c_{13}^{(m)}e_{31}^{(m)}e_{33}^{(m)} + c_{11}^{(m)}\Big(e_{33}^{(m)}\Big)^{2}\\
&\qquad + \Big[c_{11}^{(m)}c_{33}^{(m)} - \Big(c_{13}^{(m)}\Big)^{2}\Big]\chi_{33}^{(m)} \quad (m = 1, 2).
\end{aligned} \tag{2.7}$$

For material 1—PZT-4 and material 2—PZT-5H, under condition $D_{3}^{\infty} = 0$, the relation (2.7) takes the form

$$\sigma_{33}^{\infty} = 9.418\sigma_{11}^{(1)\infty} - 10.029\sigma_{11}^{(2)\infty}.$$

Table 2.1 presents several values of $\sigma_{11}^{(m)\infty}/\sigma_{33}^{\infty}$ $(m = 1, 2)$ corresponding to the above relationship.

Under additional requirement of continuity of the electric field component E_3, the relation (2.7) takes the form

$$\sigma_{11}^{(1)\infty} = 0.954\sigma_{33}^{\infty}, \qquad \sigma_{11}^{(2)\infty} = 0.797\sigma_{33}^{\infty}.$$

2.3 Numerical Realization

Simulation of the piezoelectric bimaterial with interfacial crack was carried out using FEM, the most effective computational tool for solving a wide range of continuum mechanics problems [9]. Modeling and calculations were conducted in the framework of Abaqus software. The infinite plane was approximated by a finite region $|x_1| \leqslant b$, $|x_3| \leqslant b$, where $b = 20l$. As was shown by numerical experiments, this provides sufficient agreement with available analytical results for the infinite domain.

Electromechanical boundary conditions at infinity were applied at the boundaries of the finite region. The components of the electric field were expressed through the voltage difference between two opposite sides of the finite region:

$$E_1^{\infty} = -\Delta\varphi/(2b).$$

Two springs—horizontal and vertical—were added at each corner of the outer boundary to stabilize the body from the rigid motion. The stiffness of each spring was chosen to be several orders of magnitude less than the stiffness of bimaterial. Numerous studies showed that such choice does not significantly affect the stress state. Quadrangular 8-node piezoelectric elements with second-order polynomial approximation were used to discretize the region. The components of displacements u_1 and u_3 and electric potential φ were determined at each node. The inhomogeneous finite element mesh was refined in the vicinity of the crack and became more coarse while approaching the outer boundary of the body. Special care was taken to properly model the vicinities of the crack tips where concentric finite element mesh was used. The length of the radial sides of this mesh gradually decreased from $0.3560l$ to $0.0012l$, and the arc measure was $\alpha = 11.25°$ (Fig. 2.2). The computational model had approximately 10^4 finite elements with $3 \cdot 10^4$ nodes, which guaranteed the practical invariance of results obtained with more refined meshes. In the degenerate triangular elements adjacent to the tips of the crack, the middle nodes were shifted by 1/4 of the length along the radial sides toward the crack tip. This provided modeling of asymptotics of degree 0.5 for displacements and electric potential. Possible contact between the crack faces was modeled by creating a contact pair under the conditions of mutual non-penetration of contacting sides and frictionless contact.

The boundary conditions on the electrically conductive parts of the crack are as follows:

$$\varphi = 0, \tag{2.8}$$

and on the insulated parts of the crack, they can be written as

$$D_3 = 0. \tag{2.9}$$

The finite element solution of the boundary value problems was carried out in geometrically linear formulation under static loading.

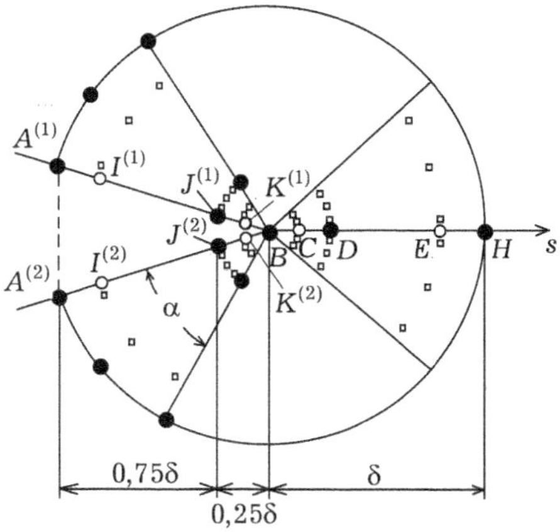

Fig. 2.2 Discretization scheme of the vicinity of the crack tip B

Since Abaqus cannot obtain force (stress intensity factors) or energy (J-integral, ERR) fracture characteristics for piezoelectric materials, there was a need for post-processing of FEM results to obtain fracture parameters. The ERR at the crack tips was determined using the virtual crack closure integral method [13–15], according to which the increment of potential energy during the crack length growth by δ is equal to the virtual work performed by stresses and electrical displacements acting on the interval δ in front of the crack on displacements and electric potentials acting on the interval of the same length δ behind the crack when it is closed to the length δ.

The ERR in the vicinity of the crack tip B (Fig. 2.2) is characterized by the value [3, 4, 24]

$$G(B) = G^M(B) + G^E(B), \tag{2.10}$$

where $G^M(B)$ and $G^E(B)$ are the mechanical and electrical components of the ERR, respectively:

$$\begin{aligned} G^M(B) &= -\frac{1}{2\delta}\int_0^\delta [\sigma_{13}(s)\Delta u_1(s-\delta) + \sigma_{33}(s)\Delta u_3(s-\delta)]\,ds, \\ G^E(B) &= -\frac{1}{2\delta}\int_0^\delta \big[D_3(s)\Delta\varphi(s-\delta) - D_3^{(1)}(s-\delta)\varphi^{(1)}(s-\delta) \\ &\quad + D_3^{(2)}(s-\delta)\varphi^{(2)}(s-\delta) - \Delta D_3(s-\delta)\varphi(s)\big]ds, \end{aligned} \tag{2.11}$$

$$\begin{aligned} \Delta u_i(s) &= u_i^{(1)}(s) - u_i^{(2)}(s) \quad (i = 1, 3), \\ \Delta\varphi(s) &= \varphi^{(1)}(s) - \varphi^{(2)}(s), \\ \Delta D_3(s) &= D_3^{(1)}(s) - D_3^{(2)}(s). \end{aligned} \tag{2.12}$$

In the framework of the proposed numerical approach, the displacement jumps $\Delta u_i^{(m)}(s)$ on the crack surfaces are approximated by the method of least squares using finite element values at nodes $A^{(m)}$, $J^{(m)}$, and B $(m = 1, 2)$:

$$\begin{aligned} \Delta u_i(s) &= \alpha_i^{\Delta u} s + \beta_i^{\Delta u} \sqrt{-s}, \quad (i = 1, 3) \\ \Delta\varphi(s) &= \alpha^{\Delta\varphi} s + \beta^{\Delta\varphi} \sqrt{-s}, \quad -\delta \leqslant s \leqslant 0 \end{aligned} \tag{2.13}$$

Potentials $\varphi(s)$ ahead of the crack front are approximated in the following form:

$$\varphi(s) = \alpha^{\varphi} + \beta^{\varphi} s + \gamma^{\varphi} \sqrt{s} \quad (0 \leqslant s \leqslant \delta) \tag{2.14}$$

with the use of appropriate values at the nodes B, D, and H. The stresses $\sigma_{i3}(s)$ and potentials $D_3(s)$ ahead of the crack front are approximated in the form

$$\begin{aligned} \sigma_{i3}(s) &= \alpha_i^{\sigma} + \beta_i^{\sigma} s + \gamma_i^{\sigma} \sqrt{s^{-1}} \quad (i = 1, 3) \\ D_3(s) &= \alpha^{D} + \beta^{D} s + \gamma^{D} \sqrt{s^{-1}} \quad (0 \leqslant s \leqslant \delta) \end{aligned} \tag{2.15}$$

with the use of values at the points C, D, and E. These values are obtained by averaging corresponding data from the two nearest integration points—nodes of the Gauss cubature formula of the third order. In Fig. 2.2, these nodes are denoted by squares. It should be noted that in FEM, the stresses and electrical displacements are calculated just at these points. The electrical displacements $D_3(s)$ on the crack faces are approximated in the form analogous to (2.15) (with replacement of s by $-s$) and use of the data at points $I^{(m)}$, $J^{(m)}$, and $K^{(m)}$ $(m = 1, 2)$. These data are also obtained from the nearest integration points, which does not introduce significant errors since the distances between mentioned points and the nearest integration points are small, as well as the variations of data in the direction normal to the crack faces are also small. It is worth noting that the points D, $I^{(m)}$, and $J^{(m)}$ $(m = 1, 2)$ coincide with the nodes of corresponding finite elements.

Because on all finite elements adjacent to the crack tip, the space is compressed in the radial direction to reflect the square root asymptotic behavior of displacements and electric potential, the x_1 coordinates of the points C, D, and E are equal to $0.25\delta(1 - \sqrt{0.6})^2$, 0.25δ, and $0.25\delta(1 + \sqrt{0.6})^2$, respectively, and the coordinates of the points $I^{(m)}$, $J^{(m)}$, and $K^{(m)}$ $(m = 1, 2)$ are $-0.25\delta(1 + \sqrt{0.6})^2$, -0.25δ, and $-0.25\delta(1 - \sqrt{0.6})^2$, respectively.

This scheme was implemented in the environment of Mathematica software. The comparison of solutions for test problems with available analytical solutions demonstrated satisfactory accuracy of the obtained results. The errors of numerical results in all considered examples did not exceed 0.5%.

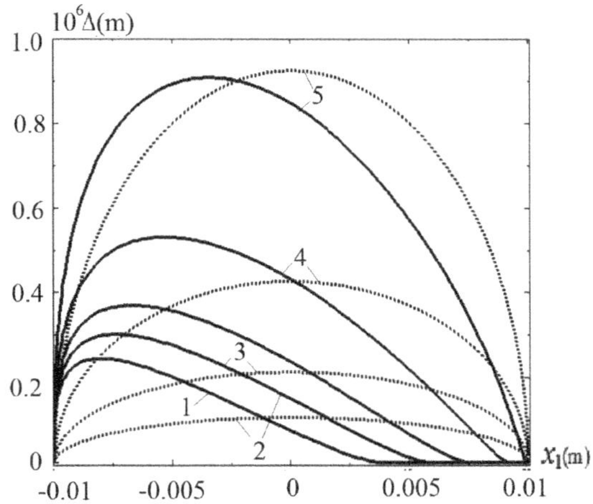

Fig. 2.3 The variation of fully conductive crack opening Δ along the crack region for $E_1^{\infty} = -2 \cdot 10^6$ V/m (solid lines) and $E_1^{\infty} = 0$ V/m (points) for different stress σ_{33}^{∞} levels: 1—$\sigma_{33}^{\infty} = 0$ MPa, 2—$\sigma_{33}^{\infty} = 0.25$ MPa, 3—$\sigma_{33}^{\infty} = 0.5$ MPa, 4—$\sigma_{33}^{\infty} = 1.0$ MPa, 5—$\sigma_{33}^{\infty} = 2.0$ MPa

2.4 Numerical Results

All the results below refer to the bimaterial PZT-4/PZT-5H. As a base case, the situation is considered where the electrode located at the interface is completely exfoliated along its entire length. Therefore, the resulting crack can be considered electrically conductive. An analytical solution for this case was given in [8]. The variation of the crack opening Δ along the crack region is presented in Fig. 2.3 for different values of σ_{33}^{∞} and E_1^{∞}. As is seen, there is a contact interaction between the crack faces due to the action of the electric field. Obtained lengths of contact zone are in good agreement with exact values in [8].

Obtained results demonstrate a significant influence of the electric field on the deformation character. This is due to the fact that free charges induced on the faces of an electrically conductive crack in a parallel electric field qualitatively change the type of the field in the vicinity of the crack (tangential component of the electric field is canceled on the crack faces). In addition, the type of stress-strain state in this zone changes drastically due to the occurrence of shear deformations caused by the piezoelectric effect.

The formation of contact zones between the faces of an electrically conductive crack takes place due to a combination of two factors—the presence of an electric field E_1^{∞} and the difference in the electromechanical properties of the piezomaterials composing the bimaterial, which causes deformations of the crack faces to be different. Figure 2.4 shows comparative crack openings for bimaterial PZT-4/PZT-5H and homogeneous material PZT-4, demonstrating no contact interaction between the faces of an electrically conductive crack in a homogeneous material.

Table 2.2 presents the variation of the ERR in the tips of a fully electrically conductive crack with respect to the levels of electromechanical load. It is worth noting that the crack opening for bimaterial takes place even in the absence of mechanical load, while in the case of a homogeneous piezomaterial under the action of only an electric

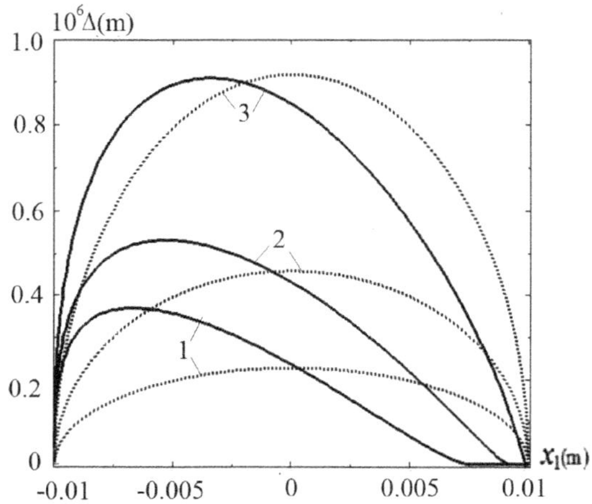

Fig. 2.4 The variation of fully conductive crack opening Δ along the crack region for homogeneous material PZT-4 (points) and PZT-4/PZT-5H bimaterial (solid lines) for different stress σ_{33}^{∞} levels and $E_1^{\infty} = -2 \cdot 10^6$ V/m: 1—$\sigma_{33}^{\infty} = 0.5$ MPa, 2—$\sigma_{33}^{\infty} = 1.0$ MPa, 3—$\sigma_{33}^{\infty} = 2.0$ MPa

Table 2.2 The variation of the ERR in the tips of a fully electrically conductive crack for some levels of electromechanical load

E_1^{∞} (V/m)	σ_{33}^{∞} (MPa)	$G^M(A)$ (N/m)	$G^E(A)$ (N/m)	$G^M(B)$ (N/m)	$G^E(B)$ (N/m)
0	0.25	0.02	0.00	0.02	0.00
0	0.5	0.08	0.00	0.08	0.00
0	1.0	0.33	0.00	0.33	0.00
0	2.0	1.34	0.00	1.34	0.00
$-2 \cdot 10^6$	0.0	6.35	1044.1	0.09	1050.1
$-2 \cdot 10^6$	0.25	7.16	1042.5	0.00	1051.2
$-2 \cdot 10^6$	0.5	8.05	1040.8	−0.10	1052.2
$-2 \cdot 10^6$	1.0	10.03	1037.4	−0.35	1054.5
$-2 \cdot 10^6$	2.0	14.61	1030.4	−1.09	1059.7

field, the crack opening is not observed—there is only mutual sliding of the crack faces. For homogeneous material PZT-4 and $E_1^{\infty} = -2 \cdot 10^6$ V/m, the ERRs have the following values: $G^M(A) = G^M(B) = 0.17$ N/m, $G^E(A) = G^E(B) = 721.0$ N/m. The data shown in Table 2.2 demonstrate a significant increase in the contribution of the electrical component to the magnitude of ERR for a fully conductive crack. There is also a large difference in the values of mechanical components of the ERR. In the case of application of the fracture criterion based on the consideration of only mechanical component G^M of the ERR [3, 4], when $E_1^{\infty} \neq 0$ the tip B is less dangerous because $G^M(B) < 0$, so that there is a deficit of potential energy required for crack growth.

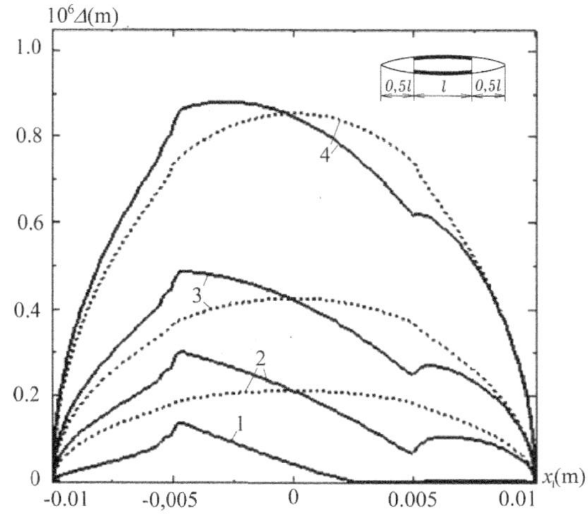

Fig. 2.5 The variation of opening Δ of the crack electroded on the interval $(-0,5l; 0,5l)$ for $E_1^\infty = -10^6$ V/m (solid lines) and $E_1^\infty = 0$ V/m (points) for different stress σ_{33}^∞ levels: 1—$\sigma_{33}^\infty = 0$ MPa, 2—$\sigma_{33}^\infty = 0.5$ MPa, 3—$\sigma_{33}^\infty = 1.0$ MPa, 4—$\sigma_{33}^\infty = 2.0$ MPa

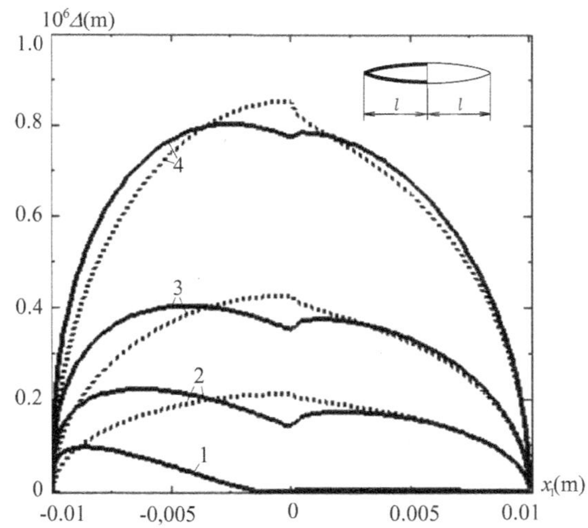

Fig. 2.6 The variation of opening Δ of the crack electroded on the interval $(-l; 0)$ for $E_1^\infty = -10^6$ V/m (solid lines) and $E_1^\infty = 0$ (points) for different stress σ_{33}^∞ levels: 1—$\sigma_{33}^\infty = 0$ MPa, 2—$\sigma_{33}^\infty = 0.5$ MPa, 3—$\sigma_{33}^\infty = 1.0$ MPa, 4—$\sigma_{33}^\infty = 2.0$ MPa

Further, some possible variants of a partially electrically conductive crack resulting from electrode delamination were considered. Figures 2.5, 2.6, 2.7, 2.8, 2.9, and 2.10 and Tables 2.3, 2.4, 2.5, 2.6, 2.7, and 2.8 show the results of these studies for deformation and fracture parameters.

As is seen from the data presented in Figs. 2.5, 2.6, 2.7, 2.8, 2.9, and 2.10, the smoothness of the crack faces is violated at the boundaries between electrically conductive and electrically insulated parts, which indicates the presence of singularities of electromechanical fields at these points. From the data presented in Tables 2.5, 2.6,

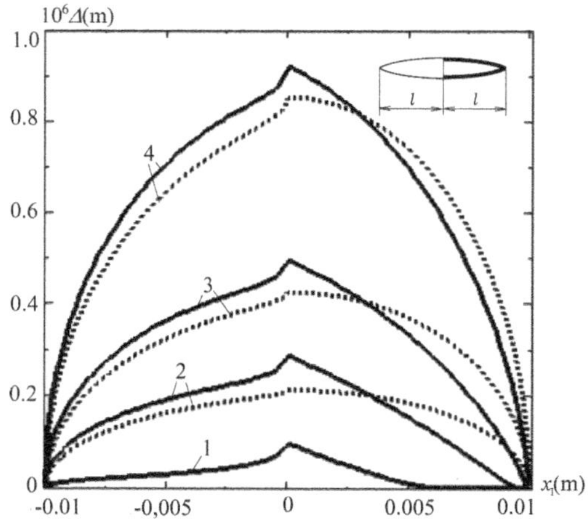

Fig. 2.7 The variation of opening Δ of the crack electroded on the interval (0; l) for $E_1^\infty = -10^6$ V/m (solid lines) and $E_1^\infty = 0$ (points) for different stress σ_{33}^∞ levels: 1—$\sigma_{33}^\infty = 0$ MPa, 2—$\sigma_{33}^\infty = 0.5$ MPa, 3—$\sigma_{33}^\infty = 1.0$ MPa, 4—$\sigma_{33}^\infty = 2.0$ MPa

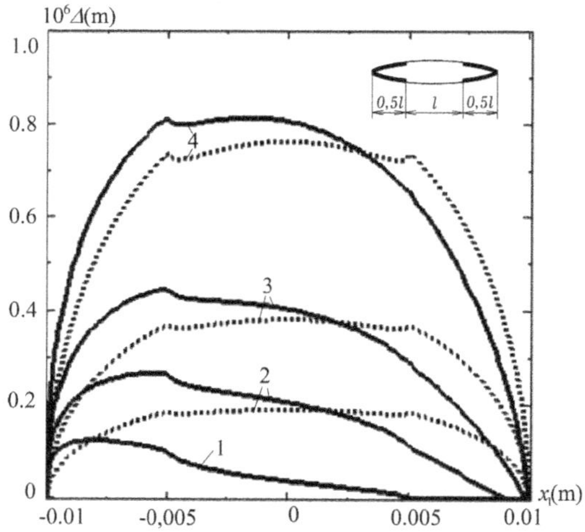

Fig. 2.8 The variation of opening Δ of the crack electroded on the intervals ($-l$; $-0.5l$) and ($0.5l$; l) for $E_1^\infty = -10^6$ V/m (solid lines) and $E_1^\infty = 0$ V/m (points) for different stress σ_{33}^∞ levels: 1—$\sigma_{33}^\infty = 0$ MPa, 2—$\sigma_{33}^\infty = 0.5$ MPa, 3—$\sigma_{33}^\infty = 1.0$ MPa, 4—$\sigma_{33}^\infty = 2.0$ MPa

2.7, and 2.8, it follows that the highest values of the electrical part of the ERR are observed when the electrode edge coincides with the crack tip. The highest values of the mechanical part of the ERR are observed when the conductivity conditions are satisfied on both sides of the crack tip.

2.5 Conclusions

The case of a multilayer electrode of finite length, completely or partially exfoliated along its length together with some part of a piezoelectric bimaterial interface, is considered. The exfoliation results in forming a partially conductive interfacial crack,

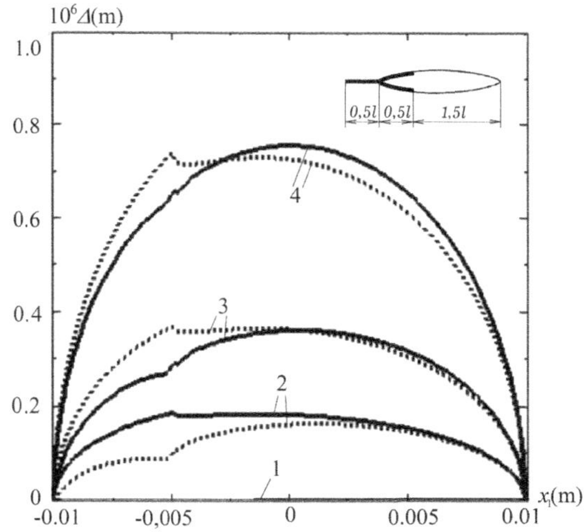

Fig. 2.9 The variation of opening Δ of the crack electroded on the interval $(-1.5l; -0.5l)$ for $E_1^\infty = -10^6$ V/m (solid lines) and $E_1^\infty = 0$ V/m (points) for different stress σ_{33}^∞ levels: 1—$\sigma_{33}^\infty = 0$ MPa, 2—$\sigma_{33}^\infty = 0.5$ MPa, 3—$\sigma_{33}^\infty = 1.0$ MPa, 4—$\sigma_{33}^\infty = 2.0$ MPa

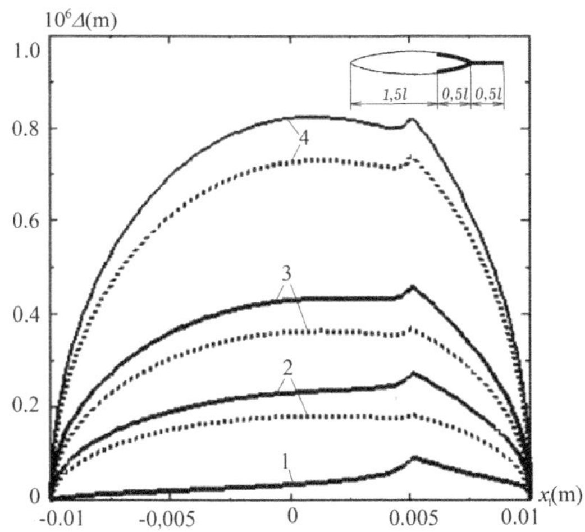

Fig. 2.10 The variation of opening Δ of the crack electroded on the interval $(0.5l; 1.5l)$ for $E_1^\infty = -10^6$ V/m (solid lines) and $E_1^\infty = 0$ V/m (points) for different stress σ_{33}^∞ levels: 1—$\sigma_{33}^\infty = 0$ MPa, 2—$\sigma_{33}^\infty = 0.5$ MPa, 3—$\sigma_{33}^\infty = 1.0$ MPa, 4—$\sigma_{33}^\infty = 2.0$ MPa

Table 2.3 The variation of the ERR in the tips of the crack electroded on the interval $(-0, 5l; 0, 5l)$ with respect to the levels of electromechanical load

E_1^∞ (V/m)	σ_{33}^∞ (MPa)	$G^M(A)$ (N/m)	$G^E(A)$ (N/m)	$G^M(B)$ (N/m)	$G^E(B)$ (N/m)
0	0.5	0.08	0.00	0.08	0.00
0	1.0	0.30	0.02	0.30	0.02
0	2.0	1.21	0.07	1.21	0.07
-10^6	0.0	0.00	0.06	0.00	−0.03
-10^6	0.5	0.11	0.26	0.05	−0.43
-10^6	1.0	0.38	0.49	0.27	−0.99
-10^6	2.0	1.39	0.95	1.17	−2.09

Table 2.4 The variation of the ERR in the tips of the crack electroded on the interval $(-l;\ 0)$ with respect to the levels of electromechanical load

E_1^∞ (V/m)	σ_{33}^∞ (MPa)	$G^M(A)$ (N/m)	$G^E(A)$ (N/m)	$G^M(B)$ (N/m)	$G^E(B)$ (N/m)
0	0.5	0.08	0.00	0.07	0.00
0	1.0	0.33	0.00	0.29	0.02
0	2.0	1.34	0.00	1.14	0.06
-10^6	0.0	−1.35	188.40	0.39	2.27
-10^6	0.5	−0.59	186.83	0.43	1.59
-10^6	1.0	0.39	185.18	0.62	0.92
-10^6	2.0	2.83	181.90	1.48	−0.40

Table 2.5 The variation of the ERR in the tips of the crack electroded on the interval $(0;\ l)$ with respect to the levels of electromechanical load

E_1^∞ (V/m)	σ_{33}^∞ (MPa)	$G^M(A)$ (N/m)	$G^E(A)$ (N/m)	$G^M(B)$ (N/m)	$G^E(B)$ (N/m)
0	0.5	0.07	0.00	0.08	0.00
0	1.0	0.29	0.02	0.33	0.00
0	2.0	1.14	0.06	1.33	0.01
-10^6	0.0	0.38	1.34	−2.49	189.57
-10^6	0.5	0.49	1.56	−2.57	190.69
-10^6	1.0	0.76	1.80	−2.70	192.02
-10^6	2.0	1.77	2.31	−3.10	195.26

Table 2.6 The variation of the ERR in the tips of the crack electroded on the intervals $(-l;\ -0.5l)$ and $(0.5l;\ l)$ with respect to the levels of electromechanical load

E_1^∞ (V/m)	σ_{33}^∞ (MPa)	$G^M(A)$ (N/m)	$G^E(A)$ (N/m)	$G^M(B)$ (N/m)	$G^E(B)$ (N/m)
0	0.5	0.08	0.00	0.08	0.00
0	1.0	0.33	0.00	0.33	0.00
0	2.0	1.34	0.00	1.33	0.00
-10^6	0.0	1.60	264.65	0.03	265.26
-10^6	0.5	2.51	261.86	−0.08	266.45
-10^6	1.0	3.65	260.02	−0.25	267.84
-10^6	2.0	6.43	256.35	−0.83	271.33

whose plane strain state under mechanical and electric loading has been analyzed using FEM.

The crack openings are presented graphically in Figs. 2.3, 2.4, 2.5, 2.6, 2.7, 2.8, 2.9, and 2.10. It follows from the obtained results that the smoothness of the crack faces is violated at the boundaries between electrically conductive and insulated zones, indicating the presence of singularities of electromechanical fields at these points. The ERRs at the crack tips are presented in Tables 2.2, 2.3, 2.4, 2.5, 2.6, 2.7,

Table 2.7 The variation of the ERR in the tips of the crack electroded on the interval $(-1.5l; -0.5l)$ with respect to the levels of electromechanical load

E_1^∞ (V/m)	σ_{33}^∞ (MPa)	$G^M(A)$ (N/m)	$G^E(A)$ (N/m)	$G^M(B)$ (N/m)	$G^E(B)$ (N/m)
0	0.5	0.08	0.00	0.07	0.00
0	1.0	0.34	0.00	0.27	0.01
0	2.0	1.35	0.00	1.10	0.04
-10^6	0.0	9.97	0.00	0.41	2.67
-10^6	0.5	10.00	0.00	0.46	2.13
-10^6	1.0	10.22	0.00	0.65	1.41
-10^6	2.0	11.16	0.00	1.49	−0.02

Table 2.8 The variation of the ERR in the tips of the crack electroded on the interval $(0.5l; 1.5l)$ with respect to the levels of electromechanical load

E_1^∞ (V/m)	σ_{33}^∞ (MPa)	$G^M(A)$ (N/m)	$G^E(A)$ (N/m)	$G^M(B)$ (N/m)	$G^E(B)$ (N/m)
0	0.5	0.07	0.00	0.08	0.00
0	1.0	0.34	0.01	0.27	0.00
0	2.0	1.35	0.04	1.10	0.00
-10^6	0.0	0.42	1.67	10.00	0.00
-10^6	0.5	0.53	1.92	10.12	0.00
-10^6	1.0	0.80	2.16	10.39	0.00
-10^6	2.0	1.79	2.68	11.43	0.00

and 2.8 for different values of external mechanical and electric fields and different positions of the points dividing the electrically conductive and insulated zones of the crack. The highest values of the electrical part of the ERR are observed when the electrode edge coincides with the crack tip. The highest values of the mechanical part of the ERR are observed when the conductivity conditions are satisfied on both sides of the crack tip.

References

1. Fett, T., Kamlah, M., Munz, D., Thun, G.: Crack resistance and fracture toughness of PZT ceramics. Proc. SPIE **4333**, 221–230 (2001)
2. Fang, D., Liu, J.: Fracture Mechanics of Piezoelectric and Ferroelectric Solids. Springer-Verlag, Berlin (2013)
3. Park, S.B., Sun, C.T.: Fracture criteria for piezoelectric ceramics. J. Am. Ceram. Soc. **78**(6), 1475–1480 (1995)
4. Park, S.B., Sun, C.T.: Effect of electric field on fracture of piezoelectric ceramics. Int. J. Fract. **70**, 203–216 (1995)
5. Herrmann, K.P., Loboda, V.V., Govorukha, V.B.: On contact zone models for an electrically impermeable interface crack in a piezoelectric bimaterial. Int. J. Fract. **111**, 203–227 (2001)

6. Beom, H.G., Atluri, S.N.: Conducting cracks in dissimilar piezoelectric media. Int. J. Fract. **118**, 285–301 (2002)
7. Lapusta, Y., Komarov, A., Labesse-Jied, F., Moutou Pitti, R., Loboda, V.: Limited permeable crack moving along the interface of a piezoelectric bi-material. Eur. J. Mech.-A/Solids **30** (5), 639–649 (2011)
8. Loboda, V., Sheveleva, A., Lapusta, Y.: An electrically conducting interface crack with a contact zone in a piezoelectric bimaterial. Int. J. Solids Struct. **51**(1), 63–73 (2014)
9. Zienkiewicz, O.C., Taylor, R.L.: The Finite Element Method for Solid and Structural Mechanics. Elsevier, Oxford (2005)
10. Kumar, S., Singh, R.N.: Crack propagation in piezoelectric materials under combined mechanical and electrical loadings. Acta Mater. **44**(1), 173–200 (1996)
11. Kumar, S., Singh, R.N.: Influence of applied electric field and mechanical boundary condition on the stress distribution at the crack tip in piezoelectric materials. Mater. Sci. Eng. **231**(1–2), 1–9 (1997)
12. Kumar, S., Singh, R.N.: Effect of the mechanical boundary condition at the crack surfaces on the stress distribution at the crack tip in piezoelectric materials. Mater. Sci. Eng. **252**(1), 64–77 (1998)
13. Kuna, M.: Finite element analyses of crack problems in piezoelectric structures. Comput. Mater. Sci. **13**(1–3), 67–80 (1998)
14. Kuna, M.: Finite element analyses of cracks in piezoelectric structures. Key Eng. Mater. **348–349**, 629–632 (2007)
15. Rao, B.N., Kuna, M.: Interaction integrals for fracture analysis of functionally graded piezoelectric materials. Int. J. Solids Struct. **45**(20), 5237–5257 (2008)
16. Gruebner, O., Kamlah, M., Munz, D.: Finite element analysis of cracks in piezoelectric materials taking into account the permittivity of the crack medium. Eng. Fract. Mech. **70**(11), 1399–1413 (2003)
17. Rybicki, E.F., Kanninen, M.F.: A finite element calculation of stress intensity factors by a modified crack closure integral. Eng. Fract. Mech. **9**(4), 931–938 (1977)
18. Benkaci, N., Maugin, G.: *J* integral computation for piezo-ceramics. Rev. Eur. Élém. Finis **10**(1), 99–128 (2001)
19. Cun-Fa, G., Minghao, Z., Pin, T., Tong-Yi, Z.: The energy release rate and the J-integral of an electrically insulated crack in a piezoelectric material. Int. J. Eng. Sci. **42**(19–20), 2175–2192 (2004)
20. McMeeking, R.M.: The energy release rate for a Griffith crack in a piezoelectric material. Eng. Fract. Mech. **71**(7–8), 1149–1163 (2004)
21. Adlucky, V.J., Loboda, V.V.: Finite element analysis of elastoplastic state of the plane with elliptic inclusion in the presence of interphase crack. Math. Mech. Phys. Fields **63**(1), 65–74 (2020)
22. Ma, P., Su, R.K.L., Feng, W.J.: Fracture analysis of an electrically conductive interface crack with a contact zone in a magnetoelectroelastic bimaterial system. Int. J. Solids Struct. **53**, 48–57 (2015)
23. Lapusta, Y., Onopriienko, O., Loboda, V.: An interface crack with partially electrically conductive crack faces under antiplane mechanical and in-plane electric loadings. Mech. Res. Commun. **81**, 38–43 (2017)
24. Parton, V.Z., Kudryavtsev, B.A.: Electromagnetoelasticity. Gordon and Breach Science Publishers, New York (1988)

Chapter 3
Asymptotic Behavior of Forced Vibrations of Layered Plates

Lenser Aghalovyan, Mher Aghalovyan, and Tatevik Zakaryan

Abstract This article delves into dynamic three-dimensional elasticity problems specifically tailored for layered plates. In contrast to classical and existing refined theories of plates and shells, which prove unsuitable for this particular class of issues, our investigation led to the discovery of fundamentally different asymptotics for stress tensor components and displacement vectors. This breakthrough enabled the formulation of an asymptotic solution to address the intricacies of the 3D problem. The study identifies specific scenarios where the derived solution attains mathematical exactness, providing a comprehensive understanding of the conditions under which the proposed approach proves particularly effective. By exploring these dynamic 3D elasticity problems with layered plates, this work contributes novel insights and solutions, extending beyond the limitations of classical plate theories.

3.1 Introduction

Several methodologies were employed to tackle dynamic elasticity problems, including the Wiener-Hopf and Smirnova-Sobolev methods (Poruchikov 1986). A numerical-analytical approach was also developed for problem-solving in plates and shells [4, 5]. A comprehensive review [12] covers experimental exploration of the dynamics of inhomogeneous shells of revolution. The asymptotic method was used to effectively solve spatial dynamic problems in plates and shells by solving singularly perturbed differential equations.

The initial applications of the asymptotic method for solving three-dimensional problems in isotropic plates and shells date back to pioneering works [6, 7, 9, 10].

L. Aghalovyan (✉) · M. Aghalovyan · T. Zakaryan
Institute of Mechanics of National Academy of Sciences of Armenia, Yerevan, Armenia
e-mail: lagal@sci.am

T. Zakaryan
e-mail: zaqaryantatevik@mail.ru

M. Aghalovyan
Armenian State University of Economics, Yerevan, Armenia

H. Altenbach et al. (eds.), *Selected Problems of Solid Mechanics and Solving Methods*,
Advanced Structured Materials 204, https://doi.org/10.1007/978-3-031-54063-9_3

The solution to singularly perturbed equations and systems comprises two distinct components: the solution to the outer problem (I^{out}) and the solution to the boundary layer (I_b)

$$I = I^{\text{out}} + I_b \tag{3.1}$$

These solutions can be independently constructed [1, 7]. The solution to the external problem not only satisfies the equations of motion (equilibrium) and elasticity relations but also adheres to the boundary conditions on the facial surfaces of plates and shells (external conditions). This solution holds beyond a certain distance from the side surface inside the plate or shell [1, 7]. The solution to the outer problem is expressed as

$$I^{\text{out}} = \varepsilon^{q_i+s} I^{(s)}, \quad s = 0, \ldots, S, \tag{3.2}$$

where s denotes summation by repeating (umbral) index s over integer values from zero to the number of approximations S (Einstein's notation).

In regularly perturbed equations, $q_i = 0$; however, in singularly perturbed equations, the value q_i depends on desired functions and the type of boundary conditions on the facial surfaces. For an isotropic plate with conditions of the first boundary value problem of elasticity theory on the facial surfaces (i.e., values of stress tensor components $\sigma_{xz}^{\pm}$, $\sigma_{yz}^{\pm}$, and $\sigma_{zz}^{\pm}$), the static problem yields [1, 7]

$$\begin{aligned} q_I &= -2 \text{ for } \sigma_{xx}, \sigma_{xy}, \sigma_{yy}, \\ q_I &= -1 \text{ for } \sigma_{xz}, \sigma_{yz}, \\ q_I &= 0 \quad \text{for } \sigma_{zz}, \\ q_I &= -2 \text{ for } u, v, \\ q_I &= -3 \text{ for } w \end{aligned} \tag{3.3}$$

This asymptotic behavior also holds for anisotropic plates and shells [1]. For a long time, only the first boundary value problem of elasticity theory for plates and shells was solved. However, it was discovered that the Kirchhoff–Love hypothesis of the classical theory of plates and shells and well-known refined theories are not applicable when conditions of the second boundary value problem (displacements are given) or mixed conditions are imposed on the facial surfaces. The asymptotic method enabled the solution of these boundary value problems. In the case of the second or mixed boundary value problems, the author established [1, 2]:

$$\begin{aligned} q_I &= -1 \quad \text{for all } \sigma_{ij}, \quad i, j = x, y, z, \\ q_I &= 0 \quad \text{for } u, v, w. \end{aligned} \tag{3.4}$$

The asymptotic (3.4), besides static problems, is also valid for solving 3D dynamic problems. Some classes of dynamic problems of anisotropic plates are solved using the asymptotic method [1–3].

In this paper, the first and mixed 3D boundary value problems for layered orthotropic plates are solved, as well as a nonclassical boundary value problem for a layered package of plates that arises in seismology. The cases when the found asymptotic solution becomes mathematically exact are indicated.

3.2 Basic Equations and the Formulation of Problems

It is required to find in the area

$$D = \{x, y, z : 0 \leqslant x \leqslant a,\ 0 \leqslant y \leqslant b,\ 0 \leqslant z \leqslant h,\ h_1 + h_2 + \cdots + h_N = h \ll l\,;\, l = \min(a, b)\}$$

which is occupied by layered plate (Fig. 3.1), the solution to equations of motion

$$\begin{aligned}
&\frac{\partial \sigma_{xx}^{(k)}}{\partial x} + \frac{\partial \sigma_{xy}^{(k)}}{\partial y} + \frac{\partial \sigma_{xz}^{(k)}}{\partial z} = \rho^k \frac{\partial^2 u^{(k)}}{\partial t^2},\\
&\frac{\partial \sigma_{xy}^{(k)}}{\partial x} + \frac{\partial \sigma_{yy}^{(k)}}{\partial y} + \frac{\partial \sigma_{yz}^{(k)}}{\partial z} = \rho^k \frac{\partial^2 v^{(k)}}{\partial t^2},\\
&\frac{\partial \sigma_{xz}^{(k)}}{\partial x} + \frac{\partial \sigma_{yz}^{(k)}}{\partial y} + \frac{\partial \sigma_{zz}^{(k)}}{\partial z} = \rho^k \frac{\partial^2 w^{(k)}}{\partial t^2}, \quad k = 1, 2, \ldots, N
\end{aligned} \tag{3.5}$$

and relations of elasticity (generalized Hooke's law) for the orthotropic body

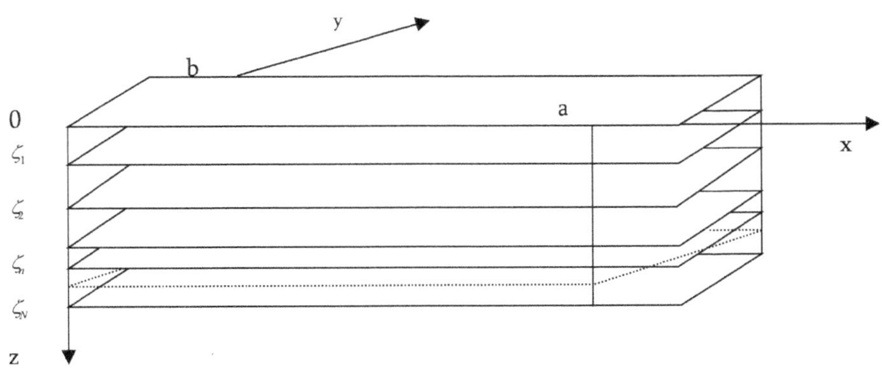

Fig. 3.1 Configuration of layered plate

$$
\begin{aligned}
\frac{\partial u^{(k)}}{\partial x} &= a_{11}^{(k)}\sigma_{xx}^{(k)} + a_{12}^{(k)}\sigma_{yy}^{(k)} + a_{13}^{(k)}\sigma_{zz}^{(k)},\\
\frac{\partial v^{(k)}}{\partial y} &= a_{12}^{(k)}\sigma_{xx}^{(k)} + a_{22}^{(k)}\sigma_{yy}^{(k)} + a_{23}^{(k)}\sigma_{zz}^{(k)},\\
\frac{\partial w^{(k)}}{\partial z} &= a_{13}^{(k)}\sigma_{xx}^{(k)} + a_{23}^{(k)}\sigma_{yy}^{(k)} + a_{33}^{(k)}\sigma_{zz}^{(k)},\\
\frac{\partial u^{(k)}}{\partial y} + \frac{\partial v^{(k)}}{\partial x} &= a_{66}^{(k)}\sigma_{xy}^{(k)},\\
\frac{\partial w^{(k)}}{\partial x} + \frac{\partial u^{(k)}}{\partial z} &= a_{55}^{(k)}\sigma_{xz}^{(k)},\\
\frac{\partial w^{(k)}}{\partial y} + \frac{\partial v^{(k)}}{\partial z} &= a_{44}^{(k)}\sigma_{yz}^{(k)},
\end{aligned}
\tag{3.6}
$$

at the following options of boundary conditions on facial surfaces of the package:

(a)

$$
\begin{aligned}
\sigma_{jz}^{I}(x, y, 0, t) &= \sigma_{jz}^{+}(\xi, \eta)\exp(i\Omega t),\\
\sigma_{jz}^{N}(x, y, h, t) &= \sigma_{jz}^{-}(\xi, \eta)\exp(i\Omega t), \quad j = x, y, z
\end{aligned}
\tag{3.7}
$$

(b)

$$
\begin{aligned}
\sigma_{jz}^{(1)}(x, y, 0, t) &= \sigma_{jz}^{+}(\xi, \eta)\exp(i\Omega t),\\
U(x, y, h, t) &= 0, \quad (U, V, W)
\end{aligned}
\tag{3.8}
$$

where $\xi = x/l$, $\eta = y/l$; Ω is the frequency of external influence; conditions of full contact between adjacent layers of the package reads

$$
\begin{aligned}
&\sigma_{\alpha z}^{(k)}(\xi, \eta, \zeta_k) = \sigma_{\alpha z}^{(k+1)}(\xi, \eta, \zeta_k), \quad \alpha = x, y, z, \quad k = 1, 2, \ldots, N-1\\
&u^{(k)}(\xi, \eta, \zeta_k) = u^{(k+1)}(\xi, \eta, \zeta_k), \quad (u, v, w)
\end{aligned}
\tag{3.9}
$$

The conditions on the lateral surface of the package have yet to be concretized. We perform them later. The solution of the formulated problem is sought in the form

$$
\begin{aligned}
&\sigma_{\alpha\beta}^{(k)}(x, y, z, t) = \sigma_{ij}^{(k)}(x, y, z)\exp(i\Omega t),\\
&\qquad \alpha, \beta = x, y, z, \quad i, j = 1, 2, 3, \quad k = 1, 2, \ldots, N\\
&\left(u^{(k)}(x, y, z, t), v^{(k)}(x, y, z, t), w^{(k)}(x, y, z, t)\right)\\
&\qquad = \left(u_x^{(k)}(x, y, z), u_y^{(k)}(x, y, z), u_z^{(k)}(x, y, z)\right)\exp(i\Omega t),
\end{aligned}
\tag{3.10}
$$

Substituting dimensionless coordinates and displacements

$$
\xi = x/l \quad \eta = y/l, \quad \zeta = z/h, \quad U = u_x/l, \quad V = u_y/l, \quad W = u_z/l \tag{3.11}
$$

into dynamic equations (3.5) and elasticity relations (3.6), we obtain the system singularly perturbed by a small parameter $\varepsilon = h/l$:

$$\begin{gathered}\frac{\partial\sigma_{11}^{(k)}}{\partial\xi}+\frac{\partial\sigma_{12}^{(k)}}{\partial\eta}+\varepsilon^{-1}\frac{\partial\sigma_{13}^{(k)}}{\partial\zeta}+\varepsilon^{-2}\rho^{(k)}\Omega_*^2U^{(k)}=0,\\ \frac{\partial\sigma_{12}^{(k)}}{\partial\xi}+\frac{\partial\sigma_{22}^{(k)}}{\partial\eta}+\varepsilon^{-1}\frac{\partial\sigma_{23}^{(k)}}{\partial\zeta}+\varepsilon^{-2}\rho^{(k)}\Omega_*^2V^{(k)}=0,\\ \frac{\partial\sigma_{13}^{(k)}}{\partial\xi}+\frac{\partial\sigma_{23}^{(k)}}{\partial\eta}+\varepsilon^{-1}\frac{\partial\sigma_{33}^{(k)}}{\partial\zeta}+\varepsilon^{-2}\rho^{(k)}\Omega_*^2W^{(k)}=0,\\ \Omega_*^2=h^2\Omega^2,\quad k=1,2,\ldots,N\end{gathered}\tag{3.12a}$$

$$\begin{aligned}\frac{\partial U^{(k)}}{\partial\xi}&=a_{11}^{(k)}\sigma_{11}^{(k)}+a_{12}^{(k)}\sigma_{22}^{(k)}+a_{13}^{(k)}\sigma_{33}^{(k)},\\ \frac{\partial V^{(k)}}{\partial\eta}&=a_{12}^{(k)}\sigma_{11}^{(k)}+a_{22}^{(k)}\sigma_{22}^{(k)}+a_{23}^{(k)}\sigma_{33}^{(k)},\\ \varepsilon^{-1}\frac{\partial W^{(k)}}{\partial\zeta}&=a_{13}^{(k)}\sigma_{11}^{(k)}+a_{23}^{(k)}\sigma_{22}^{(k)}+a_{33}^{(k)}\sigma_{33}^{(k)},\\ \frac{\partial V^{(k)}}{\partial\xi}+\frac{\partial U^{(k)}}{\partial\eta}&=a_{66}^{(k)}\sigma_{12}^{(k)},\\ \frac{\partial W^{(k)}}{\partial\xi}+\varepsilon^{-1}\frac{\partial U^{(k)}}{\partial\zeta}&=a_{55}^{(k)}\sigma_{13}^{(k)},\\ \frac{\partial W^{(k)}}{\partial\eta}+\varepsilon^{-1}\frac{\partial V^{(k)}}{\partial\zeta}&=a_{44}^{(k)}\sigma_{23}^{(k)},\end{aligned}\tag{3.12b}$$

The solution of system (3.12) consists of solutions to the external problem (I^{out}) and the boundary layer (I_b); $I=I^{\text{out}}+I_b$. The solution to the external problem we seek in the form

$$\begin{gathered}\sigma_{ij}^{(k)\text{out}}=\varepsilon^{-1+s}\sigma_{ij}^{(k,s)}(\xi,\eta,\zeta)\quad i,j=1,2,3,\quad s=0,\ldots,S,\\ (U^{(k)\text{out}},V^{(k)\text{out}},W^{(k)\text{out}})=\varepsilon^{s}(U^{(k,s)},V^{(k,s)},W^{(k,s)}),\quad k=1,\ldots,N.\end{gathered}\tag{3.13}$$

By substituting (3.13) into (3.12) and equating in each equation the coefficients at the same power ε, we will obtain the following consistent system for determining unknown functions $\sigma_{ij}^{(k,s)}$, $U^{(k,s)}$, $V^{(k,s)}$, and $W^{(k,s)}$:

$$\begin{gathered}\frac{\partial\sigma_{11}^{(k,s-1)}}{\partial\xi}+\frac{\partial\sigma_{12}^{(k,s-1)}}{\partial\eta}+\frac{\partial\sigma_{13}^{(k,s)}}{\partial\zeta}+\rho^k\Omega_*^2U^{(k,s)}=0,\\ \frac{\partial\sigma_{12}^{(k,s-1)}}{\partial\xi}+\frac{\partial\sigma_{22}^{(k,s-1)}}{\partial\eta}+\frac{\partial\sigma_{23}^{(k,s)}}{\partial\zeta}+\rho^k\Omega_*^2V^{(k,s)}=0,\\ \frac{\partial\sigma_{13}^{(k,s-1)}}{\partial\xi}+\frac{\partial\sigma_{23}^{(k,s-1)}}{\partial\eta}+\frac{\partial\sigma_{33}^{(k,s)}}{\partial\zeta}+\rho^k\Omega_*^2W^{(k,s)}=0,\end{gathered}\tag{3.14a}$$

$$
\begin{aligned}
\frac{\partial U^{(k,s-1)}}{\partial \xi} &= a_{11}^{(k)}\sigma_{11}^{(k,s)} + a_{12}^{(k)}\sigma_{22}^{(k,s)} + a_{13}^{(k)}\sigma_{33}^{(k,s)},\\
\frac{\partial V^{(k,s-1)}}{\partial \eta} &= a_{12}^{(k)}\sigma_{11}^{(k,s)} + a_{22}^{(k)}\sigma_{22}^{(k,s)} + a_{23}^{(k)}\sigma_{33}^{(k,s)},\\
\frac{\partial W^{k(s)}}{\partial \zeta} &= a_{13}^{(k)}\sigma_{11}^{(k,s)} + a_{23}^{(k)}\sigma_{22}^{(k,s)} + a_{33}^{(k)}\sigma_{33}^{(k,s)},\\
\frac{\partial V^{(k,s-1)}}{\partial \xi} + \frac{\partial U^{(k,s-1)}}{\partial \eta} &= a_{66}^{(k)}\sigma_{12}^{(k,s)},\\
\frac{\partial W^{(k,s-1)}}{\partial \xi} + \frac{\partial U^{(k,s)}}{\partial \zeta} &= a_{55}^{(k)}\sigma_{13}^{(k,s)},\\
\frac{\partial W^{(k,s-1)}}{\partial \eta} + \frac{\partial V^{(k,s)}}{\partial \zeta} &= a_{44}^{(k)}\sigma_{23}^{(k,s)},
\end{aligned}
\tag{3.14b}
$$

By utilizing (3.14), it is possible to express all stresses in terms of the displacements.

$$
\begin{aligned}
\sigma_{13}^{(k,s)} &= \frac{1}{0_{55}^{(k)}}\left(\frac{\partial U^{(k,s)}}{\partial \zeta} + \frac{\partial W^{(k,s-1)}}{\partial \xi}\right),\\
\sigma_{23}^{(k,s)} &= \frac{1}{a_{44}^{(k)}}\left(\frac{\partial V^{(k,s)}}{\partial \zeta} + \frac{\partial W^{(k,s-1)}}{\partial \eta}\right),\\
\sigma_{12}^{(k,s)} &= \frac{1}{a_{66}^{(k)}}\left(\frac{\partial U^{(k,s-1)}}{\partial \eta} + \frac{\partial V^{(k,s-1)}}{\partial \xi}\right),\\
\sigma_{11}^{(k,s)} &= \frac{1}{\Delta^{(k)}}\left(-A_{23}^{(k)}\frac{\partial W^{(k,s)}}{\partial \zeta} + A_{22}^{(k)}\frac{\partial U^{(k,s-1)}}{\partial \xi} - A_{12}^{(k)}\frac{\partial V^{(k,s-1)}}{\partial \eta}\right),\\
\sigma_{22}^{(k,s)} &= \frac{1}{\Delta^{(k)}}\left(-A_{13}^{(k)}\frac{\partial W^{(k,s)}}{\partial \zeta} - A_{12}^{(k)}\frac{\partial U^{(k,s-1)}}{\partial \xi} + A_{33}^{(k)}\frac{\partial V^{(k,s-1)}}{\partial \eta}\right),\\
\sigma_{33}^{(k,s)} &= \frac{1}{\Delta^{(k)}}\left(A_{11}^{(k)}\frac{\partial W^{(k,s)}}{\partial \zeta} - A_{23}^{(k)}\frac{\partial U^{(k,s-1)}}{\partial \xi} - A_{13}^{(k)}\frac{\partial V^{(k,s-1)}}{\partial \eta}\right),\\
Q^{(k,m)} &\equiv 0, \quad m < 0,
\end{aligned}
\tag{3.15}
$$

where

$$
\begin{aligned}
A_{11}^{(k)} &= a_{11}^{(k)}a_{22}^{(k)} - \left(a_{12}^{(k)}\right)^2, & A_{12}^{(k)} &= a_{12}^{(k)}a_{33}^{(k)} - a_{23}^{(k)}a_{13}^{(k)},\\
A_{13}^{(k)} &= a_{11}^{(k)}a_{23}^{(k)} - a_{13}^{(k)}a_{12}^{(k)}, & A_{22}^{(k)} &= a_{22}^{(k)}a_{33}^{(k)} - \left(a_{23}^{(k)}\right)^2,\\
A_{23}^{(k)} &= a_{13}^{(k)}a_{22}^{(k)} - a_{12}^{(k)}a_{23}^{(k)}, & A_{33}^{(k)} &= a_{11}^{(k)}a_{33}^{(k)} - \left(a_{13}^{(k)}\right)^2,\\
& & \Delta^{(k)} &= a_{11}^{(k)}A_{22}^{(k)} - a_{12}^{(k)}A_{12}^{(k)} - a_{13}^{(k)}A_{23}^{(k)}
\end{aligned}
\tag{3.16}
$$

By substituting functions $\sigma_{13}^{(k,s)}$, $\sigma_{23}^{(k,s)}$, and $\sigma_{22}^{(k,s)}$ into the first three equations (3.14), for determining $U^{(k,s)}$, $V^{(k,s)}$, and $W^{(k,s)}$, we obtain

$$\frac{\partial^2 U^{(k,s)}}{\partial \zeta^2} + a_{55}^{(k)} \rho^{(k)} \Omega_*^2 U^{(k,s)} = R_U^{(k,s)}, \tag{3.17}$$

$$R_U^{(k,s)} = -a_{55}^{(k)} \left(\frac{\partial \sigma_{11}^{(k,s-1)}}{\partial \xi} + \frac{\partial \sigma_{12}^{(k,s-1)}}{\partial \eta} \right) - \frac{\partial^2 W^{(k,s-1)}}{\partial \xi \partial \zeta},$$

$$\frac{\partial^2 V^{(k,s)}}{\partial \zeta^2} + a_{44}^{(k)} \rho^{(k)} \Omega_*^2 V^{(k,s)} = R_V^{(k,s)}, \tag{3.18}$$

$$R_V^{k(s)} = -a_{44}^{(k)} \left(\frac{\partial \sigma_{12}^{(k,s-1)}}{\partial \xi} + \frac{\partial \sigma_{22}^{(k,s-1)}}{\partial \eta} \right) - \frac{\partial^2 W^{(k,s-1)}}{\partial \eta \partial \zeta},$$

$$A_{11}^{(k)} \frac{\partial^2 W^{(k,s)}}{\partial \zeta^2} + \Delta^{(k)} \rho^{(k)} \Omega_*^2 W^{(k,s)} = R_W^{(k,s)}, \tag{3.19}$$

$$R_W^{(k,s)} = -\Delta^{(k)} \left(\frac{\partial \sigma_{13}^{(k,s-1)}}{\partial \xi} + \frac{\partial \sigma_{23}^{(k,s-1)}}{\partial \eta} \right) + A_{23}^{(k)} \frac{\partial^2 U^{(k,s-1)}}{\partial \xi \partial \zeta} + A_{13}^{(k)} \frac{\partial^2 V^{(k,s-1)}}{\partial \eta \partial \zeta}.$$

The solution to these equations reads

$$\begin{aligned}
U^{(k,s)} &= C_1^{(k,s)}(\xi, \eta) \sin \gamma_1^{(k)} \zeta + C_2^{(k,s)}(\xi, \eta) \cos \gamma_1^{(k)} \zeta \\
&\quad + U_\tau^{(k,s)}(\xi, \eta, \zeta), \quad \gamma_1^{(k)} = \Omega_* \sqrt{\rho^{(k)} a_{55}^{(k)}}, \\
V^{(k,s)} &= C_3^{(k,s)}(\xi, \eta) \sin \gamma_2^{(k)} \zeta + C_4^{(k,s)}(\xi, \eta) \cos \gamma_2^{(k)} \zeta \\
&\quad + V_\tau^{(k,s)}(\xi, \eta, \zeta), \quad \gamma_2^{(k)} = \Omega_* \sqrt{\rho^{(k)} a_{44}^{(k)}}, \\
W^{(k,s)} &= C_5^{(k,s)}(\xi, \eta) \sin \gamma_3^{(k)} \zeta + C_6^{(k,s)}(\xi, \eta) \cos \gamma_3^{(k)} \zeta \\
&\quad + W_\tau^{(k,s)}(\xi, \eta, \zeta), \quad \gamma_3^{(k)} = \Omega_* \sqrt{\rho^{(k)} \Delta^{(k)} / A_{11}^{(k)}},
\end{aligned} \tag{3.20}$$

where ${U_\tau}^{(k,s)}$, ${V_\tau}^{(k,s)}$, and ${W_\tau}^{(k,s)}$ are particular solutions of the corresponding equations. By substituting $U^{(k,s)}$, $V^{(k,s)}$, and $W^{(k,s)}$ in (3.15) for stresses ${\sigma_{13}}^{(k,s)}$, ${\sigma_{23}}^{(k,s)}$, and ${\sigma_{33}}^{(k,s)}$, we get

$$\begin{aligned}
\sigma_{13}^{(k,s)} &= \Omega_* \sqrt{\rho^{(k)} / a_{55}^{(k)}} \Big(C_1^{(k,s)}(\xi, \eta) \cos \gamma_1^{(k)} \zeta \\
&\quad - C_2^{(k,s)}(\xi, \eta) \sin \gamma_1^{(k)} \zeta \Big) + f_{13}^{(k,s)}(\xi, \eta, \zeta), \\
\sigma_{23}^{(k,s)} &= \Omega_* \sqrt{\rho^{(k)} / a_{44}^{(k)}} \Big(C_3^{(k,s)}(\xi, \eta) \cos \gamma_2^{(k)} \zeta \\
&\quad - C_4^{(k,s)}(\xi, \eta) \sin \gamma_2^{(k)} \zeta \Big) + f_{23}^{(k,s)}(\xi, \eta, \zeta), \\
\sigma_{33}^{(k,s)} &= \Omega_* \sqrt{\rho^{(k)} A_{11}^{(k)} / \Delta^{(k)}} \Big(C_5^{(k,s)}(\xi, \eta) \cos \gamma_3^{(k)} \zeta \\
&\quad - C_6^{(k,s)}(\xi, \eta) \sin \gamma_3^{(k)} \zeta \Big) + f_{33}^{(k,s)}(\xi, \eta, \zeta),
\end{aligned} \tag{3.21}$$

where

$$
\begin{aligned}
f_{13}^{(k,s)} &= \frac{1}{a_{55}^{(k)}}\left(\frac{\partial U_{\tau}^{(k,s)}}{\partial \zeta} + \frac{\partial W^{(k,s-1)}}{\partial \xi}\right),\\
f_{23}^{(k,s)} &= \frac{1}{a_{44}^{(k)}}\frac{\partial V_{\tau}^{(k,s)}}{\partial \zeta} + \frac{1}{a_{55}^{(k)}}\frac{\partial W^{(k,s-1)}}{\partial \eta},\\
f_{33}^{(k,s)} &= \frac{A_{11}^{(k)}}{\Delta^{(k)}}\frac{\partial W_{\tau}^{(k,s)}}{\partial \zeta} - \frac{1}{\Delta^{(k)}}\left(A_{23}^{(k)}\frac{\partial U^{(k,s-1)}}{\partial \xi} + A_{13}^{(k)}\frac{\partial V^{(k,s-1)}}{\partial \eta}\right).
\end{aligned}
\tag{3.22}
$$

By satisfying the boundary (3.7) and contact (3.9) conditions in terms of σ_{xz} and U, we will obtain an algebraic system in terms of $C_1^{(k,s)}$ and $C_2^{(k,s)}$. Solving this system yields

$$
\begin{aligned}
C_1^{(1,s)} =& \frac{1}{\Omega_*}\sqrt{a_{55}^{(1)}/\rho^{(1)}}\left(\sigma_{xz}^{+(s)} - f_{13}^{(1,s)}(\xi,\eta,0)\right),\\
C_2^{(1,s)} =& \frac{1}{\cos\gamma_1^{(1)}\zeta_1}\Big(C_1^{(2,s)}\sin\gamma_1^{(2)}\zeta_1\\
&+ C_2^{(2,s)}\cos\gamma_1^{(2)}\zeta_1 + d_2^{(1,s)} - C_1^{(1,s)}\sin\gamma_1^{(1)}\zeta_1\Big),\\
C_1^{(2,s)} =& d_4^{(1,s)}\left[b_1^{(1)}C_1^{(1,s)} + b_2^{(1)}C_2^{(2,s)} + d_3^{(1,s)}\right],\\
C_2^{(2,s)} =& \frac{1}{\cos\gamma_1^{(2)}\zeta_2 + d_4^{(1,s)}b_2^{(1)}\sin\gamma_1^{(2)}\zeta_2}\Big(C_1^{(3,s)}\sin\gamma_1^{(3)}\zeta_2\\
&+ C_2^{(3,s)}\cos\gamma_1^{(3)}\zeta_2 + d_2^{(2,s)}\\
&- d_4^{(1,s)}\left[b_1^{(1)}C_1^{(1,s)} + d_3^{(1,s)}\right]\sin\gamma_1^{(2)}\zeta_2\Big),\\
C_1^{(k,s)} =& d_4^{(k-1,s)}\Big[b_1^{(k-1)}C_1^{(k-1,s)}\\
&+ b_2^{(k-1)}C_2^{(k,s)} + d_3^{(k-1,s)}\Big], \quad k = 2,\ldots,N\\
C_2^{(k,s)} =& \frac{1}{\cos\gamma_1^{(k)}\zeta_k + d_4^{(k-1,s)}b_2^{(k-1)}\sin\gamma_1^{(k)}\zeta_k}\\
&\times\Big(C_1^{(k+1,s)}\sin\gamma_1^{(k+1)}\zeta_2 + C_2^{(k+1,s)}\cos\gamma_1^{(k+1)}\zeta_k + d_2^{(k,s)}\\
&- d_4^{(k-1,s)}\left[b_1^{(k-1)}C_1^{(k-1,s)} + d_3^{(k-1,s)}\right]\sin\gamma_1^{(k)}\zeta_k\Big),\\
&k = 2,\ldots,N-1\\
C_2^{(N,s)} =& \frac{1}{d_4^{(N-1,s)}b_2^{(N-1)}\cos\gamma_1^{(N)} - \sin\gamma_1^{(N)}}\\
&\times\Big[\frac{1}{\Omega_*}\sqrt{a_{55}^{(N)}/\rho^{(N)}}\left(\sigma_{13}^{-(s)} - f_{13}^{(N,s)}(\xi,\eta,1)\right)\\
&\left(C_1^{(N-1,s)}b_1^{(N-1)} + d_3^{(N-1,s)}\right)d_4^{(N-1,s)}\cos\gamma_1^{(N)}\Big],
\end{aligned}
\tag{3.23}
$$

$$
\begin{aligned}
d_1^{(k,s)} &= \frac{1}{\Omega_*}\sqrt{a_{55}^{(k)}/\rho^{(k)}}\Big(f_{13}^{(k+1,s)}(\xi,\eta,\zeta_k) \\
&\qquad -f_{13}^{(k,s)}(\xi,\eta,\zeta_k)\Big), \quad k=1,\ldots,N-1, \\
d_2^{(k,s)} &= u_\iota^{(k+1,s)}(\xi,\eta,\zeta_k)-u_\iota^{(k,s)}(\xi,\eta,\zeta_k), \\
d_3^{(k,s)} &= -\left(d_2^{(k,s)}\tan\gamma_1^{(k)}\zeta_k-d_1^{(k,s)}\right), \\
d_4^{(k,s)} &= \left(\sqrt{a_{55}^{(k)}\rho^{(k+1)}/\left(a_{55}^{(k+1)}\rho^{(k)}\right)}\cos\gamma_1^{(k+1)}\zeta_k\right. \\
&\qquad \left.+\tan\gamma_1^{(k)}\zeta_k\sin\gamma_1^{(k+1)}\zeta_k\right)^{-1}, \quad k=1,\ldots,N-1, \\
b_1^{(k)} &= 1/\cos\gamma_1^{(k)}\zeta_k, \\
b_2^{(k)} &= \sqrt{a_{55}^{(k)}\rho^{(k+1)}/\left(a_{55}^{(k+1)}\rho^{(k)}\right)}\sin\gamma_1^{(k+1)}\zeta_k-\sin\gamma_1^{(k)}\zeta_k.
\end{aligned}
$$

Satisfying the remaining conditions (3.7) and (3.9) leads to finding the unknown functions $C_3{}^{(k,s)}$, $C_4{}^{(k,s)}$, $C_5{}^{(k,s)}$, and $C_6{}^{(k,s)}$, corresponding formulas can be obtained from (3.23) by cyclic permutation:

$$
\Big(C_1^{(k,s)}, C_2^{(k,s)}; C_3^{(k,s)}, C_4^{(k,s)}; C_5^{(k,s)}, C_6^{(k,s)};
\qquad a_{55}^{(k)}, a_{44}^{(k)}, \Delta^{(k)}/A_{11}^{(k)}; \gamma_1^{(k)}, \gamma_2^{(k)}, \gamma_3^{(k)}\Big). \tag{3.24}
$$

In the case of problem (b), the values of constants $C_1{}^{(1,s)}$,...,$C_1{}^{(N,s)}$, $C_2{}^{(1,s)}$,..., $C_2{}^{(N-1,s)}$ can be found using (3.23), and $C_2{}^{(N,s)}$, $C_4{}^{(N,s)}$, and $C_6{}^{(N,s)}$ are determined as follows

$$
\begin{aligned}
C_2^{(N,s)} &= -\frac{1}{d_4^{(N-1,s)}b_2^{(N-1)}\sin\gamma_1^{(N)}+\cos\gamma_1^{(N)}} \\
&\times\left[u_\tau^{(N,s)}(\xi,\eta,1)+\left(b_1^{(N-1)}C_1^{(N-1,s)}+d_3^{(N-1,s)}\right)d_4^{(N-1,s)}\sin\gamma_1^{(N)}\right] \\
&\Big(C_2^{(k,s)}; C_4^{(k,s)}; C_6^{(k,s)}; a_{55}^{(k)}, a_{44}^{(k)}, \Delta^{(k)}/A_{11}^{(k)}; \gamma_1^{(k)}, \gamma_2^{(k)}, \gamma_3^{(k)}\Big).
\end{aligned} \tag{3.25}
$$

Having determined $C_j^{(k,s)}$ ($j=1,\ldots,6$) using (3.15) and (3.21), all components of the stress tensor can be determined.

As a rule, the solution to the external problem will not satisfy boundary conditions on the lateral surface of the plate. The solution for the boundary layer eliminates the arising discrepancy. The corresponding solution can be built autonomously using the method described in [1]. All components of the stress tensor and the displacement vector decrease exponentially with distance from the lateral surface into the inside of the package. Since the solution to the external problem is already known, using the solution to the boundary layer, conditions on the lateral surface of the package are satisfied. The corresponding procedure is described in [1].

In the mid-twentieth century, seismologists discovered noticeable deformations occur in the seismologically dangerous part of the Earth's crust before an earthquake [11, 13]. To determine the stress-strain states (SSS) of a layered block of Earth's crust, based on the theory of elasticity and corresponding displacement values, measuring instruments such as inclinometers and deformographs are placed at a certain depth from the surface of the seismically dangerous block. This prevents external influences, such as atmospheric factors, from affecting the true SSS.

The process of earthquake preparation can be divided into two main stages: slow and fast. The slow stage is quasistatic, while the fast stage includes foreshocks, earthquakes, and aftershocks. To determine the quasistatic state, it is necessary to solve the equilibrium equations of the theory of elasticity. This involves finding the values of the displacements of points at a certain depth, as a result of measurements, when the facial surface of the package is free. To investigate fast-flowing processes, the solution to dynamic equations must be found. The corresponding problem for the layered package of three plates (usually, the Earth's crust is considered to consist of three main layers: sedimentary, granite, and basaltic) is formulated as follows: find the solution to dynamic equations and relations, under boundary conditions, in the area D consisting of three layers, where the facial surface of the package of plates is free.

$$\sigma_{jz}^{(1)}(x, y, 0, t) = 0 \quad (j = x, y, z) \tag{3.26}$$

are known values of displacements of points of the contact surface between the second and the third layers, as data of measuring instruments

$$u^{(2)}(x, y, \zeta_2, t) = u^{(3)}(x, y, \zeta_2, t) = u^{+}(\xi, \eta), \quad (U, V, W) \tag{3.27}$$

and conditions of full contact between adjacent layers

$$\begin{aligned} \sigma_{jz}^{(1)}(x, y, \zeta_1, t) &= \sigma_{jz}^{(2)}(x, y, \zeta_1, t), \\ u^{(1)}(x, y, \zeta_1, t) &= u^{(2)}(x, y, \zeta_1, t), \\ \sigma_{jz}^{(2)}(x, y, \zeta_2, t) &= \sigma_{jz}^{(3)}(x, y, \zeta_2, t) \end{aligned} \tag{3.28}$$

Based on the general solution (3.13), (3.15), and (3.20), conditions (3.26)–(3.28) are satisfied, as a result we have

$$
\begin{aligned}
C_1^{(1,s)} &= -\frac{1}{\Omega_*}\sqrt{a_{55}^{(1)}/\rho^{(1)}}\, f_{13}^{(1,s)}(\xi,\eta,0),\\
C_2^{(1,s)} &= \frac{1}{\cos\gamma_1^{(1)}\zeta_1}\Big(C_1^{(2,s)}\sin\gamma_1^{(2)}\zeta_1\\
&\qquad + C_2^{(2,s)}\cos\gamma_1^{(2)}\zeta_1 + d_2^{(1,s)} - C_1^{(1,s)}\sin\gamma_1^{(1)}\zeta_1\Big),\\
C_1^{(2,s)} &= d_4^{(1,s)}\left[b_1^{(1)}C_1^{(1,s)} + b_2^{(1)}C_2^{(2,s)} + d_3^{(1,s)}\right],\\
C_2^{(2,s)} &= \frac{1}{\cos\gamma_1^{(2)}\zeta_2 + d_4^{(2,s)}\sin\gamma_1^{(2)}\zeta_2}\Big(U^{+(s)}(\xi,\eta)\\
&\qquad - u_\iota^{(2,s)}(\xi,\eta,\zeta_2) - \left[b_1^{(1)}C_1^{(1,s)} + d_3^{(1,s)}\right]d_4^{(1,s)}\sin\gamma_1^{(2)}\zeta_2\Big), \qquad (3.29)\\
C_1^{(3,s)} &= \frac{1}{\sin\gamma_1^{(3)}\zeta_2}\left[U^{+(s)}(\xi,\eta) - u_\iota^{(3,s)}(\xi,\eta,\zeta_3) - C_2^{(3,s)}\cos\gamma_1^{(3)}\zeta_2\right],\\
C_2^{(3,s)} &= \frac{1}{\Omega_*}\sqrt{\rho^{(3)}/a_{55}^{(3)}}\Big(U^{+(s)}(\xi,\eta) - u_\iota^{(3,s)}(\xi,\eta,\zeta_3)\Big)\cos\gamma_1^{(3)}\zeta_2\\
&\qquad - \Bigg[\frac{1}{\Omega_*}\sqrt{\rho^{(2)}/a_{55}^{(2)}}\Big(C_1^{(2,s)}\cos\gamma_1^{(2)}\zeta_2 - C_2^{(2,s)}\sin\gamma_1^{(2)}\zeta_2\Big)\\
&\qquad + f_{13}^{(3,s)}(\xi,\eta,\zeta_2) - f_{13}^{(2,s)}(\xi,\eta,\zeta_2)\Bigg]\sin\gamma_1^{(3)}\zeta_2,
\end{aligned}
$$

$$U^{+(0)} = u^+/l, \quad U^{+(s)} = 0, \quad s \neq 0$$

$$
\Big(C_1^{(k,s)}, C_2^{(k,s)}; C_3^{(k,s)}, C_4^{(k,s)}; C_5^{(k,s)}, C_6^{(k,s)};\\
a_{55}^{(k)}, a_{44}^{(k)}, \Delta^{(k)}/A_{11}^{(k)}; \gamma_1^{(k)}, \gamma_2^{(k)}, \gamma_3^{(k)}\Big)
$$

After determining $C_j^{(k,s)}$, the stresses and displacements of layers of a package are determined using (3.15) and (3.20).

When using polynomial functions u^+, v^+, w^+, and $\sigma_{\alpha z}^{\pm}$ $(\alpha = x, y, z)$ with respect to ξ and η, the iteration process stops once a certain degree of approximation is reached, which may vary based on the degree of polynomial used. This results in a mathematically precise solution to the external problem. To better understand this concept, let us consider the following examples.

1. $\sigma_{xz}^{+} = a_0 + a_1\xi + a_2\eta$, $(\sigma_{xz}^{+}, \sigma_{yz}^{+}, \sigma_{zz}^{+}; a, b, c)$, $\sigma_{\alpha z}^{-} = 0$. Iteration breaks at approximation $s = 1$. As a result, we have the mathematically exact solution

$$
\begin{aligned}
u^{(k)} &= l\left(U^{(k,0)} + \varepsilon U^{(k,1)}\right)\exp(i\Omega t), \quad (u, v, w)\\
\sigma_{\alpha\beta}^{(k)} &= \left(\varepsilon^{-1}\sigma_{\alpha\beta}^{(k,0)} + \sigma_{\alpha\beta}^{(k,1)}\right)\exp(i\Omega t), \quad \alpha, \beta = x, y, z, \quad k = 1, 2, 3\\
U^{(k,0)} &= C_1^{(k,0)}(\xi, \eta)\sin\gamma_1^{(k)}\zeta + C_2^{(k,0)}(\xi, \eta)\cos\gamma_1^{(k)}\zeta,\\
V^{(k,0)} &= C_3^{(k,0)}(\xi, \eta)\sin\gamma_2^{(k)}\zeta + C_4^{(k,0)}(\xi, \eta)\cos\gamma_2^{(k)}\zeta,\\
W^{(k,0)} &= C_5^{(k,0)}(\xi, \eta)\sin\gamma_3^{(k)}\zeta + C_6^{(k,0)}(\xi, \eta)\cos\gamma_3^{(k)}\zeta,
\end{aligned}
\tag{3.30}
$$

where

$$
\begin{aligned}
C_1^{(1,0)} &= \frac{1}{\Omega_*}\sqrt{a_{55}^{(1)}/\rho^{(1)}}\sigma_{xz}^{+(0)},\\
C_2^{(1,0)} &= \frac{1}{\cos\gamma_1^{(1)}\zeta_1}\Big(C_1^{(2,0)}\sin\gamma_1^{(2)}\zeta_1\\
&\qquad + C_2^{(2,0)}\cos\gamma_1^{(2)}\zeta_1 - C_1^{(1,0)}\sin\gamma_1^{(1)}\zeta_1\Big),\\
C_1^{(2,0)} &= d_4^{(1,0)}\left[b_1^{(1)}C_1^{(1,0)} + b_2^{(1)}C_2^{(2,0)}\right],\\
C_2^{(2,0)} &= \frac{1}{\cos\gamma_1^{(2)}\zeta_2 + d_4^{(1,0)}b_2^{(1)}\sin\gamma_1^{(2)}\zeta_2}\Big(C_1^{(3,0)}\sin\gamma_1^{(3)}\zeta_2\\
&\qquad + C_2^{(3,0)}\cos\gamma_1^{(3)}\zeta_2 - d_4^{(1,0)}b_1^{(1)}C_1^{(1,0)}\sin\gamma_1^{(2)}\zeta_2\Big),\\
C_1^{(k,0)} &= d_4^{(k-1,0)}\left[b_1^{(k-1)}C_1^{(k-1,0)} + b_2^{(k-1)}C_2^{(k,0)}\right], \quad k = 2, \ldots, N\\
C_2^{(N,0)} &= -\frac{b_1^{(N-1)}d_4^{(N-1,0)}C_1^{(N-1,0)}\cos\gamma_1^{(N)}}{d_4^{(N-1,0)}b_2^{(N-1)}\cos\gamma_1^{(N)} - \sin\gamma_1^{(N)}},
\end{aligned}
\tag{3.31}
$$

$$
\Big(C_1^{(k,s)}, C_2^{(k,s)}; C_3^{(k,s)}, C_4^{(k,s)}; C_5^{(k,s)}, C_6^{(k,s)};\\
a_{55}^{(k)}, a_{44}^{(k)}, \Delta^{(k)}/A_{11}^{(k)}; \gamma_1^{(k)}, \gamma_2^{(k)}, \gamma_3^{(k)}\Big).
$$

Stresses $\sigma_{\alpha\beta}^{k(0)}$ are determined by formulas (3.15), (3.21).
When $s = 1$, the iteration breaks (because of large volume, we don't give corresponding data). As a result, we have a mathematically exact solution.

2. Let consider the case when, in conditions (3.27), $u^+ = \text{const}$, (u, v, w)
The iteration breaks at initial approximation. As a result, according to formulas (3.15), (3.20), and (3.21), we have

$$
\begin{aligned}
u^k &= lU^{k(0)}\exp(i\Omega t), \quad (u, v, w)\\
\sigma_{\alpha\beta}^k &= \varepsilon^{-1}\sigma_{\alpha\beta}^{k(0)}\exp(i\Omega t) \quad (\alpha, \beta = x, y, z, \quad k = \text{I, II, III}),\\
U^{k(0)} &= C_1^{k(0)}(\xi, \eta)\sin\gamma_1^k\zeta + C_2^{k(0)}(\xi, \eta)\cos\gamma_1^k\zeta,\\
V^{k(0)} &= C_3^{k(0)}(\xi, \eta)\sin\gamma_2^k\zeta + C_4^{k(0)}(\xi, \eta)\cos\gamma_2^k\zeta,\\
W^{k(0)} &= C_5^{k(0)}(\xi, \eta)\sin\gamma_3^k\zeta + C_6^{k(0)}(\xi, \eta)\cos\gamma_3^k\zeta,
\end{aligned}
\tag{3.32}
$$

$$
\begin{aligned}
C_1^{(1,0)} =& 0, \\
C_2^{(1,0)} =& \frac{1}{\cos \gamma_1^{(1)} \zeta_1} \left(C_1^{(2,0)} \sin \gamma_1^{(2)} \zeta_1 + C_2^{(2,0)} \cos \gamma_1^{(2)} \zeta_1 \right), \\
C_1^{(2,0)} =& d_4^{(1,0)} b_2^{(1)} C_2^{(2,0)}, \\
C_2^{(2,0)} =& \frac{U^{+(0)}}{\cos \gamma_1^{(2)} \zeta_2 + d_4^{(1,0)} \sin \gamma_1^{(2)} \zeta_2}, \\
C_1^{(3,0)} =& \frac{U^{+(0)} - C_2^{(3,0)} \cos \gamma_1^{(3)} \zeta_2}{\sin \gamma_1^{(2)} \zeta_2}, \quad U^{+(0)} = u^+/l, \\
C_2^{(3,0)} =& \frac{1}{\Omega_*} \sqrt{\rho^{(3)}/a_{55}^{(3)}} U^{+(0)} \cos \gamma_1^{(3)} \zeta_2 \\
& - \left[\frac{1}{\Omega_*} \sqrt{\rho^{(2)}/a_{55}^{(2)}} \left(C_1^{(2,0)} \cos \gamma_1^{(2)} \zeta_2 \right. \right. \\
& \left. \left. - C_2^{(2,0)} \sin \gamma_1^{(2)} \zeta_2 \right) \right] \sin \gamma_1^{(3)} \zeta_2,
\end{aligned}
$$

$$
\begin{aligned}
\Big(C_1^{(k,0)}, C_2^{(k,0)}; C_3^{(k,0)}, C_4^{(k,0)}; C_5^{(k,0)}, C_6^{(k,0)}; \\
a_{55}^{(k)}, a_{44}^{(k)}, \Delta^{(k)}/A_{11}^{(k)}; \gamma_1^{(k)}, \gamma_2^{(k)}, \gamma_3^{(k)} \Big), \quad k = 1, 2, 3,
\end{aligned}
$$

$$
\begin{aligned}
d_4^{(1,0)} =& \left(\sqrt{a_{55}^{(1)} \rho^{(2)} / \left(a_{55}^{(2)} \rho^{(1)} \right)} \cos \gamma_1^{(2)} \zeta_1 + \tan \gamma_1^{(1)} \zeta_1 \sin \gamma_1^{(2)} \zeta_1 \right)^{-1}, \\
b_2^{(1)} =& \sqrt{a_{55}^{(1)} \rho^{(2)} / \left(a_{55}^{(2)} \rho^{(1)} \right)} \sin \gamma_1^{(2)} \zeta_1 - \sin \gamma_1^{(2)} \zeta_1.
\end{aligned}
$$

3.3 Conclusions

The asymptotic method has been successfully applied to obtain solutions for two distinct classes of 3D dynamic problems within the elasticity theory for layered plates, addressing singularly perturbed differential equations. The first class focuses on the forced vibrations of layered plates situated on an entirely rigid base, while the second class is relevant to seismology, specifically describing rapid dynamic processes such as foreshocks, earthquakes, and aftershocks. The study identifies specific cases where the derived solutions attain mathematical exactness and practical examples are provided for illustration.

Acknowledgements The work was supported by the Science Committee of RA, in the frames of the research project No 21T-2C075.

References

1. Aghalovyan, L.A.: Asymptotic Theory of Anisotropic Plates and Shells. World Scientific Publishing, Singapore (2015)
2. Aghalovyan, L.A.: On an asymptotic method for solving dynamic mixed problems of anisotropic strips and plates. Izv. Vyssh. Uchebn. Zaved. Sev.-Kavkazsk. Reg. Estestv. Nauki. **3**, 8–11 (2000). (in Russian)
3. Aghalovyan M.L., Zakaryan T.V.: Asymptotic solution of the first 3d dynamic elasticity theory problem on forced vibrations of a three-layer plate with an asymmetric structure. Mech. Compos. Mater. **55**(1), 3–18
4. Grigorenko, Y.M., Vlajkov, G., Grigorenko, A.: The numerical-analytical method for solving problems of shell mechanics based on various models. Kiev: Akademperiodika (2006)
5. Grigorenko, Y.M., Grigorenko, A.Y.: Static and dynamic problems for anisotropic inhomogeneous shells with variable parameters and their numerical solution (Review). Int. Appl. Mech. **49**(2), 123–193 (2013)
6. Friedrichs, K.O., Dressler, R.F.: A Boundary-Layer Theory for Elastic Plates. Commun. Pure Appl. Math. **14**(1), 1–31 (1961)
7. Gol'denveizer, A.L.: Derivation of an approximate theory of bending of a plate by the method of asymptotic integration of the equations of the theory of elasticity. J. Appl. Math. Mech. **26**(4), 1000–1025 (1962)
8. Gol'denveizer, A.L.: Theory of Thin Elastic Shells. Nauka, Moscow (1976). (in Russian)
9. Green, A.E.: On the linear theory of thin elastic shells. Proc. Roy. Soc. Ser. A. **266**(1325), 141–160 (1962)
10. Green, A.E.: Boundary layer equations in the linear theory of thin elastic shells. Proc. Roy. Soc. Ser. A. **2691339**, 481–491 (1962)
11. Kasahara, K.: Earthquake Mechanics. Cambridge University Press (1981)
12. Lugovoi, P.Z., Meish, V.F.: Dynamics of inhomogeneous shell systems under non-stationary loads (Review). Int. Appl. Mech. **53**(5), 3–65 (2017)
13. Rikitake, T.: Earthquake Prediction. Elsevier SPC, Amsterdam (1976)

Chapter 4
Methods of Numerical Solution of Problems of Thermo-Plasticity Taking into Account the Stress Mode

Maya Babeshko, Alexander Galishin, Vitalii Savchenko, and Nikolai Tormakhov

Abstract This study presents advanced numerical methods for analyzing the elastic–plastic stress–strain behavior of compound bodies subjected to thermo-force loading. The constitutive equations utilized in the analysis incorporate considerations for the temperature-dependent properties of isotropic materials, stress mode variations, and deformation history. Specifically, these equations capture two nonlinear dependencies involving the invariants of the stress–strain state and the stress mode angle. To enhance algorithmic clarity, the relationships between stress and strain tensor components are reformulated into a generalized form of Hooke's law, augmented with additional terms. The paper outlines a procedure for successive approximations to facilitate numerical computations. The findings of the numerical simulations are presented to provide valuable insights into the complex behavior of these compound bodies under diverse loading conditions.

4.1 Introduction

The evaluation of operational capabilities and strength of structural elements functioning under thermo-force loading conditions, beyond the limits of material elasticity, often requires numerical methods to solve thermo-plasticity problems for both massive bodies and shells that model these elements. These methods are developed based on constitutive equations that accurately depict the material properties under realistic structural load conditions. Constitutive equations, governing the relationship between stress and strain tensors, are inherently nonlinear in various thermo-plasticity theories, necessitating linearization methods when solving corresponding boundary value problems.

Among the common linearization techniques are the method of variable parameters of elasticity and the method of additional deformations [4–6]. This article

M. Babeshko (✉) · A. Galishin · V. Savchenko · N. Tormakhov
S.P. Timoshenko Institute of Mechanics, National Academy of Science of Ukraine, Kyiv, Ukraine
e-mail: bab41@ukr.net

V. Savchenko
e-mail: savchenko_vitalij@ukr.net

H. Altenbach et al. (eds.), *Selected Problems of Solid Mechanics and Solving Methods*, Advanced Structured Materials 204, https://doi.org/10.1007/978-3-031-54063-9_4

focuses on methods derived from constitutive equations [27, 28], which account for stress mode and plastic loosening in isotropic materials. The methodologies based on these equations exhibit distinctive characteristics when compared to those derived from traditional equations [21–24, 28–32]. Notably, equations [27] incorporate two experimentally determined nonlinear dependencies, considering changes in material properties not only from temperature and load history but also from the stress mode parameter.

The stress mode parameter denotes the orientation of shear stresses concerning the projection of principal stresses in the octahedral plane [14, 28]. This angle has a straightforward relationship with the Lode parameter [16], expressed in terms of the second and third deviatoric stress invariants. Experimental tests on tubular specimens under proportional loading at various stress mode angles and temperatures individualize these functions.

The constitutive equations [27, 28] have undergone experimental validation and are widely employed in solving boundary value problems. When the relationships between the first invariants of stress and strain tensors are linear, and the connections between the second invariants of their respective deviators are independent of the stress mode parameter, the constitutive equations [27, 28] transform into traditional relations [25, 26]. These traditional relations are based on the Prandtl–Reuss [18, 20] and Hencky theories [13], which pertain to the theory of deformation processes along trajectories of small curvatures and the theory of simple processes, respectively.

4.2 Formulation of the Problem of Thermo-Plasticity and Basic Relations

In this section, we present a comprehensive formulation of thermo-plasticity problems. The general problem statement is adapted to the specifics of each case, considering the geometry of the object under investigation, its material properties, and the conditions of thermo-force loading.

Let us consider a compound body composed of isotropic inelastic materials. We utilize an orthogonal coordinate system q_i $(i = 1, 2, 3)$ to describe the body, which experiences volume and surface forces along with nonuniform heating, leading to small strains in its elements. The material properties are assumed to depend on temperature and stress mode. The body undergoes simple (or nearly simple) deformation or deformation along paths of small curvature during loading and heating, accompanied by inelastic deformation and unloading. It is assumed that creep deformations are negligible compared to elastic and plastic components.

The components of the body are made of dissimilar initial isotropic materials, joined without tension, ensuring perfect mechanical and thermal contact. The stress–strain state of these solids is analyzed by determining the temperature (addressing the non-stationary heat conduction problem) and the displacement, strains, and stresses at fixed time points (addressing thermo-plastic problems). Methods for solving non-stationary thermal conductivity problems are well-established for both spatially axisymmetric bodies and shells of revolution, as detailed in works [1, 2].

In the analysis of the stress–strain state under thermo-force loading, it is assumed that the temperature field of the body is known at any moment during the process. Equilibrium equations, Cauchy relations, and linearized defining equations [27, 28] are employed in formulating boundary value problems.

The constitutive equations proposed in [27, 28] establish the relationship between stresses σ_{ij} and strains ε_{ij}. The strain tensor is represented as the sum of elastic and plastic strain tensors. The variation in elastic strains with stresses follows the generalized Hooke's law, while plastic deformation components are calculated according to the defining equations.

To account for the deformation history, the loading and heating process is divided into small time intervals, ensuring their endpoints are as close as possible to the onset of unloading of body elements. The deformation history is traced by solving the problem at each step.

The relations between components of the stress tensor σ_{ij} and components of the strain tensor ε_{ij} remain consistent irrespective of the stage of deformation. These relations, formulated as the generalized Hooke's law with additional terms, are expressed as follows:

$$\sigma_{ij} = 2G'\varepsilon_{ij} + \left(K' - 2G'\right)\varepsilon_0\delta_{ij} - \sigma^*_{ij}. \tag{4.1}$$

The coefficients of these equations G', K' and additional terms σ^*_{ij} depend on the theories of plasticity, the method of linearization, etc. If we use the equations of the theory of thermo-plastic deformation along paths of small curvature and the method of additional stresses, in (4.1) $G' = G$, $K' = K = 2G(1+\nu)/(1-2\nu)$, where G is the shear moduli and ν is Poisson's ratio.

The additional stresses read

$$\sigma^*_{ij} = 2G\left[e^{(p)}_{ij} + \frac{1+\nu}{1-2\nu}\left(\varepsilon_T + \varepsilon^{(p)}_0\right)\delta_{ij}\right], \tag{4.2}$$

where $\varepsilon_T = \alpha_T\,(T - T_0)$, α_T is the coefficient of linear thermal expansion; G, ν, and α_T depend on temperature T; T_0 is the initial temperature;

$$e^{(p)}_{ij} = \varepsilon^{(p)}_{ij} - \varepsilon^{(p)}_0\delta_{ij} \quad \text{and} \quad \varepsilon^{(p)}_0 = \tfrac{1}{3}\varepsilon^{(p)}_{ii}$$

are the components of the plastic strain deviator and plastic loosening, respectively; ${\varepsilon_{ii}}^{(p)}$ is the first invariant of the plastic strain tensor. We assume that σ^*_{ij} (4.2) are known and calculated using the values $e^{(p)}_{ij}$ and ε^p_0 obtained in the previous approximation.

At the end of the Nth stage, the components of the deviator of inelastic deformation are determined as the sum of their increments

$$e_{ij}^{(p)} = \sum_{k=1}^{N} \Delta_k e_{ij}^{(p)}. \tag{4.3}$$

We will use the dependence between the first invariants of the stress σ_{ii} and strain ε_{ii} tensors to determine $\varepsilon_0^{(p)}$

$$\sigma_0 = F_1\left(\varepsilon_0^*, T, \omega_\sigma\right), \tag{4.4}$$

$$\varepsilon_0^* = \varepsilon_0 - \varepsilon_T, \quad \varepsilon_0^{(p)} = \varepsilon_0^* - \frac{\sigma_0}{K}, \tag{4.5}$$

$$\omega_\sigma = \frac{1}{3} \arccos\left[-\frac{3\sqrt{3}}{2}\frac{I_3(D_\sigma)}{S^3}\right] \quad \left(0 \leqslant \omega_\sigma \leqslant \frac{1}{3}\pi\right) \tag{4.6}$$

where $\sigma_0 = \sigma_{ii}/3$, $\varepsilon_0 = \varepsilon_{ii}/3$ are the mean stress and strain, respectively, and ω_σ is the angle of the stress mode [14, 19], $I_3\left(D_\sigma\right) = |s_{ij}|$ is the third invariant of the stress deviator D_σ, $s_{ij} = \sigma_{ij} - \sigma_0\delta_{ij}$ are the stress deviator components, and S is the intensity of the tangential stresses

$$S = \left(s_{ij}s_{ij}/2\right)^{1/2}. \tag{4.7}$$

Note that in some works [21], instead of ω_σ, the quantity

$$\vartheta = \frac{1}{3} \arccos\left(\frac{3\sqrt{3}I_3\left(D_\sigma\right)}{2S^3}\right)$$

is used.

The increment $\Delta_k e_{ij}^{(p)}$ in (4.5) is determined by the expression

$$\Delta_k e_{ij}^{(p)} = \left\langle\frac{s_{ij}}{S}\right\rangle_k \Delta_k \Gamma^{(p)}, \tag{4.8}$$

where $\Delta_k \Gamma^{(p)}$ is the increment of the intensity of plastic shear deformations. The angle brackets in (4.8) indicate the average value of the value in them for the stage. To determine $\Delta_k \Gamma^{(p)}$, we will assume the existence of dependence

$$S = F_2\left(\Gamma, T, \omega_\sigma\right), \tag{4.9}$$

where Γ is the intensity of instantaneous shear deformations

$$\Gamma^* = \frac{S}{2G} + \Gamma^{(p)}, \quad \Gamma^{(p)} = \sum_{k=1}^{N} \Delta_k \Gamma^{(p)}. \tag{4.10}$$

The functions F_1 in (4.4) and F_2 in (4.9) are calculated based on the results of the above-mentioned basic experiments on the proportional load of tubular samples, as described in [3, 27, 28].

Note that the given constitutive equations can be used in a simplified version when, instead of (4.4), a linear dependence between the first invariants of the stress and strain tensors is assumed.

If we use the equations of the theory of small elastoplastic deformations allowing for the loading history and linearized by the method of elastic solutions, the coefficients in the constitutive equations (4.1) $G' = G_0$, $K' = K_0$ and the additional terms are as follows:

$$\sigma_{ij}^* = 2G_0\omega e_{ij} + 2G^* e_{ij}^{1p} + \left[K_0\omega_1\varepsilon_0 + K^*\left(\varepsilon_0^{1p} + \varepsilon_T\right)\right]\delta_{ij}, \tag{4.11}$$

where G_0 and K_0 are the shear and bulk moduli of the material at temperature T_0;

$$\omega = 1 - \frac{G^*}{G_0}, \quad \omega_1 = 1 - \frac{K^*}{K_0}, \quad 2G^* = \frac{S}{\Gamma}, \quad K^* = \frac{\sigma_0}{\varepsilon_0 - \varepsilon_T}, \tag{4.12}$$

${e_{ij}}^{1p}$ and ${\varepsilon_0}^{1p}$ are the plastic deviatoric strains and the mean strain at the time of unloading of an element of the body, respectively. During the loading, they are equal to zero. In the case of unloading, we have

$$e_{ij}^{1p} = e_{ij}^1 - \frac{S_{ij}^1}{2G^1}, \quad \varepsilon_0^{1p} = \varepsilon_0^1 - \varepsilon_T^1 - \frac{\sigma_0^1}{K^*}, \tag{4.13}$$

where the superscript “1” refers to the moment of unloading.

When the equations of the theory of small elastoplastic deformations are linearized by the method of variable parameters, coefficients in the constitutive equations (4.1) $G' = G^*$, $K' = K^*$ and σ_{ij}^* are expressed as

$$\sigma_{ij}^* = 2G^* e_{ij}^{1p} + K^*\left(\varepsilon_0^{1p} + \varepsilon_T\right)\delta_{ij}. \tag{4.14}$$

As in the previous cases, we believe that additional terms are known and calculated using the results obtained in the previous approximation.

Thus, the linearization of the nonlinear problem of thermo-plasticity is carried out by the method of successive approximations using the relationships (4.4) and (4.9), which correspond to the values set at the temperature stage and the angle of the stress mode calculated on the previous approximation.

4.3 Solving the Problem of Thermo-Plasticity

The constitutive equations (4.1), together with the equilibrium equations and the Cauchy relations, form a closed system of equations from which the unknown displacements, strains, and stresses can be determined. For example, to derive the governing system of equations for displacements, it is necessary to substitute the strain components (the Cauchy relations) into (4.1) and the result into the equilibrium equations. The resulting system of equations should be subjected to boundary conditions and solved by successive approximations.

In the current section, using the constitutive equations (4.1), the systems for determining the thermo-elastic–plastic state of axisymmetric spatial bodies and shells were formulated and methods of solving the corresponding boundary value problems were developed. In the case of non-axisymmetric loading, the given loads and solving functions are presented in the form of trigonometric series along the circular coordinate.

The variational Lagrange equation and the finite element method are used for spatial bodies of revolution. The problem is reduced to solving the systems of algebraic equations, the coefficients and right-hand parts of which are calculated based on the results of the previous approximation. When solving systems of algebraic equations, the Gauss method is used.

For shells of revolution, when using the hypotheses of Kirchhoff–Love or rectilinear element, in each approximation at each stage of loading, the problem is reduced to solving systems of ordinary differential equations, the coefficients and free terms of which are calculated based on the results of the previous approximation. When solving systems of differential equations, the Runge–Kutta method with discrete orthogonalization and normalization [9] according to the procedure outlined in [10–12] is used.

Let us pay attention to some details of calculations. We assume that dependencies (4.4) and (4.9) are given in the form of tables $\varepsilon_0^* \sim \sigma_0$ and $\Gamma^* \sim S$, respectively, for several values of the angle $0 \leqslant \omega_\sigma \leqslant \pi/3$ and temperature from the range under consideration. To solve the problem at the first stage, it is convenient to choose the level of loads in such a way that the body is deformed within the limits of elasticity. Let us solve the thermoelasticity problem and determine the body's stress–strain state. When constructing the process of successive approximations at the second and any of the following stages, we use the components of the stress–strain state corresponding to the end of the previous stage.

Starting the process of successive approximations at an arbitrary stage, we calculate the coefficients and additional stresses of equations (4.1) and solve the boundary value problem at the temperature, loads, and boundary conditions corresponding to this stage. Having obtained the values of the components of strains and stresses, we determine the values of $\varepsilon_0^* in$ (4.5) and ω_σ in (4.6) in the first approximation at the current stage. By linear interpolation of dependence (4.4) on temperature and angle ω_σ, we find the corresponding value $\sigma_0 = F_1(\varepsilon_0^*, T, \omega_\sigma)$ and calculate ${\varepsilon_0}^{(p)}$ in (4.5).

Then, by linear interpolation of the dependence (4.9) on temperature and angle ω_σ, we determine the value $S^{(d)}$, which corresponds to the value of the intensity of shear deformations $\Gamma = (\Gamma^{(p)})_{k-1} + S/(2G)$, where S is calculated according to (4.7). Then, using the constitutive equations of the theory of simple processes, we find K^* and $2G^* = S^d/\Gamma$ (4.12) and additional stresses (4.11) or (4.14).

In the case of the constitutive equations of the theory of thermo-plastic deformation along the trajectories of small curvature, we define

$$\Delta_k \Gamma^{(p)} = \frac{S - S^{(d)}}{2G}. \tag{4.15}$$

Using this value, we determine the increments of the component of the deviator of the plastic component deformations (4.8) and the value of these components (4.3). Now, we can determine the value of σ_{ij}^* in (4.2) and solve the boundary value problem for the new approximation. In the general case, the increment of plastic shear strain intensity at the kth step in an arbitrary Mth approximation can be calculated from the equation

$$\Delta_k \Gamma^{(p)} = \sum_{m=1}^{M-1} \Delta_m \Gamma^{(p)} + \frac{S - S^{(d)}}{2G} \tag{4.16}$$

where the shear-stress intensity S is calculated by (4.7), and $S^{(d)}$ is found from the function (4.9) corresponding to the temperature at the end of the step and to ω_σ (4.6) calculated from the previous approximation using the following expression for the strain intensity:

$$\Gamma = \Gamma_{k-1}^{(p)} + \Delta_k \Gamma^{(p)} + \frac{S}{2G}, \quad \Delta_k \Gamma^{(p)} = \sum_{m=1}^{M-1} \Delta_m \Gamma^{(p)}. \tag{4.17}$$

The process of successive approximations is terminated when

$$\left| \frac{\sigma_0 - F_1(\varepsilon_0, T, \omega_\sigma)}{K} \right| \leqslant \delta_1, \quad \left| \frac{S - F_2(\Gamma, T, \omega_\sigma)}{2G} \right| \leqslant \delta_2, \tag{4.18}$$

where δ_1 and δ_2 are measures of accuracy with which the calculated values of ω_σ, ε_0, and σ_0 satisfy function (4.4) and the values of ω_σ, S, and Γ satisfy (4.9).

The above algorithm is only used if elements of the body undergo active loading. To identify whether the process is the active loading or unloading in each element where plastic deformation occurs ($\Gamma^{(p)} > 0$) at each step, it is necessary to test the condition

$$\Delta \Gamma^{(p)} > 0 \tag{4.19}$$

after finding the first approximation. The increment $\Delta \Gamma^{(p)}$ is defined by the formula (4.15).

If this condition is satisfied, then we have the active loading, and the above algorithm can be used. Otherwise, we have the unloading. To take into account the unloading in this element of the body, it is necessary to accept $\Delta\Gamma^{(p)} = 0$ and, when determining additional stresses, use the values of the plastic component deformations corresponding to the end of the previous stage and solve the boundary value problem with these values in the following approximations at this stage. Note that the number of successive approximations at the stage required to fulfill conditions (4.18) exceeds by 20–40% the number required in the case of solving the same problem using the defining equations without taking into account the stress mode. A modification of the described methods of successive approximations, which allows to reduce the number of approximations by 30–50%, is proposed in [2].

The correctness of determining the moments of transition from active load to unloading and vice versa and choosing the size of the stages is checked by repeated calculation when the number of stages is increased. In this way, the required size of the stages is determined. The methods developed using Eqs. (4.1) found their application in the study of the thermally stressed and deformed state of structural elements in the form of spatial bodies [21] and shells [1–3, 8] of revolution made of materials, properties of which depend on the stress mode.

4.4 Numerical Results

We present the investigation results for the thermo-elastic–plastic state of a thin-walled structure, implemented using the constitutive equations of the theory of small curvature processes. The material of the structure is X18N10T alloy. The design is a cylindrical shell with spherical bottoms. A quarter of the shell meridian is shown in Fig. 4.1. The length of the cylindrical section $L = 60$ cm, the radius of the mid-surface $R = 20$ cm, the thickness of the shell $h = 1$ cm, and the initial temperature $T_0 = 20\,°\mathrm{C}$. The shell is in the process of slow heating by the external environment. As a result of heating, a temperature was established that varies linearly along the thickness from $T_0 = 20\,°\mathrm{C}$ on the inner surface to $T_0 = 120\,°\mathrm{C}$ on the outer surface

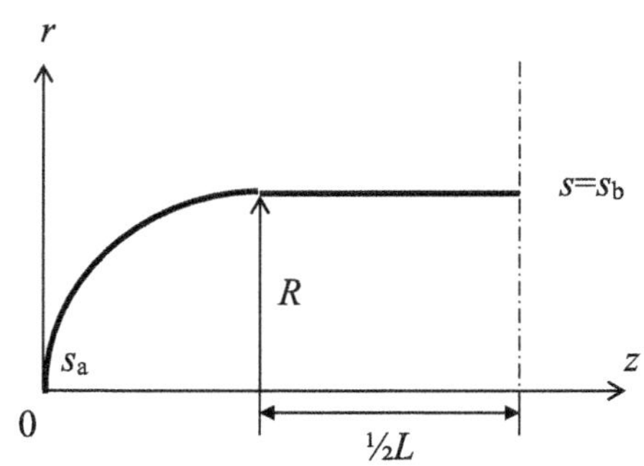

Fig. 4.1 Meridian of the shell

and does not change along the meridian. The shell experiences an escalating internal pressure attributed to the emergence of plastic strains. The analysis of the stress–strain condition of the shell is conducted after the onset of plastic deformations. The loading period is segmented into 19 steps. The following pressure values q_ς (in MPa) correspond to the following steps: $1, 2.5, 5, 7, 9, 10, 11, 12, \ldots, 23$. Due to the symmetry of the shell and the loading conditions, the problem was solved for half of the shell; at the starting point of the meridian $s = s_a = 0$ (Fig. 4.1), the conditions in the pole of the spherical type were set [11]. The symmetry conditions were set in the middle ($s = s_b$) of the cylindrical part of the shell.

The material characteristics are sourced from [3, 27, 28]. The specific values for functions (4.4) and (4.9) are compiled in Tables 4.1 and 4.2, respectively. Poisson's ratio is set at 0.27, and the coefficient of linear thermal expansion is prescribed as $\alpha_T = 0.12 \cdot 10^{-4} {}^\circ\mathrm{C}^{-1}$.

Three scenarios were considered:

1. **Stress, Pressure, and Temperature:** All factors (stress mode, pressure, and temperature) are considered.
2. **Stress and Pressure:** Stress mode and pressure are considered, while temperature is disregarded.
3. **Pressure and Temperature:** Pressure and temperature are considered, while stress mode is disregarded.

The results after the loading process are depicted in Figs. 4.2, 4.3, 4.4, 4.5 and 4.6, where solid, dashed, and dotted lines correspond to Cases 1, 2, and 3, respectively.

The analysis reveals that, at the process's end, the entire cylindrical section of the shell and the adjoining part of the caps undergo deformation beyond elasticity without unloading. In contrast, the portion of the caps near the pole ($0 \leqslant s \leqslant 20$ cm)

Table 4.1 The value of dependence (4.4)

$\varepsilon_0 - \alpha_T(T - T_0)$	σ_0, MPa					
	$T = 20\,^\circ$C			$T = 500\,^\circ$C		
	$\omega_\sigma = 0$	$\pi/6$	$\pi/3$	$\omega_\sigma = 0$	$\pi/6$	$\pi/3$
0	0	0	0	0	0	0
0.0002	88.9	88.9	88.9	71.8	71.8	71.8
0.0006	243.8	207.1	121.4	165.9	131.2	88.1
0.001	261.9	214.9	133.1	178.5	138.8	99.9
0.0024	295.7	233.4	154.2	205.3	151.4	115.5
0.003	306.0	240.4	160.7	207.4	154.0	119.4
0.004	316.6	252.0	169.0	211.4	159.0	123.9
0.006	336.6	262.0	179.5	215.4	164.0	129.3
0.008	356.6	282.0	186.2	219.0	169.0	133.0
0.02	373.0	365.0	188.0	239.8	197.0	154.0

Table 4.2 The value of dependence (4.9)

Γ	S, MPa					
	$T = 20\,°C$			$T = 500\,°C$		
	$\omega_\sigma = 0$	$\pi/6$	$\pi/3$	$\omega_\sigma = 0$	$\pi/6$	$\pi/3$
0	0	0	0	0	0	0
0.0002	32.2	32.2	32.2	26.0	26.0	26.0
0.01	166.6	166.6	166.6	104.3	104.3	104.3
0.02	188.9	168.2	191.6	128.6	128.6	128.6
0.04	220.5	201.4	223.0	149.0	134.3	164.6
0.06	243.0	222.8	247.8	173.4	145.2	187.6
0.08	261.9	235.3	268.0	178.6	152.6	200.4

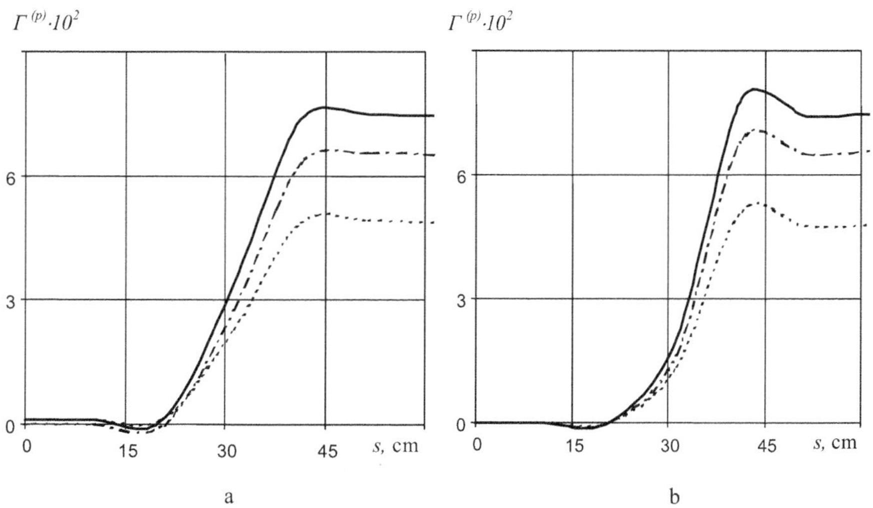

Fig. 4.2 Change of the accumulated plastic shear deformations on the inner (**a**) and outer (**b**) surfaces along the meridian

undergoes elastic deformation. Referencing Fig. 4.2, illustrating the variation in the intensity of stored plastic shear strain (4.10), we observe that the maximum intensity occurs in Case 1. In Cases 2 and 3, this quantity is 13% and 35% less than in Case 1. In the plastic range, the meridional strains ε_{ss} are significantly smaller in absolute magnitude than the hoop strains $\varepsilon_{\theta\theta}$. Figure 4.3 displays the variation in meridional strains on the inside surface (Fig. 4.3a) and outside surface (Fig. 4.3b). Figure 4.4 illustrates the meridional variation in hoop strains, constant throughout the shell's thickness. In the elastic range near the pole, strains in Cases 1 and 3 are equal. In the plastic range, the maximum hoop strains in Cases 2 and 3 are 13% and 35% less than in Case 1. Figures 4.5 and 4.6 depict the variation in meridional and hoop stresses on the inside surface (Fig. 4.5a) and outside surface (Fig. 4.5b). Notably, the

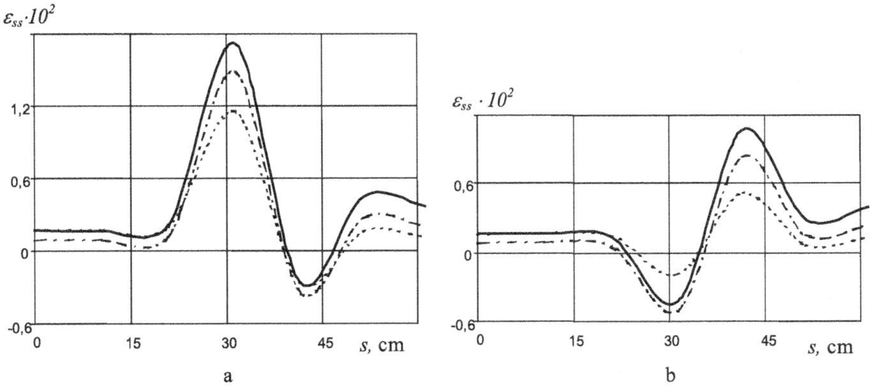

Fig. 4.3 Change of meridional deformations on the inner (**a**) and outer (**b**) surfaces along the meridian

Fig. 4.4 Change of circular deformations along the meridian

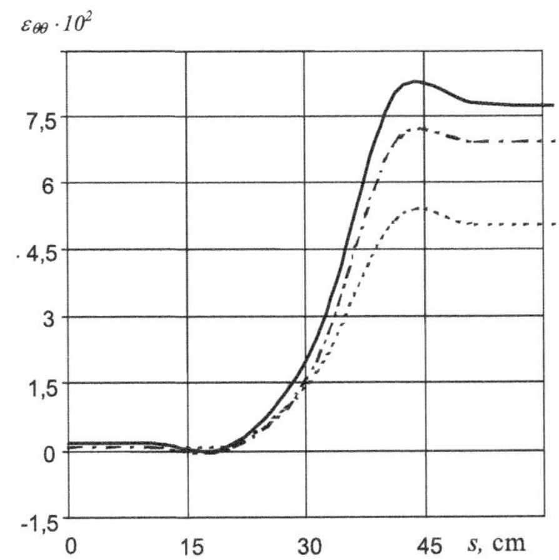

maximum meridional stresses in Cases 1 and 2 differ by approximately 50%, while hoop stresses (Fig. 4.6) vary significantly only near the pole of the spherical section. In the cylindrical section, the maximum meridional stresses in Cases 1 and 3 differ by more than 50%, while the difference between the hoop stresses is minor.

4.5 Conclusions

In this study, we have developed methods for solving thermo-plasticity problems in compound bodies and shells subjected to thermo-force loading. These methods consider the nuanced dependence of isotropic material properties on temperature, stress mode, and deformation history. Our approach is rooted in the utilization of constitu-

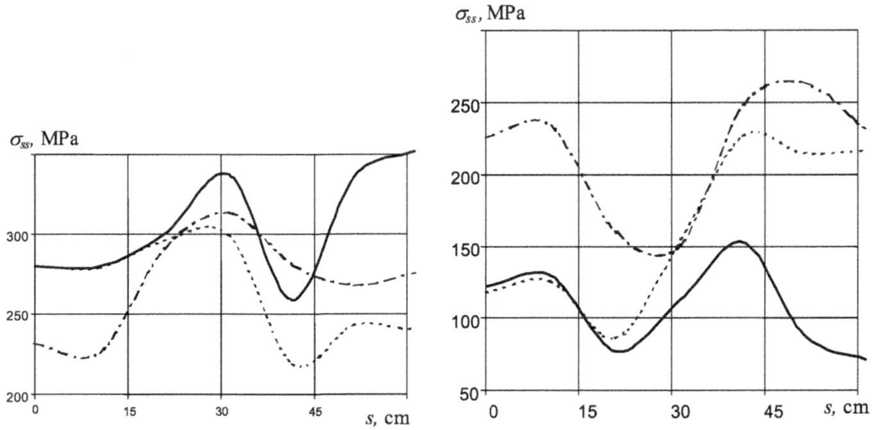

a b

Fig. 4.5 Change of meridional stresses on the inner (**a**) and outer (**b**) surfaces along the meridian

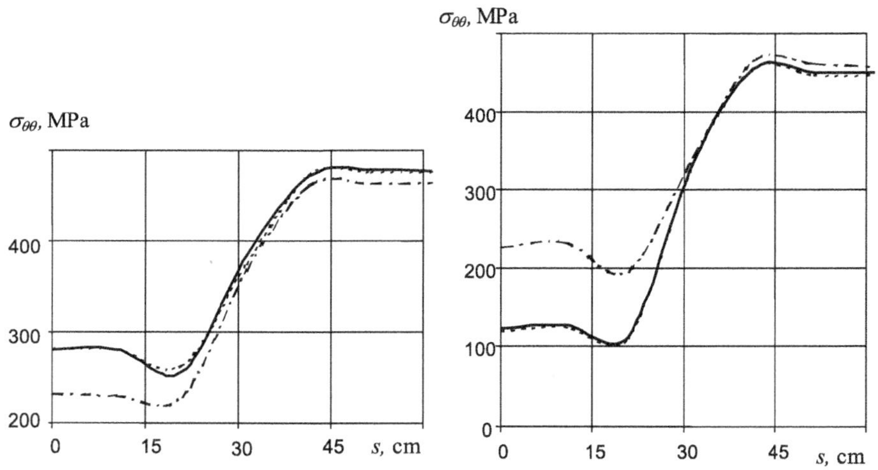

a b

Fig. 4.6 The change of circular stresses on the inner (**a**) and outer (**b**) surfaces along the meridian

tive equations that elucidate material deformation along both rectilinear trajectories and paths of small curvature.

The constitutive equations incorporate two crucial nonlinear dependencies. The first links the first invariants of stress and strain tensors, while the second associates the second invariants of their corresponding deviators. These nonlinear dependencies find their basis in experimental results obtained from proportional loading of tubular samples under different temperature conditions and stress mode parameters. The stress mode parameter, determined by the second and third invariants of the stress deviator, serves as a critical parameter. We have detailed methods for the linearization of constitutive equations, tailored for deformation processes along small curvature and rectilinear trajectories. Corresponding techniques for constructing a step-by-step process of successive approximations in determining the elastic–plastic state of the body during thermo-force loading are also outlined.

To exemplify these methods, we present a numerical example, and the results are graphically depicted. Notably, our calculations highlight the significant influence of considering the dependence of material properties on stress mode and temperature on the overall results.

This study advances our understanding of thermo-plasticity problems, offering comprehensive methods that consider the intricacies of material behavior under varying conditions. The presented numerical example demonstrates the practical application of these methods and underscores the importance of accounting for material property dependencies to obtain accurate and meaningful results in real-world scenarios.

References

1. Babeshko, M.E., Galishyn, A.Z., Savchenko, V.G.: Stress and strength of thin-wall constructions under nonisothermic repeated loading. Prikl. Mech. **58**, 6. pp. 70–83 (2022) [in Ukrainian]
2. Babeshko, M.E., Savchenko, V.G.: Technique of improvement of convergence of a successive approximation method in the boundary problems of thermoviscoplasticity. Prikl. Mech. **59**, 4. pp. 61–75 (2023) [in Ukrainian]
3. Babeshko, M.E., Shevchenko, Y.N.: Method of successive approximations for solving boundary-value problems of plasticity with allowance for the stress mode. Int. Appl. Mech. **46**(7), 744–752 (2010)
4. Birger, I.A.: Some general methods for solving problems of the theory of plasticity. Appl. Mat. and Mech. **15**, 6 pp. 765-770 (1951) [in Russian]
5. Birger, I.A.: The method of additional deformations in problems of the theory of plasticity. Izvestiya AN SSSR. Mech. Engin. **1**, 47–56 (1963). [in Russian]
6. Birger, I.A.: Methods of elastic solutions in the theory of plastic flow. Izvestiya AN SSSR. Mech. Engin. **2**, 116–118 (1963). [in Russian]
7. Bykov, D.L.: On some methods for solving problems of the theory of plasticity. Elasticity Inelasticity **4**, 119–149 (1975)
8. Galishin, A.Z., Shevchenko, Y.N.: Determining the axisymmetric elastoplastic state of thin shells with allowance for the third invariant of the stress deviator. Int. Appl. Mech. **46**(8), 869–876 (2010)
9. Godunov, S.K.: On a numerical method for solving boundary value problems for systems of linear ordinary differential equations. Uspehy Mat. Nauk **16**(3), 171–174 (1961) [in Russian]
10. Grigorenko, Y.M., Vasilenko, A.T., Bespalova, E.I.: Numerical solution of boundary value problems of statics of orthotropic shells with variable parameters. Nauk, Dumka, Kyiv (1975). [in Russian]
11. Grigorenko, Ya. M., Vasilenko, A. T.: The Theory of Shells of Variable Stiffness. Vol. 4 of the Five-Volume Series Metods of Calculation of Shells. Naukova Dumka, Kyiv (1981) [in Russian]
12. Grigorenko, Y.M., Shevchenko, Y.N., Vasilenko, A.T. et al.: Numerical Methods, Vol. 11 of the 12-Volume Series Mechanics of Composite Materials. A.S.K., Kyiv (2002) [in Russian]
13. Hencky, H.: Zur Theorie plastischer Deformationen und der hierdurch im Material hervorgerufenen Nachspannungen. ZAMM **4**(4), 323–334 (1924)

14. Kachanov, L.M.: Fundamentals of the Theory of Plasticity. Dover, New York (2004)
15. Lensky, V.S., Brovko, G.A.: The method of homogeneous linear approximations in unrelated problems of thermoradiation elasticity and plasticity. Thermal Stress. Struct. Elements. **11**, 100–113 (1971). [in Russian]
16. Lode, W.: Versuche über den Einfluß der mittleren Hauptspannung auf das Fließen der Metals - Eisen. Kupfer und Nickel. Z. Physik **36**, 913–939 (1926)
17. Malinin, N.N.: Applied Theory of Plasticity and Creep. Mashinostroenie, Moscow, p 400 (1968) [in Russian]
18. Prandtl, L.: Anwendungsbeispiele zu einem Henckyschen Satz über das plastische Gleichgewicht. ZAMM **3**(6), 401–406 (1923)
19. Rabotnov, Y.N.: Mechanics of a Deformable Solid Body. Nauka, Moscow (1979). [in Russian]
20. Reuss, A.: Berücksichtigung der elastischen Formänderung in der Plastizitätstheorie. ZAMM **10**(3), 266–274 (1930)
21. Savchenko, V.G.: A Method to Study the Nonaxisymmetric Plastic Deformation of Solids of Revolution with Allovance for the Stress Mode. Int. Appl. Mech. **44**(9), 975–981 (2008)
22. Shevchenko, Y.N.: Approximate methods for solving problems of thermo-plasticity taking into account the history of loading. Thermal Stress. Struct. Elements **7**, 37–48 (1967). [in Russian]
23. Shevchenko, Y.N.: The method of elastic solutions in the theory of plastic flow under non-isothermal loading processes. Thermal Stress. Struct. Elements **15**, 45–49 (1975). [in Russian]
24. Shevchenko, Y.N.: Thermo-Plasticity Under Variable Loading. Nauk. Dumka, Kyiv (1970). [in Russian]
25. Shevchenko, Yu.N., Savchenko, V.G.: Thermoviscoplasticity, Vol. 2 of the five-volume series Mechanics of Coupled Fields in Structural Members. Naukova Dumka, Kyiv (1987) [in Russian]
26. Shevchenko, Y.N., Terekhov, R.G.: Constitutive Equations of Thermoviscoplasticity. Naukova Dumka, Kyiv (1982). [in Russian]
27. Shevchenko, Y.N., Terekhov, R.G., Tormakhov, N.N.: Constitutive equations for describing the elastoplastic deformation of elements of a body along small-curvature paths in view of the stress mode. Int. Appl. Mech. **42**(4), 421–430 (2006)
28. Shevchenko, Y.N., Tormakhov, N.N.: Constitutive equations of thermoplasticy including the third invariant. Int. Appl. Mech. **46**(6), 613–624 (2010)
29. Steblyanko, P.A., Shevchenko, Y.N.: Computational methods in stationary and nonstationary thermal – plasticity problems. In: Hetnarski, R.B., (ed.) Encyclopedia of Thermal Stresses. in 11 volumes, vol. 2, C-D, pp. 507–1084–P. 623–630. Springer, New York (2014)
30. Ugodchikov, A.G., Korotkikh, Yu.G.: Some Methods for Solving Nonlinear Problems in the Theory of Plates and Shells on the Computer. Nauk. Dumka, Kyiv, 220 p. (1971)
31. Vorovich, I.I., Krasovsky, Y.P.: On the method of elastic solutions. Report Acad. Sci. USSR **126**(4), 740–743 (1959)
32. Życzkowski, M.: Combined Loadings in the Theory of Plasticity. – PWN – Polish Scientific Publishers (1981)

Chapter 5
Analysis of the Beam Approximation Applicability in Problems on Compression of Bodies Along Closely Spaced Cracks

Viacheslav Bogdanov, Mykhailo Dovzhyk, and Volodymyr Nazarenko

Abstract A nonclassical problem of fracture mechanics on compression of bodies along closely spaced parallel cracks is investigated. Cases of a body containing near-surface penny-shaped cracks, parallel to the boundary surface of the body, two parallel penny-shaped cracks, and an array of parallel penny-shaped cracks are considered. The start of fracture process is assumed to begin with stability loss of a part of material surrounding the cracks. By using the approach based on relationships and methods of the three-dimensional linearized theory of stability of deformable bodies for finite and small subcritical strains, the critical loads corresponding to the onset of fracture in highly elastic and composite materials with cracks are calculated. The methodology employed permits the passage to the limit when the separation distance between the parallel cracks or between the near-surface crack and the body boundary tends to zero. On the basis of the results obtained here in the framework of the rigorous three-dimensional linearized approach, we assess the applicability limits of the "beam approach" ("beam approximation"), which is used by other authors and relies on studying stability loss of the respective idealized two-dimensional objects that are "as if separated" by the cracks.

5.1 Introduction

The fracture of a body under compression along cracks contained in it and located parallel to each other is one of the nonclassical problems of fracture mechanics [16–19]. With such a loading scheme, a homogeneous stress–strain state arises in the body, and there are no singular components in the solutions of corresponding problems of the linear theory of elasticity [18, 19]. In this case, the classical approaches of fracture mechanics such as Griffith energy-based fracture criterion [11] or the equivalent Irwin force-based criterion [22], Eshelby–Cherepanov–Rice concept of invariant

V. Bogdanov · M. Dovzhyk (✉) · V. Nazarenko
S.P. Timoshenko Institute of Mechanics, National Academy of Science of Ukraine, Kyiv, Ukraine
e-mail: medved_mik@ukr.net

V. Bogdanov
e-mail: bogdanov@nas.gov.ua

H. Altenbach et al. (eds.), *Selected Problems of Solid Mechanics and Solving Methods*, Advanced Structured Materials 204, https://doi.org/10.1007/978-3-031-54063-9_5

J-integral [7], Wells criterion of critical crack opening displacement [32], as well as their generalizations [10], do not work.

In the situation considered, the start of fracture in a cracked body is determined by the phenomenon of the local loss of equilibrium stability of the part of material surrounding cracks [13, 14, 17, 18]. Currently, to calculate the critical values of compressive load corresponding to the local stability loss of the material, two fundamentally different approaches are employed.

The *first* of those was originally proposed by Obreimoff [26] and consists in the use of approximate calculation schemes and approximate theories. In the framework of this approach, the "beam approximation" [6, 23–25, 27] is applied most frequently: in it the part of the material situated between neighboring cracks (or between a near-surface crack and the body surface) is conditionally substituted with a thin-walled member (beam, plate, or shell) which is studied with reliance on the applied theories of stability of thin-walled systems. The instability of this thin-walled member in the Eulerian sense is investigated for the chosen a priori fastening on its edges. Despite its simplicity and ease of use, the method has significant drawbacks: it is necessary to carry out separate investigations to determine the possibility of its application, depending on the separation distance between the cracks (or between the crack and the body boundary surface), but even after determining this distance, there remains the question of the adequate choice of conditions for fixing the idealized thin-walled member.

The *second* approach relies on the relationships and methods of the three-dimensional linearized theory of the stability of deformable bodies [12, 15] to investigate the stability of local equilibrium of the material adjacent to cracks. This method was proposed in [13, 14] and was used to calculate the critical parameters in problems on compression of elastic and elastoplastic materials containing various arrays of parallel cracks [15–20]. It should be noted that the latter approach is not associated with the introduction of additional simplifying hypotheses and provides the precision accepted in the mechanics of a deformable solid body. Besides, the results of investigating the problems on compression of bodies along the flat cracks situated in parallel planes in the rigorous linearized formulation make it possible to assess the limits of the "beam approximation" applicability to these problems.

The results of investigating the fracture problems on compression of bodies along parallel penny-shaped cracks are presented below. Bodies with a near-surface crack, parallel to the boundary surface of the body, two parallel cracks, and an array of parallel cracks are considered. Calculation results for highly elastic isotropic materials and laminated composites are given.

To study highly elastic materials, the relationships of the three-dimensional linearized theory of stability of deformable bodies for finite (large) subcritical strains [12, 15] are employed. Due to the sufficient rigidity of the composite materials considered, we use the second variant of the three-dimensional linearized theory of stability of deformable bodies for small subcritical strains in investigating them, when the initial stress–strain state is determined by geometrically linear theory [12, 15]. Besides, we assume that crack sizes are essentially larger than the sizes of composite's structure elements (i.e., macrocracks are considered). The cases of cracks location in the interfaces of composite's structure elements are not investigated. Under such assumptions for laminated composites with isotropic layers, which are considered below, a continual model of the composite with the reduced (averaged) characteristics of a transverse isotropic body is to be used.

The proposed methodology makes it possible to investigate the problems on compression of bodies along internal cracks located closely to each other or to the boundary surface of the body and permits the passage to the limit when the separation distance between the parallel cracks tends to zero. By using it, the applicability limits of the first approach ("beam approximation") to studying such problems is analyzed.

5.2 A Body with a Near-Surface Penny-Shaped Crack

We consider a semi-infinite solid body occupying the upper half-space $x_3 \geq -h$ and containing a flat penny-shaped crack of radius a, which is located in the plane $x_3 = 0$ with its center on Ox_3-axis. We assume that the body is isotropic or transversely isotropic (in the case of the transversely isotropic body, Ox_3-axis is the axis of material's properties isotropy) and is loaded by compressive forces $S_{11}^0 = S_{22}^0$ that operate along the crack. These forces cause a uniform subcritical (initial) stress–strain state in the body determined by the following expression for the components of stresses and displacement:

$$\begin{gathered} S_{33}^0 = 0, \quad S_{11}^0 = S_{22}^0 \neq 0, \\ u_m^0 = \delta_{jm}(\lambda_j - 1)x_j, \quad \lambda_1 = \lambda_2 \neq \lambda_3, \quad \lambda_j = \text{const}. \end{gathered} \tag{5.1}$$

In (5.1), λ_j denote the contraction ratio along the axis x_j are the Lagrangian coordinates coinciding in the undeformed state with the Cartesian ones, ${S_{ij}}^0$ are the components of the symmetric stress tensor in the subcritical (initial) state, ${u_j}^0$ are components of the displacement vector corresponding to the initial stresses, and δ_{ij} is Kronecker's symbol. We assume that the crack faces and the boundary plane $x_3 = -h$ are free of stresses. We will investigate the axisymmetric loss of material stability in the local area near the crack.

In the framework of the *second* approach, the axisymmetric problem is reduced, using the methodology of [30], to the solution of a system of Fredholm integral equations of the second kind with a side condition (for more details, see [2, 5, 20]):

$$\begin{aligned} f(\xi) + \frac{1}{\pi k}\int_0^1 M_1(\xi,\eta) f(\eta) d\eta - \frac{2}{\pi k}\int_0^1 N_1(\xi,\eta) g(\eta) d\eta &= 0, \\ g(\xi) + \frac{1}{\pi k}\int_0^1 M_2(\xi,\eta) g(\eta) d\eta - \frac{2}{\pi k}\int_0^1 N_2(\xi,\eta) f(\eta) d\eta + \tilde{C}_1 &= 0, \\ \int_0^1 g(\xi) d\xi &= 0, \end{aligned} \tag{5.2}$$

with $0 \leqslant \xi, \eta \leqslant 1$. Equation (5.2) are given in dimensionless form (all functions and variables are related to the crack radius). The kernels of integral equation (5.2) are of the form

$$
\begin{aligned}
M_1(\xi,\eta) &= R_1(\eta+\xi) - R_1(1+\xi) + R_1(\eta-\xi) - R_1(1-\xi),\\
N_1(\xi,\eta) &= S_1(\eta+\xi) + S_1(\eta-\xi),\\
M_2(\xi,\eta) &= S_2(\eta+\xi) + S_2(\eta-\xi),\\
N_2(\xi,\eta) &= R_2(\eta+\xi) - R_2(1+\xi) + R_2(\eta-\xi) - R_2(1-\xi).
\end{aligned}
\tag{5.3}
$$

For the case of equal roots ($n_1^0 = n_2^0$) of the characteristic equation [15], the quantities involved in (5.3) are

$$
\begin{aligned}
R_1(\zeta) &= -2\left[\frac{k}{2}L_0(\zeta) + \frac{1}{k}L_2(\zeta) + L_1(\zeta)\right], \quad R_2(\zeta) = -\frac{1}{k}L_1(\zeta),\\
S_1(\zeta) &= -\frac{1}{k}L_3(\zeta), \quad S_2(\zeta) = -2\left[\frac{k}{2}L_0(\zeta) + \frac{1}{k}L_2(\zeta) - L_1(\zeta)\right],\\
L_0(\zeta) &= \frac{\beta_1}{\beta_1^2+\zeta^2}, \quad L_1(\zeta) = \frac{\beta_1}{2}\frac{\beta_1^2-\zeta^2}{(\beta_1^2+\zeta^2)^2},\\
L_2(\zeta) &= \frac{\beta_1^3}{2}\frac{4\beta_1^2-3\zeta^2}{(\beta_1^2+\zeta^2)^3}, \quad L_2(\zeta) = \frac{\beta_1}{2}\frac{\zeta^4-6\beta_1^2\zeta^2+\beta_1^4}{(\beta_1^2+\zeta^2)^4},\\
\beta &= ha^{-1}, \quad \beta_1 = 2(n_1^0)^{-1/2}\beta \quad k = \frac{d_2^0\,(l_1^0-l_2^0)}{d_1^0 l_1^0}.
\end{aligned}
\tag{5.4}
$$

For the case of non-equal roots ($n_1^0 \neq n_2^0$) of the characteristic equation [15]

$$
\begin{aligned}
R_1(\zeta) &= 2k_1\left\{2\frac{k_2}{k}I_0(\beta_1+\beta_2,\zeta) - \frac{1}{2}\frac{k_1+k_2}{k}\left[\frac{k_2}{k_1}I_0(2\beta_2,\zeta) + I_0(2\beta_1,\zeta)\right]\right\},\\
S_1(\zeta) &= k_1\frac{k_1+k_2}{k}\left\{I_1(\beta_1+\beta_2,\zeta) - \frac{1}{2}[I_1(2\beta_1,\zeta) + I_1(2\beta_2,\zeta)]\right\},\\
S_2(\zeta) &= 2k_2\left\{2\frac{k_1}{k_2}I_0(\beta_1+\beta_2,\zeta) - \frac{1}{2}\frac{k_1+k_2}{k}\left[\frac{k_1}{k_2}I_0(2\beta_2,\zeta) + I_0(2\beta_1,\zeta)\right]\right\},\\
R_2(\zeta) &= k_2\frac{k_1+k_2}{k}\{I_{-1}(\beta_1+\beta_2,\zeta) - \frac{1}{2}[I_{-1}(2\beta_1,\zeta) + I_{-1}(2\beta_2,\zeta)]\},\\
I_0(\rho,\zeta) &= \rho\left(\zeta^2+\rho^2\right)^{-1}, \quad I_{-1}(\rho,\zeta) = -\frac{1}{2\beta}\log\left(\zeta^2+\rho^2\right),\\
I_1(\rho,\zeta) &= \beta\left(\rho^2-\zeta^2\right)\left(\zeta^2+\rho^2\right)^{-2}, \quad \beta = ha^{-1}, \quad \beta_i = \beta(n_i^0)^{-1/2},\\
k_1 &= l_1^0\left(n_2^0\right)^{-1/2}, \quad k_2 = l_2^0\left(n_1^0\right)^{-1/2}, \quad k = k_1 - k_2
\end{aligned}
\tag{5.5}
$$

($i = 1, 2$). In (5.4) and (5.5), the values n_i^0, l_i^0, and d_i^0 ($i = 1, 2$) are determined by the choice of material's model and are related to components of elasticity tensor [15, 20].

To investigate the system of Fredholm integral equations of the second kind (5.2), we employ the procedure based on the Bubnov–Galerkin method (see [21]). A system of power functions is used as coordinate functions. The representation of the required functions in terms of the first N coordinate functions is of the form

$$f(x) = \sum_{i=0}^{N} F_i x^i, \quad g(x) = \sum_{i=0}^{N} G_i x^i. \tag{5.6}$$

For solving system (5.2), after substitution of coordinate functions (5.6) the procedure which allows the analytical calculation of integrals for the chosen system of coordinate functions using a package of character evaluations is employed. That has allowed a higher accuracy of evaluations in further numerical calculations to be achieved due to avoiding the necessity of numerical integration. For accelerating integrals evaluations, the recurrence relations are employed. The following functions have been used in further calculations:

$$L(n) = \int_0^1 \frac{x^n}{(a^2+(x+y)^2)^2}dx, \quad V(n) = \int_0^1 \frac{x^n}{(a^2+(x-y)^2)^2}dx, \tag{5.7}$$

$$L_1(n) = \int_0^1 \frac{x^n}{(a^2+(x+y)^2)^4}dx, \quad V_1(n) = \int_0^1 \frac{x^n}{(a^2+(x-y)^2)^4}dx. \tag{5.8}$$

For calculating integrals (5.7) and (5.8), it is not difficult to obtain recurrence relations:

$$\begin{aligned}
L(n) &= \frac{1}{n-3}\left(\frac{1}{a^2+(1+y)^2} - 2y(n-2)L(n-1)\right.\\
&\qquad \left. - (a^2+y^2)(n-1)L(n-2)\right) \quad (n \neq 3),\\
V(n) &= \frac{1}{n-3}\left(\frac{1}{a^2+(1-y)^2} + 2y(n-2)V(n-1)\right.\\
&\qquad \left. - (a^2+y^2)(n-1)V(n-2)\right) \quad (n \neq 3),\\
L_1(n) &= \frac{1}{n-5}\left(\frac{1}{(a^2+(x+y)^2)^3} - 2y(n-4)L_1(n-1)\right.\\
&\qquad \left. - (a^2+y^2)(n-1)L_1(n-2)\right) \quad (n \neq 5),\\
V_1(n) &= \frac{1}{n-5}\left(\frac{1}{(a^2+(1-y)^2)^3} + 2y(n-4)V_1(n-1)\right.\\
&\qquad \left. - (a^2+y^2)(n-1)V_1(n-2)\right) \quad (n \neq 5).
\end{aligned} \tag{5.9}$$

Knowing the values of integrals (5.7) for $n = 0, 1, 3$ and (5.8) for $n = 0, 1, 5$, and using recurrent relations (5.9), it is possible to calculate the integrals from functions (5.4) and (5.5). After substituting (5.6) in the Fredholm integral equation (5.2) and using (5.9) for integration of kernels (5.3), the system of $2N + 3$ equations with the same number of unknown quantities F_i, G_i, and $\tilde{C}_1$ ($i \in [0, N]$) has been obtained

$$\begin{gathered} \sum_{i=0}^{N} F_i F_{1ji} + \sum_{i=0}^{N} G_i G_{1ji} = 0, \\ \sum_{i=0}^{N} F_i F_{2ji} + \sum_{i=0}^{N} G_i G_{2ji} + \tilde{C}_1 = 0, \\ \sum_{i=0}^{N} \frac{1}{i+1} G_i = 0, \end{gathered} \tag{5.10}$$

where $0 \leqslant j \leqslant N$, and F_{kji} and G_{kji} are the exact expressions which depend on the geometric parameter of the problem $\beta = ha^{-1}$ (the dimensionless distance between the crack and the body boundary surface), material's characteristics as well as contraction ratio along the axis λ_j.

As an example, we consider numerical results for highly elastic materials with *Bartenev–Khazanovich* and with *Treloar* elastic potential, as well as those for a laminated composite material with isotropic layers. For a material with *Bartenev–Khazanovich elastic potential* [1] (an incompressible body, the case of equal roots of the characteristic equation [15]), the parameters appearing in (5.2), (5.4), and (5.5) take the form

$$\begin{aligned} n_1^0 = n_2^0 = \lambda_1^3, \quad k = \left(3\lambda_1^3 1\right)\left(1 + \lambda_1^3\right)^{-1}, \\ \beta_1 = 2\lambda_1^{-3/2}\beta, \quad S_{11}^0 = 2\mu\left(\lambda_1^3 - 1\right)\lambda_1^{-4}, \end{aligned} \tag{5.11}$$

where μ is a material constant.

By the substitution of expressions (5.11) into (5.10), we obtain the system with coefficients F_{kji} and G_{kji} depending on parameters β and λ_1. Numerical investigation of this system allows us to calculate for different values of the geometric parameter (distance ratio) $\beta = ha^{-1}$ the smallest values of critical contraction ratio λ_1, and from (5.11)—the critical compressive stress S_{11}^0 at which the system loses its stability. Using 20 coordinate functions, the dependence of critical values of relative contraction $\varepsilon = 1 - \lambda_1$ on distance ratio $\beta = ha^{-1}$ has been found. Figure 5.1 shows this dependence for small (interval $0.01 \leqslant \beta \leqslant 0.1$, the left-hand picture in Fig. 5.1) and large (interval $0.1 \leqslant \beta \leqslant 3.0$, the right-hand picture in Fig. 5.1) values of the geometric parameter β.

For a material with *Treloar elastic potential* [29] (an incompressible body, the case of non-equal roots of the characteristic equation [15]), the parameters appearing in (5.2), (5.4), and (5.5) take the form

$$\begin{aligned} n_1^0 = 1, \ n_2^0 = \lambda_1^6, \quad k_1 = 2\left(1 + \lambda_1^6\right)^{-1}, \\ k_2 = \tfrac{1}{2}\lambda_1^{-3}\left(1 + \lambda_1^6\right), \quad S_{11}^0 = 2c_{10}\left(1 - \lambda_1^{-6}\right), \end{aligned} \tag{5.12}$$

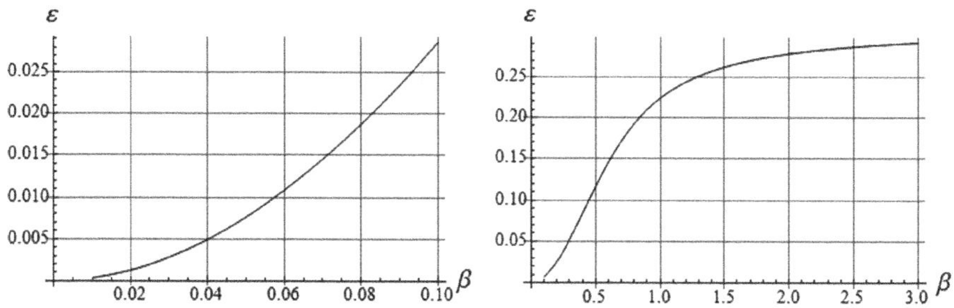

Fig. 5.1 Dependence of the critical values of relative contraction $\varepsilon = 1 - \lambda_1$ on distance ratio $\beta = ha^{-1}$ for a material with *Bartenev–Khazanovich potential* (a near-surface crack)

where c_{10} is a material constant. Figure 5.2 illustrates for this material the dependence of critical values of relative contraction $\varepsilon = 1 - \lambda_1$ on the dimensionless distance between the crack and the body boundary surface $\beta = ha^{-1}$(the left-hand and right-hand pictures in Fig. 5.2 present, respectively, the results for small and large values of β).

As an example of composite material, we considered *a laminated two-component composite with isotropic layers* (see [9]). Dependence of critical dimensionless compressive stress σ_{11}^0/E (σ_{11}^0 is the component of the symmetric stress tensor in the case of small subcritical (initial) strains; E is the reduced elasticity modulus in the plane of isotropy) on the ratio of the dimensionless distance between the crack and the free surface β are shown in Fig. 5.3. The results are given for equal values of Poisson's ratios of the layers of the composite $\nu^{(1)} = \nu^{(2)} = 0.3$ and volume concentration of the layers with elasticity modulus $E^{(1)}$ is $c_1 = 0.7$. The reduced mechanical characteristics of the composite as a transversely isotropic medium are as follows: $\nu = 0.3$, $\nu' = 0.3$, $G'/E = 0.1$, and $E'/E = 0.1$.

As can be seen from Figs. 5.1, 5.2, and 5.3, the mutual influence of the near-surface crack and the body boundary at quite small values of the distance between the crack and the half-space boundary results in a significant decrease of the critical values of

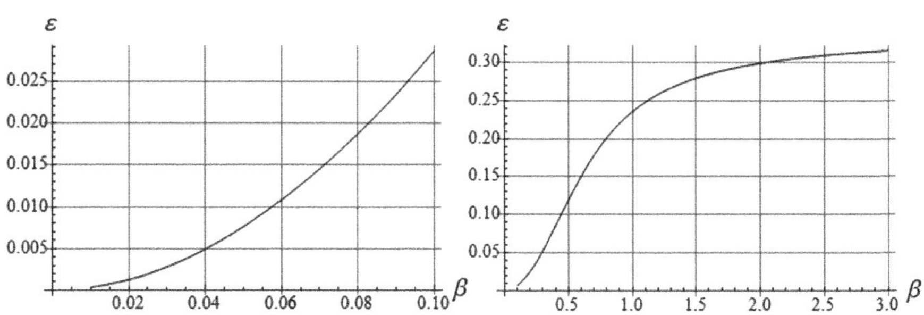

Fig. 5.2 Dependence of critical values of relative contraction $\varepsilon = 1 - \lambda_1$ on distance ratio $\beta = ha^{-1}$ for a material with *Treloar* potential (a near-surface crack)

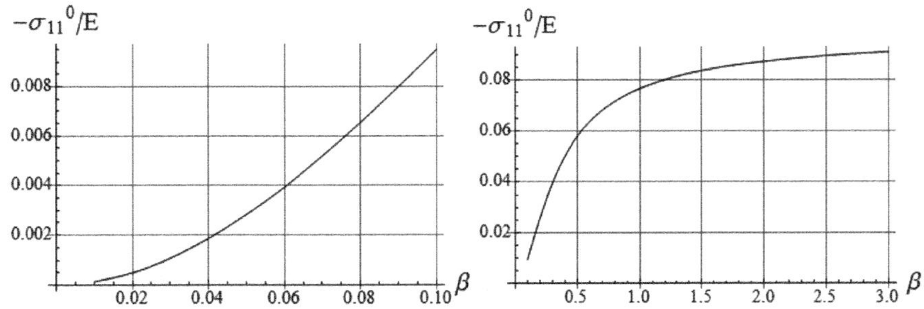

Fig. 5.3 Dependence of critical values of dimensionless compressive stress σ_{11}^{0}/E on distance ratio $\beta = ha^{-1}$ for a laminated composite material (a near-surface crack)

relative contraction $\varepsilon = 1 - \lambda_1$ (and, respectively, the critical values of compressive load) by two and more orders of magnitude as compared with the case of an isolated crack in an unbounded body (see [20]).

To assess the applicability limits of the first approach ("beam approximation") in studying the axisymmetric problem on compression of the half-space with a near-surface penny-shaped crack, it is necessary to compare the values of critical compressive stresses for closely spaced cracks, which have been obtained in the framework of the rigorous three-dimensional approach (the procedure of numerical analytical approach has been described above) and the values of Eulerian critical compressive force that are obtained within "beam approximation". It should be noted here that to investigate the above-mentioned problem by applying the "beam approximation", an idealized object in the form of a flat circular plate of radius a and thickness h is used. In this case, the critical compressive force for a linearly elastic body is determined from such expressions [28, 31]:

- for the condition of *rigid fixing* of the plate edge

$$\left(S_{11}^{0}\right)_{\mathrm{Eu}}^{(1)} = A_{\mathrm{Eu}}^{(1)} \frac{E}{\left(1 - \nu^2\right)} \beta^2, \quad A_{\mathrm{Eu}}^{(1)} = -1.2233, \tag{5.13}$$

- for the condition of *hinged supporting* of the plate edge

$$\left(S_{11}^{0}\right)_{\mathrm{Eu}}^{(2)} = A_{\mathrm{Eu}}^{(2)} \frac{E}{\left(1 - \nu^2\right)} \beta^2, \quad A_{\mathrm{Eu}}^{(2)} = -0.3497, \tag{5.14}$$

where $\beta = ha^{-1}$, E is elasticity modulus, and ν is Poisson's ratio.

As we see from (5.13) and (5.14), a variation in the condition of fastening of the "conceptionally separated" edge of the circular plate from rigid fixing to hinged supporting produces a 3.5-time change of the Eulerian critical compressive stress.

When using the "beam approximation", the Eulerian critical values of compressive forces for a flat circular plate of radius a and the plate thickness h are calculated, taking into account (5.13) and (5.14), for the material with *Bartenev–Khazanovich potential* by the following formulas:

- for the condition of *rigid fixing* of the plate edge

$$\begin{aligned} \left(S_{11}^{0}\right)_{\mathrm{Eu}}^{(1)} &= 2A_{\mathrm{Eu}}^{(1)}\mu\beta^2, \\ A_{\mathrm{Eu}}^{(1)} &= -2.4467; \end{aligned} \tag{5.15}$$

- for the condition of *hinged supporting* of the plate edge

$$\begin{aligned} \left(S_{11}^{0}\right)_{\mathrm{Eu}}^{(2)} &= 2A_{\mathrm{Eu}}^{(2)}\mu\beta^2, \\ A_{\mathrm{Eu}}^{(1)} &= -0.6991. \end{aligned} \tag{5.16}$$

By employing the procedure described above, the expression for critical compressive stress derived in the framework of the latter approach (rigorous three-dimensional approach) for the material with *Bartenev–Khazanovich potential* can be represented in the form similar to (5.15), (5.16), viz. [18]:

$$\left(S_{11}^{0}\right)_{\mathrm{B}} = 2\mu\left[(1-\varepsilon_{\mathrm{B}})^3\right](1-\varepsilon_{\mathrm{B}})^{-4} = 2A_{\mathrm{B}}\mu\beta^2, \quad A_{\mathrm{B}} = A_{\mathrm{B}}(\beta), \tag{5.17}$$

where ε_{B} is the critical relative contraction for the material with *Bartenev–Khazanovich* potential. The values of coefficient A_{B} are calculated for each value of β using the methodology described above.

For this material, Table 5.1 shows dimensionless values of critical compressive stresses $(S^0{}_{11})_{\mathrm{B}}/\mu$ obtained within the rigorous three-dimensional approach (see (5.17)), as well as the values of critical compressive stresses obtained with reliance on the "beam approximation" ($(S^0{}_{11})^{(1)}{}_{\mathrm{Eu}}/\mu$—for the idealized model of a circular plate with the rigid fixing edge (see (5.15)), $(S^0{}_{11})^{(2)}{}_{\mathrm{Eu}}/\mu$—for a circular plate with the hinged supporting edge (see (5.16))). In the fourth and fifth columns of the table, in parentheses, the percentage of deviation of the values obtained using the approximate approach from the values obtained using the rigorous three-dimensional approach is indicated.

As can be seen from Table 5.1, the approximate computational scheme in the form of a circular plate cannot be used for investigating the problem on the compression of a semi-bounded body containing a near-surface penny-shaped crack, with the values of relative distance between the crack and the body boundary exceeding 1/20 of the crack radius. Indeed, at $\beta = 0.05$, the approximate computational scheme in the form of a circular plate with the rigid fixing edge gives a substantially higher (by 5.2%) value of the critical compressive stress $(S^0{}_{11})^{(1)}{}_{\mathrm{Eu}}$ as compared with the value of $(S^0{}_{11})_{\mathrm{B}}$ obtained with the use of the rigorous three-dimensional approach; for the value of $\beta = 0.1$ the excess is 9.2%, while for $\beta = 1.0$ it amounts to 67.1%. When the approximate computational scheme in the form of a circular plate with the hinged supporting edge is employed, the obtained critical values of compressive stresses $(S^0{}_{11})^{(2)}{}_{\mathrm{Eu}}$ prove to be significantly smaller (by 52.3–71.1%) in comparison with the precise value of $(S^0{}_{11})_{\mathrm{B}}$ in the entire analyzed range of β change (at $0.01 \leqslant \beta \leqslant 1.0$).

Table 5.1 Values of critical values of relative contractions ε_B and compressive stresses in a material with Bartenev–Khazanovich potential

β	ε_B	$(S^0{}_{11})_B/\mu$	$(S^0{}_{11})^{(1)}{}_{Eu}/\mu$	$(S^0{}_{11})^{(2)}{}_{Eu}/\mu$
0.01	$8.0573 \cdot 10^{-5}$	$-4.8355 \cdot 10^{-4}$	$-4.8933 \cdot 10^{-4}$ (1.2%)	$-1.3981 \cdot 10^{-4}$ (−71.1%)
0.02	$3.1845 \cdot 10^{-4}$	$-1.9125 \cdot 10^{-3}$	$-1.9573 \cdot 10^{-3}$ (2.3%)	$-5.5924 \cdot 10^{-4}$ (−70.8%)
0.03	$7.0812 \cdot 10^{-4}$	$-4.2577 \cdot 10^{-3}$	$-4.4040 \cdot 10^{-3}$ (3.4%)	$-1.2583 \cdot 10^{-3}$ (−70.4%)
0.04	$1.2444 \cdot 10^{-3}$	$-7.4944 \cdot 10^{-3}$	$-7.8293 \cdot 10^{-3}$ (4.5%)	$-2.2369 \cdot 10^{-3}$ (−70.2%)
0.05	0.0019	−0.0116	−0.0122 (5.2%)	−0.0035 (−69.5%)
0.1	0.0073	−0.0448	−0.0489 (9.2%)	−0.0140 (−68.8%)
0.25	0.039	−0.2638	−0.3058 (15.9%)	−0.0874 (−66.9%)
0.5	0.117	−1.0187	−1.2234 (20.1%)	−0.3495 (−65.7%)
1.0	0.224	−2.9305	−4.8933 (67.1%)	−1.3981 (−52.3%)

Table 5.2 Values of critical relative contractions (critical stresses) and coefficients A_B, A_T, and A_C for a small distance between the crack and the boundary surface in material with *Bartenev–Khazanovich potential* (ε_B, A_B), material with *Treloar potential* (ε_T, A_T), and laminated composite material (σ_{11}^0/E, A_C) (a near-surface crack)

β	ε_B	A_B	ε_T	A_T	σ_{11}^0/E	A_C
$9 \cdot 10^{-2}$	$5.9607 \cdot 10^{-3}$	−2.24766	$5.96120 \cdot 10^{-3}$	−8.83140	$-7.997 \cdot 10^{-3}$	−0.987
$8 \cdot 10^{-2}$	$4.7608 \cdot 10^{-3}$	−2.26383	$4.76105 \cdot 10^{-3}$	−8.92798	$-6.553 \cdot 10^{-3}$	−1.024
$7 \cdot 10^{-2}$	$3.6850 \cdot 10^{-3}$	−2.28126	$3.68511 \cdot 10^{-3}$	−9.02475	$-5.197 \cdot 10^{-3}$	−1.061
$6 \cdot 10^{-2}$	$2.7374 \cdot 10^{-3}$	−2.30003	$2.73747 \cdot 10^{-3}$	−9.12490	$-3.951 \cdot 10^{-3}$	−1.097
$5 \cdot 10^{-2}$	$1.9224 \cdot 10^{-3}$	−2.32024	$1.92242 \cdot 10^{-3}$	−9.22762	$-2.834 \cdot 10^{-3}$	−1.134
$4 \cdot 10^{-2}$	$1.2444 \cdot 10^{-3}$	−2.34200	$1.24441 \cdot 10^{-3}$	−9.33309	$-1.871 \cdot 10^{-3}$	−1.169
$3 \cdot 10^{-2}$	$7.0812 \cdot 10^{-4}$	−2.36541	$7.08118 \cdot 10^{-4}$	−9.44157	$-1.084 \cdot 10^{-3}$	−1.204
$2 \cdot 10^{-2}$	$3.1845 \cdot 10^{-4}$	−2.39062	$3.18445 \cdot 10^{-4}$	−9.55334	$-4.947 \cdot 10^{-4}$	−1.237
$1 \cdot 10^{-2}$	$8.0573 \cdot 10^{-5}$	−2.41776	$8.05725 \cdot 10^{-5}$	−9.66870	$-1.268 \cdot 10^{-4}$	−1.268
$1 \cdot 10^{-3}$	$8.1458 \cdot 10^{-7}$	−2.44373	$8.14575 \cdot 10^{-7}$	−9.77490	$-1.290 \cdot 10^{-6}$	−1.290
$1 \cdot 10^{-4}$	$8.1506 \cdot 10^{-9}$	−2.44519	$8.15068 \cdot 10^{-9}$	−9.78076	$-1.287 \cdot 10^{-8}$	−1.287
$1 \cdot 10^{-5}$	$8.1476 \cdot 10^{-11}$	−2.44427	$8.14758 \cdot 10^{-11}$	−9.77710	$-1.285 \cdot 10^{-10}$	−1.285
$1 \cdot 10^{-6}$	$8.1461 \cdot 10^{-13}$	−2.44382	$8.14606 \cdot 10^{-13}$	−9.77527	$-1.284 \cdot 10^{-12}$	−1.284
$1 \cdot 10^{-9}$	$8.1458 \cdot 10^{-19}$	−2.44373	$8.14575 \cdot 10^{-19}$	−9.77490	$-1.284 \cdot 10^{-18}$	−1.284

For the material with *Bartenev–Khazanovich potential*, the values of critical relative contraction ε_B and coefficients $A_B = A_B(\beta)$ in the case of a small distance between the crack and the boundary surface in material are given, respectively, in columns 2 and 3 of Table 5.2.

In a similar way, the Eulerian critical values of compressive forces for the material with *Treloar potential* are calculated by the following formulas:

- for the condition of *rigid fixing* of the plate edge

$$\begin{aligned}\left(S^0_{11}\right)^{(1)}_{\text{Eu}} &= A^{(1)}_{\text{Eu}} c_{10}\beta^2,\\ A^{(1)}_{\text{Eu}} &= -9.7867;\end{aligned} \tag{5.18}$$

- for the condition of *hinged supporting* of the plate edge

$$\begin{aligned}\left(S^0_{11}\right)^{(2)}_{\text{Eu}} &= A^{(2)}_{\text{Eu}} c_{10}\beta^2,\\ A^{(1)}_{\text{Eu}} &= -2.7973.\end{aligned} \tag{5.19}$$

The expression for critical compressive stress derived in the framework of the rigorous three-dimensional approach for the material with *Treloar potential* can be represented in the form similar to (5.18), (5.19), viz. [18]:

$$\left(S^0_{11}\right)_{\text{T}} = 2c_{10}\left[1 - (1 - \varepsilon_{\text{T}})^{-6}\right] = A_{\text{T}} c_{10}\beta^2, \quad A_{\text{T}} = A_{\text{T}}(\beta), \tag{5.20}$$

where ε_{T} is critical relative contraction for the material with *Treloar potential*. For this potential, the values of critical relative contraction ε_{T} and coefficient $A_{\text{T}} = A_{\text{T}}(\beta)$ are given, respectively, in columns 4 and 5 of Table 5.2.

For a *laminated two-component composite with isotropic layers*, the dimensionless compressive stress σ^0_{11}/E at small values of geometric parameter β can be represented in the form similar to the expression of Eulerian critical compressive force for a penny-shaped plate (5.13), (5.14), viz.:

$$\sigma^0_{11}/E = A_{\text{C}}\beta^2. \tag{5.21}$$

The critical values of σ^0_{11}/E and the values of coefficient $A_{\text{C}} = A_{\text{C}}(\beta))$ for composite are given, respectively, in columns 6 and 7 of Table 5.2 (it is assumed that Poisson's ratios of the layers of the composite $\nu^{(1)} = \nu^{(2)} = 0.3$ and volume concentration of the layers with elasticity modulus $E^{(1)}$ is $c_1 = 0.7$).

Now, by examining the asymptotic behavior of critical compressive stresses when $\beta \to 0$, we need to find the limit values of coefficients A_{B}, A_{T}, A_{C} and compare these coefficients with the values of $A^{(i)}{}_{\text{Eu}}$ $(i = 1, 2)$. It follows from Table 5.2 that when $\beta < 0.01$ the difference of coefficients A_{B}, A_{T} and coefficient $A^{(1)}{}_{\text{Eu}}$, which corresponds to the *rigid fixing* of the edge of thin penny-shaped plate, is less than 1%, while for $\beta = 1 \cdot 10^{-9}$ this difference is 0.12%. Similar conclusions are obtained for the laminated composite. Thus, for materials with *Bartenev–Khazanovich* and *Treloar* elastic potentials, as well as for the laminated composite in the axisymmetric problem on the biaxial uniform compression of a body with a near-surface crack, the results of calculating critical compression parameters in the framework of the rigorous three-dimensional approach prove to be very close to the results obtained within the "beam approximation", when a circular plate with *rigid fixing* of its edge is used as an approximate scheme.

5.3 A Body Containing Two Parallel Penny-Shaped Cracks

We will consider a solid containing two coaxial penny-shaped cracks of the same radius a which are located in the planes $x_3 = 0$ and $x_3 = -2h$, with centers on Ox_3-axis. As before, it is assumed that the body is isotropic or transversely isotropic (in the case of the transversely isotropic body, axis Ox_3 is the axis of material's properties isotropy) and is loaded by compressive forces $S_{11}^0 = S_{22}^0$ directed along the planes where the cracks are located. The crack faces are assumed to be free of stress. By using the relation of the three-dimensional linearized theory of stability of deformable bodies, we reduce the problem to the solution of a system of Fredholm integral equations of the second kind (5.2); the kernels of integral equations are of the form (5.3) (see more details in [3, 8, 20]).

For the case of equal roots ($n_1^0 = n_2^0$) of the characteristic equation [15], they are represented by formulas

$$
\begin{aligned}
R_1(\zeta) &= \beta_1 \frac{\beta_1^2(1+k) + (k-1)\zeta^2}{(\beta_1^2+\zeta^2)^2}, \\
R_2(\zeta) &= \frac{\beta_1}{\beta_1^2+\zeta^2}, \\
S_1(\zeta) &= \frac{\beta_1^3}{2}\frac{\beta_1^2 - 3\zeta^2}{(\beta_1^2+\zeta^2)^3}, \\
S_2(\zeta) &= \beta_1 \frac{\beta_1^2(1-k) - (k+1)\zeta^2}{(\beta_1^2+\zeta^2)^2}.
\end{aligned}
\tag{5.22}
$$

In (5.22), β_1 and k are determined from (5.4).

For the case of non-equal roots ($n_1^0 \neq n_2^0$) of the characteristic equation [15],

$$
\begin{aligned}
R_1(\zeta) &= \frac{2k_2\beta_1}{4\beta_1^2+\zeta^2} - \frac{2k_1\beta_2}{4\beta_2^2+\zeta^2}, \\
R_2(\zeta) &= \frac{k_2}{2\beta}\left[\ln(4\beta_2^2+\zeta^2) - \ln(4\beta_1^2+\zeta^2)\right], \\
S_1(\zeta) &= k_1\beta\left[\frac{4\beta_1^2-\zeta^2}{4\beta_1^2+\zeta^2} - \frac{4\beta_2^2-\zeta^2}{4\beta_2^2+\zeta^2}\right], \\
S_2(\zeta) &= \frac{2k_1\beta_1}{4\beta_1^2+\zeta^2} - \frac{2k_2\beta_2}{4\beta_2^2+\zeta^2}.
\end{aligned}
\tag{5.23}
$$

In (5.23), β_i, k_i ($i = 1, 2$) are determined from (5.5).

To calculate the values of critical contraction ratio λ_1 and critical compressive stress S_{11}^0 corresponding to the local loss of equilibrium of the material in the vicinity of cracks at the initial stage of fracture, we employ the procedure analogous to that used in the study of the problem on body compression along a near-surface crack. After substituting expressions (5.6) in the integral equation (5.2) and using (5.9) for integral equations kernels (5.3), with taking into account (5.22), (5.23), we obtain the system of $2N + 3$ Eq. (5.10).

Figures 5.4, 5.5, and 5.6 show, respectively, the dependence of critical relative contraction $\varepsilon = 1 - \lambda_1$ on the dimensionless half-distance between the cracks $\beta = ha^{-1}$ for highly elastic materials with *Bartenev–Khazanovich* and with *Treloar potentials*, and for *two-component laminated composite with isotropic layers* (parameters of these materials are given in Sect. 5.2). The left-hand pictures in Figs. 5.4, 5.5, and 5.6 correspond to a small distance between the cracks and the right-hand pictures—to a large one.

As can be seen from Figs. 5.4, 5.5, and 5.6, when the distance between the cracks decreases, the critical values of relative contraction $\varepsilon = 1 - \lambda_1$ (and, correspondingly, the critical values of compressive load) significantly decrease as compared with the case of an isolated crack in an unbounded body (see [20]).

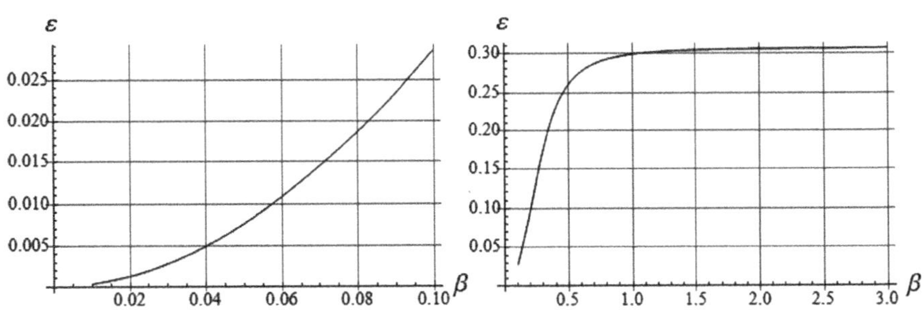

Fig. 5.4 Dependence of critical values of relative contraction $\varepsilon = 1 - \lambda_1$ on geometric parameter $\beta = ha^{-1}$ for a material with *Bartenev–Khazanovich potential* (two parallel cracks)

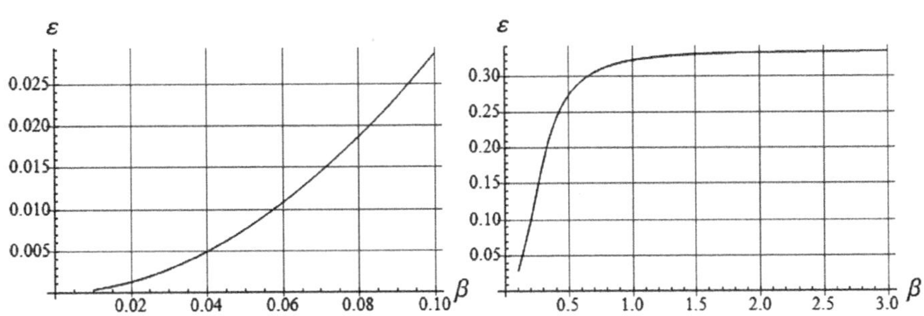

Fig. 5.5 Dependence of critical values of relative contraction $\varepsilon = 1 - \lambda_1$ on geometric parameter $\beta = ha^{-1}$ for a material with *Treloar potential* (two parallel cracks)

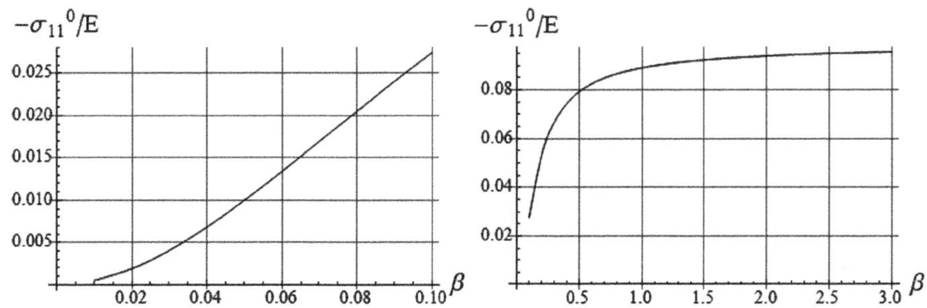

Fig. 5.6 Dependence of the critical values of dimensionless compressive stress σ_{11}^0/E on geometric parameter $\beta = ha^{-1}$ for a laminated composite material (two parallel cracks)

When the "beam approximation" is employed in the problem on compression of a body with two parallel coaxial penny-shaped cracks, the scheme in the form of a circular plate of radius a and thickness $2h$ is used, and conditions of fastening the plate edge are *rigid fixing* or *hinged supporting*. To determine the values of Eulerian critical compressive force, we can use $(5.15)_1$, $(5.16)_1$ (for the material with *Bartenev–Khazanovich potential*), (5.18), (5.19) (for the material with *Treloar potential*), and (5.13), (5.14) (for the laminated composite) if we replace β by 2β in these formulas. Coefficients $A^{(i)}{}_{\mathrm{Eu}}$ $(i = 1, 2)$ are determined by expressions $(5.15)_2$ and $(5.16)_2$ (for the material with *Bartenev–Khazanovich potential*), (5.18) and (5.19) (for the material with *Treloar potential*), and (5.13), (5.14) (for the laminated composite).

The expression for critical compressive stress derived in the framework of rigorous three-dimensional approach can be represented for *Bartenev–Khazanovich potential* in the form (5.17), for *Treloar potential*—in the form (5.20), and for the laminated composite—in the form (5.21) if we replace β by 2β in these formulas.

Table 5.3 shows the results of calculating the values A_{B}, A_{T}, and A_{C} involved, respectively, in expressions (5.17), (5.20), and (5.21) for different values of the geometric parameter β. It follows from Table 5.3 that when $\beta < 0.01$, the difference of coefficients A_{B}, A_{T} from coefficient $A^{(1)}{}_{\mathrm{Eu}}$, which corresponds to the *rigid fixing* of the edge of a thin penny-shaped plate, is less than 1.5%. Therefore, the use of a circular plate with boundary condition of *rigid fixing* on its edge as the idealized model within the "beam approximation" has been substantiated for the problem on compression of a body with two parallel coaxial cracks located at the distance $2h \leqslant 0.02a$ from each other.

Table 5.3 Values of critical relative contractions (critical stresses) and coefficients A_B, A_T, A_C for a small distance between the cracks in a material with Bartenev–Khazanovich potential (ε_B, A_B), a material with Treloar potential (ε_T, A_T), and a laminated composite material (σ_{11}^0/E, A_C) (two parallel cracks)

β	ε_B	A_B	ε_T	A_T	σ_{11}^0/E	A_C
$9 \cdot 10^{-2}$	$2.34269 \cdot 10^{-2}$	−2.32948	$2.34532 \cdot 10^{-2}$	−8.68639	$-2.40491 \cdot 10^{-2}$	−0.74226
$8 \cdot 10^{-2}$	$1.87479 \cdot 10^{-2}$	−2.32564	$1.87615 \cdot 10^{-2}$	−8.79447	$-2.05282 \cdot 10^{-2}$	−0.80188
$7 \cdot 10^{-2}$	$1.45380 \cdot 10^{-2}$	−2.32532	$1.45445 \cdot 10^{-2}$	−8.90477	$-1.69604 \cdot 10^{-2}$	−0.86533
$6 \cdot 10^{-2}$	$1.08187 \cdot 10^{-2}$	−2.32875	$1.08214 \cdot 10^{-2}$	−9.01782	$-1.34195 \cdot 10^{-2}$	−0.93191
$5 \cdot 10^{-2}$	$7.61086 \cdot 10^{-3}$	−2.33624	$7.61180 \cdot 10^{-3}$	−9.13416	$-1.00051 \cdot 10^{-2}$	−1.00051
$4 \cdot 10^{-2}$	$4.93540 \cdot 10^{-3}$	−2.34808	$4.93566 \cdot 10^{-3}$	−9.25436	$-6.84424 \cdot 10^{-3}$	−1.06941
$3 \cdot 10^{-2}$	$2.81367 \cdot 10^{-3}$	−2.36463	$2.81372 \cdot 10^{-3}$	−9.37906	$-4.09503 \cdot 10^{-3}$	−1.13626
$2 \cdot 10^{-2}$	$1.26786 \cdot 10^{-3}$	−2.38630	$1.26786 \cdot 10^{-3}$	−9.50896	$-1.91641 \cdot 10^{-3}$	−1.24780
$1 \cdot 10^{-2}$	$3.21497 \cdot 10^{-4}$	−2.41355	$3.21497 \cdot 10^{-4}$	−9.64490	$-5.01367 \cdot 10^{-4}$	−1.25342
$1 \cdot 10^{-3}$	$3.25751 \cdot 10^{-6}$	−2.44315	$3.25751 \cdot 10^{-6}$	−9.77252	$-5.16309 \cdot 10^{-6}$	−1.29077
$1 \cdot 10^{-4}$	$3.26050 \cdot 10^{-8}$	−2.44537	$3.26050 \cdot 10^{-8}$	−9.78149	$-5.15527 \cdot 10^{-8}$	−1.28881
$1 \cdot 10^{-5}$	$3.25919 \cdot 10^{-10}$	−2.44439	$3.25919 \cdot 10^{-10}$	−9.77756	$-5.14160 \cdot 10^{-10}$	−1.28540
$1 \cdot 10^{-6}$	$3.25854 \cdot 10^{-12}$	−2.44391	$3.25854 \cdot 10^{-12}$	−9.77563	$-5.13770 \cdot 10^{-12}$	−1.28443
$1 \cdot 10^{-9}$	$3.25833 \cdot 10^{-18}$	−2.44375	$3.25833 \cdot 10^{-18}$	−9.77499	$-5.13672 \cdot 10^{-18}$	−1.28418

5.4 A Body Containing a Periodic Array of Parallel Penny-Shaped Cracks

We consider an isotropic or transversely isotropic solid body with an infinite array of coaxial penny-shaped cracks of the same radius a, which are located in the parallel planes $x_3 = -2hn$ (where $n = 0, \pm 1, \pm 2, \ldots$), with centers on Ox_3-axis. Compressive forces $S_{11}^0 = S_{22}^0$ act along the cracks; the crack surfaces are free of stresses.

By using the approach in the framework of the three-dimensional linearized theory of stability of deformable bodies, the problem is reduced to the solution of Fredholm integral equation of the second kind (for more details, see [4, 20]) in the form

$$f(\xi) - \frac{1}{\pi}\int_0^1 M(\xi, \eta) f(\eta) d\eta = 0, \quad 0 \leqslant \xi, \eta \leqslant 1. \tag{5.24}$$

The kernel of the integral equation (5.24) is of the form

$$M_1(\xi, \eta) = R(\xi - \eta) - R(\eta + \xi). \tag{5.25}$$

For the case of equal roots ($n_1^0 = n_2^0$) of the characteristic equation [15], we have

$$R(\zeta) = \frac{1}{k\beta_1}\left[(k-1)\mathrm{Re}\,\psi\left(1 + \frac{i\zeta}{2\beta_1}\right) + \frac{\zeta}{2\beta_1}\mathrm{Im}\,\psi_1\left(1 + \frac{i\zeta}{2\beta_1}\right)\right], \tag{5.26}$$

where β_1 and k are determined from (5.4). For the case of non-equal roots ($n_1^0 \neq n_2^0$) of the characteristic equation [15]:

$$R(\zeta) = \frac{1}{k}\left[\frac{k_2}{\beta_1}\mathrm{Re}\,\psi\left(1 + \frac{i\zeta}{2\beta_1}\right) - \frac{k_1}{\beta_2}\mathrm{Re}\,\psi\left(1 + \frac{i\zeta}{2\beta_2}\right)\right]. \tag{5.27}$$

In (5.26), (5.27)

$$\psi(z) = \frac{d}{dz}\ln\Gamma(z),$$

where $\Gamma(z)$ is the gamma-function; ψ_1 is the derivative of $\psi(z)$; β_i and k_i ($i = 1, 2$) and k are determined from (5.5).

To investigate the system of Fredholm integral equations of the second kind (5.24), we use the procedure based on the Bubnov–Galerkin method. A system of power functions is used as coordinate functions. The representation of the required functions in terms of the first N coordinate functions is of the form

$$f(x) = \sum_{i=0}^{N} F_i x^i. \tag{5.28}$$

Similar to the procedure described in Sect. 5.2, after substituting (5.28) into (5.24), the Fredholm integral equation (5.24) is transformed into the system of equations

$$\sum_{i=0}^{N} F_i F_{ji} = 0, \quad 0 \leqslant j \leqslant N. \tag{5.29}$$

As an example, we consider numerical results for highly elastic materials with *Bartenev–Khazanovich* and *Treloar* elastic potentials. Figures 5.7 and 5.8 show, respectively, the dependence of critical relative contraction $\varepsilon = 1 - \lambda_1$ on the dimensionless half-distance between the cracks $\beta = ha^{-1}$ for highly elastic materials with *Bartenev–Khazanovich* and with *Treloar potentials.* These figures demonstrate that the mutual influence of the cracks at quite small distances between them results in a significant decrease of the critical values of relative contraction $\varepsilon = 1 - \lambda_1$ (and, correspondingly, the critical values of compressive load) as compared with the case of an isolated crack in an unbounded body (see [20]).

Further we determine critical values of Eulerian compressive force for the circular plate of radius a and thickness $2h$, which within the "beam approximation" models the part of material separated by neighboring cracks. For determining Eulerian compressive force, we can use representations (5.15), (5.16) (for the material with *Bartenev–Khazanovich potential*), and (5.18), (5.19) (for the material with *Treloar potential*) if we replace β by 2β in these formulas. Coefficients $A^{(i)}{}_{\mathrm{Eu}}$ ($i = 1, 2$) are determined by expressions $(5.15)_2$, $(5.16)_2$ (for the material with *Bartenev–Khazanovich potential*), (5.18), (5.19) (for the material with *Treloar potential*). For example, in the case of highly elastic material with *Treloar potential*, we have

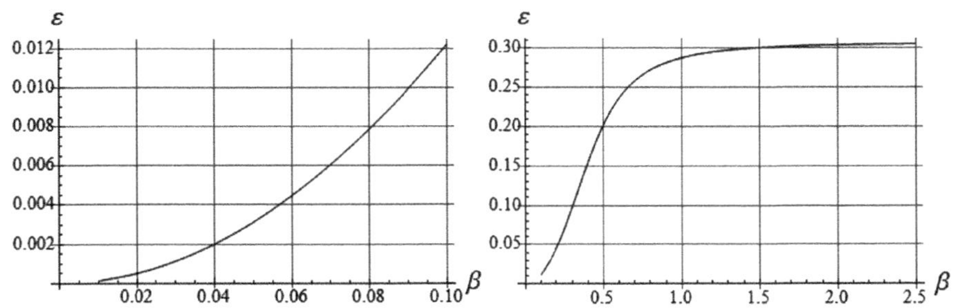

Fig. 5.7 Dependence of the critical values of relative contraction $\varepsilon = 1 - \lambda_1$ on geometric parameter $\beta = ha^{-1}$ for a material with *Bartenev–Khazanovich potential* (an array of parallel cracks)

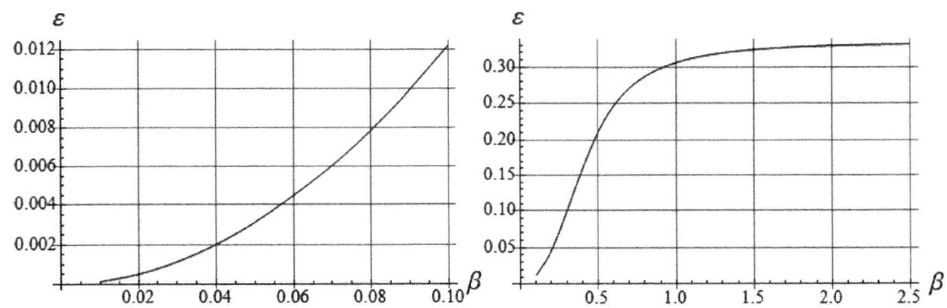

Fig. 5.8 Dependence of the critical values of relative contraction $\varepsilon = 1 - \lambda_1$ on geometric parameter $\beta = ha^{-1}$ for a material with *Treloar potential* (an array of parallel cracks)

- for the condition of *rigid fixing* of the plate edge

$$\begin{aligned} \left(S_{11}^{0}\right)_{\mathrm{Eu}}^{(1)} &= 4A_{\mathrm{Eu}}^{(1)} c_{10}\beta^2, \\ A_{\mathrm{Eu}}^{(1)} &= -9.7867; \end{aligned} \tag{5.30}$$

- for the condition of *hinged supporting* of the plate edge

$$\begin{aligned} \left(S_{11}^{0}\right)_{\mathrm{Eu}}^{(2)} &= 4A_{\mathrm{Eu}}^{(2)} c_{10}\beta^2, \\ A_{\mathrm{Eu}}^{(1)} &= -2.7973. \end{aligned} \tag{5.31}$$

The expression for critical compressive stress derived in the framework of the rigorous three-dimensional approach for the material with *Treloar potential* can be represented in the form similar to (5.18), (5.19), viz. [18]:

$$\left(S_{11}^{0}\right)_{\mathrm{T}} = 4A_{\mathrm{T}} c_{10}\beta^2, \quad A_{\mathrm{T}} = A_{\mathrm{T}}(\beta). \tag{5.32}$$

In (5.30), (5.31), and (5.32) we use geometric parameter $\beta = ha^{-1}$.

Table 5.4 Values of critical relative contractions and coefficients A_B, A_T for a small distance between the cracks in materials with *Bartenev–Khazanovich potential* (ε_B, A_B) and with *Treloar potential* (ε_T, A_T) (an array of parallel cracks)

β	ε_B	A_B	$A_B/A^{(2)}{}_{Eu}$	ε_T	A_T	$A_T/A^{(2)}{}_{Eu}$
$9 \cdot 10^{-2}$	$9.92168 \cdot 10^{-3}$	−0.946601	1.351	$9.92344 \cdot 10^{-3}$	−3.67535	1.313
$8 \cdot 10^{-2}$	$7.87937 \cdot 10^{-3}$	−0.945562	1.350	$7.88047 \cdot 10^{-3}$	−3.69397	1.319
$7 \cdot 10^{-2}$	$6.06377 \cdot 10^{-3}$	−0.945230	1.350	$6.06406 \cdot 10^{-3}$	−3.71269	1.326
$6 \cdot 10^{-2}$	$4.47813 \cdot 10^{-3}$	−0.945596	1.351	$4.47813 \cdot 10^{-3}$	−3.73177	1.333
$5 \cdot 10^{-2}$	$3.12607 \cdot 10^{-3}$	−0.946676	1.352	$3.12617 \cdot 10^{-3}$	−3.75141	1.340
$4 \cdot 10^{-2}$	$2.01133 \cdot 10^{-3}$	−0.948523	1.355	$2.01133 \cdot 10^{-3}$	−3.77124	1.347
$3 \cdot 10^{-2}$	$1.12740 \cdot 10^{-3}$	−0.951077	1.359	$1.13711 \cdot 10^{-3}$	−3.79036	1.354
$2 \cdot 10^{-2}$	$5.08301 \cdot 10^{-4}$	−0.954519	1.364	$5.08203 \cdot 10^{-4}$	−3.81152	1.361
$1 \cdot 10^{-2}$	$1.27734 \cdot 10^{-4}$	−0.958375	1.369	$1.27734 \cdot 10^{-4}$	−3.83203	1.369
$1 \cdot 10^{-3}$	$1.28418 \cdot 10^{-6}$	−0.963138	1.376	$1.28125 \cdot 10^{-6}$	−3.84375	1.373
$1 \cdot 10^{-4}$	$1.28613 \cdot 10^{-8}$	−0.964600	1.378	$1.28516 \cdot 10^{-8}$	−3.85547	1.377
$1 \cdot 10^{-5}$	$1.28711 \cdot 10^{-10}$	−0.965332	1.379	$1.28516 \cdot 10^{-10}$	−3.85547	1.377
$1 \cdot 10^{-6}$	$1.28711 \cdot 10^{-12}$	−0.965332	1.379	$1.28516 \cdot 10^{-12}$	−3.85547	1.377
$1 \cdot 10^{-9}$	$1.28809 \cdot 10^{-18}$	−0.966064	1.380	$1.28516 \cdot 10^{-18}$	−3.85547	1.377

Table 5.4 shows the values of critical relative contractions ε_B (for *Bartenev–Khazanovich potential*) and ε_T (for *Treloar potential*) as well as for coefficients A_B (for Bartenev–Khazanovich potential) and A_T (for *Treloar potential*) for geometric parameter $\beta = ha^{-1}$ changing in interval $1 \cdot 10^{-9} \leqslant \beta \leqslant 9 \cdot 10^{-2}$. For comparison, here we give the relations of value A, obtained within the rigorous three-dimensional approach) to value $A^{(1)}{}_{Eu}$ (obtained within the "beam approximation" for the condition of *rigid fixing* of the plate edge) and to value $A^{(2)}{}_{Eu}$ (obtained within the "beam approximation" for the condition of *hinged supporting* of the plate edge).

From the comparison of the results shown in Table 5.4 for $A_{B,T}/A^{(1)}{}_{Eu}$ and for $A_{B,T}/A^{(2)}{}_{Eu}$, it can be seen that the values of Eulerian critical compressive force used within the "beam approximation" both in respect of the idealized model in the form of a circular plate with *rigid fixing* of its edge and in respect of the model as a circular plate with *hinged supporting* of its edge are substantially different from the value of critical compressive stress obtained within the rigorous three-dimensional approach (the approach relying on the relationships of the three-dimensional linearized theory of the stability of deformable bodies). It should be noted here that the results for values A_B and A_T in Table 5.4 corresponding to the rigorous approach are closer to value $A^{(2)}{}_{Eu}$ that corresponds within the approximate approach to Eulerian critical compressive force for a circular plate with the boundary conditions of *hinged supporting* of its edge rather than to value $A^{(1)}{}_{Eu}$ corresponding to critical compressive force for a circular plate with the boundary conditions of *rigid fixing* of its edge. But as it follows from the analysis of the above results, even for quite long cracks, when the results of the approximate approach (the "beam approximation") are used, for *hinged supporting* in the entire interval of parameter β change there is a permanent error (in excess of 30%); when the approximate scheme in the form of a plate with *rigid fixing* of its edge is used, the above-mentioned error exceeds 140%.

Thus, we can conclude that the application of the approximate scheme in the form of a plate both with *hinged supporting* and with the *rigid fixing* of its edge is not sufficiently substantiated for the analysis of the problem on fracture of bodies with periodic arrays of coaxial cracks under compression along the cracks. To study them, the rigorous approach in the framework of the three-dimensional linearized theory of the stability of deformable bodies should be applied.

5.5 Conclusions

In the framework of the approaches of the three-dimensional linearized theory of stability of deformable bodies [12–15, 18], spatial axisymmetric problems on the fracture of bodies under compression along the circular cracks contained in them have been studied. The cases of parallel coaxial cracks located close to each other and near-surface cracks located close to the boundary surface of the body are considered. The methodology provides for the reduction of initial problems to eigenvalue problems to systems of Fredholm integral equations of the second kind and the numerical analytical study of these equations using the Bubnov–Galerkin method and the package of applied programs of symbolic mathematics for calculating integrals.

For highly elastic materials with *Bartenev–Khazanovich* and *Treloar potentials* and for two-component laminated composites with isotropic layers, the values of critical fracture parameters, which correspond to the local loss of equilibrium stability of the part of material surrounding the cracks as the start of fracture process are obtained. The results are analyzed for a fairly wide interval of the change of the relative (normalized to the crack radius) distances between parallel cracks (or between a near-surface crack and the body boundary) $\beta = ha^{-1}$ $(1 \cdot 10^{-9} \leqslant \beta \leqslant \infty)$, which also practically includes passage to the limit when $\beta \to 0$. This makes it possible to evaluate the applicability limits of the approximate approach (the so-called "beam approximation") [6, 23–28], which is popular in the studies of such problems. According to this approach, the material containing parallel cracks is substituted with a circular plate as if "separated" by cracks (or by a near-surface crack and the body boundary), and then its instability is studied with reliance on the applied theories of stability of thin-walled systems.

The following general conclusions can be made from the results obtained above:

1. For the problems investigated, at quite small values of parameter β (when $\beta \to 0$), one can obtain a decrease of the critical value of relative contraction (and, correspondingly, of critical compressive stress) by two, three, and more orders of magnitude as compared with the case of a single isolated crack.

2. The "beam approximation", which is based on substituting the part of the material between cracks (or between a crack and a body boundary) with a circular plate with a chosen a priori idealized condition of fixing its edge, seems to be absolutely inapplicable in determining critical compression parameters when separation distances between the cracks (or between the crack and the body boundary) exceed 1/20 of the crack radius. It results from the fact that in this case, when calculating the idealized model in the form of a circular plate with the rigid fixing edge, the values of critical compressive stresses obtained within the "beam approximation" are larger than the precise values by 5.2% and more, and when the computational scheme in the form of a circular plate with the hinged supporting edge is used, this difference is in excess of 50%.
3. For the problems of compression of a semi-infinite body with a near-surface penny-shaped crack and an unbounded body with two penny-shaped cracks of equal radii located in parallel planes

   - it has been ascertained that the use of boundary conditions of *rigid fixing* of the edge of the "as if separated thin circular plate" is valid in the investigation within the "beam approximation" for very small values of the relative (normalized to the crack radius) distances between parallel cracks (or between the near-surface crack and the body boundary) (when $\beta < 0.01$). The validity is achieved due to the fact that when $\beta \to 0$ the results obtained in the framework of the rigorous approach relying on the relationships of the three-dimensional linearized theory of the stability of deformable bodies transform into the results obtained within the above-mentioned "beam approximation";
   - the applicability limits of the "beam approximation" have been determined. When $\beta = ha^{-1} < 0.01$, the results obtained within the above-mentioned approximate and the rigorous approaches nearly coincide for the cases examined. For example, for $\beta = ha^{-1} < 0.05$, the results for the "beam approximation" and the rigorous three-dimensional approach coincide to within 5%.

4. For the problem of compression of an unbounded body containing a periodic array of parallel penny-shaped cracks, the applicability of the approximate design diagrams in the form of "as if separated thin plate" with boundary conditions as *rigid fixing* or *hinged supporting* of its edge cannot be substantiated within the "beam approach". This conclusion is based on the fact that when $\beta = ha^{-1} \to 0$ the results obtained within the rigorous three-dimensional approach do not transform into those obtained within the "beam approximation" either for the boundary conditions of *hinged supporting* or for the boundary conditions of *rigid fixing* of the edge of the "as if separated thin plate". It should be noted that in the case of a periodic array of cracks, the results obtained in the framework of the rigorous approach, when $\beta \to 0$, are closer to those obtained within the "beam approximation" for the boundary conditions of *hinged supporting* of the edge of "as if separated thin plate". Yet, even for quite thin "as if separated thin plates" (at $1 \cdot 10^{-9} \leqslant \beta \leqslant 9 \cdot 10^{-2}$) with the *hinged supporting* of its edge, there is a permanent error (in excess of 30%); when the approximate

scheme in the form of a circular plate with *rigid fixing* of its edge is used, the error exceeds 140%.

References

1. Bartenev, G.M., Khazanovich, T.N.: The law of high elastic deformation of polymers. Vysokomolekulyarnyye Soedineniya **2**(1), 21–28 (1960)
2. Bogdanov, V.L.: On a circular shear crack in a semiinfinite composite with initial stresses. Mater. Sci. **43**(3), 321–330 (2007)
3. Bogdanov, V.L.: Influence of initial stresses on fracture of composite materials containing interacting cracks. J. Math. Sci. **165**(3), 371–384 (2010)
4. Bogdanov, V.L.: Influence of initial stresses on the stressed state of a composite with a periodic system of parallel coaxial normal tensile cracks. J. Math. Sci. **186**(1), 1–13 (2012)
5. Bogdanov, V.L., Nazarenko, V.M.: Study of the compressive failure of a semi-infinite elastic material with a harmonic potential. Int. Appl. Mech. **30**(10), 760–765 (1994)
6. Bolotin, V.V.: Stability Problems in Fracture Mechanics. Wiley, New York (1994)
7. Chrepanov, G.P.: Mechanics of Brittle Fracture. McGraw-Hill, New York (1979)
8. Dovzhyk, M., Bogdanov, V., Nazarenko, V.: Fracture of composite materials at compression along two parallel cracks. In: Gdoutos, E. (ed.) Proceedings of the First International Conference on Theoretical, Applied and Experimental Mechanics, ICTAEM 2018, Structural Integrity, vol. 5, pp.190–193. Springer, Cham (2018)
9. Dovzhyk, M., Bogdanov, V., Nazarenko, V.: Fracture of composite material at compression along near-surface crack. In: Gdoutos, E. (ed.) Proceedings of the Second International Conference on Theoretical, Applied and Experimental Mechanics, ICTAEM 2019, Structural Integrity, vol. 8, pp. 114–118. Springer, Cham (2019)
10. Gdoutos, E.E.: Fracture Mechanic: An Introduction. Springer, Dordrecht (2005)
11. Griffith, A.A.: The phenomenon of rupture and flow in solids. Phil. Trans. Roy. Soc., Ser. A. **211**(2), 163–198 (1920)
12. Guz, A.N.: Stability of Elastic Bodies Under Finite Strains. Naukova Dumka, Kyiv (in Russian) (1973)
13. Guz, A.N.: A criterion of solid body destruction during compression along cracks. (Two-dimensional problem). Dokl. Akad. Nauk SSSR **259**(6), 1315–1318 (1981)
14. Guz, A.N.: A criterion of solid body destruction under compression along cracks. (A 3-dimensional problem). Dokl. Akad. Nauk SSSR **259**(6), 1315–1318 (1981)
15. Guz, A.N.: Fundamentals of the Three-Dimensional Theory of Stability of Deformable Bodies. Springer, Berlin-Heidelberg-New York (1999)
16. Guz, A.N.: On some nonclassical problems of fracture mechanics taking into account the stresses along cracks. Int. Appl. Mech. **40**(8), 937–942 (2004)
17. Guz, A.N.: On study of nonclassical problems of fracture and failure mechanics and related mechanisms. Int. Appl. Mech. **45**(1), 3–40 (2009)
18. Guz, A.N.: Establishing the foundations of the mechanics of fracture of materials compressed along cracks (Review). Int. Appl. Mech. **50**(1), 1–57 (2014)
19. Guz, A.N.: Eight Non-Classical Problems of Fracture Mechanics. In: Advanced Structure Materials, vol. 138. Springer, Cham (2021)
20. Guz, A.N., Bogdanov, V.L., Nazarenko, V.M.: Three-dimensional problems on loading of bodies along cracks. In: Guz, A.N., Bogdanov, V.L., Nazarenko, V.M. (eds.) Fracture of Materials Under Compression Along Cracks, pp. 249–439. Springer, Cham (2020)
21. Guz, A.N., Dovzhik, M.V., Nazarenko, V.M.: Fracture of a material compressed along a crack located at a short distance from the free surface. Int. Appl. Mech. **47**(6), 627–635 (2011)
22. Irvin, G.R.: Analysis of stresses and strains near the end of a crack traversing a plate. J. Appl. Mech. **24**(3), 361–364 (1957)

23. Kachanov, L.M.: Delamination Buckling of Composite Materials. Kluwer Academic Publications, Boston (1988)
24. Kienzler, R., Herrman, G.: Mechanics in Material Space with Applications to Defect and Fracture Mechanics. Springer, Berlin (2000)
25. Mikhailov, A.M.: Certain problems in the theory of cracks in beam approximation. J. Appl. Mech. Tech. Phys. **8**(5), 83–86 (1967)
26. Obreimoff, I.W.: The splitting strength of mica. Proc. Roy. Soc. London, metric converter Product ID 127 A127 A, 290–297 (1930)
27. Polilov, A.N., Rabotnov, Y.M.: Development of delamination in compressed composites. Izv. Akad. Nauk SSSR **2**, 166–171 (1983)
28. Slepyan, L.I.: Mechanics of Cracks. Sudostroyeniye, Leningrad (in Russian) (1990)
29. Treloar, L.R.G.: Large elastic deformations in rubber-like materials, IUTAM Colloquium. Madrid 208–217 (1955)
30. Uflyand, Y.S.: Integral Transformations in Problems of the Theory of Elasticity. Nauka, Leningrad (in Russian) (1967)
31. Volmir, A.S.: Stability of Deformable Systems. Nauka, Moscow (in Russian) (1967)
32. Wells, A.A.: Application of fracture mechanics at and beyond general yielding. Brit. Weld. J. **10**(11), 563–570 (1963)

Chapter 6
Analysis of Fracture of an Orthotropic Plate with a Crack Under Biaxial Loading

Olga Bogdanova and Anatoly Kaminsky

Abstract A generalization of the model of a crack with a process zone to the case of orthotropic materials is proposed. It is assumed that the material in the process zone satisfies the strength condition of an arbitrary type. Using the proposed crack model, the fracture of an orthotropic plate with a crack under biaxial loading conditions was studied. The crack is located along one of the axes of the anisotropy, and the external loads act in parallel and perpendicularly to it. Within the framework of the criterion of critical crack opening and the critical value of the J-integral, the influence of the biaxiality of external loading on the critical state of a plate with a crack is analyzed. A numerical solution was obtained for the von Mises–Hill and Gol'denblatt–Kopnov strength criteria. The theoretical results obtained are compared with experimental data obtained by testing specimens made of structural metals.

6.1 Introduction

As is known, some problems in fracture mechanics cannot be studied within the framework of classical approaches. One of these problems is to evaluate the influence of an external load applied along the crack axis on the cracking process. Thus, when considering an unbounded body containing a mode I crack within the framework of the Griffiths, Barenblatt, Leonov–Panasyuk, and Dugdale crack models, we find that the values of the stress intensity factor K_I, J-integral, and crack opening displacement δ do not depend on the load component applied parallel to the crack. For this reason, the fracture-toughness characteristics calculated using these classical models of cracks do not depend on the loading biaxiality [1, 16, 26], which does not correspond to experimental data (see, for example, [2, 14, 25, 31]). This stipulates the interest of researchers in such problems of fracture mechanics, which

O. Bogdanova (✉) · A. Kaminsky
S.P. Timoshenko Institute of Mechanics, National Academy of Science of Ukraine, Kyiv, Ukraine
e-mail: o.bogdanova@i.ua

A. Kaminsky
e-mail: fract@inmech.kiev.ua

H. Altenbach et al. (eds.), *Selected Problems of Solid Mechanics and Solving Methods*, Advanced Structured Materials 204, https://doi.org/10.1007/978-3-031-54063-9_6

is confirmed by numerous publications on this topic (see, for example, [4, 14, 16, 19, 26, 27]).

To study problems of this kind, several non-classical approaches have been developed that take into account the influence of stresses acting along a crack on the characteristics of fracture toughness. These approaches are as follows:

1. To study the fracture of bodies compressed along the parallel cracks they contain, Guz [16] proposed a criterion of the local loss of stability in the part of material adjacent to the crack as the fracture criterion. Some results in such formulation for different loading schemes and crack location are given in [10, 18].
2. To investigate the fracture of materials with initial (residual) stresses acting along cracks when initial stresses are significantly larger as compared to additional (operational) stresses, in [17] A. N. Guz formulated the approach within the linearized mechanics of deformable solid bodies. The results of studying some problems of the brittle fracture mechanics of materials with initial (residual) stresses, which were obtained with reliance on this approach, were presented in [7–9, 18].
3. The approach based on a more complete allowance for distribution of stresses near the mode I crack edge, and not just the singular part, was proposed by Larsson and Carlsson [29]. According to this approach, it is assumed to introduce the second parameter T (in addition to K_I or J_I), which is a nonregular term in the expansion of stresses in the vicinity of a crack tip. This approach is known as the $K_I - T$ theory. At present many researchers use this approach to calculate T-stresses in the case of specimens with different geometry and loading schemes [4, 12, 32, 33, 35].
4. The theoretical investigations (within the framework of nonlinear approach) of J-integral values and crack opening displacement are given in works of Jansson [19] and O'Dowd et al. [34, 35] for the center-cracked plates under the biaxial tensions and materials which are assumed to obey the power-law relations.
5. Another approach to solving the elastoplastic problem for thin plates made of weakly hardening materials, based on the well-known Dugdale model [13], is presented in [3, 15, 22]. According to this approach, the material in the area of the crack tip is modeled by a thin, ideally plastic strip. This approach was used to solve problems of fracture of isotropic and orthotropic plates made of ideal elastoplastic material under biaxial loading [15, 22–26].

As an analysis of experiments [6, 20, 21, 27, 28, 38] shows a process zone origins before a crack front. For different materials, the structure of this zone is different: for polymers it is known as the craze zone, for composites it is a zone filled with half-destroyed reinforcing fibers, and for metals it is a zone of destruction. In many cases, process zones in thin plates have the form of narrow wedge-shaped regions located on the continuation of the crack and are filled with loosened and partially destroyed material [6, 20]. This situation makes it possible to model process zones, by analogy with the Leonov–Panasyuk and Dugdale models, by cuts along the extension of the crack, on the edges of which self-equilibrated compressive stresses act. However, in contrast to these models, the parameters of the model we propose are determined based on the criterion of the strength of the material in the process zone.

As an advantage of the approach proposed, it can be noted that the crack model is constructed without smallness restriction imposed on the size of a process zone. Also, there is no necessity to break up the stress field near a crack tip (as it holds in the $K_I - T$ theory) into singular and regular components.

It should be noted that for an isotropic material, when applying the von Mises–Hill criterion, the model can be significantly simplified. The corresponding analysis, as well as references to the results obtained for this model by other authors, can be found in [26].

This work synthesizes and generalizes the authors' results on the construction and substantiation of a model of a crack in an orthotropic body [11, 22–25]. These results are based on some principles of the Leonov–Panasyuk and Dugdale models and strength criteria in a general form. The effects of loading biaxiality on the critical fracture-toughness characteristics are studied, as applied to the von Mises–Hill and Gol'denblatt–Kopnov strength criteria [30], in the case when the crack is directed along one of the orthotropy axes. The theoretical results are also compared with those obtained experimentally.

6.2 The Crack Model and Problem Formulation

Let us consider a thin orthotropic plate containing a crack of length $2l$, which is located along the Ox-orthotropy axis. The plate is under the action of normal loads $\sigma_x^\infty = q$, $\sigma_y^\infty = p > 0$, which are applied at infinity.

The destruction of the defect-free material of the plate is described by the strength criterion in the general form

$$F\left(\sigma_1, \sigma_2, C_i\right) = 0, \tag{6.1}$$

where σ_1 and σ_2 are the principal stresses, and C_i is the material constants.

In constructing the crack model, the following assumptions are used:

1. Process zones arising near a crack tip take the form of a narrow wedge-shaped section on the crack continuation. In modeling, they can be replaced by slits of length d whose faces are acted upon by uniformly distributed coordinate-independent compressive stresses σ_y^0 (Fig. 6.1).
2. Stress components σ_x^0 and σ_y^0 in the process zone satisfy the strength criterion (6.1) and continuity condition on the crack front. These stress components are obtained in the course of the problem solving.
3. The stress components are finite everywhere over the whole region.

Based on these hypothesis, a boundary-value problem for an orthotropic plate with a slit of length $2L = 2(l + d)$ under the following boundary conditions can be presented:

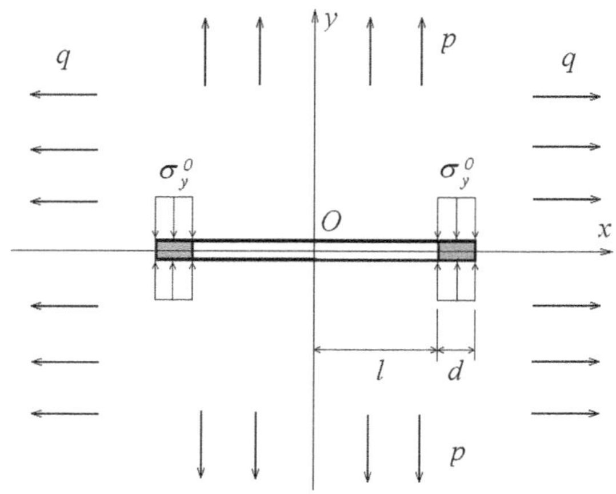

Fig. 6.1 A crack in unbounded body under biaxial loading

$$\begin{array}{l}
\sigma_y = \begin{cases} 0, & |x| < l \\ \sigma_y^0, & l \leqslant |x| \leqslant L \end{cases}, \quad y = 0, \\
\sigma_x = \sigma_x^0, \quad l \leqslant |x| \leqslant L, \quad y = 0, \\
F\left(\sigma_x^0, \sigma_y^0, C_i\right) = 0, \quad l \leqslant |x| \leqslant L, \quad y = 0, \\
\tau_{xy} = 0, \quad -\infty < x < \infty, \quad y = 0, \\
v = 0, \quad |x| \geqslant L, \quad y = 0, \\
\sigma_x = q, \quad \sigma_y = p, \quad \tau_{xy} = 0, \quad |x + iy| \to \infty.
\end{array} \tag{6.2}$$

6.3 Basic Parameters of the Model

6.3.1 The Stress State of a Process Zone

The boundary-value problem defined by (6.2) is solved by reducing to the Riemann–Gilbert conjugation problem for complex potentials $\omega_1(z_1)$, $\omega_2(z_2)$, where $z_1 = x + S_1 y$; $z_2 = x + S_2 y$; S_1, S_2, $\bar{S}_1$, $\bar{S}_2$ are the roots of a characteristic equation [37]

$$a_{11}S^4 + (2a_{12} + a_{66})S^2 + a_{22} = 0, \tag{6.3}$$

where a_{ij} are the elastic constants entering into the generalized Hooke law

$$\varepsilon_i = \sum_{j=1}^{6} a_{ij}\sigma_{ij}, \quad i = 1, \ldots 6. \tag{6.4}$$

Note that for majority of orthotropic materials whose symmetry of elastic properties coincides with coordinate axes, the following equalities hold:

$$S_1 = i\beta_1, \quad S_2 = i\beta_2, \quad \beta_1, \beta_2 \in R. \tag{6.5}$$

In what follows, it is assumed that condition (6.5) is satisfied.

The complex potentials $\omega_1(z_1)$ and $\omega_2(z_2)$ in the case under consideration are expressed as follows [25]:

$$\begin{aligned}\omega_1(z_1) = \frac{1}{\beta_2 - \beta_1}\Bigg\{ & i\frac{\beta_2\sigma_y^0 z_1}{\pi\sqrt{L^2 - z_1^2}} \arccos\frac{l}{L} - i\frac{\beta_2 p z_1}{2\sqrt{L^2 - z_1^2}} \\ & + \frac{\beta_2\sigma_y^0}{2\pi i}\ln\frac{L^2 - z_1 l + \sqrt{(L^2 - l^2)(L^2 - z_1^2)}}{L^2 + z_1 l + \sqrt{(L^2 - l^2)(L^2 - z_1^2)}} \\ & + \frac{\beta_2\sigma_y^0}{2\pi i}\ln\frac{l + z_1}{l - z_1} - |M_2|\Bigg\}\end{aligned} \tag{6.6}$$

$$\begin{aligned}\omega_2(z_2) = \frac{1}{\beta_1 - \beta_2}\Bigg\{ & i\frac{\beta_1\sigma_y^0 z_2}{\pi\sqrt{L^2 - z_2^2}} \arccos\frac{l}{L} - i\frac{\beta_1 p z_2}{2\sqrt{L^2 - z_2^2}} \\ & + \frac{\beta_1\sigma_y^0}{2\pi i}\ln\frac{L^2 - z_2 l + \sqrt{(L^2 - l^2)(L^2 - z_2^2)}}{L^2 + z_2 l + \sqrt{(L^2 - l^2)(L^2 - z_2^2)}} \\ & + \frac{\beta_1\sigma_y^0}{2\pi i}\ln\frac{l + z_2}{l - z_2} - |M_2|\Bigg\},\end{aligned}$$

where

$$|M_2| = \frac{\beta_1\beta_2 p - q}{2(\beta_1 + \beta_2)}.$$

Expressing stresses σ_x in terms of the complex potentials $\omega_1(z_1), \omega_2(z_2)$ and applying the stress continuity condition on the crack front, we find

$$\sigma_x^0 = \beta_1\beta_2\left(\sigma_y^0 - p\right) + q. \tag{6.7}$$

Besides, the stresses σ_x^0 and σ_y^0 satisfy the strength criterion (6.1). This equation together with (6.7) forms the closed system of equations for the unknown stress components σ_x^0 and σ_y^0. It is evident that stresses within the process zone will be dependent on the elastic constants of a material regardless of the form of the strength criterion (6.1).

6.3.2 Length of the Process Zone

Since the stresses σ_x^0 and σ_y^0 are expressed linearly in terms of the complex potentials $\omega_1(z_1)$ and $\omega_2(z_2)$, it is evident that the necessary condition for restriction of the stresses over all the plane and thus at $|x| = L$ with account for (6.6) can be written as

$$\frac{l}{L} = \frac{l}{l+d} = \cos\frac{\pi p}{2\sigma_y^0}. \tag{6.8}$$

This makes it possible to express the length of the process zone d in terms of the crack length l, external load p, and stress σ_y^0 within the process zone as

$$d = l\left(\sec\frac{\pi p}{2\sigma_y^0} - 1\right). \tag{6.9}$$

It is evident that dimension of the process zone depends not only on the crack length and external load, but also on elastic orthotropy constants, regarding that the stress σ_y^0 equally depends on these constants.

6.3.3 Opening Displacement in the Crack Tip

The displacement $v(x)$ on the interval $[l, L]$ in accordance with (6.6) and (6.12) is defined by

$$\begin{aligned} v(x) = &\frac{\sigma_y^0\left(|q_1|\beta_2 - |q_2|\beta_1\right)}{\pi\left(\beta_2 - \beta_1\right)} \\ &\times\left\{x\ln\frac{\left(L^2 - xl + \sqrt{\left(L^2 - x^2\right)\left(L^2 - l^2\right)}\right)(l+x)}{\left(L^2 + xl + \sqrt{\left(L^2 - x^2\right)\left(L^2 - l^2\right)}\right)(l-x)}\right. \\ &\left.\qquad - 2l\ln\frac{\sqrt{L^2 - x^2} + \sqrt{L^2 - l^2}}{\sqrt{x^2 - l^2}}\right\} \end{aligned} \tag{6.10}$$

where

$$q_1 = \frac{a_{22} - a_{12}\beta_1^2}{i\beta_1}, \quad q_2 = \frac{a_{22} - a_{12}\beta_2^2}{i\beta_2}.$$

Then, the opening displacement in the crack tip $|x| = l$, $y = 0$ with allowance for (6.8) can be written as follows:

$$\delta(l) = 2v(l) = \frac{\sigma_y^0 l\left(|q_1|\beta_2 - |q_2|\beta_1\right)}{\pi(\beta_2 - \beta_1)} \ln\cos\frac{\pi p}{2\sigma_y^0}. \tag{6.11}$$

Having expressed q_1, q_2, β_1, and β_2 in terms of elastic constants of an orhotropic body, (6.11) becomes

$$\delta(l) = \frac{4T_0\sigma_y^0 l}{\pi} \ln\sec\frac{\pi p}{2\sigma_y^0}, \quad T_0 = \frac{1}{\sqrt{E_1E_2}}\sqrt{2\left(\sqrt{\frac{E_1}{E_2}} - \nu_{12}\right) + \frac{E_1}{G_{12}}}, \tag{6.12}$$

where E_1 and E_2 are the elastic moduli along the orthotropy axes, ν_{12} is Poisson's ratio, and G_{12} is the shear modulus. If $E_1 = E_2 = E$, $\nu_{12} = \nu$ (isotropic body), and $\sigma_y^0 = \sigma_0$ (the Treska strength condition), the constitutive equations of the Leonov–Panasyuk model [36] with (6.8) and (6.12) can be derived as

$$\frac{l}{L} = \cos\frac{\pi p}{2\sigma_0}, \quad \delta(l) = -\frac{8\sigma_0 l}{\pi E}\ln\cos\frac{\pi p}{2\sigma_0}. \tag{6.13}$$

6.3.4 *J*-Integral

The value of J-integral can be obtained as

$$J = -\int_0^\delta \sigma_y d\delta = \sigma_y^0\delta(l) = \frac{4T_0\left(\sigma_y^0\right)^2 l}{\pi}\ln\sec\frac{\pi p}{2\sigma_y^0}. \tag{6.14}$$

6.4 Limit Equilibrium of a Cracked Orthotropic Plate

Let us consider a limiting state of an orthotropic plate using the δ_c- and J_c- criteria.

6.4.1 The δ_c-Criterion

In accordance with the δ_c-criterion, a crack starts when the crack tip opening displacement reaches some ultimate value δ_c, i.e., if

$$\delta(l) = \delta_c. \tag{6.15}$$

Then, basing on (6.12), the field of ultimate loads (p_*, q_*) can be defined by

$$\frac{4T_0\sigma_y^0(p_*, q_*)\, l}{\pi} \ln \sec \frac{\pi\, p_*}{2\sigma_y^0(p_*, q_*)} = \delta_c, \tag{6.16}$$

where $\sigma_y^0(p_*, q_*)$ is determined by solving the system of (6.1) and (6.7) for $p = p_*,\ q = q_*$.

If the plate is subjected to uniaxial tension ($q = 0$), the limiting state is defined by

$$\frac{4T_0\sigma_y^0\left(p_*^{(0)}, 0\right) l}{\pi} \ln \sec \frac{\pi p_*^{(0)}}{2\sigma_y^0\left(p_*^{(0)}, 0\right)} = \delta_c, \tag{6.17}$$

where $p_*^{(0)}$ is the ultimate load in uniaxial tension. By comparing (6.16) and (6.17), we get

$$\frac{\sigma_y^0(p_*, q_*)}{\sigma_y^0\left(p_*^{(0)}, 0\right)} \ln \cos \frac{\pi\, p_*}{2\sigma_y^0(p_*, q_*)} = \ln\ \cos \frac{\pi p_*^{(0)}}{2\sigma_y^0\left(p_*^{(0)}, 0\right)}. \tag{6.18}$$

Note that (6.18) characterizes the field of ultimate loads (p_*, q_*) depending on the ultimate load value under uniaxial loading. In this case, the change in the value of ${p_*}^{(0)}$ from 0 to σ_{0y} (σ_{0y} is the ultimate strength in the y-direction) corresponds to the change in the crack length from infinity to zero, i.e., it covers all the range of variation in the crack length.

6.4.2 The J_c-Criterion

In accordance with the J_c-criterion a crack starts when $J(l) = J_c$. Then, basing on (6.14), the field of ultimate loads (p_*, q_*) can be defined by

$$\frac{4T_0\left(\sigma_y^0(p_*, q_*)\right)^2 l}{\pi} \ln \sec \frac{\pi\, p_*}{2\sigma_y^0(p_*, q_*)} = J_c \tag{6.19}$$

or

$$\left[\frac{\sigma_y^0(p_*, q_*)}{\sigma_y^0\left(p_*^{(0)}, 0\right)}\right]^2 \ln \cos \frac{\pi\, p_*}{2\sigma_y^0(p_*, q_*)} = \ln\ \cos \frac{\pi\, p_*^{(0)}}{2\sigma_y^0\left(p_*^{(0)}, 0\right)}. \tag{6.20}$$

Relation (6.20) characterizes the field of ultimate loads (p_*, q_*) depending on the ultimate load value under uniaxial loading.

6.5 Numerical Analysis

6.5.1 The von Mises–Hill Strength Criterion

Let us adopt the von Mises–Hill criterion as the strength condition (6.1). This criterion in the case of a plane stress state is described by

$$\frac{\sigma_x^2}{\sigma_{0x}^2} + \frac{\sigma_y^2}{\sigma_{0y}^2} - \frac{\sigma_x \sigma_y}{\sigma_{0x}\sigma_{0y}} = 1, \tag{6.21}$$

where σ_{0x} and σ_{0y} are the ultimate strengths in the x- and y-axis directions, respectively.

Solving the system of (6.7) and (6.21) simultaneously, stress components in the process zone are found as follows:

$$\sigma_y^0 = \frac{b + \sqrt{b^2 - 4ac}}{2a}, \quad \sigma_x^0 = \beta \sigma_y^0 - Q, \tag{6.22}$$

where $\beta = \sqrt{E_1/E_2}$, $Q = \beta p - q$,

$$a = \frac{\beta^2}{\sigma_{0x}^2} + \frac{1}{\sigma_{0y}^2} - \frac{\beta}{\sigma_{0x}\sigma_{0y}}, \quad b = \frac{Q}{\sigma_{0x}}\left(\frac{2\beta}{\sigma_{0x}} - \frac{1}{\sigma_{0y}}\right), \quad c = \frac{Q^2}{\sigma_{0x}^2} - 1.$$

Figure 6.2 shows limiting strength diagrams calculated by (6.21) and fracture diagrams, constructed on the basis of the δ_c-criterion (6.18). Figure 6.3 shows analogous diagrams determined using the J_c-criterion (6.20).

Figure 6.4 shows fracture diagrams, determined using the δ_c-criterion (6.18) and the J_c-criterion (6.20) for $\beta = 0.3$. Here the limiting strength diagrams correspond to ultimate stresses in a defect-free material and fracture diagrams to ultimate stresses in a material with longitudinal stresses.

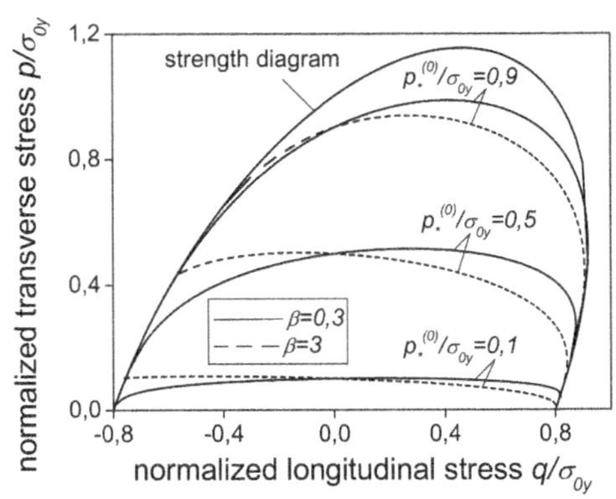

Fig. 6.2 Limiting diagrams calculated by the δ_c-criterion

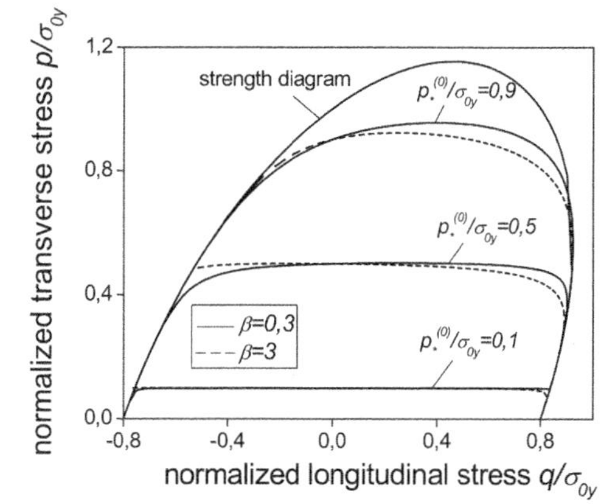

Fig. 6.3 Limiting diagrams calculated by the J_c-criterion

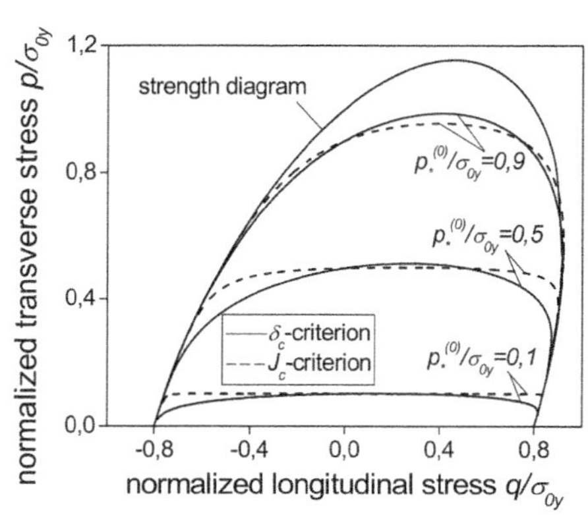

Fig. 6.4 Limiting diagrams calculated by the δ_c- and J_c- criteria

6.5.2 The Gol'denblatt–Kopnov Strength Criterion

The Gol'denblatt–Kopnov strength criterion [30] takes into account the dependence of the material strength on the sign of shear stresses and difference in resistance to tensile and compressive loads. In the case of a plane stress state, this criterion is written in the form

$$\frac{\sigma_{0x}^p-\sigma_{0x}^s}{2\sigma_{0x}^p\sigma_{0x}^s}\sigma_x+\frac{\sigma_{0y}^p-\sigma_{0y}^s}{2\sigma_{0y}^p\sigma_{0y}^s}\sigma_y+\frac{1}{2}\left\{\left(\frac{\sigma_{0x}^p+\sigma_{0x}^s}{\sigma_{0x}^p\sigma_{0x}^s}\right)^2\sigma_x^2\right.$$
$$+\left(\frac{\sigma_{0y}^p+\sigma_{0y}^s}{\sigma_{0y}^p\sigma_{0y}^s}\right)^2\sigma_y^2-\left[\left(\frac{\sigma_{ox}^p+\sigma_{0x}^s}{\sigma_{0x}^p\sigma_{0x}^s}\right)^2+\left(\frac{\sigma_{0y}^p+\sigma_{0y}^s}{\sigma_{0y}^p\sigma_{0y}^s}\right)^2\right.$$
$$\left.\left.-\left(\frac{\tau_{45}^{+}+\tau_{45}^{-}}{\tau_{45}^{+}\tau_{45}^{-}}\right)^2\right]\sigma_x\sigma_y+\frac{4\tau_{12}^2}{\tau_0^2}\right\}^{1/2}=1. \quad (6.23)$$

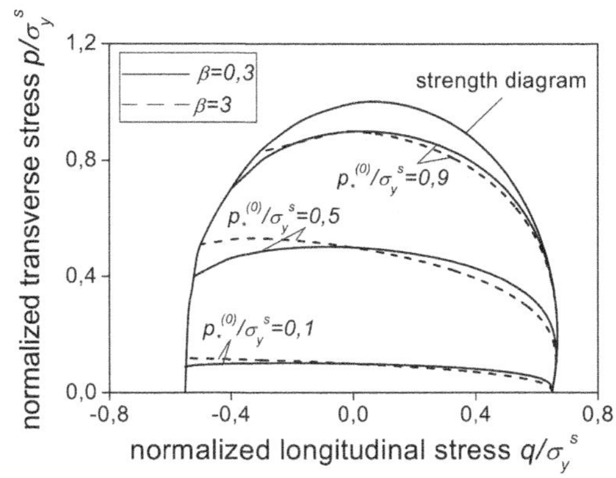

Fig. 6.5 Limiting diagrams for materials satisfying the Gol'denblatt–Kopnov strength criterion

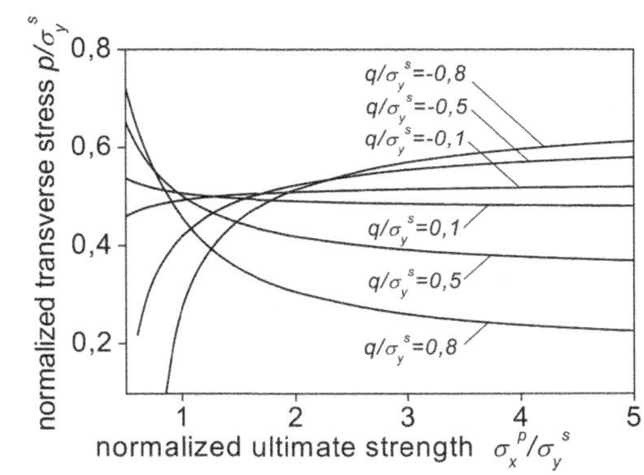

Fig. 6.6 The ultimate load p_*/σ^s_{0y} versus $\sigma^p_{0y}/\sigma^s_{0y}$

The stress components within the process zone can be determined solving jointly the system of equations (6.7) and (6.23).

Figure 6.5 shows strength diagrams calculated by (6.20) for $\sigma^p{}_{0x}/\sigma^s{}_{0y} = 0.84$, $\sigma^p{}_{0y}/\sigma^s{}_{0y} = 0.63$, $\sigma^s{}_{0x}/\sigma^s{}_{0y} = 0.65$, $\tau^+_{45}/\tau^-_{45} = 0.87$, and fracture diagrams obtained by (6.18).

The Gol'denblatt–Kopnov strength criterion permits to investigate the effect of different ultimate strengths under tension and compression on the values of ultimate loads. Taking into account this effect is important to study the fracture in composites, concretes, and rocks. Figure 6.6 shows the ultimate load $p_*/\sigma^s{}_y$ versus $\sigma^p{}_{0y}/\sigma^s{}_{0y}$. Figure 6.7 shows the ultimate load $p_*/\sigma^s{}_y$ versus $\sigma^p{}_{0x}/\sigma^s{}_{0y}$. As is seen the difference between material ultimate strengths under tension and compression brings considerable influence on the ultimate state of a cracked orthotropic plate with crack.

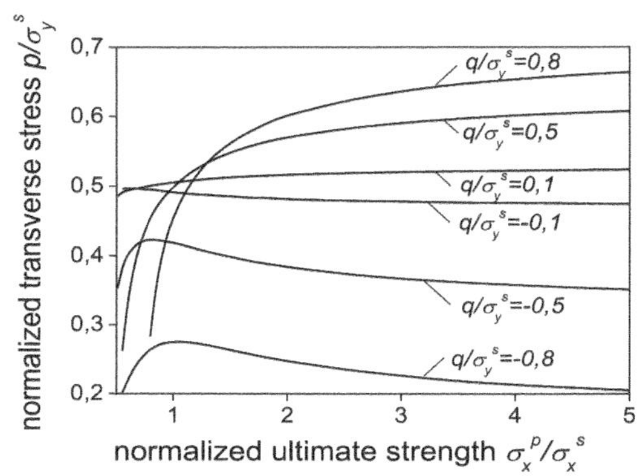

Fig. 6.7 The ultimate load p_*/σ_{0y}^s versus $\sigma_{0x}^p/\sigma_{0y}^s$

6.6 Macrocrack and Domain of Quasi-Brittle Fracture

Consider the case where process zones are small ($d \ll l$). From (6.8) it follows that in this case $p \ll \sigma_y^0$. Then, in view of (6.9) and (6.12), it can be obtained as

$$d = \frac{\pi^2 p^2 l}{8\left(\sigma_y^0\right)^2} = \frac{\pi K_I^2}{8\left(\sigma_y^0\right)^2}, \tag{6.24}$$

$$\delta(l) = \frac{T_0 \pi p^2 l}{2\sigma_y^0} = \frac{T_0 K_I^2}{2\sigma_y^0}, \tag{6.25}$$

$$J = \frac{T_0 \pi p^2 l}{2} = \frac{T_0 K_I^2}{2}, \tag{6.26}$$

where $K_I = p\sqrt{\pi l}$ is the stress intensity factor for the mode I crack. Hence, the model parameters of the macrocrack can be expressed in terms of K_I. Then, to study the fracture process, the linear fracture mechanics apparatus can be applied.

From (6.25) and (6.15), it can be concluded that the ultimate load $p_*^{(m)}$ for the macrocrack can be expressed as

$$\frac{p_*^{(m)}}{\sigma_{0y}} = \sqrt{2\frac{d_*}{l}\frac{\sigma_y^0}{\sigma_{0y}}}, \quad d_* = \frac{\delta_c}{T_0 \pi \sigma_{0y}}. \tag{6.27}$$

As is seen, the ultimate load $p_*^{(m)}$, as it follows from the postulates of linear fracture mechanics, increases beyond all bounds as $l \to 0$, whereas the value of p_* obtained from the exact solution is limited. Such situation corresponds to physical meaning and is one of the model requirements.

Let us estimate the domain within which the relations expressed by (6.24)–(6.26) can be employed. It is assumed that the material satisfies the von Mises–Hill strength criterion (6.21). Then the condition $p \ll \sigma_y^0$ becomes equivalent to the condition

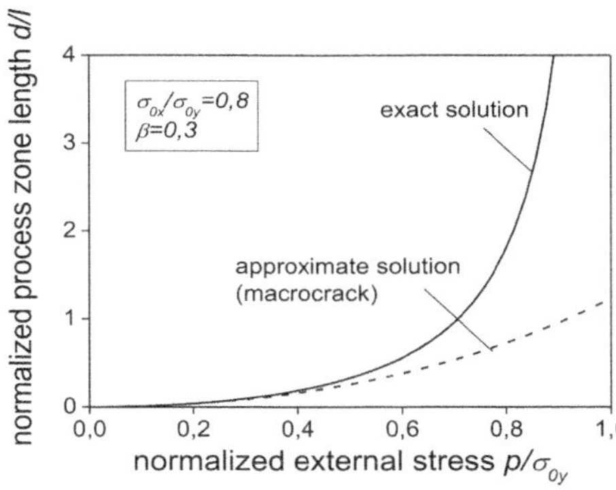

Fig. 6.8 The length of process zone depending on the external loading

$$\frac{q^2}{\sigma_{0x}^2} + \frac{p^2}{\sigma_{0y}^2} - \frac{pq}{\sigma_{0x}\sigma_{0y}} \ll 1 \tag{6.28}$$

which is valid also for $q = 0$.

Figure 6.8 shows the dependency of the process zone length d/l on the load p/σ_y^0. This dependency is defined using the exact solution of (6.8) and approximate one for (6.24). As is seen, the discrepancy between the curves beginning with $p/\sigma_y^0 \approx 0.4$ exceeds 10%. Consequently, the domain of quasi-brittle fracture for the given material is confined by the semiellipse

$$\frac{q^2}{\sigma_{0x}^2} + \frac{p^2}{\sigma_{0y}^2} - \frac{pq}{\sigma_{0x}\sigma_{0y}} = 0.16. \tag{6.29}$$

6.7 Experimental Verification

Verification of the fracture criteria in the form of (6.18) and (6.20) was carried out by testing tubular specimens (their gauge length, diameter, and wall thickness are equal to 100, 30, and 0.7 mm, respectively) made of VT-14 titanium alloy (4.6% Al; 2.8% Mo; 0.6% V) and St3 low-carbon steel (0.2%C) [5]. To test the specimens, which were manufactured of bars with a circular cross-section, a crack was simulated for both materials by a through slit of width 0.3 mm and length 3 mm. The slit was marked by the electrosparking method and oriented in the longitudinal direction. Prior to testing, the specimens were hermetized by a thin rubber patch that makes it possible to load them by axial force (tensile or compressive) and internal pressure under various ratios. In this case a biaxial stress state arises in the specimen wall characterized by longitudinal σ_z and circumferential σ_θ stresses.

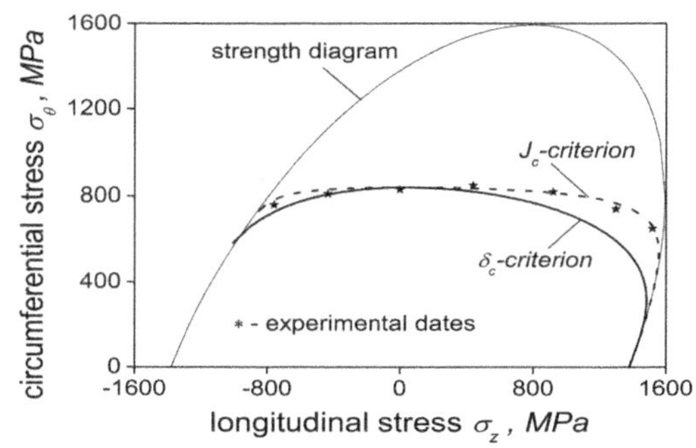

Fig. 6.9 Limiting stress diagrams for VT-14 titanium alloy

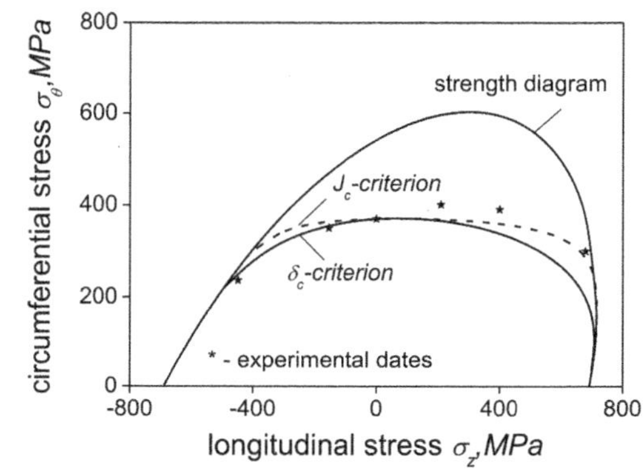

Fig. 6.10 Limiting stress diagrams for ST3 steel

The ultimate stresses σ_z^* and σ_θ^*, determined with specimens of the VT-14 alloy with crack for various ratios $n_1 = \sigma_z/\sigma_\theta$, are shown by asterisks in Fig. 6.9. The limiting strength diagram was calculated by the Hencky–von Mises criterion for a defect-free material

$$\sigma_\theta^2 + \sigma_\theta\sigma_z + \sigma_z^2 = \sigma_0^2. \tag{6.30}$$

Figure 6.10 shows analogous data for St 3 steel. Here the limiting strength diagram corresponds to the strength criterion obtained by generalization of the Hencky–von Mises criterion for isotropic materials on materials with deformation anisotropy [5]

$$\frac{2}{3}\left(\sigma_z^2 + \sigma_\theta^2 - \sigma_z\sigma_\theta\right) + 225\sigma_\theta - 3176 = 0. \tag{6.31}$$

As is seen, the J-integral criterion is in better agreement with the experimental data.

6.8 Conclusions

Based on the results obtained, we can note the following:

1. As $l \to 0$, fracture diagrams tend to a limiting strength diagram for a defect-free material.
2. The stress component acting parallel to a crack in the domain of quasi-brittle fracture influences weakly on the ultimate state of a plate, so there is no deviation from the principles of linear fracture mechanics.
3. As applied to a certain crack length, there are such domains of biaxial loading, where fracture of a cracked plate occurs by brittle mechanism, namely by fracture of the total cross-section on continuation of the crack.
4. The load q acting along a crack may both increase the ultimate load p_* and decrease it in comparison with uniaxial tension (in perpendicular direction). So, if two materials have similar strength characteristics but differ in anisotropy level of elastic properties, the crack resistance in the domain of biaxial tension will be higher for the material with a lower ratio E_1/E_2. The result for the tension-compression domain is opposite to previous one.
5. Unlike the models of Barenblatt, Leonov–Panasyuk, and Dugdale, the proposed model allows to predict the strength of orthotropic constructional materials under biaxial loading.

References

1. Adams, N.J.I.: Some comments on the effect of biaxial stress on fatigue crack growth and fracture. Eng. Fract. Mech. **5**, 983–989 (1973)
2. Ainsworth, R.A., Sharples, J.K., Smit, S.D.: Effects of residual stresses on fracture behaviour—experimental results and assessment methods. J. Strain Anal. Eng. Design **35**(4), 307–316 (2000)
3. Alpa, G., Bozzo, E., Gambarotta, L.: Validity limits of the Dugdale model for thin cracked plates under biaxial loading. Eng. Fract. Mech. **12**, 523–529 (1979)
4. Andrews, R.M., Garwood, S.F.: An analysis of fracture under biaxial loading using the non-singular T-stress. Fatigue Fract. Eng. Mater. Struct. **24**, 53–62 (2001)
5. Bastun, V.N.: Plasticity condition of orthotropic metallic materials produced by pressure treatment. Int. Appl. Mech. **29**(2), 149–152 (1993)
6. Bastun, V.N.. Kaminsky, A.A.: Applied problems of mechanics of processes of strain-hardening the engineering materials. Int. Appl. Mech. **41**(10), 1092–1129 (2005)
7. Bogdanov, V.L.: On a circular shear crack in a semiinfinite composite with initial stresses. Mater. Sci. **43**(2), 321–33 (2007)
8. Bogdanov, V.L.: Influence of initial stresses on the stressed state of a composite with a periodic system of parallel coaxial normal tensile cracks. J. Math. Sci. **186**(1), 1–13 (2012)
9. Bogdanov, V.L.: On one approach in fracture mechanics of composites with parallel cracks under the action of initial (residual) stresses. Mech. Comp. Mater. **59**(2), 239–262 (2023)
10. Bogdanov, V.L., Nazarenko, V.M.: Study of the compressive failure of a semiinfinite elastic material with a harmonic potential. Int. Appl. Mech. **30**(10), 760–765 (1994)
11. Bogdanova, O.S., Kaminsky, A.A.: On one model in the mesomechanics of fracture of orthotropic material with different strengths limits under tension and compression. Dopovidi NAN Ukrainy **2**, 64–68 (2008). [in Russian]

12. Chen, Y.Z.: Closed form solutions of T-stress in plane elasticity crack problem. Int. J. Solids Struct. **37**(1), 1763–1783 (2000)
13. Dugdale, D.S.: Yielding of steel sheets containing slits. J. Mech. Phys. Solids **8**(2), 100–108 (1960)
14. Eftis, J., Jones, I., Liebowith, H.: Load biaxiality and fracture: synthesis and summary. Eng. Frac. Mech. **36**(4), 537–574 (1990)
15. Galatenko, G.V., Degtyaryeva, O.S., Kaminsky, A.A.: Elastoplastic failure of an orthotropic plate with a crack under biaxial loading. Sov. Appl. Mech. **26**(8), 794–799 (1990)
16. Guz, A.N.: A criterion of solid body destruction during compression along cracks (Two-dimensional problem). Dokl. AN SSSR **259**(6), 1315–1318 (1981)
17. Guz, A.N.: Theory of cracks in elastic bodies with initial stress—formulation of problems, tear cracks. Sov. Appl. Mech. **16**(12), 1015–1024 (1980)
18. Guz, A.N., Bogdanov, V.L., Nazarenko, V.M.: Two-dimensional problems on fracture of bodies under compression along cracks. In: Guz, A.N., Bogdanov V.L., Nazarenko V.M. (eds.) Fracture of Materials under Compression along Cracks, pp. 149–248. Springer, Cham (2020)
19. Jansson, S.: Fully plastic plane stress solutions for biaxially loaded center-cracked plates. J. Appl. Mech. **53**(3), 555–560 (1986)
20. Kaminsky, A.A.: Subcritical crack growth in polymer composite materials. In: Cherepanov, G.P. (ed.) Fracture: A Topical Encyclopedia of Current Knowledge, pp. 758–765. Krieger Publishing Company, Malabar (1998)
21. Kaminsky, A.A.: Analyzing the laws of stable subcritical growth of cracks in polymeric materials on the basis of fracture mesomechanics models: theory and Experiment. Int. Appl. Mech. **40**(8), 829–846 (2004)
22. Kaminsky, A.A., Bogdanova, O.S.: Long-term crack resistance of an orthotropic viscoelastic plate with a crack under biaxial loading. Int. Appl. Mech. **31**(9), 747–753 (1995)
23. Kaminsky, A.A., Bogdanova, O.S.: Modelling the failure of orthotropic materials subject to biaxial loading. Int. Appl. Mech. **32**(10), 813–820 (1996)
24. Kaminsky, A.A., Bogdanova, O.S.: A mesomechanical fracture model for an orthotropic material with different tensile and compressive strengths. Int. Appl. Mech. **45**(3), 290–296 (2009)
25. Kaminsky, A.A., Bogdanova, O.S., Bastun, V.N.: On modeling cracks in orthotropic plates under biaxial loading: synthesis and summary. Fatigue Fract. Eng. Mat. Struct. **34**(5), 345–355 (2011)
26. Kaminsky, A.A., Galatenko, G.V.: Two-parametric model of the normal-rupture crack in an Elasto-plastic body at the plane strain. Int. Appl. Mech. **41**(6), 621–630 (2005)
27. Kaminsky, A.A., Kurchakov, E.E., Gavrilov, G.V.: Formation of a plastic zone in an anisotropic body under loads acting along a crack. Int. Appl. Mech. **43**(5), 475–490 (2007)
28. Kaminsky, A.A., Nizhnik, S.B.: Study of the laws governing the change in the plastic zone at the crack tip and characteristics of the fracture toughness of metallic materials in relation to their structure (survey). Int. Appl. Mech. **31**(10), 777–798 (1995)
29. Larsson, S.G., Carlsson, A.J.: Influence of non-singular stress terms and specimen geometry on small-scale yielding at crack tips in elastic plastic materials. J. Mech. Phys. Solids **21**, 17–26 (1973)
30. Lebedev, A.A., Koval'chuk, B.I., Giginjak, F.F., Lamashewsky, V.P.: Handbook of Mechanical Properties of Structural Materials at a Complex Stress State. Begell House Inc., New-York (2001)
31. Lee, E.U., Taylor, R.E.: Fatigue behavior of aluminium alloys under biaxial loading. Eng. Fract. Mech. **78**(8), 1555–1564 (2011)
32. Leevers, P.S., Radon, J.: Inherent stress biaxiality in various fracture specimen geometries. Int. J. Fracture **19**, 311–325 (1982)
33. Nikishkov, G.P., Brückner-Foit, A., Munz, D.: Calculation of the second fracture parameter for finite cracked bodies using a three-term elastic-plastic asymptotic expansion. Eng. Fract. Mech. **52**, 685–701 (1995)
34. O'Dowd, N.P., Kolednik, O., Naumenko, V.P.: Elastic-plastic analysis of biaxially loaded center-cracked plates. Int. J. Solids Struct. **36**(36), 5639–5661 (1999)

35. O'Dowd, N.P., Shih, C.F.: Family of crack-tip fields characterized by a triaxiality parameter—I: structure of fields. J. Mech. Phys. Solids **39**, 989–1015 (1991)
36. Panasyuk, V.V.: Limiting Equilibrium of Brittle Solids with Fractures. Management Information Services, Detroit-Michigan (1969)
37. Savin, G.N.: Stress Concentration Around Holes. Pergamon Press, Oxford-London (1961)
38. Savin, G.N., Kaminsky, A.A.: The crack growth in fracture of hard polymers. Prikl. Mech. **3**(3), 33–39 (1967). [in Russian]

Chapter 7
Two-Parametric Analysis of a Semi-Infinite Three-Layered High-Contrast Elastic Strip Under Antiplane Shear Deformation

Illia Chernomorets, Julius Kaplunov, and Danila Prikazchikov

Abstract Antiplane shear of a symmetric three-layered elastic semi-strip under the self-equilibrated edge load is studied. The arbitrary ratios of thickness and stiffness are considered, with emphasis on high-contrast scenario; in doing so, two aforementioned dimensionless problem parameters are assumed mutually independent. A simple explicit condition, supporting a slowly decaying solution of the equilibrium equations, is derived. Leading order estimates for the decay rate and amplitude of this solution are presented, highlighting its dominant contribution to displacement field. The derivation involves an asymptotic formula ensuring elimination of the boundary layer localised near the edge of the semi-strip. The peculiarities of the latter are also addressed. Numerical results, including comparisons of the derived shortened and full expressions for the sought for slowly decaying solutions, are presented.

7.1 Introduction

Layered structures have been an important subject of studies in solid mechanics since long ago; see, e.g., monographs by [1–5] and many more. More recently, substantial interest was drawn to high-contrast laminates, inspired by fresh engineering applications in photovoltaic panels and laminated glass, as well as advanced sandwich-type panels; see [6–10] to name a few.

Multiparametric dynamic analysis of high-contrast thin three-layered elastic plates subject to plane-strain deformation was seemingly first carried out in [10]. This paper was restricted to several practically motivated scenarios of contrast in stiffness, thickness and density of the layers. One of the key findings was emergence of an

I. Chernomorets · J. Kaplunov (✉) · D. Prikazchikov
School of Computer Science and Mathematics, Keele, UK
e-mail: j.kaplunov@keele.ac.uk

I. Chernomorets
e-mail: i.chernomorets@keele.ac.uk

D. Prikazchikov
e-mail: d.prikazchikov@keele.ac.uk

H. Altenbach et al. (eds.), *Selected Problems of Solid Mechanics and Solving Methods*, Advanced Structured Materials 204, https://doi.org/10.1007/978-3-031-54063-9_7

additional low-frequency shear mode, along with the canonical fundamental bending mode. In the following contributions [11, 12], two-mode long-wave low-frequency equations of motion were derived for antiplane shear of three-layered plates. Moreover, a natural link between the low-frequency shear mode and the slowly decaying quasi-static particular solution was revealed in [12]. The above-mentioned particular solution is in fact associated with the degenerated Saint-Venant's type boundary layer in sandwich structures, discovered in earlier publications, e.g., see [13, 14].

In this paper, we consider the simplest example of a scalar static problem in antiplane elasticity for a three-layered semi-strip, in which the two dimensionless contrast parameters, including relative thickness and stiffness, are mutually independent. A transcendental equation for exponentially decaying solutions is derived and analysed. Explicit conditions for the parameter range allowing a slowly decaying particular solution are determined. The one-dimensional approximate boundary value problem governing this solution is formulated using the asymptotic results in [12, 15]. In particular, the asymptotic decay condition adapting the classical Saint-Venant's principle for high-contrast plates is implemented.

The paper is organised as follows. The problem is formulated in Sect. 7.2. The slowly decaying particular solution of the equilibrium equations is analysed in Sect. 7.3. Numerical results testing the derived leading order approximations for the slow decay rate are displayed in Sect. 7.4, along with elucidation of the solutions corresponding to the boundary layers. Then, the approximate one-dimensional boundary value problem is discussed in Sect. 7.5, including a model example of a self-equilibrated polynomial load prescribed at the edge of the structure.

7.2 Formulation of the Problem

Consider equilibrium of a semi-infinite three-layered symmetric elastic laminate, with core layer occupying the domain $\{x \geqslant 0,\ |y| \leqslant h_1\}$, and the two skin layers satisfying $\{x \geqslant 0,\ h_1 \leqslant |y| \leqslant h_1 + h_2\}$; see Fig. 7.1.

Fig. 7.1 Problem statement

Throughout this contribution we restrict our attention to antiplane shear, within which the vector problem in elasticity reduces to a scalar one, with the only non-zero displacement being an out of plane component $u = u(x, y)$. In view of the symmetry, we will only consider the problem for the upper half of the structure; in doing so, for the sake of simplicity throughout the paper we restrict ourselves to an antisymmetric problem for which the displacement satisfies $u(x, -y) = -u(x, y)$.

The conventional equation of equilibrium within the framework of antiplane elasticity is the Laplace equation

$$\frac{\partial^2 u}{\partial x^2} + \frac{\partial^2 u}{\partial y^2} = 0. \tag{7.1}$$

The outer skin face $y = h_1 + h_2$ is assumed traction-free

$$\left.\frac{\partial u}{\partial y}\right|_{y=h_1+h_2} = 0, \tag{7.2}$$

with continuity of displacement and traction on the interface $y = h_1$ implying

$$u\big|_{y=h_1-0} = u\big|_{y=h_1+0}, \tag{7.3}$$

and

$$\mu_1 \left.\frac{\partial u}{\partial y}\right|_{y=h_1-0} = \mu_2 \left.\frac{\partial u}{\partial y}\right|_{y=h_1+0}. \tag{7.4}$$

Here μ_1 and μ_2 denote the shear moduli of the core and skin, respectively. At the edge $x = 0$ a prescribed antiplane shear load $P(y)$ is imposed, i.e.,

$$P(y) = \begin{cases} \mu_1 \left.\dfrac{\partial u}{\partial x}\right|_{x=0}, & 0 \leqslant y \leqslant h_1, \\ \mu_2 \left.\dfrac{\partial u}{\partial x}\right|_{x=0}, & h_1 \leqslant y \leqslant h_1 + h_2. \end{cases} \tag{7.5}$$

In addition, we assume a decay at infinity of the traction entering the formula above, i.e., $\partial u/\partial x \to 0$ as $x \to \infty$.

Let us also introduce the dimensionless quantities

$$h = \frac{h_1}{h_2}, \quad \mu = \frac{\mu_1}{\mu_2}. \tag{7.6}$$

The main aim of this consideration is to investigate the combinations of problem parameters leading to slowly decaying solutions of the equilibrium equations, as well as to determine the amplitudes of these solutions from the boundary conditions. In this case, in contrast to earlier developments on the subject, e.g., see [11, 12], the two problem parameters (7.6) are assumed to be independent.

7.3 Slowly Decaying Solution

On employing the ansatz $u(x, y) = \exp(-kx)U(y)$, dependence on transverse coordinate is found from (7.1) as

$$U(y) = \begin{cases} A\sin(ky), & 0 \leqslant y \leqslant h_1, \\ B\sin(ky) + C\cos(ky), & h_1 \leqslant y \leqslant h_1 + h_2, \end{cases} \tag{7.7}$$

where k is longitudinal scale, and A, B and C are arbitrary constants. Then, on substituting the results into the boundary and continuity conditions (7.2)–(7.4), we arrive at

$$\begin{aligned} B\cos K - C\sin K &= 0 \\ (A - B)\sin\frac{Kh}{h+1} - C\cos\frac{Kh}{h+1} &= 0 \\ (\mu A - B)\cos\frac{Kh}{h+1} + C\sin\frac{Kh}{h+1} &= 0, \end{aligned} \tag{7.8}$$

where $K = k(h_1 + h_2)$ and the dimensionless parameters h and μ are given by (7.6). The solvability of the algebraic system (7.8) implies vanishing of the associated determinant, yielding the secular equation

$$\tan\frac{Kh}{h+1}\tan\frac{K}{h+1} = \mu. \tag{7.9}$$

The related eigensolution (7.7) takes the form

$$U(\zeta) = \begin{cases} C\cos\frac{K}{h+1}\sec K\csc\frac{Kh}{h+1}\sin K\zeta, & 0 \leqslant \zeta \leqslant \frac{h}{h+1} \\ C\left(\tan K\sin K\zeta + \cos K\zeta\right), & \frac{h}{h+1} \leqslant \zeta \leqslant 1, \end{cases} \tag{7.10}$$

where $\zeta = y/(h_1 + h_2)$ denotes the dimensionless transverse coordinate.

The sought for small solution $K = K_0$ of (7.9) at leading order is given by

$$K_0 = \sqrt{\frac{\mu}{h}}\,(h+1) \ll 1, \tag{7.11}$$

provided that

$$\mu \ll \frac{h}{(h+1)^2}. \tag{7.12}$$

The estimate (7.11) coincides with that in [13]; see also references therein. Therefore, the relative shear modulus μ satisfies $\mu \lesssim h$ for $h \ll 1$, whereas $\mu \lesssim h^{-1}$ for $h \gg 1$. It is worth noting that the minimal stiffness contrast supporting the small solution K_0

occurs at h or order unity, when $\mu \ll 1$. In this case, the leading order approximation of the formulae (7.10) takes the polynomial form

$$U(\zeta) = \begin{cases} C\dfrac{h+1}{h}\zeta, & 0 \leqslant \zeta \leqslant \dfrac{h}{h+1} \\ C, & \dfrac{h}{h+1} \leqslant \zeta \leqslant 1. \end{cases} \tag{7.13}$$

Similar piecewise linear displacement variation was previously observed in the low-frequency longitudinal vibration of high contrast multi-span rods, see [16, 17], as well as for the low-frequency antiplane shear motion of a three-layered laminate [11, 12].

7.4 Numerical Examples

Let us first discuss and illustrate the transcendental equation (7.9). Several first branches showing variation of the decay rate K with respect to the relative thickness h at $\mu = 0.1$ are shown below in Fig. 7.2. It may be shown from (7.9) that the limiting values of K as $h \to 0$ and $h \to \infty$ correspond to zeros of cosine function $\cos K = 0$, i.e., $K = (2m+1)\pi/2, \ m \in \mathbb{N}$.

Let us now focus on the lowest curve in Fig. 7.2, associated with a slowly decaying solution at $K \ll 1$, studied in the previous section. It is presented in the next Fig. 7.3

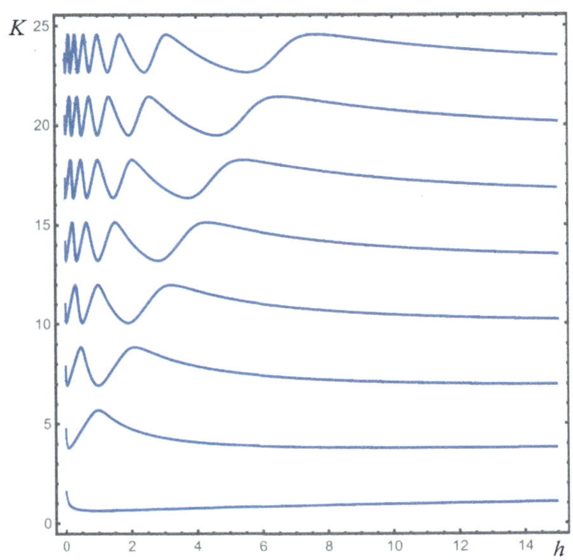

Fig. 7.2 The decay rate K versus the relative thickness h according to the secular Eq. (7.9) at $\mu = 0.1$

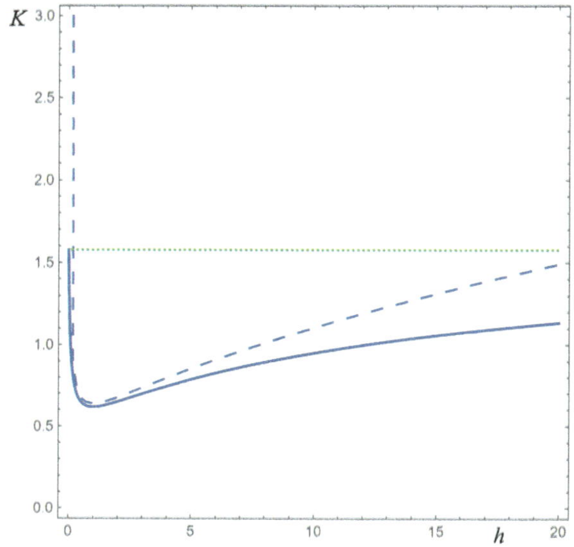

Fig. 7.3 The smallest root of (7.9) associated with slowly decaying solution, together with its approximation (7.11), and the corresponding limiting value $K = \pi/2$ for $\mu = 0.1$

by solid line, along with the approximation (7.11) displayed by dashed line and the limiting value $K = \pi/2$ depicted by dotted line ($\mu = 0.1$).

Another important observation following from (7.9) is related to the fact that its right-hand side μ is small for high-contrast structures ($\mu \ll 1$), hence, implying approximate solutions corresponding to zero tan functions at leading order, namely

$$K = n\pi \left(1 + \frac{1}{h}\right) \tag{7.14}$$

and

$$K = n\pi(h + 1). \tag{7.15}$$

Several of the higher order roots associated with rapidly decaying eigensolutions are shown in Fig. 7.4. Here, exact solutions are depicted by solid black curves, with approximations (7.14) and (7.15) displayed by blue and red curves, respectively. As might be expected, the asymptotic behaviours break around the point intersection of blue and red curves.

Finally, we present a comparison between the exact slowly decaying eigensolution (7.10) and its piecewise linear approximation (7.13), depicted by solid and dashed lines, respectively (Fig. 7.5), showing the variation of the scaled displacement U/C on the transverse coordinate ζ at $\mu = 0.05$.

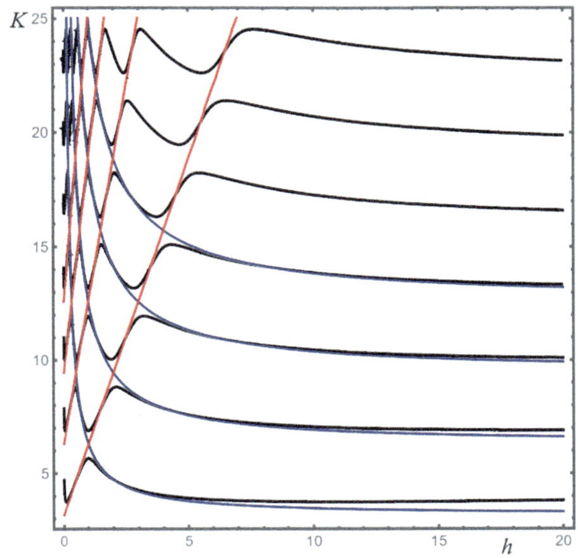

Fig. 7.4 Higher roots of (7.9) corresponding to rapidly decaying solutions, and their approximations (7.14) and (7.15) for $\mu = 0.1$

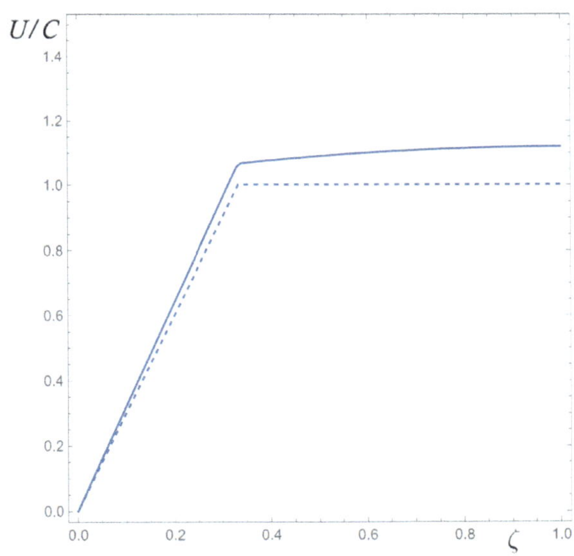

Fig. 7.5 Slowly decaying eigensolution (7.10) and its approximation (7.13) at $\mu = 0.05$

7.5 One-Dimensional Boundary Value Problem

Let us demonstrate that the slow decaying solution studied in Sect. 7.3 can be also deduced from a one-dimensional boundary value problem, corresponding to the long-scale approximation of the original two-dimensional problem. Using the asymptotic methodology developed in [11, 12], we have at leading order

$$(h_1 + h_2)^2 \frac{d^2 u_0}{dx^2} - K_0^2 u_0 = 0, \tag{7.16}$$

where

$$u(x, y) \approx \begin{cases} \dfrac{y}{h_1} u_0(x), & 0 \leqslant y \leqslant h_1 \\ u_0(x), & h_1 \leqslant y \leqslant h_1 + h_2. \end{cases} \tag{7.17}$$

The slowly decaying solution of this equation for a semi-strip takes the form

$$u_0 = C \exp\left(-\frac{K_0 x}{h_1 + h_2}\right). \tag{7.18}$$

To determine the unknown constant C we adapt the generalisation of the Saint-Venant principle oriented to high-contrast laminates, restricting ourselves to the scenario of $h \sim 1$, e.g., see [12, 15]. The implementation of Saint-Venant's principle for thin plates and shells was the subject of several insightful contributions, including [18–21]. In this case, we have the condition

$$\frac{1}{h_1} \int_0^{h_1} y \left(P(y) - \mu_1 \frac{\partial u}{\partial x}\bigg|_{x=0} \right) dy + \int_{h_1}^{h_1+h_2} \left(P(y) - \mu_2 \frac{\partial u}{\partial x}\bigg|_{x=0} \right) dy = 0. \tag{7.19}$$

This formula guarantees a strong decay of the boundary layer (over the domain of strip thickness $x/(h_1 + h_2) \sim 1$), arising because of the discrepancy of the prescribed edge load $P(y)$ and stress field for the long-scale mode of interest, having a simple polynomial variation across the thickness. On substituting (7.17) into (7.19) and neglecting asymptotically secondary terms, we finally obtain

$$C = -\frac{h_1 + h_2}{\mu_2 K_0 h_2} \left(\frac{1}{h_1} \int_0^{h_1} y\, P(y)\, dy + \int_{h_1}^{h_1+h_2} P(y)\, dy \right). \tag{7.20}$$

Thus, the contribution of the slowly varying mode into the overall displacement field dominates over that of the modes forming the boundary layers localised over a narrow vicinity of the edge. Indeed, all of the roots ($K \gtrsim 1$) of the transcendental relation (7.9) are much greater than the small root K_0 in the denominator of the last formula.

It could be easily verified that (7.20) corresponds to the Neumann boundary condition for Eq. (7.16), given by

$$\frac{du_0}{dx}\bigg|_{x=0} = -\frac{K_0 C}{h_1 + h_2}. \tag{7.21}$$

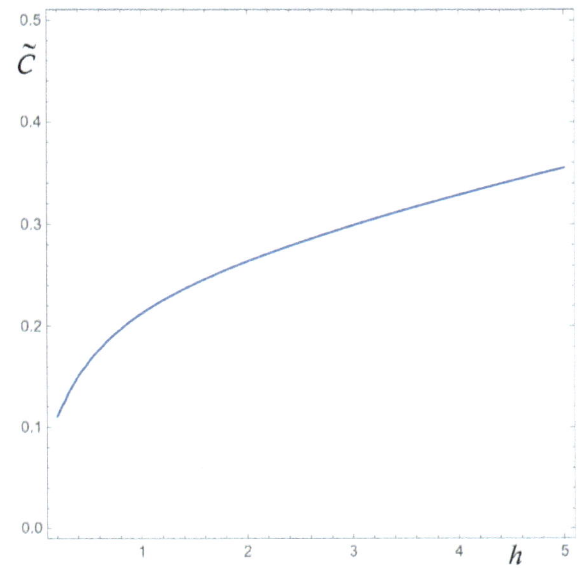

Fig. 7.6 Change of the scaled amplitude $\tilde{C}$ with the relative thickness h

As an example, consider a self-equilibrated load with polynomial variation across the thickness, given by

$$P(y) = \frac{\mu_2 \, y}{h_1 + h_2} \left[1 - \left(\frac{y}{h_1 + h_2} \right)^2 \right]. \tag{7.22}$$

For the latter (7.20) implies

$$C = -(h_1 + h_2) \sqrt{\frac{h}{\mu} \frac{8h^4 + 40h^3 + 80h^2 + 60h + 15}{60(h+1)^4}}, \tag{7.23}$$

where the dimensionless parameters h and μ are given by (7.6).

As an illustration, we display in Fig. 7.6 the variation of dimensionless amplitude $\tilde{C}(h) = -C\sqrt{\mu}/(h_1 + h_2)$ on the relative thickness h. It is observed that the dimensionless amplitude $\tilde{C}$ demonstrates a monotonous behaviour over the range of interest $h \sim 1$.

7.6 Concluding Remarks

Two-parametric analysis developed in the paper has resulted in explicit formula (7.12) ensuring the existence of a slowly decaying mode, with the decay rate given by (7.11) which has been also previously established in [13]. It is important that the contribution of the above-mentioned particular solution into the displacement field of a semi-infinite strip subjected to prescribed edge loading is dominant. Its amplitude is calculated using the so-called asymptotic decay condition, see (7.19), enabling elimination of the effect of the boundary layer.

The presented numerical results illustrate the accuracy of the approximate formula (7.11) through its comparison with the numerical solution of the full transcendental Eq. (7.9). In addition, the example for a self-equilibrated polynomial load demonstrates the effect of the relative thickness h on the amplitude of the slowly varying solution; see (7.22) and (7.23).

The prospects of extending of the obtained results to more general formulations including non-symmetric structures and dynamic problems appear to be rather limited. The point is that taking into consideration extra problem parameters hardly allows their fully independent analysis, i.e., extra constraints will be needed, e.g., similar to [10], where several practical setups have been studied.

References

1. Grigorenko, Y.M.: Isotropic and Anisotropic Layered Shells of Revolution of Variable Thickness. Naukova Dumka, Kiev (1973)
2. Grigorenko, Y.M., Kryukov, N.N.: Numerical Solution of Static Problems of Flexible Layered Shells with Varying Parameters. Naukova Dumka, Kiev (1988)
3. Reddy, J.N.: Mechanics of Laminated Composite Plates and Shells: Theory and Analysis. CRC Press, Boca Raton (2003)
4. Aghalovyan, L.A.: Asymptotic Theory of Anisotropic Plates and Shells. World Scientific, Singapore (2015)
5. Mikhasev, G.I., Altenbach, H.: Thin-Walled Laminated Structures. Springer, New York (2019)
6. Altenbach, H., Eremeyev, V.A., Naumenko, K.: On the use of the first order shear deformation plate theory for the analysis of three-layer plates with thin soft core layer. ZAMM-J. Appl. Math. Mech./Zeit. Angew. Math. Mech. **95**(10), 1004–1011 (2015)
7. Eisenträger, J., Naumenko, K., Altenbach, H., Köppe, H.: Application of the first-order shear deformation theory to the analysis of laminated glasses and photovoltaic panels. Int. J. Mech. Sci. **96**, 163–171 (2015)
8. Boutin, C., Viverge, K.: Generalized plate model for highly contrasted laminates. Europ. J. Mech. A/Solids **55**, 149–166 (2016)
9. Boutin, C., Viverge, K., Hans, S.: Dynamics of contrasted stratified elastic and viscoelastic plates-application to laminated glass. Compos. Part B: Eng. **212**, 108551 (2021)
10. Kaplunov, J., Prikazchikov, D.A., Prikazchikova, L.A.: Dispersion of elastic waves in a strongly inhomogeneous three-layered plate. Int. J. Solids Struct. **113**, 169–179 (2017)
11. Prikazchikova, L., Ece Aydın, Y., Erbaş, B., Kaplunov, J.: Asymptotic analysis of an anti-plane dynamic problem for a three-layered strongly inhomogeneous laminate. Math. Mech. Solids **25**(1), 3–16 (2020)
12. Kaplunov, J., Prikazchikova, L., Alkinidri, M.: Antiplane shear of an asymmetric sandwich plate. Cont. Mech. Thermodyn. **33**(4), 1247–1262 (2021)
13. Baxter, S.C., Horgan, C.O.: End effects for anti-plane shear deformations of sandwich structures. J. Elast. **40**, 123–164 (1995)
14. Horgan, C.: Saint-Venant end effects for sandwich structures. In: Fourth International Conference on Sanwich Construction, vol. 1, pp. 191–200. EMAS Publishing, UK (1998)
15. Prikazchikova, L.: Decay conditions for antiplane shear of a high-contrast multi-layered semi-infinite elastic strip. Symmetry **14**, 1697 (2022)

16. Kaplunov, J., Prikazchikov, D.A., Sergushova, O.: Multi-parametric analysis of the lowest natural frequencies of strongly inhomogeneous elastic rods. J. Sound Vib. **366**, 264–276 (2016)
17. Kaplunov, J., Prikazchikov, D.A., Prikazchikova, L., Sergushova, O.: The lowest vibration spectra of multi-component structures with contrast material properties. J. Sound Vib. **445**, 132–147 (2019)
18. Goldenveizer, A.L.: Theory of Thin Elastic Shells. Izdatel'stvo Nauka, Moskva (1976). (in Russian)
19. Goldenveizer, A.L.: The boundary conditions in the two-dimensional theory of shells: the mathematical aspect of the problem. J. Appl. Math. Mech. **62**(4), 617–629 (1998)
20. Gregory, R.D., Wan, F.Y.: Decaying states of plane strain in a semi-infinite strip and boundary conditions for plate theory. J. Elast. **14**(1), 27–64 (1984)
21. Gregory, R.D., Wan, F.Y.: On plate theories and Saint-Venant's principle. Int. J. Solids Struct. **21**(10), 1005–1024 (1985)

Chapter 8
Optimization of Thermal Treatment Modes of Bodies Made of Functionally Graded Materials

Bogdan Drobenko and Evgeniy Irza

Abstract A general approach to optimal control of thermal modes of heat treatment processes of deformable elastic and viscoelastic bodies made of functionally graded materials is developed. The minimum duration of the heat treatment process is the optimality criterion at restrictions imposed on the stress–strain state of the body and the technological parameters of heating–cooling. An original method for the numerical solution of a formulated optimization problem, which, after its spatiotemporal discretization, is reduced to the parametric minimization of a function of one variable at each step in time, has been developed. Solutions to the problems of optimal control of thermal regimes in the processes of heating and cooling of finite thermosensitive cylinders made of elastic and viscoelastic functionally graded materials were obtained.

8.1 Introduction

The structures made of functionally graded materials are often subjected to conditions related to their heating or cooling. As a result of the action of temperature fields upon these structures, we observe the appearance of temperature stresses that may exceed admissible values and can lead to the destruction of structures or the deterioration of their operating characteristics. At the same time, modern high requirements for the productivity of heating–cooling processes, which are used in the manufacture of structural elements and providing them with the appropriate operational characteristics, dictate their conditions regarding the duration of these processes, which are significantly energetically consuming. Therefore, there is a practical need for the development of effective mathematical models and methods for optimizing the modes of heat treatment in such structures according to the productivity and energy

B. Drobenko (✉) · E. Irza
Pidstryhach Institute for Applied Problems of Mechanics and Mathematics, National Academy of Science of Ukraine, Lviv, Ukraine
e-mail: drobenko@ukr.net

E. Irza
e-mail: evgen_irza@ukr.net

H. Altenbach et al. (eds.), *Selected Problems of Solid Mechanics and Solving Methods*, Advanced Structured Materials 204, https://doi.org/10.1007/978-3-031-54063-9_8

consumption under restrictions imposed on the stress state of the analyzed structure and the technological conditions of heating and cooling.

General approaches, methods, and results for optimizing thermal regimes in the processes of technological heating–cooling of structures are quite fully covered in monographs (Hachkevych et al. [4]; Kushnir et al. [5]), which contain an extensive bibliography. In the vast majority of applied works devoted to describing and optimizing thermomechanical processes in structural elements under thermal loads, simple homogeneous structures are considered with temperature-independent characteristics and, as a rule, in a spatially one-dimensional formulation. Using such models in cases when complex structures are heated to high temperatures can lead to significant errors in estimating their thermomechanical state parameters since the properties of bodies at elevated temperatures and in their natural state are different. This is especially important for designing optimal heating–cooling modes of thermosensitive structures made of functionally graded materials since the accuracy of solving direct problems of researching the deformation processes of such structures under intense temperature load affects the optimal mode significantly.

In view of the above, there is a need to develop effective models and methods for studying thermomechanical processes in structures of a fairly general appearance subjected to a non-stationary temperature load, taking into account the dependence of material properties on temperature and spatial coordinates, as well as at the design on this basis, optimal modes of their heating–cooling according to various criteria. The statement of the problem is given, and the stress–strain state of elastic bodies of canonical and non-canonical shape under force loads and temperature fields is studied (Grigorenko and Vasilenko [2]; Grigorenko, Vasilenko, Emelyanov et al. [3]) taking into account the anisotropy, heterogeneity of material and temperature dependence of mechanical characteristics.

In this work, a general approach to optimal control of heat treatment modes of deformable elastic and viscoelastic bodies made of thermosensitive functionally graded materials is developed. The optimality criterion is chosen as a minimum duration of the heat treatment process at restrictions on the stress–strain state of the body and the technological parameters of heating–cooling.

8.2 Statement of the Optimization Problem and Its Solution

Consider a thermosensitive inhomogeneous body that occupies a region Ω of Euclidean space R^3 with the Lipschitz continuous surface Γ (physical and mechanical characteristics of the body depend on spatial coordinates and temperature). The body is referred to the curvilinear coordinate system $Ox^1x^2x^3$. On a part Γ_u of the surface Γ a displacement $\mathbf{u} = \mathbf{u}_0$ is set, and on another part Γ_σ a force load, which is characterized by vector $\mathbf{p}$, is given; $\mathbf{r} = [x^1, x^2, x^3]$ is the radius-vector of the point of the body in the coordinate system $Ox^1x^2x^3$; $\Gamma_u \cup \Gamma_\sigma = \Gamma$. The body is subjected to high-temperature heat treatment (heating–cooling).

The temperature field and the strain–stress state of the body in the heat treatment process are determined by the following state parameters: a temperature $t\,(\mathbf{r},\,\tau)$; a stress tensor $\hat{\sigma}\,(\mathbf{r},\,\tau)$; a strain tensor $\hat{\varepsilon}\,(\mathbf{r},\,\tau)$; a displacement vector $\mathbf{u}\,(\mathbf{r},\,\tau)$. The state parameters are related to each other, to the physical and geometric parameters of the body, to external force $\mathbf{p}$ and to the heat load function $h\,(\tau)$ (control function), by a system of equations

$$L_i\left(\mathbf{r},\tau,t,\hat{\sigma},\hat{\varepsilon},\mathbf{p},\mathbf{u},h\right)=0\quad(i=1,\ldots,n_0),\tag{8.1}$$

where nonlinear integral-differential operators are denoted by L_i, τ is time.

Note that this system of equations generally includes heat conduction equations, equations of motion, geometric relations between deformations and displacements, phenomenological relations between the components of tensors of stresses and strains, compatibility equations of deformations, and initial and boundary conditions.

In general, the system of Eq. (8.1) is a system of nonlinear integral-differential equations. The problem of determining state parameters from this system of equations for a given control function $h(\tau)$ is called a direct problem. Taking into account the phenomenological and geometric relations, the direct problem can be written in terms of temperature and displacements. Then the system of Eq. (8.1) takes the form

$$\bar{L}_j\,(\mathbf{r},\tau,t,\mathbf{u},\mathbf{p},h)=0\quad(i=1,\ldots,n_0).\tag{8.2}$$

We will formulate the problem of optimal control of the thermal regimes of heating and cooling of the body under consideration. For the optimization function, we will choose the duration of the heat treatment process, i.e.

$$J=\tau^*(h).\tag{8.3}$$

The minimum value of the functional (8.3) is chosen as the optimality criterion, i.e.

$$I=\min_h \tau^*.\tag{8.4}$$

The control function $h(\tau)$ is selected based on the technological possibilities of controlling the existing physical and mechanical processes in a specific heat treatment technology. In particular, the control function can be ambient temperature, heat flow, heat transfer coefficient, etc.

The requirements for the quality of products and the parameters of technological heat treatment processes lead to certain additional restrictions imposed on the state variables and technological conditions of the heat treatment process. These constraints consist of inequalities

$$\phi_j\left(\mathbf{r},\tau,t,\hat{\sigma},\hat{\varepsilon},\mathbf{p},\mathbf{u},h\right)\leqslant 0\quad(j=1,\ldots,m).\tag{8.5}$$

Here ϕ_j are the given functions. The selection of restrictions on state parameters and technological conditions is carried out taking into account the goals of heat treatment. In the formulation under consideration, the task of optimizing thermal regimes in heat treatment processes consists in finding a control function $h(\tau)$ that ensures the minimum duration of the heating–cooling process (8.4) under constraints (8.2) and (8.5).

The solution of the formulated extremal problem is based on the principle of phased parametric optimization, according to which thermomechanical processes in a body undergoing heat treatment, as in a system with distributed parameters, are first described by a discrete model, and then the optimization problem is solved as a problem of nonlinear mathematical programming. Since the system of Eq. (8.2) is nonlinear, and the geometric configuration of the body is often quite complex, when solving a direct problem, there is a need to use universal computational methods. One of these methods is the finite element method.

As a result of standard finite element discretization, the original direct problem (8.2) is reduced to the system of equations

$$\sum_{e=1}^{n_e} \int_{\Omega^{(e)}} \bar{L}_j\left(\tau, T^{(e)}, U^{(e)}, h(\tau)\right) d\Omega = 0 \quad (j = 1, \ldots, n) \tag{8.6}$$

with unknown temperature $T^{(e)}$ and displacements $U^{(e)}$ at the nodes of the finite-element mesh (Bathe [1]; Altenbach and Sakharov [7]; Zienkiewicz [8]); n_e is a total number of finite elements Ω_e. Note that there are no longer spatial derivatives in (8.6); the nodal values of temperature and displacements depend only on time.

To find the solution of this system of equations, a unified set of single-step algorithms (Zienkiewicz et al. [9]) is used, according to which the entire time interval of the heat treatment process is sequentially passed with a constant step $\Delta\tau$ (the sought values of the state parameters at the end of each time interval $[\tau_i, \tau_i + \Delta\tau]$ $(i = 0, 1, \ldots)$ are determined using their values at the beginning of the interval). Within the framework of the proposed approach, the control function on the time interval $\left[\tau_i, \tau_{i+1}\right]$ is approximated by a linear function. The value of the control function at the moment of time τ_{i+1} is given by the relation

$$h_{i+1} = h_i + a_i \Delta\tau, \tag{8.7}$$

where a_i is the maximum possible rate of increase (decrease) of the control function during heating (cooling), taking into account constraints (8.2) and (8.5).

As a result of the performed time discretization and the application of the unified set of single-step algorithms to determine the desired values of temperature and displacements of the body at the nodes of the finite element mesh at the moment of time τ_{i+1} (at the end of the considered time interval), a system of algebraic equations is obtained

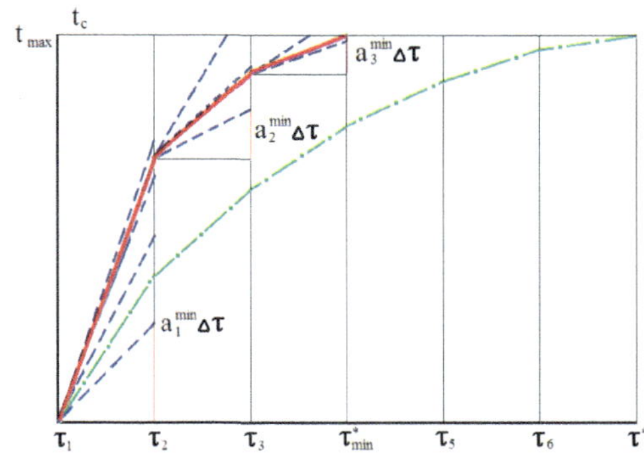

Fig. 8.1 Illustration of finding of the optimal solution ($\tau^*_{\min} = 3\Delta\tau$, $\tau^* = 6\Delta\tau$)

$$\sum_{e=1}^{n_e} \int_{\Omega^{(e)}} \tilde{L}_j \left(\{T^{(e)}\}_{i+1}, \ldots, \{U^{(e)}\}_{i+1}, a_i\right) d\Omega = 0 \quad (j = 1, \ldots, n). \tag{8.8}$$

Here $\tilde{L}_j$ are algebraic operators obtained from the corresponding operators $\bar{L}_j$, $\{T^{(e)}\}_{i+1}$, $\{U^{e}\}_{i+1}$ are the values of the state parameters (temperature and displacements) at the nodes of the e-th finite element at the moment of time τ_{i+1}.

In the general case, the system of algebraic Eq. (8.8) is nonlinear and is solved based on iterative approaches. At each step of the iterative process, based on the obtained approximations of displacements at the end of the corresponding time interval, the corresponding deformations and stresses are determined, using geometrical and phenomenological relations, and checked for the fulfillment of constraint (8.5), which, as a result of the performed discretization in the time interval $\left[\tau_i, \tau_{i+1}\right]$, acquire the corresponding form

$$\phi_j \left(\{T\}_{i+1}, \ldots, \{U\}_{i+1}, \{\sigma_{i+1}\}, a_i\right) \leqslant 0, \quad (j = 1, \ldots, l). \tag{8.9}$$

Thus, according to the proposed approach, the optimization task for the time interval $\left[\tau_i, \tau_{i+1}\right]$ is reduced to the task of finding the maximum rate of change of the control function, and practically—of one parameter a_i, the value of which is determined by direct search from a set of admissible values of its change based on information about stress, deformation, and displacement in body, as well as given restrictions on the considered step. In other words, instead of the problem of finding the trajectory of the change of the control function, which belongs to the set of the power of continuum, a piecewise linear function is sought, the values of which are determined at the nodal points of the time discretization of the considered optimization problem.

The process of finding the optimal solution in the case when the temperature of the external environment is selected as the control function $t_c(\tau)$ is illustrated in Fig. 8.1. The search for the maximum possible rate of change of the control function a_i at each time interval $\left[\tau_i, \tau_{i+1}\right]$ is performed with a step Δa_i: the direct problem is solved at $a_i = \Delta a_i$, the stress–strain state of the body is determined, and the fulfillment of constraints (8.9) is evaluated. If all the constraints are fulfilled, the rate of change of the control function increases and the direct problem is solved at $a_i = 2\Delta a_i$ with further evaluation of the fulfillment of the constraints. Calculations in this time interval according to this scheme are continued until constraints (8.9) are met. If at least one constraint is violated, there is a return to the last value of the parameter a_i, at which all constraints were fulfilled, and the parameter variation step Δa_i is halved. The search for the maximum possible rate of change of the control function at the *i*th time step continues with this calculation procedure until the increment of the step Δa_i becomes small enough.

Note that in order to optimize the calculations, the initial step Δa_i on the next time interval is selected taking into account the number of solutions of the direct problem on the previous step. If the number of such solutions increases, the initial step Δa_i is doubled. If with such step on the considered time interval, there is an immediate departure from the set of admissible rates of change of the control function, the step of changing the parameter a_i is halved.

Thus, at each step of the extremum search, direct problems are solved using the finite element method in combination with the unified set of single-step algorithms. This approach makes it possible to solve a wide class of problems for bodies of complex geometric shape and structure; moreover, the search for the optimal value of the control function does not require zero approximation and is reduced at each time step to the parametric minimization of the function of one variable.

8.3 Solving of Direct Problems

The direct problem of determining the parameters of the temperature and the stress–strain state in the body is formulated in a quasi-static setting. Methods of solving direct problem occupy a key place in the construction of the optimal heat treatment mode. As a result of the standard finite-element discretization, the heat conduction problem is reduced to a system of ordinary differential equations (Altenbach and Sakharov [7])

$$[C_t] \frac{d\{T\}}{d\tau} + [K_t]\{T\} = \{f_t\} \tag{8.10}$$

with respect to the unknown temperature values at the nodes of the finite-element mesh, where $[C_t]$, $[K_t]$, $\{f_t\}$ are general matrices of heat capacity, of thermal conductivity and of heat load vector, obtained by summing up the appropriate matrix–vector characteristics of individual finite elements.

The system of ordinary differential Eq. (8.10) is solved using a unified set of single-step algorithms (Zienkiewicz, Wood and Nine [9]). As a result of the standard finite-element discretization of the system of equations, which describe the processes of deformation of the body, we obtain a system of algebraic equations (Altenbach and Sakharov [7])

$$[K_u]\{U(\tau_{i+1})\} = \{F_0(\tau_{i+1})\} \tag{8.11}$$

for elastic body and Volterra integral equations of the second kind

$$[K_u]\{U(\tau)\} = \{F_0(\tau)\} + \int_0^\tau [F(\tau,\xi)]\{U(\xi)\}\,d\xi \tag{8.12}$$

for viscoelastic body (Stashchuk and Irza [6]). Here, $[K_u]$ and $\{F_0\}$ are respectively the stiffness matrix and the load vector, obtained by summing up the appropriate matrix–vector characteristics of individual finite elements. In particular, in the case of an elastic body, these characteristics are as follows (Altenbach and Sakharov [7]):

$$[K_u^{(e)}]=\int_{\Omega^{(e)}}[B_u^{(e)}]^T[D^{(e)}][B_u^{(e)}]d\Omega,\ \{F_0^{(e)}\}=\int_{\Omega^{(e)}}[B_u^{(e)}]^T[D^{(e)}]\{\varepsilon_t^{(e)}\}d\Omega, \tag{8.13}$$

where $[B_u^{(e)}]$ is the matrix of geometrical relations of the theory of elasticity; $[D^{(e)}]$ is the matrix of elastic constants; $\{\varepsilon_t\}$ is the vector of components of the temperature strain tensor.

In the case of viscoelastic body (Stashchuk and Irza [6]),

$$\left[F^{(e)}(\tau,\xi)\right] = \int_{\Omega^{(e)}} R(\tau,\xi)\left[B_u^{(e)}\right]^T[D_0]\left[B_u^{(e)}\right]d\Omega, \tag{8.14}$$

$$\begin{aligned}\left\{F_0^{(e)}\right\} &= \int_{\Omega^{(e)}}\left[B_u^{(e)}\right]^T\left[D^{(e)}\right]\left\{\varepsilon_t^{(e)}\right\}d\Omega \\ &-\int_{\Omega^{(e)}} R_0\left[B_u^{(e)}\right]^T[D_0]\{\sigma^{\text{res}}\}d\Omega + \int_{\Omega^{(e)}} 2GR_0\left[B_u^{(e)}\right]^T[D_0]\{\varepsilon^{\text{res}}\}d\Omega,\end{aligned} \tag{8.15}$$

where

$$R(\tau,\upsilon) = \frac{2G(\tau)G(\upsilon)}{\eta(\upsilon)}\exp\left\{-\int_\upsilon^\tau \frac{G}{\eta}du\right\},\quad R_0(\tau) = \exp\left\{-\int_{\tau_0}^\tau \frac{G}{\eta}du\right\}$$

(G is shear modulus; η is dynamic viscosity).

The system of Eq. (8.12) is solved by the method of stepwise integration over time. To do this, we enter a discrete set of points τ_0, τ_1, ... on the time scale, with a step $\Delta\tau$. At each time step, we use the trapezium method to calculate the integral. Using the same discretization step for all time intervals, we obtain the following algorithm for determining the values of displacements $\{U\}$ at the nodes of the finite element mesh at the instant of time τ_n:

$$\begin{aligned}&\left([K_u] - \frac{1}{2}\Delta\tau[F(\tau_n,\tau_n)]\right)\{U(\tau_n)\} \\ &= \{F_0\} + \Delta\tau\left(\frac{1}{2}[F(\tau_n,0)]\{U(0)\} + \sum_{i=1}^{n-1}[F(\tau_n,\tau_i)]\{U(\tau_i)\}\right).\end{aligned} \tag{8.16}$$

The temperature- and coordinate-dependent properties of the body are approximated by interpolation splines, reconstructed from the points of known experimental dependencies, which describe the thermomechanical behavior of materials in a wide range of temperatures.

8.4 Heat Treatment Modes of Elastic Cylinder

As an example, the optimal mode of the heat treatment of an elastic hollow finite cylinder $\Omega = \{(r, z) : r_v \leqslant r \leqslant r_z;\ |z| \leqslant L\}$ made of functionally graded material was developed. The cylinder Ω is heated from the initial temperature t_0 to a certain temperature $t_{\max}$, after which it is cooled to the initial temperature. The duration of the heating–cooling process is chosen as the optimization function. The control function is the temperature $t_c(\tau)$ on the outer surface of the cylinder, and thermal insulation conditions are set on the inner surface and ends of the cylinder. During heating–cooling, the technological restraints must be observed:

$$t_0 \leqslant t\,(r, z, \tau) \leqslant t_{\max}, \quad \frac{dt}{d\tau} \leqslant v_{\max}, \quad \sigma_{\phi\phi} \leqslant \sigma_d\,(r, z, t)\,. \tag{8.17}$$

According to the developed concept, we will perform finite-element discretization of the area Ω using isoparametric biquadratic finite elements (Zienkiewicz [8]) and represent the time interval in the form of a union of intervals of length $\Delta\tau$. After the discretization of the optimization problem, we begin to vary the control function on the first time interval according to the methodology developed. For each such variation, we determine the distributions of the temperature and mechanical fields in the cylinder from a direct problem, and check whether the existing restrictions are fulfilled. Having determined the value of the control function at the end of the considered time step, at which we reach the limit of one of the constraints (8.17) with a given accuracy, we proceed to the variations of the control function at the next time interval.

Computational experiments were performed for the cylinder ($r_v = 0.206$ m, $r_z = 0,216$ m, $L = 0,1$ m) with the following parameters:

$$t_0 = 20\,^\circ\mathrm{C}, \quad t_{\max} = 310\,^\circ\mathrm{C}, \quad v_{\max} \leqslant 0.5\,^\circ\mathrm{C}/s, \quad \Delta a_0 = 0.5\,^\circ\mathrm{C}/s.$$

Dependencies of material properties on temperature in three typical sections of the inhomogeneous cylinder along the thickness are presented on the Fig. 8.2 (curves 1, 2, 3 correspond to sections $r_1 = 0.216$ m, $r_2 = 0.211$ m, $r_3 = 0.206$ m). The optimal mode is shown on the Fig. 8.3 (curve 1), and corresponding stress is shown on the Fig. 8.4 (curves 1, 2, 3 reflect stresses at points ($r_1 = 0.216$ m, $z_1 = 0$ m), ($r_2 = 0.211$ m, $z_2 = 0$ m), and ($r_3 = 0.206$ m, $z_3 = 0$ m) respectively).

A computational experiment was also performed for the case when the allowable stress σ_d are assumed to be constant (the material is still heterogeneous along thick-

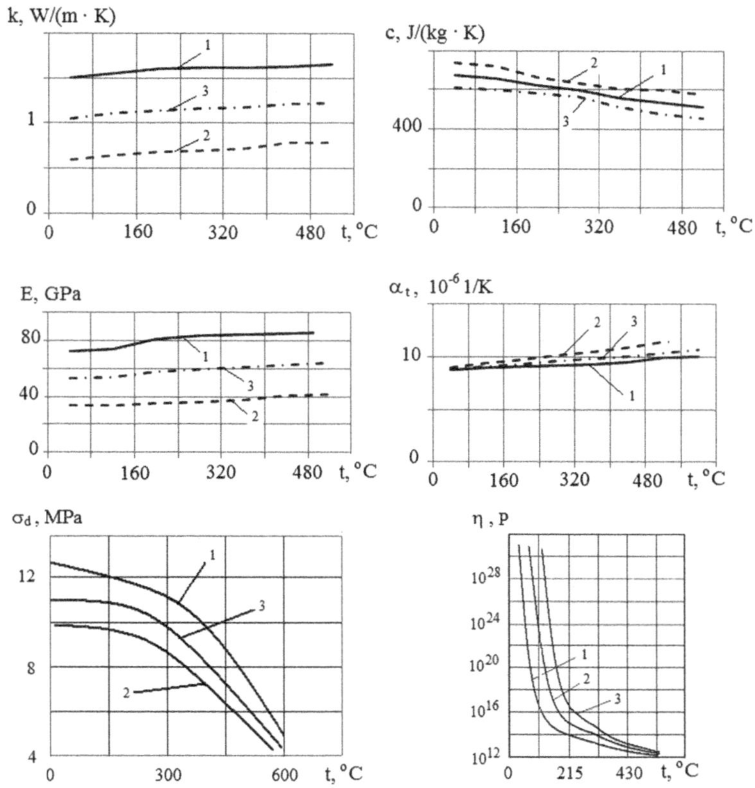

Fig. 8.2 Dependencies of material properties on temperature (k is thermal conductivity; c is specific heat capacity; E is modulus of elasticity; α_t is linear temperature expansion; σ_d is allowable stress; η is viscosity)

Fig. 8.3 The optimal mode:
1—$\sigma_d = \sigma_d(r, t)$;
2—$\sigma_d = \sigma_d(r)$

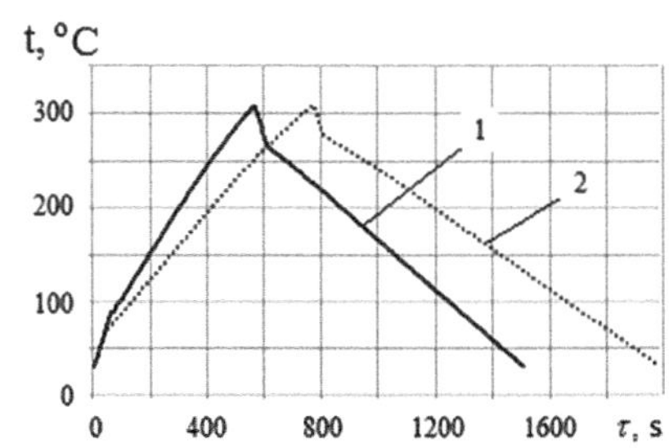

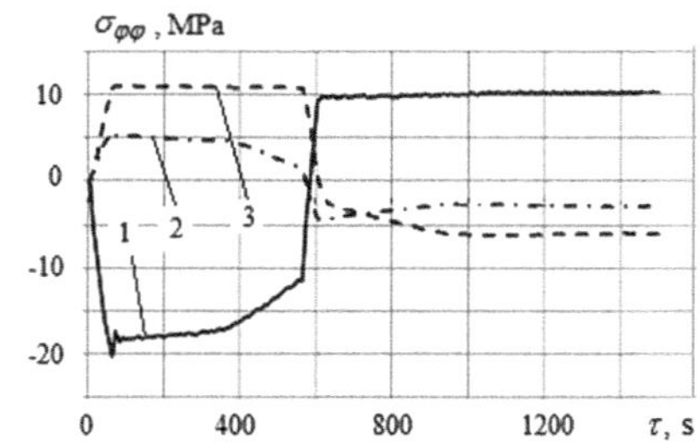

Fig. 8.4 Changes of circumferential stress in time

ness of the cylinder). The obtained optimal mode in this case is shown on Fig. 8.3 (curve 2). The analysis of the obtained results indicates the need to take into account the dependence of allowable stresses on temperature, since the durations of the modes for these two cases differ by 32%.

The convergence of the obtained optimal solutions was studied. For this, computational experiments were performed with different time steps (from a second to a minute), the order of approximation of the temperature and displacement fields (bilinear, biquadratic, bicubic finite elements) when solving direct problems, as well as with different sizes of finite elements (from five to twenty equilateral finite elements along thickness of the cylinder). A practically exact solution was obtained already at a time step of 15 s on meshes with ten equilateral biquadratic elements per thickness of the cylinder. With further reduction of the time step and the size of finite elements, the obtained solutions coincide with an accuracy of 1%.

8.5 Heat Treatment Modes of Viscoelastic Cylinder

The optimal mode of the heat treatment of a viscoelastic thermosensitive hollow finite cylinder $\Omega = \{(r, z) : r_v \leqslant r \leqslant r_z, |z| \leqslant L\}$ made of functionally graded material also was developed in order to remove residual stresses. The heat treatment of the cylinder takes place in three stages. On the first stage, the cylinder is heated from the initial temperature t_0 to temperature $t_{\max}$, on the second stage, it is kept for a certain time at the temperature $t_{\max}$ (in order to relax the residual stresses to a given level), and on the third stage, it is cooled to the initial temperature. The temperature on the outer surface of the cylinder is selected by the control function $t_c(\tau)$, and the thermal insulation conditions are set on the inner surface and the cylinder ends.

The technological restrictions (8.17) must be met during the heat treatment on the first and third stages, and restrictions on the temperature and the residual stresses

$$t_c = t_{\max}, \quad \max\{\sigma_{rr}^{\text{res}}, \sigma_{zz}^{\text{res}}, \sigma_{\phi\phi}^{\text{res}}, \sigma_{rz}^{\text{res}}\} = \sigma_*^{\text{res}} \tag{8.18}$$

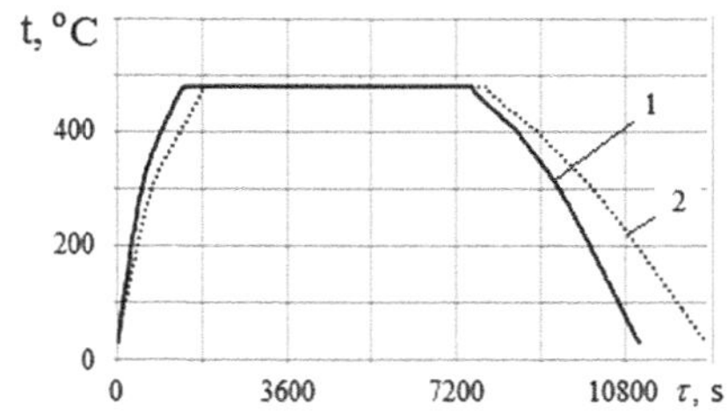

Fig. 8.5 Optimal mode:
1—$\sigma_d = \sigma_d(r, t)$;
2—$\sigma_d = \sigma_d(r)$

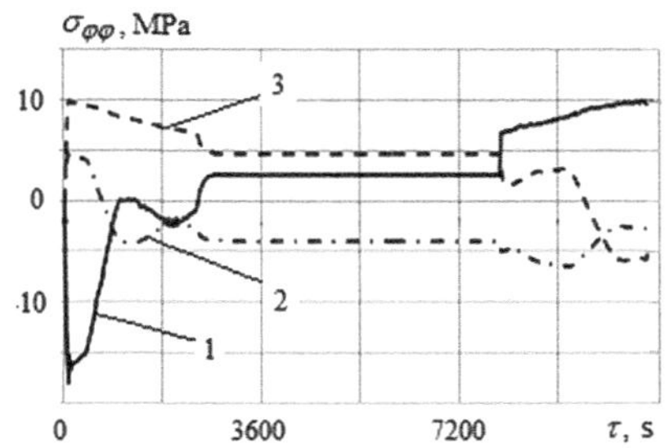

Fig. 8.6 Changes of circumferential stress in time

must be fulfilled in the second stage. Computational experiments were performed for the cylinder $r_v = 0.206$ m, $r_z = 0.216$ m, $L = 0.1$ m, with the following parameters:

$$t_0 = 20\,^\circ\text{C},\quad t_{\max} = 480\,^\circ\text{C},\quad v_{\max} \leqslant 0.5\,^\circ\text{C}/s,\quad \Delta a_0 = 0.5\,^\circ\text{C}/s,\quad \sigma_*^{\text{res}} = 0.5\text{ MPa}.$$

Dependencies of material properties of functionally graded material on temperature in three typical sections of the cylinder are presented on Fig. 8.2 (curves 1, 2, 3 correspond to sections $r_1 = 0.216$ m, $r_2 = 0.211$ m, $r_3 = 0.206$ m).

Figure 8.5 shows a graph of the optimal heat treatment mode (curve 1). Figure 8.6 shows the corresponding change in time of circumferential stress of a cylinder along the r coordinate in section $z_3 = 0$ (curves 1, 2, 3 reflect stress at $r_1 = 0.216$ m, $r_2 = 0.211$ m, and $r_3 = 0.206$ m, respectively). Figure 8.7 illustrates the relaxation of residual stress in time. Figure 8.5 also shows the optimal mode obtained without taking into account the dependence of allowable stresses on temperature (curve 2). As you can see, the durations of the modes in the cases $\sigma_d = \sigma_d(r, t)$ and $\sigma_d = \sigma_d(r)$ differ by 28%.

Thus, the models and methods proposed in the work as a whole make it possible to more adequately (compared to existing models) predict the real thermo-mechanical behavior of heterogeneous elastic and viscoelastic bodies in a wide temperature range and to design on this basis the optimal regimes for their heat treatment for practically any amount technological limitations, as well as take them into account when automating processes of high-temperature heat treatment of products made of elastic and viscoelastic materials, the physical and mechanical characteristics of which depend on temperature and spatial coordinates.

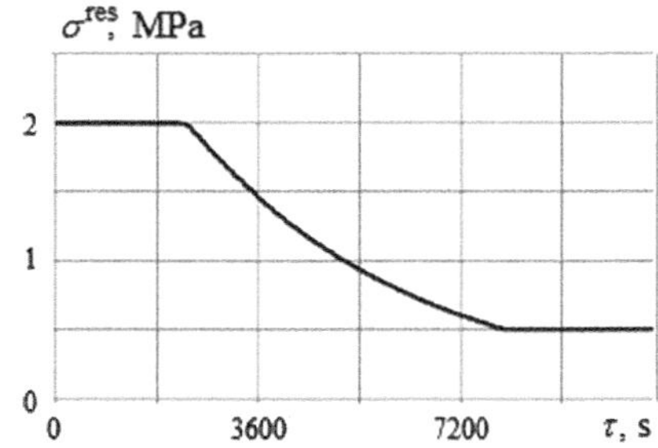

Fig. 8.7 The relaxation of residual stress in time

8.6 Conclusions

A general formulation of the problem of optimizing the thermal modes of heating–cooling of bodies, the characteristics of which depend both on spatial coordinates and on temperature is proposed. An original method of numerical solution of formulated optimization problems based on the principle of stepwise parametric optimization has been developed. The use of the finite element method for solving direct problems in the process of iterative construction of optimal heating–cooling regimes makes it possible to solve a wide class of problems for bodies of complex geometric shape. Problem the search for the optimal value of the control function does not require zero approximation and is reduced at each step in time to the parametric minimization of the function of one variable. The step of variation of the control function in the current time interval is selected taking into account the number of iterations in the previous one, it significantly accelerates the process of calculation.

Solutions of optimal control problems of thermal regimes in the processes of heating and cooling of finite thermosensitive cylinders made of elastic and viscoelastic functionally graded materials were obtained, taking into account the relaxation of residual stresses at elevated temperatures. The need to take into account the dependence of the physical and mechanical characteristics of the material on temperature is shown. In particular, it was found that the durations of the optimal modes of high-temperature heat treatment of the cylinder, obtained taking into account the dependence of allowable stresses on temperature and without such consideration, differ by a third.

References

1. Bathe, K.J.: Finite Element Procedures Analysis. Prentice Hall, Englewood Cliffs, N.J. (1995)
2. Grigorenko, YaM., Vasilenko, A.T.: Determination of the stress-strain state of anisotropic thermosensitive cylinders. J. Math. Sci. **97**, 3826–3829 (1999)
3. Grigorenko, Ya.M., Vasilenko, A.T., Emelyanov, I.G. et al.: Statics of structural elements. In: Guz, A.N. (Eds.) Mechanics of Composites, vol. 8. Institute of Mechanics of NAS of Ukraine. Nauk. Dumka, Kyiv (1999). (In Rus.)

4. Hachkevych, O.R., Hachkevych, M.G., Budz, S.F.: Stress optimization of heating regimes of glass piecewise homogeneous shells. Pidstryhach Institute for Applied Problems of Mechanics and Mathematics of the NAS of Ukraine, Lviv (2014). (In Ukr.)
5. Kushnir, R.M., Popovych, V.S., Yasinskyy, A.V.: Optimization and identification in termomechanics of inhomogeneous solids in Modeling and optimization. In: Chif Burak, Y.J., Kushnir, R.M. (Eds.) Thermomechanics of electroconductive heterogeneous bodies, vol. 5. SPOLOM, Lviv (2011). (In Ukr.)
6. Stashchuk, M.H., Irza, E.M.: Thermal stressed states of the bodies of revolution made of functionally graded materials. Mater. Sci. **55**(3), 311–319 (2019)
7. Altenbach, von.J., Sakharov, A.S.: The Finite Element Method in Solid Mechanics. Vyshcha Shkola, Kyiv (1982). (in Rus.)
8. Zienkiewicz, O.C.: The Finite Element Method in Engineering Science. McGraw-Hill, London (1971)
9. Zienkiewicz, O.C., Wood, W.L., Nine, N.W.: A unified set of single step algorithm. Part 1: general formulation and applications. Int. J. Num. Meth. Eng. **20**, 1529–1552 (1984)

Chapter 9
Numerical Analysis of Contact Between Elastic Bodies in the Presence of Thin Coating and Nonlinear Winkler Surface Layers

Ivan I. Dyyak, Ihor I. Prokopyshyn, Ivan A. Prokopyshyn, and Andriy O. Styahar

Abstract We consider the problem of unilateral contact through a nonlinear Winkler surface layer of a massive elastic body and a composite body consisting of a thin coating in the form of a Timoshenko-type shell and a massive elastic base connected with the coating via the other nonlinear Winkler layer. We present a weak formulation of this problem in the form of a nonlinear variational equation. To solve this equation, we propose a parallel Robin-type domain decomposition method that reduces the original contact problem to the iterative solution of independent linear variational equations corresponding to elasticity problems for massive bodies and a problem of the Timoshenko-type shell theory for the thin coating. With the help of finite element approximations, the developed method is used to study the contact between two massive elastic bodies with surface groove in the presence of a thin coating and the nonlinear Winkler-type surface layers. We analyze the dependence of the contact and interface stresses on the height of the coating and the Winkler layer parameter. The results obtained for the coatings modeled using the Timoshenko-type shell theory and the classical elasticity theory are compared.

I. I. Dyyak · I. A. Prokopyshyn · A. O. Styahar
Ivan Franko National University of Lviv, Lviv, Ukraine
e-mail: ivan.dyyak@lnu.edu.ua

I. A. Prokopyshyn
e-mail: ivan.prokopyshyn@lnu.edu.ua

A. O. Styahar
e-mail: andriy.styahar@lnu.edu.ua

I. I. Prokopyshyn (✉)
Pidstryhach Institute for Applied Problems of Mechanics and Mathematics,
National Academy of Science of Ukraine, Lviv, Ukraine
e-mail: ihor84@gmail.com

H. Altenbach et al. (eds.), *Selected Problems of Solid Mechanics and Solving Methods*,
Advanced Structured Materials 204, https://doi.org/10.1007/978-3-031-54063-9_9

9.1 Introduction

The increase in requirements for the reliability of elements and units of modern machinery, the functioning of which is affected by contact deformations, has led to intensive research into the contact interaction of bodies with the fullest possible consideration of their surface properties, in particular inhomogeneities in the form of thin coatings, surface layers, and micro-irregularities that arise both as a result of treatment of surfaces using various technologies and during operation.

Thin elastic coatings are used to improve the strength and bearing capacity of machine parts, to protect the materials against the action of corrosive media, high temperatures, and negative influence of the wear particles, friction, and shock, and to decrease the level of stress concentration in the main body.

A brief survey of investigations of contact problems for bodies with thin coatings can be found in Martynyak et al. [12] and the works cited in it.

The analysis of the stress–strain state of bodies with thin coatings and inclusions under the conditions of perfect contact between the body and the coating (inclusion) was carried out by Grigorenko et al. [8], Liashenko et al. [9], Lopata et al. [10], Pasternak [14], Pasternak and Sulym [15], Savula and Kossak [25], Solovykh et al. [26], Sulym [27], and Vynnytska and Savula [28].

An asymptotic analysis of the spatial contact problem for an elastic layer is important for the substantiation of applied coating theories. In particular, such analysis can be found in Alblas and Kuipers [1, 2].

Mathematical models and methods for the solution of contact problems for elastic bodies with thin coatings and interface layers were developed by Pelekh et al. [17].

The methods based on the integral equations for solving the problems of contact interaction between infinite bodies with thin coatings were developed by Martinyak and Shvets' [11] and Mkhitaryan et al. [13]. The numerical methods for solving the contact problems for finite bodies in the presence of thin coatings based on the variational formulations and the finite element method (FEM) were investigated by Blokh and Orobinskii [3] and Podgornyi et al. [17].

The contact interaction between bodies with discontinuous coatings and rigid punches was studied by Gong and Konvopoulos [5, 6] and Ramachandra and Ovaert [23].

As an efficient approach to the investigation of the stress–strain state of multiscale systems consisting of many bodies and coatings under the conditions of their contact interaction, we mention the use of domain decomposition methods (algorithms) (DDMs). These methods allow reducing the solution of boundary-value problems for complex structures to the solution of sequences of simpler problems for its individual elements, which makes it possible to efficiently combine different mathematical models (3D theory, shell theory) and methods.

The Dirichlet–Neumann-type DDMs based on the combination of the FEM with the direct boundary element method were proposed by Dyyak et al. [4] for the solution of the problems of plane deformation of an elastic body with thin coating in the form of a Timoshenko-type shell under the conditions of perfect contact between the body and the coating.

Parallel iterative Robin-type DDMs were proposed for the solution of the problems of contact between several linear and nonlinear elastic bodies by Grigorenko et al. [7] and Prokopyshyn et al. [19]. Martynyak et al. [12] generalized these methods to the case of contact problems for elastic bodies with nonlinear Winkler surface layers. Prokopyshyn et al. [20] used a Robin-type DDM to study the axisymmetric problem of contact between two elastic bodies, one of which has a discontinuous cylindrical elastic coating. For the last problem, the model of the axisymmetric elastic body was used to describe the coating.

Prokopyshyn and Styahar [22] developed a Robin-type domain decomposition algorithm to solve the contact problem for two massive elastic bodies in the presence of a discontinuous thin elastic coating in the form of a Timoshenko-type shell between the bodies. Prokopyshyn and Styahar [21] generalized Robin-type DDMs for solving the problem of unilateral contact interaction between a massive elastic body and a composite body consisting of a continuous thin elastic coating in the form of Timoshenko shell and a massive elastic base connected with the coating through a nonlinear Winkler layer.

In the present paper, we consider the contact problem similar to that in Prokopyshyn and Styahar [21]. However, compared with that problem, an additional nonlinear Winkler surface layer is imposed between the massive and composite bodies. It means that we impose the conditions of unilateral frictionless contact through the nonlinear layer of Winkler type on the interface between the massive elastic body and the coating with the unknown actual contact zone, while the condition of bilateral contact through the other nonlinear Winkler layer with the known contact zone holds on the interface between the coating and the massive elastic base. As in Prokopyshyn and Styahar [21, 22], the stress–strain state of the massive elastic body and the massive elastic base is described by equations of the classical elasticity theory, whereas the Timoshenko-type shell theory is used to model the thin coating.

We perform the weak formulation of this problem in the form of a nonlinear variational equation. For solving this variational equation, we develop a parallel iterative Robin-type domain decomposition algorithm aimed at reduction of the procedure of solution of the contact problem to the simultaneous solution, in each iteration, of three independent linear variational equations, one of which corresponds to the boundary-value problem of the Timoshenko-type shell theory for a thin coating and the other two correspond to the boundary-value elasticity problems for massive bodies with Robin conditions on the common boundaries.

With the help of the proposed algorithm and the finite element approximations, we investigate the contact interaction between two elastic bodies with surface groove in the presence of a thin coating and nonlinear Winkler surface layers and analyze the influence of the Winkler parameter and the height of the coating on the contact and interface stresses.

9.2 Statement of the Problem

Consider the problem of imperfect contact of an elastic body $\Omega_1 \subset \mathbb{R}^2$ and a thin body $\Omega_2 \subset \mathbb{R}^2$, which is the coating of the other elastic body $\Omega_3 \subset \mathbb{R}^2$ (Fig. 9.1). We assume that on the interface between the elastic body Ω_1 and the thin coating Ω_2 of the body Ω_3 the conditions of unilateral frictionless contact through a nonlinear Winkler layer are imposed, whereas on the interface between the body Ω_3 and its coating Ω_2 the conditions of bilateral contact through the other nonlinear Winkler layer are satisfied. Suppose that the boundaries $\Gamma_\alpha = \partial\Omega_\alpha$ $(\alpha = 1, 2, 3)$, of all bodies are Lipschitz. In $\mathbb{R}^2$, we introduce a Cartesian coordinate system x_1, x_2. At a point $\mathbf{x} = (x_1, x_2)^{\mathrm{T}}$, the stress-strain state of each massive body Ω_α $(\alpha = 1, 3)$, is determined by the displacement vector $\mathbf{u}_\alpha(\mathbf{x}) = u_{\alpha i}(\mathbf{x})\mathbf{e}_i$, the strain tensor $\widehat{\boldsymbol{\varepsilon}}_\alpha(\mathbf{x}) = \varepsilon_{\alpha ij}(\mathbf{x})\mathbf{e}_i \otimes \mathbf{e}_j$, and the stress tensor $\widehat{\boldsymbol{\sigma}}_\alpha(\mathbf{x}) = \sigma_{\alpha ij}(\mathbf{x})\mathbf{e}_i \otimes \mathbf{e}_j$ whose components satisfy the following equations of the classical theory of elasticity under the conditions of plane deformation:

$$\sum_{k=1}^{2} \frac{\partial \sigma_{\alpha ik}(\mathbf{x})}{\partial x_k} = 0, \tag{9.1}$$

$$\sigma_{\alpha ij}(\mathbf{x}) = \delta_{ij}\lambda_\alpha \left(\varepsilon_{\alpha 11}(\mathbf{x}) + \varepsilon_{\alpha 22}(\mathbf{x})\right) + 2\,\mu_\alpha \varepsilon_{\alpha ij}(\mathbf{x}), \tag{9.2}$$

$$\varepsilon_{\alpha ij}(\mathbf{x}) = \frac{1}{2}\left(\frac{\partial u_{\alpha i}(\mathbf{x})}{\partial x_j} + \frac{\partial u_{\alpha j}(\mathbf{x})}{\partial x_i}\right), \tag{9.3}$$

where $i, j = 1, 2$, $\mathbf{x} \in \Omega_\alpha$, $\alpha = 1, 3$, $\delta_{ij} = \{1, i = j\} \vee \{0, i \neq j\}$ is the Kronecker symbol and $\lambda_\alpha(\mathbf{x})$, $\mu_\alpha(\mathbf{x})$ are Lamé parameters.

In order to describe the stress–strain state of the coating Ω_2, we use the equations of Timoshenko-type shell theory (Pelekh, [16]). For this purpose, in the thin body Ω_2, we introduce a local curvilinear coordinate system ξ_1, ξ_2, ξ_3 related to its middle surface Ω_2^*. Assume that the coating Ω_2 is a cylindrical shell infinite in the direction ξ_2. By v_{21}, w_2, and γ_{21}, we denote the tangential displacement, the normal displacement, and the angle of rotation, respectively. Further, we denote the strains by ε_{211}, ε_{213},

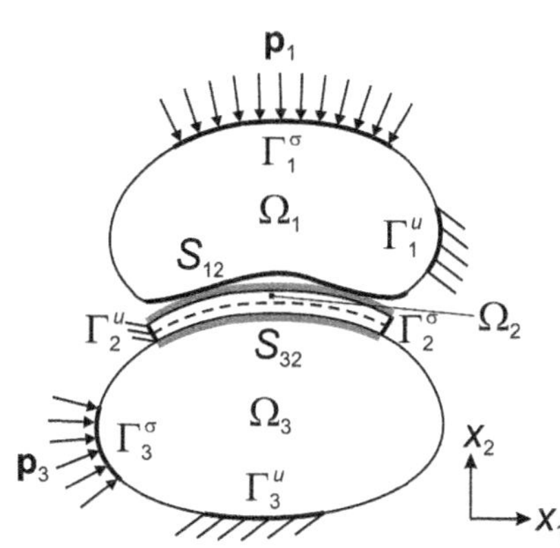

Fig. 9.1 Contact of the bodies in the presence of a coating and Winkler layers

and χ_{211}. The forces and moment in the shell are denoted by T_{211}, T_{213}, and M_{211}, respectively.

The equations of Timoshenko-type shell theory for the thin body Ω_2 have the form (Pelekh, [16])

$$-\frac{1}{A_{21}}\frac{dT_{211}}{d\xi_1}-k_{21}T_{213}=p_{21},$$

$$-\frac{1}{A_{21}}\frac{dT_{213}}{d\xi_1}+k_{21}T_{211}=p_{23},\quad \xi_{1b}\leqslant\xi_1\leqslant\xi_{1e}, \tag{9.4}$$

$$-\frac{1}{A_{21}}\frac{dM_{211}}{d\xi_1}+T_{213}=m_{21},\quad \xi_{1b}\leqslant\xi_1\leqslant\xi_{1e}, \tag{9.5}$$

$$T_{211}=\frac{E_2h_2\varepsilon_{211}}{1-\nu_2^2},\quad T_{213}=k_2'\mu_2h_2\varepsilon_{213},\quad M_{211}=\frac{E_2h_2^3\chi_{211}}{12\,(1-\nu_2^2)}, \tag{9.6}$$

$$\varepsilon_{211}=\frac{1}{A_{21}}\frac{dv_{21}}{d\xi_1}+k_{21}w_2,\quad \varepsilon_{213}=\frac{1}{A_{21}}\frac{dw_2}{d\xi_1}+\gamma_{21}-k_{21}v_{21},$$

$$\chi_{211}=\frac{1}{A_{21}}\frac{d\gamma_{21}}{d\xi_1}, \tag{9.7}$$

where h_2 is the thickness of the shell; $A_{21}=A_{21}(\xi_1)$ and $k_{21}=k_{21}(\xi_1)$ are the Lamé parameter and the curvature of the middle line of the shell, respectively; $E_2=\frac{1}{2}\lambda_2/(\lambda_2+\mu_2)$ is Young's modulus, $\nu_2=\mu_2(3\lambda_2+2\mu_2)/(\lambda_2+\mu_2)$ is Poisson's ratio, and $k_2'=5/6$ is the shear coefficient.

The functions of loading of the shell take the form

$$p_{21}=\left(1+k_{21}\frac{h_2}{2}\right)\sigma_{213}^{+}+\left(1-k_{21}\frac{h_2}{2}\right)\sigma_{213}^{-}, \tag{9.8}$$

$$p_{23}=\left(1+k_{21}\frac{h_2}{2}\right)\sigma_{233}^{+}-\left(1-k_{21}\frac{h_2}{2}\right)\sigma_{233}^{-}, \tag{9.9}$$

$$m_{21}=\frac{h_2}{2}\left(\left(1+k_{21}\frac{h_2}{2}\right)\sigma_{213}^{+}-\left(1-k_{21}\frac{h_2}{2}\right)\sigma_{213}^{-}\right), \tag{9.10}$$

where σ_{2ij}^{+} and σ_{2ij}^{-} $(i,j=1,3)$, are the components of the vector of surface forces on the outer ($\xi_3=h_2/2$) and inner ($\xi_3=-h_2/2$) surfaces of the shell, and m_{21} is the moment of the external load.

On each boundary $\Gamma_\alpha=\partial\Omega_\alpha$ $(\alpha=1,3)$ of massive bodies, we introduce a local orthonormal basis $\boldsymbol{\tau}_\alpha$, $\mathbf{n}_\alpha$, where $\boldsymbol{\tau}_\alpha$ is the unit tangent vector and $\mathbf{n}_\alpha$ is the unit vector of external normal. On this basis, the vectors of displacements and stresses on Γ_α can be represented as follows:

$$\mathbf{u}_\alpha=u_{\alpha\tau}\boldsymbol{\tau}_\alpha+u_{\alpha n}\mathbf{n}_\alpha,\quad \boldsymbol{\sigma}_\alpha=\widehat{\boldsymbol{\sigma}}_\alpha\cdot\mathbf{n}_\alpha=\sigma_{\alpha\tau}\boldsymbol{\tau}_\alpha+\sigma_{\alpha n}\mathbf{n}_\alpha\quad(\alpha=1,3).$$

Suppose that the boundaries of the massive bodies consist of three disjoint parts: $\Gamma_1 = \Gamma_1^u \cup \Gamma_1^\sigma \cup S_{12}$, $\Gamma_3 = \Gamma_3^u \cup \Gamma_3^\sigma \cup S_{32}$; and the boundary of the thin body consists of four disjoint parts: $\Gamma_2 = \Gamma_2^u \cup \Gamma_2^\sigma \cup S_{21} \cup S_{23}$.

On the parts Γ_α^u of the boundaries of each body, we impose kinematic boundary conditions. At the same time, static boundary conditions are imposed on the parts Γ_α^σ. For the massive bodies Ω_1 and Ω_3, these conditions take the following form:

$$\begin{aligned} &\mathbf{u}_\alpha(\mathbf{x}) = 0, \quad \mathbf{x} \in \Gamma_\alpha^u \quad (\alpha = 1, 3), \\ &\boldsymbol{\sigma}_\alpha(\mathbf{x}) = \mathbf{p}_\alpha(\mathbf{x}), \quad \mathbf{x} \in \Gamma_\alpha^\sigma \quad (\alpha = 1, 3), \end{aligned} \tag{9.11}$$

where $\mathbf{p}_\alpha = p_{\alpha\tau}\boldsymbol{\tau}_\alpha + p_{\alpha n}\mathbf{n}_\alpha$ are given boundary forces. For the thin coating Ω_2, we use the following kinematic and static boundary conditions:

$$\begin{aligned} &v_{21} = 0, \quad w_2 = 0, \quad \gamma_{21} = 0, \quad \xi_1 = \xi_{1e}, \\ &T_{211} = 0, \quad T_{213} = 0, \quad M_{211} = 0, \quad \xi_1 = \xi_{1b}, \end{aligned} \tag{9.12}$$

where the edge $\xi_1 = \xi_{1e}$ corresponds to the boundary Γ_2^u and $\xi_1 = \xi_{1b}$ corresponds to the boundary Γ_2^σ.

Assume that $S_{12} \subset \Gamma_1$ is the possible contact zone between the body Ω_1 and the coating Ω_2, and that $S_{21} \subset \Gamma_2$ is the possible contact zone between the coating Ω_2 and the body Ω_1. We suppose that these zones are sufficiently close ($S_{12} \approx S_{21}$). Therefore, $\mathbf{n}_1(\mathbf{x}) \approx -\mathbf{n}_2(\mathbf{x}')$, where $\mathbf{x}' = P(\mathbf{x})$ is the projection of the point $\mathbf{x} \in S_{12}$ onto S_{21}. The normal distance between Ω_1 and Ω_2 prior to contact is denoted by

$$d_n(\mathbf{x}) = ||\mathbf{x} - \mathbf{x}'|| = \sqrt{\sum_{i=1}^{2} (x_i - x_i')^2}.$$

The zones S_{12} and S_{21} have nonlinear Winkler-type surface layers. The total compression z_{12n} of these layers is described by a known function of the normal contact stresses

$$z_{12n}(\mathbf{x}) = \tilde{g}_{12n}(\sigma_{1n}(\mathbf{x})) = \tilde{g}_{12n}(-\overset{+}{\sigma}_{233}(\mathbf{x}')), \quad \mathbf{x} \in S_{12}, \quad \mathbf{x}' = P(\mathbf{x}) \in S_{21}.$$

The inverse dependence can be represented in the form

$$\sigma_{1n}(\mathbf{x}) = -\overset{+}{\sigma}_{233}(\mathbf{x}') = g_{12n}(z_{12n}(\mathbf{x})), \quad \mathbf{x} \in S_{12}, \quad \mathbf{x}' = P(\mathbf{x}) \in S_{21}.$$

On the boundaries S_{12} and S_{21} we impose the following conditions of unilateral contact through the nonlinear Winkler layer:

$$\sigma_{1\tau}(\mathbf{x}) = \sigma_{213}^{+}(\mathbf{x}') = 0, \mathbf{x} \in S_{12}, \mathbf{x}' = P(\mathbf{x}) \in S_{21}, \quad (9.13)$$

$$\sigma_{1n}(\mathbf{x}) = -\sigma_{233}^{+}(\mathbf{x}') = g_{12n}(z_{12n}(\mathbf{x})) \leqslant 0, \mathbf{x} \in S_{12}, \mathbf{x}' = P(\mathbf{x}) \in S_{21}, \quad (9.14)$$

$$u_{1n}(\mathbf{x}) + w_2(\mathbf{x}') + z_{12n}(\mathbf{x}) \leqslant d_n(\mathbf{x}), \mathbf{x} \in S_{12}, \mathbf{x}' = P(\mathbf{x}) \in S_{21}, \quad (9.15)$$

$$[u_{1n}(\mathbf{x}) + w_2(\mathbf{x}') + z_{12n}(\mathbf{x}) - d_n(\mathbf{x})]\sigma_{1n}(\mathbf{x}) = 0, \mathbf{x} \in S_{12}, \mathbf{x}' = P(\mathbf{x}) \in S_{21}. \quad (9.16)$$

In addition, the thin coating Ω_2 is connected with the main body Ω_3 along the common boundary $S_{23} = S_{32}$ through one more nonlinear Winkler layer. Thus, on the boundary $S_{23} = S_{32}$, the following bilateral contact conditions are satisfied:

$$-\sigma_{213}^{-}(\mathbf{x}) = \sigma_{3\tau}(\mathbf{x}) = g_{23\tau}\left(v_{21}(\mathbf{x}) - \tfrac{1}{2}h_2\gamma_{21}(\mathbf{x}) - u_{3\tau}(\mathbf{x})\right), \mathbf{x} \in S_{23}, \quad (9.17)$$

$$-\sigma_{233}^{-}(\mathbf{x}) = \sigma_{3n}(\mathbf{x}) = g_{23n}(-w_2(\mathbf{x}) - u_{3n}(\mathbf{x})), \mathbf{x} \in S_{23}, \quad (9.18)$$

where $g_{23\tau}(z)$ and $g_{23n}(z)$ are nonlinear functions that describe the relationship between the jumps $z_{23\tau} = v_{21} - \frac{1}{2}h_2\gamma_{21} - u_{3\tau}$ and $z_{23n} = -w_2 - u_{3n}$ of tangential and normal displacements and the corresponding contact stresses.

Note that the contact problem (9.1)–(9.18) is nonlinear because the real contact zones between the body Ω_1 and the coating Ω_2 are unknown and the boundary conditions (9.14), (9.16), (9.17), and (9.18) are nonlinear.

9.3 Weak Formulation and Domain Decomposition Algorithm

For the massive bodies Ω_α ($\alpha = 1, 3$), we introduce Sobolev spaces $V_\alpha = [H^1(\Omega_\alpha)]^2$ ($\alpha = 1, 3$), whose elements are vectors of generalized displacements $\mathbf{u}_\alpha = (u_{\alpha 1}, u_{\alpha 2})^{\mathrm{T}}$ ($\alpha = 1, 3$). Furthermore, for the coating Ω_2, we introduce a Sobolev space $V_2 = [H^1(\Omega_2^*)]^3$ with elements $\mathbf{u}_2 = (v_{21}, w_2, \gamma_{21})^{\mathrm{T}}$, where v_{21}, w_2, and γ_{21} are, respectively, the generalized displacements and the angle of rotation of the shell. We now consider closed subspaces $V_\alpha^0 = \{\mathbf{u}_\alpha \in V_\alpha : \mathbf{u}_\alpha = 0 \text{ on } \Gamma_\alpha^u\}$ ($\alpha = 1, 2, 3$), of these spaces and consider a reflexive Banach space

$$V_0 = V_1^0 \times V_2^0 \times V_3^0 = \{\mathbf{u} = (\mathbf{u}_1, \mathbf{u}_2, \mathbf{u}_3)^{\mathrm{T}} : \mathbf{u}_\alpha \in V_\alpha^0, \alpha = 1, 2, 3\}$$

as the direct product of the spaces V_α^0, $\alpha = 1, 2, 3$.

In the space V_0, we define a bilinear form $A(\mathbf{u}, \mathbf{u}^*)$ such that $A(\mathbf{u}, \mathbf{u})$ corresponds to the total energy of elastic deformation of the massive bodies and the shell:

$$A(\mathbf{u}, \mathbf{u}^*) = \sum_{\alpha=1}^{3} a_\alpha(\mathbf{u}_\alpha, \mathbf{u}_\alpha^*), \quad \mathbf{u}, \mathbf{u}^* \in V_0,$$

$$a_\alpha(\mathbf{u}_\alpha, \mathbf{u}_\alpha^*) = \int_{\Omega_\alpha} \left(\lambda_\alpha \sum_{i=1}^{2} \varepsilon_{\alpha ii}(\mathbf{u}_\alpha) \sum_{j=1}^{2} \varepsilon_{\alpha jj}(\mathbf{u}_\alpha^*) + 2\mu_\alpha \sum_{i,j=1}^{2} \varepsilon_{\alpha ij}(\mathbf{u}_\alpha)\, \varepsilon_{\alpha ij}(\mathbf{u}_\alpha^*) \right) d\Omega \quad (\alpha = 1, 3),$$

$$a_2(\mathbf{u}_2, \mathbf{u}_2^*) = \left(T_{211}, \frac{dv_{21}^*}{d\xi_1} \right)_{L_2(\Omega_2^*)} + \left(A_{21} k_{21} T_{213}, v_{21}^* \right)_{L_2(\Omega_2^*)} + \left(T_{213}, \frac{dw_2^*}{d\xi_1} \right)_{L_2(\Omega_2^*)} + \left(A_{21} k_{21} T_{211}, w_2^* \right)_{L_2(\Omega_2^*)} + \left(M_{211}, \frac{d\gamma_{21}^*}{d\xi_1} \right)_{L_2(\Omega_2^*)} + (A_{21} T_{213}, \gamma_{21}^*)_{L_2(\Omega_2^*)},$$

$$(u, u^*)_{L_2(\Omega_2^*)} = \int_{\xi_{1b}}^{\xi_{1e}} u u^* \, d\xi_1,$$

and a linear form $L(\mathbf{u})$ corresponding to the work of all external forces applied to the massive bodies and the shell:

$$L(\mathbf{u}) = \sum_{\alpha=1}^{3} \ell_\alpha(\mathbf{u}_\alpha), \quad \mathbf{u} \in V_0,$$

$$\ell_\alpha(\mathbf{u}_\alpha) = \int_{\Gamma_\alpha^\sigma} \mathbf{p}_\alpha \cdot \mathbf{u}_\alpha dS, \quad (\alpha = 1, 3),$$

$$\ell_2(\mathbf{u}_2) = (A_{21} p_{21}, v_{21})_{L_2(\Omega_2^*)} + (A_{21} p_{23}, w_2)_{L_2(\Omega_2^*)} + (A_{21}, m_{21}, \gamma_{21})_{L_2(\Omega_2^*)}.$$

Moreover, in the space V_0, we also define a nonquadratic functional $J(\mathbf{u})$ that corresponds to the deformation energy of all nonlinear Winkler layers between the shell and the massive bodies

$$J(\mathbf{u}) = \int_{S_{12}} \left[\int_0^{d_n - u_{1n} - w_2} g_{12n}^{-}(z)\, dz \right] dS + \int_{S_{23}} \left[\int_0^{v_{21} - \gamma_{21} h_2/2 - u_{3\tau}} g_{23\tau}(z)\, dz + \int_0^{-w_2 - u_{3n}} g_{23n}(z)\, dz \right] dS, \quad \mathbf{u} \in V_0,$$

where $g_{12n}^{-}(z) = \{0, z \geqslant 0\} \vee \{g_{12n}(z), z < 0\}$. The functional $J(\mathbf{u})$ is Gâteaux differentiable:

$$J'(\mathbf{u}, \mathbf{u}^*) = -\int_{S_{12}} g_{12n}^-(d_n - u_{1n} - w_2)[u_{1n}^* + w_2^*]dS$$
$$-\int_{S_{23}} g_{23\tau}\left(v_{21} - \frac{1}{2}h_2\gamma_{21} - u_{3\tau}\right)\left[-v_{21}^* + \frac{1}{2}h_2\gamma_{21}^* + u_{3\tau}^*\right]dS$$
$$-\int_{S_{23}} g_{23n}(-w_2 - u_{3n})[w_2^* + u_{3n}^*]dS, \quad \mathbf{u}, \mathbf{u}^* \in V_0.$$

The Gâteaux differential $J'(\mathbf{u}, \mathbf{u}^*)$ is linear with respect to $\mathbf{u}^*$ but nonlinear with respect to $\mathbf{u}$.

By using the results of Dyyak et al. [4], Martynyak et al. [12], Prokopyshyn and Styahar [21], we have shown that the original contact problem (9.1)–(9.18) is equivalent in weak sense to the problem of minimization of the nonquadratic functional

$$F(\mathbf{u}) = \frac{1}{2}A(\mathbf{u}, \mathbf{u}) + J(\mathbf{u}) - L(\mathbf{u}) \to \min_{\mathbf{u} \in V_0} \tag{9.19}$$

in the space V_0 with subsequent determination of the quantity z_{12n} by the formula

$$z_{12n} = \tilde{g}_{12n}(\sigma_{1n}), \quad \sigma_{1n} = g_{12n}^-(d_n - u_{1n} - w_2), \tag{9.20}$$

and, in addition, the minimization problem (9.19) is equivalent to the nonlinear variational equation:

$$F'(\mathbf{u}, \mathbf{u}^*) = A(\mathbf{u}, \mathbf{u}^*) + J'(\mathbf{u}, \mathbf{u}^*) - L(\mathbf{u}^*) = 0 \quad \forall \mathbf{u}^* \in V_0, \quad \mathbf{u} \in V_0. \tag{9.21}$$

Thus, the solution of the original contact problem is reduced to the solution of the nonlinear variational equation (9.21) in the space V_0.

To solve the nonlinear variational equation (9.21) of the contact problem (9.1)–(9.18), we use the following implicit nonstationary iterative method:

$$G^k(\mathbf{u}^{k+1}, \mathbf{u}^*) = G^k(\mathbf{u}^k, \mathbf{u}^*)$$
$$- \rho^k\left[A(\mathbf{u}^k, \mathbf{u}^*) + J'(\mathbf{u}^k, \mathbf{u}^*) - L(\mathbf{u}^*)\right] \quad \forall \mathbf{u}^* \in V_0, \quad k = 0, 1, \ldots, \tag{9.22}$$

where $G^k : V_0 \times V_0 \to \mathbb{R}$ are bilinear forms in the space V_0, $\rho^k > 0$ are iterative parameters, and $\mathbf{u}^k \in V_0$ are the kth approximations to the exact solution of the variational equation (9.21).

Note that, in the general case, the iterative method (9.22) applied to solve (9.21) does not lead to the decomposition of the problem into subdomains. The required decomposition can be realized only as a result of the proper choice of the bilinear forms G^k in this method.

In what follows, we assume that the functions $g_{12n}(z)$, $g_{23\tau}(z)$, and $g_{23n}(z)$ have generalized derivatives denoted by $g_{12n}'(z)$, $g_{23\tau}'(z)$, and $g_{23n}'(z)$, respectively.

We choose the bilinear forms G^k in method (9.22) as follows:

$$G^k(\mathbf{u}, \mathbf{u}^*) = A\,(\mathbf{u}, \mathbf{u}^*) + X^k(\mathbf{u}, \mathbf{u}^*), \quad \mathbf{u}, \mathbf{u}^* \in V_0, \tag{9.23}$$

where $X^k : V_0 \times V_0 \to \mathbb{R}$ are the following bilinear forms:

$$\begin{aligned} X^k(\mathbf{u}, \mathbf{u}^*) = &\int_{S_{12}} \psi_{12}^k g_{12n}'(d_n - u_{1n}^k - w_2^k)[u_{1n}u_{1n}^* + w_2 w_2^*]dS \\ &+ \int_{S_{23}} \psi_{23}^k\, g_{23\tau}'\left(v_{21}^k - \tfrac{1}{2}h_2\gamma_{21}^k - u_{3\tau}^k\right) \\ &\times \left[\left(-v_{21} + \tfrac{1}{2}h_2\gamma_{21}\right)\left(-v_{21}^* + \tfrac{1}{2}h_2\gamma_{21}^*\right) + u_{3\tau}u_{3\tau}^*\right]dS \\ &+ \int_{S_{23}} \psi_{23}^k\, g_{23n}'(-w_2^k - u_{3n}^k)[w_2 w_2^* + u_{3n}u_{3n}^*]dS, \quad \mathbf{u}, \mathbf{u}^* \in V_0. \end{aligned}$$

Here $\psi_{\alpha\beta}^k(\mathbf{x}) = \{0, \mathbf{x} \in S_{\alpha\beta}\backslash S_{\alpha\beta}^k\} \vee \{1, \mathbf{x} \in S_{\alpha\beta}^k\}$ are the characteristic functions specifying some given subboundaries $S_{\alpha\beta}^k \subseteq S_{\alpha\beta}$, $\{\alpha, \beta\} = \{1, 2\}, \{2, 3\}$. In particular, these functions can be chosen as follows (Prokopyshyn and Styahar, [21]):

$$\psi_{12}^k = \chi_{12}^k = \chi_{12}(\mathbf{u}^k) = -[\mathrm{sgn}(d_n - u_{1n}^k - w_2^k)]^-, \quad \psi_{23}^k \equiv 1. \tag{9.24}$$

We now show that, as a result of the indicated choice of the forms G^k, we obtain the decomposition of the problem into subdomains. Let us introduce the notation $\tilde{\mathbf{u}}^{k+1} = [\mathbf{u}^{k+1} - (1 - \rho^k)\,\mathbf{u}^k]/\rho^k$. Then the iterative method (9.22) with the bilinear forms (9.23) can be written in the following equivalent form:

$$\begin{aligned} &A(\tilde{\mathbf{u}}^{k+1}, \mathbf{u}^*) + X^k(\tilde{\mathbf{u}}^{k+1}, \mathbf{u}^*) \\ &= L(\mathbf{u}^*) + X^k(\mathbf{u}^k, \mathbf{u}^*) - J'(\mathbf{u}^k, \mathbf{u}^*) \quad \forall \mathbf{u}^* \in V_0, \end{aligned} \tag{9.25}$$

$$\mathbf{u}^{k+1} = \rho^k \tilde{\mathbf{u}}^{k+1} + (1 - \rho^k)\,\mathbf{u}^k, \quad k = 0, 1, \ldots. \tag{9.26}$$

Since the quantities common for the subdomains are known from the previous iteration, the variational equation (9.25) splits into three independent variational equations for the subdomains (bodies) Ω_α. Hence, method (9.25), (9.26) is equivalent to the following iterative method:

$$\begin{aligned} a_1(\tilde{\mathbf{u}}_1^{k+1}, \mathbf{u}_1^*) &+ \int_{S_{12}} \psi_{12}^k g_{12n}'(d_n - u_{1n}^k - w_2^k)\, \tilde{u}_{1n}^{k+1} u_{1n}^* dS \\ &= \ell_1(\mathbf{u}_1^*) + \int_{S_{12}} \psi_{12}^k g_{12n}'(d_n - u_{1n}^k - w_2^k) u_{1n}^k u_{1n}^* dS \\ &\quad + \int_{S_{12}} g_{12n}^-(d_n - u_{1n}^k - w_2^k) u_{1n}^* dS \quad \forall \mathbf{u}_1^* \in V_1^0, \end{aligned} \tag{9.27}$$

$$
\begin{aligned}
&a_2(\tilde{\mathbf{u}}_2^{k+1}, \mathbf{u}_2^*) + \int_{S_{21}} \psi_{12}^k g_{12n}'(d_n - u_{1n}^k - w_2^k)\tilde{w}_2^{k+1} w_2^* dS \\
&+ \int_{S_{23}} \psi_{23}^k g_{23\tau}' \left(v_{21}^k - \tfrac{1}{2}h_2\gamma_{21}^k - u_{3\tau}^k\right)\left(-\tilde{v}_{21}^{k+1} + \tfrac{1}{2}h_2\tilde{\gamma}_{21}^{k+1}\right)\left(-v_{21}^* + \tfrac{1}{2}h_2\gamma_{21}^*\right) dS \\
&\qquad + \int_{S_{23}} \psi_{23}^k g_{23n}'(-w_2^k - u_{3n}^k)\tilde{w}_2^{k+1} w_2^* dS \\
&\qquad = \ell_2(\mathbf{u}_2^*) + \int_{S_{21}} \psi_{12}^k g_{12n}'(d_n - u_{1n}^k - w_2^k) w_2^k w_2^* dS \\
&+ \int_{S_{23}} \psi_{23}^k g_{23\tau}' \left(v_{21}^k - \tfrac{1}{2}h_2\gamma_{21}^k - u_{3\tau}^k\right)\left(-v_{21}^k + \tfrac{1}{2}h_2\gamma_{21}^k\right)\left(-v_{21}^* + \tfrac{1}{2}h_2\gamma_{21}^*\right) dS \\
&+ \int_{S_{23}} \psi_{23}^k g_{23n}'(-w_2^k - u_{3n}^k) w_2^k w_2^* dS + \int_{S_{21}} g_{12n}^-(d_n - u_{1n}^k - w_2^k) w_2^* dS \\
&\qquad + \int_{S_{23}} g_{23\tau}\left(v_{21}^k - \tfrac{1}{2}h_2\gamma_{21}^k - u_{3\tau}^k\right)\left(-v_{21}^* + \tfrac{1}{2}h_2\gamma_{21}^*\right) dS \\
&\qquad + \int_{S_{23}} g_{23n}(-w_2^k - u_{3n}^k) w_2^* dS \quad \forall \mathbf{u}_2^* \in V_2^0,
\end{aligned} \tag{9.28}
$$

$$
\begin{aligned}
&a_3(\tilde{\mathbf{u}}_3^{k+1}, \mathbf{u}_3^*) + \int_{S_{32}} \psi_{23}^k g_{23\tau}' \left(v_{21}^k - \tfrac{1}{2}h_2\gamma_{21}^k - u_{3\tau}^k\right)\tilde{u}_{3\tau}^{k+1} u_{3\tau}^* dS \\
&\qquad + \int_{S_{32}} \psi_{23}^k g_{23n}'(-w_2^k - u_{3n}^k)\tilde{u}_{3n}^{k+1} u_{3n}^* dS \\
&\qquad = \ell_3(\mathbf{u}_3^*) + \int_{S_{32}} \psi_{23}^k g_{23\tau}' \left(v_{21}^k - \tfrac{1}{2}h_2\gamma_{21}^k - u_{3\tau}^k\right) u_{3\tau}^k u_{3\tau}^* dS \\
&+ \int_{S_{32}} \psi_{23}^k g_{23n}'(-w_2^k - u_{3n}^k) u_{3n}^k u_{3n}^* dS + \int_{S_{32}} g_{23\tau}\left(v_{21}^k - \tfrac{1}{2}h_2\gamma_{21}^k - u_{3\tau}^k\right) u_{3\tau}^* dS \\
&\qquad + \int_{S_{32}} g_{23n}(-w_2^k - u_{3n}^k) u_{3n}^* dS \quad \forall \mathbf{u}_3^* \in V_3^0,
\end{aligned} \tag{9.29}
$$

$$
\mathbf{u}_\alpha^{k+1} = \rho^k \tilde{\mathbf{u}}_\alpha^{k+1} + (1 - \rho^k)\mathbf{u}_\alpha^k, \quad \alpha = 1, 2, 3, \quad k = 0, 1, \ldots. \tag{9.30}
$$

In each kth iteration of method (9.27)–(9.30), it is necessary to solve three independent linear variational equations (9.27)–(9.29) in the separate bodies Ω_α ($\alpha = 1, 2, 3$). Two of these equations [i.e. (9.27) and (9.29)] correspond to the problems of the theory of elasticity for massive bodies, while the other equation [equation (9.28)] corresponds to the problem of Timoshenko-type shell theory for a thin body with Robin boundary conditions on the common boundaries. Therefore, the iterative method (9.27)–(9.30) belongs to parallel Robin domain decomposition algorithms.

9.4 Numerical Results

We have performed software implementation of the proposed domain decomposition algorithm (9.27)–(9.30) with the use of the FEM to solve the problems in separate bodies.

The developed DDM has been used to study the plane problem of unilateral contact through a nonlinear Winkler layer between a massive elastic body Ω_1 containing a boundary groove and a composite body $\Omega_2 \cup \Omega_3$ consisting of a thin elastic coating Ω_2 and a massive elastic base Ω_3 connected through the other nonlinear Winkler layer (Fig. 9.2).

A compressive normal load with constant intensity q acts upon the top boundary of the body Ω_1 and the bottom boundary of the body Ω_3. We impose symmetry conditions on the right-hand boundary of each of the bodies. The left-hand boundaries of all bodies are free of loads. The massive bodies Ω_1 and Ω_3 have identical heights $h_1 = h_3 = h$ and the height of the thin coating Ω_2 is equal to h_2 ($h_2 \ll h$). The length of all bodies is equal to ℓ. The height of the groove is described by the function $r(x_1) = r_0[1 - (x_1 - \ell)^2/b^2]^{3/2}$, $x_1 \in [\ell - b, \ell]$, where $r_0 = r(\ell)$ is the maximum height of the groove and b is the length of the groove. Young's moduli of the materials of massive bodies Ω_1 and Ω_3 are identical $E_1 = E_3 = E$. At the same time, Young's modulus of the material of thin coating Ω_2 is equal to E_2. Poisson's ratios of the materials of all bodies are also identical: $\nu_1 = \nu_2 = \nu_3 = \nu$. The possible contact zone between the body Ω_1 and the coating Ω_2 of the body Ω_3 is equal to $S_{12} = \{\mathbf{x} : x_1 \in [0, \ell], x_2 = h + h_2\}$ and the interface boundary between the coating Ω_2 and the base Ω_3 is equal to $S_{23} = \{\mathbf{x} : x_1 \in [0, \ell], x_2 = h\}$. The distance between the bodies Ω_1 and Ω_2 before the contact is equal to

$$d_n(\mathbf{x}) = \mathrm{sgn}[(x_1 - \ell + b)^+]r(x_1) = r_0\left\{[1 - (x_1 - \ell)^2/b^2]^+\right\}^{3/2}, \quad \mathbf{x} \in S_{12},$$

where $z^+ = \max\{0, z\}$.

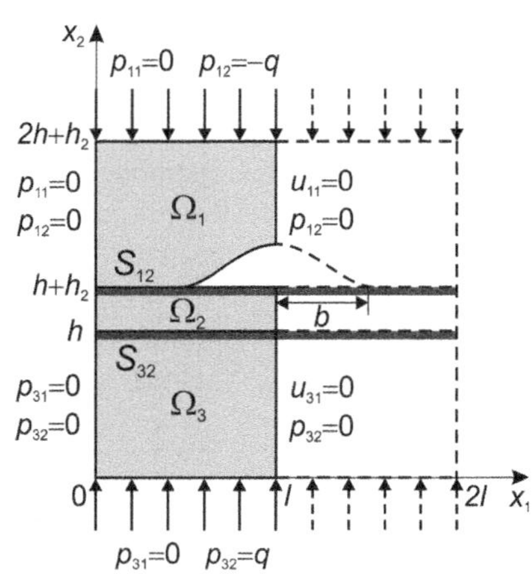

Fig. 9.2 Scheme of contact between the bodies

The functions g_{12n}, g_{23n}, and $g_{23\tau}$ that describe the relationships between the stresses and the jumps of displacements of the nonlinear Winkler layers are taken as follows:

$$g_{12n}(z_{12n}(\mathbf{x})) = \operatorname{sgn}(z_{12n}(\mathbf{x}))\,|z_{12n}(\mathbf{x})|^{1/a_{12n}}\,/B_{12n}^{1/a_{12n}},\quad \mathbf{x} \in S_{12},$$
$$g_{23n}(z_{23n}(\mathbf{x})) = \operatorname{sgn}(z_{23n}(\mathbf{x}))\,|z_{23n}(\mathbf{x})|^{1/a_{23n}}\,/B_{23n}^{1/a_{23n}},\quad \mathbf{x} \in S_{23},$$
$$g_{23\tau}(z_{23\tau}(\mathbf{x})) = z_{23\tau}(\mathbf{x})/B_{23\tau},\quad \mathbf{x} \in S_{23}.$$

The investigation of the problem in question is performed for the following physical and geometric parameters: $q = 10$ MPa, $h = 8b$, $h_2 \in [b/16, b]$, $\ell = 4b$, $b = 1$ cm, $r_0 = r(\ell) = 5 \cdot 10^{-4}$ cm, $E = 2.1 \cdot 10^5$ MPa, $E_2 = E/4$, $\nu = 0.3$, $B_{12n} \in [10^{-5}\ \text{cm/(MPa)}^{a_{12n}}, 5 \cdot 10^{-3}\text{cm/(MPa)}^{a_{12n}}]$, $a_{12n} = 0.5$, $B_{23n} = 5 \cdot 10^{-5}$ cm/(MPa)$^{a_{23n}}$, $a_{23n} = 0.5$, and $B_{23\tau} = 5 \cdot 10^{-5}$ cm/MPa.

To solve this problem, we have used the DDM (9.27)–(9.30). The numerical solutions to the equations (9.27) and (9.29) have been obtained using a two-dimensional FEM with linear triangular elements. At the same time, a one-dimensional FEM with bubble basis functions of the fourth order has been used for the solution to equation (9.28). For each body Ω_1 and Ω_3, we have used a uniform finite element mesh with 8192 linear triangular elements. For this mesh each boundary S_{12} and S_{32} contains 33 finite element nodes. For the thin coating Ω_2, we have used a one-dimensional mesh with 32 finite elements whose nodes coincide with nodes of the two-dimensional meshes for the massive bodies Ω_1 and Ω_3.

In order to check and compare the numerical results, this problem has also been solved by the DDM obtained according to the model in which the stress–strain state of the thin coating Ω_2 is described by the equations of the classical theory of elasticity. The algorithm of this method differs from the domain decomposition algorithm (9.27)–(9.30) by the fact that in each iteration k, instead of the variational equation (9.28) of the shell theory, it is necessary to solve the following variational equation:

$$\begin{aligned}
\tilde{a}_2\big(\tilde{\mathbf{u}}_2^{k+1}, \mathbf{u}_2^*\big) &+ \int_{S_{21}} \psi_{12}^k\, g_{12n}'(d_n - u_{1n}^k - u_{2n}^k)\tilde{u}_{2n}^{k+1} u_{2n}^* dS \\
&+ \sum_{m=\tau,n} \int_{S_{23}} \psi_{23}^k\, g_{23m}'(-u_{2m}^k - u_{3m}^k)\, \tilde{u}_{2m}^{k+1} u_{2m}^* dS \\
&= \ell_2(\mathbf{u}_2^*) + \int_{S_{21}} \psi_{12}^k g_{12n}'(d_n - u_{1n}^k - u_{2n}^k) u_{2n}^k u_{2n}^* dS \\
&+ \sum_{m=\tau,n} \int_{S_{23}} \psi_{23}^k g_{23m}'(-u_{2m}^k - u_{3m}^k) u_{2m}^k u_{2m}^* dS \\
&+ \int_{S_{21}} g_{12n}^{-}(d_n - u_{1n}^k - u_{2n}^k) u_{2n}^* dS \\
&+ \sum_{m=\tau,n} \int_{S_{23}} g_{23m}(-u_{2m}^k - u_{3m}^k) u_{2m}^* dS \quad \forall \mathbf{u}_2^* \in \tilde{V}_2^0,
\end{aligned} \tag{9.31}$$

where

$$\tilde{V}_2^0 = \left\{ \mathbf{u}_2 = (u_{21}, u_{22})^{\mathrm{T}} \in \tilde{V}_2 : \mathbf{u}_2 = 0 \text{ on } \Gamma_2^u \right\}, \quad \tilde{V}_2 = [H^1(\Omega_2)]^2,$$
$$\tilde{a}_2(\mathbf{u}_2, \mathbf{u}_2^*) = \int_{\Omega_2} \left(\lambda_2 \sum_{i=1}^{2} \varepsilon_{2ii}(\mathbf{u}_2) \sum_{j=1}^{2} \varepsilon_{2jj}(\mathbf{u}_2^*) + 2\mu_2 \sum_{i,j=1}^{2} \varepsilon_{2ij}(\mathbf{u}_2)\varepsilon_{2ij}(\mathbf{u}_2^*) \right) d\Omega,$$
$$\ell_2(\mathbf{u}_2) = \int_{\Gamma_2^\sigma} \mathbf{p}_2 \cdot \mathbf{u}_2 dS,$$

which corresponds to the plane boundary-value elasticity problem for the body Ω_2 with Robin conditions on the boundaries S_{21} and S_{23}. In what follows, we denote this method by (9.27), (9.29), (9.30), and (9.31).

To solve all variational equations in method (9.27), (9.29), (9.30), and (9.31), we have applied the FEM with linear triangular elements. For the massive bodies Ω_1 and Ω_3, we have chosen the same finite element mesh as for these bodies in method (9.27)–(9.30). To solve problem (9.31) in the thin body Ω_2 with height $h_2 = b/8$, $b/4$, $b/2$, b ($b = 1$ cm), we have used meshes with 256, 512, 1024, and 2048 finite elements, respectively. For these meshes, 32 finite elements are located on each side of the boundaries S_{21} and S_{23}.

The initial approximations for displacements and the termination criterion for the iterative process are chosen in the same way as in (Prokopyshyn and Styahar, [21]).

The iterative parameters ρ^k in both methods have been set equal in each iteration, i.e., $\rho^k = \rho \in [0.4,\ 0.5]$ ($k = 0, 1, \ldots$), and the characteristic functions ψ_{12}^k and ψ_{23}^k have been chosen in the form (9.24). In this case, the DDM (9.27)–(9.30) attains the relative accuracy $\varepsilon_u = 10^{-3}$ for displacements after 31–110 iterations, while the DDM (9.27), (9.29), (9.30), and (9.31) attains the same accuracy after 48–359 iterations depending on the parameters of the problem.

For the formulated problem, we have studied the influence of the height h_2 of the coating and the parameter B_{12n} of nonlinear Winkler layer between the top body Ω_1 and the thin coating Ω_2 of body Ω_3 on the contact and interface stresses obtained by the DDM (9.27)–(9.30) (with the use of the Timoshenko-type shell theory) and by the algorithm (9.27), (9.29), (9.30), and (9.31) (based on the model of elastic body).

In Fig. 9.3(a–d), we present the plots of normal contact stress σ_{12n} on the interface S_{12} between the top body Ω_1 and the coating Ω_2 for different heights of the coating: $h_2 = b$, $b/2$, $b/4$, and $b/8$ ($b = 1$ cm). Curves 1–6 in this figure are the plots of the stress for the Winkler layer parameters: $B_{12n} = 5 \cdot 10^{-3}$, 10^{-3}, $5 \cdot 10^{-4}$, 10^{-4}, $5 \cdot 10^{-5}$, and 10^{-5}cm/(MPa)$^{a_{12n}}$ ($a_{12n} = 0.5$). The solid curves correspond to the numerical results obtained by DDM (9.27)–(9.30), whereas the dashed curves correspond to the results obtained using algorithm (9.27), (9.29), (9.30), and (9.31).

In this figure, we see that the difference between the distributions of normal contact stress obtained using both analyzed methods is more significant for the case of lower values of parameter B_{12n} and higher values of height h_2. As the height h_2 decreases, the plots of σ_{12n} obtained by both methods approach each other because the accuracy of the method based on the shell theory becomes higher. Moreover, a decrease in the height h_2 leads to a decrease in the peak of normal contact stress σ_{12n}, which

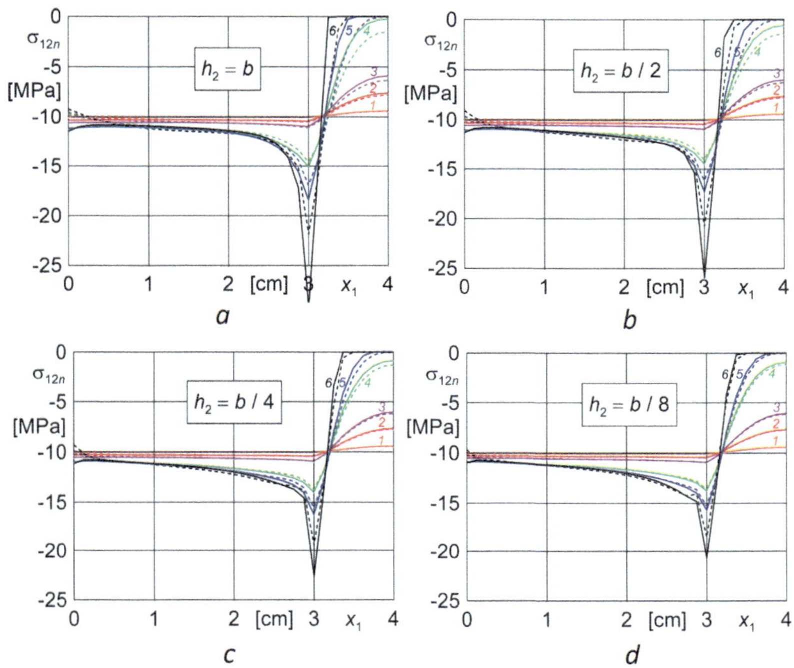

Fig. 9.3 Normal contact stress σ_{12n} for the heights of the coating $h_2 = b$ (*a*), $b/2$ (*b*), $b/4$ (*c*), $b/8$ (*d*) and Winkler layer parameters $B_{12n} = 5 \cdot 10^{-3}$, 10^{-3}, $5 \cdot 10^{-4}$, 10^{-4}, $5 \cdot 10^{-5}$, and 10^{-5} cm/(MPa)$^{0.5}$ (curves 1–6). The solid lines correspond to the numerical solutions obtained by the DDM (9.27)–(9.30), while the dashed lines correspond to the DDM (9.27), (9.29), (9.30), and (9.31)

is attained near the left edge of the groove, and an increase in the size of the actual contact zone, whereas a decrease in the parameter B_{12n} leads to an increase in this peak and a decrease in the size of the actual contact zone. The higher the values of the parameter B_{12n}, the smoother the distributions of normal contact stress.

We have also performed the comparison of the distributions of normal contact stress σ_{12n} on the boundary S_{12} and normal interface stress σ_{23n} on the interface S_{23} between the bottom body Ω_3 and its coating Ω_2 obtained for different values of the Winkler layer parameter and different heights of the coating.

In Fig. 9.4(a–d), we show the plots of contact stress σ_{12n} (solid curves) and interface stress σ_{23n} (dashed curves) for different heights of the coating $h_2 = b/2$, $b/4$, $b/8$, and $b/16$ ($b = 1$ cm) obtained by the DDM (9.27)–(9.30) based on the Timoshenko-type shell theory for the Winkler layer parameters $B_{12n} = 5 \cdot 10^{-3}$, $10^{-3}, 5 \cdot 10^{-4}$, 10^{-4}, $5 \cdot 10^{-5}$, and 10^{-5}cm/(MPa)$^{a_{12n}}$ ($a_{12n} = 0.5$) (curves 1–6).

It follows from this figure that as the height h_2 of the coating decreases, the distributions of normal interface stress σ_{23n} become closer to the distributions of normal contact stress σ_{12n}. This can be explained by the fact that sufficiently thin coatings ($h_2 \leqslant b/16$) can no longer exert significant influence on the stress–strain state of the system of contacting bodies.

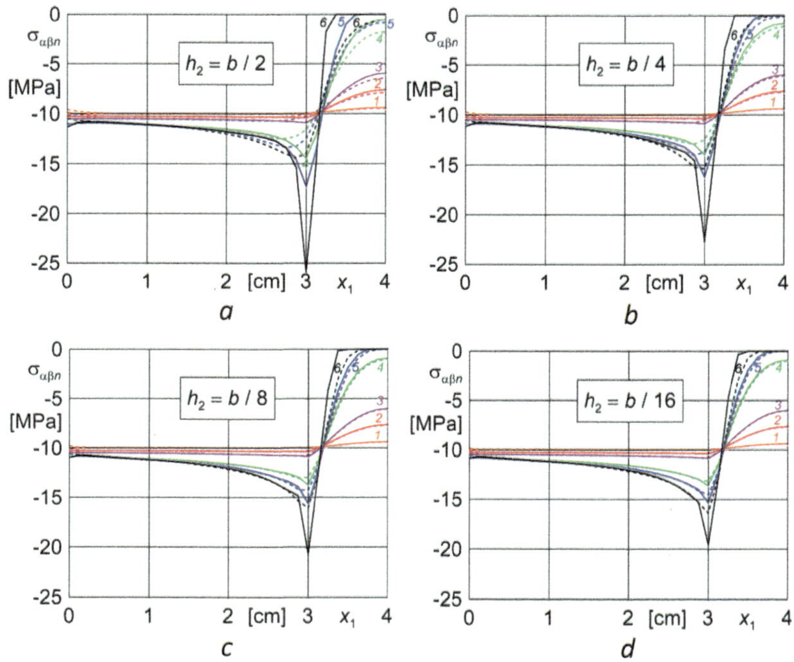

Fig. 9.4 Contact stress σ_{12n} (solid curves) and interface stress σ_{23n} (dashed curves) for the heights of the coating $h_2 = b/2$ (*a*), $b/4$ (*b*), $b/8$ (*c*), $b/16$ (*d*) and Winkler layer parameters $B_{12n} = 5 \cdot 10^{-3}$, 10^{-3}, $5 \cdot 10^{-4}$, 10^{-4}, $5 \cdot 10^{-5}$, and 10^{-5}cm/(MPa)$^{0.5}$ (curves 1–6)

9.5 Conclusions

We considered the problem of unilateral contact through a nonlinear Winkler layer of a massive elastic body and a thin elastic coating in the form of Timoshenko-type shell connected with the massive elastic base via another nonlinear Winkler layer. We obtained a weak formulation of this problem in the form of a nonlinear variational equation in the Hilbert space. To solve the nonlinear variational equation of the contact problem, we developed a parallel iterative Robin-type domain decomposition algorithm, which reduces the procedure of its solution in each iteration to the simultaneous solution of two independent linear variational equations corresponding to the linear elasticity problems for massive elastic bodies and a linear variational equation corresponding to a problem of the Timoshenko-type shell theory for the thin coating with the Robin boundary conditions on the contact surfaces. We also performed software implementation of the algorithm using the FEM with triangular elements to solve the elasticity problems in massive bodies and the one-dimensional FEM with bubble basis functions of high order to solve the problem of the Timoshenko-type shell theory.

The proposed method was used for the numerical investigation of the problem of contact interaction between two elastic bodies with surface groove in the presence of a thin elastic coating and the nonlinear Winkler layers. We performed the comparison of the numerical solutions obtained by the domain decomposition algorithm based on the use of the Timoshenko-type shell theory to model the stress–strain state of the thin coating with the solutions obtained by the DDM in which the classical elasticity theory was used to model the coating. The influence of the height of the coating and the parameter of the Winkler layer on the contact and the interface stresses was studied. It was established that as the height of the coating decreases, the results obtained using both models become closer, and the distributions of interface stress on the surface between the coating and its base converge to the distributions of contact stress on the boundary between the top elastic body and the coating.

References

1. Alblas, J.B., Kuipers, M.: Contact problems of a rectangular block on an elastic layer of finite thickness - Part I: The thin layer. Acta. Mech. **8**(3–4), 133–145 (1969)
2. Alblas, J.B., Kuipers, M.: On the two dimensional problem of a cylindrical stamp pressed into a thin elastic layer. Acta. Mech. **9**(3–4), 292–311 (1970)
3. Blokh, M.V., Orobinskii, A.V.: Modification of the finite-element method to solve two-dimensional elastic and plastic contact problems. Streng. Mater. **15**(3), 613–620 (1983)
4. Dyyak, I., Savula, Ya., Styahar, A.: Numerical investigation of a plain strain state for a body with thin cover using domain decomposition. J. Numer. Appl. Math. **109**(3), 23–33 (2012)
5. Gong, Z.-Q., Komvopoulos, K.: Effect of surface patterning on contact deformation of elastic-plastic layered media. J. Tribol. **125**(1), 16–24 (2003)
6. Gong, Z.-Q., Komvopoulos, K.: Mechanical and thermomechanical elastic-plastic contact analysis of layered media with patterned surfaces. J. Tribol. **126**(1), 9–17 (2004)
7. Grigorenko, A.Ya., Dyyak, I.I., Matysyak, S.I., Prokopyshyn, I.I.: Domain decomposition methods applied to solve frictionless-contact problems for multilayer elastic bodies. Int. Appl. Mech. **46**(4), 388–399 (2010)
8. Grigorenko, Ya.M., Savula, Ya.G., Kossak, O.S.: Stress-strain analysis of elastic bodies on the basis of a heterogeneous mathematical model. Int. Appl. Mech. **36**(12), 1595–1604 (2000)
9. Liashenko, B.A., Volkov, Y.V., Solovykh, Ye.K., Lopata, L.A.: Elevation of the wear resistance of parts of the ship cars and mechanisms by coatings of discrete structure. Technological support of coatings with discrete structures by electrocontact sintering. Probl. Friction Wear **67**, 110–126 (2015)
10. Lopata, L.A., Liashenko, B.A., Kalinichenko, V.I., Volkov, Y.V., Lopata, T.V.: Obtaining of wear resistant discrete coatings by electrocontact sintering. Probl. Friction Wear **51**, 139–148 (2009)
11. Martinyak, R.M., Shvets', R.M.: A mathematical model of mechanical contact of bodies across a thin inhomogeneous layer. J. Math. Sci. **90**(2), 2000–2002 (1998)
12. Martynyak, R.M., Prokopyshyn, I.A., Prokopyshyn, I.I.: Contact of elastic bodies with nonlinear Winkler surface layers. J. Math. Sci. **205**(4), 535–553 (2015)
13. Mkhitaryan, V.P., Shekyan, A.L., Shekyan, L.A.: Indentation of a round punch into a rough elastic half-space. Mech. Solids. **44**(5), 729–736 (2009)
14. Pasternak, Ia.M.: Plane problem of elasticity theory for anisotropic bodies with thin elastic inclusions. J. Math. Sci. **186**(1), 31–47 (2012)
15. Pasternak, Ia., Sulym, H.: Stress state of solids containing thin elastic crooked inclusions. J. Eng. Math. **78**(1), 167–180 (2013)
16. Pelekh, B.L.: Generalized Theory of Shells (in Russ.). Vyshcha Shkola, Lviv (1978)

17. Pelekh, B.L., Maksymuk, A.V., Korovaichuk, I.M.: Contact Problems for Layered Elements of Structures and Bodies with Coatings (in Russ.). Naukova Dumka, Kyiv (1988)
18. Podgornyi, A.N., Gontarovskii, P.P., Kirkach, B.N., Matyukhin, Yu.I., Khavin, G.L.: Problems of Contact Interaction of Structural Elements (in Russ.). Naukova Dumka, Kyiv (1989)
19. Prokopyshyn, I.I., Dyyak, I.I., Martynyak, R.M., Prokopyshyn, I.A.: Penalty Robin-Robin domain decomposition schemes for contact problems of nonlinear elasticity. Lect. Notes Comput. Sci. Eng. **91**, 647–654 (2013)
20. Prokopyshyn, I.I., Dyyak, I.I., Prokopyshyn, I.A.: Algorithms of domain decomposition for the axisymmetric problem of contact of elastic bodies. Appl. Probl. Mech. Math. **17**, 68–81 (2019)
21. Prokopyshyn, I.I., Styahar, A.O.: Investigation of contact between elastic bodies one of which has a thin coating connected with the body through a nonlinear Winkler layer by the domain decomposition methods. J. Math. Sci. **258**(4), 477–506 (2021)
22. Prokopyshyn, I.I., Styahar, A.O.: Numerical analysis of contact of the elastic bodies one of which has a discontinuous thin coating. Mater. Sci. **57**(5), 734–744 (2022)
23. Ramachandra, S., Ovaert, T.C.: Effect of coating geometry on contact stresses in two-dimensional discontinuous coatings. J. Tribol. **122**(4), 665–671 (2000)
24. Riederer, K., Duenser, C., Beer, G.: Simulation of the linear inclusions with the BEM. Eng. Anal. Bound. Elem. **33**(7), 959–965 (2009)
25. Savula, Ya.G., Kossak, O.S.: Numerical simulation of free oscillations of elastic bodies with a thin coating. J. Math. Sci. **109**(1), 1295–1302 (2002)
26. Solovykh, Ye.K., Liashenko, B.A., Kalinichenko, V.I.: Wear resistant discontinuous coatings of frame type. Probl. Friction Wear **54**, 31–46 (2010)
27. Sulym, H.T.: Fundamentals of the Mathematical Theory of Thermoelastic Equilibrium of Deformable Solids with Thin Inclusions (in Ukr.). NTSh Research and Publishing Center, Lviv (2007)
28. Vynnytska, L., Savula, Ya.: Mathematical modeling and numerical analysis of elastic body with thin inclusion. Comput. Mech. **50**(5), 533–542 (2012)

Chapter 10
Weight Optimization of Non-homogeneous Rotation Shells by Methods of Optimal Processes Theory

Anatoliy Dzyuba, Petr Dzyuba, Larisa Levitina, and Volodymyr Sirenko

Abstract This study presents mathematically rigorous and mechanically accurate models that apply the theory of optimal processes to reduce the amount of material used in elastic inhomogeneous shells of rotation with variable stiffness. The research also includes the results of weight optimization for shell structures.

10.1 Introduction

Shell and plate elements are widely and quite efficiently used in various branches of engineering. These are the bodies of rockets, ships, docks, reservoirs, cisterns, bunkers, airplane fuselages, wings, etc. In this case, the shell structure can often be reinforced with ribs in one or both directions, have hatches, various openings, variable wall thickness in one or two directions, or other non-homogeneities. It is also necessary to take into account the variety of external influences acting on the structure, the peculiarities of the structure, and the properties of the material from which they are made. The high requirements for the ratio of strength and lightness (material consumption) of such rather complex structural elements are also obvious in modern conditions.

These strength and weight requirements are very important for the rocket and space industry (Degtyarev [4]; Mossakovsky et al. [19]; Degtjarev et al. [5]; Hudramovich and Dzyuba [17]), which determine the development of new and more advanced mathematical models, sufficiently reliable and low-cost methods for their

A. Dzyuba (✉) · P. Dzyuba · L. Levitina · V. Sirenko
Oles Honchar Dnipro National University, Dnipro, Ukraine
e-mail: dzb@ua.fm

P. Dzyuba
e-mail: avatarrr@ua.fm

L. Levitina
e-mail: ldlora@i.ua

V. Sirenko
e-mail: vsirenko@i.ua

H. Altenbach et al. (eds.), *Selected Problems of Solid Mechanics and Solving Methods*, Advanced Structured Materials 204, https://doi.org/10.1007/978-3-031-54063-9_10

direct calculation, effective techniques and special algorithms for optimal design that would provide as many opportunities as possible to determine an often quite significant number of variable parameters and take into account various constraints on the state parameters and features of the structure's behavior with minimal computational costs for finding the optimal design.

Optimization parameters for individual non-homogeneous structures can be either discrete (finite-dimensional) or piecewise continuous variable functions. The use of generalized discrete optimization methods in conjunction with the finite element method is often due to the large dimensionality of the arising optimization problem, the difficulty of rebuilding the finite element mesh at each step of the search algorithm due to the current change in optimization parameters, in particular, wall thickness or other geometry of thin-walled structural elements, and the need to satisfy a significant number of constraints, which creates a number of fundamental difficulties for solving the problem of weight optimization of real composite structures of new equipment.

The modern formulation of optimal control problems has emerged from the request to take into account (while meeting certain specified criteria) various kinds of constraints imposed on the control parameters and state functions of the optimization object described by a system of differential equations.

The approach presented in this research paper is based on the joint use of methods of the theory of optimal processes (Pontryagin et al. [21]; Bryson and Ho [2]), algorithms of finite-dimensional optimization (Rekleitis et al. [22]; Himmelblau [16]), decomposition and synthesis of substructures and their application to reduce the material consumption of elastic non-homogeneous shells of rotation of variable stiffness (Grigorenko [11]; Grigorenko et al. [12]; Grigorenko and Vasilenko [15]).

10.2 Necessary Conditions for Optimality

The problem of weight optimization of non-homogeneous shells of rotation is to find the optimal control $\bar{\delta}(x) \in D_\delta$ from the condition for a minimum of the functional

$$\int_{x_0}^{x_L} V\left(\bar{u}, \bar{\delta}, x\right) dx \tag{10.1}$$

for the processes described by the system of differential equations

$$\frac{d\bar{u}_i}{dx} = \varphi_0\left(\bar{u}, \bar{\delta}, x\right) \quad (i = 1, \ldots n) \tag{10.2}$$

with boundary conditions

$$\bar{\Psi}\left(\bar{u}\left(x_0\right)\right) = 0, \quad \bar{u}\left(x_0\right) \in U_0, \quad \bar{\Psi}\left(\bar{u}\left(x_L\right)\right) = 0, \quad \bar{u}\left(x_L\right) \in U_L. \tag{10.3}$$

In the problems of optimizing the parameters of mechanical engineering structures, Eq. (10.2) describe the state of shell and load-bearing rod elements and are linear (or reduced to them) in phase variables $\bar{u}\,(E)$ (which can be displacements, internal forces, deformations, stresses, etc.) and nonlinear in terms of control $\bar{\delta}(x)$; boundary conditions (10.3) are the conditions of fixation or interaction of substructures; $\bar{\psi}\,[\bar{u}\,(x_e)]$ are functions of boundary conditions at the left x_0 $(e = 0)$ and right x_L $(e = L)$ ends of the trajectory; p_e is the number of boundary conditions at x_0 or x_L, $e = 0 \vee e = L$.

The components of the optimal control vector $\bar{\delta}(x) = (\delta_1(x), \delta_2(x), \ldots, \delta_m(x))$ can be the wall thickness or shape of the median surface of the shells and/or the geometric dimensions of their cross-section, which are variable in length of the rod load-bearing elements, for non-homogeneous composite materials–the physical characteristics of the material, etc. Limitations $\bar{\delta}(x) \in D_\delta$ and $\bar{u} \in D_u$ are the conditions of strength, stiffness, stability, design, and technological requirements, the main of which are as follows

$$\bar{f_1}\left(\bar{u}, \bar{\delta}, x\right) \leqslant 0, \tag{10.4a}$$

$$\int_{x_0}^{x_L} \bar{f_2}\left(\bar{u}, \bar{\delta}, x\right) dx \leqslant 0, \tag{10.4b}$$

$$\bar{f_3}\,(\bar{u}, x) \leqslant 0. \tag{10.4c}$$

The necessary conditions for optimality are formulated in the form of the maximum principle of LS Pontryagin (Pontryagin et al. [21]; Bryson and Ho [2]). The extended Hamiltonian operator and the system for conjugate functions with boundary conditions of transversality are as follows

$$H^* = H + \bar{\xi} \cdot \bar{g}, \tag{10.5a}$$

$$\frac{d\bar{\lambda}_i}{dx} = -\frac{\partial H}{\partial u_i} - \sum_{j=1}^{z} \xi_j(x) \cdot \frac{\partial g_j}{\partial u_i} \quad (i = 1, \ldots n), \tag{10.5b}$$

$$\bar{\lambda}\,(x_e) = \sum_{j=1}^{P_e} c_j \operatorname{grad} \Psi_j(\bar{u}(x_e)), \quad e = 0 \vee e = L, \tag{10.5c}$$

where $H = \sum_{i=0}^{n} \lambda_i \varphi_i$; $\bar{\lambda}(x)$ are conjugate functions satisfying Eq. (10.5b) with boundary conditions of transversality (10.5c); the components of the column vector $\bar{g}$ are the components of the z generalized constraints (10.4) (for the case (10.4c), this is the first of the higher derivatives of the constraint, where control is first explicitly introduced (Bryson and Ho [2]); $\bar{c}$ and $\bar{\xi}\,(x)$ are Lagrange multipliers and functions. The optimal control $\bar{\delta}^*(x)$ is found from the condition for the maximum of the Hamiltonian operator:

$$H^*\left(\bar{u}^*(x), \bar{\lambda}^*(x), \bar{\delta}^*(x), x\right) = \sup_{\bar{\delta}\in D_m} H^*\left(\bar{u}^*(x), \bar{\lambda}^*(x), \bar{\delta}(x), x\right), \tag{10.6}$$

which in principle establishes the dependence of $\bar{\delta}^*(x) = \bar{\delta}^*\left(\bar{u}^*(x), \bar{\lambda}^*(x), x\right)$ and, for certain initial values of $\bar{\delta}(x)$, allows us to determine $\bar{u}^*(x)$, $\bar{\lambda}^*(x)$ and then the optimal control $\bar{\delta}^*(x)$ for all $x_0 \leqslant x \leqslant x_L$.

It is proposed to satisfy the necessary conditions for the optimality of the maximum principle in the case of constraints (10.4a) containing control and phase variables according to the scheme of the method of successive approximations:

$$\bar{\delta}^{k-1} \to \bar{u}^k \to \bar{\lambda}^k \to \sup_{\bar{\delta}} H \to \bar{\delta}^k.$$

The calculation of $\bar{u}^k$ and $\bar{\lambda}^k$ is carried out by integrating the corresponding boundary value problems (10.2), (10.3) and (10.5b), (10.5c) by the sweep method with orthogonalization according to S.K. Godunov (Grigorenko and Mukoed [13]; Biderman [1]).

In the presence of phase constraints, the boundary value problem for conjugate variables has its peculiarities since the trajectory, in this case, consists of segments, some of which are on the boundaries of the constraints, and others are inside them, and the conjugate system (10.5b) may differ in such intervals by the presence or absence of components with the corresponding Lagrange multipliers (functions) $\xi_j(x)$ (Golshtein and Tretyakov [10]). Depending on the exit (descent) of the phase trajectory on the corresponding constraints (10.4) or their violation at the kth step of approximations at an arbitrary nodal point $x = x_i$ under the requirements of (10.6) and

$$g_t^k\left(\bar{\delta}^k(E_V)\right) < 0, \quad \xi_t^k = 0 \vee g_t^k\left(\bar{\delta}^k(E_V)\right) = 0, \quad \xi_t^k < 0, \quad t = 1, \ldots, z \tag{10.7}$$

the calculation of generalized Lagrange multipliers to determine the free terms of the system (10.5b) is carried out in the form

$$\xi_t^{k,s+1}(x_i) = \begin{cases} 0, & g_t\left(\bar{\delta}^{k,s}, x_i\right) < 0 \\ \xi_t^{k,s}(x_i), & \left(\bar{\delta}^{k,s}, x_i\right) = 0 \\ \omega\xi_t^{k,s}(x_i), & \left(\bar{\delta}^{k,s}, x_i\right) > 0, \quad \xi_t^{k,s} \neq 0 \\ \omega\varepsilon, & \left(\bar{\delta}^{k,s}, x_i\right) > 0, \quad \xi_t^{k,s} = 0 \end{cases} \tag{10.8}$$

where

$$\omega = \min_{\delta_j^k}\left\{1 - \left|\frac{\partial H\left(\bar{u}^k, \bar{\lambda}^k, \bar{\delta}^{k,s}, x_i\right)/\partial\delta_j^{k,s}}{\sum_e \xi_e^{k,s}\partial g_e/\partial\delta_j^{k,s}}\right|\right\}$$

$(j = 1, \ldots, m, i = 0, \ldots, N, e \in \{1, \ldots, z\})$.

Thus, if the tth constraint ($t = 1, \ldots, z$) is violated, the corresponding Lagrange multiplier must be reduced (see (10.7)). If the trajectory is on the boundary of the constraint, it should be left unchanged. When the constraint is not violated, the Lagrange multiplier can obviously be omitted, i.e., increased to zero. Here j and t are the numbers of the corresponding components of the vectors $\bar{\delta}$, $\bar{g}$, e correspond to the indices t ($e \in \{1, \ldots, z\}$) of the violated constraints (10.4), $|\varepsilon| \ll 1$.

Suppose that before the kth step of the method of successive approximations for solving the optimal control problem, the components of the Lagrange multipliers $\bar{\xi}^{k,s}$ and the control $\bar{\delta}^{k,s}$ are known, where s is the number of the internal iteration of the search $\bar{\xi}\,(E)$ and $\bar{\delta}\,(E)$. Then, with known $\bar{u}^k$, $\bar{\lambda}^k$, $\bar{\xi}^{k,s}$, the Lagrange $\bar{\xi}_t^{k,s+1}(x_i)$ and control $\bar{\delta}^{k,s+1}$ multipliers are determined according to (10.8) and the maximum condition

$$\sup_{\delta^{k,s+1}\in D'_m} H^* = \sup_{\delta^{k,s+1}\in D'_m} \Big\{ - V\big(\bar{u}^k, \bar{\delta}^{k,s+1}\big) + \bar{\lambda}^{T^k}\big(\bar{\delta}^{k,s+1}, x_i\big)\bar{\varphi}\big(\bar{u}^k, \bar{\delta}^{k,s+1}, x_i\big)\Big\} + \bar{\xi}^{T^{k,s}}\bar{g}\big(\bar{u}^k, \bar{\delta}^{k,s+1}, x_i\big) \quad (10.9)$$

by solving the corresponding auxiliary problems of nonlinear programming for fixed (nodal) points x_i ($i = 0, \ldots, N$) of the integration interval.

For this purpose, the methods of nonlinear programming (Rekleitis et al. [22]; Himmelblau [16]) for finding the extremum of (10.9) can be used. In many cases, the use of one-dimensional optimization methods is sufficient.

In the general case of reaching the phase constraints (10.4c), the resulting multipoint boundary value problem for conjugate variables is reduced to a sequence of two-point problems and solved using an algorithm of residuals of additional conditions at the internal points of the trajectory (Dzyuba and Torskyy [9]).

Thus, the application of the approach makes it possible to reduce the multidimensional value problem of nonlinear programming of determining discrete variable parameters (the number of which often coincides with the number of nodal points of integration of the original system (10.2), (10.3) or finite elements and, as a rule, is too large for thin-walled structures) to two main stages of a joint iterative process: solving boundary value problems for the main (10.2), (10.3) and conjugate (10.5b), (10.5c) systems in order to find $\bar{u}^*(x)$, $\bar{\lambda}^*(x)$ and a sequence of auxiliary nonlinear programming problems, usually of low dimensionality, maximizing the Hamiltonian operator (10.6) with respect to the variables $\bar{\delta}(x)$ at fixed (nodal) points x_i of the integration interval.

10.3 Mathematical Models of Rotation Shells with Wall Thickness Variation

The equations of the moment theory of shells of rotation with an arbitrary meridian shape in general case and a wall thickness variable in 2 directions $h = h\,(s, \varphi)$ under asymmetric loading are given in Grigorenko and Vasilenko [15], assuming that they are homogeneous, isotropic, thin and elastic, for which the Kirchhoff hypotheses and the conditions of small deformations and angles of rotation compared to unit are fulfilled.

After appropriate transformations, these equations can be reduced to a system of eight partial-derivative differential equations:

$$
\begin{aligned}
\frac{\partial u}{\partial s} &= -\mu\frac{\cos\theta}{r}u - \left(\mu\frac{\sin\theta}{r} + \frac{1}{R_1}\right)w + \frac{1}{Kr}\left(N_1 r\right) - \mu\frac{1}{r}\frac{\partial v}{\partial\varphi} \qquad (10.10)\\
\frac{\partial v}{\partial s} &= \frac{\cos\theta}{r}v + \frac{2}{K(1-\mu)}\frac{1}{r}\left(S^{*}r\right) - \left(\frac{1}{r} + \frac{4D}{KR_2}\frac{\sin\theta}{r^2}\right)\frac{\partial u}{\partial\varphi}\\
&\quad + \frac{4D}{KR_2}\frac{\cos\theta}{r^2}\frac{\partial w}{\partial\varphi} + \frac{4D}{KR_2}\frac{\partial\vartheta_1}{r\partial\varphi}\\
\frac{\partial w}{\partial s} &= \frac{u}{R_1} - \vartheta_1\\
\frac{\partial\vartheta_1}{\partial s} &= -\mu\frac{\cos\theta}{r}\vartheta_1 + \frac{1}{Dr}\left(M_1 r\right) - \mu\frac{\sin\theta}{r^2}\frac{\partial v}{\partial\varphi} + \mu\frac{1}{r^2}\frac{\partial^2 w}{\partial\varphi^2},\\
\frac{\partial\left(N_1 r\right)}{\partial s} &= K\left(1-\mu^2\right)\frac{\cos^2\theta}{r}u - K\left(1-\mu^2\right)\frac{\cos\theta\sin\theta}{r}w\\
&\quad + \mu\frac{\cos\theta}{r}\left(N_1 r\right) - \frac{1}{R_1}\left(Q_1^{*}r\right) + \frac{2(1-\mu)}{R_2}\frac{\sin\theta}{r^2}\frac{\partial D}{\partial\varphi}\frac{\partial u}{\partial\varphi}\\
&\quad + K\left(1-\mu^2\right)\frac{\cos\theta}{r}\frac{\partial v}{\partial\varphi} - \frac{2(1-\mu)}{R_2}\frac{\cos\theta}{r^2}\frac{\partial D}{\partial\varphi}\frac{\partial w}{\partial\varphi}\\
&\quad - \frac{2(1-\mu)}{R_2}\frac{1}{r}\frac{\partial D}{\partial\varphi}\frac{\partial\vartheta_1}{\partial\varphi} + \frac{2D(1-\mu)}{R_2}\frac{\sin\theta}{r^2}\frac{\partial^2 u}{\partial\varphi^2}\\
&\quad - \frac{2D(1-\mu)}{R_2}\frac{\cos\theta}{r^2}\frac{\partial^2 w}{\partial\varphi^2} - \frac{2D(1-\mu)}{R_2}\frac{1}{r}\frac{\partial^2\vartheta_1}{\partial\varphi^2}\\
&\quad - \frac{1}{r}\frac{\partial}{\partial\varphi}\left(S^{*}r\right) - rq_1\\
\frac{\partial\left(S^{*}r\right)}{\partial s} &= -\frac{\partial K}{\partial\varphi}\left(1-\mu^2\right)\frac{\cos\theta}{r}u - \frac{\partial K}{\partial\varphi}\left(1-\mu^2\right)\frac{\sin\theta}{r}w\\
&\quad - \frac{\partial D}{\partial\varphi}\left(1-\mu^2\right)\frac{\sin\theta\cos\theta}{r^2}\vartheta_1 - \frac{\cos\theta}{r}\left(S^{*}r\right)\\
&\quad - K\left(1-\mu^2\right)\frac{\cos\theta}{r}\frac{\partial u}{\partial\varphi} - \left(1-\mu^2\right)\left(\frac{1}{r}\frac{\partial K}{\partial\varphi} + \frac{\sin^2\theta}{r^3}\frac{\partial D}{\partial\varphi}\right)\frac{\partial v}{\partial\varphi}\\
&\quad - K\left(1-\mu^2\right)\frac{\sin\theta}{r}\frac{\partial w}{\partial\varphi} - D\left(1-\mu^2\right)\frac{\sin\theta\cos\theta}{r^2}\frac{\partial\vartheta_1}{\partial\varphi}\\
&\quad - \mu\frac{1}{r}\frac{\partial\left(N_1 r\right)}{\partial\varphi} - \mu\frac{\sin\theta}{r^2}\frac{\partial\left(M_1 r\right)}{\partial\varphi}
\end{aligned}
$$

$$-\left(1-\mu^2\right)\left(K\frac{1}{r}-D\frac{\sin^2\theta}{r^3}\right)\frac{\partial^2 v}{\partial\varphi^2}+\frac{\sin\theta}{r^3}\left(1-\mu^2\right)\frac{\partial D}{\partial\varphi}\frac{\partial^2 w}{\partial\varphi^2}$$
$$+\left(1-\mu^2\right)D\frac{\sin\theta}{r^3}\frac{\partial^3 w}{\partial\varphi^3}-rq_2.$$

$$\frac{\partial(Q_1^* r)}{\partial s}=\left(\frac{1}{R_1}+\mu\frac{\sin\theta}{r}\right)(N_1 r)+K\left(1-\mu^2\right)\frac{\cos\theta\sin\theta}{r}u$$
$$+K\left(1-\mu^2\right)\frac{\sin^2\theta}{r}w-\frac{\partial^2 D}{\partial\phi^2}\left(1-\mu^2\right)\frac{\cos\theta}{r^2}\theta_1$$
$$+\frac{\partial D}{\partial\phi}(1-\mu)\frac{2\cos\theta\sin\theta}{r^3}\frac{\partial u}{\partial\phi}-\left(1-\mu^2\right)\frac{\sin\theta}{r}\left(\frac{\partial^2 D}{\partial\phi^2}\frac{1}{r^2}-K\right)\frac{\partial v}{\partial\phi}$$
$$-\frac{\partial D}{\partial\phi}(1-\mu)\frac{2\cos^2\theta}{r^3}\frac{\partial w}{\partial\phi}-\frac{\partial D}{\partial\phi}\frac{2\cos\theta}{r^2}\left[(1-\mu)+\left(1-\mu^2\right)\right]\frac{\partial\theta_1}{\partial\phi}$$
$$+D(1-\mu)\frac{2\cos\theta\sin\theta}{r^3}\frac{\partial^2 u}{\partial\phi^2}-\frac{\partial D}{\partial\phi}\left(1-\mu^2\right)\frac{2\sin\theta}{r^3}\frac{\partial^2 v}{\partial\phi^2}$$
$$-\left(D(1-\mu)\frac{2\cos^2\theta}{r^3}-\frac{\partial^2 D}{\partial\phi^2}\left(1-\mu^2\right)\frac{1}{r^3}\right)\frac{\partial^2 w}{\partial\phi^2}$$
$$-D(1-\mu)\frac{2\cos\theta}{r^2}\frac{\partial^2\theta_1}{\partial\phi^2}-\frac{\mu}{r^2}\frac{\partial^2(M_1 r)}{\partial\phi^2}-D\left(1-\mu^2\right)\frac{\sin\theta}{r^3}\frac{\partial^3 v}{\partial\phi^3}$$
$$+\frac{\partial D}{\partial\phi}\left(1-\mu^2\right)\frac{2}{r^3}\frac{\partial^3 w}{\partial\phi^3}+D\left(1-\mu^2\right)\frac{1}{r^3}\frac{\partial^4 w}{\partial\phi^4}-rq_3$$
$$\frac{\partial(M_1 r)}{\partial s}=D\left(1-\mu^2\right)\frac{\cos^2\theta}{r}\vartheta_1+\left(Q_1^* r\right)+\mu\frac{\cos\theta}{r}(M_1 r)$$
$$+2(1-\mu)\frac{\partial D}{\partial\varphi}\frac{\sin\theta}{r^2}\frac{\partial u}{\partial\varphi}+D\left(1-\mu^2\right)\frac{\cos\theta\sin\theta}{r^2}\frac{\partial v}{\partial\varphi}$$
$$-2(1-\mu)\frac{\partial D}{\partial\varphi}\frac{\cos\theta}{r^2}\frac{\partial w}{\partial\varphi}-2(1-\mu)\frac{\partial D}{\partial\varphi}\frac{1}{r}\frac{\partial\vartheta_1}{\partial\varphi}+2D(1-\mu)\frac{\sin\theta}{r^2}\frac{\partial^2 u}{\partial\varphi^2}$$
$$-\left(2D(1-\mu)\frac{\cos\theta}{r^2}+D\left(1-\mu^2\right)\frac{\cos\theta}{r^2}\right)\frac{\partial^2 w}{\partial\varphi^2}-2D(1-\mu)\frac{1}{r}\frac{\partial^2\vartheta_1}{\partial\varphi^2}.$$

Here, for the force factors, the generally accepted notations N_1, N_2, S, M_1, M_2, M, Q_1, and Q_2 are introduced (Grigorenko and Mukoed [13]; Grigorenko et al. [12]); q_1, q_2, and q_3 are the meridional, circumferential, and normal components of the external load intensity, respectively.

As the main variables with respect to which the system of equation is written, we choose four variables u, v, w, and ϑ_1 (Fig. 10.1), which characterize the displacements and the corresponding four force factors N_1, S^*, Q_1^*, and M_1, where

$$S^*=S+\frac{2M}{R_2},\quad Q_1^*=Q_1+\frac{1}{r}\frac{\partial M}{\partial\varphi}$$

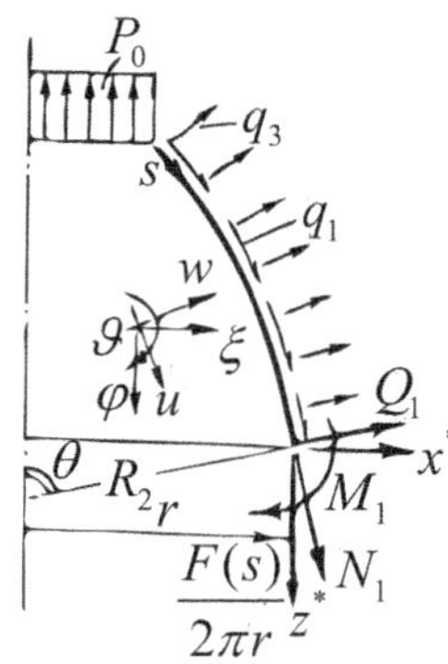

Fig. 10.1 Forces and displacements in the shell

are the reduced forces; R_1 and R_2 are the radiuses of curvature of the surface; $D = Eh^3/(12(1-\mu^2))$ is the cylindrical stiffness; $K = Eh/(1-\mu^2)$ is the tensile stiffness; E and μ are the elastic modulus and Poisson's ratio, respectively;

$$F(s) = P_0 + \int_{S_0}^{S_L} (q_3 \cos\theta - q_1 \sin\theta)\, 2\pi r ds;$$

$\theta(s)$ is the angle between the normal to the median surface and the axis of rotation of the shell; $r(s)$ is the radius of a parallel circle.

In the case when the thickness of the shell is variable in two directions, it is proposed to use the idea of the method of straight lines (Grigorenko and Mukoed [13]) to solve system (10.10).

Assuming that the parameters of the shell state are sufficiently smooth in the circumferential direction, the partial derivatives of the variable $\varphi = \varphi_m$ ($m = 1, 2, \ldots, M$) of Eq. (10.10) are replaced by finite differences of the fourth order of accuracy:

$$\begin{aligned}
y' &= \left(y_{m-2} - 8y_{m-1} + 8y_{m+1} - y_{m+2}\right) / (12\Delta),\\
y'' &= \left(-y_{m-2} + 16y_{m-1} - 30y_m + 16y_{m+1} - y_{m+2}\right) / \left(12\Delta^2\right),\\
y''' &= \left(y_{m-3} - 8y_{m-2} + 13y_{m-1} - 13y_{m+1} + 8y_{m+2} - y_{m+3}\right) / \left(8\Delta^3\right),\\
y^{\text{iv}} &= \left(-y_{m-3} + 12y_{m-2} - 39y_{m-1} + 56y_m - 39y_{m+1} + 12y_{m+2} - y_{m+3}\right) / \left(6\Delta^4\right),
\end{aligned}$$

where $\Delta = \Delta\varphi_m = 2\pi/M$ is the step of the difference grid in the circumferential direction.

It should be noted that, depending on the method of replacing the derivatives with finite differences, different systems of the method of straight lines are possible, which may differ in the accuracy of the approximating relations for the corresponding derivatives.

This approach makes it possible to obtain a set of one-dimensional boundary value problems for each mth meridian relative to the components of the stress-strain vector

$$\bar{Y}^m = \left(u^m, v^m, w^m, (\vartheta_1)^m, (N_1 r)^m, (S_1^* r)^m, (Q_1^* r)^m, (M_1 r)^m\right)^{\mathrm{T}}.$$

In general, this makes it possible to obtain a boundary value problem for a system of $8 \times M$ ordinary differential equations for calculating a shell with a stiffness variable in two directions:

$$\frac{d\bar{Y}^m}{ds} = \bar{A}^m\left(h_m\left(s, \varphi_m\right), s\right)\bar{Y}^m + \bar{B}^m\left(h_m\left(s, \varphi_m\right), s\right) \quad (m = 1, \ldots, M) \quad (10.11)$$

with boundary conditions for fixing the ends of the shell in the form of (10.3).

For the case of a cylindrical shell ($\sin\theta = 1$, $\cos\theta = 0$, $R_1 = \infty$, and $R_2 = R$), the system of such equations is given in Dzyuba et al. [7].

The anchoring conditions on the meridional edges of the shell ($s = 0$, $s = L$, where L is the length of the shell) are taken into account when solving boundary value problems along the meridian s_m by the sweep method with orthogonalization according to S.K. Godunov (Grigorenko and Mukoed [14]; Biderman [1]).

For the case of a change in the shell wall thickness only in the meridional direction $h = h(s)$, the use of the Fourier method (Grigorenko et al. [12]; Emel'yanov [3]; Biderman [1]) allows us to reduce Eq. (10.10) to boundary value problems for systems of ordinary differential equations with respect to the coefficients of the expansions of the corresponding functions in trigonometric sequences

$$f = \sum_{t=0}^{\infty} f_t^c \cos t\varphi + \sum_{t=1}^{\infty} f_t^s \sin t\varphi, \quad \psi = \sum_{t=1}^{\infty} \psi_t^c \sin t\varphi - \sum_{t=0}^{\infty} \psi_t^s \cos t\varphi,$$

respectively for symmetric and oblique components of displacements, internal forces, and loads. Here f_t^A, f_t^s, ψ_t^c, and ψ_t^s are the coefficients of such expansions into trigonometric sequences.

Since the forces and displacements in the equations of state of the shell are referred to a local coordinate system associated with its normal and tangent to the meridian, the system coefficients may have discontinuities when the shell meridian consists of several sections with angular points between them. This leads to the need to construct equations of compatibility of deformation of different sections.

These difficulties, according to (Biderman [1]), can be circumvented by switching to global coordinates. To do this, internal forces and displacements are projected not on the tangent and normal to the meridian but on the normal to the symmetry axis of the shell and on the axis itself. In this case, instead of displacements u and w, displacements ξ and ζ are introduced, and instead of forces N_1 and Q_1^*, forces X^* and Z^* (Fig. 10.1) as follows:

$$\xi = u\cos\theta + w\sin\theta, \qquad \zeta = u\sin\theta - w\cos\theta,$$
$$X^* = N_1\cos\theta + Q_1^*\sin\theta, \quad Z^* = N_1\sin\theta - Q_1^*\cos\theta.$$

The coefficients of the expansions u_t, w_t, N_{1t}, and Q^*_{1t} in the Fourier series of the corresponding functions and their derivatives through ξ_t, ζ_t, X_t, and Z_t, as well as the radial and axial components of the load, will be related to each other by the same dependencies:

$$q_{xt} = q_{1t}\cos\theta + q_{3t}\sin\theta, \quad q_{zt} = q_{1t}\sin\theta - q_{3t}\cos\theta.$$

After such transformations, the coefficients of the resulting system of equations do not contain the curvature of the meridian $1/R_1$, so the main unknowns referred to the fixed coordinate system remain continuous even for shells with surface fractures, which allows us not to write conjugation equations for such cases. As for the force unknowns, $(X^*r)_t$, $(Z^*r)_t$, $(S^*r)_t$, and $(M_1r)_t$ undergo discontinuities of a known value in advance only where the shells are subjected to concentrated forces on a specific load parallel.

The system $(8 \times t^*)$ of linear differential equations obtained in this way is given in the form:

$$\frac{d\overline{Y}_t}{dS} = \bar{A}_t\left(h\left(s\right), s\right)\bar{Y}_t + \bar{B}_t\left(h\left(s\right), s\right) \quad (t = 1, 2, \ldots), \tag{10.12}$$

where t^* denotes the number of Fourier series components used to approximate the functions being sought; the components of the state vector $\bar{Y}_t = \{y_{it}\}$ are the corresponding coefficients f_t^A, f_t^s, ψ_t^c, and ψ_t^s of such expansions for the functions $\xi, \zeta, v, \vartheta_1, Xr, Zr, S^*r, M_1r$, and the elements of the matrix $\bar{A}_t$ and the load column vector $\bar{B}_t$ are the corresponding variable coefficients obtained after substituting the expansions of the main variables into trigonometric sequences into system (10.10) (Biderman [1]; Dzyuba et al. [7]).

10.4 Optimization of Variable Stiffness of Shells of Rotation

The problem of weight optimization of arbitrarily loaded shells of rotation with variable stiffness in two directions is to find the wall thickness $h\left(s, \varphi\right)$ that would provide the minimum volume of material

$$V = \int_0^{2\pi}\int_{s_0}^{s_N} r\left(s\right) h\left(s, \varphi\right) ds d\varphi \tag{10.13}$$

while the equations of state (10.10) are fulfilled with boundary conditions for fixing the ends and the presence of strength, stability, stiffness, and design requirements, respectively:

$$\begin{gathered}\max_{-h/2\leqslant z\leqslant h/2}\sigma_{\text{eq}}\left(s, \varphi, z\right) \leqslant \left[\sigma\right], \quad \left|\sigma_-\right| \leqslant \sigma_{\text{kr}}, \\ \bar{u}\left(s, \varphi\right) \leqslant \bar{u}_{\max}, \quad h_1 \leqslant h\left(s, \varphi\right) \leqslant h_2.\end{gathered} \tag{10.14}$$

Here, to calculate the equivalent stress, the dependence

$$\sigma_{\text{eq}} = \sqrt{\frac{1}{2}\left(\sigma_1^2 + \sigma_2^2 - \sigma_1\sigma_2 + 3\tau_{12}^2 + 3\tau_{1z}^2 + 3\tau_{2z}^2\right)},$$

σ_{kr} is determined in accordance with Mossakovsky et al. [19], Lizin and Pyatkin [18]; $\bar{u}(s, \varphi)$ is the displacement vector; $\sigma_1, \sigma_2, \tau_{12}, \tau_{1z}$, and τ_{2z} is the corresponding normal and tangential stresses in the shell sections.

Following the differential-difference approach, the functional (10.13) is given by replacing the integration over φ with summation as follows

$$\begin{aligned} V &= \int_0^{2\pi} \int_{s_0}^{s_n} r(s)\, h(s, \varphi)\, ds d\varphi \\ &\approx \sum_{m=1}^{M} \Delta \int_{s_0}^{s_n} r(s)\, h(s, \varphi_m)\, ds \\ &= \frac{2\pi}{M}{}_1 \int_{s_0}^{s_n} r(s) \sum_{m=1}^{M} h(s, \varphi_m)\, ds \end{aligned}$$

and, thus, the problem of optimal design is reduced to finding the minimum of the functional

$$V^* = \int_{s_0}^{s_n} r(s) \sum_{m=1}^{M} h(s, \varphi_m)\, ds \tag{10.15}$$

for the object described by the system of $8 \times M$ Eq. (10.11), in the presence of constraints (10.14). Representing criterion (10.13) in the form of (10.15) and system (10.10) in the form of (10.11) allows us to apply the principle of Pontryagin's maximum in its usual one-dimensional form to the original two-dimensional problem. In this case, the Hamiltonian operator of the problem takes the form

$$\begin{aligned} H(s) = \sum_{m=1}^{M} \Big[& - r(s)\, h^m(s, \varphi_m) \\ & + \sum_{i=1}^{8} \lambda_i^m(s, \varphi_m) \sum_{j=1}^{8} \left(A_{ij}^{(m)}\left(h^m(s)\right) Y_j^{(m)} + B_i^{(m)}\left(h^m(s)\right) \right) \Big]. \end{aligned} \tag{10.16}$$

The constraints (10.14) using Lagrange functions $\bar{\xi}(s, \varphi_m)$ are added to (10.16) similarly to (10.5a), the boundary value problem for the conjugate functions λ_i^m ($i = 1, \ldots, 8$, $m = 1, \ldots, M$) is given in the form (10.5b), and the optimal control problem is solved in accordance with the algorithm outlined in Dzyuba and Torskyy [9].

The determination of Y_i^m ($i = 1, \ldots, 8$, $m = 1, \ldots, M$) by integrating system (10.11) and solving the analogous equations for λ_i^m obtained in accordance with

(10.5b) can be carried out in the form of simultaneous integration of the boundary value problems of the corresponding systems by $8 \times M$ equations.

At the same time, when the number of M bands of the differential-difference representation is increased in order to improve the calculation accuracy, the difference in the values of the coefficients of the rows of the matrix $\bar{A}$ on the adjacent $m - 1, m, m + 1$ lines will decrease, which can lead to its weak conditionality, an increase in inaccuracy and, as a result, to the loss of stability of the corresponding numerical solution of systems of linear algebraic equations in the sweep method. In addition, the simultaneous solution of $8 \times M$ equations as a single system leads to the need to find the maximum of the Hamiltonian operator (10.16) as a problem of $\acute{I}$-dimensional optimization of finding variables $h^m(s, \varphi_m)$ for each of the nodal points $s_0 \leqslant s_e \leqslant s_L$, $e = 0, \ldots, N$.

In the present work, it is proposed to solve these problematic aspects by sequentially solving M systems of 8 equations. In this case, with a known control of the $k - 1$ approximation step $h^{m,k-1}(s_e, \varphi_m)$, the integration of the defining systems for the shell and the calculation of conjugate functions at N nodal points $(e = 0, \ldots, N)$ is carried out sequentially for each of the $m = 1, \ldots, M$ conditional lines by the sweep method, where the results of previous iterations are used to determine $Y_i^t(s_e)$ for neighboring lines $(t \neq m)$. For the initial approximation $(k = 0)$, $h^{m,0}(s_e, \varphi_m) = \text{const}$ is assumed and the corresponding components of the vectors $\bar{Y}_i{}^{m,0}$ $(i = 1, \ldots, 8)$ are then determined. If it is impossible (or difficult) to solve this problem, $\bar{Y}_i{}^{m,0} = 0$ is assumed, and then a sequence of M boundary value problems is solved for eight variables $u^m, v^m, w^m, (\vartheta_1)^m, (N_1 r)^m, (S_1^* r)^m, (Q_1^* r)^m$, and $(M_1 r)^m$ and the corresponding components of the vector λ_i^m $(i = 1, \ldots, 8, m = 1, \ldots, M)$ and the main parameters of the stress-strain state and constraints (10.14) are determined.

The optimal control $h^*(s_e, \varphi_m)$ of the current step of approximations is found by solving $N \times M$ one-dimensional maximum search problems (10.16), taking into account the constraints (10.14) at the nodal points (s_e, φ_m) $(e = 1, \ldots, N, m = 1, \ldots,$
$M)$. In this case, it is considered that convergence is achieved if one of the conditions is met

$$\max_{s_e, \varphi_m} \frac{\left|h_e^{m,k+1} - h_e^{m,k}\right|}{\left|h_e^{m,k+1}\right|} \leqslant \varepsilon_1, \qquad 0 < \varepsilon_1 \ll 1,$$

$$\sum_{e=1}^{N} \sum_{m=1}^{M} \frac{\left|h_e^{m,k+1} - h_e^{m,k}\right|^2}{\left|h_e^{m,k+1}\right|^2} \leqslant \varepsilon_2, \quad 0 < \varepsilon_2 \ll 1.$$

To accelerate the convergence of such a process, the algorithms presented in Dzyuba et al. [8] proved to be effective.

Figure 10.2 shows the results of finding the optimal variable in two directions (longitudinal in Fig. 10.2b; in the transverse direction in Fig. 10.2c) for strengthening a hinged steel (X18H9H) cylindrical shell of radius $R = 0.057$ m, length

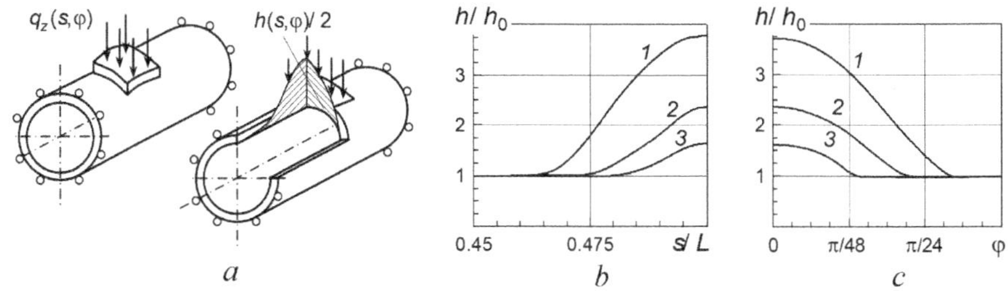

Fig. 10.2 Designing the optimal shell wall thickening

$L = 2R$, with wall thickness $h_0 = 2.8 \cdot 10^{-3}$ m at the site of a local radial load of $1.5N$. Lines 1, 2, and 3 correspond to the values $[\sigma]$ in constraints (10.14) equal to 100, 150, 200 MPa, respectively. Compared to the case of a rectangular lining of constant thickness $h^* = h_0$ (Fig. 10.2a), the stresses σ_{eq} were almost two times lower for the same volume of material.

For the case of the shell wall thickness varying only in the radial direction, the optimality criterion (10.1) is taken as follows:

$$V = \min 2\pi \int_{s_0}^{s_L} r(s)\, h(s)\, ds.$$

The defining equations and constraints are given in the form of (10.12) and (10.14), respectively. The algorithm for solving this problem is similar to the previous one.

It should be noted that shells of optimal variable stiffness usually have a rather complex configuration of their material distribution. Therefore, despite their obvious advantage in terms of material consumption when ensuring the conditions of strength, stability, and stiffness, their practical use is often limited to comparison with a reference, which makes it possible to assess the degree of perfection of the actual structure. There is an obvious need to meet the requirements of cost and manufacturability additionally.

To this end, the approach formulates auxiliary finite-dimensional optimization problems for the best piecewise constant approximation of continuous controls that realize the satisfaction of technological and design requirements in the form:

$$\begin{gathered} V = \min_{h_i, l_i \in D_k} \left(V_{\text{var}} - V_{\text{c}}\right)^2, \qquad V_{\text{c}} = 2\pi \sum_{i=1}^{k} l_i h_i r_i, \\ h_i \geqslant \max_{s \in [s_i, s_{i+1}]} h_{\text{var}}(s) \geqslant h_0, \quad \sum_{i=1}^{k} l_i = s_L - s_0, \quad l_i \geqslant 0. \end{gathered} \tag{10.17}$$

Here V_{var} and V_{c} are the volumes of the shell material with the optimal continuously variable $h_{\text{var}}(s)$ and piecewise constant thickness h_i of the shell wall with the corresponding length l_i.

This approach makes it possible to obtain a rational, more technological distribution of the thickness (continuous, stepped, two- and one-sided) of the shell wall. It is used as a fragment of the considered optimal control problem to finalize the design of variable stiffness according to technological requirements.

In situations where the curvature of the meridian changes, or when connecting multiple shells, or due to local external loading, there may be a sudden change in the thickness (thickening) of the shell wall. As a result, it becomes necessary to reinforce such shell structures with a frame. This was noted by Myachenkov and Grigoryev in [20], as well as by Hudramovich and Dzyuba in [17].

The problem of rational material distribution of such thickening is formulated as a problem of weight optimization of the geometric shape and dimensions of the power rod element (frame), which is under the influence of external (usually local) loads and forces of interaction with the attached shells, which consists in finding the variable geometric dimensions δ_k cross section of the frame, providing a minimum volume of its material, while fulfilling the strength limitations imposed at dangerous and (or) specified points of its cross section, on the displacements (linear, angular) of some points of the cross section and a number of structural (technological) requirements for the geometric parameters of the cross section of the frame.

In contrast to traditional approaches to solving the problem of weight optimization of a rod element by determining its certain geometric dimensions of a cross-section of a given configuration, the present work uses an algorithm that allows, by varying the coordinates δ_k of the corner points of the section, to create a "whole mosaic" of configurations and identify the best among them (in terms of material consumption (10.18a)), taking into account the constraints (10.18b, 10.18c) on the stresses $\sigma_{\max}$ and displacements $\bar{u}$ of the contour points (Dzyuba [6])

$$V_{\text{opt}} = \min_{\delta_k} F\left(\delta_k\right) R_c, \tag{10.18a}$$

$$\sigma_{\max_j}\left(\delta_k\right) \leqslant [\sigma] \quad (j = 1, \ldots, n) \tag{10.18b}$$

$$\bar{u}(\delta_k) \leqslant [\bar{u}], \tag{10.18c}$$

$$f_t(\delta_k) \leqslant 0, \quad (k = 1, \ldots, p) \tag{10.18d}$$

Here, F is the cross-sectional area; R_c is the radius of the centerline of the frame.

The introduction of a system of geometric relationships between the coordinates of points in the form of a moving, rigid, connected "slide" in the form of (10.18d) and design conditions $\delta_k^{=} \leqslant \delta_k \leqslant \delta_k^2$ $(k = 1, \ldots, M)$, allows to significantly reduce the number of discrete variable parameters δ_k of the problem and determine both the geometric dimensions and the optimal configuration of the frame as a whole. This problem is successfully solved using finite-dimensional optimization algorithms (Himmelblau [16]; Rekleitis et al. [22]).

Figure 10.3 shows the configurations of the rod element at separate iterations of the process of transforming the original rectangular spar to a rational configuration of the power spar under complex loading by moments M_x and M_y.

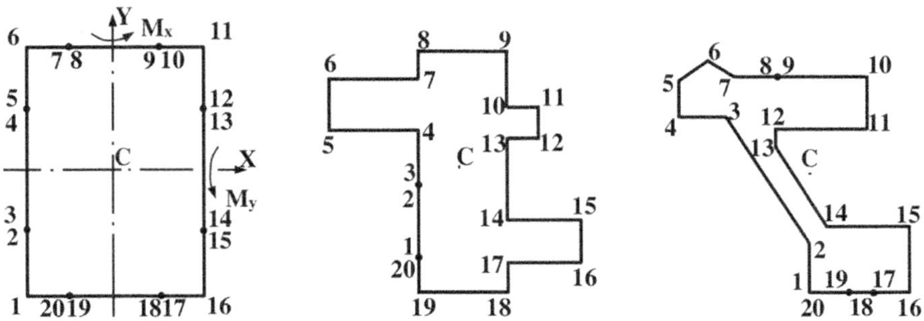

Fig. 10.3 Changing the shape of the frame cross-section in the process of optimization

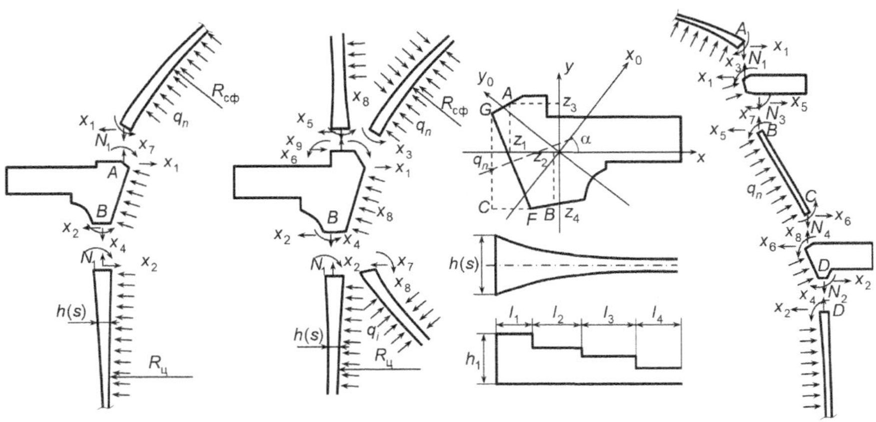

Fig. 10.4 Design scheme and optimization results

The previous sections' approaches to weight optimization of irregular parameters of structural elements can be applied to reduce material consumption of complex structures.One such problem is the problem of optimal design of the shells of rotation connected by a power frame bundle under asymmetric loading. Figure 10.4 shows the calculation schemes and qualitative (due to the significant amount of input data description) results of weight optimization as a rocket and space technology design fragment.

The structure can have several frames of different configurations (Fig. 10.3c), and each of the shell substructures can have an arbitrary meridian shape, physical and mechanical characteristics, anchoring conditions, etc.

The task is to find the variable parameters (shell wall thickness, configuration and geometric dimensions of the cross-section of the frames) that ensure minimum material consumption when the equations of state with boundary conditions of fixation and constraints imposed on stresses, deformations (displacements), structural parameters, etc. are fulfilled. Such a problem is solved by decomposing it into separate substructures in the form of reinforcing rings and shells of variable rotation along the thickness meridian $h(s)$, further optimizing their parameters and synthesizing the structure as a whole using iterative refinement of the interaction forces of individual elements.

Since these forces depend on the geometric dimensions sought (stiffness ratios of shells and frames), the forces of their interaction, taking into account the conditions of joint deformation of the structure, may change as a result of solving the corresponding optimization problems for individual substructures. The iterative refinement of interaction forces (synthesis) of individual substructures is carried out by the method of successive approximations, taking into account the results of solving the direct problem of calculating the stress-strain state of a composite structure. In the case of a bundle of shells of rotation connected by a frame under asymmetric loading, to determine the interaction forces of substructures, the equation of state of the frame is presented as a system of linear algebraic equations with respect to the coefficients of the expansion of displacements along the line of the frame and shells joint into Fourier series (Myachenkov and Grigoryev [20]):

$$\bar{G}_k \cdot \bar{\Delta}_k = \bar{\theta}_k \left(\bar{f}_{0k} + \sum_{t=1}^{m_P} \bar{\psi}_t^P \cdot \bar{Q}_{tk}^{*P} \frac{1}{R} - \sum_{t=1}^{m_L} \bar{\psi}_t^L \cdot \bar{Q}_{tk}^{*L} \frac{1}{R} \right).$$

These equations are supplemented by the conditions of continuity of displacements of the lines of the shells and frames joint, which establish the connection between the generalized edge displacements $\bar{W}^{*P(L)}{}_{tk}$ of the ends of the contour of the tth shell element with the edge forces $\bar{Q}_{0(L)k}{}^t$ at these ends and the displacements of the points of their connections with the frame and further displacements Δ_k of the frame centerline for the kth harmonic of the Fourier series

$$\bar{W}_{tk}^{*P(L)}(s_e) = \bar{c}_{tk}^{P(L)} \cdot \bar{\Delta}_k,$$

where $\bar{\psi}_t{}^{P(L)}$ and $\bar{c}_{tk}{}^{P(L)}$ are the transformation matrices.

This approach, in combination with the application of the above discrete-continuum model of direct calculation of shells, optimization of parameters of individual substructures, decomposition and synthesis of the composite structure, and the use of the proposed algorithms for accelerating convergence, has significant advantages over the direct use of finite-dimensional optimization algorithms.

10.5 Conclusions

- The mathematically and mechanically correct models and algorithms for applying the methods of the theory of optimal processes to reduce the material consumption of elastic non-homogeneous shells of variable stiffness have been constructed.
- The application of the necessary optimality conditions of L.S. Pontryagin for weight optimization of shells of rotation with variable stiffness parameters allows

simplifying the process of finding optimization parameters by fragmenting the problem into simpler components, the possibility of using analytical and analytical-numerical research methods at certain stages, and thus significantly reducing the computational costs of finding an optimal design as a whole.

- The proposed algorithm of step-by-step optimization allows to find the optimal parameters of sufficiently complex components of shell structures by dividing them into separate substructures and further synthesizing the structure with optimal parameters by iterative means, which allows to fully use the capabilities of both methods of the theory of optimal processes and finite-dimensional optimization, simplify the consideration of real design requirements and significantly reduce the cost of solving the problem of reducing the material consumption of structures when creating samples of new equipment.

References

1. Biderman, V.L.: Mechanics of thin-walled structures. Mashinostroenie, Moscow (1977). [In Russian]
2. Bryson, A.E., Ho, Y.-S.: Applied theory of optimal control. Mir, Moscow (1972). [In Russian]
3. Emel'yanov, I.G.: Application of discrete Fourier series to the stress analysis of shell structures. Comput. Contin. Mech. **8**(3), 245–253 (2015)
4. Degtyarev, A.V.: Problems and prospects. ART-PRESS, Dnipro, Missile technology (2014). [In Russian]
5. Degtyarev, M.A., Dzyuba, A.P., Avramov, K.V., Sirenko, V.N.: Method of reducing the material consumption of the tail compartments of launch vehicles. J. Mech. Eng. **23**(3), 24–36 (2020). [In Russian]
6. Dzyuba, A.P.: Algorithm for the synthesis of the optimal cross section configuration of a rod element with a complex bend, Problems of. computational mechanics and strength of structures. Iss. **1**, 37–45 (1997)
7. Dzyuba, A.A., Dzyuba, A.P., Levitina, L.D., Safronova, I.A.: Mathematical simulation of deformation for the rotation shells with variable wall thickness. J. Optim. Differ. Equ. Their Appl. **29**(1), 79–95 (2021)
8. Dzyuba, A.P., Safronova, I.A., Levitina, L.D.: Algorithm for computational costs reducing in problems of calculation of asymmetrically loaded shells of rotation. Strength Mater. Theory Struct. 105, 107–117 (2020)
9. Dzyuba, A., Torskyy, A.: Algorithm of the successive approximation method for optimal control problems with phase restrictions for mechanics tasks. J. Math. Model. Comput. **9**(3), 734–749 (2022)
10. Golshtein, E.G., Tretyakov, N.V.: Modified Lagrange Functions. Theory and Methods of Optimization. Nauka, Moscow (1989). [In Russian]
11. Grigorenko Y.M.: Isotropic and anisotropic layered shells of rotation of variable stiffness. Naukova dumka, Kyiv (1973). [In Russian]
12. Grigorenko, Y.M., Vlaikov, G.G., Grigorenko, A.Y.: Numerical-analytical solution of shell mechanics problems based on various models. Academicperiodica, Kiev (2006). [In Russian]
13. Grigorenko Y.M., Mukoed, A.P.: Solving nonlinear problems in the theory of shells on a computer. Vishcha Shkola, Kiev (1983). [In Russian]
14. Grigorenko Y.M., Mukoed, A.P.: Solving the problems of shell theory on a computer. Vishcha shkola, Kiev (1979). [In Russian]

15. Grigorenko, Y.M., Vasilenko, A.T.: Methods for calculating shells. In: Theory of Shells of Variable Rigidity, vol.4. Naukova Dumka, Kyiv (1981). [In Russian]
16. Himmelblau, D.: Applied Nonlinear Programming. Mir, Moscow (1975). [In Russian]
17. Hudramovich, V.S., Dzyuba, A.P.: Contact interaction and optimization of locally loaded shell structures. J. Math. Sci. **162**, 231–245 (2009). Springer Science + Business Media
18. Lizin, V.T., Pyatkin, V.A.: Design of thin-walled structures. Mashinostroenie, Moscow (1985). [In Russian]
19. Mossakovsky, V.I., Makarenkov, A.G., Nikitin, P.I., et al.: Strength of rocket structures. Vysshaya Shkola, Moscow (1990). [In Russian]
20. Myachenkov, V.I., Grigoryev, I.V.: Calculation of compound shell structures on a computer: Reference. Mashinostroenie, Moscow (1981). [In Russian]
21. Pontryagin, L.S., Bolteanskii, V.G., Gamkrelidze, R.V., Mishchenko, E.F.: The Mathematical Theory of Optimal Processes. Interscience, New York. NY, USA (1962)
22. Rekleitis, G., Reyvindran, A., Ragsdell, K.: Optimization in technology: in 2 books. Mir, Moscow (1986). [In Russian]

Chapter 11
On M-Integral in Linear Micropolar Elasticity with Symmetric Stress Tensor

Victor Eremeyev

Abstract In this note, we discuss a M-integral and related conservation law within a particular case of linear micropolar elasticity. Here we restrict ourselves to the case of decoupled constitutive equations with symmetric stress tensor. In this case, it is possible to formulate an additional conservation law as in the classic Cauchy elasticity.

11.1 Introduction

Conservation laws form a basis of mechanics and physics of solids. In addition to classic balance laws such as the balance of mass, energy, momentum, and moment of momentum, it is worth to mention other nontrivial conservation laws discussed by [5, 6, 9, 13] in more detail. These additional conservation laws have found various applications in mechanics of fracture. In particular, using them various path-independent integrals could be derived such as J-, L-, and M-integrals.

In addition to classic elasticity conservation laws, path-independent integrals were also considered for micropolar media. This model of continuum originated in works by Cosserat brothers and possesses both rotational and rotational degrees of freedom. As a counterpart of enhanced kinematics, there exist stresses and couple stresses. The Cosserat (micropolar) model has found recently various applications in modelling of media with complex inner structures such as foams, granular media, masonries, and composites [2, 4]. Using Noether's theorem, conservation laws were introduced by [11, 16] within nonlinear micropolar elasticity, whereas the case of small deformations was analyzed by [7]. A similar analysis was provided for micropolar shells by [1]; see also further references therein.

Within Noether's approach, nontrivial conservation laws can be related to certain variational symmetries, i.e. the invariance of Lagrangian under some transformations. Unlike classic elasticity, i.e. elasticity of simple materials, the case of micropolar solids is more complex, in general. For example, it was noted by [7] that M-integral

V. Eremeyev (✉)
University of Cagliari, Cagliari, Italy
e-mail: victor.eremeev@unica.it

H. Altenbach et al. (eds.), *Selected Problems of Solid Mechanics and Solving Methods*, Advanced Structured Materials 204, https://doi.org/10.1007/978-3-031-54063-9_11

cannot be derived, in general. This fact reflects the absence of invariance under the scaling transformation. The aim of this paper is to discuss a particular class of micropolar solids for which this conservation law exists.

The remainder of the paper is organized as follows. In Sect. 11.2, we briefly discuss constitutive equations of micropolar solids under small deformations. Here we also present equilibrium equations. In Sect. 11.3, we introduce the Eshelby tensor and corresponding conservation law. Finally, considering transformations of the Eshelby tensor, in Sect. 11.4, we introduce another conservation law related to the M-integral. Note that its existence as a path-independent integral is based on the symmetry of the stress tensor.

11.2 Governing Equations

In the following, we restrict ourselves to small deformations. Following [2, 4, 12], we introduce the vector of displacements $\mathbf{u}$ and the vector of infinitesimal rotations $\boldsymbol{\varphi}$ as kinematical descriptors

$$\mathbf{u} = \mathbf{u}(\mathbf{x}), \quad \boldsymbol{\varphi} = \boldsymbol{\varphi}(\mathbf{x}), \tag{11.1}$$

where $\mathbf{x}$ is the position vector. The corresponding strain measures are given by the formulae

$$\mathbf{e} = \nabla\mathbf{u} - \mathbf{I} \times \boldsymbol{\varphi}, \quad \mathbf{k} = \nabla\boldsymbol{\varphi}, \tag{11.2}$$

where ∇ is the 3D nabla-operator defined as in [3, 8], $\mathbf{I}$ is the unite tensor, and $\times$ stands for the cross product.

The constitutive equations take the form

$$W = W(\mathbf{e}, \mathbf{k}), \quad \mathbf{T} = \frac{\partial W}{\partial \mathbf{e}}, \quad \mathbf{M} = \frac{\partial W}{\partial \mathbf{k}}, \tag{11.3}$$

where W is the strain energy density and $\mathbf{T}$ and $\mathbf{M}$ are the stress and couple stress, respectively.

Equilibrium equations take the form

$$\nabla \cdot \mathbf{T} = \mathbf{0}, \quad \nabla \cdot \mathbf{M} + \mathbf{T}_\times = \mathbf{0}, \tag{11.4}$$

where the subindex $\times$ stands for the vectorial invariant of a second-order tensor and $\cdot$ denotes the dot product [3]. Hereinafter we do not consider mass forces and mass couples.

The general expression of W in the case of small deformations takes form

$$W = \frac{1}{2}\mathbf{e} : \mathbf{C} : \mathbf{e} + \mathbf{e} : \mathbf{B} : \mathbf{k} + \frac{1}{2}\mathbf{k} : \mathbf{D} : \mathbf{k} \tag{11.5}$$

with the following relations for **T** and **M**:

$$\mathbf{T} = \mathbf{C} : \mathbf{e} + \mathbf{A} : \mathbf{k}, \quad \mathbf{M} = \mathbf{D} : \mathbf{k} + \mathbf{e} : \mathbf{B}, \tag{11.6}$$

where **C**, **A**, and **D** are fourth-order tensors of elastic moduli. For a linear isotropic micropolar solid, W takes the form

$$\begin{aligned} W =& \frac{1}{2}\lambda \operatorname{tr}^2 \mathbf{e} + \frac{1}{2}(\mu + \kappa)\mathbf{e} : \mathbf{e} + \frac{1}{2}\mu \mathbf{e} : \mathbf{e}^T \\ &+ \frac{1}{2}\alpha \operatorname{tr}^2 \mathbf{k} + \beta \mathbf{k} : \mathbf{k}^T + \gamma \mathbf{k} : \mathbf{k}, \end{aligned} \tag{11.7}$$

where λ, μ, κ, α, β, and γ are elastic moduli. Hereinafter tr is the trace operator and : is the double-dot product. Here **T** and **M** are given by

$$\mathbf{T} = \lambda \mathbf{I} \operatorname{tr} \mathbf{e} + (\mu + \kappa)\mathbf{e} + \mu \mathbf{e}^T, \quad \mathbf{M} = \alpha \mathbf{I} \operatorname{tr} \mathbf{k} + 2\gamma \mathbf{k} + 2\beta \mathbf{k}^T. \tag{11.8}$$

From the positive definiteness of W, it follows that

$$3\lambda + 2\mu + \kappa > 0, \quad 2\mu + \kappa > 0, \quad \kappa > 0, \tag{11.9}$$

$$3\alpha + 2\beta + 2\gamma > 0, \quad \beta + \gamma > 0, \quad \gamma - \beta > 0. \tag{11.10}$$

Let us note that **T** and **M** are nonsymmetric, in general [10] discussed the case $\kappa = 0$ where **T** becomes symmetric, $\mathbf{T} = \mathbf{T}^T$. In what follows, we discuss this particular case of constitutive equations with symmetric stress tensor.

11.3 Eshelby Tensor

For micropolar solids, the following Eshelby tensor can be introduced:

$$\mathbf{B} = W\mathbf{I} - \mathbf{T} \cdot \nabla \mathbf{u}^T - \mathbf{M} \cdot \mathbf{k}^T. \tag{11.11}$$

Using straightforward calculations, one can easily check that for a homogeneous solid the following conservation law is fulfilled:

$$\nabla \cdot \mathbf{B} = \mathbf{0}. \tag{11.12}$$

From (11.12), it follows the integral identity

$$\iint_S \mathbf{n} \cdot \mathbf{B} \, \mathrm{d}S = \mathbf{0}, \tag{11.13}$$

where S is a closed smooth enough surface and $\mathbf{n}$ is the unit vector of outer normal to S. For two-dimensional problems, it brings us the famous J-integral for the micropolar continuum.

11.4 M-integral

In order to derive another conservation law, let us use the following identity:

$$\nabla \cdot (\mathbf{B} \cdot \mathbf{X}) = \operatorname{tr} \mathbf{B}, \tag{11.14}$$

where $\mathbf{X}$ is the position vector. It is valid for any divergence-free tensor $\mathbf{B}$, so we used (11.12). As $\operatorname{tr} \mathbf{B}$ has the form

$$\operatorname{tr} \mathbf{B} = 3W - \mathbf{T} : \nabla \mathbf{u} - \mathbf{M} : \nabla \boldsymbol{\varphi}, \tag{11.15}$$

we come to the integral identity

$$\iint\limits_S \mathbf{n} \cdot \mathbf{B} \cdot \mathbf{X} \, \mathrm{d}S = \iiint\limits_V (3W - \mathbf{T} : \nabla \mathbf{u} - \mathbf{M} : \nabla \boldsymbol{\varphi}) \, \mathrm{d}V, \tag{11.16}$$

where V is such a volume that $\partial V = S$. Integrating by parts in the right part of (11.16), we get

$$\begin{aligned} &\iint\limits_S \mathbf{n} \cdot \mathbf{B} \cdot \mathbf{X} \, \mathrm{d}S + \iint\limits_S \mathbf{n} \cdot (\mathbf{T} \cdot \mathbf{u} + \mathbf{M} \cdot \boldsymbol{\varphi}) \, \mathrm{d}S \\ &= \iiint\limits_V \left[3W + (\nabla \cdot \mathbf{T})\mathbf{u} + (\nabla \cdot \mathbf{M}) \cdot \boldsymbol{\varphi})\right] \mathrm{d}V. \end{aligned} \tag{11.17}$$

Using Eqs. (11.4), we get

$$\iint\limits_S \left[\mathbf{n} \cdot \mathbf{B} \cdot \mathbf{X} + \mathbf{n} \cdot (\mathbf{T} \cdot \mathbf{u} + \mathbf{M} \cdot \boldsymbol{\varphi})\right] \mathrm{d}S = 3 \iiint\limits_V W \, \mathrm{d}V - \iiint\limits_V \mathbf{T}_\times \cdot \boldsymbol{\varphi} \, \mathrm{d}V. \tag{11.18}$$

The presence of integrals over V prevents us from considering (11.18) as a conservation law. Nevertheless, assuming symmetry of $\mathbf{T}$, we can transform the right part of (11.18) to a surface integral. First, let us note that for a symmetric tensor $\mathbf{T}$ its vectorial invariant vanishes, $\mathbf{T}_\times = \mathbf{0}$. Then, in the linear case, W can be written as

$$W = \frac{1}{2}\mathbf{T} : \mathbf{e} + \frac{1}{2}\mathbf{M} : \mathbf{k}. \tag{11.19}$$

Assuming symmetry of $\mathbf{T}$, we transform W into the form

$$W = \frac{1}{2}\mathbf{T} : \nabla\mathbf{u} + \frac{1}{2}\mathbf{M} : \nabla\boldsymbol{\varphi}. \tag{11.20}$$

Indeed, we have $\mathbf{T} : \mathbf{e} = \mathbf{T} : (\nabla\mathbf{u} - \mathbf{I} \times \boldsymbol{\varphi}) = \mathbf{T} : \nabla\mathbf{u}$ as $\mathbf{I} \times \boldsymbol{\varphi}$ is a skew-symmetric tensor and so it is orthogonal to any symmetric tensor.

As a result, we can integrate again by parts and obtain the formula

$$\iiint\limits_V 3W\,\mathrm{d}V = \frac{3}{2}\iint\limits_S \mathbf{n}\cdot(\mathbf{T}\cdot\mathbf{u} + \mathbf{M}\cdot\boldsymbol{\varphi})\,\mathrm{d}S. \tag{11.21}$$

Finally, using (11.21), we get

$$\iint\limits_S \left[\mathbf{n}\cdot\mathbf{B}\cdot\mathbf{X} - \frac{1}{2}\mathbf{n}\cdot(\mathbf{T}\cdot\mathbf{u} + \mathbf{M}\cdot\boldsymbol{\varphi})\right]\mathrm{d}S = 0. \tag{11.22}$$

As a result, from (11.22), we obtain another conservation law

$$\nabla\cdot\left[\mathbf{B}\cdot\mathbf{X} - \frac{1}{2}(\mathbf{T}\cdot\mathbf{u} + \mathbf{M}\cdot\boldsymbol{\varphi})\right] = 0. \tag{11.23}$$

It is a straightforward analog of a similar conservation law in elasticity of simple elasticity; see [6].

For a two-dimensional case, our derivations should be slightly modified. In the 2D case instead of (11.15), we have

$$\operatorname{tr}\mathbf{B} = 2W - \mathbf{T} : \nabla\mathbf{u} - \mathbf{M} : \nabla\boldsymbol{\varphi}, \tag{11.24}$$

so we can derive the final formulae

$$\iint\limits_L \mathbf{n}\cdot\mathbf{B}\cdot\mathbf{X}\,\mathrm{d}s = 0, \quad \nabla\cdot[\mathbf{B}\cdot\mathbf{X}] = 0, \tag{11.25}$$

where L is a closed plane curve. Equation $(11.25)_1$ is the M-integral in plane linear micropolar elasticity.

11.5 Conclusions

We provided a derivation of a new conservation law for linear micropolar solids with symmetric stress tensor. This conservation law results in the so-called M-integral that is useful in mechanics of fracture. Let us note that symmetry of $\mathbf{T}$ means decoupling of equilibrium equations for isotropic solids, whereas for anisotropic solid it is not the case, in general. In particular, this conservation law is valid for the model proposed by [10]. Let us note that one can apply to equilibrium equations expressed in translations and rotations the same technique as presented by [14, 15] for classic linear elasticity. Moreover, the same results could be derived for micropolar shells having a base surface as a minimal surface [1].

Acknowledgements The author acknowledges the support within the European Union's Horizon 2020 research and 105 innovation programme under the Marie Sklodowska-Curie grant agreement EffectFact No 101008140.

References

1. Eremeyev, V.A.: Minimal surfaces and conservation laws for bidimensional structures. Math. Mech. Solids **28**(1), 380–393 (2023)
2. Eremeyev, V.A., Lebedev, L.P., Altenbach, H.: Foundations of Micropolar Mechanics. SpringerBriefs in Applied Sciences and Technologies. Springer, Heidelberg (2013)
3. Eremeyev, V.A., Cloud, M.J., Lebedev, L.P.: Applications of Tensor Analysis in Continuum Mechanics. World Scientific, New Jersey (2018)
4. Eringen, A.C.: Microcontinuum Field Theory. I. Foundations and Solids. Springer, New York (1999)
5. Gurtin, M.E.: Configurational Forces as Basic Concepts of Continuum Physics. Springer, Berlin (2000)
6. Kienzler, R., Herrmann, G.: Mechanics in Material Space with Applications to Defect and Fracture Mechanics. Springer, Berlin (2000)
7. Lubarda, V.A., Markenscoff, X.: Conservation integrals in couple stress elastic-ity. J. Mech. Phys. Solids **48**(3), 553–564 (2000)
8. Lurie, A.I.: Nonlinear Theory of Elasticity. North-Holland, Amsterdam (1990)
9. Maugin, G.A.: Configurational Forces: Thermomechanics, Physics, Mathemat-ics, and Numerics. CRC Press, Boca Raton (2011)
10. Neff, P.: The Cosserat couple modulus for continuous solids is zero viz the linearized Cauchy-stress tensor is symmetric. ZAMM **86**(11), 892–912 (2006)
11. Niktitin, E., Zubov, L.M.: Conservation laws and conjugate solutions in the elas-ticity of simple materials and materials with couple stress. J. Elast. **51**(1), 1–22 (1998)
12. Nowacki, W.: Theory of Asymmetric Elasticity. Pergamon Press, Oxford (1986)
13. Olver, P.: Applications of Lie Groups to Differential Equations, 2nd edn. Springer, New Yourk (1993)
14. Olver, P.J.: Conservation laws in elasticity. I. General results. Arch. Rat. Mech. Anal. **85**, 111–129 (1984)
15. Olver, P.J.: Conservation laws in elasticity. II. Linear homogeneous isotropic elastostatics. Arch. Rat. Mech. Anal. **85**(2), 131–160 (1984b)
16. Zubov, L.M.: Variational principles and invariant integrals for nonlinearly elastic bodies with couple stress. Mech. Solids **25**(2), 7–13 (1990)

Chapter 12
Nonlinear Dynamic Analysis of the Progressive Collapse on Multi-core Computers

Sergiy Fialko

Abstract The method of analysis of structures for progressive collapse, based on the integration of nonlinear equations of motion, is considered. A finite element library has been developed that includes both bar and flat shell finite elements that take into account transverse shear deformations. For reinforced concrete structural elements, discrete placement of reinforcement is taken into account. The behavior of concrete is described by the plastic flow theory when using the yield surface in the form of a non-circular paraboloid, and the behavior of reinforcement is described by von Mises's plastic flow theory with the use of kinematic hardening. The destruction of concrete and the rupture of reinforcement are simulated using the damage parameter. Geometric nonlinearity is taken into account. The method is focused on shared memory multi-core computers.

12.1 Introduction

Analysis of the stability of buildings and structures to progressive collapse arising from extreme influences such as an explosion of household gas, a sabotage act, and a vehicle collision is of great interest. Reviews of articles devoted to this problem are given in [1–3]. Among the many definitions of progressive and disproportionate collapse given by various authors, it is emphasized that in this type of analysis, the initial damage to the structure must be local in nature and lead to a disproportionate increase in the number of damaged structural elements and the degree of their damage in the final state.

Typically, in this type of analysis, a group of structural elements is specified, which, after applying dead and live loads, are instantly removed at specified moments in time. The so-called scenario of initial destruction is being formed. The removal of selected structural elements is modeled either by the addition of initiating loads that should nullify the impact of the removed finite elements on their neighbors or by the action of the removed elements is replaced by their reactions, which are zeroed

S. Fialko (✉)
Tadeusz Kościuszko Cracow University of Technology, Kraków, Poland
e-mail: sergiy.fialko@pk.edu.pl

H. Altenbach et al. (eds.), *Selected Problems of Solid Mechanics and Solving Methods*, Advanced Structured Materials 204, https://doi.org/10.1007/978-3-031-54063-9_12

out in a given time moment. In the mentioned publications, both static and dynamic methods of analysis are considered, and among the latter, linear dynamic analysis, as well as nonlinear elastoplastic, is used. The analysis of progressive destruction is associated with the creation of design models in which a number of elements get large deformations and accumulate significant damages. Therefore, for reliable modeling of the structure behavior, it is necessary to take into account the nonlinear properties of materials. The second significant factor is to take into account the dynamic nature of the ongoing processes [5–9].

Modern design models of multistory buildings contain a large number of different structural elements, and their dimensionality reaches hundreds of thousands of degrees of freedom, which for nonlinear dynamic analysis is a difficult computational problem. Therefore, in the vast majority of studies, significant simplifications of the calculation models are usually resorted to. Most studies consider design models of a highly simplified geometric shape, and often the design models contain only the rod finite elements (for instance, [1, 4–8]). The next type of simplification is the use of lumped plasticity in the rod elements (for example, [8]), when all plastic deformations are concentrated in the so-called plastic hinges, the position of which must be guessed along the length of the rod.

In the proposed approach, three types of finite elements are used to simulate the behavior of the entire design model, taking into account both geometric and physical nonlinearity: the rod finite element, as well as triangular and quadrilateral isoparametric finite elements for medium-thickness shells and plates. These finite elements contain inclusions—reinforcing rods that allow us to describe the behavior of reinforced concrete elements. In the case of a homogeneous material (for example, steel), there are no inclusions. To describe the behavior of concrete and reinforcement, the plastic flow theory is used, and the destruction of concrete of stretched and compressed zones, as well as the rupture of reinforcement, is modeled using damage parameters. For both concrete and reinforcement, geometric nonlinearity is taken into account. It is shown that the simultaneous accounting of plastic deformations and geometric nonlinearity in some cases makes it possible to present the sequence of destruction in more detail and to predict more accurately whether or not a progressive collapse will occur in this scenario of initial destruction. In addition, the computational model can contain any number and any type of linear finite elements.

The removal of a group of elements in the proposed approach is carried out by excluding a contribution of each finite element being deleted at a given time in the algorithm for calculating internal forces. If the Newton–Raphson method is used to solve a nonlinear problem, then the contributions of finite elements to be removed are excluded in the algorithm for assembling a tangent stiffness matrix. This approach is much more convenient than the methods mentioned above and makes it possible to easily remove both rod elements (columns) and shell elements (fragments of bearing walls), which is fundamentally important in the analysis of modern multistory buildings. We emphasize that in all the mentioned works, the initial destruction is associated with the sudden removal of columns, which significantly narrows the class of the problems under consideration.

The proposed formulation of the problem is based on the methods and approaches used in solid mechanics, to which the generally accepted hypotheses of the theory of Mindlin–Reissner plates and shells and the theory of S.P. Timoshenko beams are subsequently applied.

To analyze real design models on widely spread computers (desktops, laptops, and workstations with shared memory), such formulation of the problem in a conventional implementation would require an unreasonably long computing time. Therefore, in the proposed approach, much attention is paid to the development of unique parallel algorithms focused on modern multi-core computers with shared memory, allowing acceleration of the solution of such problems tenfold.

12.2 Problem Formulation

12.2.1 Models of Materials

The behavior of concrete is described by the equations of the plastic flow theory when the yield surface is represented by both a circular and a non-circular paraboloid proposed in [10] as a strength surface:

$$f = 3\bar{\alpha}^2 b I_1 + \frac{3\sqrt{3}}{2}\beta\,(a - bI_1)\,J_3 J_2^{-3/2} + 3J_2 - \sigma_0, \tag{12.1}$$

where $a = \bar{\sigma}_c\bar{\sigma}_t, b = \bar{\sigma}_c + \bar{\sigma}_t, \bar{\alpha} \in (0.531, \sqrt{3}]$ defines the deviation of the paraboloid from circular shape, $\beta = 1 - 3\bar{\alpha}^2, \sigma_0 = 3a\bar{\alpha}^2$, I_1 is a first invariant of the stress tensor, J_2 is a second invariant of the stress deviator D_σ, $J_3 = \det(\mathrm{D}_\sigma)$ is the third invariant of the stress deviator, $\bar{\sigma}_c$, and $\bar{\sigma}_t$ are the compressive and tensile strength of concrete, changing during evolution of the plastic strains [11]:

$$\bar{\sigma}_t = \begin{cases} \sigma_t + H_t\varepsilon_{ps}, & \sigma_t + H_t\varepsilon_{ps} > \alpha\sigma_t \\ \alpha\sigma_t, & \sigma_t + H_t\varepsilon_{ps} \leqslant \alpha\sigma_t \end{cases}, \quad t \leftrightarrow c, \tag{12.2}$$

where ε_{ps} is a measure of the plastic strain, $H_t = E_t/(1 - E_t/E)$. So, when the image point moves along stress surface (12.1), this surface expands ($H_t > 0$) or compresses ($H_t < 0$) [11].

When deformation exceeds ε_d or is less than ε_u, the damage parameter

$$d(\varepsilon) = 1 - p(\varepsilon),$$

where $p(\varepsilon) \in [1, \alpha]$ [12], is used to simulate the destruction of concrete (Fig. 12.1). Note that in the stretched zone of concrete is usually taken $\varepsilon_d = \varepsilon_t$. Parameter α takes into account the residual strength of concrete. The isotropic hardening is taken. The bilinear diagram, von Mises yield criterion, and kinematic hardening are taken

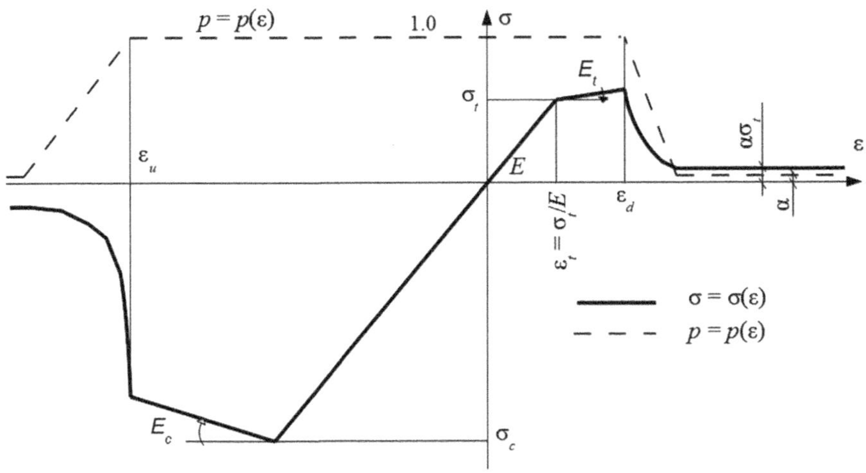

Fig. 12.1 σ–ε diagram for concrete

for steel. To ensure the lack of slipping of the reinforcement in concrete, kinematic coupling conditions for the reinforcement and concrete are applied:

$$\varepsilon_s = \varepsilon_c, \tag{12.3}$$

where ε_s is a longitudinal strain of the reinforcement rod and ε_c is a strain of concrete in the same point reduced to the direction of the longitudinal axis of the rod.

12.2.2 Flat Shell Finite Elements

Quadrilateral and triangular flat shell finite elements (FE) are presented in Fig. 12.2. To simplify the input of source data and take into account that for actual finite element meshes several reinforcement rods are in the domain of the reinforcement layer, we smear the rods in the plane of finite element and create the reinforcement layers. Each reinforcement layer consists of rods of the same cross-section, material properties, and direction. In addition, the steps between neighborhood rods are the same. The number of reinforcement layers is not restricted. In the direction across the thickness of the FE, the discreteness of rebar accounting is preserved. Here $Oxyz$ is a local coordinate system, s_1–s_4 are the axes of reinforcement layers (their direction coincides with the direction of rebar's rods and can be not parallel to the Ox and Oy axes of the local coordinate system) and z_{s1}–z_{s4} are the distances between the corresponding reinforcement layer and the middle surface.

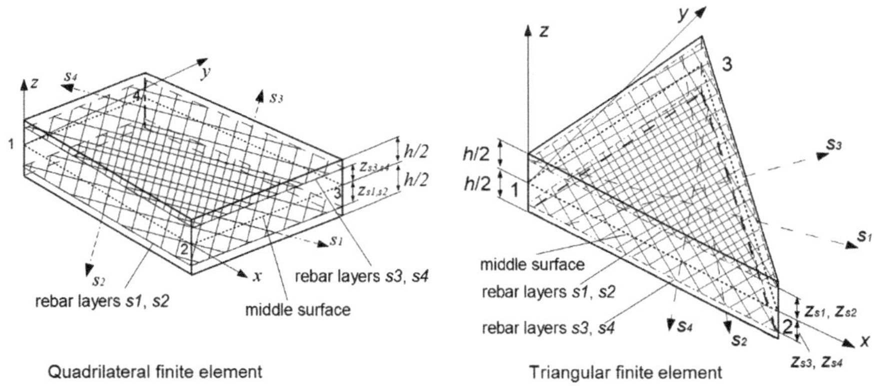

Fig. 12.2 Quadrilateral and triangular flat shell finite elements

Reinforcement rods work not only for tension/compression but also for transverse shear, which significantly increases the computational stability of the method in cases where concrete is completely destroyed in the cross-section.

The principle of virtual work is applied to obtain both the tangent stiffness matrix of FE and the vector of internal forces:

$$\iiint_V \boldsymbol{\sigma}_c : \delta\boldsymbol{\varepsilon}_c dV + \sum_s \frac{A_s}{h_s} \iint_\Omega \boldsymbol{\sigma}_s : \delta\boldsymbol{\varepsilon}_s d\Omega - \delta A_{\text{ext}} = 0, \tag{12.4}$$

where $\boldsymbol{\sigma}_c$ and $\boldsymbol{\varepsilon}_c$ are the stress and strain tensors for concrete, $\boldsymbol{\sigma}_s$ and $\boldsymbol{\varepsilon}_s$ are the stress and strain tensors for rebar rod in the layer s, A_{ext} is a virtual work of external forces, A_s and h_s are cross-sectional area and step between rods of layer s, Ω is an area of FE and V is a volume. Summation by s covers all reinforcement layers. According to the static and kinematic hypothesis of the Mindlin–Reissner shell theory, stress and strain tensors are as follows

$$\boldsymbol{\sigma}_c = \begin{pmatrix} \sigma_x & \tau_{xy} & \tau_{xz} \\ \tau_{yx} & \sigma_y & 0 \\ \tau_{zx} & 0 & 0 \end{pmatrix}, \quad \boldsymbol{\varepsilon}_c = \begin{pmatrix} \varepsilon_x & 0.5\gamma_{xy} & 0.5\gamma_{xz} \\ 0.5\gamma_{yx} & \varepsilon_y & 0 \\ 0.5\gamma_{zx} & 0 & \varepsilon_z \end{pmatrix}, \tag{12.5}$$

where $\varepsilon_z = -\frac{\nu}{1-\nu}\left(\varepsilon_x + \varepsilon_y\right)$ and ν is Poisson's ratio. The stress and strain tensors for reinforcement rods are

$$\boldsymbol{\sigma}_s = \begin{pmatrix} \sigma_s & \tau_{sn} & \tau_{sz} \\ \tau_{ns} & 0 & 0 \\ \tau_{zs} & 0 & 0 \end{pmatrix}, \quad \boldsymbol{\varepsilon}_s = \begin{pmatrix} \varepsilon_s & 0.5\gamma_{sn} & 0.5\gamma_{sz} \\ 0.5\gamma_{ns} & -\nu_s\varepsilon_s & 0 \\ 0.5\gamma_{zn} & 0 & -\nu_s\varepsilon_s \end{pmatrix}, \tag{12.6}$$

where ν_s is Poisson's ratio for reinforcement rod. The local coordinate system for the rod belonging to reinforcement layer s is shown in Fig. 12.3a.

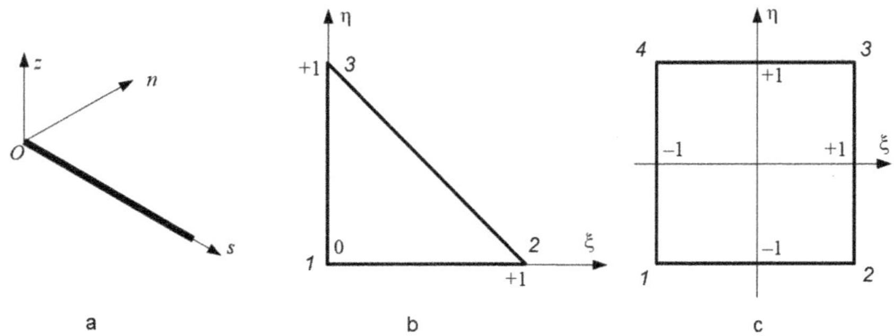

Fig. 12.3 Rebar rod in the local coordinate system $Osnz$ (**a**), the isoparametric coordinate system for triangular (**b**), and quadrilateral finite element (**c**)

The direction of the Oz axis coincides with the direction of the Oz axis of the finite element local coordinate system, and the On axis is parallel to the plane Oxy (Fig. 12.2). Taking into account the quadratic approximation of the angles of rotation, the Cauchy relations are as follows:

$$\begin{aligned}
\varepsilon_x &= \frac{\partial u}{\partial x} + \frac{1}{2}\left(\frac{\partial u}{\partial x}\right)^2 + \frac{1}{2}\left(\frac{\partial v}{\partial x}\right)^2 + \frac{1}{2}\left(\frac{\partial w}{\partial x}\right)^2 + z\kappa_x \\
\varepsilon_y &= \frac{\partial v}{\partial y} + \frac{1}{2}\left(\frac{\partial u}{\partial y}\right)^2 + \frac{1}{2}\left(\frac{\partial v}{\partial y}\right)^2 + \frac{1}{2}\left(\frac{\partial w}{\partial y}\right)^2 + z\kappa_y \\
\gamma_{xy} &= \frac{\partial u}{\partial y} + \frac{\partial v}{\partial x} + \frac{\partial u}{\partial x}\frac{\partial u}{\partial y} + \frac{\partial v}{\partial x}\frac{\partial v}{\partial y} + \frac{\partial w}{\partial x}\frac{\partial w}{\partial y} + 2z\chi
\end{aligned} \tag{12.7}$$

where u, v, and w are the displacements of the middle surface, κ_x, κ_y, and χ are its changes curvatures and torsion.

The shape functions for triangular FE (Fig. 12.3b) are $N_1(\xi, \eta) = 1 - \xi - \eta$, $N_2(\xi, \eta) = \xi$, $N_3(\xi, \eta) = \eta$ and for quadrilateral FE (Fig. 12.3c) are $N_i(\xi, \eta) = (1 \pm \xi)(1 \pm \eta)/4$ $(i = 1, \ldots, 4)$. For the numerical calculation of the volume integral in (12.4), the finite element is divided by thickness by equidistant layers parallel to the middle surface, and then in the direction of the Oz axis, the trapezium method is used. For each layer, when integrating into the plane of a finite element, the Gauss–Legendre method is applied with a 2×2 integration scheme for a quadrilateral FE, and one Gauss point is taken for a triangular FE. The components of stress and strain tensors, as well as stress and strain deviators, are calculated in each Gauss point of each layer in concrete and reinforcement layers.

To avoid the shear locking effect, which is typical for Mindlin–Reissner plates and shells, the mixed interpolation of tensorial components (MITC) [13] is applied for quadrilateral FE and discrete shear gap approach [14] for triangular one. The details of implementation are in [11, 15].

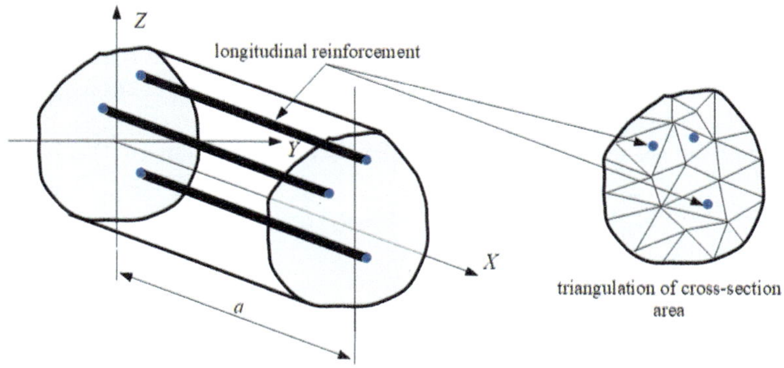

Fig. 12.4 Bar finite element

12.2.3 Bar Finite Element

Figure 12.4 presents the two-node bar finite element. The bar cross-section can have an arbitrary shape and arbitrary holes. S.P. Timoshenko shear theory is applied.

The principle of the virtual works is used to obtain a tangent stiffness matrix as well as a vector of internal forces:

$$\iiint_V \boldsymbol{\sigma}_c : \delta\boldsymbol{\varepsilon}_c dV + \sum_s A_s \int_0^a \boldsymbol{\sigma}_s : \delta\boldsymbol{\varepsilon}_s dx - \delta A_{\text{ext}} = 0, \tag{12.8}$$

where

$$\boldsymbol{\sigma}_c = \begin{pmatrix} \sigma_x & \tau_{xy} & \tau_{xz} \\ \tau_{yx} & 0 & 0 \\ \tau_{zx} & 0 & 0 \end{pmatrix}, \quad \boldsymbol{\varepsilon}_c = \begin{pmatrix} \varepsilon_x & 0.5\gamma_{xy} & 0.5\gamma_{xz} \\ 0.5\gamma_{yx} & -\nu\varepsilon_x & 0 \\ 0.5\gamma_{zx} & 0 & -\nu\varepsilon_x \end{pmatrix}, \tag{12.9}$$

and tensors $\boldsymbol{\sigma}_s$ and $\boldsymbol{\varepsilon}_s$ are presented by (12.6). The kinematic coupling conditions $\varepsilon_s = \varepsilon_x, \gamma_{sn} = \gamma_{xy}, \gamma_{sz} = \gamma_{xz}$ ensure no slip behavior of reinforcement rods in concrete. In the coordinate system associated with the non-deformable position of the bar finite element, the longitudinal deformation is presented as

$$\varepsilon_x = \sqrt{1 + a^2} - 1, \quad a^2 = 2\left[\frac{\partial u}{\partial x} + \frac{1}{2}\left(\frac{\partial u}{\partial x}\right)^2 + \frac{1}{2}\left(\frac{\partial v}{\partial x}\right)^2 + \frac{1}{2}\left(\frac{\partial w}{\partial x}\right)^2\right], \tag{12.10}$$

where u, v, and w are the displacements of the central axis of bar. To describe the initial post-buckling behavior of the bar, we retain up to 18 members of the Taylor expansion. It may be important when the structure has significant destructions.

The triangulation of the bar's cross-section is produced (Fig. 12.4), and the body of the bar is divided into triangular prisms. Then the Gauss points are taken in the gravity centers of these prisms, and numerical integration by volume is performed (the first integral in (12.8)). The summation by s covers all longitudinal reinforcement rods, which are taken into account fully discretely. Integral in the second term of (12.8) is the virtual work of internal forces in sth reinforcement rod, and the single Gauss point is taken at the mid of this rod. The linear shape functions are applied: $N_1(x) = 1 - x/a$ and $N_2(x) = x/a$.

To avoid the shear locking effect, the shear strains are taken as

$$\gamma_{xy}(x, z) = \frac{1}{2}\left(\gamma_{xy}(0, z) + \gamma_{xy}(a, z)\right), \quad y \leftrightarrow z, \tag{12.11}$$

Details are in [11, 16].

12.3 Numerical Realization

In the proposed formulation of the problem, we operate with thin-walled plates and shells, as well as bars with three-dimensional bodies. However, we use the kinematic and static hypotheses of the Mindlin–Reissner plates and the shells and the rods of S.P. Timoshenko. The components of stress and strain tensors for the shell finite elements are calculated and stored at the Gauss points of each layer of concrete and each reinforcement layer and for the rod finite elements—at the Gauss points of each triangular prism that arose as a result of the triangulation of the cross-section, and in the middle of each reinforcing rod. This allows us to naturally realize the relations of the flow plastic theory, avoiding the need to create a yield surface in terms of generalized forces (longitudinal and transverse forces, bending and torsion moments, etc.). On the other hand, each finite element requires processing and storage in the RAM of a computer for a large amount of data. In order to create an effective method that allows us to analyze the real-life design models of buildings and structures in an acceptable time, it is necessary to use modern achievements of computer science: parallelization and vectorization of calculations, as well as maximum support of the processor's pipelines.

The finite element discretization turns out the following Cauchy problem:

$$\begin{aligned} &\mathbf{M}\ddot{\mathbf{u}} + \mathbf{C}\dot{\mathbf{u}} + N(\mathbf{u}) = \mathbf{f}_{\text{ext}}, \\ &\mathbf{u}(0) = \mathbf{u}_0, \quad \dot{\mathbf{u}}(0) = \mathbf{v}_0, \end{aligned} \tag{12.12}$$

where $\mathbf{M}$ is a mass matrix, $\mathbf{C}$ is a dissipation matrix, and N is nonlinear operator, determining internal force vector $\mathbf{f}_{\text{int}}$, $\mathbf{f}_{\text{ext}}$ is an external force vector, $\mathbf{u} = \mathbf{u}(t)$ is a nodal displacement vector, $\mathbf{u}_0$ and $\mathbf{v}_0$ are initial displacement and velocity vectors. In the case of proportional damping, dissipation matrix is taken as $\mathbf{C} = \alpha\mathbf{M} + \beta\mathbf{K}$, where parameters α and β are determined using the modal damping parameters, and $\mathbf{K}$ is a tangent stiffness matrix in initial time moment $t = t_0$. The proposed approach allows us to use so-called material damping for the simulation of the structural behavior in the case when different parts of the structure have different dissipation parameters [17].

The solution to this problem consists of two stages. In the first stage, we apply the dead load as well as the exploitation load, solve a static nonlinear problem $N(\mathbf{u}^{\text{stat}}) = \mathbf{f}_{\text{ext}}^{\text{stat}}$, and determine the stress–strain state caused by static loads. In the second stage, we solve the Cauchy problem (12.12), taking $\mathbf{u}_0 = \mathbf{u}^{\text{stat}}$ and $\mathbf{v}_0 = 0$. In addition, for nonlinear finite elements, the final stress–strain stage of the previous static analysis is taken as an initial stress–strain stage for dynamic analysis. The motion of structure begins when some finite elements are removed according to the accepted scenario.

The implicit predictor–corrector approach [18] is applied. The predictor stage uses a Newmark method, and the corrector stage is based on the Newton–Raphson method. The details of our implementation are presented in [19]. Here we only mention the principal stages of the corrector's iterations because the Newmark-like predictor works very fast. The considered problems demonstrate a strong nonlinearity, and this fact requires re-assembly and factorizing the tangent stiffness matrix at each iteration of the Newton–Raphson method. The most time-consuming stages are:

1. calculation of internal forces vector;
2. assembling of tangent stiffness matrix;
3. factoring of tangent stiffness matrix;
4. forward-back substitutions for obtaining of solution.

12.3.1 Calculation of Internal Forces Vector

Evaluation of the internal forces vector takes almost all the time when the residual vector of the Newton–Raphson method is evaluated. The parallel algorithm is based on parallelization of the loop covering all finite elements except removed finite elements according to the scenario of initial destruction: $\forall \text{ele} \in \text{e}$ calculate a contribution of nodal reaction vector to the internal force vector $\mathbf{f}_{\text{int}} += \mathbf{r}_{\text{ele}}(\mathbf{u}_{\text{ele}})$. Here e is a set of finite elements taking part in the calculation of $\mathbf{f}_{\text{int}}$, $\mathbf{f}_{\text{ele}}$ and $\mathbf{u}_{\text{ele}}$ are the nodal reaction vector and the nodal displacements vector for the finite element ele. This approach is simple and obvious for parallelization since for each finite element the calculation of the vector of nodal reactions does not depend on the calculation of the vectors of nodal reactions of other finite elements. The problem of synchronization when adding the components of the vector of nodal reactions to the components of the vector of internal forces is solved as follows: each thread performs this operation storing results in its own vector, and at the end of the parallel region covering the loop on the finite elements, these vectors are summed up. The details of the parallel algorithm are presented in [19].

12.3.2 Assembling of Tangent Stiffness Matrix

Since the non-zero structure of a sparse tangent stiffness matrix can occupy a significant amount of RAM, we cannot create multiple such structures whose number is equal to the number of threads. Therefore, the basic idea of a parallel assembly algorithm is based on finding groups of finite elements that do not have a neighborhood with each other simultaneously. In this case, the finite element tangent stiffness matrices belonging to each of these groups can be calculated on different threads and simultaneously added to the non-zero structure of the sparse global matrix. In the first approach, the structure of levels of the adjacency graph for finite elements is applied to obtain the groups of non-neighborhood finite elements [19]. The creation of these groups is performed before the parallel loop covering all finite elements except those removed due to the initial destruction scenario begins. The static mapping of the computational tasks onto threads is used.

However, a more efficient approach using a dynamic mapping of computational tasks onto threads is presented in [20], Algorithm 4. The initial queue containing all aggregated finite elements is created. Then in the parallel region in the critical section, the nearest finite element is retrieved, and all neighbor finite elements are marked as "locked". As soon as the "locked" finite element is retrieved from the queue by another thread, this FE is pushed at the end of the queue and will be retrieved afterward. So, only unmarked FE can be put to aggregation, and in a parallel region simultaneously are only non-neighborhood finite elements. As soon as selected finite elements proceeded, all their neighbors are unmarked.

12.3.3 Factoring of Tangent Stiffness Matrix and Forward-Back Substitutions

The highly parallelized sparse direct solver PARFES is used [20]. The parallelization of the forward and back substitutions is presented in [21]. The dynamic mapping of computing tasks onto threads is used in both: factorization stage and the stage of forward-back substitutions. The efficiency of all stages of the Newton–Raphson method is demonstrated in [20].

12.4 Numerical Results

The reliability of the results obtained by the proposed approach is confirmed by comparing them with the experimental results of other authors, as well as with numerical results that inspire confidence. Testing of triangular and quadrilateral finite elements under the action of static loads is given in [11, 15], as well as testing of bar finite elements is presented in [11, 16].

In [22], a comparison of the proposed numerical approach with experimental results [23] is presented. The compressed column subjected to imposed cyclic lateral displacement of the top end is considered. This test proves the applicability of the proposed approach for numerical modeling of bars finite elements under the action of cyclic loads.

12.4.1 Progressive Collapse Analysis Test

A reinforced concrete slab with a size in the plan of 2.1 × 4.1 m and a thickness of 0.08 m is considered [24], Fig. 12.5. The construction is loaded with a dead weight of 0.00188 MN/m^2, as well as sandbags, creating a superimposed load with an intensity of 0.0068 MN/m^2. The sudden removal of the corner support is simulated.

For concrete, $E = 28{,}000$ MN/m^2, $\nu = 0.2$, $\sigma_c = 26.7$ MN/m^2, $\sigma_t = 2$ MN/m^2, and for steel, $E = 200{,}000$ MN/m^2, $\sigma_y = 500$ MN/m^2. The height of the sandbags significantly exceeds the thickness of the structure. Therefore, in order to take into account the inertial characteristics of the rotational motion of the attached masses, the upper elastic membrane with a fictitiously small modulus of elasticity $E = 0.01$ MN/m^2 is introduced into the design model, which practically does not affect the stiffness characteristics of the system, but has a real mass equivalent to the mass of sandbags. Eccentricity $e = 0.24$ m. The details of the test are in [24].

Figure 12.6 presents the comparison of the numerical results obtained by the proposed approach and experimental results taken from the mentioned article. Taking into account the geometric nonlinearity (solid curve) allows us to get essentially closer to the results of the experiment with a case when the geometric nonlinearity does not account (dash curve). In addition, Fig. 12.6 demonstrates an acceptable agreement of our numerical results with experimental ones. The peak value of controlled displacement is practically the same as in the experiment. The final displace-

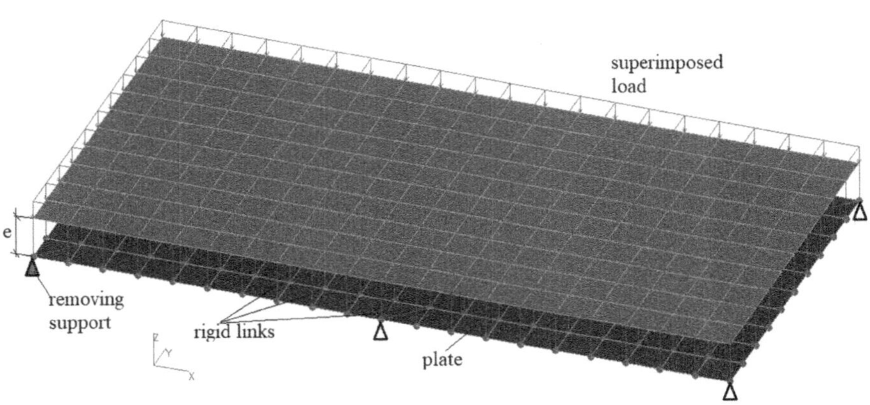

Fig. 12.5 Reinforced concrete slab with suddenly removed corner support

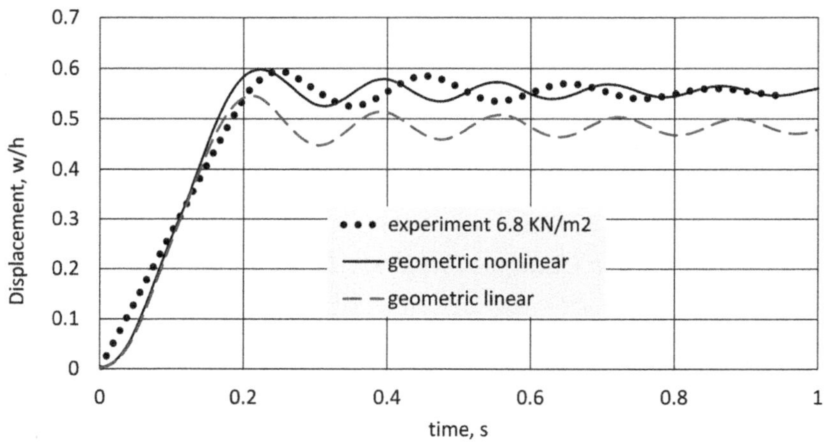

Fig. 12.6 Vertical displacement of plate corner under removing support

ment after the dissipation of vibration is the same too. So, the proposed approach demonstrates reliable results.

12.4.2 Progressive Collapse Analysis of Multistory Building

Analysis of a large number of works showed that when analyzing progressive destruction, they are usually limited to the scenario of removing a particular column of the lower floor. The author could not find works in which the instantaneous removal of load-bearing walls would be considered. However, modern multi-story buildings contain not only load-bearing columns and pylons but also staircase-elevator blocks, in which the bearing structures are reinforced concrete walls. Therefore, it is fundamentally important to develop methods that can simulate the removal of not only rods but also fragments of plates and shells.

Figure 12.7 demonstrates a typical design model of the multi-story building. The paper [25] considers a simulation of the destruction process when the corner column has been suddenly removed. Here we consider the sudden removal of the bottom part of the staircase–elevator blocks (Fig. 12.8) in the same design model. In the article mentioned above, it has been shown that the installation of outrigger stiffeners (Fig. 12.9) is an inexpensive and, at the same time, effective way to increase the vitality of the structure. This conclusion is also confirmed by the results given in this work.

Figure 12.10 shows a destruction scheme of the structure without outrigger stiffeners and taking into account both: plasticity, destruction of concrete as well as reinforcement, and geometrical nonlinearity of floor slabs, walls as well as columns. In the lower parts of the columns surrounding the staircase–elevator block, there

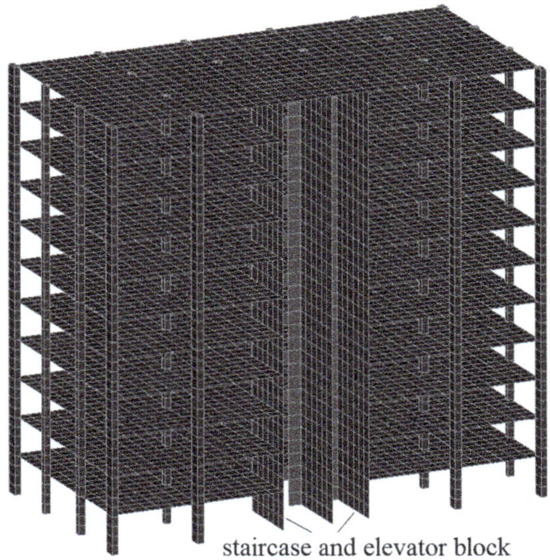

Fig. 12.7 Finite element model of multi-story building

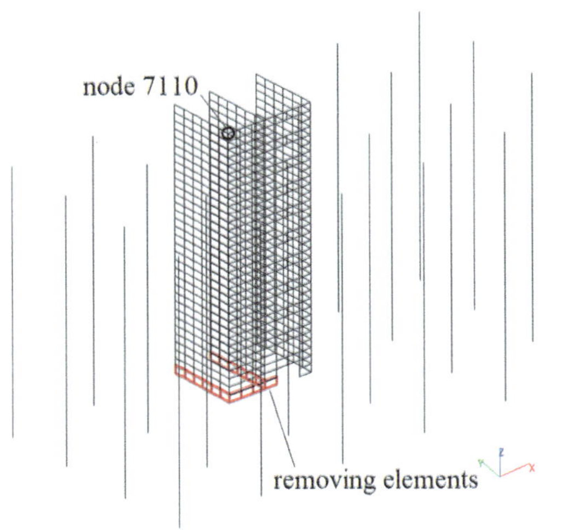

Fig. 12.8 Finite element model of the vertical bearing structures

was an exhaustion of the bearing capacity of compressed concrete, as well as the destruction of reinforcement rods due to the buckling of the reinforcement after the destruction of the concrete. The solid curve in Fig. 12.11 demonstrates the increasing nature of the displacement of control node No. 7110 in the direction of OX. The non-taking into account the geometric nonlinearity gives a distorted picture of the behavior of the structure, from which it would be possible to conclude that the movement of the controlled displacement in time is limited and the progressive destruction does not occur—dash curve in Fig. 12.11.

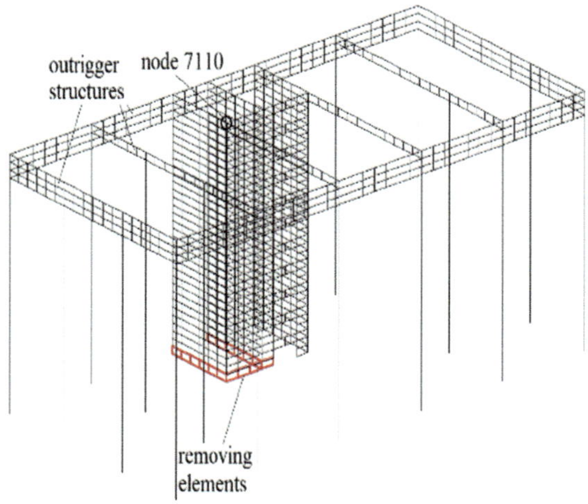

Fig. 12.9 Outrigger stiffeners

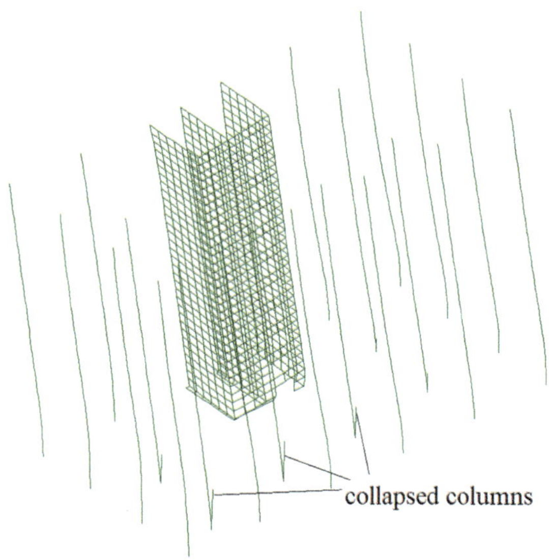

Fig. 12.10 Collapse of the vertical bearing structures. No outrigger stiffeners

The addition of stiffening outriggers (Fig. 12.9) radically changes the behavior of the design model—the movement of the controlled node is oscillatory in nature (Fig. 12.12), both taking into account geometric nonlinearity and without it. This means that after the sudden removal of the noted elements (Fig. 12.9), there is no progressive collapse.

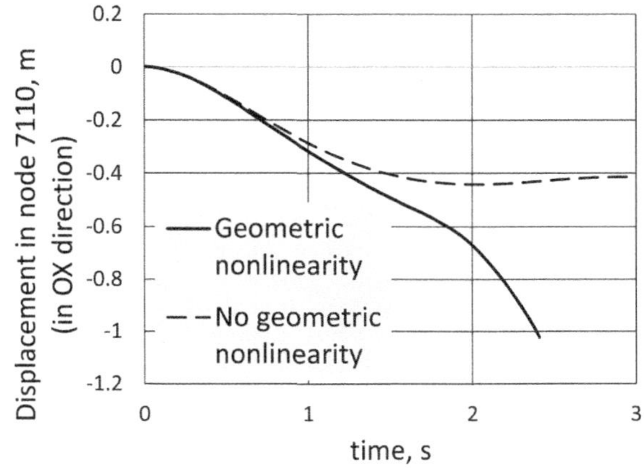

Fig. 12.11 Displacement of the controlled node when no outrigger stiffeners

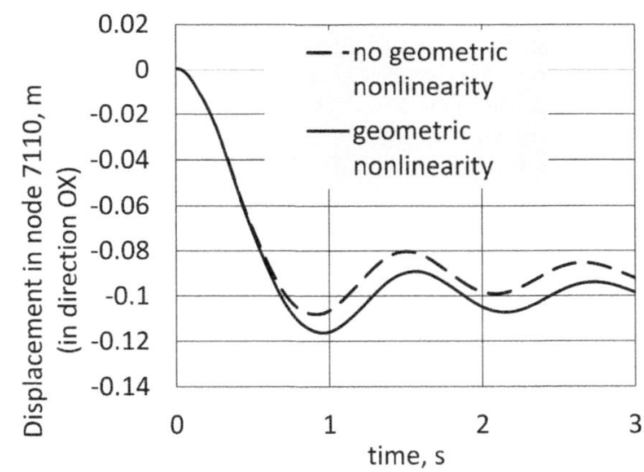

Fig. 12.12 Displacement of the controlled node when outrigger stiffeners are presented

12.5 Conclusions

The proposed approach allows us to perform an analysis of the progressive destruction of the entire design model of a multi-story building for initial local destruction, initiated by the instantaneous removal of the shell finite elements of the lower part of the staircase–elevator unit with and without an outrigger floor. It is shown that simultaneous consideration of geometric and physical nonlinearity in a structure without outrigger stiffeners allows detecting the progressive collapse while taking into account only one physical nonlinearity does not reveal a progressive collapse. The presence of outrigger stiffeners leads to a redistribution of the internal forces of the structure not only on columns surrounding the damaged staircase–elevator unit but on far columns; therefore, there is no progressive collapse.

References

1. Adam, J.M., Parisi, F., Sagaseta, J., Lu, X.: Research and practice on progressive collapse and robustness of building structures in the 21st century. Eng. Struct. **173**, 122–149 (2018)
2. Wang, H., Zhang, A., Li, Y., Yan, W.: A review on progressive collapse of building structures. Open Civil Eng. J. **8**, 183–192 (2014)
3. Kiakojouri, F., Sheidaii, M.R., De Biagi, V., Chiaia, B.: Progressive collapse of structures: a discussion on annotated nomenclature. Structures **29**, 1417–1423 (2021)
4. Kwasniewski, L.: Nonlinear dynamic simulations of progressive collapse for a multistory building. Eng. Struct. **32**, 1223–1235 (2010)
5. Izzuddin, B.A., Vlassis, A.G., Elghazouli, A.Y., Nethercot, D.A.: Progressive collapse of multistorey buildings due to sudden column loss—Part I: simplified assessment framework. Eng. Struct. **30**, 1308–1318 (2008)
6. Fu, F.: 3-D nonlinear dynamic progressive collapse analysis of multi-storey steel composite frame buildings–parametric study. Eng. Struct. **32**, 3974–3980 (2010)
7. Li, J., Hao, H.: Numerical study of structural progressive collapse using substructure technique. Eng. Struct. **52**, 101–113 (2013)
8. Kokot, S., Anthoine, A., Negro, P., Solomos, G.: Static and dynamic analysis of a reinforced concrete flat slab frame building for progressive collapse. Eng. Struct. **40**, 205–217 (2012)
9. Sadek, F., Main, J.A., Lew, H.S., Bao, Y.: Testing and analysis of steel and concrete beam-column assemblies under a column removal scenario. J. Struct. Eng. **137**(9), 881–892 (2011)
10. Geniev, G.A., Kissyuk, V.N., Tyupin, G.A.: The Theory of Plasticity of Concrete and Reinforced Concrete. Stroyizdat, Moscow (1974)
11. Fialko S., Yu.: Application of finite element method to analysis of strength and bearing capacity of thin-walled concrete structures, taking into account the physical nonlinearity. SCAD Soft, ASV, Moscow (2018)
12. Criesfield MA (2000) Nonlinear finite element analysis for solids and structures (vol. 2: Advanced topics). John Wiley & Sons Ltd Chichester, New York
13. Bathe, K.J.: Finite Element Procedures. Prentice Hall, NJ (1996)
14. Bletzinger, K.U., Bischoff, M., Ramm, E.: A unified approach for shear-locking-free triangular and rectangular shell finite elements. In: Idelsohn, S., Oñate, E., Dvorkin, E. (eds.) Computational Mechanics, pp. 1–22. CIMNE, New Trends and Applications Barcelona, Spain (1998)
15. Fialko, S.: Quadrilateral finite element for analysis of reinforced concrete floor slabs and foundation plates. Appl. Mech. Mater. **725–726**, 820–835 (2015)
16. Fialko, S., Karpilovskyi, V.: Spatial thin-walled reinforced concrete structures taking into account physical nonlinearity in SCAD software. Rod finite element. In: 13th International Conference "Modern Building Materials, Structures and Techniques", 16–17 May Vilnius Lithuania, pp. 728–735 (2019)
17. Fialko, S., Karpilovskyi, V.: Time history analysis formulation in SCAD FEA software. J. Measur. Eng. **6**(4), 173–180 (2018)
18. Hughes, T.J.R., Belytschko, T.: Nonlinear Finite Element Analysis. Course Notes Sept. 4–8 Munich, Germany (1995)
19. Fialko, S., Karpilovskyi, V.: Multithreaded parallelization of the finite element method algorithms for solving physically nonlinear problems. In: Proceedings of the Federated Conference on Computer Science and Information Systems ACSIS, vol. 15, pp. 311–318 (2018)
20. Fialko, S.: Parallel finite element solver for multi-core computers with shared memory. Comput. Math. Appl. **94**, 1–14 (2021)
21. Fialko, S.: Parallel algorithms for forward and back substitution in linear algebraic equations of finite element method. J. Telecommun. Inf. Technol. **4**, 20–29 (2019)
22. Fialko, S.: Dynamic analysis of the elasto-plastic behaviour of buildings and structures in the SCAD ++ software package. IOP Conf. Ser.: J. Phys.: Conf. Series 1425 012041 (2020)
23. Ohno, T., Nishioka, T.: An experimental study of energy absorption capacity of columns in reinforced concrete structures. Proc. JSCE Struct. Eng./Earthquake Eng. **1**, 23–33 (1984)

24. Russell, J.M., Owena, J.S., Hajirasouliha, I.: Experimental investigation on the dynamic response of RC flat slabs after a sudden column loss. Eng. Struct. **99**, 28–41 (2015)
25. Fialko, S.Y., Kabantsev, O.V., Perelmuter, A.V.: Elasto-plastic progressive collapse analysis based on the integration of the equations of motion. Mag. Civil Eng. **102**(2), Article No. 10214 (2021)

Chapter 13
Fatigue Endurance of Thin-Walled Cylindrical Shells Under Biaxial Combined Loading

Vladyslav Golub

Abstract The problem of determining the number of cycles to failure of thin-walled cylindrical shells due to multicycle fatigue under combined uniaxial tension-compression/torsion and bending/torsion loads is solved. Solutions are built based on the concept of equivalent stresses, the structure of which is given based on classical failure criteria. The criterion of maximum normal stresses, the criterion of maximum tangential stresses, and the criterion of the specific strain energy due to distortion were used as fracture criteria. The results of the calculations were tested experimentally.

13.1 Introduction

For a reliable assessment of the efficiency of structural elements in power engineering, aviation, rocketry, and space engineering, it is necessary to take into account the impact of cyclic stress changes when such a phenomenon as multicycle (classical) fatigue can develop [1–3]. The number of stress change cycles during the entire service life of the structure can reach tens and hundreds of millions and cause progressive destruction under conditions when the stress is below the material's yield point.

One of the most important elements of modern structures, which can be destroyed due to multicycle fatigue, can be attributed to thin-walled cylindrical shells. The problem of the destruction of shells due to fatigue under conditions of biaxial cyclic loading, when tensile-compressive/torsional or bending/torsional stresses are acting, is particularly relevant. In this case, the solution to strength problems concerning fatigue of shells and other typical structural elements is built based on the analysis of experimental data using various empirical relations [1, 4–6].

In this work, the fatigue life of thin-walled cylindrical shells under biaxial combined loading is determined based on equivalent stresses using classical failure criteria.

V. Golub (✉)
S. P. Timoshenko Institute of Mechanics, National Academy of Science of Ukraine, Kyiv, Ukraine
e-mail: creep@inmech.kiev.ua

H. Altenbach et al. (eds.), *Selected Problems of Solid Mechanics and Solving Methods*, Advanced Structured Materials 204, https://doi.org/10.1007/978-3-031-54063-9_13

13.2 Statement of the Problem. Basic Relations

Fatigue failure of thin-walled cylindrical shells (Fig. 13.1) is considered under the conditions of combined loading with (i) symmetric cyclic tension-compression by force $\tilde{N}$ and symmetric cyclic torsion by moment $\tilde{M}_\tau$, (ii) symmetric cyclic bending by moment $\tilde{M}_\varphi$ and symmetric cyclic torsion by moment $\tilde{M}_\tau$ applied at the end-walls of the shells.

We denote the diameter of the shell's middle surface by D and the wall thickness by $h \ll D$ so that the radial stress can be neglected. We also assume that the shell is long enough and its middle zone is in a momentless statically determined stress state. Cyclic axial $\tilde{\sigma}$, cyclic longitudinal bending $\tilde{\sigma}_\varphi$, and cyclic tangential $\tilde{\tau}$ stresses are determined by the following formulas:

$$\begin{aligned}
\tilde{\sigma} &= \frac{\tilde{N}_x}{\pi Dh} = \frac{N_{xa}}{\pi Dh}\sin 2\pi n = \sigma_{xa}\sin 2\pi n,\\
\tilde{\sigma}_\varphi &= \frac{2\tilde{M}_\varphi}{\pi Dh} = \frac{2M_{\varphi a}}{\pi Dh}\sin 2\pi n = \sigma_{\varphi a}\sin 2\pi n,\\
\tilde{\tau} &= \frac{2\tilde{M}_\tau}{\pi D^2 h} = \frac{2M_{\tau a}}{\pi D^2 h}\sin 2\pi n = \tau_a \sin 2\pi n,
\end{aligned} \tag{13.1}$$

where N_{xa}, $M_{\varphi a}$, $M_{\tau a}$, σ_{xa}, $\sigma_{\varphi a}$, and τ_a are the amplitude values of the corresponding load and stress components; n is the number of load cycles.

The maximum normal stress criterion, the maximum tangential stress criterion, and the specific deformation energy criterion are considered as fatigue failure criteria of thin-walled cylindrical shells under the conditions of biaxial combined loading (13.1), which determine the endurance of the shells. It is believed that the failure criteria should satisfy the critical value of the fatigue characteristics, as the bounded limits of fatigue $\sigma_n(n_R)$ under conditions of uniaxial tensile-compressive loading are considered.

The criterion of maximum normal stresses (Rankine criterion) specifies the strength condition for fatigue under biaxial combined loading in the form

$$\sigma_{1a} = \frac{\sigma_a}{2} + \sqrt{\left(\frac{\sigma_a}{2}\right)^2 + \tau_a^2} = \sigma_n(n_R), \tag{13.2}$$

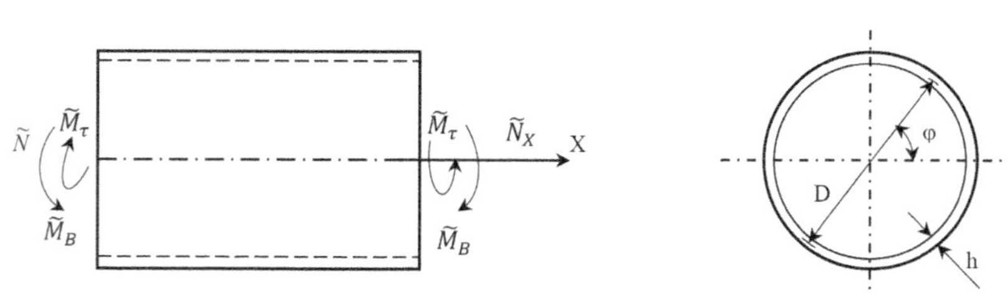

Fig. 13.1 Geometry of thin-walled cylindrical shells under biaxial combined loading

the criterion of maximum tangential stresses (Tresca criterion) has the form

$$\sigma_{1a} - \sigma_{3a} = 2\sqrt{\left(\frac{\sigma_a}{2}\right)^2 + \tau_a^2} = \sigma_n(n_R) \tag{13.3}$$

and the criterion of the specific strain energy for distortion (von Mises criterion) reads

$$\frac{1}{\sqrt{2}}\sqrt{2\sigma_a^2 + 6\tau_a^2} = \sigma_n(n_R), \tag{13.4}$$

where σ_{1a} and σ_{3a} are the amplitude values of the principal stresses.

The function $\sigma_n(n_R)$ that specifies the relationship between the bounded fatigue limit σ_n and the number of cycles to failure n_R under uniaxial tension-compression loading is specified as

$$n_R = \frac{1}{(1+q_\sigma)D_\sigma(\sigma_n)^{q_\sigma}} \rightarrow \sigma_n = \frac{1}{[(1+q_\sigma)D_\sigma n_R]^{1/q_\sigma}}, \tag{13.5}$$

which satisfies the power law of fatigue damage accumulation.

The task consists in the development of a calculation method and in solving the problems of calculating the number of cycles to failure of thin-walled cylindrical shells due to multicycle fatigue under the conditions of biaxial combined tension-compression and torsion, bending and torsion, and in the experimental approval of the calculation results.

13.3 Basic Resolving Equations of the Fatigue Endurance Calculation Method

Based on the failure criteria, a system of equations establishing the dependence of the number of cycles to failure of thin-walled cylindrical shells on the amplitudes of the normal and tangential stresses of the biaxial combined load is formulated.

13.3.1 Equations Based on the Rankine Criterion

Let us present Eq. (13.2), which specifies the strength condition due to fatigue under conditions of biaxial loading based on the criterion of maximum normal stresses, in the form of the ratio

$$\frac{\sigma_a}{\sigma_n(n_R)} + \left[\frac{\tau_a}{\sigma_n(n_R)}\right]^2 = 1, \tag{13.6}$$

from which, taking into account (13.5) for dependence n_R on σ_a, we obtain the equation

$$n_R = \frac{1}{(1+q_\sigma)D_\sigma(\sigma_a)^{q_\sigma}}\left\{1-\left\{\ [(1+q_\sigma)D_\sigma n_R]^{1/q_\sigma}\ \nu\sigma_a\right\}^2\right\}^{q_\sigma}, \tag{13.7}$$

and for n_R dependence on τ_a–equation

$$n_R = \frac{1}{(1+q_\sigma)D_\sigma\left(\frac{\tau_a}{\nu}\right)^{q_\sigma}}\left\{1-\left\{\ [(1+q_\sigma)D_\sigma n_R]^{1/q_\sigma}\ \tau_a\right\}^2\right\}^{q_\sigma}, \tag{13.8}$$

where it is accepted that $\nu = \tau_a/\sigma_a$.

13.3.2 Equations Based on the Tresca Criterion

We will submit (13.3), which sets the condition of strength due to fatigue under conditions of biaxial loading on the basis of the criterion of maximum tangential stresses in the form of a ratio

$$\left[\frac{\sigma_a}{\sigma_n(n_R)}\right]^2+\left[\frac{2\tau_a}{\sigma_n(n_R)}\right]^2=1, \tag{13.9}$$

from which, taking into account (13.5) for dependence n_R on σ_a, we obtain the equation

$$n_R = \frac{1}{(1+q_\sigma)D_\sigma(\sigma_a)^{q_\sigma}}\left\{1-\left\{\ [(1+q_\sigma)\,D_\sigma n_R]^{1/q_\sigma}\ 2\nu\sigma_a\right\}^2\right\}^{q_\sigma/2}, \tag{13.10}$$

and for dependence n_R on τ_a–equation

$$n_R = \frac{1}{(1+q_\sigma)D_\sigma\left(\frac{\tau_a}{\nu}\right)^{q_\sigma}}\left\{1-\left\{[(1+q_\sigma)D_\sigma n_R]^{1/q_\sigma}\ 2\tau_a\right\}^2\right\}^{q_\sigma/2}, \tag{13.11}$$

where it is accepted that $\nu = \tau_a/\sigma_a$.

13.3.3 Equations Based on the von Mises Criterion

We submit (13.4), which sets the condition of strength due to fatigue under conditions of biaxial loading on the basis of the criterion of specific strain energy due to distortion in the form of a ratio

$$\left[\frac{\sigma_a}{\sigma_n(n_R)}\right]^2 + \left[\frac{\sqrt{3}\tau_a}{\sigma_n(n_R)}\right]^2 = 1, \tag{13.12}$$

from which, taking into account (13.5), for the dependence n_R on σ_a, we obtain the equation

$$n_R = \frac{1}{(1+q_\sigma)D_\sigma(\sigma_a)^{q_\sigma}}\left\{1 - \left\{[(1+q_\sigma)D_\sigma n_R]^{1/q_\sigma}\sqrt{3}\nu\sigma_a\right\}^2\right\}^{q_\sigma/2}, \tag{13.13}$$

and for dependence n_R on τ_a–equation

$$n_R = \frac{1}{(1+q_\sigma)D_\sigma\left(\frac{\tau_a}{\nu}\right)^{q_\sigma}}\left\{1 - \left\{[(1+q_\sigma)D_\sigma n_R]^{1/q_\sigma}\sqrt{3}\tau_a\right\}^2\right\}^{q_\sigma/2}, \tag{13.14}$$

where it is accepted that $\nu = \tau_a/\sigma_a$.

13.4 Fatigue Endurance of Thin-Walled Cylindrical Shells

The problems of calculating the number of cycles to failure of thin-walled cylindrical shells under conditions of combined tension-compression and torsion, bending and torsion are solved.

13.4.1 Materials of Shells. Material Constants

Steel 45, JIS SNCM8 steel, steel 20, and steel St 460, which are widely used in energy and aviation, are considered shell materials. The values of the material constants q_σ and D_σ, which are included in (13.7)–(13.14), are determined based on the results of approximation of the experimental data of fatigue tests of thin-walled tubular specimens under conditions of uniaxial tension-compression. The task of determining the constants q_σ and D_σ is reduced to the minimization of the functional

$$\Phi(q_\sigma D_\sigma) = \sum_{j=1}^{s}\left\{n_{Rj}(\sigma_{aj}) - \frac{1}{(1+q_\sigma)D_\sigma(\sigma_a)^{q_\sigma}}\right\}^2, \tag{13.15}$$

where (13.5) is used as an approximating function. Here σ_{aj} and n_{Rj} are a set of discrete values of the amplitudes of normal cyclic tension-compression stresses and the corresponding numbers of cycles to failure.

The values of the material constants q_σ and D_σ for the shell materials, calculated according to the method described above using condition (13.15), are given in

Table 13.1 The value of material constants

Material constants	Materials of shells			
q_σ	Steel 45	Steel JIS SNCM8	Steel 20	Steel St 460
	15.87	12.19	3.10	5.46
D_σ, MPa^{-q_σ}, cycle^{-1}	$2.24 \cdot 10^{-45}$	$6.27 \cdot 10^{-40}$	$2.99 \cdot 10^{-13}$	$2.49 \cdot 10^{-19}$
$\mu = \tau_n(n_R)/\sigma_n(n_R)$	0.68	0.67	1.2–1.8	0.72

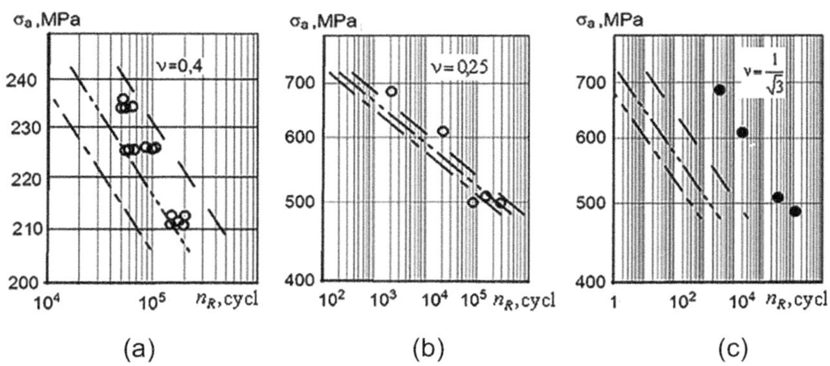

Fig. 13.2 Calculated (lines) and experimental (markers) fatigue life values of thin-walled cylindrical shells made of steel 45 **a** and steel JIS SNCM8 **b**, **c** under combined cyclic tension-compression and cyclic torsion loading: (——)–calculation according to Eq. (13.7); (—•—)–calculation according to Eq. (13.10); (—••—)–calculation according to Eq. (13.13)

Table 13.1. The experimental data used in determining the material constants of the shells are taken from [4–6].

13.4.2 Combined Loading by Cyclic Tension-Compression and Cyclic Torsion

The dependence of the number of cycles to failure of shells n_R on the amplitude of cyclic normal stresses σ_a at $\nu = \text{const}$ is calculated using (13.7), (13.10), and (13.13). The materials of the shells are steel 45 and steel JIS SNCM8. The results of calculations fulfilled using the values of material constants given in the table are compared in Fig. 13.2 with experimental data. Calculation results are plotted with lines, and experimental data are plotted with markers. Equations (13.7), (13.10), and (13.13) are solved numerically by the iteration method. Experimental data are taken from [4, 5].

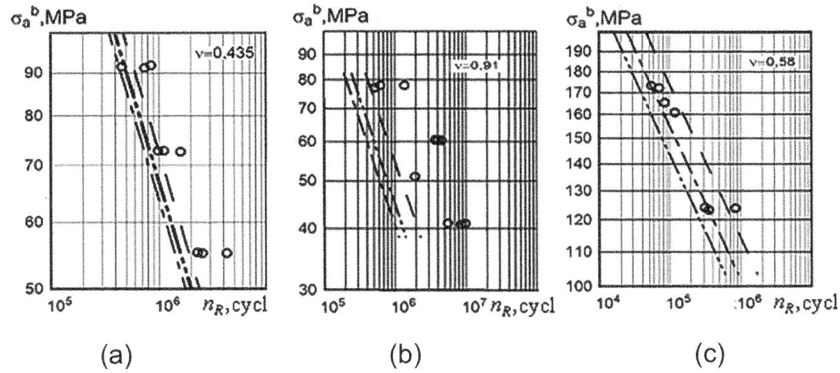

Fig. 13.3 Calculated (lines) and experimental (markers) fatigue life values of thin-walled cylindrical shells made of steel 20 **a**, **b** and steel StE 460 **c** under combined cyclic bending and cyclic torsion loading: (——)–calculation according to Eq. (13.7); (—•—)–calculation according to Eq. (13.10); (—••—)–calculation according to Eq. (13.13)

13.4.3 Combined Loading by Cyclic Bending and Cyclic Torsion

The dependence of the number of cycles failure of shells n_R on the amplitude of bending cyclic stresses σ_a at $\nu = \text{const}$ is calculated using (13.7), (13.10), and (13.13). The materials of the shells are steel 20 and steel St 460. The results of calculations fulfilled using the values of material constants given in the table are compared in Fig. 13.3 with experimental data. Calculation results are plotted with lines, and experimental data are plotted with markers. Equations (13.7), (13.10), and (13.13) are solved numerically by the iteration method. Experimental data are taken from [5, 6].

13.5 Analysis of the Results

The solution to the problems of fatigue endurance calculation of thin-walled cylindrical shells under the conditions of biaxial combined loading is constructed in the work based on the concept of equivalent stresses. Classical failure criteria are considered as equivalent stresses, including the criterion of maximum normal stresses, the criterion of maximum tangential stresses, and the criterion of the specific strain energy due to distortion. The results of the calculations are compared with the experimental data taken from the literature.

The effectiveness of using classical failure criteria for solving the fatigue problems of thin-walled cylindrical shells under conditions of biaxial combined loading is analyzed on the problems of calculating of the number of cycles to failure of shells under the conditions of the combined action of cyclic tension-compression and cyclic torsion, cyclic bending, and cyclic torsion. In general, as evidenced by the data shown in Figs. 13.2 and 13.3, a completely satisfactory agreement of the calculation results with the experiment was obtained.

The best agreement with experimental data was obtained for calculations fulfilled according to the criterion of the specific strain energy due to distortion regardless of the nature of the load (shells made of steels 45, JIS SNCM8, St 460). In some cases (shells made of steel 20), a satisfactory agreement with the experimental data and the calculation results was obtained, which was fulfilled in accordance with the criterion of maximum normal stresses. Worse agreement with experimental data was obtained for calculations fulfilled according to the criterion of maximum tangential stresses, regardless of the nature of the combined load.

13.6 Conclusions

The solution to the problems of the calculation of fatigue endurance of thin-walled cylindrical shells under conditions of biaxial combined loading can be built on the basis of classical failure criteria. The accuracy of the calculations depends, first of all, on the degree of compliance of the fatigue failure criterion with the initial plastic properties of the shell material. These properties are given by the average values of the coefficient $\mu = \tau_n(n_R)/\sigma_n(n_R)$, which for the studied materials varies from 0.67 to 1.2–1.8. For plastic materials, the best agreement with experimental data was obtained for the criterion of the specific strain energy due to distortion. The coefficient μ for such materials can vary from 0.577 to 0.80. For brittle materials, the best agreement with experimental data is obtained for the maximum normal stress criterion. The coefficient μ for such materials can vary from 0.8 to 1.2.

References

1. Heywood, R.B.: Designing Against Fatigue. Chapman and Hall Ltd., London (1962)
2. Serensen, S.V., Kogaev, V.P., Schnejderovich, R.M.: The Supporting Power and the Structural Design of Machine Elements. Machinostrojenije, Moscow (1975). [in Russian]
3. Bolotin, V.V.: The Service Life of Machines and Construction. Machinostrojenije, Moscow (1990). [in Russian]
4. Tanaka, K., Matsuoka, S.: The strength of JIS SNCM8 steel under combined alternating stresses. In: Advances in Research on the Strength and Fracture of Materials, vol. 28, pp. 1161–1168. Pergamon Press, New York (1978)
5. Panfilov, Y.A.: Description of the limit state in variable biaxial stress and its application in calculations communication 1 simple loading. Strength Mater. **13**, 51–55 (1981)
6. Sonsino, C.M., Kueppers, M.: Multiaxial fatigue of welded joints under constant and variable amplitude loading. Fatigue Fract. Eng. Mater. Struct. **24**(5), 309–327 (2001)

Chapter 14
Stress State of Corrugated Shells with Oblique Cuts

Alexander Grigorenko, Mykola Kryukov, Wolfgang H. Müller, and Serhii Yaremchenko

Abstract An approach to solving problems of the stress-strain state of corrugated cylindrical shells with oblique cuts is proposed. The equations of the shell theory based on the straight line hypothesis were chosen. The approach to the solution is based on the parameterization of the surface of the shell, the reduction of the two-dimensional boundary value problem to the one-dimensional one by the spline collocation method, and the solution of the latter by the stable numerical method of discrete orthogonalization. The influence of geometric parameters on the displacement fields of corrugated shells with beveled sections is analyzed.

14.1 Introduction

The analysis of corrugated shells poses an intriguing problem in mechanics, given their widespread application in construction and engineering [9]. Including corrugations in shells induces notable changes in their structural characteristics, leading to increased bending stiffness and the ability to reduce shell thickness. As a result, extensive research has been conducted on the stress-strain state, stability, and natural vibrations of shells with corrugations oriented in various directions [2, 4, 6, 7, 10, 12].

This article investigates the stress-deformed state of cylindrical shells featuring corrugations in the cross-section with beveled sections. The analysis is based on the

A. Grigorenko (✉) · S. Yaremchenko
S.P. Timoshenko Institute of Mechanics, National Academy of Science of Ukraine, Kyiv, Ukraine
e-mail: ayagrigorenko1991@gmail.com

S. Yaremchenko
e-mail: yaremch@gmail.com

M. Kryukov
State University of Infrastructure and Technologies, Kyiv, Ukraine
e-mail: mmkryukov@ukr.net

W. H. Müller
Institute of Mechanics, Technische Universität Berlin, Berlin, Germany
e-mail: wolfgang.h.mueller@tu-berlin.de

H. Altenbach et al. (eds.), *Selected Problems of Solid Mechanics and Solving Methods*, Advanced Structured Materials 204, https://doi.org/10.1007/978-3-031-54063-9_14

equations derived from the refined theory of shells, specifically the Timoshenko type. The shell's shape is precisely defined in non-orthogonal coordinates. The equations are transformed into a rectangular domain through a variable substitution [3, 5, 8]. This transformation facilitates the application of the spline collocation method to solve the resulting two-dimensional boundary value problem, ultimately reducing it to a more manageable one-dimensional problem. The solution to the one-dimensional boundary value problem is performed using the method of discrete orthogonalization. This approach enhances the clarity and efficiency of the analysis, providing valuable insights into the stress-deformed behavior of cylindrical shells with beveled corrugations.

14.2 Basic Relations

Let us consider corrugated along directrix cylindrical shells, the cross-section of which is defined in polar coordinates by the formula:

$$r(\theta) = r_0 + \alpha \cos(m\theta), \tag{14.1}$$

where r_0 is base circle radius, θ is polar angle, α is amplitude of corrugation, and m is corrugation pace. The midsurface of corrugated cylindrical shells with oblique cuts (Fig. 14.1) is given in Cartesian coordinates as

$$\begin{cases} x_1 = r\cos(\theta), \\ x_2 = r\sin(\theta), \\ x_3 = \xi L\left[L + r(\cos\theta - 1)(\tan\beta_1 + \tan\beta_2)\right] + r(1 - \cos\theta)\tan\beta_1, \end{cases} \tag{14.2}$$

$0 \leqslant \xi \leqslant L, 0 \leqslant \theta \leqslant 2\pi, \beta_1; \beta_2$ are the oblique cuts angles.

Let the shell be subjected to the influence of uniform pressure q. Let us investigate the stress-strain state of such a shell using refined shell theory based on straight line hypothesis. We obtain linear law for displacement distribution along the thickness according to the Timoshenko hypothesis:

$$\begin{aligned} \tilde{u}_1(\alpha_1, \alpha_2, \alpha_3) &= u_1(\alpha_1, \alpha_2) + \alpha_3\psi_1, \\ \tilde{u}_2(\alpha_1, \alpha_2, \alpha_3) &= u_2(\alpha_1, \alpha_2) + \alpha_3\psi_2, \\ \tilde{u}_3(\alpha_1, \alpha_2, \alpha_3) &= w(\alpha_1, \alpha_2). \end{aligned} \tag{14.3}$$

Here u_1, u_2, and w are displacements of points of coordinate surface in orthogonal curvilinear coordinates α_1, α_2, and α_3, respectively; ψ_1 and ψ_2 are total angles of rotation of the rectilinear element.

Expressions for deformations are

$$\begin{aligned}
e_1(\alpha_1,\alpha_2,\alpha_3) &= \varepsilon_1(\alpha_1,\alpha_2)+\alpha_3\varkappa_1(\alpha_1,\alpha_2),\\
e_2(\alpha_1,\alpha_2,\alpha_3) &= \varepsilon_2(\alpha_1,\alpha_2)+\alpha_3\varkappa_2(\alpha_1,\alpha_2),\\
e_{12}(\alpha_1,\alpha_2,\alpha_3) &= \varepsilon_{12}(\alpha_1,\alpha_2)+\alpha_3 2\varkappa_{12}(\alpha_1,\alpha_2),\\
e_{13}(\alpha_1,\alpha_2,\alpha_3) &= \gamma_1(\alpha_1,\alpha_2),\\
e_{23}(\alpha_1,\alpha_2,\alpha_3) &= \gamma_2(\alpha_1,\alpha_2).
\end{aligned} \tag{14.4}$$

The first fundamental form coefficient $A_1 = 1$ and curvature $k_1 = 0$ for noncircular cylinders; that is why

$$\begin{aligned}
&\varepsilon_1=\frac{\partial u_1}{\partial\alpha_1},\quad \varepsilon_2=\frac{1}{A_2}\frac{\partial u_2}{\partial\alpha_2}+k_2w,\quad \varepsilon_{12}=\frac{\partial}{\partial u_2\alpha_1}+\frac{1}{A_2}\frac{\partial u_1}{\partial\alpha_2},\\
&\varkappa_1=\frac{\partial\psi_1}{\partial\alpha_1},\quad \varkappa_2=\frac{1}{A_2}\frac{\partial\psi_2}{\partial\alpha_2}-k_2\left(\frac{1}{A_2}\frac{\partial u_2}{\partial\alpha_2}+k_2w\right),\quad 2\varkappa_{12}=\frac{1}{A_2}\frac{\partial\psi_1}{\partial\alpha_2}+\frac{\partial\psi_2}{\partial\alpha_1}-\frac{k_2}{A_2}\frac{\partial u_1}{\partial\alpha_2},\\
&\gamma_1=\psi_1+\frac{\partial w}{\partial\alpha_1},\quad \gamma_2=\psi_2+\frac{1}{A_2}\frac{\partial w}{\partial\alpha_2}-k_2u_2.
\end{aligned} \tag{14.5}$$

Quantities included in (14.5) have such geometrical sense: ε_1 and ε_2 characterize tension (compression) of midsurface in directions α_1 and α_2; ε_{12} is shear deformation; $\varkappa_1$, $\varkappa_2$, and $\varkappa_{12}$ are bending and torsion deformations; γ_1 and γ_2 are angles of rotation of rectilinear element obtained as a result of transverse shears. Coefficient of first fundamental form A_2 and curvature k_2 depend on polar radius:

$$A_2=\sqrt{r^2+(r')^2},\quad k_2=\frac{r^2+2(r')^2-rr''}{[r^2+(r')^2]^{3/2}}.$$

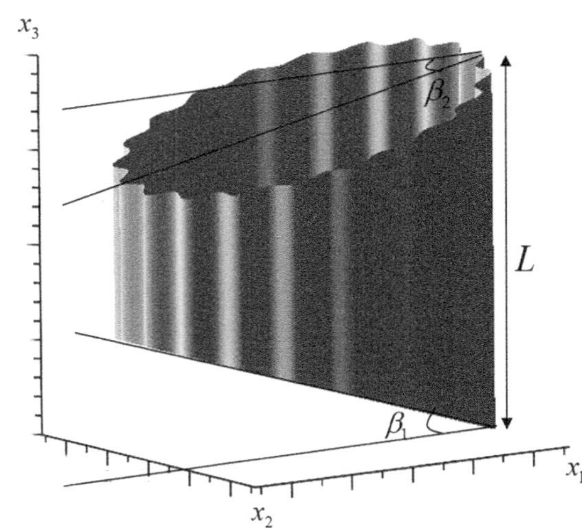

Fig. 14.1 Midsurface of corrugated shell with oblique cuts

Equations of equilibrium:

$$
\begin{gathered}
\frac{\partial N_1}{\partial \alpha_1}+\frac{1}{A_2}\frac{\partial N_{21}}{\partial \alpha_2}=0,\quad \frac{1}{A_2}\frac{\partial N_2}{\partial \alpha_2}+\frac{\partial N_{12}}{\partial \alpha_1}+k_2Q_2=0,\\
\frac{\partial Q_1}{\partial \alpha_1}+\frac{1}{A_2}\frac{\partial Q_2}{\partial \alpha_2}-k_2N_2+q=0,\quad \frac{\partial M_1}{\partial \alpha_1}+\frac{1}{A_2}\frac{\partial M_{21}}{\partial \alpha_2}-Q_1=0,\\
\frac{1}{A_2}\frac{\partial M_2}{\partial \alpha_2}+\frac{\partial M_{12}}{\partial \alpha_1}-Q_2=0,
\end{gathered}
\tag{14.6}
$$

where N_1 and N_2 are normal tangent forces; N_{12} and N_{21} are shear tangent forces; Q_1 and Q_2 are transverse forces; M_1 and M_2 are bending moments; M_{12} and M_{21} are torsional moments.

Elastic relations for orthotropic shells, when the coordinate surface is midsurface, are the following:

$$
\begin{aligned}
&N_1=C_{11}\varepsilon_1+C_{12}\varepsilon_2, && N_2=C_{12}\varepsilon_1+C_{22}\varepsilon_2,\\
&N_{12}=C_{66}\varepsilon_{12}+2k_2D_{66}\varkappa_{12}, && N_{21}=C_{66}\varepsilon_{12},\\
&M_1=D_{11}\varkappa_1+D_{12}\varkappa_2, && M_2=D_{12}\varkappa_1+D_{22}\varkappa_2,\\
&M_{21}=M_{12}=2D_{66}\varkappa_{12}, && Q_1=K_1\gamma_1,\quad Q_2=K_2\gamma_2,
\end{aligned}
\tag{14.7}
$$

where

$$
\begin{aligned}
&C_{11}=\frac{E_1h}{1-\nu_1\nu_2},\quad C_{12}=\nu_2C_{11},\quad C_{22}=\frac{E_2h}{1-\nu_1\nu_2},\\
&C_{66}=G_{12}h,\quad D_{11}=\frac{E_1h^3}{12(1-\nu_1\nu_2)},\quad D_{12}=\nu_2D_{11},\\
&D_{22}=\frac{E_2h^3}{12(1-\nu_1\nu_2)},\quad D_{66}=\frac{1}{12}G_{12}h^3,\quad K_1=\frac{5}{6}hG_{13},\quad K_2=\frac{5}{6}hG_{23}.
\end{aligned}
$$

Here C_{ij}, K_i, and D_{ij} are stiffness characteristics of shell defined by thickness h and elastic properties of material E_i, ν_i, and G_{ij}.

14.3 Resolving Equations

Let us choose as desired function displacements u_1, u_2, w and total rotation angles ψ_1, ψ_2. Substituting (14.5) into elastic relations (14.7) and then substituting obtained expressions into (14.6), we obtain

$$
\begin{aligned}
&C_{11}\partial^2 u_1 \partial\alpha_1^2 - \frac{C_{66}A_2'}{A_2^3}\frac{\partial u_1}{\partial\alpha_2} + \frac{C_{66}}{A_2^2}\frac{\partial^2 u_1}{\partial\alpha_2^2} + \frac{C_{66}+C_{12}}{A_2}\frac{\partial^2 u_2}{\partial\alpha_1\partial\alpha_2} + C_{12}k_2\frac{\partial w}{\partial\alpha_1} = 0,\\
&\frac{C_{66}+C_{12}-k_2^2 D_{66}}{A_2}\frac{\partial^2 u_1}{\partial\alpha_1\partial\alpha_2} - k_2^2 K_2 u_2 + C_{66}\frac{\partial^2 u_2}{\partial\alpha_1^2} - \frac{C_{22}A_2'}{A_2^3}\frac{\partial^2 u_2}{\partial\alpha_1\partial\alpha_2}\\
&\qquad + \frac{C_{22}}{A_2^2}\frac{\partial^2 u_2}{\partial\alpha_2^2} + \frac{C_{22}k_2'}{A_2}w\frac{k_2(K_2+C_{22})}{A_2}\frac{\partial w}{\partial\alpha_2}\\
&+\frac{C_{22}}{A_2^2}\frac{\partial^2 u_2}{\partial\alpha_2^2} + \frac{C_{22}}{A_2^2}\frac{\partial^2 u_2}{\partial\alpha_2^2} + \frac{k_2 D_{66}}{A_2}\frac{\partial^2\psi_1}{\partial\alpha_1\partial\alpha_2} + k_2 K_2\psi_2 + k_2 D_{66}\frac{\partial^2\psi_2}{\partial\alpha_1^2} = 0,\\
&\qquad -k_2 C_{12}\frac{\partial u_1}{\partial\alpha_1} - \frac{k_2' K_2}{A_2}u_2 - \frac{k_2(K_2+C_{22})}{A_2}\frac{\partial u_2}{\partial\alpha_2} - k_2^2 C_{22} w\\
&\qquad +K_1\frac{\partial^2 w}{\partial\alpha_1^2} - \frac{K_2 A_2'}{A_2^3}\frac{\partial w}{\partial\alpha_2} + \frac{K_2}{A_2^2}\frac{\partial^2 w}{\partial\alpha_2^2} + K_1\frac{\partial\psi_1}{\partial\alpha_1} + \frac{K_2}{A_2}\frac{\partial\psi_2}{\partial\alpha_2} + q = 0,\\
&\frac{k_2 D_{66}A_2' - k_2' D_{66}A_2}{A_2^3}\frac{\partial u_1}{\partial\alpha_2} - \frac{k_2 D_{66}}{A_2^2}\frac{\partial^2 u_1}{\partial\alpha_2^2} - \frac{k_2 D_{12}}{A_2}\frac{\partial^2 u_2}{\partial\alpha_1\partial\alpha_2} - (k_2^2 D_{12}+K_1)\frac{\partial w}{\partial\alpha_1}\\
&\quad -K_1\psi_1 + D_{11}\frac{\partial^2\psi_1}{\partial\alpha_1^2} - \frac{D_{66}A_2'}{A_2^3}\frac{\partial\psi_1}{\partial\alpha_2} + \frac{D_{66}}{A_2^2}\frac{\partial^2\psi_1}{\partial\alpha_2^2} + \frac{D_{12}+D_{66}}{A_2}\frac{\partial^2\psi_2}{\partial\alpha_1\partial\alpha_2} = 0,\\
&\quad -\frac{k_2 D_{66}}{A_2}\frac{\partial^2 u_1}{\partial\alpha_1\alpha_2} + k_2 K_2 u_2 + \frac{k_2 D_{22}A_2' - k_2' D_{22}A_2}{A_2^3}\frac{\partial u_2}{\partial\alpha_2} - \frac{k_2 D_{22}}{A_2}\frac{\partial^2 u_2}{\partial\alpha_2^2}\\
&\qquad -\frac{2k_2 k_2' D_{22}}{A_2}w - \frac{k_2^2 D_{22}+K_2}{A_2}\frac{\partial w}{\partial\alpha_2} + \frac{D_{12}+D_{66}}{A_2}\frac{\partial^2\psi_1}{\partial\alpha_1\partial\alpha_2}\\
&\qquad -K_2\psi_2 + D_{66}\frac{\partial^2\psi_2}{\partial\alpha_1^2} - \frac{D_{22}A_2'}{A_2^3}\frac{\partial\psi_2}{\partial\alpha_2} + \frac{D_{22}}{A_2^2}\frac{\partial^2\psi_2}{\partial\alpha_2^2} = 0.
\end{aligned}
\tag{14.8}
$$

Then, to describe the stress-strain state of corrugated shells with oblique cuts according to (14.2), we transform variables in Eq. (14.8). New coordinates ξ, θ are related to α_1, α_2 as follows:

$$
\alpha_1 = \frac{\xi}{L}\left[L + r(\cos\theta - 1)(\tan\beta_1 + \tan\beta_2)\right] + r(1-\cos\theta)\tan\beta_1, \quad \alpha_2 = \theta. \tag{14.9}
$$

In new coordinates, let us represent Eq. (14.8) as

$$
\begin{aligned}
\frac{\partial^2 u_1}{\partial\theta^2} &= a_{11}\frac{\partial u_1}{\partial\xi}. + a_{12}\frac{\partial^2 u_1}{\partial\theta} + a_{13}\frac{\partial u_1}{\partial\theta} + a_{14}\frac{\partial^2 u_1}{\partial\theta\partial\xi} + a_{15}\frac{\partial u_2}{\partial\xi}\\
&\quad + a_{16}\frac{\partial^2 u_2}{\partial\xi^2} + a_{17}\frac{\partial^2 u_2}{\partial\theta\partial\xi} + a_{18}\frac{\partial w}{\partial\xi},\\
\frac{\partial^2 u_2}{\partial\theta^2} &= a_{21}\frac{\partial u_1}{\partial\xi} + a_{22}\frac{\partial^2 u_1}{\partial\xi^2} + a_{23}\frac{\partial^2 u_1}{\partial\theta\partial\xi} + a_{24}u_2 + a_{25}\frac{\partial u_2}{\partial\xi} + a_{26}\frac{\partial^2 u_2}{\partial\xi^2}\\
&\quad + a_{27}\frac{\partial u_2}{\partial\theta} + a_{28}\frac{\partial^2 u_2}{\partial\theta\partial\xi} + a_{29}w + a_{2,10}\frac{\partial w}{\partial\xi} + a_{2,11}\frac{\partial w}{\partial\theta} + a_{2,12}\frac{\partial\psi_1}{\partial\xi}\\
&\quad + a_{2,13}\frac{\partial^2\psi_1}{\partial\theta} + a_{2,14}\frac{\partial^2\psi_1}{\partial\theta\partial\xi} + a_{2,15}\psi_2 + a_{2,16}\frac{\partial^2\psi_2}{\partial\xi^2},
\end{aligned}
$$

$$
\begin{aligned}
\frac{\partial^2 w}{\partial \theta^2} &= a_{31}\frac{\partial u_1}{\partial \xi} + a_{32}u_2 + a_{33}\frac{\partial u_2}{\partial \xi} + a_{34}\frac{\partial u_2}{\partial \theta} + a_{35}w + a_{36}\frac{\partial w}{\partial \xi} + a_{37}\frac{\partial^2 w}{\partial \xi^2} \\
&\quad + a_{38}\frac{\partial w}{\partial \theta} + a_{39}\frac{\partial^2 w}{\partial \theta \partial \xi} + a_{3,10}\frac{\partial \psi_1}{\partial \xi} + a_{3,11}\frac{\partial \psi_2}{\partial \xi} + a_{3,12}\frac{\partial \psi_2}{\partial \theta} + a_{3,13}q, \\
\frac{\partial^2 \psi_1}{\partial \theta^2} &= a_{41}\frac{\partial u_1}{\partial \xi} + a_{42}\frac{\partial^2 u_1}{\partial \xi^2} + a_{43}\frac{\partial u_1}{\partial \theta} + a_{44}\frac{\partial u_2}{\partial \xi} + a_{45}\frac{\partial^2 u_2}{\partial \xi^2} + a_{46}\frac{\partial^2 u_2}{\partial \theta \partial \xi} \\
&\quad + a_{47}\frac{\partial w}{\partial \xi} + a_{48}\psi_1 + a_{49}\frac{\partial \psi_1}{\partial \xi} + a_{4,10}\frac{\partial^2 \psi_1}{\partial \xi^2} + a_{4,11}\frac{\partial \psi_1}{\partial \theta} + a_{4,12}\frac{\partial^2 \psi_1}{\partial \xi \partial \theta} \\
&\quad + a_{4,13}\frac{\partial \psi_2}{\partial \xi} + a_{4,14}\frac{\partial^2 \psi_2}{\partial \xi^2} + a_{4,15}\frac{\partial^2 \psi_2}{\partial \xi \partial \theta}, \\
\frac{\partial^2 \psi_2}{\partial \theta^2} &= a_{51}\frac{\partial u_1}{\partial \xi} + a_{52}\frac{\partial^2 u_1}{\partial \xi^2} + a_{53}\frac{\partial^2 u_1}{\partial \theta \partial \xi} + a_{54}u_2 + a_{55}\frac{\partial u_2}{\partial \xi} + a_{56}\frac{\partial^2 u_2}{\partial \xi^2} \\
&\quad + a_{57}\frac{\partial u_2}{\partial \theta} + a_{58}w + a_{59}\frac{\partial w}{\partial \xi} + a_{5,10}\frac{\partial w}{\partial \theta} + a_{5,11}\frac{\partial \psi_1}{\partial \xi} + a_{5,12}\frac{\partial^2 \psi_1}{\partial \xi^2} + a_{5,13}\frac{\partial^2 \psi_1}{\partial \theta \partial \xi} \\
&\quad + a_{5,14}\psi_2 + a_{5,15}\frac{\partial \psi_2}{\partial \xi} + a_{5,16}\frac{\partial^2 \psi_2}{\partial \xi^2} + a_{5,17}\frac{\partial \psi_2}{\partial \theta} + a_{5,18}\frac{\partial^2 \psi_2}{\partial \theta \partial \xi}.
\end{aligned}
\tag{14.10}
$$

Coefficients a_{ij} depend on ξ and θ [5].

Let us add boundary conditions of clamped contours $\xi = 0$ and $\xi = L$ to (14.10):

$$u_1 = u_2 = w = 0, \quad \psi_1 = \psi_2 = 0, \tag{14.11}$$

and symmetry condition at $\theta = 0$ and $\theta = \pi$:

$$\psi_2 = 0, \quad u_2 = 0, \quad \frac{\partial u_1}{\partial \theta} = \frac{\partial w}{\partial \theta} = 0, \quad \frac{\partial \psi_1}{\partial \theta} = 0. \tag{14.12}$$

The two-dimensional boundary value problem is reduced to one-dimensional by the spline-collocation method [11] along ξ-coordinate. The one-dimensional boundary value problem is solved by the discrete orthogonalization method [1].

14.4 Numerical Results

Based on the proposed approach, an analysis of the stress-strain state of isotropic shells with the following parameters was carried out:

$$r_0 = 20, \quad L = 60, \quad h = 1, \quad \alpha = 0.1, \quad \nu = 0.3.$$

Figure 14.2 shows the distribution of displacements of shell with cut angles $\beta_1 = \beta_2 = 30°$ at different values of the corrugation pace in the section $\xi = L/2$ at $m = 4$ (dashed line), $m = 8$ (dotted line), $m = 12$ (dashed-dotted line), and $m = 16$ (solid line). We can see that at $m = 4$ the maximum displacement at the top of the corrugation exceeds the minimum in the cave by more than seven times. When the corrugation pace is increased, the differences between the values of deflections in

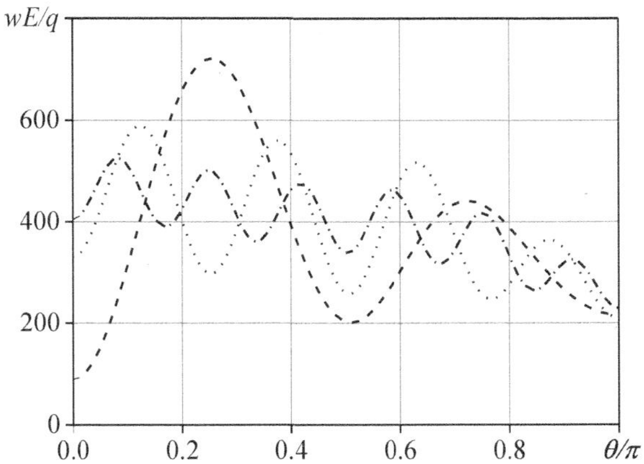

Fig. 14.2 Shell displacements at various values of the pace of corrugation

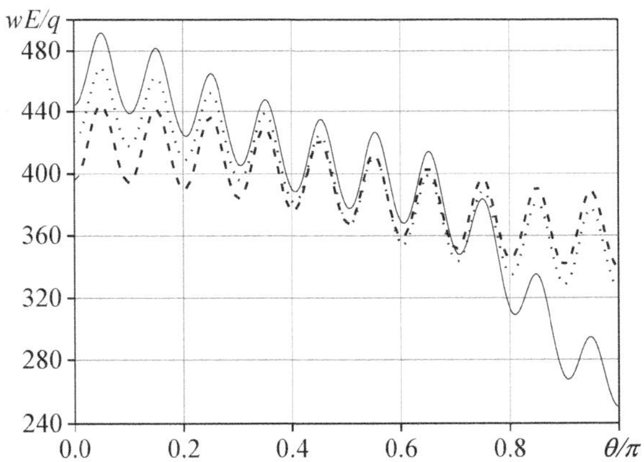

Fig. 14.3 Displacement distribution at various values of cut angles

the peaks are significantly reduced as the shell becomes more rigid. For $m = 8$, the values at the peaks differ by a factor of two, and the difference between them does not exceed 20% for $m = 16$.

Figure 14.3 shows the distribution of displacements of the shell with the corrugation step $m = 20$ in the section $\xi = L/2$ at different values of the cut angles $\beta_1 = \beta_2 = 10°$ (dashed line), $\beta_1 = \beta_2 = 20°$ (dotted line), and $\beta_1 = \beta_2 = 30°$ (solid line). We can see on the graphs for all cut angles in the neighborhood of $\theta = 0$ that the differences between the displacements depending on the cut angles are insignificant. For $\theta = \pi$, a sharp decrease in displacements by almost two times can be observed for angles of 30°, while the graphs for deflections at angles of 10° and 20° differ slightly.

14.5 Conclusions

By employing the refined theory of Timoshenko-type shells, we derived resolving equations to assess the stress-strain state of cylindrical corrugated shells subjected to a uniformly distributed load. We implemented a coordinate substitution to simplify the problem domain into a rectangular form. The resulting resolving equations enable the application of methods such as spline collocation and discrete orthogonalization for solving the associated two-dimensional boundary value problem.

This developed approach was effectively employed to scrutinize the impact of cut angles and corrugation spacing within the shell. Notably, an increase in corrugation spacing in shells with oblique cuts led to a reduction in maximum displacements and the disparity between displacements at the corrugation's top and bottom. Conversely, an elevation in cut angles resulted in a slight increase in displacements in the uncut section of the shell. Within the cut region, displacements exhibited a significant decrease with rising cut angles. These findings contribute valuable insights into the behavior of corrugated shells, offering a nuanced understanding of their response to varying geometric parameters.

References

1. Bellman, R., Kalaba, R.: Quasilinearization and Nonlinear Boundary-Value Problems. American Elsevier Publishing Company, New York (1965)
2. Bespalova, O.I., Boreiko, N.P.: Stability of shells of revolution of alternating gaussian curvature. Int. Appl. Mech. **58**(1), 53–62 (2022). https://doi.org/10.1007/s10778-022-01134-5
3. Civalek, O., Gurses, M.: Frequency analysis of trapezoidal plates and membrane using discrete singular convolution. Asian J. Civil Eng. (Build. Housing) **9**(6), 593–605 (2009)
4. Grigorenko, O.Y., Borisenko, M.Y., Boichuk, O.V.: Free vibrations of a corrugated closed cylindrical shell. Int. Appl. Mech. **58**(1), 43–52 (2022). https://doi.org/10.1007/s10778-022-01133-6
5. Grigorenko, Ya.M., Grigorenko, O.Y., Kryukov N.N., Yaremchenko S.N.: Stress-strain state of elliptic cross-section cylindrical shells with beveled cuts. Rep. NAS Ukraine **6**, 21–29. (2020). https://doi.org/10.15407/dopovidi2020.06.021
6. Grigorenko, Y.M., Rozhok, L.S.: Stress analysis of circumferentially corrugated hollow orthotropic cylinders. Int. Appl. Mech. **42**(12), 1389–1397 (2006). https://doi.org/10.1007/s10778-006-0209-6
7. Grigorenko, Y.M., Zakhariichenko, L.I.: Design of corrugated cylindrical shells under different end conditions. Int. Appl. Mech. **35**(9), 897–905 (1999). https://doi.org/10.1007/BF02682285
8. Kornishin, M.S., Paimyshin, V.N., Snegirev, V.F.: Computational Geometry in Problems of Shell Mechanics. Nauka, Moscow (1989). [in Russian]
9. Lai, M., Eugster, S.R., Reccia, E., Spagnuolo, M., Cazzani, A.: Corrugated shells: an algorithm for generating double-curvature geometric surfaces for structural analysis. Thin-Walled Struct. **173**, 109019 (2022). https://doi.org/10.1016/j.tws.2022.109019
10. Nemish, Y.N., Chernopiskii, D.I.: Elastic Equilibrium of Corrugated Solids. Naukova Dumka, Kyiv (1983). [in Russian]
11. Schumaker, L.L.: Spline Functions: Basic Theory. Cambridge University Press, Cambridge (2007)
12. Semenyuk, N.P., Zhukova, N.B., Babich, I.Y.: Stability of transversely corrugated cylindrical shells. Theor. Appl. Mech. **45**, 106–113 (2009). [in Russian]

Chapter 15
Numerical Analysis of Free Vibration Frequencies of Hexagonal Plate

Alexander Grigorenko, Maksym Borysenko, Olena Boychuk, and Nataliia Boreiko

Abstract Free vibrations of the isotropic hexagonal plates of the different thicknesses with the free edges are considered based on two different approaches. The approach to calculating the frequencies of free vibrations of isotropic square plates by the Rayleigh–Ritz method has been extended to the calculation of the frequencies of free vibrations of hexagonal plates. The coefficients of the modes and the boundary conditions in the reduced formula are calculated, depending on the ratio of the thickness to the side of the plate. The limits of the application of the formula are established. Frequencies and forms of free oscillations of the plates of this class are calculated by the finite element method (FEM). The frequencies calculated by the FEM were compared with the frequencies calculated using the reduced formula, and the accuracy of the calculations by the two methods was established. The modes of the vibrations obtained based on the FEM are compared with the modes of vibrations obtained numerically and experimentally by other authors. The topology of the forms of the free vibrations for hexagonal, pentagonal, quadrangular, and triangular plates with free edges is established. The approaches implemented in the work make it possible to study the dynamic characteristics of plates of other configurations and can be used to assess the accuracy of other approaches.

15.1 Introduction

Thin-walled constructions and engineering structures, including modern buildings, rockets, and spacecraft, commonly use plates of various shapes. When designing these structures, it is essential to evaluate the load-bearing elements under sudden

A. Grigorenko (✉) · M. Borysenko · N. Boreiko
S.P. Timoshenko Institute of Mechanics, National Academy of Science of Ukraine, Kyiv, Ukraine
e-mail: ayagrigorenko1991@gmail.com

M. Borysenko
e-mail: mechanics530@gmail.com

O. Boychuk
Mykolayiv National Agrarian University, Mykolayiv, Ukraine

H. Altenbach et al. (eds.), *Selected Problems of Solid Mechanics and Solving Methods*, Advanced Structured Materials 204, https://doi.org/10.1007/978-3-031-54063-9_15

dynamic loads of different types. One critical aspect of ensuring the reliability of these plates is to determine the frequencies and forms of free vibrations with high accuracy, taking into account the properties of the material and boundary conditions. This is a significant issue in applied mathematics and mechanics.

For the last two centuries, questions about free vibrations of plates have been widely considered. Chladni did some of the first work on vibrations of thin isotropic rectangular plates [1], and the calculation of the dynamic characteristics of a square plate by the Ritz method was first presented in 1909 [2]. Classical theory for determining the frequencies and forms of free vibrations uses many different methods [3–9].

Meleshko and Papkov presented the problem of vibrations of a square plate, investigated the accuracy of the fulfillment of homogeneous boundary conditions, and analyzed the theoretical and experimental data [10]. An interesting examination of free vibrations of polygonal and rounded polygonal plates using an improved Ritz method on the class of homotopic forms was carried out in [11]. Axisymmetric resonant vibrations of an elastic round sandwich plate under local periodic surface loads of rectangular, sinusoidal, and parabolic shapes were presented in 2003 [12]. Axisymmetric forced vibrations of a circular sandwich plate on an elastic base were demonstrated in 2010 [13]. In both papers, the broken normal hypothesis is used by authors to describe the kinematics of the plate, an analytical solution to the problem is obtained, and numerical results are analyzed.

One modern method for studying the frequencies and modes of free vibrations is the finite element method (FEM), which is the basis of many programs for engineering calculations, such as FEMAP with the NX Nastran solver. By FEM is implemented in FEMAP, and the frequencies and modes of free vibrations of a thin isotropic triangular plate with a hole in the center are determined for different variations of rigid fastening at the edges and at the hole [14]. The frequencies and modes of free vibrations of a thin, rigidly fixed square plate of constant thickness are determined based on the FEM and the Rayleigh–Ritz method [15].

The free vibrations of thin triangular, quadrangular, and pentagonal plates with free edges and various physical and mechanical characteristics are studied using the FEM [16]. A comparative analysis of the calculated frequencies of free vibrations is performed, and dependence of the corresponding frequencies on the physical and mechanical characteristics of the material is established.

The transverse vibrations of a thin elastic plate of regular hexagonal shape were investigated by FEM in [17], where some frequencies and forms of free vibrations of a thin plate with free edges and rigidly fixed edges were presented, and the obtained results were compared with the results obtained by other authors.

The main method of verifying the results of the calculations is by physical experiment. Numerical analysis is only a simulation of the real construction. The compliance with the results of experimental investigations depends on the fidelity of the mathematical model describing the construction and on the numerical process used to solve the equations on the model. There are various experimental methods for studying the frequencies and forms of free vibrations of plates [18–21].

Some of the most significant difficulties in the theoretical study of free structural vibrations are associated with the need to find numerical solutions for partial differential equations with variable coefficients; solutions of such equations are often plagued by computational instability and by the loss of the expected order of accuracy. In particular, problems of this sort are magnified in the free vibration analysis of hexagonal plates, whereby the numerical analysis involves additional complexity associated with the use of a non-orthogonal coordinate system.

The purpose of this work is to apply the Rayleigh–Ritz method approach for calculating the frequencies of free vibrations of isotropic square plates [15] to the calculation of the frequencies of free vibrations of isotropic hexagonal plates of different thicknesses with free edges and to apply the finite element method to solve problems of this class.

To calculate frequencies with a formula derived from the Rayleigh–Ritz method, it is necessary to calculate the coefficients of the mode and boundary conditions using an additional FEM calculation; both methods must be applied. FEM is a proven method for a large class of problems, so its calculations can be a reference for establishing the accuracy of calculations by formulas. The proposed two methods are completely different. The finite element method is a purely numerical method and requires the use of special skills. The second method is a calculation by an analytical formula with a table of numerically calculated factors of the modes and the boundary conditions; it is simple and convenient to use.

Studies are carried out for non-thin plates of various thicknesses within the ratio of the plate thickness to its side length $h/a \leqslant 1/8$, which is a relevant and new task because known publications are devoted to the research of thin plates.

15.2 Basic Relations

The Rayleigh–Ritz method can be used to calculate the frequencies of free vibrations of an isotropic plate of constant thickness in an orthogonal curvilinear coordinate system. This method uses beam functions [22] describing the deflections of a beam with free ends in the approximate determination of the forms of free vibrations

$$w_{m_1 m_2} = c_{m_1 m_2} F_{m_1}(x_1) F_{m_2}(x_2). \tag{15.1}$$

Applying the Rayleigh–Ritz formula [22], we obtain the first approximation

$$\omega = \pi^2 \sqrt{\frac{D}{\rho h} \left(\frac{A_m^4}{a_1^4}\right) + \left(\frac{A_n^4}{a_2^4}\right) + \frac{2}{a_1^2 a_2^2} \left[\nu B_m B_n + (1 - \nu) C_m C_n\right]}, \tag{15.2}$$

where

$$D = \frac{Eh^3}{12\left(1 - \nu^2\right)}; \tag{15.3}$$

$$A_m = \begin{cases} 0 & (m = 0), \\ 1.506 & (m = 1), \\ m + 0.5 & (m \geqslant 2); \end{cases} \qquad B_m = \begin{cases} 0 & (m = 0), \\ 1.248 & (m = 1), \\ A_m\left(A_m - \frac{2}{\pi}\right) & (m \geqslant 2); \end{cases}$$
$$C_m = \begin{cases} \frac{12}{\pi^2} & (m = 0), \\ 5.017 & (m = 1), \\ A_m\left(A_m + \frac{6}{\pi}\right) & (m \geqslant 2). \end{cases} \tag{15.4}$$

Note that (15.2) determines the cyclic frequency.

Since the plate is square:

$$\begin{aligned} a_1 &= a_2 = a, \\ A_m &= A_n = A, \\ B_m &= B_n = B, \\ C_m &= C_n = C. \end{aligned} \tag{15.5}$$

Substituting (15.3) and (15.5) into (15.2), performing some mathematical transformations and dividing by 2π, we get:

$$f = \frac{h}{a^2}\sqrt{\frac{E}{\rho\left(1 - \nu^2\right)}}\left[\frac{\pi}{2\sqrt{6}}\sqrt{A^4 + \nu B^2 + (1 - \nu)\,C^2}\right], \tag{15.6}$$

where f is the frequency in Hz.

As a result, we obtain a reduced form of writing formula (15.2) for calculating natural frequencies by the Rayleigh–Ritz method, a feature of which is the grouping of multipliers into coefficients. Let us introduce the following notation:

$$G = \frac{h}{a^2}, \quad M = \sqrt{\frac{E}{\rho\left(1 - \nu^2\right)}}, \quad F = \frac{\pi}{2\sqrt{6}}\sqrt{A^4 + \nu B^2 + (1 - \nu)\,C^2}, \tag{15.7}$$

where G is geometry factor, M is material factor; F is coefficient of mode and boundary conditions.

After substituting (15.7) into (15.6), we obtain

$$f = GMF = \frac{h}{a^2}\sqrt{\frac{E}{\rho\left(1 - \nu^2\right)}}F. \tag{15.8}$$

The resulting formula will be called the reduced formula (RF) (15.8), the application of which was extended to the calculation of the frequencies of free vibrations of triangular plates with free edges [23] and square plates with rigidly fixed edges [15].

The coefficients of modes and boundary conditions are calculated by the formula

$$F_i = \frac{f_i^{\mathrm{FEM}}}{h} a^2 \sqrt{\frac{\rho\left(1-\nu^2\right)}{E}}, \tag{15.9}$$

where f_i^{FEM} represents the pre-calculated frequencies of the same plate by the FEMAP program.

When calculating dynamic processes using the FEM, the motion of a mechanical system with a finite number of degrees of freedom in the absence of external forces is described by a system of Lagrange equations of the second kind:

$$\frac{d}{dt}\left(\frac{\partial L}{\partial \dot{q}_j}\right) - \frac{\partial L}{\partial q_j} = 0, \quad j = 1, 2, \ldots, s, \tag{15.10}$$

where $L = T - U$. Using the discrete form of the kinetic energy and deformation energy functionals, respectively

$$T = \frac{1}{2}\, \dot{\overline{\Phi}}_i^T \mathbf{M}_i\, \dot{\overline{\Phi}}_i, \quad U = \frac{1}{2}\, \dot{\overline{\Phi}}_i^T \mathbf{K}_i\, \dot{\overline{\Phi}}_i,$$

where $\mathbf{K}_i$ and $\mathbf{M}_i$ are the stiffness matrix and the mass matrix of the i-th finite element, and $\overline{\Phi}_i$ is the vector of nodal displacements of the i-th element. From the Lagrange equation (15.10), we obtain the following equations of motion in the absence of damping

$$\mathbf{K}\, \overline{\Phi}_j + \mathbf{M}\, \ddot{\overline{\Phi}}_j = 0, \tag{15.11}$$

where $\mathbf{K}$ and $\mathbf{M}$ are the stiffness matrix and the mass matrix of the mechanical system and $\overline{\Phi}_j$ is the displacement vector of system nodes, which corresponds to the j-th degree of freedom, and reproduces the j-th mode of vibration.

With free vibrations of the mechanical system, all nodal points carry out harmonic vibrations as a function of time

$$\overline{\Phi}_j\,(t) = \overline{\Phi}_j\, \sin\left(\omega_j t\right). \tag{15.12}$$

After substituting the functions (12) into the equation of motion (11), the determination of the natural frequencies and vibration modes is reduced to solving the system of algebraic equations

$$\mathbf{K}\overline{\Phi}_j - \omega_j^2 \mathbf{M}\overline{\Phi}_j = 0, \quad j = 1, 2, \ldots, s, \tag{15.13}$$

where ω_j is the cyclic frequency of the harmonic vibrations.

For determining the natural modes and vibration frequencies when energy dissipation and damping are not considered, NX Nastran primarily uses the Lanczos method, which requires fewer resources than other methods. The Lanczos method allows one to determine n necessary eigenvalues and modes, while the results can be considered practically accurate for this discrete model since the error is

$$\frac{\left\| \overline{\Phi}_j - \omega_j^2 \mathbf{K}^{-1} \mathbf{M} \overline{\Phi}_j \right\|}{\left\| \overline{\Phi}_j \right\|} \leqslant 10^{-7}.$$

The Lanczos method uses reduction to transform the matrix $\mathbf{T}$ to the tridiagonal form

$$\mathbf{T} = \mathbf{Q}_k^T \mathbf{M} \mathbf{K}^{-1} \mathbf{M} \mathbf{Q}_k, \tag{15.14}$$

where $\mathbf{Q}_k = \{\overline{q}_1, \overline{q}_2, \ldots, \overline{q}_k\}$ is a rectangular matrix with elements $N_{\text{eq}} \times k$; N_{eq} is the number of equations; k is a step number according to the Lanczos method; $\overline{q}_k$ is the k-th vector of Lanczos.

The formula

$$\beta_{k+1}\, \overline{q}_{k+1} = \mathbf{K}^{-1} \mathbf{M}\, \overline{q}_k - \alpha_k\, \overline{q}_k - \beta_k\, \overline{q}_k, \tag{15.15}$$

generates the next Lanczos vector and determines the current row of the matrix $\mathbf{T}$

$$\mathbf{T} = \begin{vmatrix} \alpha_1 & \beta_2 & & & \\ \beta_2 & \alpha_2 & \beta_3 & & \\ & \beta_3 & \alpha_3 & \beta_4 & \\ \cdot & \cdot & \cdot & \cdot & \cdot \\ & & & \beta_k & \alpha_k \end{vmatrix}.$$

Thus, we get the eigenvalue problem:

$$\mathbf{T}\, \overline{s}_h^k - \lambda_h^k\, \overline{s}_h^k = 0, \quad h = 1, 2, \ldots, k, \tag{15.16}$$

$$\left(\omega_h^k\right)^2 = 1/\lambda_h^k,$$

where ω_h^k is the k-th approximation of the circular frequency ω_h, $h = 1, 2, \ldots, n$; n is the number of eigen pairs.

The algorithm continues the calculations (with increasing k, the step number of the Lanczos procedure) until the specified accuracy is achieved with all the necessary eigenvalues.

The selective orthogonalization procedure maintains the required level of orthogonalization of Lanczos' vectors q_k, which ensures the reliability and stability of the numerical calculation process. In this case, economical methods are used to implement the selective orthogonalization procedure and to determine the eigenvalues (15.16) by using double QR iterations with shifts. The output eigenvectors are determined by the formula

$$\overline{\Phi}_h^k = Q_k\, \overline{s}_h^k = 0, \quad h = 1, 2, \ldots, n. \tag{15.17}$$

15.3 Building a Finite Element Model

We will solve the problem of the calculation of the frequencies and the modes of the free vibrations of the plate based on the FEM. It is necessary to solve the test problem, correctly build a finite element calculation model, and correctly select the type of finite element and its size while taking into account the optimal calculation time and accuracy of the results obtained.

The study of the selection of the type and the size of the finite element (plate, brick, tetrahedral (Fig. 15.1)) will be carried out on two aluminum plates in the form of a regular hexagon with the following parameters: plate side $a = 0.24$ m, thickness $h = 0.004$ m or $h = 0.024$ m, Young's modulus $E = 71$ GPa, Poisson's ratio $\nu = 0.33$, and density $\rho = 2710\ \text{kg/m}^3$. The sizes of the elements will be halved until the relative differences of the previous and current frequencies become $\varepsilon \leqslant 1\%$ for at least one of the types of elements.

The calculated frequencies for the plate $h/a = 1/60$ meshed with different types and sizes of finite elements are shown in Table 15.1, and the relative differences in frequencies are shown in Table 15.2. Figure 15.2 shows the convergence of the first and tenth frequencies in cases of meshes of different element types.

Table 15.1 shows the frequencies of free vibrations of hexagonal plates $h/a = 1/60$ meshed with different finite elements of different sizes: $4h$, $2h$, and h. The table also shows the number of elements (Elements), the number of nodes (Nodes), and the relative time of the computer calculation (Time).

Analyzing Table 15.1 and Fig. 15.2, we can see the convergence of calculations of the frequency of free vibrations for a thin plate meshed with different elements. In the cases of meshes with plate elements and brick elements of size $4h$, we have lower frequencies than in the cases of meshes with the same elements of size $2h$. Meshes of tetra-elements give higher frequencies and take higher relative computer calculation

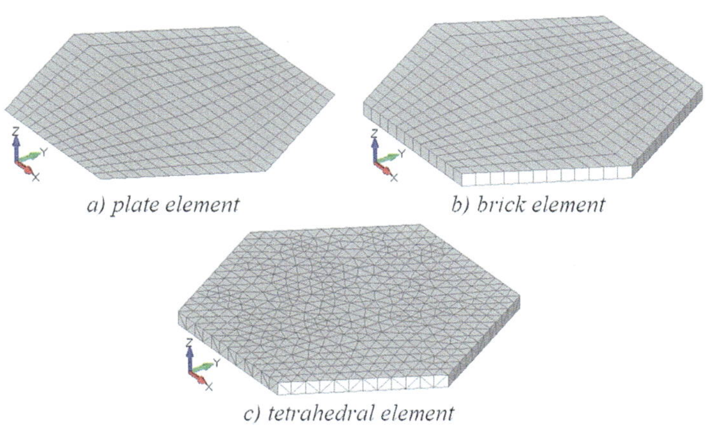

Fig. 15.1 Finite element meshing

Table 15.1 Frequencies of free vibrations of a hexagonal plate $h/a = 1/60$

	f, Hz								
No	Plate			Brick			Tetrahedral		
	$4h$	$2h$	h	$4h$	$2h$	h	$4h$	$2h$	h
1	107.27	108.04	108.21	107.59	108.12	108.22	108.37	108.27	108.24
2	107.85	108.17	108.23	107.94	108.19	108.22	108.38	108.27	108.24
3	184.07	185.48	185.83	184.12	185.39	185.71	186.16	185.90	185.91
4	228.62	230.30	230.58	229.81	230.71	230.72	231.88	230.88	230.66
5	270.66	273.19	273.81	271.35	273.19	273.63	274.93	273.98	273.95
6	400.67	408.76	410.81	403.07	409.05	410.54	413.49	411.52	411.42
7	408.23	410.73	411.32	409.35	410.70	410.97	413.76	411.52	411.44
8	427.74	434.28	435.73	430.86	434.97	435.72	439.75	436.39	436.03
9	431.22	435.00	435.76	433.52	435.54	435.73	440.02	436.40	436.04
10	672.13	683.02	685.52	677.60	684.12	685.25	695.61	686.94	686.24
Time	1	1	5	1	1	10	5	20	70
Elements	450	1800	7381	450	1800	7200	3802	14507	50992
Nodes	496	1891	7200	992	3782	14762	7877	29572	99407

Table 15.2 Relative frequency differences of a hexagonal plate $h/a = 1/60$

	ε, %					
No	Plate		Brick		Tetrahedral	
	$\mid 1 - f_{2h}/f_{4h} \mid$	$\mid 1 - f_h/f_{2h} \mid$	$\mid 1 - f_{2h}/f_{4h} \mid$	$\mid 1 - f_h/f_{2h} \mid$	$\mid 1 - f_{2h}/f_{4h} \mid$	$\mid 1 - f_h/f_{2h} \mid$
1	0.72	0.16	0.49	0.09	0.09	0.03
2	0.30	0.06	0.23	0.03	0.10	0.03
3	0.77	0.19	0.69	0.17	0.14	0.01
4	0.73	0.12	0.39	0.00	0.43	0.10
5	0.93	0.23	0.68	0.16	0.35	0.01
6	2.02	0.50	1.48	0.36	0.48	0.02
7	0.61	0.14	0.33	0.07	0.54	0.02
8	1.53	0.33	0.95	0.17	0.76	0.08
9	0.88	0.17	0.47	0.04	0.82	0.08
10	1.62	0.37	0.96	0.17	1.25	0.10
Max	2.02	0.50	1.48	0.36	1.25	0.10

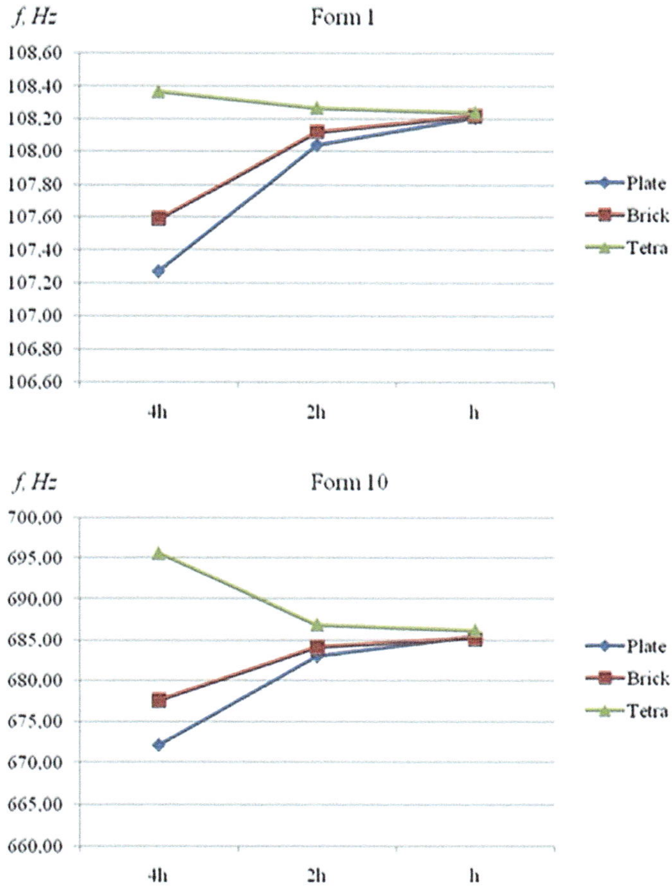

Fig. 15.2 Convergence of the first and tenth frequencies for different types of elements of meshing for a plate $h/a = 1/60$

time due to the large number of elements. A good convergence of the results and the fulfillment of the condition $\varepsilon \leqslant 1\%$ are observed for the element size h.

The calculated frequencies for the plate $h/a = 1/10$ meshed with different types and sizes of finite elements are shown in Table 15.3, and the relative differences in frequencies are shown in Table 15.4. Figure 15.3 shows the convergence of the first and tenth frequencies in cases of meshes of different element types.

Table 15.3 shows the frequencies of free vibrations of hexagonal plates meshed with different finite elements of different sizes: $2h$, h, $h/2$. Analyzing Table 15.3 and Fig. 15.3, we can see a large discrepancy between the frequencies for non-thin plates in cases of meshes of brick elements and a tetra-element with size $2h$, and one can observe a good convergence of results in cases of meshes of elements with $h/2$ and fulfillment of the condition $\varepsilon \leqslant 1\%$ only in the case of mesh of $h/2$ tetra-elements.

Table 15.3 Frequencies of free vibrations of a hexagonal plate $h/a = 1/10$

No	f, Hz								
	Plate			Brick			Tetrahedral		
	$2h$	h	$h/2$	$2h$	h	$h/2$	$2h$	h	$h/2$
1	591.46	629.00	638.11	599.40	624.88	635.54	638.30	637.86	637.20
2	620.73	636.38	639.80	615.42	628.90	636.34	638.46	637.88	637.31
3	1014.06	1080.76	1099.20	1001.34	1057.65	1087.38	1094.19	1094.40	1094.95
4	1232.26	1311.67	1328.17	1259.71	1302.41	1323.59	1339.61	1328.78	1322.34
5	1457.06	1574.28	1606.26	1450.60	1535.96	1585.51	1600.99	1596.44	1596.83
6	1946.29	2259.82	2361.16	2003.52	2210.48	2330.32	2382.99	2365.82	2358.70
7	2153.69	2357.27	2387.33	2212.93	2284.98	2351.02	2384.20	2366.53	2362.38
8	2231.71	2408.47	2481.36	2213.15	2361.98	2454.81	2503.63	2477.98	2467.23
9	2279.56	2447.87	2489.98	2289.80	2388.17	2459.82	2504.28	2478.20	2467.77
10	3094.33	3710.01	3832.69	3330.93	3590.76	3769.90	3879.56	3819.71	3796.70
Time	1	1	1	1	1	1	1	5	10
Elements	50	231	861	50	462	2583	429	1770	27267
Nodes	66	200	800	132	200	1600	947	3635	16588

Table 15.4 Relative frequency differences of a hexagonal plate $h/a = 1/10$

No	ε, %					
	Plate		Brick		Tetrahedral	
	$\lvert 1 - f_h/f_{2h} \rvert$	$\lvert 1 - f_{0.5h}/f_h \rvert$	$\lvert 1 - f_h/f_{2h} \rvert$	$\lvert 1 - f_{0.5h}/f_h \rvert$	$\lvert 1 - f_h/f_{2h} \rvert$	$\lvert 1 - f_{0.5h}/f_h \rvert$
1	6.35	1.45	4.25	1.71	0.07	0.10
2	2.52	0.54	2.19	1.18	0.09	0.09
3	6.58	1.71	5.62	2.81	0.02	0.05
4	6.44	1.26	3.39	1.63	0.81	0.48
5	8.04	2.03	5.88	3.23	0.28	0.02
6	16.11	4.48	10.33	5.42	0.72	0.30
7	9.45	1.28	3.26	2.89	0.74	0.18
8	7.92	3.03	6.72	3.93	1.02	0.43
9	7.38	1.72	4.30	3.00	1.04	0.42
10	19.90	3.31	7.80	4.99	1.54	0.60
Max	19.90	4.48	10.33	5.42	1.54	0.60

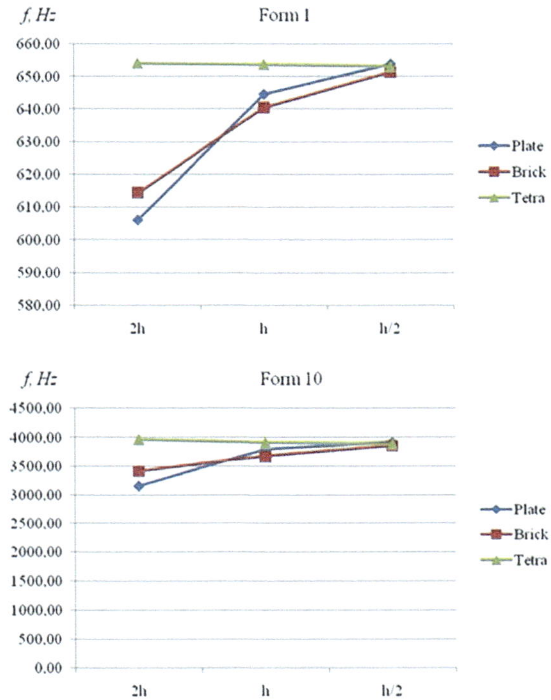

Fig. 15.3 Convergence of the first and tenth frequencies for different types of elements of meshing for a plate $h/a = 1/10$

We can conclude that for thin plates it is better to use plate elements or brick elements; the difference is only in the time of computer calculation and for non-thin and thick plates it is better to use brick elements. Below, we will mesh plates of different thicknesses with brick elements. The value of the element will be selected for each plate thickness individually, depending on the plate thickness.

15.4 Solution of the Test Problem

As a test problem, we will calculate the frequencies of free vibrations by the FEM in the FEMAP program of an aluminum square plate with free edges [19] with the following parameters: side of the plate $a = 0.08$ m, thickness $h = 0.001$ m, Young's modulus $E = 70$ GPa, Poisson's ratio $\nu = 0.33$, and density $\rho = 2700\ \text{kg/m}^3$. Table 15.5 shows 26 frequencies of free vibrations calculated by the FEM and determined numerically and experimentally by other authors, as well as the given discrepancy in percentage compared to the experimental results ε_A and calculations by the FEM ε_F. Figure 15.4 shows a comparison of the first ten modes of free vibrations obtained in FEMAP and by other authors.

Table 15.5 Comparison of frequencies of free vibrations of a square plate

	f, Hz										
No	AF-ESPI	FEM	FEM	ε_A, %	ε_F, %	No	AF-ESPI	FEM	FEM	ε_A, %	ε_F, %
	[19]	[19]					[19]	[19]			
1	494	510	509	3.04	0.20	14	5934	6163	6122	3.17	0.67
2	735	745	744	1.22	0.13	15	6294	6458	6421	2.02	0.57
3	890	945	944	6.07	0.11	16	7400	7628	7586	2.51	0.55
4	1271	1324	1321	3.93	0.23	17	7550	7810	7758	2.75	0.67
5	2330	2357	2351	0.90	0.25	18	8104	8221	8157	0.65	0,78
6	2345	2424	2416	3.03	0.33	19	9290	9296	9225	0.70	0.76
7	2630	2634	2624	0.23	0.38	20	10560	10709	10606	0.44	0.96
8	2930	2974	2965	1.19	0.30	21	10790	11125	11014	2.08	1.00
9	3895	4021	4001	2.72	0.50	22	10920	11267	11182	2.40	0.75
10	4402	4499	4481	1.79	0.40	23	11100	11342	11462	3.26	1.06
11	4630	4731	4714	1.81	0.36	24	11160	11403	11301	1.26	0.89
12	4920	5048	5024	2.11	0.48	25	12440	12857	12733	2.36	0.96
13	5654	5827	5791	2.42	0.62	26	12780	13206	13089	2.42	0.89

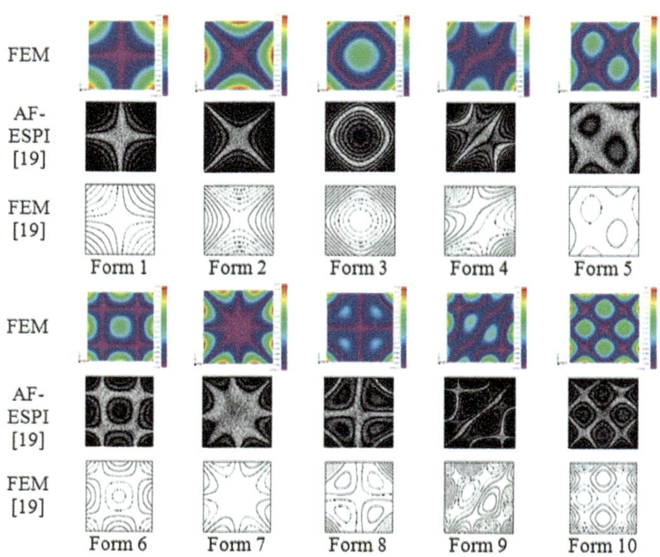

Fig. 15.4 Comparison of ten modes of free vibrations of a square plate

Analyzing the results presented in Table 15.5 and in Fig. 15.4, we can conclude that it is expedient to use the FEM to calculate the frequencies and forms of free vibrations of the plates.

15.5 Results of Numerical Calculations

To calculate the coefficients of the modes of vibrations and the boundary conditions of the RF (15.8), we will calculate the frequencies of the FEM of an aluminum hexagonal plate with free edges with the parameters: side of the plate $a = 0.24$ m, thickness $h = 0.004$ m ($h/a = 1/60$), Young's modulus $E = 71$ GPa, Poisson's ratio $\nu = 0.33$, and density $\rho = 2710\ \mathrm{kg/m^3}$. We will calculate the first ten coefficients of mode and boundary conditions using (15.9). The obtained coefficients for hexagonal plates with free edges are shown in Table 15.6.

Using the FEM and RF (15.8) with the coefficients from Table 15.6, we calculate the frequencies of free vibrations of aluminum hexagonal plates with a side $a = 0.24$ m and the following ratios of plate thickness to plate side h/a: 1/240, 1/120, 1/60, 1/30, 1/20, 1/ 15, 1/12, 1/10, 1/5 (the corresponding ratios for triangular plates with free edges are considered in [23]). The results obtained, namely, the first ten frequencies, are presented in Table 15.7, where ε are the percentage deviations between the frequencies obtained in the calculations by the two methods.

Analyzing the results presented in Table 15.7, it is possible to establish the limits of application of the RF (15.8) for hexagonal plates with free edges. Namely: for the first ten frequencies at $h/a \leq 1/30$, the deviation between the calculated frequencies using the RF (15.8) compared with the frequencies calculated by the FEM is $\varepsilon < 1\%$. The deviation at $h/a \leqslant 1/10$ is $\varepsilon < 9\%$, and for the first three frequencies at $h/a \leqslant 1/5$, the deviation is $\varepsilon < 7\%$. In mechanics, the first frequencies of free vibrations are considered the most dangerous since they have the largest vibration amplitudes. Therefore, the RF (15.8) with the coefficients from Table 15.6 is accurate enough to calculate the first three vibration frequencies of plates with a size ratio within $h/a \leqslant 1/5$.

Let us compare the maximum deviations between the calculated frequencies by the two methods for a triangular plate taken from [23] and a hexagonal plate with free edges taken from Table 15.7. The deviations are presented in Table 15.8 and in Fig. 15.5.

Table 15.6 Calculated coefficients according to formula (15.9) at $h/a = 1/60$

i	F_i	i	F_i
1	0.287390	6	1.090806
2	0.287402	7	1.091462
3	0.493323	8	1.156977
4	0.612552	9	1.157157
5	0.726832	10	1.819838

Table 15.7 The first ten frequencies (Hz) of free vibrations of a hexagonal plate calculated by two methods

No	1/240		ε, %	1/120		ε, %	1/60		ε, %
	RF (8)	FEM		RF (8)	FEM		RF (8)	FEM	
1	27.05	27.10	0.18	54.11	54.17	0.11	108.22	108.22	0.00
2	27.06	27.10	0.15	54.11	54.17	0.11	108.22	108.22	0.00
3	46.44	46.49	0.11	92.88	92.96	0.09	185.76	185.76	0.00
4	57.66	57.89	0.40	115.33	115.65	0.28	230.65	230.65	0.00
5	68.42	68.54	0.18	136.84	137.04	0.15	273.69	273.69	0.00
6	102.68	102.97	0.28	205.37	205.83	0.22	410.74	410.74	0.00
7	102.75	103.00	0.24	205.49	205.88	0.19	410.99	410.99	0.00
8	108.91	109.39	0.44	217.83	218.52	0.32	435.66	435.66	0.00
9	108.93	109.39	0.42	217.86	218.54	0.31	435.72	435.72	0.00
10	171.31	172.26	0.55	342.63	344.04	0.41	685.25	685.25	0.00
Max			0.55			0.41			0.00
No	1/30		ε, %	1/20		ε, %	1/15		ε, %
	RF (8)	FEM		RF (8)	FEM		RF (8)	FEM	
1	216.43	215.91	0.24	324.65	322.80	0.57	432.86	428.76	0.96
2	216.44	215.92	0.24	324.66	322.81	0.57	432.88	428.78	0.96
3	371.52	370.90	0.17	557.28	554.97	0.42	743.04	737.41	0.76
4	461.31	458.46	0.62	691.96	681.86	1.48	922.62	900.60	2.45
5	547.37	545.88	0.27	821.06	815.41	0.69	1094.74	1081.17	1.26
6	821.48	817.80	0.45	1232.22	1218.62	1.12	1642.96	1610.69	2.00
7	821.97	818.64	0.41	1232.96	1219.86	1.07	1643.95	1612.31	1.96
8	871.31	865.38	0.69	1306.97	1285.29	1.69	1742.62	1693.56	2.90
9	871.45	865.51	0.69	1307.17	1285.39	1.69	1742.89	1693.58	2.91
10	1370.51	1358.21	0.91	2055.76	2010.20	2.27	2741.02	2636.62	3.96
Max			0.91			2.27			3.96
No	1/12		ε, %	1/10		ε, %	1/5		ε, %
	RF (8)	FEM		RF (8)	FEM		RF (8)	FEM	
1	541.08	533.58	1.41	649.29	637.20	1/90	1298.59	1225.30	5.98
2	541.10	533.65	1.40	649.32	637.31	1.88	1298.64	1225.92	5.93
3	928.80	917.25	1.26	1114.55	1094.95	1.79	2229.11	2084.67	6.93
4	1153.27	1114.49	3.48	1383.93	1322.34	4.66	2767.85	2435.04	13.67
5	1368.43	1341.51	2.01	1642.12	1596.83	2.84	3284.23	2973.61	10.45
6	2053.70	1990.22	3.19	2464.44	2358.70	4.48	4928.87	4246.01	16.08
7	2054.93	1993.34	3.09	2465.92	2362.38	4.38	4931.84	4258.69	15.81
8	2178.28	2087.84	4.33	2613.94	2467.23	5.95	5227.87	4388.38	19.13
9	2178.62	2088.14	4.33	2614.34	2467.77	5.94	5228.68	4392.22	19.04
10	3426.27	3232.35	6.00	4111.52	3796.70	8.29	8223.05	6477.74	26.94
Max			6.00			8.29			26.94

Table 15.8 Maximum deviations between frequencies calculated by two methods

	ε_{max}, %								
Plates	1/240	1/120	1/60	1/30	1/20	1/15	1/12	1/10	1/5
Triangular plate [23]	1.29	1.04	0	3.31	7.85	13.56	20.3	27.9	90.36
Hexagonal plate	0.55	0.41	0	0.91	2.27	3.96	6.00	8.29	26.94

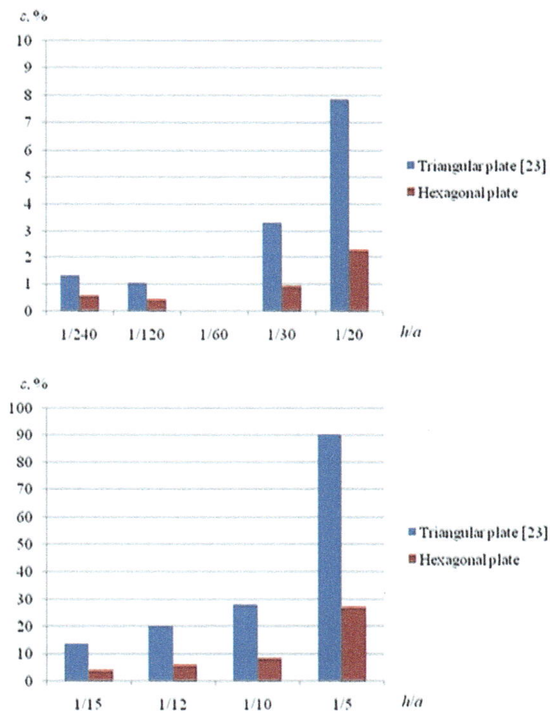

Fig. 15.5 Maximum deviations between frequencies calculated by two methods

Analyzing the results presented in Table 15.8 and on the histograms (Fig. 15.5), one can observe a significantly smaller discrepancy between the frequencies calculated using the RF (15.8) with the frequencies calculated by the FEM for hexagonal plates compared to the discrepancies for calculating the frequencies of triangular plates, provided that both coefficients were determined at the same ratio, $h/a = 1/60$.

Additionally, the coefficients of mode and boundary conditions were determined according to formula (15.9) for hexagonal plates with the ratios $h/a \geqslant 1/10$, taking as a basis the frequencies calculated by the FEM for the plate with the ratio $h/a = 1/8$, results in the coefficients presented in Table 15.9.

We calculate by FEM and using the RF (15.8) with the coefficients from Table 15.9 the frequencies of free vibrations of aluminum hexagonal plates with a side $a =$

Table 15.9 Calculated coefficients according to formula (15.9) at $h/a = 1/8$

i	F_i	i	F_i
1	0.279780	6	1.022306
2	0.279815	7	1.023312
3	0.480245	8	1.064616
4	0.574549	9	1.064810
5	0.696966	10	1.622272

Table 15.10 The first ten frequencies (Hz) of free vibrations are calculated by two methods, taking into account the coefficients of Table 15.9

No	1/10		ε, %	1/8		ε, %	1/5		ε, %
	RF (8)	FEM		RF (8)	FEM		RF (8)	FEM	
1	632.10	637.20	0.80	790.13	790.13	0.00	1264.20	1225.30	3.17
2	632.18	637.31	0.80	790.23	790.22	0.00	1264.36	1225.92	3.14
3	1085.01	1094.95	0.91	1356.26	1356.26	0.00	2170.02	2084.67	4.09
4	1298.07	1322.34	1.84	1622.58	1622.59	0.00	2596.13	2435.04	6.62
5	1574.64	1596.83	1.39	1968.30	1968.30	0.00	3149.28	2973.61	5.91
6	2309.68	2358.70	2.08	2887.09	2887.09	0.00	4619.35	4246.01	8.79
7	2311.95	2362.38	2.13	2889.94	2889.94	0.00	4623.90	4258.69	8.58
8	2405.27	2467.23	2.51	3006.58	3006.58	0.00	4810.53	4388.38	9.62
9	2405.70	2467.77	2.52	3007.13	3007.13	0.00	4811.41	4392.22	9.54
10	3665.17	3796.70	3.46	4581.46	4581.46	0.00	7330.33	6477.74	13.16
Max			3.46			0.00			13.16

0.24 m and ratios of plate thickness to the side of the plate: 1/10, 1/8, 1/5. The results obtained, namely the first ten frequencies, are presented in Table 15.10.

Analyzing the results presented in Table 15.10, we can see that the corrections made to the coefficients significantly improved the calculations using RF (15.8). At the same time, for the first ten frequencies at $h/a \leqslant 1/10$, the deviation between the calculated frequencies using RF (15.8) in comparison with the frequencies calculated by the FEM is $\varepsilon < 3.5\%$, which indicates a twofold increase in result accuracy and for frequencies at $h/a \leqslant 1/5$ deviation is $\varepsilon < 13.2\%$, which also indicates a twofold increase in result accuracy. For the first three frequencies at $h/a \leqslant 1/5$, the deviation is $\varepsilon < 4.1\%$.

RF (15.8), with tables of coefficients of modes and boundary conditions, can be used for a quick engineering calculation of the first ten frequencies and modes of free vibrations of hexagonal plates with free edges, as shown in Table 15.11.

Let us compare the vibration modes obtained for hexagonal plates with free edges with the vibration modes obtained by other authors using different methods (Figs. 15.6 and 15.7).

Table 15.11 Generalization of using RF (15.8)

$$f = \frac{h}{a^2}\sqrt{\frac{E}{\rho(1-\nu^2)}}F_i$$

i	F_i, $h/a < 1/10$ ($\varepsilon < 8\%$)	F_i, $1/10 \leqslant h/a \leq 1/5$ ($\varepsilon < 13.2\%$)	Form	i	F_i, $h/a < 1/10$ ($\varepsilon < 8\%$)	F_i, $1/10 \leqslant h/a \leq 1/5$ ($\varepsilon < 13.2\%$)	Form
1	0.287390	0.279780	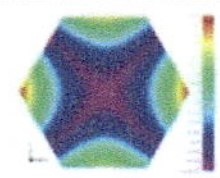	6	1.090806	1.022306	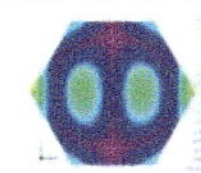
2	0.287402	0.279815	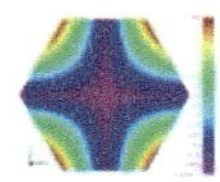	7	1.091462	1.023312	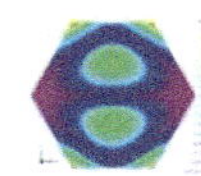
3	0.493323	0.480245	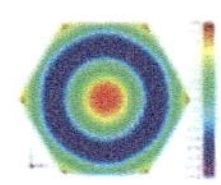	8	1.156977	1.064616	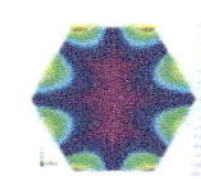
4	0.612552	0.574549	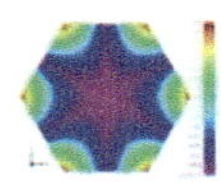	9	1.157157	1.064810	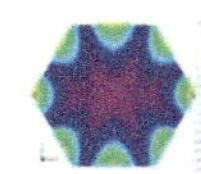
5	0.726832	0.696966	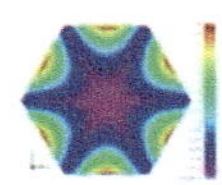	10	1.819838	1.622272	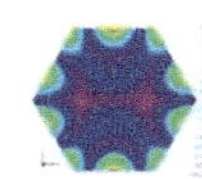

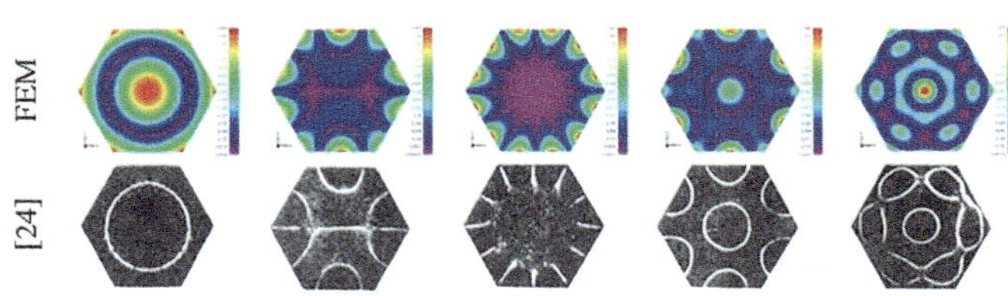

Fig. 15.6 Comparison of five vibration modes obtained by the FEM and experimentally [24]

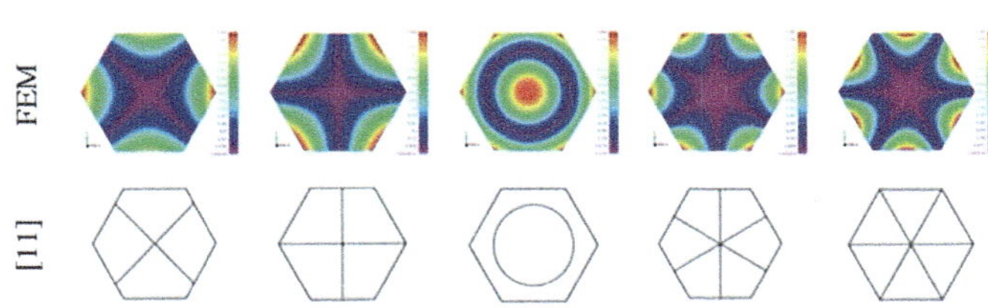

Fig. 15.7 Comparison of five modes of vibrations obtained by the FEM and Ritz method [11]

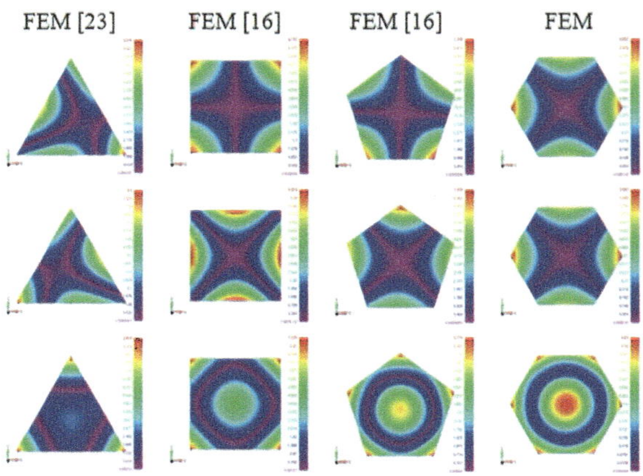

Fig. 15.8 Topology of the first three vibration modes of different plates with free edges

We establish the topology of vibration modes between triangular, quadrangular, pentagonal, and hexagonal plates with free edges (Fig. 15.8), which shows the presence of the same vibration modes for plates of different shapes.

15.6 Conclusions

The paper extends the approach for calculating the frequencies of free vibrations of isotropic square plates using the Rayleigh–Ritz method in order to calculate the frequencies of free vibrations of hexagonal plates of different thicknesses with free edges and calculates the coefficients of modes and boundary conditions depending on the ratio of thickness to the side of the plate. The FEM is further used to determine the frequencies and forms of free vibrations of plates of the specified class.

The following concrete results were obtained during the study:

- The FEM solved the test problem of calculating the frequencies and modes of free vibrations of a thin isotropic plate with free edges. Small discrepancies between the calculated frequencies and the frequencies obtained by other authors numerically (no more than 1.1%) and in the frequencies obtained experimentally (no more than 4%), also the identical modes of vibrations at the corresponding frequencies indicate the expediency of using the FEM for solving problems of this class;
- The problem of free vibrations of a hexagonal plate with different thicknesses and free edges was solved and the coefficients of the form and boundary conditions of the reduced formula were refined for plates $1/10 \leqslant h/a \leqslant 1/5$;
- A generalization of the use of the reduced formula was carried out, according to which the error in calculating the first ten frequencies of free vibrations according to the formula in comparison with the calculation using the FEM does not exceed 8% for $h/a < 1/10$ and 13.2% for $1/10 \leqslant h/a \leq 1/5$, and for the first three frequencies does not exceed 4.1% for all considered plate thicknesses;

- It was found that the frequencies calculated using the reduced formula for hexagonal plates have a smaller discrepancy with the frequencies calculated by the FEM compared with the results for a triangular plate, provided that the coefficients were refined at the same ratio $h/a = 1/60$;
- A comparison of the modes of vibrations obtained by the FEM with the modes of vibrations obtained by the Ritz method and experimentally for hexagonal plates was carried out;
- The topology of the modes of free vibrations for hexagonal, pentagonal, quadrangular, and triangular plates with free edges has been established, showing the presence of the same vibration modes for plates of different shapes.

The approaches implemented in the paper demonstrate that it is possible to study the dynamic characteristics of plates of other configurations and can be used to assess the accuracy of other approaches, as well as to analyze the behavior of specific structural elements.

References

1. Chladni, E.F.F.: Die Akustik. Breitkopf und Hartel, Leipzig (1802)
2. Ritz, W.: Theorie der Transversalschwingungen einer quadratischen Platte mit freien Randern. Ann. Phys. **333**(4), 737–786 (1909). https://doi.org/10.1002/andp.19093330403
3. Grigorenko, A.Y., Efimova, T.L.: Spline-approximation method applied to solve natural-vibration problems for rectangular plates of varying thickness. Int. Appl. Mech. **41**(10), 1161–1169 (2005). https://doi.org/10.1007/s10778-006-0022-2
4. Hadian, J., Nayfeh, A.H.: Free vibration and buckling of shear-deformable cross-ply laminated plates using the state-space concept. Comput. Struct. **48**(4), 677–693 (1993). https://doi.org/10.1016/0045-7949(93)90261-B
5. Karunasena, W., Kitipornchai, S., Al-Bermani, F.G.A.: Free vibration of cantilevered arbitrary triangular Mindlin plates. Int. J. Mech. Sci. **38**(4), 431–442 (1996). https://doi.org/10.1016/0020-7403(95)00060-7
6. Lam, K.Y., Liew, K.M., Chow, S.T.: Free vibration analysis of isotropic and orthotropic triangular plates. Int. J. Mech. Sci. **32**(5), 455–464 (1990). https://doi.org/10.1016/0020-7403(90)90172-F
7. Leissa, A.W., Jaber, N.A.: Vibrations of completely free triangular plate. Int. J. Mech. Sci. **34**(8), 605–616 (1992). https://doi.org/10.1016/0020-7403(92)90058-O
8. Leissa, A.W.: Vibration of Plates. NASA SP-160, Washington (1969)
9. Liew, K.M., Xiang, Y., Kitipornchai, S.: Research on thick plate vibration: a literature survey. J. Sound Vib. **180**(1), 163–176 (1995). https://doi.org/10.1006/jsvi.1995.0072
10. Meleshko, V.V., Papkov, S.O.: Flexural vibration of elastic rectangular plates with free edges: from Chladni (1809) and Ritz (1990) to the present day. Akust. Visnyk, 12, 4, 34–51 (2009) [in Russian]
11. Wang, C.Y.: Vibrations of completely free rounded regular polygonal plates. Int. J. Acoust. Vib. **20**(2), 107 (2015)
12. Starovoitov, E.I., Leonenko, D.V., Yarovaya, A.V.: Vibrations of circular sandwich plates under resonance loads. Int. Appl. Mech. **39**(12), 1458–1463 (2003). https://doi.org/10.1023/B:INAM.0000020831.16802.4a
13. Starovoitov, E.I., Leonenko, D.V.: Resonant effects of local loads on circular sandwich plates on an elastic foundation. Int. Appl. Mech. **46**(1), 86–93 (2010). https://doi.org/10.1007/s10778-010-0285-5

14. Grigorenko, O.Y., Borisenko, M.Y., Boichuk, O.V., Vasileva, L.Y.: Free Vibrations of Triangular Plates with a Hole. Int. Appl. Mech. **57**(5), 534–542 (2021). https://doi.org/10.1007/s10778-021-01104-3
15. Grigorenko, A.Ya., Borysenko, M.Yu., Boychuk, O.V., Novytskyi, V.S.: Numerical Analysis of Free Vibrations of Rectangular Plates Based on Different Approaches. Visnyk Z.N.U. Physic. Math. Sci. **1**, 33–41 (2019) [in Ukrainian]. https://doi.org/10.26661/2413-6549-2019-1-05
16. Borysenko, M., Zavhorodnii, A., Skupskyi, R.: Numerical analysis of frequencies and forms of own collars of different forms with free zone. J. Appl. Math. Comput. Mech. **18**(1), 5–13 (2019). https://doi.org/10.17512/jamcm.2019.1.01
17. Laura, P.A.A., Rossi, R.E.: Transverse vibrations of a thin, elastic plate of regular hexagonal shape. J. Sound Vib. **256**(2), 367–372 (2002). https://doi.org/10.1006/jsvi.2001.4053
18. Grigorenko, A.Ya., Borysenko, M.Yu., Boychuk, O.V., Novytskyi, V.S.: Application of Experimental and Numerical Methods to the Study of Free Oscillations of Rectangular Plates. Prob. Comput. Mech. Streng. Struct. **29**, 103–112 (2019) [in Ukrainian]. https://doi.org/10.15421/4219009
19. Ma, C.C., Huang, C.H.: Experimental whole-field interferometry for transverse vibration of plates. J. Sound Vib. **271**(3–5), 493–506 (2004). https://doi.org/10.1016/S0022-460X(03)00276-1
20. Karlash, V.L.: Resonant electromechanical vibrations of piezoelectric plates. Int. Appl. Mech. **41**(7), 709–747 (2005). https://doi.org/10.1007/s10778-005-0140-2
21. Karlash, V.L.: Planar electroelastic vibrations of piezoceramic rectangular plate and half-disk. Int. Appl. Mech. **43**(5), 547–553 (2007). https://doi.org/10.1007/s10778-007-0053-3
22. Birger, I.A., Panovko, Y.G.: Strength. Stability. Vibrations. Vol. 3. Mashinostroenie, Moscow (1968) [in Russian]
23. Grigorenko, O.Ya., Borysenko, M.Yu., Boychuk, O.V.: Numerical Evaluation of Frequencies and the Modes of Free Vibrations of Isosceles Triangular Plates with Free Edges. J. Math. Sci. **273**, 27–43 (2023). https://doi.org/10.1007/s10958-023-06481-3
24. Waller, M.D.: Vibrations of free plates: line symmetry; corresponding modes. Proc. R. Soc. Lond. A. Math. Physic. Sci. **211**(1105), 265–276 (1952). https://doi.org/10.1098/rspa.1952.0038

Chapter 16
Selected Aspects of Thermomechanics of Ferrite Bodies Under Electromagnetic Actions

Oleksandr Hachkevych, Roman Ivas'ko, Roman Kushnir, and Anida Stanik-Besler

Abstract Physical and mathematical problems related to the construction of the proposed model of quantitative description and methods of studying the thermally stressed state of ferrite bodies caused by the action of low external quasi-steady electromagnetic fields (EMF) of high carrier frequency are discussed. The model is based on the general theory of the interaction of EMF and material continuum (which can be simultaneously magnetized and polarized in certain parts of the frequency spectrum of quasi-steady EMFs), in which the effect of electromagnetic radiation is taken into account by heat dissipations (associated with both remagnetization and repolarization, as well as Joule) and ponderomotive forces (at the statistical description of the power factors of the impact of radiation on the considered body), and also known from the literature for such materials' experimentally determined characteristics—complex magnetic and dielectric permeabilities and corresponding tangents of the loss angles. Solutions of new problems on determination of electromagnetic, temperature, and mechanical fields in a layer made from ferrite material under the influence of EMF caused by a given quasi-steady electric current, uniformly as well as periodically distributed along the spatial coordinate in the current-carrying plane parallel to the base of the layer, are obtained. A number of new regularities of the mechanical behavior of ferrite bodies subjected to external electromagnetic loads, in particular related to resonance phenomena, have been revealed.

O. Hachkevych · R. Ivas'ko (✉) · R. Kushnir
Pidstryhach Institute for Applied Problems of Mechanics and Mathematics,
National Academy of Science of Ukraine, Lviv, Ukraine
e-mail: roman_ivasko@ukr.net

O. Hachkevych
e-mail: dept13@iapmm.lviv.ua

A. Stanik-Besler
Opole University of Technology, Opole, Poland
e-mail: a.stanik-besler@po.edu.pl

H. Altenbach et al. (eds.), *Selected Problems of Solid Mechanics and Solving Methods*,
Advanced Structured Materials 204, https://doi.org/10.1007/978-3-031-54063-9_16

16.1 Introduction

In engineering practice, electrical and electronic devices with ferrite elements of various functional purposes, operated in external electromagnetic fields (EMF), including quasi-steady (the amplitudes of the electric and magnetic field strengths which change little over a period) are widely used. Often in various production processes, the technology of manufacturing such elements, in particular the annealing technology, involves the use of the above-mentioned fields. The influence of EMF on the ferrite body causes a number of interrelated processes of different physical nature (thermal, mechanical, electromagnetic, etc.) to occur in it, which significantly affect the strength and functional parameters of the corresponding ferrite elements. This is due to the fact that ferrites (due to their manufacturing method and mechanical properties) belong to the ceramic materials, for products of which even small stresses can cause significant changes in their electromagnetic functional parameters. Strictional phenomena are also inherent to them to varying degrees—the interconnectedness of deformation and magnetization. Therefore, it is important to study physical and mechanical processes in deformable solid ferrite bodies subjected to EMF to predict the functional properties and strength of ferrite elements of devices for various technical purposes.

16.2 State of the Problem

Studies of mechanical behavior of both ferromagnetic and dielectric bodies under complex loads (mechanical, electromagnetic, thermal); resonance oscillations and dissipative heating of piezoelectric and piezomagnetic ceramic bodies (using in the description of EMF the formalisms of forced electro- or magnetostatic approximations used in this type of problems); interconnection of arising processes and phenomena of various physical nature are widely known in the literature (Burak et al. [2]; Hachkevych et al. [10, 11]; Hachkevych and Drobenko [6]; Hachkevych and Kushnir [9]; Hachkevych, Ivas'ko and Stanik-Besler [8], etc.). At the same time, it is important to note several fundamental works dedicated to solving various physical and mathematical issues related to the formulation and methods of solving problems in this topic (Grigorenko and Molchenko [4, 5]; Grigorenko [3] among others). These works provide a refined formulation and a more complete account of the deformation features of shell- and plate-type elements widely used in technology caused by electromagnetic influence.

The researches of the interaction of EMF with the ferrite medium (exhibiting simultaneous magnetization and polarization) presented in the literature concern mainly the optimization (using the results of experimental studies) of operating modes of ferrite products subjected to EMF and the prediction of the initial characteristics of individual electrical and electronic devices. They are aimed at minimizing hysteresis losses in the material. Known studies are related to the manufacture of composites employing induction heating method using adhesive mixtures with ferrite fillers. However, they concern only the study of the process of heating ferrite materials with the help of EMF (e.g. Zhang et al. [20]).

The relationship between mechanical, thermal, and electromagnetic processes in the ferrite material body subjected to EMF, as well as its influence on the parameters of these processes at different amplitude-frequency characteristics of external EMF in the literature, are not sufficiently studied. The ability of the ferrite medium to both magnetize and polarize simultaneously in certain intervals of the frequency spectrum is not taken into account. Therefore, there is a need to build a mathematical model to quantitatively describe and study the stress state of ferrite bodies caused by the influence of quasi-steady EMF (QSEMF), taking into account the relationship of electrical conductivity, thermal conductivity, and deformation, as well as the characteristics of electromagnetic material properties depending on the frequency of EMF. This model is a theoretical basis for the rational design and development of electromagnetic devices with ferrite elements, as well as their operating modes and modes of processing such elements using EMF, while ensuring the functionality of both these elements and devices as a whole.

16.3 Research Results

We give a brief description of the state of the selected research in this problem and the works on thermomechanics of electrically conductive bodies capable of polarization and magnetization under external electromagnetic influences, which are close in direction, and we also consider the issues associated with the formulation of problems of thermomechanics of electrically conductive bodies and methods of their solution. It is established that at the present time individual studies relating to the problems of targeted fabrication of ferrite elements of devices for various technical applications are known. There are also studies of temperature fields and stresses in ferromagnetic bodies under selected external electromagnetic influences (Landau and Lifshits [14] and Starodubtsev [16], among others). However, studies of the thermal stress state of ferrite bodies due to the influence of QSEMF (taking into account the relationship of electrical conductivity, thermal conductivity, and deformation processes) for the rational design and development of electromagnetic devices with ferrite elements, as well as their operation and processing modes using EMF, while ensuring the functional ability of both these elements and devices in general are not covered enough in the literature.

A number of works are devoted to the description of physical and mechanical properties of ferrite materials subjected to QSEMF (e.g. Tsutaoka [19]; Hachkevych, Ivas'ko and Stanik-Besler [8]); their classification by magnetic and electrical properties (magnetically soft, electrically soft; magnetically hard, electrically hard; magnetostrictive, electrostrictive; materials with a rectangular hysteresis loop, etc.) has been made. According to their mechanical properties, they belong to the brittle materials. The electromagnetic effect is considered (at the characteristic parameters

$H_m < 10^4 \text{A/m}, \omega_m < 10^{10}\,\text{s}^{-1}$, where H_m is the highest value of the magnetic field strength in the ferrite body, and ω_m is the maximum carrier circular frequency of the EMF), in which it can be assumed that displacements, deformations, and their velocities in the studied brittle bodies are so small that the assumptions of linear elasticity theory are satisfied, and the influence of motion on EMF characteristics can be neglected. We consider ferrites, for which electromechanical and thermoelectric effects are insignificant. It is assumed that the functional properties of ferrite products are preserved at maximum temperatures lower than their Curie temperatures and stresses lower than those permitted by thermal strength. The medium is considered isotropic, in which the magnetization and polarization vectors are parallel to the vectors of magnetic and electric field strengths, respectively. During cyclic remagnetization and repolarization, the dependences between inductions and field strengths have the form of dynamic hysteresis loops, which are characterized by saturation. Loop shapes for specific materials depend on the frequency and amplitude of the external field. In the case of relatively low fields and high frequencies, the magnetization and polarization curves are elliptical. For elliptic dependence between inductions and strengths at harmonic strengths of magnetic and electric fields, complex permeabilities are used in the electrical engineering literature.

Ferritic material is a magnetic ceramic and in terms of its strength characteristics, it belongs to the brittle materials, for which there is a linear dependence between deformations and stresses up to the strength limit. Under typical operating conditions of products made of ferrite materials, there are low fields ($H_{max} < 10^3 \text{A/m}$) at high carrier frequencies ($\omega \geqslant 10^5\,\text{s}^{-1}$) containing a range where ferrite has the properties of both a ferromagnet and a dielectric. The same parameter values have the fields used in many known heat treatment technologies for ferrite products.

A variant of a mathematical model and a method of quantitative description of interrelated processes of electric conductivity, thermal conductivity, and deformation occurring in ferrite bodies subjected to EMF created by a system of quasi-steady electric currents of the radio frequency range with amplitude modulation given in the external environment ($V \subset E^3$) are developed. The density of these currents is described by the dependence:

$$\mathbf{J}_{**}^{(0)}(\mathbf{r}, t) = \text{Re}\left[\mathbf{J}^{0*}(\mathbf{r}, t)\right],$$
$$\text{div}\mathbf{J}_{**}^{(0)}(\mathbf{r}, t) = 0 \quad (\mathbf{r} \in V,\ t \in [0, \infty)),$$

where $\mathbf{J}^{0*}(\mathbf{r}, t) = \mathbf{J}_{*}^{(0)}(\mathbf{r}, t)e^{i\omega t}$ is complex vector of current density; $\mathbf{r}$ is radius-vector of considered point of space; t is time; $\mathbf{J}_{*}^{(0)}(\mathbf{r}, t) = J_0(t)\mathbf{J}_0(\mathbf{r})$; ω is circular frequency; $\mathbf{J}_0(\mathbf{r})$ is complex amplitude of carrying signal currents; $J_0(t)$ is function changing slowly over a period $f_* = 2\pi/\omega$ of electromagnetic oscillations, satisfying the condition of quasi-steady approximation (e.g. Biessonov [1]; Hachkevych, Ivas'ko and Stanik-Besler [8])

$$\left|\frac{dJ_0(t)}{dt}\right| \ll \omega\,|J_0(t)|\,. \tag{16.1}$$

Under the conditions mentioned above, which are adopted in the experimental characterization of ferrite materials and the derivation of material equations, the influence of QSEMF on the processes of heat conduction and deformation in the ferrite body, as in the known models for ferromagnetic and dielectric bodies, is accounted for through the Joule and hysteresis (associated with remagnetization and repolarization) heat dissipations and ponderomotive forces due to this field. Ponderomotive moments, due to the assumed parallelism of induction vectors and magnetic and electric field strengths, are equal to zero (Tamm [17]; Landau and Lifshits [14]; Hutter et al. [13]; Hachkevych et al. [12]; Hachkevych, Ivas'ko and Stanik-Besler [8], among others). In this approximation and at constant material characteristics, the initial relations for quantitative description of the parameters of thermal, mechanical, and electromagnetic processes in ferrite bodies subjected to QSEMF are formulated in two stages. At the first stage, based on Maxwell's equations, we wrote down the problems for determining EMF parameters and expressions for heat dissipations and ponderomotive forces (in their statistical description) as functions of electromagnetic parameters. At the second stage, based on the relations of the dynamic problem of thermoelasticity in stresses, we obtained the dependences describing the thermal and mechanical parameters at given initial and boundary conditions for the temperature T and stress tensor $\hat{\sigma}$ components, in which the heat sources and bulk forces are heat dissipations and ponderomotive forces found at the first stage (e.g. Siedov [15]; Hachkevych and Ivas'ko [7]; Hachkevych, Ivas'ko and Stanik-Besler [8]).

The case of low EMF at high frequencies, important for engineering applications, is considered, for which data on the values of magnetic and dielectric permeabilities and the corresponding loss angle tangents and their frequency dependences for different types of ferrite materials are available in the scientific and technical literature. An approximate method of investigating the physical and mechanical fields that exist under such electromagnetic influences has been proposed. Elliptic shape of hysteresis loops in low EMFs of high frequency allows to linearize the initial problem of electrodynamics by introducing approximate complex representations of electric and magnetic field strength vectors for such fields using the method of complex amplitudes (when describing EMF parameters in the first harmonic approximation):

$$\mathbf{A}_{**}(\mathbf{r},t) = \mathrm{Re}\left[\mathbf{A}^{*}(\mathbf{r},t)\right], \quad \mathbf{A} = \{\mathbf{E},\mathbf{H}\}, \tag{16.2}$$

where

$$\mathbf{A}^{*}(\mathbf{r},t) = \mathbf{A}_{*}(\mathbf{r},t)e^{i\omega t} = J_0(t)\mathbf{A}(\mathbf{r})e^{i\omega t}, \tag{16.3}$$

$\mathbf{A}_{*}(\mathbf{r},t)$—modulated complex amplitudes satisfying the condition defining the quasi-steady field character (Biessonov [1]; Hachkevych, Ivas'ko and Stanik-Besler [8], among others)

$$\left|\frac{\partial \mathbf{A}_{*}(\mathbf{r},t)}{\partial t}\right| \ll \omega\left|\mathbf{A}_{*}(\mathbf{r},t)\right|$$

(which follows approximately from (16.1)).

To describe the elliptic dependence between inductions $\mathbf{D}_{**}$, $\mathbf{B}_{**}$ and the corresponding field strengths $\mathbf{E}_{**}$, $\mathbf{H}_{**}$, the frequency-dependent complex permeabilities:

$$\varepsilon'_*(\omega) = \varepsilon'(\omega) - i\varepsilon''(\omega), \quad \mu'_*(\omega) = \mu'(\omega) - i\mu''(\omega)$$

and the corresponding tangents of the angles of hysteresis losses:

$$\tan\delta_\mathrm{d} = \frac{\varepsilon''}{\varepsilon'}, \quad \tan\delta_\mathrm{m} = \frac{\mu''}{\mu'}$$

known from the literature are introduced. In this case, the complex amplitudes of induction and electric and magnetic field strengths are related by dependences (e.g. Tamm [17]; Biessonov [1]; Hachkevych, Ivas'ko and Stanik-Besler [8]):

$$\mathbf{D}_*(\mathbf{r},t) = \varepsilon_0\,\varepsilon'_*(\omega)\mathbf{E}_*(\mathbf{r},t), \quad \mathbf{B}_*(\mathbf{r},t) = \mu_0\,\mu'_*(\omega)\mathbf{H}_*(\mathbf{r},t). \tag{16.4}$$

Taking into account the complex representations (16.2) and (16.3) of the field vectors and the phenomenological relations (16.4), the expressions for the power density of heat dissipations Q and the ponderomotive force $\mathbf{F}$ are obtained as a sum of two terms:

$$Q = Q^{(1)} + Q^{(2)}, \quad \mathbf{F} = \mathbf{F}^{(1)} + \mathbf{F}^{(2)}, \tag{16.5}$$

where $Q^{(1)}$ and $\mathbf{F}^{(1)}$ are slowly changing in time (coinciding with the bulk density of the corresponding value averaged over the period of EMF oscillations); $Q^{(2)}$ and $\mathbf{F}^{(2)}$ are quasi-periodic (whose total power in the period is zero) (Hachkevych, Ivas'ko and Stanik-Besler [8], among others).

Based on the structure of expressions for heat dissipations and ponderomotive forces (16.5), the temperature and stress tensor are represented as a sum of slowly time-varying $T^{(1)}$, $\hat{\sigma}^{(1)}$ and quasi-steady $T^{(2)}$, $\hat{\sigma}^{(2)}$ components:

$$T = T^{(1)} + T^{(2)}, \quad \hat{\sigma} = \hat{\sigma}^{(1)} + \hat{\sigma}^{(2)}.$$

In this case, the first of them are determined in the quasi-static formulation, neglecting the influence of the coupling of the strain and temperature fields, and the second are determined in the quasi-steady approximation (neglecting the influence of the initial stage of the process) (Hachkevych, Ivas'ko and Stanik-Besler [8] among others).

In order to obtain the regularities of mechanical behavior of ferrite bodies subjected to external QSEMFs, the parameters of the thermal stress state of a ferrite elastic layer of thickness h under the influence of EMF induced by the system of quasi-steady electric currents given in the external environment and uniformly distributed over a thin current-carrying layer (modeled by a plane parallel to the layer base) were determined based on the proposed method (Hachkevych, Ivas'ko and Stanik-Besler [8] among others) and using analytical and numerical methods (e.g. Tikhonov and Samarskii [18]). The plane is at a distance l from the surface of the considered ferrite layer. In this case, in order to carry out a broad analysis of the results, the solution of the corresponding coupled problem of thermomechanics was found approximately using a cubic approximation of the distributions of the quasi-steady components of temperature and stress along the thickness coordinate.

Numerical studies were performed for a layer coupled to a rigid half-space at $z = 0$ (displacement $u_z = 0$). The temperature fields and stresses in a layer of magnetically soft nickel-zinc ferrite 1000HH were quantitatively investigated at thicknesses $h = 0.03, 0.15, 0.30$ m and $l = 10^{-3}$ m. The frequency ω of the external EMF was selected from a list of amplitude-modulated waves (radio frequency range, ultra-short waves, and microwave television range) authorized for use in industry. The dielectric and magnetic permeabilities, as well as the tangents of the angles of hysteresis losses, were taken into account as a function of the field frequency. The upper base $z = h$ is free from mechanical loads. It has a convective heat exchange with the external environment (whose temperature is constant and equal to the initial temperature of the layer) under the Bio criterion $\mathrm{Bi} = 0.2$, and the lower base $z = 0$ is insulated.

Extensive studies of the thermomechanical behavior of the layer depending on material characteristics and field parameters have been carried out, in particular, with available resonance phenomena (Hachkevych and Ivas'ko [7]; Hachkevych, Ivas'ko and Stanik-Besler [8], among others).

Figure 16.1 shows thickness coordinate distributions in the layer ($h = 0.03$ m) of amplitudes of quasi-periodic stress components for the EMF resonance frequencies $\omega_{\text{res}}^{\text{I}} \approx 1.5 \cdot 10^5\ \text{s}^{-1}$ (Fig. 16.1a) and $\omega_{\text{res}}^{\text{II}} \approx 5.3 \cdot 10^5\ \text{s}^{-1}$ (Fig. 16.1b), where the temperature and stress levels increase significantly due to quasi-periodic in time components of heat dissipations and ponderomotive forces (when their frequency 2ω is close to one of the natural frequencies of thermoelastic (thickness) vibrations of the layer). The symbol σ_{F} marks the quasi-periodic components of the stresses due to the action of the ponderomotive force only ("force stresses"). The values of quasi-periodic stress components obtained at resonant frequencies can exceed the strength limits of the material and cause the destruction of the product.

To estimate the influence of inhomogeneity of external electric current distribution on the parameters of technological heating of flat ferrite elements, taking into account the stress state (strength characteristics), the thermomechanical behavior of the ferrite layer subjected to QSEMF created by the current flowing in the current-carrying plane, whose density has a sinusoidal character of change in coordinate y was determined and studied (Hachkevych, Ivas'ko and Stanik-Besler [8] among others).

According to the results given above, for a stable dependence of the heating process on the frequency of electromagnetic influence, it is assumed that the circular carrier frequency ω lies outside the vicinity of the EMF resonant frequencies $\omega_{\text{res}}^{\text{I}}$ and $\omega_{\text{res}}^{\text{II}}$. In this case, the quasi-static formulation was used to determine the stress state. The solution is obtained in three stages, which are reduced to the successive determination of the parameters describing the electromagnetic, temperature, and mechanical fields.

In order to build rational modes of the mentioned technological heating (in particular for technological training of ferrite products, low-temperature annealing to remove structural, mounting or other residual stresses, etc.) using EMF, quasi-static

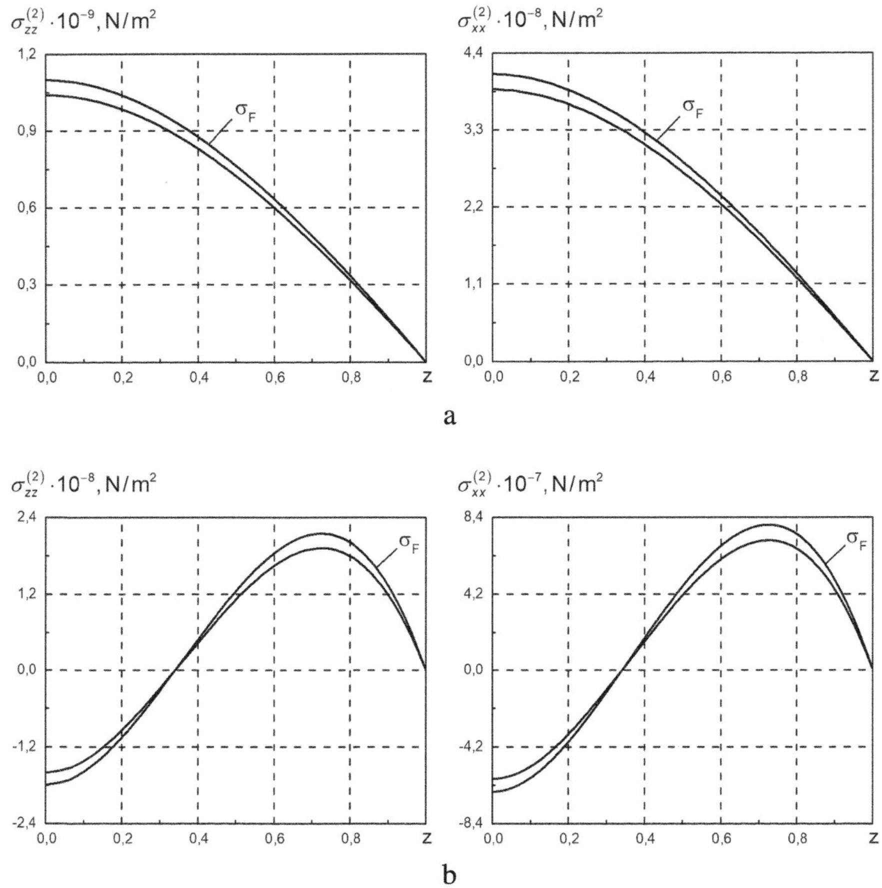

Fig. 16.1 Distributions of quasi-periodic stress components at resonance

components of temperature and stresses in the layer at a uniform current in the current-carrying plane were studied.

Selected issues related to the possibility of modeling and investigating the influence of striction effects on the thermomechanical behavior of ferrite bodies under electromagnetic loading were discussed (Hachkevych, Ivas'ko and Stanik-Besler [8] among others).

A variant of the mathematical model for quantitative description of mechanical and electromagnetic properties of ferrite striction bodies subjected to slowly changing in time EMF (at slowly changing in time magnetic field strength) was considered. In the framework of the quasi-static approximation, the determination of the parameters of the magnetoelastic state of such bodies is reduced to solving a system of coupled differential equations (in partial derivatives) of continuum mechanics and magnetostatics (taking into account the linearized material equations of magnetostriction).

The approaches of computer simulation of physico-mechanical processes in homogeneous bodies made of ferrite striction materials under external slowly changing in time magnetic loading are considered. Within the framework of such approaches, the parameters of electromagnetic and mechanical processes in a non-linear rod magnetostrictive transducer were investigated.

16.4 Conclusions

The article considers selected aspects of the performed studies of thermal and mechanical properties of ferrite bodies subjected to QSEMF to analyze the occurring physical and mechanical processes in order to predict the functional capability of ferrite elements of devices for various technical purposes (by permissible stresses and Curie temperatures). The main results obtained on this problem are as follows:

- a mathematical model for the quantitative description of the thermomechanical behavior of ferrite bodies subjected to QSEMF is constructed, taking into account the relatedness of the processes of electrical conductivity, thermal conductivity, and deformation. It is a development of known models for ferromagnetic materials subjected to EMF and is based on the general theory of interaction between EMF and material continuum (capable of simultaneous magnetization and polarization in certain parts of the frequency spectrum of EMF), in which the influence of electromagnetic radiation on physical and mechanical processes is accounted for through heat dissipations (associated with both remagnetization and repolarization, as well as Joule) and ponderomotive forces (at the statistical description of the power factors of the impact of radiation on the considered body as an external action in relation to it), as well as the dependences of magnetic and electric field inductions on the corresponding field strengths available for such materials;
- an approximate method for studying the thermal stress state of such bodies at moderate magnetic field strength and high frequencies ($H_{\mathrm{m}} < 10^3\,\mathrm{A/m}$, $\omega \geqslant 10^5\,\mathrm{s}^{-1}$) has been developed, based on the experimentally determined characteristics of specific ferrite materials—complex magnetic and dielectric permeabilities and corresponding tangents of loss angles—known from literature. This allowed us to introduce the formalism of approximate complex representations for EMF parameters with further reduction of the initial problem for determining these parameters to the problem on complex amplitudes of magnetic or electric field strengths (in the approximation of the basic harmonics of these strengths);
- using the proposed methodology, we obtained solutions to new complex problems of determining the electromagnetic, temperature, and mechanical fields in an elastic layer of ferrite material subjected to EMF induced by a given quasi-steady electric current with uniform, as well as sinusoidal, distribution along the coordinate in the current-carrying plane parallel to the layer base;

- the analysis of the solutions found revealed a number of new regularities in the mechanical behavior of ferrite bodies and peculiarities of their functional ability under the influence of external QSEMF. The main ones are
  - unlike ferromagnetic bodies, in ferrite bodies (as well as in bodies of low electrical conductivity (BLEC)) the type (linear or wave) of the amplitude distributions of EMF strengths, heat dissipations, and ponderomotive forces under the influence of an external EMF depends on the ratio between the characteristic size of the body in the direction of EMF propagation and its length in the ferrite medium, determined by both the electrical and magnetic properties of the material;
  - parametric resonance occurs in ferrite bodies (as well as in BLEC): significant increase of electric and magnetic field strengths in the vicinity of certain (resonance) EMF frequencies, caused by the appearance of stationary (standing) electromagnetic waves. The occurrence of standing waves depends on the hysteresis losses, which for ferrite bodies depend significantly on the EMF frequency and the magnetic (rather than electrical, as for BLEC) properties of the material;
  - as for linear relative to electric and magnetic properties of bodies, ferromagnetic bodies and BLEC, resonance occurs in ferrite bodies due to the periodic nature of changes in the time of heat dissipations and ponderomotive forces. For it, each of the resonant EMF frequencies is practically equal to half of the corresponding natural frequency of body vibrations (at the values of the coupling parameter ε_* of the strain and temperature fields are small—$\varepsilon_* < 10^{-2}$). In this case, approximate values of the first two frequencies of the resonant oscillations by thickness can be more easily obtained using a cubic approximation of the distributions of the quasi-steady components of temperature and stresses on the spatial variables;
  - the effect of ponderomotive forces on the stressed state of ferrite bodies is negligible with the exception of the vicinities of EMF frequencies, where resonance phenomena may occur. Then the influence of ponderomotive forces is decisive (in contrast to BLEC, where the influence of heat dissipations is decisive).

The constructed mathematical model and the proposed method for predicting the thermomechanical behavior of ferrite elements of electromagnetic devices make it possible to find solutions to new problems of electromagnetothermomechanics of ferrite bodies of a specific geometric configuration subjected to given types of QSEMF. The results obtained with its help can be used in engineering practice to estimate the dependence of stress levels in products on the characteristics of operating loads in order to predict the parameters of electrical and electronic devices with ferrite elements at their design.

The basic mechanical and magnetic relations of magnetoelasticity of magnetostrictive bodies under complex slowly changing in time mechanical and magnetic influences, often used as technological factors, are considered. Within the framework of the quasi-static approximation, the determination of the parameters of the magnetoelastic state of such bodies is reduced to the solution of a system of coupled partial differential equations of continuum mechanics and magnetostatics taking into account linearized material equations of magnetostriction.

The possibility of using modern computer engineering systems for modeling physical and mechanical processes occurring in magnetostrictive ferrite bodies subjected to slowly changing in time magnetic loading is discussed. Using the finite-element software COMSOL Multiphysics and basic mechanical and magnetic relations of magnetoelasticity of magnetostrictive bodies, the model problem of studying the parameters of electromagnetic and mechanical processes in a nonlinear rod magnetostrictive transducer is considered.

The presented results of scientific research were obtained mainly within the framework of international scientific cooperation of the Pidstryhach Institute for Applied Problems of Mechanics and Mathematics of the National Academy of Sciences of Ukraine and the Opole University of Technology (Poland).

References

1. Biessonov, L.A.: Theoretical Foundations of Electrical Engineering. Electromagnetic Field (in Russ.). Yurayt, Moscow (2018)
2. Burak, Ya., Hachkevych, O., Terlets'kyi, R.: Thermomechanics of Multicomponent Bodies of Low Electrical Conductivity (in Ukrain.). In: Burak, Ya., Kushnir, R. (Eds.) Modeling and Optimization in Thermomechanics of Electrically Conductive Inhomogeneous Bodies, vol 1. Spolom, L'viv (2006)
3. Grigorenko, Ya.: Approaches to the numerical solution of linear and nonlinear problems in shell theory in classical and refined formulations. Int. Appl. Mech. **32**(6), 409–442 (1996)
4. Grigorenko, Ya., Molchenko, L.: Basics of the Theory of Plates and Shells (in Ukrain.). Lybid', Kyiv (1993)
5. Grigorenko, Ya., Molchenko, L.: Basics of the Theory of Plates and Shells with Elements of Magnetoelasticity. Textbook (in Ukrain.). PPC Kyiv University, Kyiv (2010)
6. Hachkevych, O., Drobenko, B.: Thermomechanics of Magnetizable Electrically Conductive Thermosensitive Bodies (in Ukrain.). In: Burak, Ya., Kushnir, R. (Eds) Modeling and Optimization in Thermomechanics of Electrically Conductive Inhomogeneous Bodies, vol 4. Spolom, L'viv (2010)
7. Hachkevych, O., Ivas'ko, R.: Thermoelastic state of a ferritic layer in a quasistationary electromagnetic field. Mater. Sci. **41**(3), 376–387 (2005)
8. Hachkevych, O., Ivas'ko, R., Stanik-Besler, A.: Selected Mathematical Problems of Thermomechanics of Ferrite Solids (in Russ.). Pidstryhach Institute for Applied Problems of Mechanics and Mathematics NAS of Ukraine, L'viv (2022)
9. Hachkevych, O., Kushnir, R.: Selected problems of the mechanics of coupled fields. J. Math. Sci. **229**(2), 115–132 (2018)
10. Hachkevych, O., Kushnir, R., Terletskii, R.: Mathematical problems of thermomechanics for deformable bodies subjected to thermal irradiation. Ukrainian Math. J. **73**(10), 1522–1536 (2022)
11. Hachkevych, O., Terlets'kyi, R., Kournyts'kyi, T.: Mechanothermodiffusion in Partially Transparent Bodies (in Ukrain.). In: Burak, Ya., Kushnir, R. (Eds.) Modeling and Optimization in Thermomechanics of Electrically Conductive Inhomogeneous Bodies, vol 2. Spolom, L'viv (2007)
12. Hachkevych, O., Terletskii, R., Ivas'ko, R.: Modeling of electromagnetic, thermal, and mechanical processes in magnetic media with regard for the moment factors. J. Math. Sci. **256**(4), 497–517 (2021)

13. Hutter, K., van de Ven, A.A.F., Ursescu, A.: Electromagnetic Field-Matter Interactions in Thermoelastic Solids and Viscous Fluids. Lecture Notes in Physics, vol. 710. Springer, Berlin-Heidelberg (2006)
14. Landau, L.D., Lifshits, Ye.M.: Electrodynamics of Continuous Media (in Russ.). Fizmatlit, Moscow (2005)
15. Sedov, L.I.: Continuum Mechanics: In 2 vols (in Russ.). Lan', St. Petersburg (2004)
16. Starodubtsev, Yu.N.: Magnetically Soft Materials: Encyclopedic Dictionary-Reference Book (in Russ.). Tekhnosfera, Moscow (2011)
17. Tamm, I.Ye.: Fundamentals of Electricity Theory (in Russ.). Fizmatlit, Moscow (2003)
18. Tikhonov, A.N., Samarskii, A.A.: Equations of Mathematical Physics (in Russ.). Moscow State University Press, Moscow (2004)
19. Tsutaoka, T.: Frequency dispersion of complex permeability in Mn-Zn and Ni-Zn spinel ferrites and their composite materials. J. Appl. Phys. **93**(5), 2789–2796 (2003)
20. Zhang, X.K., Li, Y.F., Xiao, J.Q., Wetzel, E.D.: Theoretical and experimental analysis of magnetic inductive heating in ferrite materials. J. Appl. Phys. **93**(10), 7124–7126 (2003)

Chapter 17
The Influence of the Temperature Dependence of Thermomechanical Characteristics of FGM on the Thermostressed State of a Hollow Sphere

Halyna Harmatii, Bohdan Kalynyak, and Myroslav Kutniv

Abstract The uncoupled quasi-static problem of thermoelasticity for a hollow sphere was solved, taking into account the temperature dependence of the thermomechanical characteristics of the components of the functional gradient material (FGM). The temperature field has been determined, taking into account the complex heat exchange with media of time-varying temperature. The nonlinear non-stationary boundary value problem of thermal conductivity was solved by the numerical method of the lines. The corresponding problem of thermoelasticity was solved by its reduction to integral Fredholm equations of the second kind and by replacing the integral with a quadrature formula. For various models of homogenization of the properties of the selected metal-ceramic two-component material, the influence of the temperature dependence of thermophysical and mechanical characteristics on the thermostressed state of the hollow sphere has been investigated.

17.1 Introduction

Components of constructions of various shapes, such as pipes, burners, and pressure tanks, are often exposed to the destructive effects of the external environment. Transient thermal and mechanical processes in this case can cause the destruction of construction materials up to their destruction before reaching a steady state. For brittle materials, the main threat is rapid destruction, which is due, in particular, to small values of critical stresses compared to the modulus of elasticity. It is essential

H. Harmatii (✉) · B. Kalynyak · M. Kutniv
Pidstryhach Institute for Applied Problems of Mechanics and Mathematics, National Academy of Science of Ukraine, Lviv, Ukraine
e-mail: galynaharmatiy@gmail.com

B. Kalynyak
e-mail: b-kalynyak@litech.net

M. Kutniv
Rzeszow University of Technology, Rzeszow, Poland

H. Altenbach et al. (eds.), *Selected Problems of Solid Mechanics and Solving Methods*, Advanced Structured Materials 204, https://doi.org/10.1007/978-3-031-54063-9_17

to analyze the primary thermal processes and stresses caused by them while designing structural elements. Nonhomogeneous materials can be used to reduce critical values of the stresses. Paper [26] provides a solution for the transient thermal fields that arise in thick-walled vessels if there is convective heat exchange from the inner surface with an exponential change in the medium-superficial temperature over time.

Models and analytical-numerical methods for calculating the stress state in heterogeneous and anisotropic bodies of complex shapes are proposed, particularly in [6–8] and others. An overview of FGM manufacturing technologies is given in [5]. An analysis of publications of such studies up to 2010 is provided in [31]. Given the significant increase in research activity on FGM, the work of [1] is an attempt to highlight the most relevant topics and publications related to FGM. The tasks of determining the thermoelastic state and residual stresses in bodies of canonical shape, made of functional gradient materials with and without taking into account thermosensitivity, are considered in [1, 3, 4, 9, 15, 17, 18, 21–24, 27, 28]. These studies highlight the significance of examining the overall effects of time-varying thermal loads on structural elements made of FGM. They also emphasize the influence of temperature and coordinates on the properties of heterogeneous materials.

The paper proposes a method for determining the thermal stress state of a hollow sphere made of FGM, taking into account the temperature dependence of the material characteristics and time-varying temperatures of the environment on the surfaces. The calculated values of the thermal stress state of hollow spheres with FGM, the characteristics of which were determined based on the homogenization models of a simple mixture (see [13, 23]) and a modified model of a simple mixture (see [20]) with and without taking into account thermal sensitivity of the material, have been compared.

As the thermal conductivity boundary value problem for determining the distribution of the temperature field in the FGM of a hollow sphere under the conditions of complex heat exchange is nonlinear, numerical methods, namely the method of lines, were used to solve it. In this approach, the solution process consists of spatial discretization and numerical integration over time. The differential operator on the spatial variable has a divergent form. Therefore, similar to [11], we used the integro-interpolational method for the spatial difference approximation. For the numerical solving of the Cauchy problem for the system of ordinary differential equations, backward differentiation formulas from the first to the sixth orders of accuracy, which are known to be $A(\alpha)$-stable (see, e.g., [10]), were used. The stress state resulting from temperature and force loads is determined by reducing elasticity problems to integral equations using the direct integration method (see [2, 14, 30]).

17.2 Problem Statement

Let us consider a hollow sphere assigned to the spherical coordinate system (r, φ, ψ) with an inner radius r_0 and an outer radius R, manufactured of functionally graded material (FGM) in the radial direction with temperature-dependent characteristics. The inner and outer surfaces of the hollow sphere are subjected to time τ dependent force loads $p_1(\tau)$ and $p_2(\tau)$. Through them, convective heat exchange occurs with the media, the temperatures of which accordingly $t_{c0}(\tau)$ and $t_{cR}(\tau)$ are functions of time τ. The temperature t and thermal stress state of the sphere depend only on the radial variable r and time τ. The thermomechanical characteristics of the FGM hollow sphere are the coefficient of thermal conductivity $\lambda = \lambda(r, t)$, volumetric heat capacity $c = c_v(r, t)$, elasticity modulus $E = E(r, t)$, and Poisson's ratio $\nu = \nu(r, t)$; the coefficient of linear thermal expansion $\alpha = \alpha(r, t)$ are functions of the radial coordinate and temperature. It is necessary to determine the thermal stress state in such an inhomogeneous thermosensitive hollow sphere.

The stresses in the hollow thermosensitive sphere caused by the temperature field $t(r, \tau)$ and force loads $F_r(r)$ will be determined based on the model of unbound quasi-static thermoelasticity in terms of stresses, which includes (see [14])

- equilibrium equation

$$\frac{d}{dr}(r^3\sigma_r) = 2r^2\sigma(r) - r^3 F_r(r), \quad \sigma = \frac{1}{2}\sigma_r + \sigma_\varphi, \tag{17.1}$$

- strain-stress relations

$$\begin{aligned} e_r &= \frac{1+\nu(r,t)}{E(r,t)}\sigma_r - 2\frac{\nu(r,t)}{E(r,t)}\sigma + \Phi(r,t), \\ e_\varphi &= \frac{1-\nu(r,t)}{E(r,t)}\sigma - \frac{1}{2}\frac{1+\nu(r,t)}{E(r,t)}\sigma_r + \Phi(r,t), \end{aligned} \tag{17.2}$$

- compatibility equation $r\dfrac{de_\varphi}{dr} = e_r - e_\varphi$, presented in terms of stresses as

$$\frac{d}{dr}\left[\frac{1-\nu(r,t)}{E(r,t)}\sigma + \Phi(r,t)\right] = \frac{\sigma_r}{2}\frac{d}{dr}\left(\frac{1+\nu(r,t)}{E(r,t)}\right) - \frac{1+\nu(r,t)}{E(r,t)}\frac{F_r(r)}{2}, \tag{17.3}$$

- boundary conditions on surfaces

$$\sigma|_{r=r_0} = -p_1(\tau), \quad \sigma|_{r=R} = -p_2(\tau). \tag{17.4}$$

Here

$$\Phi(r,t) = \int_{t_p}^{t} \alpha(r, T)dT,$$

σ is the total stresses, σ_r and σ_φ are the radial and circumferential components of the stress, respectively, and t_p is the reference temperature at which there are no stresses. The temperature field $t = t(r, \tau)$ is defined as the solution of a nonlinear non-stationary boundary value problem of thermal conductivity

$$\frac{1}{r^2}\frac{\partial}{\partial r}\left(r^2\lambda(r,t)\frac{\partial t}{\partial r}\right) = c(r,t)\frac{\partial t}{\partial \tau}, \quad r_0 < r < R, \tag{17.5}$$

$$\left[\lambda(r,t)\frac{\partial t}{\partial r} - \beta_0(t - t_{c0}(\tau))\right]\Big|_{r=r_0} = 0, \tag{17.6}$$

$$\left[\lambda(r,t)\frac{\partial t}{\partial r} + \beta_R(t - t_{cR}(\tau))\right]\Big|_{r=R} = 0, \tag{17.7}$$

$$t|_{\tau=0} = t_p, \tag{17.8}$$

where β_0 and β_R are heat exchange coefficients from surfaces $r = r_0$ and $r = R$ correspondingly.

17.3 Solving the Heat Transfer Problem

We reduce the nonlinear boundary value problem (17.5), (17.6), (17.7), and (17.8) to the Cauchy problem for the system of ordinary differential equations by the method of lines.

For discretization along the radial coordinate, we will use the integro-interpolation method (see, e.g., [24]). For this, a uniform grid

$$\bar{\omega}_h = \{r_i = r_0 + ih,\ i = 0, 1, \ldots, n\}, \quad h = (R - r_0)/n$$

and the notations

$$r_{i-1/2} = r_i - 1/2h, \quad r_{i+1/2} = r_i + 1/2h, \quad w(r) = r^2\lambda(r,t)\frac{\partial t}{\partial r}$$

are introduced on the segment $[r_0, R]$.

Let us integrate equation (17.5) over the segment $r_{i-1/2} \leqslant r \leqslant r_{i+1/2}$. Then we will get the heat balance equation

$$w_{i+1/2} - w_{i-1/2} = \int_{r_{i-1/2}}^{r_{i+1/2}} r^2 c(r,t)\frac{\partial t}{\partial \tau} dr, \tag{17.9}$$

where $w_{i\pm 1/2} = w(r_{i\pm 1/2})$.

Let us replace the integral included in the balance equation with an approximate equality

$$\int_{r_{i-1/2}}^{r_{i+1/2}} r^2 c(r,t)\frac{\partial t}{\partial \tau} dr \approx h r_i^2 c(r_i, t_i)\frac{\partial t(r_i,\tau)}{\partial \tau},$$

where $t_i = t_i(\tau) = t(r_i, \tau)$. Then, instead of (17.9), we get the equation

$$\frac{w_{i+1/2} - w_{i-1/2}}{h r_i^2} = c(r_i, t_i)\frac{\partial t(r_i,\tau)}{\partial \tau}. \tag{17.10}$$

Let's express $w_{i\pm 1/2}$ in terms of the values $t(r, \tau)$ in the grid nodes. To do this, let us integrate the equality

$$\frac{\partial t}{\partial r} = \frac{w(r)}{r^2\lambda(r,t)}$$

on the segment $r_{i-1} \le r \le r_i$:

$$t_i - t_{i-1} = \int_{r_{i-1}}^{r_i} \frac{w(r)}{r^2\lambda(r,t)} dr \approx \frac{w_{i-1/2}}{r_{i-1/2}^2} \int_{r_{i-1}}^{r_i} \frac{dr}{\lambda(r,t)}. \tag{17.11}$$

Let

$$\left(\frac{1}{h}\int_{r_{i-1}}^{r_i} \frac{dr}{\lambda(r,t)}\right)^{-1} \approx \lambda\left(r_{i-1/2}, \frac{t_{i-1}+t_i}{2}\right) = a_i(t), \tag{17.12}$$

then equality

$$w_{i-1/2} \approx r_{i-1/2}^2 a_i(t) \frac{t_i - t_{i-1}}{h}$$

follows from (17.11). By substituting the expression $w_{i\pm 1/2}$ in (17.10), we obtain a system of ordinary differential equations

$$\frac{dt_i}{d\tau} = \frac{1}{hr_i^2 c(r_i, t_i)} \left(r_{i+1/2}^2 a_{i+1}(t) \frac{t_{i+1} - t_i}{h} - r_{i-1/2}^2 a_i(t) \frac{t_i - t_{i-1}}{h}\right),$$

$$i = 1, \ldots, n-1, \tag{17.13}$$

where $a_i(t)$ is determined by the equality (17.12) or

$$a_i(t) = 0.5\left[\lambda(r_{i-1}, t_{i-1}) + \lambda(r_i, t_i)\right].$$

Let us construct a differential analog of the boundary condition (17.6). To do this, according to the integro-interpolation method, we will integrate equation (17.5) on the segment $r_0 \le r \le r_{1/2}$:

$$w_{1/2} - w(r_0) = \int_{r_0}^{r_{1/2}} c(r,t) r^2 \frac{\partial t}{\partial \tau} dr.$$

Considering equalities

$$\int_{r_0}^{r_{1/2}} c(r,t) r^2 \frac{\partial t}{\partial r} dr \approx \frac{h}{2} r_0^2 c(r_0, t_0) \frac{\partial t(r_0, \tau)}{\partial \tau},$$

$$w_{1/2} \approx r_{1/2}^2 a_1(t) \frac{t_1 - t_0}{h}, \quad w(r_0) = r^2 \lambda(r,t) \frac{\partial t}{\partial r}\bigg|_{r=0} = r_0^2 \beta_0 (t_0 - t_{c0}(\tau)),$$

we get

$$\frac{dt_0}{d\tau} = \frac{2}{hr_0^2 c(r_0, t_0)} \left(r_{1/2}^2 a_1(t) \frac{t_1 - t_0}{h} - r_0^2 \beta_0 (t_0 - t_{c0}(\tau)) \right).$$

Similarly, we construct a semi-discrete analog of the boundary condition (17.7). Let us integrate equation (17.5) over the segment $[r_{n-1/2}, R]$:

$$w(r_n) - w_{n-1/2} = \int_{r_{n-1/2}}^{R} c(r, t) r^2 \frac{\partial t}{\partial \tau} dr,$$

where

$$\int_{r_{n-1/2}}^{R} c(r, \tau) r^2 \frac{\partial t}{\partial r} dr \approx \frac{h}{2} R^2 c(R, t_n) \frac{\partial t(r_n, \tau)}{\partial \tau},$$

$$w_{n-1/2} \approx r_{n-1/2}^2 a_n(t) \frac{t_n - t_{n-1}}{h},$$

$$w(r_n) = r^2 \lambda(r, \tau) \left. \frac{\partial t}{\partial r} \right|_{r=R} = -R^2 \beta_R (t_n - t_{cR}(\tau)).$$

We get

$$\frac{dt_n}{d\tau} = \frac{2}{hR^2 c(R, t_n)} \left(-R^2 \beta_R (t_n - t_{cR}(\tau)) - r_{n-1/2}^2 a_n(t) \frac{t_n - t_{n-1}}{h} \right).$$

Therefore, the nonlinear boundary value problem of heat conduction (17.5), (17.6), (17.7), and (17.8) is reduced to the Cauchy problem for the system of ordinary differential equations

$$\begin{aligned} \frac{dt_0}{d\tau} &= \frac{2}{hr_0^2 c(r_0, t_0)} \left(r_{1/2}^2 a_1(t) \frac{t_1 - t_0}{h} - r_0^2 \beta_0 (t_0 - t_{c0}(\tau)) \right), \\ \frac{dt_i}{d\tau} &= \frac{1}{hr_i^2 c(r_i, t_i)} \left(r_{i+1/2}^2 a_{i+1}(t) \frac{t_{i+1} - t_i}{h} - r_{i-1/2}^2 a_i(t) \frac{t_i - t_{i-1}}{h} \right), \\ & \qquad i = 1, \ldots, n-1, \\ \frac{dt_n}{d\tau} &= \frac{2}{hR^2 c(R, t_n)} \left(-R^2 \beta_R (t_n - t_{cR}(\tau)) - r_{n-1/2}^2 a_n(t) \frac{t_n - t_{n-1}}{h} \right), \\ t_i(0) &= t_p, \quad i = 0, \ldots, n. \end{aligned} \tag{17.14}$$

For large n, this Cauchy problem is stiff. Therefore, to solve the problem (17.14), backward differentiation formulas from the first to the sixth orders of accuracy, which are known to be $A(\alpha)$-stable, were used.

17.4 Determination of Stressed State

To determine the thermally stressed state of the sphere, we reduce problem (17.1) to (17.4) to Fredholm's integral equation of the second kind with respect to radial stress using the direct integration method (see [2, 30]). After integrating the equilibrium equation (17.1), written through the relative coordinate $\rho = r/R$, we obtain the relationship between the radial and total stresses

$$\rho^3 \bar{\sigma}_r(\rho, \tau) - \rho_1^3 \bar{\sigma}_r(\rho_1, \tau) = \int_{\rho_1}^{\rho} \eta^2 \left(2\bar{\sigma}(\eta, \tau) - \eta \bar{F}_r(\eta)\right) d\eta, \tag{17.15}$$

where $\rho_1 = r_0/R$, $\bar{F}_r(\rho) = F_r(\rho R)$, $\bar{\sigma}(\rho, \tau) = \sigma(\rho R, t(\rho, \tau))$, $\bar{\sigma}_r(\rho, \tau) = \sigma_r(\rho R, t(\rho, \tau))$. When $\rho = R/R = 1$, from the expression (17.15), we obtain the integral condition that must be satisfied by the total stresses $\bar{\sigma}(\eta, \tau)$

$$\int_{\rho_1}^{1} \eta^2 \left[2\bar{\sigma}(\eta, \tau) - \eta \bar{F}_r(\eta)\right] d\eta = \rho_1^3 p_1(\tau) - p_2(\tau). \tag{17.16}$$

As a result of integrating the compatibility equation (17.3), we obtain the dependence between the total $\bar{\sigma}(\eta, \tau)$ and radial stresses $\bar{\sigma}_r(\eta, \tau)$:

$$\begin{aligned}\bar{\sigma}(\rho, \tau) - \frac{1}{2} \frac{\bar{E}(\rho, \tau)}{1 - \bar{\nu}(\rho, \tau)} \int_{\rho_1}^{\rho} \bar{\sigma}_r(\eta, \tau) \varphi'(\eta, \tau) d\eta + \\ = \frac{\bar{E}(\rho, \tau)}{1 - \bar{\nu}(\rho, \tau)} \left[B(\tau) - \frac{1}{2} \int_{\rho_1}^{\rho} \bar{F}_r(\eta) \varphi(\eta, \tau) d\eta - \bar{\Phi}(\rho, \tau) \right].\end{aligned} \tag{17.17}$$

Here

$$\varphi(\rho, \tau) = \frac{1 + \bar{\nu}(\rho, \tau)}{\bar{E}(\rho, \tau)}, \quad \varphi'(\rho, \tau) = \frac{d}{d\rho} \left(\frac{1 + \bar{\nu}(\rho, \tau)}{\bar{E}(\rho, \tau)} \right),$$

$$\bar{\nu}(\rho, \tau) = \nu(\rho R, t(\rho, \tau)), \quad P(r, t) = P(\rho R, t(\rho R, \tau)) = \bar{P}(\rho, \tau)$$

are thermomechanical characteristics of FGM,

$$\bar{\Phi}(\rho, \tau) = \int_{t_p}^{t(\rho R, \tau)} \alpha(\rho R, T) dT,$$

$B(\tau)$ is a time function determined from the integral condition (17.16).

From the expression (17.15) and the equilibrium equation (17.1), we have

$$\frac{1}{2\rho^2}\frac{d}{d\rho}(\rho^3\bar{\sigma}_r(\rho,\tau))+\frac{\rho\bar{F}_r(\rho)}{2}=\frac{1}{2}\frac{\bar{E}(\rho,\tau)}{1-\bar{\nu}(\rho,\tau)}\int_{\rho_1}^{\rho}\bar{\sigma}_r(\eta,\tau)\varphi'(\eta)d\eta$$
$$+\frac{\bar{E}(\rho,\tau)}{1-\bar{\nu}(\rho,\tau)}B(\tau)-F(\rho,\tau)-\frac{\bar{E}(\rho,\tau)}{1-\bar{\nu}(\rho,\tau)}\bar{\Phi}(\rho,\tau), \tag{17.18}$$

where

$$F(\rho,\tau)=\frac{1}{2}\frac{\bar{E}(\rho,\tau)}{1-\bar{\nu}(\rho,\tau)}\int_{\rho_1}^{\rho}\bar{F}_r(\eta)\varphi(\eta,\tau)d\eta.$$

As a result of integration of the integro-differential equation (17.18) in the range from ρ_1 to ρ along the radial variable and transformation of repeated integrals using the formula of integration by parts, we obtain the integral equation with respect to the radial stresses $\bar{\sigma}_r(\rho,\tau)$:

$$\bar{\sigma}_r(\rho,\tau)-\frac{1}{\rho^3}\int_{\rho_1}^{\rho}[V(\rho,\tau)-V(\eta,\tau)]\varphi'(\eta,\tau)\bar{\sigma}_r(\eta,\tau)d\eta$$
$$=2B(\tau)\frac{V(\rho,\tau)}{\rho^3}+H(\rho,\tau)+Q(\rho,\tau)-\frac{\rho_1^3}{\rho^3}p_1(\tau), \tag{17.19}$$

where

$$V(\rho,\tau)=\int_{\rho_1}^{\rho}\eta^2\frac{\bar{E}(\rho,\tau)}{1-\bar{\nu}(\rho,\tau)}d\eta,$$
$$H(\rho,\tau)=-\frac{1}{\rho^3}\int_{\rho_1}^{\rho}\left(2\eta^2F(\eta,\tau)+\eta^3\bar{F}_r(\eta)\right)d\eta,$$
$$Q(\rho,\tau)=-\frac{2}{\rho^3}\int_{\rho_1}^{\rho}\eta^2\frac{\bar{E}(\rho,\tau)}{1-\bar{\nu}(\rho,\tau)}\bar{\Phi}(\eta,\tau)d\eta.$$

Substituting the value $\rho=1$ into the expression (17.19) and taking into account the boundary conditions (17.4), we determine $2B(\tau)$:

$$2B(\tau)=-\frac{1}{2}\int_{\rho_1}^{1}\left[1-\frac{V(\eta,\tau)}{V(1,\tau)}\right]\varphi'(\eta,\tau)\bar{\sigma}_r(\eta,\tau)d\eta+$$
$$-\frac{H(1,\tau)}{V(1,\tau)}-\frac{Q(1,\tau)}{V(1,\tau)}+\frac{\rho_1^3p_1(\tau)}{V(1,\tau)}-\frac{p_2(\tau)}{V(1,\tau)}. \tag{17.20}$$

Thus, the integral equation for determining the radial stresses, obtained after substituting (17.20) into (17.19) and summing similar terms has the form:

$$\bar{\sigma}_r(\rho,\tau)+\int_{\rho_1}^{1}K(\rho,\eta,\tau)\bar{\sigma}_r(\eta,\tau)d\eta=\Psi(\rho,\tau), \tag{17.21}$$

where

$$\Psi(\rho,\tau) = H(\rho,\tau) - \frac{H(1)}{\rho^3}\frac{V(\rho,\tau)}{V(1,\tau)} + Q(\rho,\tau) - \frac{Q(1)}{\rho^3}\frac{V(\rho,\tau)}{V(1,\tau)} + \frac{\rho_1^3 p_1(\tau)}{\rho^3}\left[\frac{V(\rho,\tau)}{V(1,\tau)} - 1\right] - \frac{p_2(\tau)}{\rho^3}\frac{V(\rho,\tau)}{V(1,\tau)}, \tag{17.22}$$

$$K(\rho,\eta,\tau) = \begin{cases} V(\eta,\tau)\left[1 - \dfrac{V(\rho,\tau)}{V(1,\tau)}\right]\dfrac{\varphi'(\eta,\tau)}{2\rho^3}, & \eta \leqslant \rho, \\ V(\rho,\tau)\left[1 - \dfrac{V(\eta,\tau)}{V(1,\tau)}\right]\dfrac{\varphi'(\eta,\tau)}{2\rho^3}, & \eta > \rho. \end{cases} \tag{17.23}$$

The integral equation (17.21) is a Fredholm equation of the second kind, from the solution of which it is possible to determine the radial stresses $\bar{\sigma}_r(\rho,\tau)$ taking into account the mechanical conditions (17.4) at the boundaries and the defined temperature field. The kernel (17.23) of the integral equation is continuous and depends only on the physical characteristics of the material and their derivatives with respect to the radial variable. The right-hand part (17.22) of the integral equation (17.21) contains terms $H(\rho,\tau)$ and $Q(\rho,\tau)$, which take into account the influence of mass forces and the temperature field on the radial stresses, respectively. The total stresses can be determined by formulas (17.17), (17.20), circumferential stresses by the formula $\bar{\sigma}_\varphi(\rho,\tau) = \bar{\sigma}(\rho,\tau) - 1/2\bar{\sigma}_r(\rho,\tau)$, and deformations by the formulas (17.2), and radial displacements $\bar{u}_r(\rho,\tau)$ by formulas $e_\varphi(r, t(r,\tau)) = u_r(r,\tau)/r$, where $\bar{u}_r(\rho,\tau) = u_r(\rho R, t(\rho R,\tau))$.

To determine the solution of the integral equation (17.21), let us divide the sphere into n thin concentric layers with conditions of ideal mechanical contact between them. We will assume that the thickness of the kth layer is such that the trapezoid formula is satisfied for the integrand function

$$\int_{\rho_1}^{1} f(\eta)d\eta \approx \frac{1}{2}\sum_{i=1}^{n}\left[\rho_{i+1} f(\rho_{i+1}) + \rho_i f(\rho_i)\right]\Delta\eta_i, \quad \Delta\eta_i = \rho_{i+1} - \rho_i.$$

Then the integral equation (17.21) is given in the form:

$$\bar{\sigma}_r(\rho,\tau) + \sum_{k=1}^{n}\Big[\frac{\rho_{k+1} - \rho_k}{2}\big(K(\rho,\rho_k,\tau)\bar{\sigma}_r(\rho_k,\tau) + K(\rho,\rho_{k+1},\tau)\bar{\sigma}_r(\rho_{k+1},\tau)\big)\Big] = \Psi(\rho,\tau). \tag{17.24}$$

From this equality, written for all values of $\rho = \rho_i$ $(i = 1,\dots,n+1)$ in the case of a uniform grid $\omega_h = \{\rho_i = (i-1)h,\ i = 1,\dots,n+1,\ h = (\rho_{n+1} - \rho_1)/n\}$, the following system of linear algebraic equations with respect to $n+1$ unknowns $\bar{\sigma}_r(\rho_i,\tau)$ $(i = 1,\dots,n+1)$ is obtained:

$$
\begin{aligned}
&\left(1 + \tfrac{1}{2} h K_{11}(\tau)\right) \bar{\sigma}_r(\rho_1, \tau) + h K_{12}(\tau) \bar{\sigma}_r(\rho_2, \tau) \\
&\qquad + \cdots + \tfrac{1}{2} h K_{1,n+1}(\tau) \bar{\sigma}_r(\rho_{N+1}, \tau) = \Psi(\rho_1, \tau), \\
&\tfrac{1}{2} h K_{21}(\tau) \bar{\sigma}_r(\rho_1, \tau) + \left(1 + h K_{22}(\tau)\right) \bar{\sigma}_r(\rho_2, \tau) \\
&\qquad + \cdots + \tfrac{1}{2} h K_{2,n+1}(\tau) \bar{\sigma}_r(\rho_{n+1}, \tau) = \Psi(\rho_2, \tau), \\
&\cdots\cdots\cdots\cdots\cdots\cdots\cdots\cdots\cdots\cdots\cdots\cdots \quad \cdots\cdots\cdots \\
&\tfrac{1}{2} h K_{n+1,1}(\tau) \bar{\sigma}_r(\rho_1, \tau) + h K_{12}(\tau) \bar{\sigma}_r(\rho_2, \tau) \\
&\qquad + \cdots + \left(1 + \tfrac{1}{2} h K_{n+1,n+1}(\tau)\right) \bar{\sigma}_r(\rho_{n+1}, \tau) = \Psi(\rho_{n+1}, \tau),
\end{aligned}
\tag{17.25}
$$

where $K_{ij}(\tau) = K(\rho_i, \rho_j, \tau)$ and ρ_{n+1} is the radius of the outer surface of the sphere.

The system of linear algebraic equations (17.25) is solved by the Gaussian method.

In the case of a thin hollow sphere (spherical shell), instead of the system of equations (17.25), one equation is obtained, from which the radial stress is determined by the approximate formula

$$
\bar{\sigma}_r(\rho, \tau) = \frac{\Psi(\rho, \tau) + \dfrac{\rho - \rho_1}{2} K(\rho, \rho_1, \tau) p_1(\tau) + \dfrac{\rho_2 - \rho}{2} K(\rho, \rho_2, \tau) p_2(\tau)}{1 + \dfrac{\rho_2 - \rho_1}{2} K(\rho, \rho, \tau)}.
$$

17.5 Numerical Investigations

As a numerical example, a hollow sphere with an inner radius $r_0 = 0.01$ m and an outer radius $R = 0.03$ m is considered. Through the surfaces and convective heat exchange occurs with the media with temperature dependencies on time on the surfaces $t_{c0}(\tau) = (t_c - t_p)(1 - e^{-k_0\tau})$ and $t_{cR}(\tau) = (t_c - t_p)(1 - e^{-k_R\tau})$, where $t_p = 293$ K, $t_c = 873$ K, and $k_0 = k_R = 1000$. The calculation of the temperature distribution was carried out at the given values of the heat exchange coefficients $\beta_0 = \beta_R = 750\,\mathrm{W/(m \cdot K)}$. The materials of the sphere are two-component metal-ceramic FGM, the characteristics of which are described by two different models—a simple mixture model (Voigt model, from now on referred to as Model 1) (see [24]) and a generalized simple mixture model (hereinafter Model 2) (see [20]). The temperature dependencies of the characteristics of the jth ($j = 1, 2$) component of the FGM $P_j(t)$ are given in the form (see [13, 24]):

$$
P_j(t) = P_{j0}\left(\frac{P_{j,-1}}{t} + 1 + P_{j1}t + P_{j2}t^2 + P_{j3}t^3\right), \tag{17.26}
$$

where $P_{j,-1}$, P_{j0}, P_{j1}, P_{j2}, and P_{j3} are coefficients of the temperature dependence of the physical and mechanical characteristics of the jth component of the material.

Then, the characteristics $P(r,t)$ (see [13, 19]) of the FGM will be written as

$$P(r,t) = P_1(t)S_1(r) + P_2(t)S_2(r), \tag{17.27}$$

for Model 1 and

$$\lambda = \lambda_1\left[1 + \frac{3(\lambda_2-\lambda_1)S_2}{3\lambda_1 + (\lambda_2-\lambda_1)S_1}\right], \quad E = \frac{E_1\left[E_1 + (E_2-E_1)S_2^{2/3}\right]}{E_1 + (E_2-E_1)\left(S_2^{2/3} - S_2\right)}, \tag{17.28}$$

$$\alpha = \frac{\alpha_2 S_2 E_2/(1-\nu_2) + \alpha_1 S_1 E_1/(1-\nu_1)}{S_2 E_2(1-\nu_2) + S_1 E_1/(1-\nu_1)}, \quad \nu = \nu_2 S_2 + \nu_1 S_1, \tag{17.29}$$

$$\gamma = \gamma_2 S_2 + \gamma_1 S_1, \quad c_p = \frac{c_2\gamma_2 S_2 + c_1\gamma_1 S_1}{\gamma_2 S_2 + \gamma_1 S_1}, \quad c = c_p\gamma \tag{17.30}$$

for Model 2, where c_1, c_2, λ_1, λ_2, E_1, E_2, α_1, α_2, ν_1, ν_2 are temperature-dependent functions (17.26); $S_1(r)$ and $S_2(r)$ are the volume fractions of the components ($S_1(r) + S_2(r) = 1$), γ_1 and γ_2 are, accordingly, the density of the metal and ceramics. We believe that $S_1(r) = ((r-r_0)/(R-r_0))^n$, $n = 2$. In our case, index one corresponds to metal, and index two to ceramics.

Let us determine the thermal stress state of the sphere for the above-described models of thermomechanical characteristics with components titanium alloy Ti_6Al_4V (component 1) and zirconium dioxide ZrO_2 (component 2), the temperature dependencies of which are taken from [29]:

- for titanium alloy

$$\begin{aligned}
&\lambda_1(t) = 1.1 + 0.017t\ \mathrm{W/(m\cdot K)}, \quad \mathrm{c_1(t) = c_{p1}(t)}\gamma_1\mathrm{(t)},\\
&c_{p1}(t) = 3.5\cdot 10^2 + 0.878t - 9.74\cdot 10^{-4}t^2 + 4.43\cdot 10^{-7}t^3\ \mathrm{J/kg\cdot K},\\
&\gamma_1(t) = 4420.0/[1.0 + \alpha_1(t)(t-300.0)]^3\ \mathrm{kg/m^3},\\
&\alpha_1(t) = 7.43\cdot 10^{-6} + 5.56\cdot 10^{-9}t - 2.69\cdot 10^{-12}t^2\ 1/\mathrm{K},\\
&E_1(t) = 122.7 - 0.0565t\ \mathrm{GPa}, \quad \nu_1(t) = 0.2888 + 32.0\cdot 10^{-6}t,
\end{aligned} \tag{17.31}$$

- for ceramics

$$\begin{aligned}
&\lambda_2(t) = 1.71 + 0.21\cdot 10^{-3}t + 0.116\cdot 10^{-6}t^2\ \mathrm{W/(m\cdot K)},\\
&c_{p2}(t) = 2.74\cdot 10^2 + 0.7.95t - 6.19\cdot 10^{-4}t^2 + 1.71\cdot 10^{-7}t^3\ \mathrm{J/kg\cdot K},\\
&\gamma_2(t) = 3657/[1.0 + \alpha_2(t)(t-300.0)]^3\ \mathrm{kg/m^3}, \quad c_2(t) = c_{p2}(t)\gamma_2(t),\\
&\alpha_2(t) = 13.31\cdot 10^{-6} - 18.9\cdot 10^{-10}t + 12.7\cdot 10^{-12}t^2\ 1/\mathrm{K},\\
&E_2(t) = 132.2 - 50.3\cdot 10^{-3}t - 8.1\cdot 10^{-6}t^2\ \mathrm{GPa}, \quad \nu_2(t) = 0.333.
\end{aligned} \tag{17.32}$$

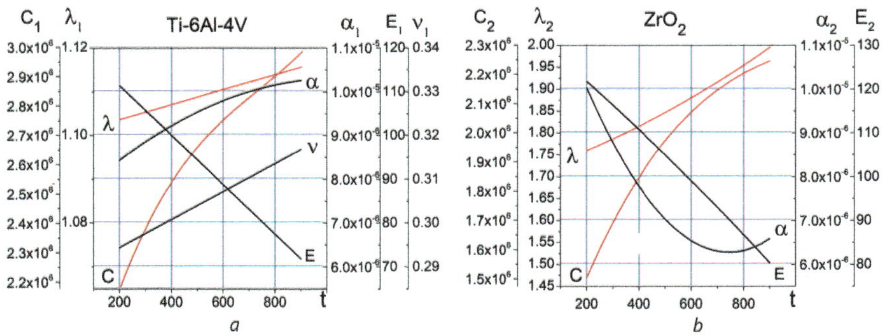

Fig. 17.1 Temperature-dependent characteristics of ceramics and metal (**a**—for metal, **b**—for ceramics)

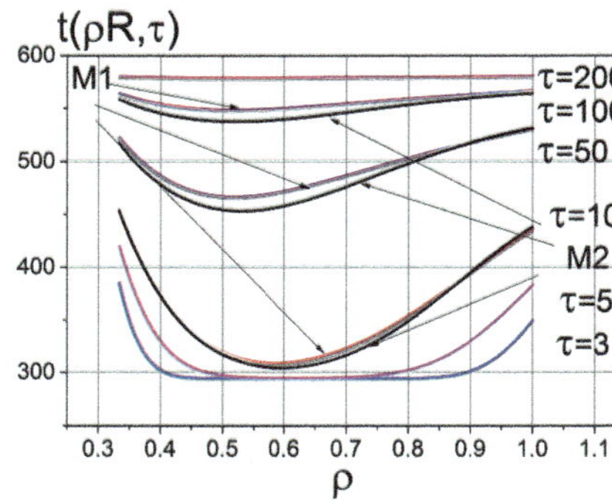

Fig. 17.2 Temperature distribution at different time moments for different homogenization models

Temperature dependencies of FGM components, taking into account thermal sensitivity, calculated according to formulas (17.31) and (17.32), are presented in Fig. 17.1. All characteristics are given in the SI system, modulus of elasticity in Gigapascals. The constant Poisson's ratio of ceramics is not shown in Fig. 17.1b. The results of the calculated distributions of the temperature field based on formulas (17.14) for the dependence on temperature and coordinates of material characteristics (17.26)–(17.32) are shown in Fig. 17.2. Designations M1 and M2 correspond to homogenization Models 1 and 2. The differences between the temperature values for the time moments $\tau = 3, 5, 200$ s in both homogenization models do not exceed 5 K. Therefore, the graphs for Model 2 are not shown in the figure.

Temperature distributions for a nonthermosensitive ($t = 300$ K) sphere are not shown in the figure because they slightly differ from those given. Figure 17.3 shows the coordinate distributions of the characteristics of the sphere material, which correspond to the temperature values in Fig. 17.2 for the moments of time $\tau = 10, 50, 200$ s with and without taking into account the temperature dependence of the FGM properties for different homogenization models. In all the figures, dashed lines show dependencies corresponding to the components' temperature-independent characteristics.

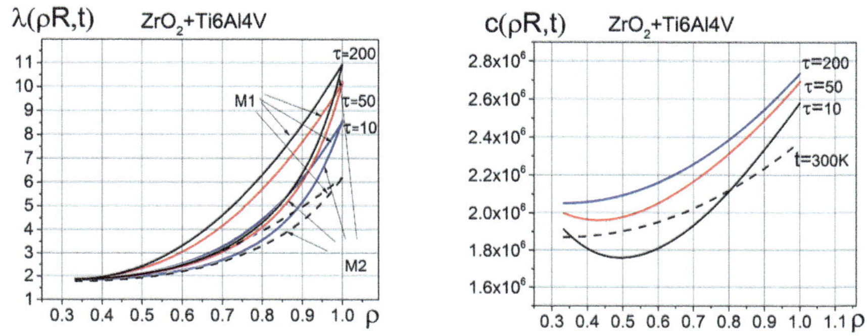

Fig. 17.3 Radial dependence of heat transfer coefficient and volume heat for different time moments and homogenization models

Fig. 17.4 Radial dependence of linear thermal expansion coefficient, elasticity modulus, Poisson ratio for different time moments at temperature 300 K and taking into account temperature dependence of material characteristic

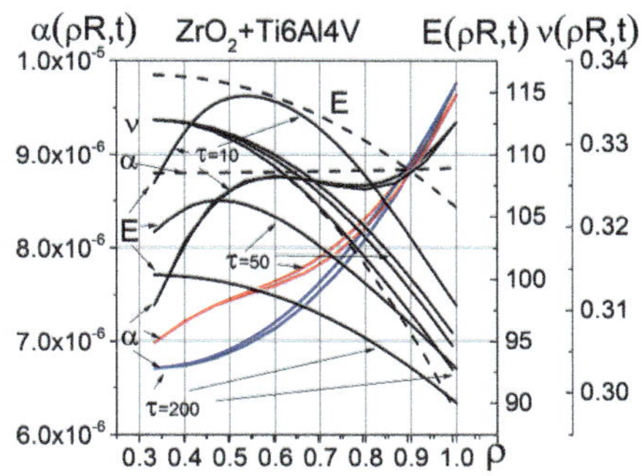

Volumetric heat capacity and Poisson's ratio in Fig. 17.3 are calculated according to the same formulas in both models, so they are presented only for the homogenization Model 1. The numerical values of the coefficient of linear thermal expansion and the modulus of elasticity calculated according to the formulas (17.27), (17.28), (17.29), and (17.30) are shown in Fig. 17.4 for the case of their different numerical values.

From Figs. 17.2, 17.3, and 17.4, it can be seen that taking into account the temperature dependence of the characteristics of the components has a much more significant effect on the properties of FGM compared to the choice of the homogenization model. From Fig. 17.4, it can be seen that the joint consideration of the temperature dependence, the heterogeneity of the FGM characteristics, and the temperature field lead to specific changes in the FGM characteristics in terms of the thickness of the sphere over time.

From Fig. 17.5, it follows that taking into account the temperature dependence of the FGM characteristics for the selected pair of its components has a much more significant effect on the stress distribution compared to the choice of the homogenization model.

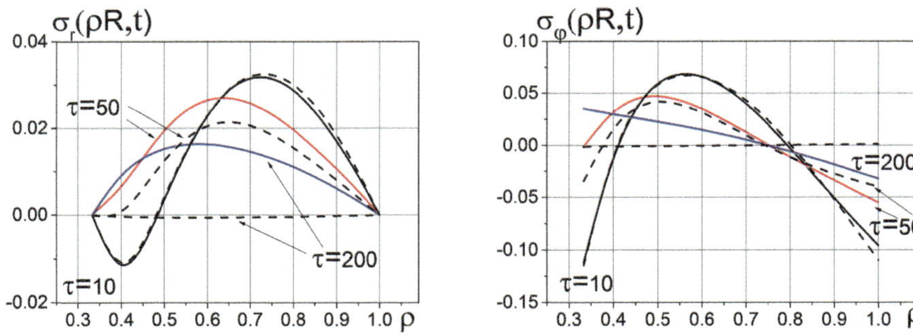

Fig. 17.5 Distribution of radial and circumferential stresses for different time moments at temperature 300 K and with taking into account temperature dependence of material characteristic

The effect of taking into account the temperature dependence of the characteristics of FGM components compared to a nonthermosensitive material can be more than 25% for circumferential stresses.

17.6 Conclusions

Analytical-numerical methods for solving the quasi-static problem of thermoelasticity are proposed, taking into account the dependence of material characteristics on coordinates and temperature. The nonlinear heat conduction problem is reduced by the integro-interpolation method with respect to the radial coordinate to the Cauchy problem for the system of ordinary differential equations with respect to time. The resulting system is solved by backward differentiation formulas that are $A(\alpha)$-stable. The corresponding problem of elasticity at a defined temperature field is solved by the method of reduction to the Fredholm integral equation of the second kind, which is solved using a quadrature scheme.

The relatively large difference between the corresponding temperature dependencies of the components of the selected FGM almost does not affect the distribution of temperature and stresses in different models of homogenization. The effect of the temperature dependence on the physical and mechanical characteristics of the material has a much more significant influence on the distribution of temperature and, especially, stresses. The most significant difference between the stresses calculated at a constant temperature and taking into account the temperature dependence of the FGM characteristics was obtained at times that correspond to the stationary temperature distribution. The largest voltage values are obtained during transient processes at the initial moments of time. In these temperature limits, the thermal stresses do not reach values commensurate with FGM's compressive and tensile strength limits.

References

1. Akbarzadeh, A.H., Chen, Z.T.: Transient heat conduction in functionally graded hollow cylinders and spheres. In: Proceedings of the ASME 2012 Pressure Vessels and Piping Conference, PVP2012, July 15–19, Toronto, Ontario, Canada, pp. 41–47 (2012). https://doi.org/10.1115/PVP2012-78617
2. Artemyuk, V.Yu., Kalynyak, B.M.: Integral equation for the radial stresses in a radially inhomogeneous hollow sphere. J. Math. Sci. **223**(2), 132–144 (2017). https://doi.org/10.1007/s10958-017-3343-2
3. Ching, H.K., Chen, J.K.: Thermal stress analysis of functionally graded composites with temperature-dependent material properties. J. Mech. Mater. Struct. **2**(4), 633–653 (2007)
4. Dhitala, S., Rokayaa, A., Kaizer, M.R., Zhang, Yu., Kima, J.: Accurate and efficient thermal stress analyses of functionally graded solids using incompatible graded finite elements. Compos. Struct. **222**(110909), 1–13 (2019). https://doi.org/10.1016/j.compstruct.2019.110909
5. El-Galy, I.M., Saleh, B.I., Ahmed, M.H.: Functionally graded materials classifications and development trends from industrial point of view. SN Appl. Sci. **1**(1378), 1–23 (2019). https://doi.org/10.1007/s42452-019-1413-4
6. Grigorenko, Y.M., Rozhok, L.S.: Equilibriun of elastic hollow inhomogeneous cylinders with cross sections in the form of convex semicorrugations. J. Math. Sci. **220**(2), 133–148 (2017). https://doi.org/10.1007/s10958-016-3172-8
7. Grigorenko, Y.M., Grigorenko, A.Y., Zakhariichenko, L.I.: Analysis of influence of the geometrical parameters of elliptic cylindrical shells with variable thickness on their stress-strain state. Int. Appl. Mech. **54**, 155–162 (2018). https://doi.org/10.1007/s10778-018-0867-1
8. Grigorenko, Y.M., Vasilenko, A.T.: The effect of inhomogeneity of elastic properties on the stress state in composite cylindrical panels. Mech. Compos. Mater. **37**(2), 85–90 (2001). https://doi.org/10.1023/A:1010663916881
9. Gupta, A., Talha, M.: Recent development in modeling and analysis of functionally graded materials and structures. Prog. Aerosp. Sci. **79**, 1–14 (2015). https://doi.org/10.1016/j.paerosci.2015.07.001
10. Hairer, E., Wanner, G.: Solving Ordinary Differential Equations II: Stiff and Differential-Algebraic Problems. Springer, Berlin, New York (2002)
11. Harmatiy, G.Yu., Kalynyak, B.M., Kutniv, M.V.: Uncoupled quasistatic problem of thermoelasticity for a two-layer hollow thermally sensitive cylinder under the conditions of convective heat exchange. J. Math. Sci. **96**(4), 439–454 (2021). https://doi.org/10.1007/s10958-021-05437-9
12. Jha, D.K., Kant, T., Singh, R.K.: A critical review of recent research on functionally graded plates. Compos. Struct. **96**, 833–849 (2013)
13. Kiani, Y., Eslami, M.R.: Thermal postbuckling of imperfect circular functionally graded material plates: examination of voigt, mori-tanaka, and self-consistent schemes. J. Press. Vessel. Technol. **137**(021201), 1–11 (2015)
14. Kushnir, R.M., Kalynyak, B.M., Horoshko, V.O.: Determination of the thermo-elastic state of the three-layer hollow sphere at completed heat exchange. Bull. Taras Shevchenko Natl. Univ. Kyiv Ser.: Phys. Math. **3**, 111–114 (2017)
15. Łatka, L., Pawłowski, L., Winnicki, M., Sokołowski, P., Małachowska, A., Kozersk, S.: Review of functionally graded thermal sprayed coatings. Appl. Sci. **10**(15), 5153 (2020). https://doi.org/10.3390/app10155153
16. Manthena, V.R., Kedar, G.D.: Mathematical modeling of thermoelastic state of a functionally graded thermally sensitive thick hollow cylinder with internal heat generation. Int. J. Thermodyn. (IJoT) **21**(4), 202–212 (2018). https://doi.org/10.5541/ijot.434180
17. Manthena, V.R., Srinivas, V.B., Kedar, G.D.: Analytical solution of heat conduction of a multilayered annular disk and associated thermal deflection and thermal stresses. J. Therm. Stress. **43**(5), 563–578 (2020). https://doi.org/10.1080/01495739.2020.1735975
18. Noda, N.: Thermal stresses in functionally graded materials. J. Therm. Stress. **22**(4–5), 477–512 (1999). https://doi.org/10.1080/014957399280841

19. Noda, N., Ishihara, M., Yamamoto, N.: Two-crack propagation paths in a functionally graded material plate subjected to thermal loadings. J. Therm. Stress. **27**(6), 457–469 (2004). https://doi.org/10.1080/01495730490451422
20. Obata, Y., Noda, N.: Steady thermal stresses in a hollow circular cylinder and hollow sphere of a functionally gradient material. J. Therm. Stress. **17**, 471–487 (1994)
21. Pawar, S.P., Deshmukh, K.C., Kedar, G.D.: Thermal stresses in functionally graded hollow sphere due to non-uniform internal heat generation. Appl. Appl. Math. **10**(1), 552–569 (2015)
22. Popovych, V.S., Sulym, H.T.: Centrally symmetric quasistatic problem of thermoelasticity for a temperature sensitive body. Mater. Sci. **40**, 365–375 (2004). https://doi.org/10.1007/PL00022000
23. Reddy, J.N.: Thermomechanical behavior of functionally graded materials. Final Report for AFOSR Grant F49620-95-1-0342. CML Report, pp. 98–101
24. Reddy, J.N., Chin, C.D.: Thermomechanical analysis of functionally graded cylinders and plates. J. Therm. Stress. **21**(6), 593–626 (1998). https://doi.org/10.1080/01495739808956165
25. Samarskii, A.A.: Theory of Difference Schemes. Marcel Dekker Inc, New York (2001)
26. Segall, A.E.: Thermoelastic analysis of thick-walled vessels subjected to transient thermal loading. ASME J. Press. Vessel. Technol. **123**(1), 146–149 (2001). https://doi.org/10.1115/1.1320818
27. Shao, Z.S., Wang, T.J., Ang, K.K.: Transient thermo-mechanical analysis of functionally graded hollow circular cylinders. J. Therm. Stress. **30**(1), 81–104 (2007). https://doi.org/10.1080/01495730600897211
28. Sharma, J.N., Sharma, P.K., Mishra, K.C.: Dynamic response of functionally graded cylinders due to time-dependent heat flux. Meccanica **51**, 139–154 (2016). https://doi.org/10.1007/s11012-015-0191-3
29. Tanigawa, Y., Akai, T., Kawamura, R., Oka, N.: Transient heat conduction and material thermal stress problems of a nonhomogeneous plate with temperature-dependent properties. J. Therm. Stress. **19**(1), 77–102 (1996)
30. Tokovyy, Y., Ma, C.C.: The Direct Integration Method for Elastic Analysis of Nonhomogeneous Solids. Cambridge Scholars Publishing (2021)
31. Pasternak, Y.M., Sulym, G.T., Pasternak, R.M.: Studies of stressed states in bodies from functionally-graded materials: review of publications up to 2010. Opir Materialiv I teoriya sporud [Resistance of Materials and Theory of Structures], vol. 95, pp. 35–80 (2015)

Chapter 18
Stress Concentration Around a Circular Hole in Thin Plates and Cylindrical Shells with a Radially Inhomogeneous Inclusion

Vadym Hudramovich, Eteri Hart, and Bohdan Terokhin

Abstract Computer simulation and FEM analysis of the stress-strain state of thin plates and thin-walled cylindrical shells with a circular hole in the presence of a surrounding ring inclusion of a functionally graded material (FGM) is carried out. The influence of the dimensions of the FGM-inclusion and the law of change of its elastic modulus on the concentration of the parameters of the stress-strain state of plates and shells in the vicinity of the hole is studied. The distribution fields of stress and strain intensities of plate-shell structural elements in the zones of local stress concentration are obtained. It has been established that when using a radially inhomogeneous ring inclusion with certain mechanical properties and geometric parameters, it is possible to reduce the stress concentration factor and the corresponding strain intensities in the vicinity of the hole by more than 35%. The law of change in the modulus of elasticity of the FGM-inclusion and the width of the inclusion have a significant effect not only on the concentration of the parameters of the stress-strain state of the plate and shell but also on the nature of the stress distribution over their surfaces. The results of a series of large-scale computational experiments show that the use of an FGM annular inclusion makes it possible to reduce the intensity of both stresses and deformations around the hole, which opens up prospects for finding rational parameters of inclusions from the point of view of the maximum possible reduction in local stress concentration.

V. Hudramovich
Institute of Technical Mechanics, National Academy of Sciences of Ukraine, State Space Agency of Ukraine, Dnipro, Ukraine
e-mail: hudramovich@i.ua

E. Hart (✉) · B. Terokhin
Oles Honchar Dnipro National University, Dnipro, Ukraine
e-mail: hart@ua.fm

H. Altenbach et al. (eds.), *Selected Problems of Solid Mechanics and Solving Methods*, Advanced Structured Materials 204, https://doi.org/10.1007/978-3-031-54063-9_18

18.1 Introduction

Thin-walled structures, which are plates and shells, are widely used in various branches of technology, particularly rocket and space, oil and gas, energy, and construction, due to the combination of significant strength and relatively low weight. In most cases, plate-shell structural elements have holes, the presence of which, as a rule, leads to a sharp increase in local stresses, which, in turn, affect the strength and reliability of the structure as a whole [3, 4, 12, 16, 17]. The issue of stress concentration around holes in structurally inhomogeneous plate-shell elements of thin-walled structures is an important problem of modern mechanics of new technology, in particular, the rocket and space industry. Bays with reinforced openings are typical of launch vehicle designs. Research, which consists of choosing rational types of reinforcements that provide minimum stress concentration factors, is relevant in the design of modern launch vehicles.

Many scientific papers have been devoted to the study of the stress-strain state (SSS) of thin-walled structures with holes. The works of well-known domestic scientists G N Savin, AN Guz, and others are fundamental in this direction. Thus, in [17], extensive results of systematic studies of stress concentration around various holes in thin plates and shells are presented using the complex variable theory. In [6], information is given on the theory of thin-walled cylindrical shells weakened by one and several holes; the formulations of the corresponding problems, methods, and results of their solution in the case of isotropic and composite materials, including taking into account plastic deformations, are considered; solutions are obtained for the cases of small and large holes, taking into account the rigidity of the reinforcing elements.

The presence of reinforcing elements or inclusions with certain mechanical properties helps to reduce the stress concentration around the holes [5, 7–11]. The use of functionally graded materials (FGM), which have been increasingly used in recent years, also makes it possible to influence the stress-strain state (SSS) of thin-walled structures with holes in order to reduce the stress concentration factor (SCF). FGM are the latest materials characterized by a smooth spatial change in composition and properties. The property gradient allows you to influence the SSS of structural elements. The feature of FGM makes them very useful for reducing interfacial stresses, and therefore they are widely used in aerospace, power engineering, electronics, chemical, optical, and biomedical engineering. FGM mechanics has aroused great interest in the last two decades, and a lot of work has been done on theoretical, numerical, and experimental studies of FGM. Thus, in [1], mixed axisymmetric problems for functionally graded media were considered, and their analytical solutions were obtained. In [18], using an analytical method, the stress distribution in an FGM plate with a circular hole was studied; in [14], the stress concentration in multi-wedge systems with functionally graded wedges was estimated. In [13], using the finite element method (FEM), using various isoparametric finite elements, the SCF was determined in the vicinity of a circular cut in an inhomogeneous plate under the action of a uniaxial tensile load; in [15], the SCF was determined around a circular cut in an FGM plate under the action of a biaxial tension and shift.

Taking into account the heterogeneity of the FGM significantly increases the complexity of the mathematical model of the problem and obtaining its solution. Analytical methods for solving problems of deformation of FGM structures can be used only in some special cases; therefore, when studying SSS structures with various inhomogeneities (holes, cutouts, grooves), it is more expedient to use numerical methods of mechanics, which, in contrast to analytical, are quite universal and effective for solving a wide class of problems.

The search for opportunities to reduce the stress concentration in thin-walled structures is an urgent problem in the mechanics of a deformable solid body. The presented work reviews a series of works by the authors on computer simulation of the behavior of elastic thin plates and thin-walled cylindrical shells with a circular hole in the presence of a radially inhomogeneous inclusion (FGM-inclusion) around it. Based on the application of the FEM, an analysis is made of the influence of the dimensions of the FGM-inclusion and the law of change in its elastic modulus on the SSS parameters of plates and shells in the zones of their local concentration.

18.2 Statement of the Problem

SSS of thin elastic homogeneous isotropic plates and cylindrical shells with a centrally located circular hole and an annular FGM-inclusion is considered. The dimensions of the plates are $a \times b$, thickness h, hole radius R, and inclusion radius R_1 (Fig. 18.1a); the length of the shells L, diameter d, thickness, radius of the hole, and the radially inhomogeneous inclusion are the same as those of the plate (Fig. 18.1b). The inclusion has a thickness h (located in the plane of the plate/shell), and rigid adhesion conditions are set on its boundary with the matrix. The value R_1 varies. A uniform uniaxial tensile load $p = \text{const}$ acts on the side faces of the plates and the ends of the shells, which does not lead to the appearance of plastic deformations.

In numerical calculations, several types of model materials for inclusion from FGM were selected, which have radial arbitrary elastic properties (according to linear and nonlinear laws of change in the elastic modulus), with the same Poisson's ratio $\nu_0 = 0.25$ and a variable elastic modulus $E_i(r)$ $(i = 1, \dots 8)$:

$$E_1(r) = 0.5E_0\left(2 - \exp\left\{-5\frac{r-R}{R_1-R}\right\}\right), \tag{18.1}$$

$$E_2(r) = 0.5E_0\left(1 + \frac{r-R}{R_1-R}\right); \tag{18.2}$$

$$E_3(r) = 0.5E_0\left(1 + \exp\left\{-5\left(1 - \frac{r-R}{R_1-R}\right)\right\}\right), \tag{18.3}$$

$$E_4(r) = E_0\left(1 + 0.5\exp\left\{-5\frac{r-R}{R_1-R}\right\}\right); \tag{18.4}$$

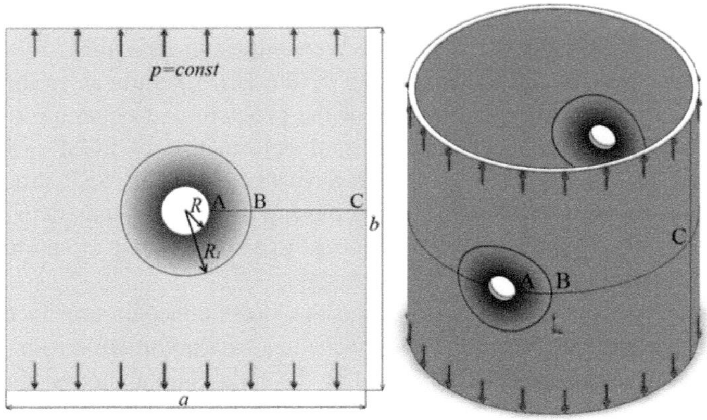

a *b*

Fig. 18.1 Geometry and loading pattern of a plate (**a**) and a shell (**b**) with a radially inhomogeneous inclusion

$$E_5(r) = E_0\left(1.5 - 0.5\frac{r-R}{R_1-R}\right), \tag{18.5}$$

$$E_6(r) = E_0\left(1.5 - 0.5\exp\left\{-5\left(1-\frac{r-R}{R_1-R}\right)\right\}\right), \tag{18.6}$$

$$E_7 = \begin{cases} E_0(1+l), & l \in [0; 0.5] \\ E_0(2-l), & l \in [0.5; 1] \end{cases}, \tag{18.7}$$

$$E_8(r) = \begin{cases} E_0\left(1+\frac{1}{\tilde{h}_1}l\right), & l \in [0; \tilde{h}_1] \\ 2E_0, & l \in [\tilde{h}_1; \tilde{h}_1+\tilde{h}_2] \\ E_0\left(2-\frac{l-\left(\tilde{h}_1+\tilde{h}_2\right)}{\tilde{h}_3}\right), & l \in [\tilde{h}_1+\tilde{h}_2; 1], \end{cases} \tag{18.8}$$

where $E_0 = 100$ GPa is the modulus of elasticity of the plate/shell, $0 \leqslant l \leqslant 1$ is the normalized parametric distance in the radial direction from the edge of the hole (point A, Fig. 18.1) along the width of the inclusion $AB = h_{\text{incl}} = R_1 - R$:

$$l = (r-R)/(R_1-R), \tag{18.9}$$

r is the distance from the center of the hole to an arbitrary point of inclusion, R and R_1 are the radii of the hole and the inclusion, respectively, $\tilde{h}_i = h_i/h_{\text{incl}}$ $(i = 1, 2, 3)$ is the width of each of the three characteristic zones of the FGM-inclusion, and $h_{\text{incl}} = \sum_{i=1}^{3} h_i$ (Fig. 18.2d).

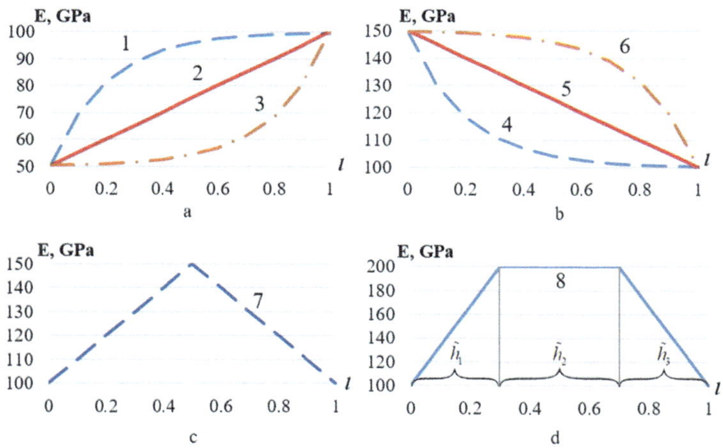

Fig. 18.2 The laws of change in the modulus of elasticity of FGM-inclusions: "soft" (**a**), "hard" (**b**–**d**)

The form of dependences for model materials (18.1) and (18.4) is similar to [18]. Dependences (18.1)–(18.3) correspond to a "soft", and (18.4)–(18.8) correspond to a "harder" inclusion than the main material of the plate/shell.

Figure 18.2 shows the corresponding graphical representations of the laws of change in the modulus of elasticity of the FGM-inclusion for "soft" (Fig. 18.2a) and "hard" (Fig. 18.2b, c, d) inclusions. Lines 1–8 correspond to dependencies (18.1)–(18.8). Here and below in the figures, the abscissa shows the normalized parametric distance (18.9), $0 \leqslant l \leqslant 1$, in the radial direction from the edge of the hole along the width of the inclusion.

For definiteness, it was assumed that the plate is square, and the length of the shells L is equal to the side of the plate. On Fig. 18.2d, we have a graphical representation of three characteristic zones of the law of change in the modulus of elasticity of the FGM-inclusion of the form (18.8): (1) an increase zone, the width of which is $\tilde{h}_1$; (2) a zone of constant (fixed) width $\tilde{h}_2$; (3) a zone of decrease in the modulus of elasticity of the width $\tilde{h}_3$.

18.3 Mathematical Model of the Problem

In the variational formulation, the original problem for a cylindrical shell leads to the minimization of the functional of the total potential energy of the deformation of the system [2]:

$$F=\sum_{s=1}^{n+1}\left\{\frac{1}{2}\int_{\Omega_s}\frac{E_s(x,y)h}{(1-\nu_s^2)}\left[\left(\frac{\partial u}{\partial x}\right)^2+\left(\frac{\partial v}{\partial y}+\frac{w}{\tilde{R}}\right)^2\right.\right.$$
$$\left.+2\nu_s\left(\frac{\partial u}{\partial x}\right)\left(\frac{\partial v}{\partial y}+\frac{w}{\tilde{R}}\right)+\frac{1-\nu_s}{2}\left(\frac{\partial u}{\partial y}+\frac{\partial v}{\partial x}\right)^2\right]\mathrm{d}x\mathrm{d}y$$
$$+\frac{1}{2}\int_{\Omega_s}\frac{E_s(x,y)h^3}{12(1-\nu_s^2)}\left[\left(\frac{\partial^2 w}{\partial x^2}\right)^2+\left(\frac{\partial^2 w}{\partial y^2}+\frac{w}{\tilde{R}}\right)^2+\right.$$
$$\left.\left.+2\nu_s\left(\frac{\partial^2 w}{\partial x^2}\right)\left(\frac{\partial^2 w}{\partial y^2}+\frac{w}{\tilde{R}}\right)+2(1-\nu_s)\left(\frac{\partial^2 w}{\partial x\partial y}\right)^2\right]dxdy\right\}$$
$$-\int_{\gamma}(p_x u+p_y v+p_z w)\,\mathrm{d}x\mathrm{d}y,$$

where $u(x,y)$, $v(x,y)$ and $w(x,y)$ are the projections of the displacement vector on the axes Ox, Oy, and Oz, respectively, h is shell thickness, $\tilde{R}$ is the radius of the shell, $E_s(x,y)$ and ν_s are modulus of elasticity and Poisson's ratio of the shell Ω_1 (matrix) ($s=1$) material and inclusion Ω_s ($s=2,\ldots,n+1$; n is the number of inclusions), $\Omega=\cup_{s=1}^{n+1}\Omega_s$ is domain of definition of variables x and y, and γ is the boundary of the region Ω along which the external intensity load

$$P(x,y)=(p_x(x,y),p_y(x,y),p_z(x,y))^{\mathrm{T}}$$

is applied. In the case of a uniaxial tensile load,

$$p_x(x,y)=p_z(x,y)=0,\quad p_y(x,y)=p=\text{const}.$$

In the case of a plate, we arrive at the problem of minimizing the functional of the total potential strain energy of a system of this type [19]:

$$F=\sum_{s=1}^{n+1}\left\{\frac{1}{2}\int_{\Omega_s}\frac{E_s(x,y)h}{(1-\nu_s^2)}\left[\left(\frac{\partial u}{\partial x}\right)^2+\left(\frac{\partial v}{\partial y}\right)^2+2\nu_s\left(\frac{\partial u}{\partial x}\right)\left(\frac{\partial v}{\partial y}\right)\right.\right.$$
$$\left.\left.+\frac{1-\nu_s}{2}\left(\frac{\partial u}{\partial y}+\frac{\partial v}{\partial x}\right)^2\right]dxdy\right\}-\int_{\gamma}(p_x u+p_y v)\,\mathrm{d}x\mathrm{d}y. \tag{18.10}$$

The stated variational problems were solved using the FEM [20] using isoparametric triangular six-node Lagrangian finite elements of the second degree, while the unknown displacement functions inside each finite element were approximated by a quadratic polynomial. In the areas of stress concentration, an adaptive mesh was used (Fig. 18.3) with a refinement factor of 10. The convergence of the FEM in the case of

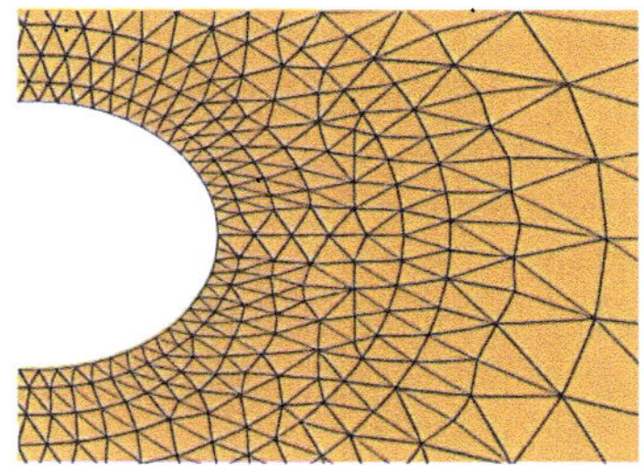

Fig. 18.3 A fragment of an adaptive finite element mesh

using flat finite elements for thin-walled shells is ensured here by mesh thickening. The use of a finer grid leads to an increase in the accuracy of the approximation of the shell surface by the geometry of the inscribed polyhedron.

18.4 Numerical Analysis

Computational experiments were carried out on a PC with an Intel Core i7-10700F processor, a clock frequency of 2.9-4.8 GHz, 32 GB RAM, x64 system capacity. The average number of finite elements in the calculation of plates is 2126, and the number of nodes is 4408; when calculating shells, these numbers are 6372 and 13002, respectively.

Numerical results were obtained for 1) square plates with the following geometric parameters: $h = 0.005$ m, $a = b = 0.2$ m, $R = a/20$; 2) cylindrical shells with parameters: $L = d = a$, $h = 0.005$ m, $R = d/10$. Tensile load in both cases $p = 10$ MPa.

The radius R_1 of the annular FGM-inclusion around the hole was varied with a step R, while the width of the FGM-inclusion was $h_{\text{incl}} = R, 2R, 3R, \ldots, 9R$.

18.4.1 The Plate

For the purpose of comparative analysis, a calculation was made for a homogeneous plate with a circular hole without inclusion. Received SCF = 3.05; the maximum value of the strain intensity in this case $\varepsilon_i^{\max} = 2.13 \times 10^{-4}$, which is in good agreement with the results of [17]. As a result of the computational experiments, using the FEM, the distribution of stress and strain intensities in the plate was obtained, and the SCF was calculated for uniaxial tension of the plate with "soft" and "hard" FGM-inclusions with the inclusion width R and $2R$. The results for "soft" inclusions (Fig. 18.2) are shown in Table 18.1, and for "hard" inclusions (Fig. 18.2b, c) in Table 18.2.

Table 18.1 The stress concentration factor and the corresponding deformations in a plate with a "soft" FGM-inclusion at $h_{incl} = 2R$

The problem	SCF	δ_1, %	$\varepsilon_i^{\max} \times 10^4$	δ_2, %
Homogeneous inclusion, $E_{\text{incl}} = 0.5E_0$	2.25	−26.4	3.14	+47.4
FGM-inclusion 1	2.05	−33.1	2.54	+19.3
FGM-inclusion 2	2.03	−33.7	2.75	+29.1
FGM-inclusion 3	2.11	−30.9	2.92	+37.1

Table 18.2 The stress concentration factor and the corresponding deformations in a plate with a "hard" FGM-inclusion at $h_{incl} = 2R$

The problem	SCF	δ_1, %	$\varepsilon_i^{\max} \times 10^4$	δ_2, %
Homogeneous inclusion, $E_{\text{incl}} = 1.5E_0$	3.51	15.1	1.63	−23.5
FGM-inclusion 4	3.90	27.9	1.89	−11.3
FGM-inclusion 5	3.79	24.3	1.77	−16.9
FGM-inclusion 6	3.62	18.7	1.68	−21.1
FGM-inclusion 7	2.84	-6.9	1.93	−9.4

Figure 18.4 illustrates the nature of the distribution of relative stresses σ_y/p in the plate along the width of the inclusion in the section AB with $h_{\text{incl}} = 2R$ in the case of "soft" type (18.1)–(18.3) (Fig. 18.4a) and "hard" type (18.4)–(18.7) (Fig. 18.4b, c) of radially inhomogeneous inclusions.

By the nature of the stress distribution (Fig. 18.4), the presence of FGM-inclusions leads to a redistribution of stresses in the cross-section of the AB plate compared to the case of inclusions from the corresponding homogeneous material: in the case of "soft" FGM-inclusions, the stresses decrease in the vicinity of the hole and some increase further along the width of the inclusion (Fig. 18.4a), and in the case of "hard" FGM-inclusions, the opposite trend is observed (Fig. 18.4b). As can be seen from Fig. 18.4c, in the case of a "hard" FGM-inclusion of the form (18.7), the maximum stresses in the vicinity of the hole and the relative stresses σ_y/p along the section AB in the interval $l \in [0.7, 1]$ along the width of the inclusion turn out to be smaller than in the plate without inclusions. As an example, Fig. 18.5 shows the distribution patterns of stress and strain intensities in a plate with an FGM-inclusion 7.

Therefore, the use of FGM-inclusions around the hole leads to a smooth distribution of stresses in the matrix without jump-like perturbations, in contrast to a homogeneous inclusion. The width of FGM-inclusions affects the nature of the stress distribution: the larger the width of the inclusion, the smoother the redistribution of stresses in the matrix.

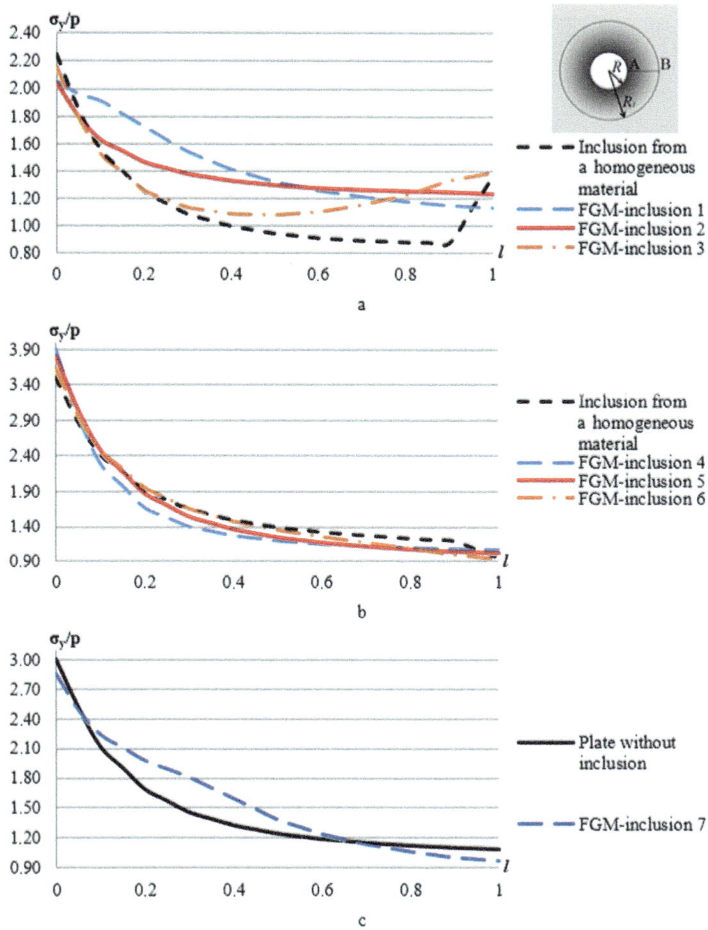

Fig. 18.4 Distribution of relative stresses σ_y/p in the plate along the width of the inclusion in the section AB with $h_{\text{incl}} = 2R$ in the case of "soft" type (18.1)–(18.3) (**a**) and "hard" type (18.4)–(18.7) (**b**, **c**) FGM-inclusions

The best option among those considered (Fig. 18.2a, b, c) turned out to be the case of an FGM-inclusion, in which the elastic modulus increases radially from the edge of the hole in the direction of the periphery, and then decreases from the middle of the inclusion (Fig. 18.2c). In this case, not only does the strength of the plate as a whole increase, but it is also possible to reduce the concentration of both stresses and strains around the hole by ~7% and ~9%, respectively. This confirms the expediency of using FGM-inclusions of this type and opens up prospects for further searches for their rational parameters. Here, δ_1 and δ_2 are the deviation of the SCF and the maximum value of the strain intensity $\varepsilon_i^{\max}$ from the corresponding value for a plate with a circular hole without an inclusion [17].

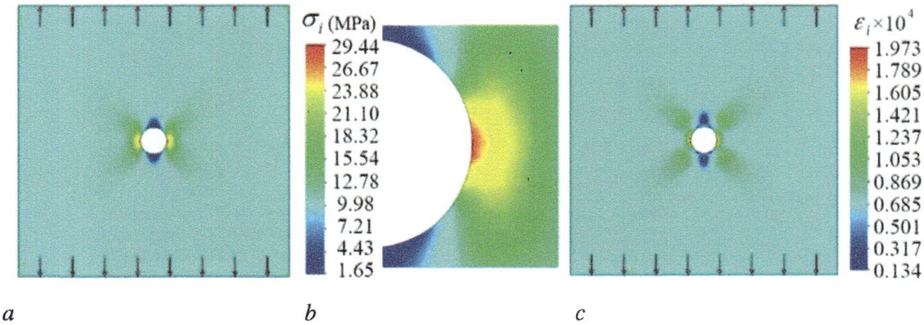

Fig. 18.5 The SSS components of a plate with FGM-inclusion 7 at $h_{\text{incl}} = R$: stress intensity in the plate (**a**) and in the vicinity of the hole (**b**), strain intensity (**c**)

Table 18.3 Variants of FGM-inclusions depending on the width of the zones h_i

Inclusion type	h_1	h_2	h_3
FGM-inclusion 1	R	R	R
FGM-inclusion 2	R	2R	R
FGM-inclusion 3	R	3R	R
FGM-inclusion 4	R	4R	R
FGM-inclusion 5	R	5R	R
FGM-inclusion 6	R	6R	R
FGM-inclusion 7	R	7R	R

From Table 18.1, it can be seen that in the case of a radially inhomogeneous inclusion of width $2R$ from FGM of the form (18.1)–(18.3), the maximum deformations and SCF in a plate with a hole are less than in the presence of a "soft" homogeneous inclusion: for FGM-inclusion 2, the difference in maximum deformations is by ~1.6 times, for FGM-inclusion 1 by ~2.5 times; the decrease in SCF is ~7% in both cases. Consequently, when using "soft" inclusions, the SCF decreases.

The use of "hard" inclusions increases the strength of the structure as a whole. In the case of FGM-inclusions of the form (18.4)–(18.6), there is some increase in the SCF in the vicinity of the hole, but in the case of FGM-inclusions of the form (18.7) both stresses and strains around the hole decrease compared to a plate without inclusions (Table 18.2). The results of studies carried out in [8] for a similar problem indicate that the width h_2 of the central zone of the FGM-inclusion of the form (18.8) has the greatest influence on the SCF value (Fig. 18.2d).

Let us find out further how the SCF changes for different variants of the value (Table 18.3) for fixed values h_1, h_3 and a variable value h_2 of the FGM-inclusion of the form (18.8).

In the case of an FGM-inclusion 7 ($h_{\text{incl}} = 9R$), its radius R_1 is equal to half the width of the plate. In this case, it was assumed that the entire plate was made of FGM.

Table 18.4 The stress concentration factor and the corresponding deformations in a plate with an FGM-inclusion of the form (18.8)

The problem	SCF	δ_1, %	$\varepsilon_i^{\max} \times 10^4$	δ_2, %
FGM-inclusion 1	2.39	−21.6	1.60	−24.9
FGM-inclusion 2	2.23	−26.9	1.50	−29.6
FGM-inclusion 3	2.10	−31.1	1.41	−33.8
FGM-inclusion 4	1.99	−34.8	1.34	−37.1
FGM-inclusion 5	1.93	−36.7	1.30	−39.0
FGM-inclusion 6	1.90	−37.7	1.28	−39.9
FGM-inclusion 7	1.93	−36.7	1.30	−39.0

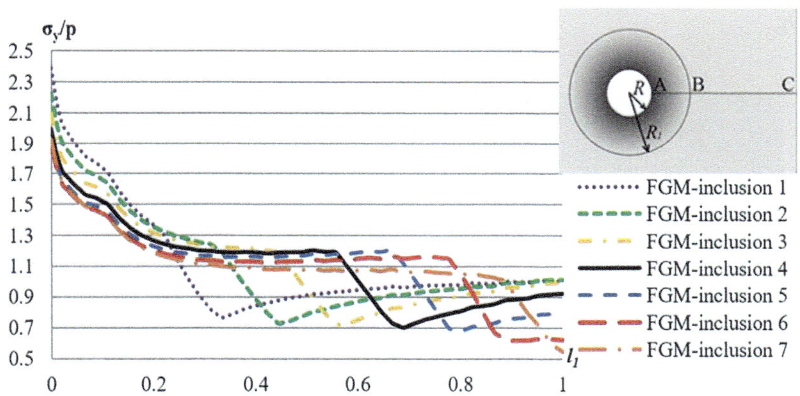

Fig. 18.6 Distribution of relative stresses σ_y/p in a plate with an FGM-inclusion in the cross section AC in the case of $h_2 = R, 2R, \ldots, 7R$

The results of a series of computational experiments using the FEM for a plate with FGM-inclusions of the form (18.8) of different widths ($h_{\text{incl}} = 3R, 4R, \ldots, 9R$) are summarized in Table 18.4.

The presence of an annular FGM-inclusion with a given law of change in the elastic modulus (18.8) makes it possible to reduce the SCF value in the plate by ~21–38%, and the maximum deformations by 25–40% (Table 18.4). Analyzing the results of calculations in the presence of FGM-inclusions of different widths, we find that the larger the width of the inclusion, the smaller the values of SCF and deformations in the plate. The smallest value of the SCF among the variants under consideration was obtained in the case of the FGM-inclusion 6 ($h_1 = h_3 = R$, $h_2 = 6R$).

Figure 18.6 shows the graphs of the distribution of relative stresses σ/p in the characteristic section AC of a plate with an FGM-inclusion of the form (18.8) for various options for the width h_2 of the second zone of the FGM-inclusion $h_2 = 4R, 5R, 6R, 7R$ (see Table 18.3). The abscissa shows the normalized parametric distance $0 \leqslant l_1 \leqslant 1$ in the radial direction from the edge of the hole (point A, Fig. 18.1a)

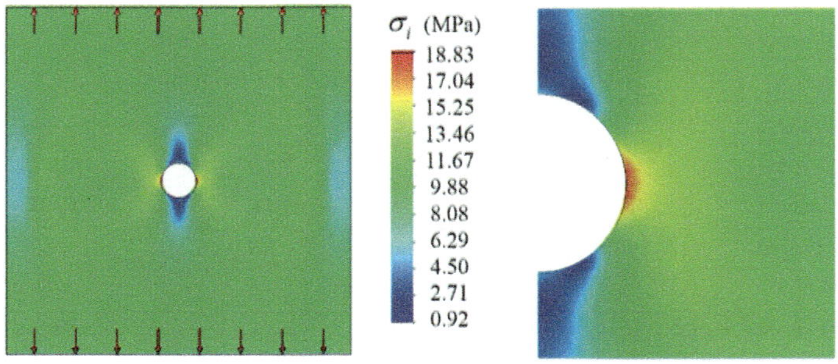

a b

Fig. 18.7 Distribution of stress intensity σ_i in a plate with an FGM-inclusion 6 (**a**) and around the hole (**b**)

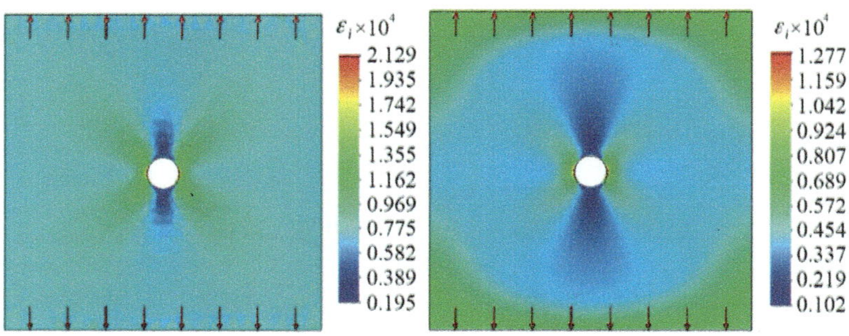

a b

Fig. 18.8 Distribution of strain intensity ε_i in a homogeneous plate (**a**) and a plate with an FGM-inclusion 6 of the form (18.8) (**b**)

along the plate cross-section $AC = (a - 2R)/2$: $l_1 = 2(r - R)/(a - 2R)$, r is the distance from the center of the hole to an arbitrary point of the segment AC.

As can be seen from Fig. 18.6, when using FGM-inclusions of different widths, a redistribution of stresses in the section C occurs, associated with the width h_2 of the central zone of the FGM-inclusion: the larger the value h_2, the lower the stress σ/p in the section AC; in the transition from the second to the third zone, for each FGM-inclusion 4–7, the stresses σ_y/p decrease. Therefore, by setting a certain law of change in the elastic modulus of the inclusion, it is possible to influence the magnitude of the SCF and, in general, the distribution of stresses in the plate.

As an example, Fig. 18.7 shows the distribution of stress intensity σ_i in a plate with a circular hole and FGM-inclusion 6 ($h_1 = h_3 = R, h_2 = 6R$). Figure 18.8 illustrates the qualitative nature of the difference in the distribution of strain intensities in a homogeneous plate and in a plate with a radially inhomogeneous inclusion of radius $R_1 = 9R$ ($h_1 = h_3 = R, h_2 = 6R$).

Table 18.5 The stress concentration factor and corresponding deformations in a cylindrical shell with FGM-inclusion

The problem	SCF	δ_1, %	$\varepsilon_i^{max} \times 10^4$	δ_2, %
FGM-inclusion 1	2.49	−23.9	1.66	−27.2
FGM-inclusion 2	2.32	−29.1	1.55	−32.0
FGM-inclusion 3	2.20	−32.7	1.46	−36.0
FGM-inclusion 4	2.11	−35.5	1.41	−38.2
FGM-inclusion 5	2.07	−36.7	1.39	−39.0
FGM-inclusion 6	2.06	−37.0	1.38	−39.5
FGM-inclusion 7	2.11	−35.5	1.41	−38.2

18.4.2 The Shell

Next, we consider the effect of the presence of FGM-inclusions around a circular hole in a thin-walled cylindrical shell on the SCF value.

The results of the calculations for a shell with FGM-inclusions of the form (18.8) of different widths (see Table 18.3) with a given law of change in the elastic modulus (Fig. 18.2d) are given in Table 18.5.

Here, δ_1 and δ_2 are the deviation of the SCF and the maximum value of the strain intensity ε_i^{max} from the corresponding values for a thin-walled cylindrical shell with a circular hole without an inclusion [6]. From Table 18.5, it can be seen that the presence of a radially inhomogeneous annular inclusion with a given law of change in the elastic modulus leads to a decrease in the SCF value in the shell by ~24–37%, and maximum deformations by ~27–39%. As in the case with plates, the FGM-inclusion 6 turned out to be the best of the considered options for the shell in terms of reducing the SCF.

As can be seen from Fig. 18.9, the distribution of relative stresses σ_y/p over the surface of the shell (in section AC) for FGM-inclusions of different widths is similar to the case for plates. The abscissa shows the normalized parametric distance $0 \leqslant l_2 \leqslant 1$ from the edge of the hole (point A, Fig. 18.1b) along the arc $AC = (\pi d - 4R)/4$.

Figures 18.10 and 18.11 illustrate a qualitative picture of the distribution of stress σ_i and strain ε_i intensities in a cylindrical shell with a circular hole and an annular FGM-inclusion 6 ($h_1 = h_3 = R$, $h_2 = 6R$), respectively.

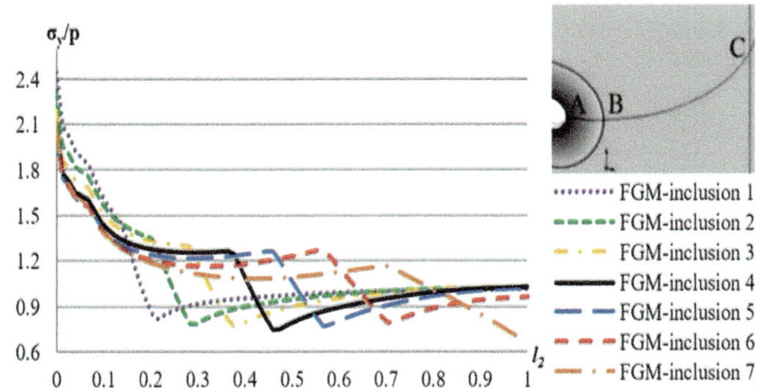

Fig. 18.9 Distribution of relative stresses σ_y/p in a shell with an FGM-inclusion in the cross-section AC in the case of $h_2 = R,\ 2R, \ldots 7R$

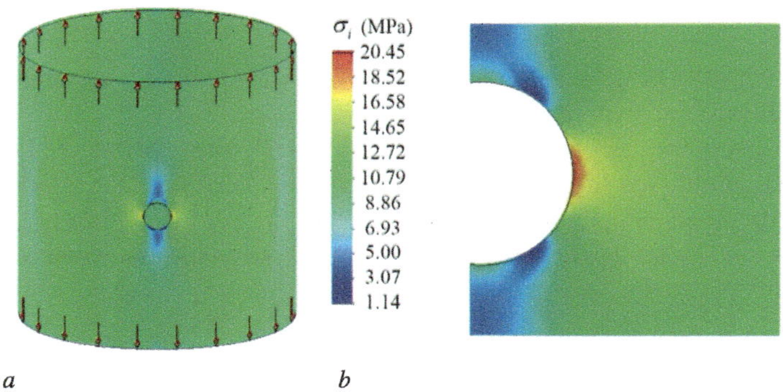

Fig. 18.10 Stress intensity σ_i distribution in the shell with FGM-inclusion 6 (**a**) and around the hole (**b**)

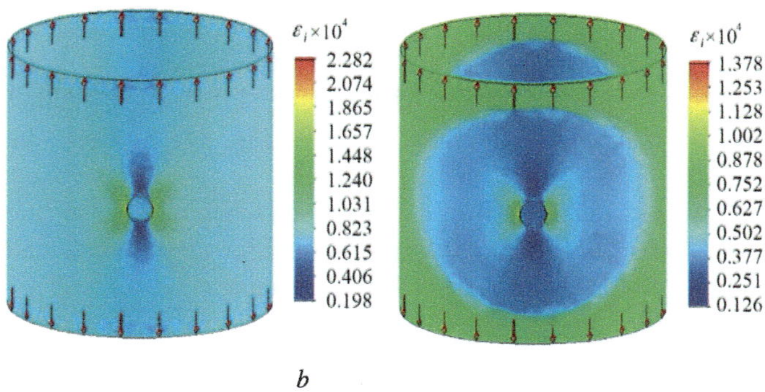

Fig. 18.11 Distribution of strain intensity ε_i in a homogeneous shell (**a**) and a shell with an FGM-inclusion 6 (**b**)

18.5 Conclusions

Based on a series of large-scale computational experiments using the FEM, modeling and analysis of the influence of a radially inhomogeneous annular inclusion on the stress concentration around a circular hole in thin plates and cylindrical shells was carried out.

The results of computer simulation and numerical study of the effect of the geometric characteristics of the FGM-inclusion and the law of variation of its elastic modulus in the radial direction on the concentration of SSS parameters around the hole of the plate-shell structural elements showed that in the presence of FGM-inclusions with certain mechanical properties and geometric characteristics, it is possible to reduce the SCF and the corresponding strain intensity in the vicinity of the hole by more than 35%.

Therefore, the use of annular reinforcements of circular holes in the form of FGM-inclusions in plates and cylindrical shells with holes is advisable. It allows us to influence not only the nature of the distribution but also the magnitude of stress and strain intensities in the zones of local concentration of their SSS parameters.

It is promising to search for rational parameters of FGM-inclusions, to find their types and configurations from the point of view of the influence on the decrease in the concentration of the SSS parameters of plates and shells with different holes.

References

1. Aizikovich, S.M., Aleksandrov, V.M., Vasil'ev, A.S. et al.: Analytic Solutions of Mixed Axisymmetric Problems for Functionally Graded Media (in Russ.). Fizmatlit, Moscow (2011)
2. Avdonin, A.S.: Applied Methods for Calculating Shells and Thin-Walled Structures (in Russ.). Nauka, Moscow (1969)
3. Grigorenko, A.Y., Muller, W.H., Grigorenko, Y.M., Vlaikov, G.G.: Recent Developments in Anisotropic Heterogeneous Shell Theory. Springer, Berlin (2016)
4. Grigorenko, Y.M., Vlaikov, G.G., Grigorenko, A.Y.: Numerical-Analytical Solution of Shell Mechanics Problems Based on Various Models (in Russ.). Akademperiodika, Kyiv
5. Gudramovich, V.S., Gart, É.L., Strunin, K.À.: Modeling of the behavior of plane-deformable elastic media with elongated elliptic and rectangular inclusions. Mater. Sci. **52**(6), 768–774 (2017)
6. Guz, A.N., Chernyshenko, I.S., Chekhov, V.I.N., Chekhov V.N., Shnerenko, K.I.: Cylindrical Shells with Cutouts (in Russ.). Naukova Dumka, Kyiv (1974)
7. Hart, E.L., Hudramovich, V.S.: Computer simulation of the stress-strain state of plates with reinforced elongate rectangular holes of various orientations. Strength Mater. Theory Struct. **108**, 77–86 (2022)
8. Hart, E.L., Hudramovich, V.S., Terokhin, B.I.: Effect of a functionally graded material inclusion on the stress concentration in thin plates and cylindrical shells with a circular opening (in Ukrainian). Tech. Mech. **4**, 67–78 (2022)
9. Hart, E.L., Terokhin, B.I.: Choice of rational parameters of reinforcement elements in computer simulation of behavior of a cylindrical shell with two rectangular holes (in Ukrainian). Probl. Comput. Mech. Strength Struct. **30**, 19–32 (2019)

10. Hart, E.L., Terokhin, B.I.: Computer simulation of the stress-strain state of the plate with circular hole and functionally graded inclusion. J. Optim. Diff. Equ. Their Appl. **29**(1), 42–53 (2021)
11. Hudramovich, V.S., Hart, E.L., Marchenko, O.A.: Reinforcing inclusion effect on the stress concentration within the spherical shell having an elliptical opening under uniform internal pressure. Strength Mater. **52**(6), 832–842 (2020)
12. Kosmodamianskii, A.S.: Plane Problem of the Theory of Elasticity for Plates with Holes, Cutouts, and Protrusions (in Russ). Vishcha Shkola, Kyiv (1975)
13. Kubair, D.V., Bhanu-Chandar, B.: Stress concentration factor due to a circular hole in functionally graded panels under uniaxial tension. Int. J. Mech. Sci. **50**(4), 732–742 (2008)
14. Linkov, A., Rybarska-Rusinek, L.: Evaluation of stress concentration in multi-wedge systems with functionally graded wedges. Int. J. Eng. Sci. **61**, 87–93 (2012)
15. Mohammadi, M., Dryden, J.R., Jiang, L.: Stress concentration around a hole in a radially inhomogeneous plate. Int. J. Solids Struct. **48**(3–4), 483–491 (2011)
16. Peterson, R.: Stress Concentration Factors (in Russ.). Mir, Moscow (1977)
17. Savin, G.N.: Stress Distribution Around Holes (in Russ.). Naukova Dumka, Kyiv (1968)
18. Yang, Q., Gao, C.F., Chen, W.: Stress analysis of a functional graded material plate with a circular hole. Arch. Appl. Mech. **80**, 895–907 (2010)
19. Washizu, K.: Variational Methods in Elasticity and Plasticity. Pergamon Press, Oxford, New York (1982)
20. Zienkiewicz, O.C., Taylor, R.L.: The Finite Element Method for Solid and Structural Mechanics. Elsevier, New York (2005)

Chapter 19
Solving Incorrect Problems of the Elasticity Theory

Oleksandr Khimich and Alexandr Popov

Abstract The possibilities of effective research and calculation of mathematical models described by problems with a unique solution on the subspace are substantiated by the example of the first main problem of the elasticity theory. A technique for obtaining a single solution of a variation problem on a subspace—a normal generalized solution—using a discrete finite element model on the entire space is proposed. The three-stage regularization method is used to find the normal generalized solution (or the weighted normal generalized solution) of the discrete problem—a system of linear algebraic equations with a symmetric semidefinite matrix. The three-stage regularization method allows us to approximate these solutions with a predetermined accuracy. This method is effectively implemented on modern high-performance computers.

19.1 Introduction

Mathematical modeling and related computer experiments are currently the main means of studying real physical phenomena and processes. At the same time, various numerical methods are widely used, which are implemented in computer programs to quickly obtain simulation results in order to optimize the parameters of processes or phenomena. When choosing numerical modeling methods, it is essential to obtain reliable solutions with estimates of the accuracy of the results.

In many cases, the solution of systems of linear algebraic equations (SLAE) is the basis for studying a mathematical model. For example, SLAEs arise when discretizing boundary value problems by the projection-difference method (finite differences, finite elements) [9, 10, 16, 17, 20]. Moreover, as a rule, solving the SLAE or problems related to the SLAE takes a significant part (50% or more) of the time to solve the entire problem. In some cases, the SLAE matrix turns out to be semidefinite. Such SLAEs arise, for example, as a result of discretization in the calculation of the strength of loose structures, that is, when numerically solving the

O. Khimich (✉) · A. Popov
V.M. Glushkov Institute of Cybernetics, National Academy of Science of Ukraine, Kyiv, Ukraine
e-mail: khimich505@gmail.com

H. Altenbach et al. (eds.), *Selected Problems of Solid Mechanics and Solving Methods*,
Advanced Structured Materials 204, https://doi.org/10.1007/978-3-031-54063-9_19

first basic problem of the theory of elasticity. SLAEs with a semidefinite matrix have a single solution on a subspace—a normal generalized solution or a normal pseudo-solution, which can be found by the method of least squares or weighted least squares (allows taking into account the "weight" of individual unknowns of the problem) [6, 8, 9]. Due to the fact that the data of applied models are specified with errors, the properties of the resolving SLAE may change. For example, a positive semidefinite matrix can become positive-definite or, conversely, sign-indefinite. Therefore, it is important to obtain a reliable solution of such a SLAE on the corresponding subspace.

An important feature of SLAEs that arises during mathematical modeling in various fields is their high orders, often exceeding 10^6. The matrices of such SLAEs have a sparse structure; that is, the number of nonzero elements of the matrix is equal to kn, where n is the order of the matrix and $k \ll n$, and the numbering of the unknowns of the problem determines the structure itself. Therefore, to solve problems with sparse matrices, it is necessary to use methods and algorithms that consider the matrix structure of this problem. Taking into account the high orders of these SLAEs and, as a result, a considerable number of arithmetic operations, it is advisable to use computers with parallel organization of calculations, including hybrid architecture (with multi-core and graphic processors), to solve such problems.

In this paper, using the example of the first basic problem of the theory of elasticity, the possibilities of effective research and calculation of mathematical models described by problems with a unique solution on the subspace are substantiated. In addition, the method of obtaining a single solution on the subspace of the variation problem using a normal pseudo-solution of the SLAE of a discrete finite element model on the entire space is substantiated. To find this normal pseudo-solution, the three-stage regularization method proposed in [3, 6, 12, 14] is used. This method makes it possible to obtain an approximation to the normal generalized solution of the SLAE with a predetermined accuracy. This method is effectively implemented on modern high-performance computers [3, 5, 12, 14, 15].

19.2 Problem Formulation

The basis of modeling and calculations of objects and processes in modern mechanics is the mathematical theory of elasticity. The purpose of the mathematical theory of elasticity is to determine stresses and deformations under any loads on the boundary and inside the elastic body of any shape. The task of obtaining an exact solution in the theory of elasticity is to determine such stresses, shifts, and deformations so that at each point inside the body, the equilibrium conditions and conditions of continuity of the body are ensured, and at the boundary, the internal forces would be in equilibrium with the external forces, which act on the surface.

Let $\Omega \subset \Re^3$ be a finite region with a sufficiently smooth boundary Γ. By

$$U_0(\Omega) = (W_2^1(\Omega))^3,$$

we will denote the Sobolev spaces of vector functions

$$w = (w_1, w_2, w_3)^T.$$

We will also introduce the following notations:

- $u \in U_0(\Omega)$ is a vector of elastic displacements,
- $\varepsilon = (\varepsilon_{11}, \varepsilon_{22}, \varepsilon_{33}, \varepsilon_{12}, \varepsilon_{23}, \varepsilon_{31})^T$ is a vector of deformations,
- $\varepsilon_{lm}(u) = \varepsilon_{ml}(u) = \frac{1}{2}\left(\frac{\partial u_l}{\partial x_m} + \frac{\partial u_m}{\partial} x_l\right)$ are components of the strain tensor,
- $\sigma = (\sigma_{11}, \sigma_{22}, \sigma_{33}, \sigma_{12}, \sigma_{23}, \sigma_{31})^T$ is a vector of stresses,
- $\sigma_{ik}(u) = \sigma_{ki}(u)$ are components of the stress tensor,
- $f_\Omega = (f_{\Omega,1}, f_{\Omega,2}, f_{\Omega,3})^T \in (L_2(\Omega))^3$ is a vector of distributed volume forces,
- $f_\Gamma = (f_{\Gamma,1}, f_{\Gamma,2}, f_{\Gamma,3})^T \in (L_2(\Gamma))^3$ is vector of distributed surface forces.

We consider the case when, for each body temperature, regardless of time, there is a mutually unambiguous linear relationship between stresses and strains—Hooke's law $\sigma = C(x)\varepsilon$. For the general case of a stressed state in the theory of elasticity, it is proved that the sixth-order matrix $C(x)$ is symmetric, i.e.,

$$c_{jm}(x) = c_{mj}(x), \quad j, m = 1, \ldots, 6.$$

In the domain $\overline{\Omega} = \Omega \bigcup \Gamma$, we consider the boundary problem of the theory of elasticity [18]:

$$Lu = -\sum_{i,k=1}^{3} \frac{\partial \sigma_{ik}}{\partial x_k} e_i = f_\Omega(x), \qquad x \in \Omega, \tag{19.1}$$

$$lu = -\sum_{i,k=1}^{3} \sigma_{ik} \cos(n,\ x_k) e_i = f_\Gamma(x), \quad x \in \Gamma, \tag{19.2}$$

where e_i is the ortho of the Ox_i axis and n is the internal normal to the boundary Γ. We assume that $f_\Omega \in \left(L_2(\Omega)\right)^3$, $f_\Gamma \in \left(L_2(\Gamma)\right)^3$, and the elasticity coefficients $c_{lm}(x)$ satisfy the condition

$$\varepsilon^T \sigma = \varepsilon^T C \varepsilon \geqslant \mu \left\|\varepsilon\right\|^2, \quad \mu_0 = \text{const} > 0. \tag{19.3}$$

The problems (19.1) and (19.2) do not always have a solution. The necessary condition for its solving is that the set of forces applied to the elastic body must be statically equivalent to zero, that is,

$$\int_\Omega f_\Omega dx + \int_\Gamma f_\Gamma d\gamma = 0, \quad \int_\Omega (r \times f_\Omega) dx + \int_\Gamma (r \times f_\Gamma) d\gamma = 0, \tag{19.4}$$

where r is the radius-vector of a point in the domain Ω and the oblique cross denotes the vector product. The following conditions distinguish the only solution:

$$\int_{\Omega} u dx = 0, \quad \int_{\Omega} (r \times u) dx = 0, \tag{19.5}$$

the first of which excludes arbitrary translational transfer, and the second—arbitrary hard rotation of the body.

The norms in $\left(L_2(\Omega)\right)^3$ and in $\left(L_2(\Gamma)\right)^3$ are determined by formulas

$$\|u\|^2_{(L_2(\Omega))^3} \equiv \|u\|^2_{0,\Omega} = \int_{\Omega} \sum_{m=1}^{3} u_m^2 dx, \ \|u\|^2_{(L_2(\Gamma))^3} \equiv \|u\|^2_{0,\Gamma} = \int_{\Gamma} \sum_{m=1}^{3} u_m^2 d\gamma.$$

We define bilinear and linear forms:

$$a(u, v) = \int_{\Omega} (\varepsilon(u))^T C(x) \varepsilon(v) dx, \quad l(f, v) = \int_{\Omega} f_{\Omega}^T v dx + \int_{\Gamma} f_{\Gamma}^T v d\gamma$$

for vector functions $u, v \in U_0(\Omega)$. In the $\bar{\Omega} = \Omega \bigcup \Gamma$, we consider the boundary value problem of the theory of elasticity in the weak formulation: to find such a vector of displacements $u \in U_0(\Omega)$ so that the integral identity is fulfilled:

$$a\,(u, v) = l\,(f, v) \quad \forall v \in U_0(\Omega). \tag{19.6}$$

The problem (19.6) has a solution if the condition (19.4) is fulfilled.

In this formulation, the problem's solution (regarding displacements) is not the only one since it is possible to move the body (translational transfer and hard rotation) as a solid. The only solution is distinguished by two conditions (19.5), which exclude arbitrary translational transfer and arbitrary hard rotation of the body. These conditions determine the subspace $V_0(\Omega) \subset U_0(\Omega)$, in which a unique solution is sought [1].

19.3 The Correctness of the First Basic Problem of the Theory of Elasticity on the Subspace

To establish the correctness of the problem on the subspace [1], it is necessary to establish the existence, uniformity, and coefficient stability of the solution. The positive-definiteness of a quadratic functional on the lineal of vector functions satisfying the conditions (19.5) is shown in [8], i.e., there is an inequality

$$a(v, v) \geqslant M_0 \|v\|_{0,\Omega}, \quad \forall v \in \left(L_2(\Omega)\right)^3 \tag{19.7}$$

with a constant M_0 independent of v. The V-ellipticity of the quadratic functional is established there, that is, an estimate

$$a(v, v) \geqslant M_1 \|v\|_{1,\Omega}, \quad \forall v \in \left(V_2^1(\Omega)\right)^3 \tag{19.8}$$

is obtained.

Let us define a linear set of vector functions $\left(V_2^k(\Omega)\right)^3 \subset \left(W_2^k(\Omega)\right)^3$:

$$\left(V_2^k(\Omega)\right)^3 = \left\{v \mid v \in \left(W_2^k(\Omega)\right)^3, \ \int_\Omega v dx = 0, \ \int_\Omega r \times v dx = 0\right\}.$$

Problems (19.1) and (19.2) correspond to the variational problem on the minimum of the functional

$$J(u) = a(u, u) - l(f, u). \tag{19.9}$$

It is known that in $\left(V_2^1(\Omega)\right)^3$ this variational problem has a unique solution (see [1, 7, 8]). There is also a unique solution $u \in \left(V_2^1(\Omega)\right)^3$ of problem (19.1) and (19.2) in the weak formulation (19.6) if $v \in \left(V_2^1(\Omega)\right)^3$.

Together with problems (19.1) and (19.2), we consider the problem with perturbed coefficients and the right-hand side (here $\tilde{\sigma} = \tilde{C}(x)\varepsilon$)

$$\tilde{L}\tilde{u} = -\sum_{i,k=1}^{3} \frac{\partial \tilde{\sigma}_{ik}}{\partial x_k} e_i = \tilde{f}_\Omega(x), \quad x \in \Omega, \tag{19.10}$$

$$\tilde{l}\tilde{u} = -\sum_{i,k=1}^{3} \tilde{\sigma}_{ik} \cos(n, x_k) e_i = \tilde{f}_\Gamma(x), \quad x \in \Gamma. \tag{19.11}$$

Assume that the conditions (analogous to conditions (19.4)) for the existence of a solution to problems (19.10), (19.11) are fulfilled

$$\int_\Omega \tilde{f}_\Omega dx + \int_\Gamma \tilde{f}_\Gamma d\gamma = 0, \quad \int_\Omega (r \times \tilde{f}_\Omega) dx + \int_\Gamma (r \times \tilde{f}_\Gamma) d\gamma = 0, \tag{19.12}$$

and the only solution of the problems (19.10) and (19.11) is distinguished by the conditions of the form (19.5)

$$\int_\Omega \tilde{u} dx = 0, \ \int_\Omega (r \times \tilde{u}) dx = 0. \tag{19.13}$$

We also assume that conditions of the form (19.3) are fulfilled for the coefficient matrix $\tilde{C}(x)$. Problems (19.10)–(19.13) are met by a bilinear functional

$$\tilde{a}(u, v) = \int_\Omega \left(\varepsilon(u)\right)^T \tilde{C}(x) \varepsilon(v) dx$$

and a similar weak statement (19.9)

$$\tilde{a}\left(\tilde{u}, v\right) = l\left(\tilde{f}, v\right) \forall v \in \left(V_2^1(\Omega)\right)^3. \quad (19.14)$$

Let $z = \tilde{u} - u$. It is obvious that $z \in \left(V_2^1(\Omega)\right)^3$. Substituting $\tilde{u} = z + u$ in (19.14) and taking into account (19.6), we get

$$\tilde{a}(z, z) = \int_\Omega \left(\varepsilon(u)\right)^T \left(C(x) - \tilde{C}(x)\right)\varepsilon(z)dx + \int_\Omega (\tilde{f}_\Omega - f_\Omega)zdx + \int_\Gamma (\tilde{f}_\Gamma - f_\Gamma)zd\gamma. \quad (19.15)$$

By virtue of (19.6), under the conditions (19.3), the quadratic functional $\tilde{a}(z, z)$ is V-elliptic, so from (19.15) we have

$$M_1 \|z\|_{1,\Omega}^2 \leqslant \delta \int_\Omega \sum_{i,k,l,m=1}^3 |\varepsilon_{ik}(u)|\, |\varepsilon_{lm}(z)|\, dx + \left|\int_\Omega (\tilde{f} - f)zdx\right| + \left|\int_\Gamma (\tilde{g} - g)zd\gamma\right|, \quad (19.16)$$

where

$$\delta = \max_{i,k} \max_x |c_{ik}(x) - \tilde{c}_{ik}(x)|,$$

M_1 is a constant that does not depend on u and z. Let us evaluate each of the terms in the right-hand side of the inequality (19.16).

We assume the coefficients $c_{ik}(x)$ are bounded functions. In view of the obvious inequality

$$\int_\Omega \varepsilon_{ik}^2(w)dx \leqslant \|w\|_{1,\Omega}^2,$$

the Cauchy–Bunyakovsky inequality and the a priori estimate for the solution of problems (19.1)–(19.5), which can be obtained based on inequalities (19.3) and (19.6), the Cauchy–Bunyakovsky inequality, the ε-inequality (here and further $M_k =$ const > 0, $k = 2, 3, 4, 5, 6$) and estimate

$$M_3 \|v\|_{0,\Gamma}^2 \leqslant \|v\|_{1,\Omega}^2, \quad \forall v \in \left(W_2^1(\Omega)\right)^3, \quad (19.17)$$

which obtained in [8], we have

$$\int_\Omega \sum_{i,k,l,m=1}^3 |\varepsilon_{ik}(u)|\, |\varepsilon_{lm}(z)|\, dx \leqslant M_4(\|f_\Omega\|_{0,\Omega} + \|f_\Gamma\|_{0,\Gamma})\, \|z\|_{1,\Omega}. \quad (19.18)$$

Due to the Cauchy–Bunyakovsky inequality,

$$\left|\int_\Omega (\tilde{f}_\Omega - f_\Omega) z dx\right| \leqslant \left\|(\tilde{f}_\Omega - f_\Omega)\right\|_{0,\Omega} \|z\|_{0,\Omega} \leqslant \left\|(\tilde{f}_\Omega - f_\Omega)\right\|_{0,\Omega} \|z\|_{1,\Omega} . \tag{19.19}$$

Finally, on the basis of the Cauchy–Bunyakovsky inequality and inequality (19.17), we obtain

$$\left|\int_\Gamma (\tilde{f}_\Gamma - f_\Gamma) z d\gamma\right| \leqslant M_5 \left\|\tilde{f}_\Gamma - f_\Gamma\right\|_{0,\Gamma} \|z\|_{1,\Omega} . \tag{19.20}$$

Combining estimates (19.18)–(19.20), from (19.16), we obtain

$$\|z\|_{1,\Omega} \leqslant M_6 \left(\delta \left(\|f_\Omega\|_{0,\Omega} + \|f_\Gamma\|_{0,\Gamma}\right) + \left\|\tilde{f}_\Omega - f_\Omega\right\|_{0,\Omega} + \left\|\tilde{f}_\Gamma - f_\Gamma\right\|_{0,\Gamma}\right) . \tag{19.21}$$

Inequality (19.21) implies the coefficient stability of the first basic boundary value problem of the elasticity theory. Therefore, the following statement is proved.
Statement 1. Let u be the solution of the problems (19.1)–(19.5), and let $\tilde{u}$ be the solution of the problems (19.10)–(19.13) with perturbed coefficients, then the error of the solution $z = \tilde{u} - u$ is estimated by (19.21).

Thus, the coefficient stability for the first basic problem of elasticity theory is established. This problem is conditionally correct; that is, its solution exists on the set of functions that meet the condition (19.4), and the single solution is determined on the subspace $\left(V_2^1(\Omega)\right)^3$. This circumstance leads to some difficulties in the numerical solution of the first basic problem of elasticity theory.

19.4 Solving the First Basic Problem of the Elasticity Theory by the Finite Element Method

To numerically solve the problem, we use the finite element method (FEM) [16, 17, 20], which essentially consists in replacing the infinite-dimensional space of admissible functions with its finite-dimensional subspace. At the same time, the basis of this subspace is usually made up of vector functions, the components of which are piecewise polynomials, different from zero only on a few finite elements.

Let V^h, the finite-dimensional space of the FEM with the basis $\{\varphi_i\}$ $(i = 1, 2, \ldots, N)$, be a subspace of the space $\left(V_2{}^1(\Omega)\right)^3$; that is, $V^h \subset \left(V_2{}^1(\Omega)\right)^3$. Consider the abstract discrete problem of the finite element method: to find such a function $u^h \in V^h$ that

$$a(u^h, v^h) = l(f, v^h) \quad \forall v^h \in V^h . \tag{19.22}$$

The existence and uniqueness of the solution of the discrete problem (19.22) follows from the Lax–Milgram lemma [16]. The convergence and accuracy of the solution follows from the general theory of the finite element method (see, for example, [16, 17, 20]), since all necessary conditions for this are fulfilled on the subspace V^h.

It should be noted that in order to construct the finite element basis $\{\varphi_i\}$ of the space V^h from $\left(V_2{}^1(\Omega)\right)^3$, it is necessary to know the null-space of the operator of the original problem in an explicit form. This null-space is defined by the condition (19.5) and is stretched over vector functions

$$\begin{aligned} &u_1^{(0)} = (1, 0, 0)^T, \quad u_2^{(0)} = (0, 1, 0)^T, \quad u_3^{(0)} = (0, 0, 1)^T, \\ &u_4^{(0)} = (0, -x_3, x_2)^T, \, u_6^{(0)} = (-x_2, x_1, 0)^T, \, u_6^{(0)} = (0, -x_3, x_2)^T. \end{aligned}$$

However, the algorithm for constructing the basis V^h in the general case for problems with non-unique solutions over the entire space is unknown in the literature.

Therefore, in addition to (19.22), we consider the following discrete problem:

$$a\left(u^h, v^h\right) = l\left(f, v^h\right), \quad \forall v^h \in W^h \subset (W_2^1(\Omega))^3. \tag{19.23}$$

The solution to this problem is not unique to the whole space. Choosing the functions ψ_i $(i = 1, 2, \ldots, N)$ as the basis of the space W^h and setting

$$u^h = \sum_{i=1}^{N} x_i \psi_i,$$

we obtain from (19.23) a system of linear algebraic equations (SLAE) with a sparse positive semidefinite matrix

$$Ax = b. \tag{19.24}$$

The following theorem is valid.

Theorem 1 *If $x^\#$ is a normal generalized solution of SLAE* (19.24) *with a sparse semidefinite matrix, and $\{\psi_i\}$ $(i = 1, 2, \ldots, N)$ is a basis of $W^h \subset \left(W_2{}^1(\Omega)\right)^3$, then the function*

$$u_\#^h = \sum_{i=1}^{N} x_i^\# \psi_i \tag{19.25}$$

will be the solution to the problem (19.22).

Proof In order to prove the theorem, it is necessary to show that the function (19.25) will be orthogonal to the null-space of the operator of the original problem. In turn, we will prove this fact in two stages. First, we will show that $u_\#{}^h$ is orthogonal to eigenfunctions $u_{0,j}{}^h (j = 1, 2, \ldots, n')$ which correspond to the zero eigenvalues of the discrete eigenvalue problem: find such a vector of displacements $u^h \in U^h$ that

$$a\left(u^h, v^h\right) = \lambda^h\left(u^h, v^h\right) \quad \forall v^h \in U^h, \tag{19.26}$$

that is, let us establish that

$$\left(u^h_{\#}, u^h_{0,j}\right) = 0, \quad u^h_{\#}, u^h_{0,j} \in U^h \quad (j = 1, 2, \ldots, n'). \tag{19.27}$$

From (19.26), we obtain a generalized algebraic eigenvalue problem [10]

$$Ax = \lambda Bx \tag{19.28}$$

with a symmetric positive-definite mass matrix B. Similarly, for the definition

$$u^h_{0,j} = \sum_{i=1}^{N} x^0_{j,i}\psi_i \quad (j = 1, 2, \ldots, n'), \tag{19.29}$$

we will obtain homogeneous systems of linear algebraic equations

$$Ax^{(0)}_j = 0 \quad (j = 1, 2, \ldots, n'). \tag{19.30}$$

Thus, the eigenfunctions corresponding to the zero eigenvalues of the discrete problem (19.26) are determined by the elements of the core of the matrix A. At the same time, the core elements of the matrix A are not only orthogonal to the elements of the image of the matrix A, but according to (19.28) are also B-orthogonal (vectors u and v are called B-orthogonal if $(u, v)_B = u^T Bv = 0$).

Further, for $j = 1, 2, \ldots, n$, using the expansion $x^{\#}$ in terms of eigenvectors $z_i{}^{\#}$ $(i = 1, 2, \ldots, N - n')$ of problem (19.28), corresponding to nonzero eigenvalues, we obtain

$$\left(u^h_{\#}, u^h_{0,j}\right) = \left(\sum_{i=1}^{N} x^{\#}_i\psi_i, \ \sum_{i=1}^{N} x^0_{j,i}\psi_i\right) = \left(x^{\#}, \ x^0_j\right)_B = \sum_{i=1}^{N} \alpha_i\left(x^{\#}_i, \ x^0_j\right)_B = 0$$

due to the B-orthogonality of the eigenvectors corresponding to different eigenvalues of the problem (19.28). Therefore, orthogonality is established.

It is easy to verify that the vector functions $u_j{}^{(0)}$ $(j = 1, 2, \ldots, 6)$ are the basis (non-orthonormal) of the null-space of the operator of the original problem. So,

$$a\left(u^{(0)}_j, v^h\right) = 0 \quad \forall v^h \in U^h, \quad j = 1, 2, \ldots, n'.$$

Since $u_j{}^{(0)} \in U^h$ $(j = 1, 2, \ldots, n')$, according to (19.28), these vector functions are eigenfunctions of the discrete problem corresponding to zero eigenvalues. Thus, the theorem is proven. □

The result of the theorem is the reasonable application of the normal generalized solution of the computational problem—a system of linear algebraic equations—to obtain an approximate solution of a discrete problem on a subspace. For the practical application of this approach, knowledge of the null-space of the operator of the original problem is not required. Therefore, the components of the stress tensor can be determined in a unique way with the required accuracy.

The same approach can also be applied to the numerical solution of other problems of the theory of elasticity that have a unique solution in the subspace. For example, for one- and two-dimensional problem formulations, mixed boundary value problems, etc., as well as for the numerical solution of boundary value problems that arise during mathematical modeling in other fields of science and technology.

It is advisable to use the method of three-stage regularization to calculate the normal generalized solution of the SLAE with a sparse symmetric semidefinite matrix on a computer with a parallel organization of calculations.

19.5 Solving Incorrect SLAE with Approximate Data of a Sparse Structure

Let us consider the solving of incorrectly formulated problems with approximate data and sparse matrices. A method of three-stage regularization was proposed in [6, 13] to obtain with guaranteed accuracy an approximation to the normal generalized solution of the SLAE with a sparse symmetric semidefinite matrix. This method is effectively implemented on modern high-performance computers, including those with parallel computing.

19.5.1 Formulation of the Problem

Let us consider a system of linear algebraic equations with a sparse symmetric positive semidefinite matrix ($A = A^T$, $A \geqslant 0$) of order n and rank $k = n - (n_0 - 1)$

$$Ax = b. \tag{19.31}$$

Problem (19.31) has a solution on the subspace—a generalized least squares solution—any vector x_0 for which the Euclidean norm of the residual reaches its smallest value:

$$\min_x \|Ax - b\| = \|Ax_0 - b\|. \tag{19.32}$$

The solution of x_n, which has the smallest Euclidean norm

$$\min_{x_0} \|x\| = \|x_n\| \tag{19.33}$$

is unique and is called a normal generalized solution.

Let us also consider the SLAE with a perturbed right-hand part

$$A\tilde{x} = \tilde{b}, \tag{19.34}$$

where $\tilde{b} = b + \Delta b$ and

$$\|\Delta b\| \leqslant \varepsilon_b \|b\| \leqslant \delta. \tag{19.35}$$

Let us denote by x and $\tilde{x}$ the normal generalized solutions of problems (19.31) and (19.34), respectively,

$$\begin{aligned} b_k &= Ax, \\ \tilde{b}_k &= A\tilde{x}, \\ \Delta b_k &= \tilde{b}_k - b_k, \\ \varepsilon_b &= \|\Delta b_k\| \,/\, \|b_k\|, \end{aligned}$$

$A^{\#}$ is a pseudo-inverse matrix, I_n is a unit matrix of order n, and $h(A) = \|A\| \|A^{\#}\|$ is a condition number of the matrix A.

For the error of the normal generalized solution of the problems (19.34) and (19.35), it is easy to obtain an estimate $\|x - \tilde{x}\| \,/\, \|x\| \leqslant h(A)\varepsilon_b$ [2, 4], from which the continuous dependence of this solution on the perturbation of the right-hand part follows. The problem (19.31) will be considered well-conditioned if the properties of the problem and the error of the initial data allow us to obtain the specified accuracy ε of the solution, that is, if

$$h(A)\varepsilon_b \leqslant \varepsilon. \tag{19.36}$$

19.5.2 Solving the Problem

The normal generalized solution of the problem (19.31), i.e., the solution of the least squares problem with a symmetric semidefinite matrix, can be calculated using the singular value decomposition of the matrix, which for symmetric matrices is the decomposition by eigenvectors $A = U\Lambda U^T$ (Λ is the diagonal matrix of eigenvalues). However, the matrix of eigenvectors U is a dense matrix of the general form, which does not preserve the sparse (for example, band) structure of the SLAE matrix A. Therefore, different regularization methods are usually used for such cases.

A two-stage regularization method is proposed in [11] for solving the least squares problem. Moreover, a three-stage regularization method is proposed in [6, 13, 14]. It solves the problem of choosing the regularization parameter, and this method allows us to calculate an approximation to the generalized solution with a predetermined accuracy.

The approximation u to the normal generalized solution of problem (19.31), x is obtained by solving the following two SLAEs with an arbitrarily selected parameter $\alpha > 0$ (e.g., $\alpha = 0.01$).

$$(A+\alpha I_n)\,z=\tilde{b}_k,\quad (A+\alpha I_n)\,u=Az. \tag{19.37}$$

In [11], the convergence of u to the normal generalized solution of problem (19.31) for $\alpha \to 0$ was established. It should be noted that in practical calculations, the right-hand part b is orthogonal to the kernel of the matrix A and, therefore, coincides with b_k. The perturbed right-hand part may not be orthogonal to the kernel of the matrix A; however, due to (19.34),

$$\left\|\tilde{b}_k-\tilde{b}\right\|\leqslant\|\Delta b\|\,.$$

Therefore, in the first SLAE from (19.37), we can use $\tilde{b}$ instead of $\tilde{b}_k$.

Let us estimate the error of the approximate solution u.

Theorem 2 *If $h(A)\varepsilon_b<\varepsilon<1$, then with accuracy to the values of the second order of smallness, the error of the approximate generalized solution of the SLAE with a symmetric positive semidefinite matrix is estimated by the value*

$$\frac{\|x-u\|}{\|x\|}\leqslant\alpha\frac{2\lambda_{n_0}+\alpha}{\left(\lambda_{n_0}+\alpha\right)^2}+\frac{\lambda_n\lambda_{n_0}\varepsilon_b}{\left(\lambda_{n_0}+\alpha\right)^2}, \tag{19.38}$$

where λ_{n_0} is the minimum nonzero eigenvalue of the matrix A (eigenvalues are numbered in ascending order).

Proof To prove this, we introduce the necessary matrix norm, consistent with the Euclidean vector norm:

$$\|A\|_*=\sup_{v\in\operatorname{Im}A}\frac{\|Av\|}{\|v\|}.$$

Based on the fact that $Ax=b_k$, the first of equation (19.38) can be written in the form of two equations

$$\begin{aligned}(A+\alpha I_n)z_k&=b_k+\Delta b_k,\\ \alpha z_{n-k}&=b_{n-k}+\Delta b_{n-k},\end{aligned}$$

and the second in the form

$$(A+\alpha I_n)\,u=Az_k,$$

we have

$$\begin{aligned}(A+\alpha I_n)z_k&=Ax_k+\Delta b_k,\\ (A+\alpha I_n)(u-x)&=\Delta b_k-\alpha z_k-\alpha x.\end{aligned}$$

Then

$$u-x=-\alpha\,(A+\alpha I_n)^{-2}\,(2A+\alpha I_n)\,x+(A+\alpha I_n)^{-2}\,A\Delta b_k.$$

The norm of the error can be estimated as follows:

$$\|u - x\| \leqslant \alpha \left\|(A + \alpha I_n)^{-2}(2A + \alpha I_n)\right\|_* \|x\| + \left\|(A + \alpha I_n)^{-2} A\right\|_* \|\Delta b_k\| .$$

When calculating the spectral norm of the matrices, it is necessary to study the behavior of the functions

$$\frac{2\lambda + \alpha}{(\lambda + \alpha)^2}, \quad \frac{\lambda}{(\lambda + \alpha)^2}.$$

They go to zero at $\lambda \to \infty$. Hence, the greatest value of these functions is achieved at the smallest possible value of λ, that is, at $\lambda = \lambda_{n_0}$. Thus, the following equalities hold for the norms of the matrices

$$\left\|(A + \alpha I_n)^{-2}(2A + \alpha I_n)\right\|_* = \frac{2\lambda_{n_0} + \alpha}{(\lambda_{n_0} + \alpha)^2}, \quad \left\|(A + \alpha I_n)^{-2} A\right\|_* = \frac{\lambda_{n_0}}{(\lambda_{n_0} + \alpha)^2}.$$

Taking into account that $\|x\| \geqslant \|b_k\| / \|A\|$, we obtain inequality (19.38), which had to be proved. □

From (19.38), with accuracy to values of the second order of smallness, it is possible to obtain the condition for selecting the parameter α to calculate the solution with the specified accuracy ε

$$\alpha \leqslant \frac{\lambda_{n_0}\varepsilon - \|A\| \varepsilon_b}{2\sqrt{1 - \varepsilon}}, \tag{19.39}$$

that is, to fulfill the condition

$$\frac{\|x - u\|}{\|x\|} \leqslant \varepsilon < 1.$$

So, to choose α, it is necessary to estimate λ_{n_0}. To do this, we can use calculations that implement one step of the power method to determine the maximum (modulo) eigenvalue of the matrix $(A + \alpha I_n)^{-1}$ [10], then the approximation to $(\lambda_{n_0} + \alpha)^{-1}$ is determined by

$$\mu = \max_i |w_i| , \tag{19.40}$$

where w is the solution of the SLAE

$$(A + \alpha I_n) w = u_H \equiv \frac{u}{\max_i |u_i|}. \tag{19.41}$$

Then, if $\alpha \ll \lambda_{n_0}$ is sufficiently small, condition (19.39) can be written as follows:

$$\alpha \leqslant \frac{\varepsilon - \|A\| \mu\varepsilon_b}{2\mu\sqrt{1 - \varepsilon}}. \tag{19.42}$$

Thus, in order to obtain a reliable solution of the system (19.31), it is necessary that the initial data satisfy condition (19.36) (the problem is well conditioned). The choice of the parameter α that satisfies the estimate (19.42) will guarantee the specified accuracy of the normal generalized solution.

The three-stage regularization method consists in performing the following steps:

1. sequential solving of two SLAEs (19.37) with the same positive-definite matrix $A + \alpha I_n$;
2. calculation of the approximation μ to $\left(\lambda_{n_0} + \alpha\right)^{-1}$ according to (19.41);
3. verification of the fulfillment of the inequality

$$(2\alpha + \|A\| \, \varepsilon_b)\mu \leqslant \varepsilon.$$

If the inequality is fulfilled, then the required accuracy ϵ is achieved at the selected value $\alpha = \alpha_0$. If the inequality is not fulfilled, then from condition (19.42) the new value of the shift $\alpha = \alpha_1$ is determined by the formula

$$\alpha_1 \leqslant 0.5\,(\varepsilon/\mu - \|A\| \, \varepsilon_b)$$

and step 1 is repeated with the matrix $A + \alpha_1 I_n$—a new approximation u is calculated, which ensures the required accuracy ε of the normal pseudo-solution.

The three-stage regularization method is an alternative to both the methods of the regularizing functional

$$(A^2 + \alpha I_n)\, u_\alpha = Ab$$

and the method of discrete regularization [19], based, for example, on the SVD algorithm. The advantage of the three-stage regularization method over the first approach is the better conditionality of the problem of determining the normal generalized solution, and over the second one—an obviously smaller number of arithmetic operations. The advantages of the three-stage regularization algorithm are enhanced for matrices of a special structure, for example, band matrices. It is known that the band structure of the matrix is broken in the first approach. In the second approach, although there is a band version of the SVD algorithm, it is not possible to reduce the number of arithmetic operations significantly due to the need to calculate all singular vectors.

19.5.3 Solving the Problem of Weighted Least Squares

Along with problem (19.32) and (19.33), we consider the weighted least squares problem with symmetric positive-definite weight matrices $M = M^T$ and M^{-1}. In this case, a weighted generalized least squares solution or a weighted pseudo-solution is any vector x_0 for which the weighted norm of the residual reaches the smallest value:

$$\min_x \|Ax - b\|_{M^{-1}} = \|Ax_0 - b\|_{M^{-1}}, \tag{19.43}$$

where

$$\|y\|_K = (y^T K y)^{\frac{1}{2}}.$$

Let us consider the case when M is a sparse matrix, the set of nonzero elements (structure) of which coincides with the set of nonzero elements of the matrix A or is its subset. And the solution x_n, which has the smallest weighted norm

$$\|x_n\|_M = \min_{x_0} \|x\|_M \tag{19.44}$$

is called the weighted normal generalized solution. The condition

$$\|\Delta b\|_{M^{-1}} \leqslant \varepsilon_b \|b\|_{M^{-1}} \leqslant \delta \tag{19.45}$$

is fulfilled for the right-hand part of the perturbed SLAR (19.44).

The weighted least squares problems (19.43) and (19.44) in the case of a symmetric positive-definite matrix M can be quite easily reduced to the problems (19.32) and (19.33) with a different matrix and right-hand side. For example, as follows. Given that M is a symmetric positive-definite matrix, there exists its decomposition $L_M L_M{}^T = M$. Based on the fact that

$$\begin{aligned} \|Ax - b\|_{M^{-1}} &= \left\| L_M^{-1} A L_M^{-T} L_M^T x - L_M^{-1} b \right\| \equiv \|Cy - d\|, \\ \|x\|_M &= \left\| L_M^T x \right\| \equiv \|y\|, \end{aligned}$$

where

$$C = L_M^{-1} A L_M^{-T}, \quad d = L_M^{-1} b, \quad y = L_M^T x,$$

instead of problems (19.43) and (19.44), we will get a least squares problems of the form (19.32) and (19.33) with a symmetric semidefinite matrix C and the right-hand part d. However, the matrix C preserves the sparse structure of the matrix A only under the condition that the matrix M is diagonal. In other cases, the density of the matrix C increases, and, in general, it should be considered a dense matrix. We also note that for the perturbed SLAE (19.34), (19.35) we have the problem (where $\bar{d} = L_M{}^{-1}\bar{b}$)

$$Cy = \bar{d}, \quad \left\|\bar{d} - d\right\| \leq \varepsilon_b \|d\|. \tag{19.46}$$

To calculate the approximation to the normal generalized solution of problem (19.46), we can use the three-stage regularization method, which can be used to find the approximation to the weighted normal generalized solution of problems (19.34), (19.35). However, this approach requires the explicit calculation of the matrix C (which may lose its sparse structure) and the vector $\bar{d}$. Therefore, it is advisable to modify the three-stage regularization method to operate with the matrices A, M, and the vector $\bar{b}$ and avoid these calculations.

Therefore, with an arbitrarily chosen parameter $\alpha_0 > 0$, for example, $\alpha_0 = 0.01$, an approximation to the weighted normal pseudo-solution u is obtained by sequentially solving two SLAEs

$$A + \alpha M z = b_k, \quad A + \alpha M u = A z. \tag{19.47}$$

Further, we solve the SLAE

$$(A + \alpha M) w = M u \left(\max_i |u_i| \right)^{-1}$$

and calculate the approximation

$$\mu \approx \max_i |w_i|$$

to the maximum eigenvalue of the generalized algebraic problem

$$M v = \mu (A + \alpha M) v.$$

In essence, one step of the power method variant is performed for the generalized problem.

Similar to the above, it is possible to obtain an estimate of the form (19.38) of the accuracy of the approximate weighted normal generalized solution

$$\|x - u\|_M \,/\, \|x\|_M \leqslant \delta = (2\alpha + \left\| L_M^{-1} A L_M^{-T} \right\| \varepsilon_b)\mu.$$

Therefore, if the inequality $\delta \leqslant \varepsilon$ is satisfied, then the required accuracy ε is achieved at the chosen value of α. If the inequality is not satisfied, then a new value of the shift is determined by the formula:

$$\alpha_1 \leqslant 0.5 \left(\varepsilon / \mu - \left\| L_M^{-1} A L_M^{-T} \right\| \varepsilon_b \right)$$

and systems (19.47) with the matrix $A + \alpha_1 M$ are solved—a new approximation u is calculated, which ensures the required accuracy ε of the weighted normal generalized solution.

The proposed schemes of the three-stage regularization method are divided into several subproblems with sparse matrices, the solving of which is effectively implemented (see, for example, [5, 13, 14]) on modern high-performance computers, including those with parallel organization of calculations.

19.6 Approbation of the Method of Numerical Solution

The proposed method has been tested on several model problems. Below are the results of solving one of the problems.

We considered the problem of the plane-stressed state of a thin isotropic loose rectangular plate with dimensions $a_1 \times a_2$ and thickness δ, which is under the action of uniformly distributed volume forces of intensity $f_\Omega(x)$ kg/cm^3, parallel to the plane of the plate and directed at an angle φ to the axis Ox. That is, in (19.1) and (19.2), $\sigma_{13} = \sigma_{23} = \sigma_{33} = f_{\Omega,3} = f_{\Gamma,3} = 0$, and the problem becomes two-dimensional. When solving such a problem, the displacements are determined ambiguously, while the stresses are calculated in a uniform way.

Next, the results of numerical calculations of the plate with the following data: $a_1 = 60$ cm; $a_2 = 20$ cm; $E = 2 \times 10^{11}$ Pa, $\nu = 0.3$; $F^V \equiv 0$, therefore, $f_1 = f_2 = 0$;

$$f_{\Gamma,1}(x)=\begin{cases} 0.005(3x_1^2-(2x_2-a_2)^2)(2x_2-a_2), \; x_1=0, \;\; x_1=a, \;\; 0\leqslant x_2\leqslant a_2; \\ 0.001(-14x_1^2+7.5(2x_2-a_2)^2x_1), \; 0\leqslant x_1\leqslant a, \;\; x_2=0, \;\; x_2=a_2; \end{cases}$$

$$f_{\Gamma,2}(x)=\begin{cases} 0.001(-14x_1^2+7.5(2x_2-a_2)^2x_1), \; x_1=0, \;\; x_1=a, \;\; 0\leqslant x_2\leqslant a_2; \\ 0.001(14x_1^2-1.25(2x_2-a_2)^2)(2x_2-a_2), \; 0\leqslant x_1\leqslant a, \; x_2=0, \; x_2=a_2 \end{cases}$$

are presented. The exact solution (tension) of this problem is known:

$$\begin{aligned} \sigma_{11} &= (0.015x_1^2 - 0.005(2x_2-a_2)^2)(2x_2-a_2); \\ \sigma_{12} &= -0.014x_1^2 + 0.0075(2x_2-a_2)^2x_1; \\ \sigma_{22} &= (0.014x_1^2 + 0.00125(2x_2-a_2)^2)(2x_2-a_2). \end{aligned}$$

After the FEM discretization of the problem of the plane stress state of a thin isotropic loose plate, we obtain SLAE (19.24) with a symmetric band positive semidefinite matrix. However, when calculating the elements of the matrix A and the vector b in the computer, a problem with approximate data will be obtained

$$A\tilde{x} = \tilde{b}, \; \left\| b - \tilde{b} \right\| = \|\Delta b\| \leqslant \delta. \tag{19.48}$$

Nothing is known a priori about the properties (singularity, conditionality, positive-definiteness, rank, etc.) of the matrix and, therefore, the properties (including compatibility) of system (19.48). In particular, for this problem, the minimum eigenvalue of the computer problem is estimated by a value of the order of 10^{-9} (see, for example, σ_1, σ_2, σ_3 on the 9×15 grid in Table 19.1). That is, the matrix of the machine

Table 19.1 Singular values of the matrix of SLAE

Grid	System order	Bandwidth	σ_n	σ_4	σ_3	σ_2	σ_1
9×15	320	49	5.24×10^7	9.93×10^3	6.23×10^{-9}	4.62×10^{-9}	3.18×10^{-9}
12×20	548	61	5.30×10^7	6.06×10^3	19.2×10^{-9}	8.00×10^{-9}	3.12×10^{-9}
18×30	1178	85	5.35×10^7	2.95×10^3	30.6×10^{-9}	20.8×10^{-9}	7.75×10^{-9}
24×40	2050	109	5.36×10^7	1.74×10^3	17.3×10^{-9}	10.6×10^{-9}	5.03×10^{-9}

Table 19.2 The maximum relative errors of the calculated stresses

Grid	System order	Bandwidth	δ_{11} (%)	δ_{12} (%)	δ_{22} (%)
18×30	1 178	171	2.85	3.37	26.8
36×60	4 514	315	1.47	1.74	15.3
72×120	17 666	603	0.75	0.88	8.12
90×150	27 482	747	0.60	0.71	6.57

problem is not singular in the classical sense (its determinant is different from zero), while the three eigenvalues of the discrete problem (19.24) are equal to zero.

Table 19.1 shows the maximum and four minimum singular values (these are dimensionless quantities) of the matrix of the algebraic system in the absence of the plate fixing.

Therefore, before solving the problem, it is necessary to investigate the properties of the matrix (for example, the dimensions of the kernel and image of the matrix, their bases) of the corresponding machine problem (19.30), using the band version of the SVD algorithm. For multivariate calculation on smaller grids, it is advisable to use the three-stage regularization method, as it is more economical. The comparison of the approximations to the displacements u_1, u_2 on the 18×30 grid, which are obtained by solving an algebraic problem using the SVD algorithm and the three-stage regularization algorithm, showed that the differences between these solutions do not exceed 0.05%.

Table 19.2 shows some parameters of algebraic problems (19.24) and the maximum relative errors (in percentages, denoted by δ_{11}, δ_{12}, δ_{22}) of the corresponding calculated stresses in the centers of mass of finite elements on different grids. Table 19.3 shows some results of the numerical solving of the problem in the original formulation using the three-stage regularization method. The results are grouped by three values at the points (x_1, x_2), and the calculated values of the stresses (in Pa) are given: in the first row—σ_{11}, in the second row—σ_{12}, and in the third row—σ_{22} (x_1 and x_2 are given in cm). Table 19.4 shows the exact values of stresses (according to the scheme similar to Table 19.3).

Testing showed that the three-stage regularization method, according to time costs, used memory, and reliability of the obtained results, is the most effective method for solving problems of this type on the computer. It should also be noted that the SVD algorithm also provides a reliable solution to the problem, but with significantly greater time and memory consumption, and it is advisable to use it for researching the properties of computer problems on coarse grids.

Table 19.3 Results (stresses) of the numerical solving of the problem

x_2/x_1	0.0	15.0	30.0	45.0	60.0
0.0	20.243697	−47.378245	−250.130855	−587.934534	−1 038.96160
	−0.099154	−47.623490	−101.624858	−161.784164	−220.77310
	−9.566290	−13.941354	−18.850377	−24.162044	−26.99765
5.0	2.493502	−31.237734	−132.469540	−301.128438	−530.23829
	−0.039838	−14.402766	−35.176005	−62.108855	−82.45452
	−1.259985	−3.346201	−5.463770	−7.541295	−7.02207
10.0	0.000000	0.000000	0.000000	0.000000	0.00000
	−0.002502	−3.159099	−12.688162	−28.370445	−37.51795
	0.000000	0.000000	0.000000	0.000000	0.00000
15.0	−2.493502	31.237734	132.469540	301.128438	530.238297
	−0.039838	−14.402766	−35.176005	−62.108855	−82.454524
	1.259985	3.346201	5.463770	7.541295	7.022079
20.0	−20.243697	47.378245	250.130855	587.934534	1 038.96162
	−0.099154	−47.623490	−101.624858	−161.784164	−220.77311
	9.566290	13.941354	18.850377	24.162044	26.99765

Table 19.4 The exact values of stresses

x_2/x_1	0.0	15.0	30.0	45.0	60.0
0.0	20.000	−47.500	−250.000	−587.500	−1 060.000
	0.0	−48.150	−102.600	−163.350	−230.400
	−10.000	−14.200	−18.400	−22.600	−26.800
5.0	2.500	−31.250	−132.500	−301.250	−537.500
	0.000	−14.400	−35.100	−62.100	−95.400
	−1.250	−3.350	−5.450	−7.550	−9.650
10.0	0.0	0.0	0.0	0.0	0.0
	0.0	−3.150	−12.600	−28.350	−50.400
	0.0	0.0	0.0	0.0	0.0
15.0	−2.500	31.250	132.500	301.250	537.500
	0.0	−14.400	−35.100	−62.100	−95.400
	1.250	3.350	5.450	7.550	9.650
20.0	−20.000	47.500	250.000	587.500	1 060.000
	0.0	−48.150	−102.600	−163.350	−230.400
	10.000	14.200	18.400	22.600	26.800

19.7 Conclusions

The methods and algorithms for solving incorrect problems with elliptic operators are considered. The first main problem of the theory of elasticity has a unique solution on a subspace, but discrete models use finite-dimensional subspaces of the entire space, and therefore the solutions of these models are not unique.

The methodology for calculating the unique solution on the subspace of the first basic problem of the theory of elasticity has been developed and justified. It is proved that the normal generalized solution of the system of linear algebraic equations (which arises as a result of discretization) with a sparse symmetric semidefinite matrix allows obtaining a finite element approximation to the unique solution of the original problem.

To calculate the normal generalized solution and the weighted normal generalized solution of SLAR with a sparse symmetric semidefinite matrix with guaranteed accuracy, a three-stage regularization method was developed and substantiated. Economical parallel algorithms of the three-stage regularization method have been developed for computers with parallel organization of calculations of various architectures. Estimates of the accuracy of the approximate solutions in the case of perturbation of the right part of the SLAR were obtained.

Therefore, the proposed methodology for solving the first basic problem of elasticity theory allows for obtaining reliable solutions with guaranteed accuracy. This methodology is invariant for a wide class of problems with Neumann conditions on the boundary.

References

1. Galba, E.F., Gladkiy, A.V., Khimich, A.N., Yakovlev, M.F.: On the correctness of the first basic problem of elasticity theory on a subspace. Computer Mathematics. V.M. Glushkov Institute of Cybernetics of NAS of Ukraine, Kyiv, No. 1, pp. 54–62 (2002)
2. Khimich, A.N.: Estimates of the total error in the solution of systems of linear algebraic equations for matrices of arbitrary ranks. Computer Mathematics, No. 2, pp. 41–49 (2002)
3. Khimich, A.N., Molchanov, I.N., Popov, A.V., Chistyakova, T.V., Yakovlev, M.F.: Parallel Algorithms for the Solving Of Computational Mathematics Problems. Naukova Dumka, Kyiv (2008). (In Russian)
4. Khimich, A.: Perturbation bounds for the least squares problem. Cybern. Syst. Anal. **32**(3), 434–436 (1996). https://doi.org/10.1007/BF02366509
5. Khimich, A.N., Popov, A.V., Polyanko, V.V.: Algorithms of parallel computations for linear algebra problems with irregularly structured matrices. Cybern. Syst. Anal. **47**(6), 973–985 (2011). https://doi.org/10.1007/s10559-011-9377-4
6. Khimich, A.N., Yakovlev, M.F.: On the solution of systems with matrices of incomplete rank. Computer Mathematics, No. 1, pp. 1–15 (2003)
7. Mikhlin, S.G.: Variational Methods in Mathematical Physics. Nauka, M. (1970). (In Russian)
8. Molchanov, I.N., Galba, E.F.: Variational formulations of the static problem of the theory of elasticity for given external forces. Ukr. Math. J. **43**(2), 158–161 (1991)

9. Molchanov, I.N., Levchenko, I.S., Fedonyuk, N.N., Khimich, A.N., Chistyakova, T.V.: Numerical simulation of the stress concentration in an elastic half-space with a two-layer inclusion. Int. Appl. Mech. **38**(3), 308–314 (2002). https://doi.org/10.1023/A:1016078010683
10. Molchanov, I.N., Popov, A.V., Khimich, A.N.: Algorithm to solve the partial eigenvalue problem for large profile matrices. Cybern. Syst. Anal. **28**(3), 281–286 (1992). https://doi.org/10.1007/BF01126215
11. Morozov, V.A.: Regularization Methods for Unstable Problems. Publishing House of Moscow University, M. (1987). (In Russian)
12. Popov, A.V.: About one implementation of a parallel algorithm for solving incorrect systems of linear algebraic equations. System analysis and information technologies. In: Proceedings of the VII International Scientific and Technical Conference (June 28–July 2, 2005, Kyiv), pp. 72. National Technical University of Ukraine "KPI", K. (2005)
13. Popov, A.V., Khimich, A.N.: Research and solution of the first basic problem of the theory of elasticity. Computer Mathematics, No. 2, pp. 105–114 (2003)
14. Popov, A.V.: On an effective method for solving incorrect problems with sparse matrices. Theory of Optimal Solutions, pp. 77–81 (2013)
15. Popov, A.V.: On the parallel algorithms of sparse matrices factorization. Computer Mathematics, No. 2, pp. 115–124 (2013)
16. Siarlet, P.G.: The Finite Element Method for Elliptic Problems. North Holland (1978)
17. Strang, G., Fix, G.J.: An Analysis of the Finite Element Method. Prentice-Hall Inc, N.J. (1973)
18. Timoshenko, S.P., Goodier, J.N.: Theory of Elasticity. McGraw-Hill, N.Y. (1970)
19. Voevodin, V.V., Kuznetsov, Yu.A.: Matrices and Computing. Nauka, M. (1984). (In Russian)
20. Zienkiewicz, O.C., Morgan, K.: Finite Elements and Approximation. A Wiley-Interscience Publication, N.Y. (1983)

Chapter 20
Influence of Boundary Conditions and Dissipative Heating Onto Resonance Vibration of Shear Compliant Viscoelastic Cylindrical Shell with Piezoelectric Sensors

Ivan Kirichok, Yaroslav Zhuk, and Olga Chernyushok

Abstract A problem on forced resonant vibrations and dissipative heating of viscoelastic cylindrical shell with piezoelectric sensors with taking account for the in-plane shear strain, temperature dependence of viscoelastic characteristic of material and edge clamping conditions are considered. The influence of temperature dependence, in-plane shear strain, and heat exchange conditions on the surfaces onto amplitude and temperature-frequency characteristics of the forced vibration of the shell is studied. A special technique for the shell life prediction based on the thermal failure criterion is developed.

20.1 Introduction

Cylindrical shells made of inelastic materials are widely used as elements of modern structures. The shells usually possess complex geometrical and material characteristics such as layer-wise composition, anisotropy, and high in-plane shear compliance. They can be subjected to intensive harmonic loading with frequencies equal to or close to the resonance frequencies of the structure. Operation of the shells under that severe condition is usually accompanied by dissipative heating due to internal hysteresis losses in the inelastic materials (see, for example, [2, 3]). Furthermore, harsh working conditions under elevated temperatures resulting in high levels of

I. Kirichok
S.P. Timoshenko Institute of Mechanics, National Academy of Science of Ukraine, Kyiv, Ukraine
e-mail: term@inmech.kiev.ua

Y. Zhuk (✉)
Faculty of Mechanics and Mathematics, Taras Shevchenko National University of Kyiv, Kyiv, Ukraine
e-mail: y.zhuk@i.ua

O. Chernyushok
National University of Food Technologies, Kyiv, Ukraine
e-mail: chernyshokolqa@ukr.net

H. Altenbach et al. (eds.), *Selected Problems of Solid Mechanics and Solving Methods*, Advanced Structured Materials 204, https://doi.org/10.1007/978-3-031-54063-9_20

dynamic stresses, deflection amplitudes, and overheating temperatures can significantly affect the structural element's performance and even cause its failure. Special active (usually piezoelectric) components operating as sensors or actuators are commonly introduced to control the thermomechanical response of the structure [3, 6, 14, 16–18].

Derivation of the electro-thermomechanical models to describe the dynamic response of layered thin-wall structural members made of inelastic passive (without piezoeffect) and active materials in the framework of both classic and high-order theories (with the relaxed normality condition) along with solutions of the numerous problems are proposed in numbers of papers and books [4, 12–14]. The authors would like to mention the pioneering book [15] having academician Ya.M. Grigorenko among its contributors in this respect. All these works are among those reviewed in the original summary publications [7–10]. In particular, papers [4, 11–13, 19, 20] are devoted to the investigation of the forced resonance vibration and dissipative heating of viscoelastic cylindrical shells with embedded piezoactive sensors and actuators in the framework of classic Kirchhoff–Love theory.

The present paper addresses the problem of forced resonant vibration and dissipative heating of viscoelastic cylindrical shells with piezoelectric sensors in the framework of the Timoshenko shell theory. Special attention is paid to the accounting for temperature dependence of material properties as well as the type of boundary conditions. The influence of boundary conditions, in-plane shear strain, the temperature dependence of both passive and piezoactive constituents of the shell onto the structure deflection amplitudes, overheating temperature level, electric parameters of the sensor, and shell durability under forced resonance vibration are studied.

20.2 Problem Statement. Main Equations

As it was mentioned above, the general theory of thermoviscoelastic plates and shells with distributed sensors and actuators under monoharmonic mechanical or electrical loading was developed in [5, 6, 8, 9, 11, 14, 17]. The case of controlling the axisymmetric resonant vibrations and dissipative heating of shells of revolution with piezoelectric sensors and actuators is addressed in [8, 11] in particular. The problem of resonant vibration and heating of ring plates with piezoactuators under electromechanical loading is investigated in [10] in detail. In this paper, let us consider a flexible cylindrical three-layer shell of length l and thickness $h = h_0 + 2h_1$ relative to the orthogonal coordinate system α, θ, z with the origin of normal coordinate $z = 0$ located in the middle of the internal layer having the radius R. Surfaces of the internal passive layer $z = \pm h_0/2$ of the shell are assumed to be ideally bounded with the piezoactive layers of thickness h_1. It is also supposed that the material of the internal passive layer is transversely isotropic while the outer active layers are fabricated of the piezoceramics polarized over the thickness in the opposite (to each other) directions. Materials of all three layers are considered to demonstrate viscoelastic mechanical response. The electromechanical properties of the materials under consideration are assumed to be temperature-dependent. Let us suppose the upper ($^1/_2 h_0 \leqslant z \leqslant h_0 + h_1$) and lower ($-h_1 - {}^1/_2 h_0 \leqslant z \leqslant -{}^1/_2 h_0$) piezolayers are characterized by the piezoelectric constant $+d_{31}$ and $-d_{31}$, respectively.

The inner contact faces of the piezolayers and passive layer are covered with solid continuous infinitely thin electrodes and are kept at zero electric potential $\varphi_{1,2}(\pm{}^1/_2 h_0) = 0$. The outer surfaces $z = \pm({}^1/_2 h_0 + h_1)$ of the shell are electroded over the areas $s^{\pm} = 2\pi R \Delta_\alpha$ and are not covered with the electrodes elsewhere out of those areas.

Under these assumptions, the electrostatic boundary conditions at the outer surfaces of the piezosensor can be written in the following form:

$$\iint_{s^{\pm}} D_z^{\pm} dz = 0 \quad (\alpha_0 \leqslant \alpha \leqslant \alpha_1), \quad D_z^{\pm} = 0 \quad (0 \leqslant \alpha < \alpha_0, \ \alpha_1 < \alpha \leqslant l), \tag{20.1}$$

where $D_z^{\pm}$ is the normal component of the electric-flux density vector in the piezolayers; α stands for the shell axial coordinate; $\alpha_0 \leqslant \alpha \leqslant \alpha_1$ is the segment of the sensor electroded area, $\Delta_\alpha = \alpha_1 - \alpha_0$.

The shell is subjected to the axisymmetric outer pressure $q_z = q_z'(\alpha) \cos \omega t$ that varies harmonically in time t with amplitude q_z' at frequency ω which is close to the flexural vibration resonance of the shell. Under the loading, the electric potentials of amplitudes $\pm V_s$ are induced on the sensor $s^{\pm}$ electrodes. These values can be either calculated or determined experimentally.

As to the mechanical boundary conditions, it is supposed that the shell facets are always free along the axial direction. Along the transverse direction, both facets are assumed to be either fixed or simply supported (or one edge is simply supported while the other is fixed).

Regarding the thermal boundary condition, let us assume that the heat exchange with the environment of temperature T_c takes place on all the shell surfaces.

The Timoshenko hypotheses are supposed to be valid for the mechanical variables through the total thickness of the laminate to simulate the mechanical response of the shell. The in-plane shear deformation and rotation inertia of the cross-section are taken into account as it was adopted in [5]. For electrical variables of the problem, it is assumed the components of electric-flux density vector D_α and D_θ are negligibly small. Then the components E_α and E_θ of the electric field vector can be determined from the constitutive equations $D_\alpha = 0$ and $D_\theta = 0$. In so doing, it follows from the equations of electrostatics that normal component $D_z = \text{const}$ does not depend on the coordinate z [6].

Temperature-dependent viscoelastic material properties are described in the frame of the complex moduli approach [4, 6]. Temperature is supposed to be constant over the laminate thickness.

According to the assumptions adopted and following the derivation procedure of two-dimensional equations for the shell electro-thermomechanics theory [4, 6, 8], the problem statement of forced resonance vibration and dissipative heating of the shell under consideration with respect to the unknown complex amplitudes of the main field variables consists of the following relationships:

- harmonic vibration equation (multiplier $e^{i\omega t}$ is omitted)

$$\begin{aligned} \frac{dN_\alpha}{d\alpha} + \rho_* \omega^2 u &= 0, \\ \frac{dQ_\alpha}{d\alpha} - \frac{N_\theta}{R} + \rho_* \omega^2 w + q_z &= 0, \\ \frac{dM_\alpha}{d\alpha} - Q_\alpha + \rho_{**} \omega^2 \psi_\alpha &= 0; \end{aligned} \tag{20.2}$$

- amplitude constitutive equations of electro-viscoelasticity for forces and moments

$$\begin{aligned} &N_\alpha = C_{11}\varepsilon_\alpha + C_{12}\varepsilon_\theta, \quad N_\theta = C_{12}\varepsilon_\alpha + C_{11}\varepsilon_\theta, \\ &Q_\alpha = k_s C_{44}\varepsilon_{\alpha z}, \\ &M_\alpha = D_{11}\kappa_\alpha + M_E; \quad M_\theta = D_{12}\kappa_\alpha + M_E; \end{aligned} \tag{20.3}$$

- geometrical relations between amplitudes of strain and displacement

$$\varepsilon_\alpha = \frac{du}{d\alpha}, \quad \varepsilon_\theta = \frac{w}{R}, \quad \kappa_\alpha = \frac{d\psi_\alpha}{d\alpha}, \quad \vartheta_\alpha = -\frac{dw}{d\alpha}, \quad \varepsilon_{\alpha z} = \psi_\alpha - \vartheta_\alpha; \tag{20.4}$$

- expressions for electric-flux density in the piezosensor layers

$${}^{1,2}D_z = -b_{33}\frac{V_a}{h_1} \pm b_{31}(\varepsilon \mp \bar{h}_1 \kappa_\alpha) \quad (z \leqslant -h_0/2, \ z \geqslant h_0/2), \tag{20.5}$$

  where superscripts 1 and 2 stand for the upper and lower pizoactive layer, respectively;
- heat conduction equation averaged over the shell thickness and the vibration period

$$\frac{1}{a}\frac{\partial T}{\partial t} = \frac{\partial^2 T}{\partial \alpha^2} - \frac{2\alpha_s}{\lambda H}(T - T_c) + \frac{\omega}{2\lambda H}\langle W \rangle \tag{20.6}$$

  with the averaged dissipation rate of the form

$$\begin{aligned} \langle W \rangle =& N''_\alpha \varepsilon'_\alpha - N'_\alpha \varepsilon''_\alpha + N''_\theta \varepsilon'_\theta - N'_\theta \varepsilon''_\theta + M''_\alpha \kappa'_\alpha - M'_\alpha \kappa''_\alpha \\ &+ Q''_\alpha \varepsilon'_{\alpha z} - Q'_\alpha \varepsilon''_{\alpha z} + ({}^1D''_z + {}^2D''_z)V'_a - ({}^1D'_z + {}^2D'_z)V''_a, \end{aligned} \tag{20.7}$$

  where real and imaginary parts of any complex quantity a, $a = a' + ia''$, are marked as $(\cdot)'$ and $(\cdot)''$ correspondently, $i = \sqrt{-1}$.

As it was mentioned above, it is assumed that the mechanical boundary conditions are of the form that provides free displacement of the shell facets along the axial direction. The following cases of edge support in the transverse direction are possible:

- both facets are hinged

$$N_\alpha = 0, \quad w = 0, \quad M_\alpha = 0 \quad (\alpha = 0, l); \tag{20.8}$$

- one facet is hinged while another one is rigidly fixed

$$\begin{aligned} N_\alpha = 0, \quad w = 0, \quad M_\alpha = 0 \;\; (\alpha = 0), \\ N_\alpha = 0, \quad w = 0, \quad \psi_\alpha = 0 \;\; (\alpha = l); \end{aligned} \tag{20.9}$$

- both edges are rigidly fixed

$$N_\alpha = 0, \quad w = 0, \quad \psi_\alpha = 0 \;\; (\alpha = 0, l). \tag{20.10}$$

Boundary and initial conditions for heat conduction equation are of the form

$$\lambda \frac{\partial T}{\partial \alpha} = \pm \alpha_{0,l}(T - T_c) \quad (\alpha = 0, l), \quad T = T_0 \quad (t = 0). \tag{20.11}$$

The voltage amplitude across the sensor open-circuited electrodes induced by the shell harmonic vibration can be calculated using the solution of electro-thermomechanics problem (20.2)–(20.11) and first electrostatic boundary condition (20.1), yielding the formula

$$\frac{V_s}{h_1} = \frac{\displaystyle\int_{\alpha_0}^{\alpha_1} b_{31}(\varepsilon + \bar{h}_1 \kappa_\alpha) d\alpha}{\displaystyle\int_{\varepsilon_0}^{\alpha_1} b_{33} d\alpha}. \tag{20.12}$$

The following notations are introduced in Eqs. (20.1)–(20.12)

$$\begin{aligned}
&C_{1n} = c_{1n} h_0 + 2c_{1n}^E h_1, \quad C_{44} = G_{\alpha z} h_0 + 2c_{44}^E h_1, \\
&D_{1n} = \frac{1}{12}\left(c_{1n} h_0^3 + 2c_{1n}^E \bar{h}_{13} + 2\gamma_{33} h_1^3\right), \\
&c_{11} = E/(1 - \nu^2), \quad c_{12} = \nu c_{11}, \quad c_{11}^E = 1/\left[s_{11}^E\left(1 - \nu_E^2\right)\right], \\
&c_{12}^E = \nu_E c_{11}^E, \quad \nu_E = -s_{12}^E / s_{11}^E, \quad c_{44}^E = 1/\left(s_{44}^E - d_{15}^2/\varepsilon_{11}^T\right), \\
&b_{31} = d_{31}/\left[s_{11}^E(1 - \nu_E)\right], \quad b_{33} = \varepsilon_{33}^T\left(1 - k_p^2\right), \\
&k_p^2 = 2d_{31}^2/\left[\varepsilon_{33}^T s_{11}^E (1 - \nu_E)\right], \quad \gamma_{33} = b_{31}^2/b_{33}, \\
&\rho_* = 2\rho_1 h_1 + \rho_0 h_0, \quad \rho_{**} = \frac{1}{12}\left(2\rho_1 \bar{h}_{13} + \rho_* h_0^3\right), \\
&\bar{h}_{13} = 4h_1^3 + 6h_1^2 h_0 + 3h_1 h_0^2, \quad \bar{h}_1 = \frac{1}{2}(h_0 + h_1), \\
&H = 2h_1 + h_0, \quad M_E = -\bar{h}_1 b_{31} V_a, \quad \varepsilon = \varepsilon_\alpha + \varepsilon_\theta.
\end{aligned} \tag{20.13}$$

In (20.13),

$$s_{kk}^{E} = s_{kk}'\left(1 - i\delta_{kk}^{s}\right), \quad d_{ik} = d_{ik}'\left(1 - i\delta_{ik}^{d}\right), \quad \varepsilon_{kk}^{T} = \varepsilon_{kk}'\left(1 - i\delta_{kk}^{\varepsilon}\right)$$

are the temperature-dependent complex compliances, piezoelectric coefficients, and dielectric permeabilities for piezoceramics, respectively; $E = E' + iE''$ and $G_{\alpha z} = G_{\alpha z}' + iG_{\alpha z}''$ are the temperature-dependent Young modulus and shear modulus, respectively; $\nu = \text{const}$ is Poisson's ratio of the passive material; k_s is the shear correction factor; $w = w' + iw''$, $u = u' + iu''$, and $\psi_\alpha = \psi_\alpha' + i\psi_\alpha''$ are the complex amplitudes of deflection, longitudinal displacement, and normal rotation angle; N_α, N_θ, Q_α and M_α, M_θ are the amplitudes of the complex forces and moments; ρ_0 and ρ_1 are the specific densities of passive and piezoactive material; λ and a are the averaged coefficients of heat conductivity and thermal diffusivity; $\alpha_s = {}^1\!/\!{}_2(\alpha_+ + \alpha_-)$; $\alpha_\pm$ and $\alpha_{0,l}$ are the heat transfer coefficients at the correspondent shell surfaces and facets; T_0 is the initial temperature of the shell.

20.3 Solution Construction Technique

The problem of electro-thermomechanics (20.1)–(20.13) is coupled and nonlinear, caused by the thermal dependencies of material electromechanical characteristics. The problem's solution can be organized in the frame of a time-step procedure [6, 19]. For this purpose, the Eqs. (20.2)–(20.4) should be rewritten in terms of the sought-for complex variables u, w, ψ_α, N_α, Q_α, and M_α as a system of normal form first order differential equations

$$\begin{aligned}
&\frac{dN_\alpha}{d\alpha} = -\rho_* \omega^2 u, && \frac{dQ_\alpha}{d\alpha} = \frac{\nu_c}{R} N_\alpha + \frac{\tilde{C}_{11}}{R^2} w - \rho_* \omega^2 w - q_z, \\
&\frac{dM_\alpha}{d\alpha} = Q_\alpha - \rho_{**} \omega^2 \psi_\alpha, && \frac{du}{d\alpha} = J_c N_\alpha - \frac{\nu_c}{R} w, \\
&\frac{d\psi_\alpha}{d\alpha} = J_D M_\alpha, && \frac{dw}{d\alpha} = -\psi_\alpha + J_{SD} Q_\alpha,
\end{aligned} \tag{20.14}$$

where $J_c = 1/C_{11}$, $J_D = 1/D_{11}$, $\nu_c = C_{12}/C_{11}$, $\tilde{C}_{11} = C_{11}\left(1 - \nu^2\right)$, and $J_{SD} = 1/(k_s C_{44})$.

Separation of the real and imaginary parts in the system of equations (20.14) with boundary conditions (20.8)–(20.10) yields the boundary-value problem that can be solved at each time step with the use of discrete-orthogonalization method. Then, the standard numerical technique is used to solve the system of ordinary differential equations [15]. In the first step, the linear problem is solved with the isothermal values of material characteristics. In the second step, the dissipative function (20.7) is calculated, and the heat conduction problem (20.6) with boundary and initial conditions (20.11) is solved by the finite difference technique in the framework of the explicit numerical algorithm. Accounting for the temperature field obtained over the shell, the refined temperature-dependent values of the material characteristics (20.13) are determined. Then, solving the boundary-value problem is continued at the next time step up to the moment when the temperature steady state is reached. To implement the proposed solution procedure, the nondimensional spatial $x = \alpha/l$ and time $\tau = at/l^2$ coordinates as well as parameters $\gamma_s = \alpha_{\pm,0,l}l/\lambda$ and $\nu_E = \text{const}$ were used. Numerical calculations presented in the paper were performed for the particular case of monoharmonic loading of the shell with the external surface pressure of constant amplitude $q_z'(\alpha) = q_0$.

20.4 Numerical Results and Analysis

In the present paper, the primary attention is paid to the axisymmetric bending vibration of the shell in the vicinity of the first mode frequency. This regime is chosen because the axisymmetric bending vibration occurs mainly under loading conditions considered. Moreover, the first mode is the most significant in terms of energy characteristics and industrial applications. Therefore, the piezoelectric sensors were placed so that the electroded area midpoint coincides with the maximum deflection point of the shell cross-section to provide the most effective functioning precisely for the first vibration mode. The shell passive layer is considered to be fabricated of polymer [2] with the following temperature-dependent moduli: $E' = (1672 - 118.6T) \cdot 10^6$ Pa, $E'' = (15.01 - 1.205T) \cdot 10^6$ Pa, $G_{\alpha z} = 2(1+\nu)E$ Pa, Poisson's ratio $\nu = 0.36$ and density $\rho_0 = 929\ \text{kg/m}^3$. Piezoelectric constituents of the shell are made of PZT piezoceramics, demonstrating viscoelastic properties under harmonic loading [5, 6]. Experimentally measured thermal dependencies of electromechanical properties are approximated by the following expressions:

$$
\begin{aligned}
s_{11}' &= 12.5\left(1 - 0.3077 \cdot 10^{-3}\bar{T}\right) \cdot 10^{-12}\ \text{m}^2/\text{N},\\
s_{44}' &= 39.7\left(1 + 0.5458 \cdot 10^{-3}\bar{T}\right) \cdot 10^{-12}\ \text{m}^2/\text{N},\\
\delta_{11}^s &= 0.16\left(1 + 0.6155 \cdot 10^{-3}\bar{T} + 0.4158 \cdot 10^{-4}\bar{T}^2\right) \cdot 10^{-2},\\
\bar{T} &= T - T_0, \quad \nu_E = 0.37,\\
\delta_{44}^s &= 0.14\left(1 + 8.33 \cdot 10^{-3}\bar{T}\right) \cdot 10^{-2},\\
d_{31}' &= -1.6\left(1 + 0.219 \cdot 10^{-2}\bar{T}\right) \cdot 10^{-10}\ \text{C/m},\\
\delta_{31}^d &= 0.4\left(1 + 1.198 \cdot 10^{-2}\bar{T} + 1.823 \cdot 10^{-4}\bar{T}^2\right) \cdot 10^{-2},\\
d_{15}' &= 4.5\left(1 + 0.9722 \cdot 10^{-3}\bar{T}\right) \cdot 10^{-10}\ \text{C/m},\\
\delta_{15}^d &= 0.35\left(1 + 0.3571 \cdot 10^{-2}\bar{T}\right) \cdot 10^{-2},\\
\varepsilon_{11}' &= 18.5\varepsilon_0\left(1 + 0.4505 \cdot 10^{-2}\bar{T}\right) \cdot 10^2,\\
\delta_{11}^\varepsilon &= 0.5\left(1 + 0.015\bar{T}\right) \cdot 10^{-2},\\
\varepsilon_{33}' &= 21\varepsilon_0\left(1 + 0.111 \cdot 10^{-3}\bar{T} + 0.8426 \cdot 10^{-4}\bar{T}^2\right) \cdot 10^2,\\
\varepsilon_0 &= 8.854 \cdot 10^{-12}\ \text{F/m},\\
\delta_{33}^\varepsilon &= 0.35\left(1 + 0.0119\bar{T} + 0.119 \cdot 10^{-3}\bar{T}^2\right) \cdot 10^{-2},\\
\rho_1 &= 7520\ \text{kg/m}^3, \quad \lambda = 0.45\ \text{W/(m} \cdot {^\circ}\text{C)}.
\end{aligned}
$$

Geometrical dimension of the shell are as follows: $l = 0.2$ m, $h_0 = 0.04$ m, $h_1 = 0.1 \cdot 10^{-4}$ m, and $R = 0.2$ m. The shear correction factor $k_s = 5/6$ [5].

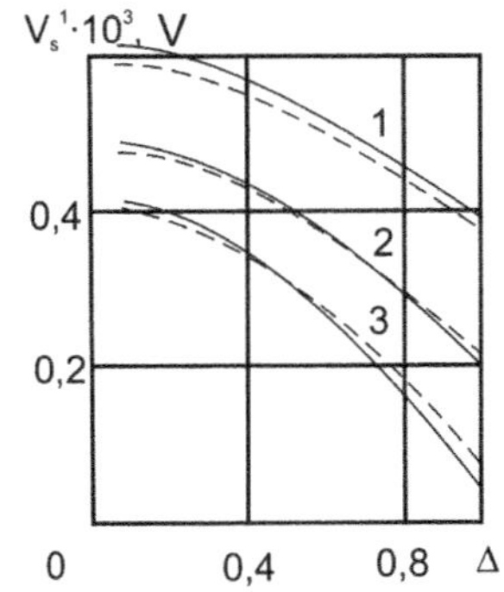

Fig. 20.1 Dependency of the amplitude of sensor reference electric voltage V_s^1 on the sensor width parameter $\Delta = \Delta_\alpha / l$ for different types of boundary conditions

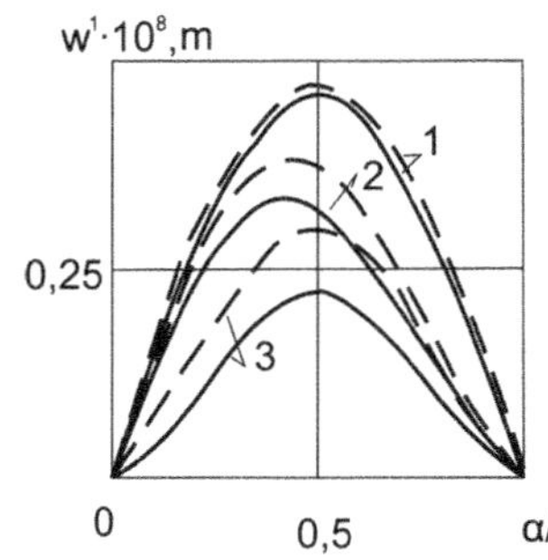

Fig. 20.2 Deflection amplitude distribution along the shell axis for different types of boundary conditions

In Fig. 20.1, curves 1, 2, and 3 illustrate dependency of the amplitude of sensor reference electric voltage V_s^1 on the parameter $\Delta = \Delta_\alpha / l$ for three types of boundary conditions (20.8), (20.9), and (20.10), respectively. The voltage V_s^1 is determined according to formula (20.12) with making use of the results of the forced vibration investigation under loading amplitude $q_0 = 1$ calculated for isothermal material properties ($T = T_0$).

Deflection amplitude distribution along the shell axis for the sensor width parameter $\Delta = 0.05$ under the same types of boundary conditions (20.8), (20.9), and (20.10) are shown in Fig. 20.2 by the curves 1, 2, and 3, respectively.

In Figs. 20.1 and 20.2, solid lines correspond to classical problem statement developed in the framework of the Kirchhoff–Love hypotheses, while dashed lines stand for the results obtained by taking into account the in-plane shear strain. At this, to build the solid curves 1, 2, and 3, the data calculated for load frequencies $\omega = (90; 107; 133) \cdot 10^2\ \text{s}^{-1}$ close to the resonance were used, while frequencies $\omega = (88; 99; 113) \cdot 10^2\ \text{s}^{-1}$ were chosen to obtain numerical results for dashed lines. Analysis of the behavior of the curves in Figs. 20.1 and 20.2 shows the stiffer shell facets constraint is, the lower both deflection amplitudes and sensor reference electric voltage are. Accounting for the in-plane shear strain (dashed lines) predicts higher amplitudes, which are even more profound in the case of rigidly fixed shell facets (curves 3 in Fig. 20.2).

Influence of the shell facet constraint conditions (20.8)–(20.10) and the accounting for the in-plane shear strain on frequency characteristics of the maximum deflection

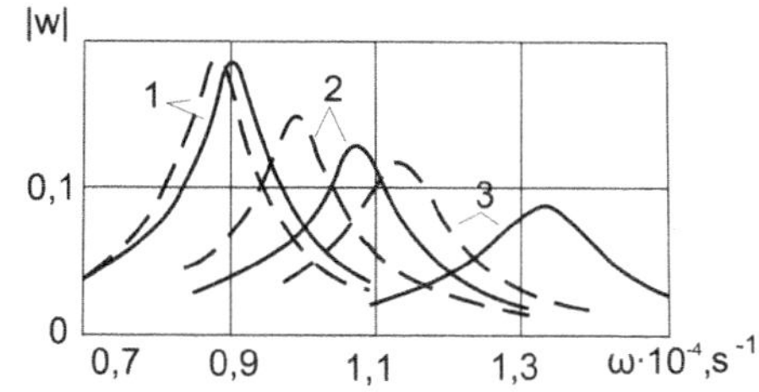

Fig. 20.3 Influence of constraint conditions on amplitude-frequency curves in the framework of the Kirchhoff–Love (solid lines) and Timoshenko (dashed lines) theories

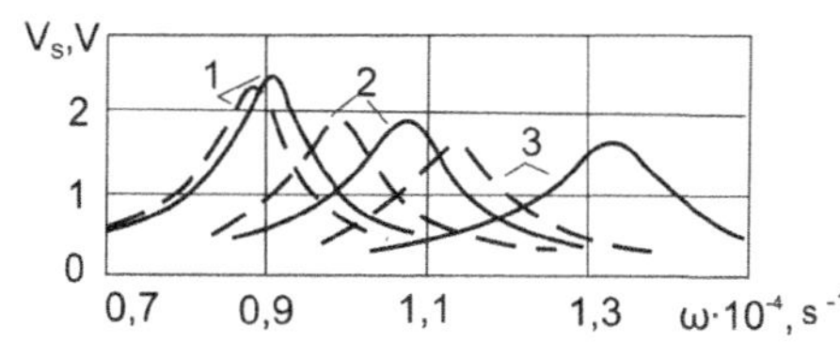

Fig. 20.4 Influence of constraint conditions on sensor electric parameter versus frequency curves in the framework of the Kirchhoff–Love (solid lines) and Timoshenko (dashed lines) theories

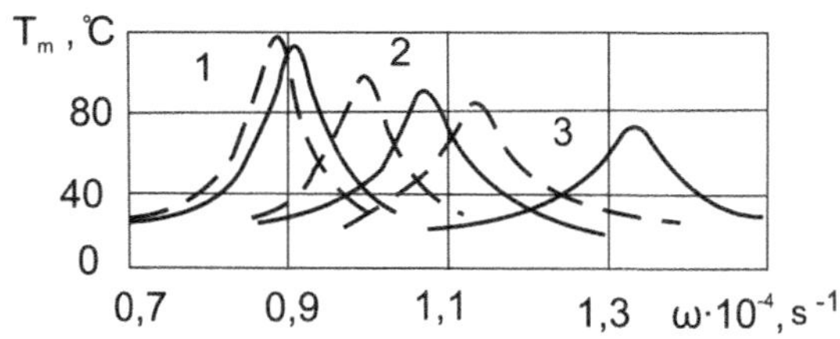

Fig. 20.5 Influence of constraint conditions on maximum overheating temperature versus frequency curves in the framework of the Kirchhoff–Love (solid lines) and Timoshenko (dashed lines) theories

amplitudes, sensor electric parameter, and maximum overheating temperature are demonstrated by curves 1, 2, and 3 in Figs. 20.3, 20.4 and 20.5, respectively. The shown results correspond to the external pressure amplitude $q_0 = 0.4 \cdot 10^4$ Pa and heat exchange parameter $\gamma_s = 0.5$ for temperature-independent moduli of the shell materials. Analysis of Figs. 20.3, 20.4 and 20.5 shows that accounting for the in-plane shear strain manifests itself in a shift of the frequency curves for maximum amplitude w, sensor electric parameter V_s and maximum overheating temperature T_m to the lower frequency region. The shift is maximal in the case of rigidly fixed shell facets (lines 3). It is worth mentioning that the stiffer constraint conditions on the shell edges are (see solid curves 2 and 3 in Figs. 20.3, 20.4 and 20.5), the greater shift of the resonance characteristics to the high-frequency region is observed. At that, the shift is accompanied by a decrease in the maximum values of the abovementioned quantities.

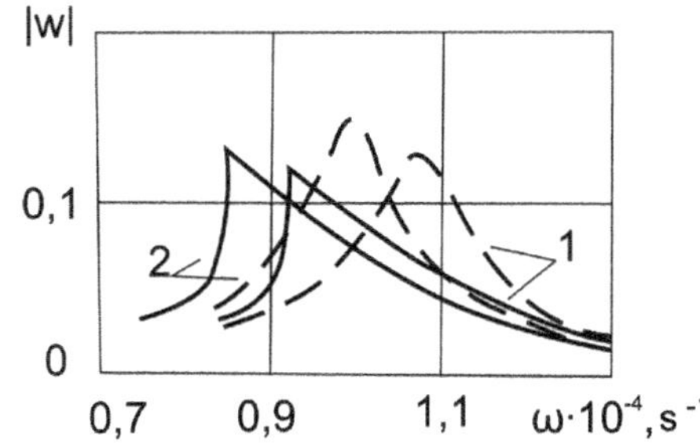

Fig. 20.6 Amplitude-frequency curves of maximum deflection for isothermal material characteristics ($T = T_0$) (dashed lines) and temperature-dependent material moduli (solid lines)

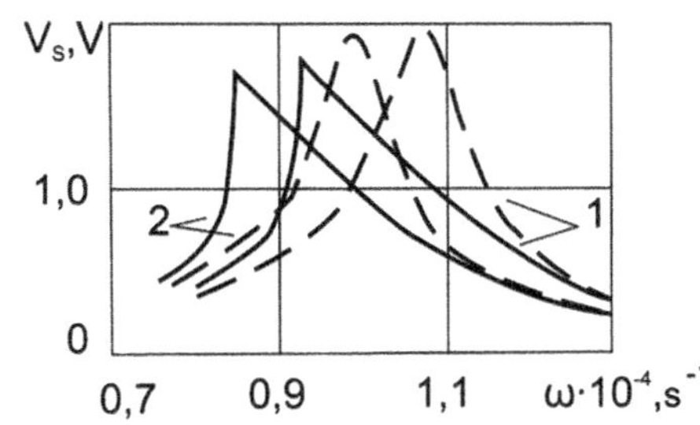

Fig. 20.7 Amplitude-frequency curves of sensor electric parameter for isothermal material characteristics ($T = T_0$) (dashed lines) and temperature-dependent material moduli (solid lines)

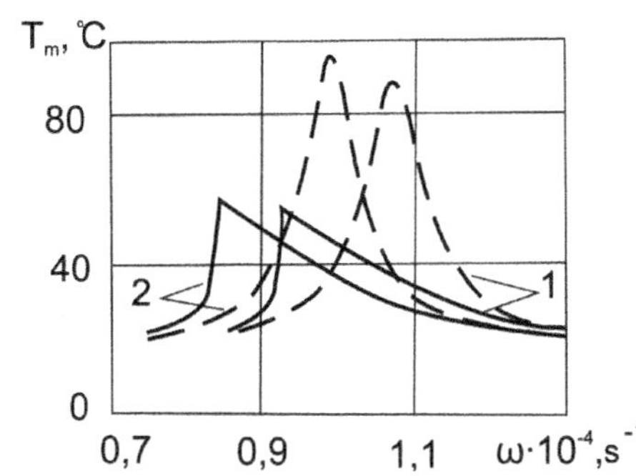

Fig. 20.8 Frequency curves of maximum heating temperature for isothermal material characteristics ($T = T_0$) (dashed lines) and temperature-dependent material moduli (solid lines)

Analogous frequency curves for maximum amplitude w, sensor electric parameter V_s, and maximum overheating temperature T_m are shown in Figs. 20.6, 20.7 and 20.8) for the shell having one facet hinged and another one rigidly fixed (correspondent boundary conditions are described by (20.9)). These data were calculated for loading parameter $q_0 = 0.4 \cdot 10^4$ Pa, $\gamma_s = 0.5$ and for both isothermal values ($T = T_0$) (dashed lines) and temperature-dependent material moduli (solid lines). Numerical data for curves 1 and 2 were computed in the classical and Timoshenko approach frameworks, respectively. Analysis of the data presented in Figs. 20.6, 20.7 and 20.8 shows that accounting for the temperature dependence of electromechanical properties of the shell materials for both classical (Kirchhoff–Love) and refined (Timoshenko) formulations induces a shift in frequency curves (see solid lines in Figs. 20.6, 20.7 and 20.8) to the low-frequency region and their transformation to nonlinear characteristics of soft type. In the vicinity of resonance frequency, a significant decrease in maximum values of deflection, sensor electrical parameter amplitude, and the maximum dissipative heating temperature occurs compared to the values obtained for the isothermal resonance case.

Under high amplitudes of harmonic loading and typical heat exchange conditions with the ambient environment, the dissipative heating temperature can reach the critical value T_k, causing failure of a structure due to degradation of electro-thermomechanical properties of constituent materials, especially in the vicinity of resonance. At this, the working efficiency of the smart structure can be diminished due to either softening of the passive layer or depolarization of the piezoactive material as soon as Curie temperature is exceeded. It should be mentioned here that degradation temperature value T_k is assumed to be found experimentally for each particular material.

Upper bounds for the permissible strain/stress amplitudes and temperature are widely used in engineering practice to assess the durability of structural members made of inelastic materials under intense harmonic loading. Those characteristics are generally considered the most adequate to describe the workability of the structural elements under investigation [14]. One of the most important techniques for estimation of element durability under harsh operation conditions is the application of the thermal criterion, according to which the maximum overheating temperature should be kept lower than the thermal degradation temperature ($T_{\max} \leqslant T_k$) (that, in turn, should be the minor of the softening temperature of the passive layer material and Curie point for piezoceramic layer). At this, the critical value of the mechanical loading amplitude q_k corresponding to the steady-state maximum overheating temperature (i.e., degradation temperature T_k) should be found. Under any loading amplitude lower than the critical amplitude $q_0 < q_k$ at which the dissipative heating temperature falls short of T_k, the durability of the structural member corresponds to some practically determined time interval. In this case, to solve the thermomechanics problem, it is reasonable to restrict ourselves by considering the steady-state heat conduction equation with a correspondent dissipative function. Under loading amplitudes exceeding the critical value $q_0 \geqslant q_k$, it is necessary to determine the critical operational time τ_k (or the number of cycles to failure) of the structure with making use of nonstationary heat conduction problem solution. To be specific, for numerical calculations, let us assume the degradation of the shell under consideration occurs at critical overheating temperature $T_m = T_k = 120\,^{\circ}\mathrm{C}$ (the safety factor is applied as well) at least at one point of the shell.

In Fig. 20.9, the relationship between maximum steady-state dissipative heating temperature T_m and loading pressure amplitude q_0 is illustrated for the shell having one facet hinged and another one rigidly fixed (corresponding boundary conditions are described by (20.9) with sensor width parameter $\Delta = 0.05$ and vibrating on the frequency $\omega = 9 \cdot 10^3\ \mathrm{s}^{-1}$. Numbers 1–3 by the lines correspond to the data calculated for the following values of the heat exchange coefficient: $\gamma_s = 0.1, 0.25$, and 0.5. Results obtained in the framework of the Timoshenko refined theory for isothermal moduli values (temperature-independent characteristics at $T = T_0$) and in the case of temperature-dependent moduli are shown with the dashed and solid lines, respectively. Values of degradation temperature T_k and correspondent critical amplitudes of the mechanical load q_k are marked with crosses on the ordinate and

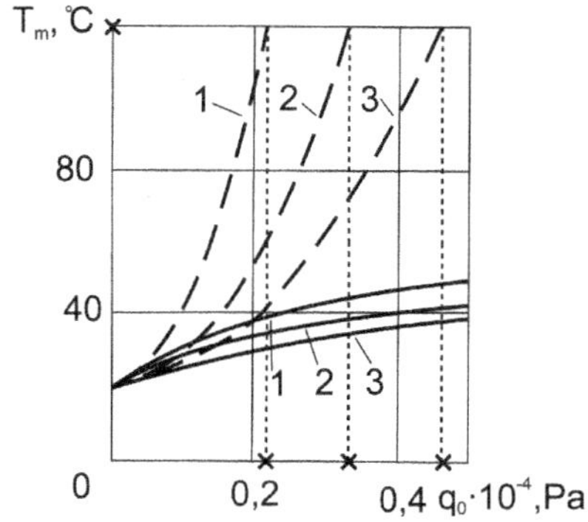

Fig. 20.9 Maximum steady-state temperature of dissipative heating T_m versus load amplitude q_0 under frequency $\omega = 9 \cdot 10^3$ s^{-1} calculated in the framework of the Timoshenko refined theory with temperature-independent (dashed lines) and temperature-dependent moduli (solid lines); 1, 2, and 3 correspond to $\gamma_s = 0.01, 0.25$, and 0.5

abscissa axis, respectively, for the three cases mentioned above. Analysis of the data and curves 1–3 presented in Fig. 20.9 for the shell under forced resonance vibration shows that the intensification of heat exchange on the shell surfaces leads to an overestimation of the critical load amplitude q_k corresponding to the degradation temperature T_k if material properties is considered to be temperature-independent. Accounting for the temperature dependence of electromechanical properties of the shell materials leads to the prediction of much lower maximum dissipative heating temperatures T_m and higher critical amplitude q_k at this particular frequency due to shift in the temperature-frequency curve to the low-frequency region.

In Fig. 20.10, evolution of maximum dissipative heating temperature in time $\tau = at/l^2$ is illustrated for the case of isothermal material properties and frequency $\omega = 9 \cdot 10^3$ s^{-1} with the heat exchange coefficient $\gamma_s = 0.25$. Curves 1–3 correspond to the following values of the loading amplitude: $q_0 = (0.30; 0.335; 0.35) \cdot 10^4$ Pa, respectively. As it was mentioned above, the degradation of the shell under consideration is assumed to occur if the maximum overheating temperature reaches the

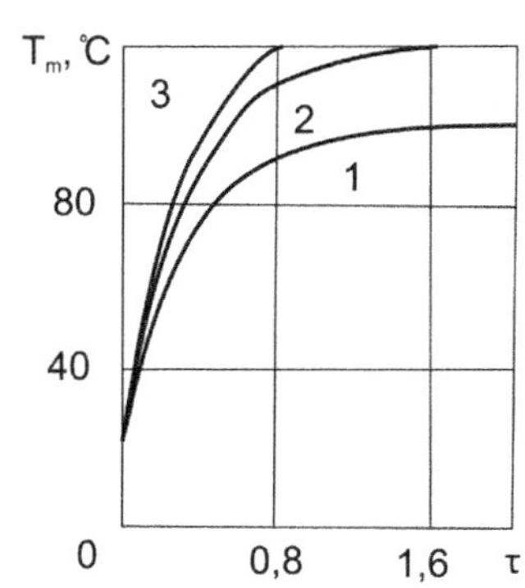

Fig. 20.10 Evolution of maximum dissipative heating temperature in time for isothermal material properties

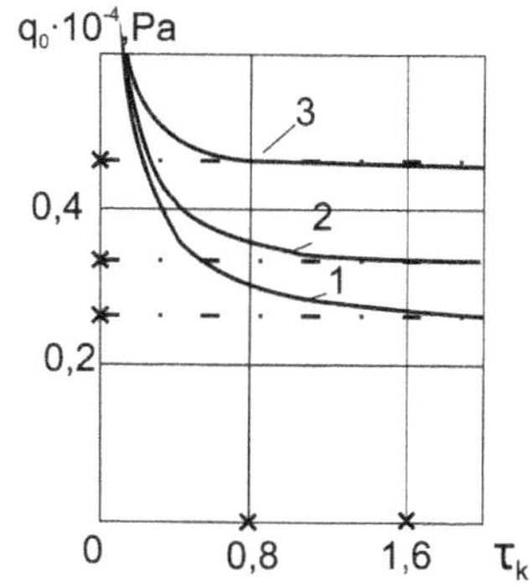

Fig. 20.11 Dependency between loading amplitude and critical operational time calculated following thermal failure criterion

critical level $T_m = T_k = 120\,^{\circ}$C. Thus, the system can preserve its workability for an infinitely long time with respect to the thermal failure criterion if the steady-state maximum overheating temperature is lower than the critical value. The relationship between loading amplitude q_0 and critical operational time τ_k (analog of the Wöhler diagram) calculated according to the thermal failure criterion is illustrated in Fig. 20.11. Values of the critical loading amplitude q_k and critical times τ_k are illustrated with the crosses on the ordinate and abscissa axis, respectively. Curves 1–3 correspond to the heat exchange coefficient $\gamma_s = 0.1, 0.25$, and 0.5, respectively. For each particular value of γ_s, a critical value q_k can be found as a horizontal asymptote. For any loading amplitude that is lower than this value $q_0 < q_k$, the vibration can last infinitely long without failure due to overheating. If loading amplitude exceeds the critical value $q_0 \geqslant q_k$, there is a critical time moment τ_k (see curves 2 and 3 in Fig. 20.11, for instance) when the overheating temperature reaches the critical level T_k, and thus the system fails due to the overheating. Nevertheless, the shell can function for the time interval $0 < \tau < \tau_k$.

20.5 Conclusions

In the framework of Timoshenko-type refined shell theory, the statement and numerical solution of the problem on prediction of coupled electro-thermomechanical response and durability of viscoelastic cylindrical three-layer shell with piezoelectric sensors under axisymmetric resonance vibration is studied for different types of the constraint condition of shell facets. For the most power-intensive first axisymmetric bending mode, the influence of the mechanical and thermal boundary conditions and the accounting for the in-plane shear strain and thermal dependencies of electromechanical properties of the shell constituent (both passive and piezoactive materials) on frequency curves for maximum deflection amplitude, sensor electric parameter, and dissipative heating temperature are investigated in detail for the case of external harmonic pressure loading. A technique for the shell life prediction using the critical value of the dissipative heating temperature is proposed. Thermal criterion is applied to estimate the shell's durability for both steady and transient regimes. An approach is developed to predict the workability time interval for forced resonance vibration of the shell under load amplitudes exceeding the critical amplitude level.

References

1. Blanguernon, A., Lene, F., Bernadou, M.: Active control a beam a piezoceramic element. Smart Mater. Struct. **8**, 116–124 (1999)
2. Bolkisev, A.M., Karlash, V.L., Shul'ga, N.A.: Temperature dependence of the properties of piezoelectric ceramics. Int. Appl. Mech. **20**, 650–653 (1984)
3. Gabbert, U., Tzou, H.S.: Smart Structures and Structronic Systems. Kluwer Academic Publishers, Dordrecht (2001)
4. Grigorenko, A.Y., Müller, W.H., Loza, I.A.: Selected Problems in the Elastodynamics of Piezoceramic Bodies. Advanced Structured Materials (STRUCTMAT, volume 154). Springer, Cham, Switzerland (2021)
5. Karnaukhov, V.G., Kirichok, I.F., Kozlov, V.I.: Thermomechanics of inelastic thin-walled structural members with piezoelectric sensors and actuators under harmonic loading (review). Int. Appl. Mech. **53**, 6–58 (2017)
6. Karnaukhov, V.G., Mikhailenko, V.V.: Nonlinear single-frequency vibrations and dissipative heating of inelastic piezoelectric bodies. Int. Appl. Mech. **38**, 521–547 (2002)
7. Karnaukhov, V.G., Kirichok, I.F.: Forced harmonic vibrations and dissipative heating-up of viscoelastic thin-walled elements (review). Int. Appl. Mech. **36**, 174–195 (2000)
8. Karnaukhov, V.G., Kirichok, I.F., Kozlov, V.I.: Electromechanical vibrations and dissipative heating of viscoelastic thin-walled piezoelements (review). Int. Appl. Mech. **37**, 182–212 (2001)
9. Karnaukhov, V.G., Kirichok, I.F., Kozlov, V.I.: Thermomechanics of inelastic thin-wall structural members with piezoelectric sensors and actuators under harmonic loading (review). Int. Appl. Mech. **53**, 6–58 (2017)
10. Katunin, A.: Criticality of the self-heating in polymers and polymer matrix composites during fatigue and their application in non-destructive testing. J. Polym. **11**(1), 19 (2019)
11. Kirichok, I.F.: Resonance axisymmetric vibrations and dissipative heating of viscoelastic cylindrical shell with piezolayers under electromechanical excitation. Int. Appl. Mech. **51**, 567–573 (2015)
12. Kirichok, I.F., Chernyushok, O.A.: Forced vibrations and vibroheating of flexible viscoelastic beam with piezoelectric sensor and actuators with allowance for the shear strains. Int. Appl. Mech. **54**, 568–576 (2018)
13. Kirichok, I.F., Karnaukhova, T.V.: Forced axisymmetrical vibrations and heating of termoviskoelastic cylindrical shells with piezoactuators. Int. Appl. Mech. **46**, 1132–1138 (2010)
14. Kirichok, I.F., Zhuk, Y.A., Kruts, S.Y.: Accounting for shear deformation in the problem of vibrations and dissipative heating of flexible viscoelastic structural element with piezoelectric sensor and actuator. In: Sadovnichiy, V.A., Zgurovsky, M.Z. (eds.) Contemporary Approaches and Methods in Fundamental Mathematics and Mechanics. Understanding Complex Systems book series (UCS), pp. 51–69. Springer (2021)
15. Kovalenko, A.D., Grigorenko, Y.M., Illin, L.A., Polischuck, T.I.: The design of conical shells subjected to antisymmetric loadings. National Aeronautics and Space Administration Technical Translation (NASA) (1969)
16. Tani, J., Takaga, T., Qiu, J.: Intelligent material systems: application of functional materials. Appl. Mech. Rev. **51**, 505–521 (1998)
17. Tzou, H.S.: Piezoelectric Shells (Distributed Sensing and Control of Continua). Kluwer Academic Publishers, Dordrecht-Boston-London (1993)

18. Tzou, H.S., Bergman, L.A.: Dynamics and Control of Distributed Systems. Cambridge University Press, New York (1998)
19. Zhuk, Y.A., Guz, I.A.: Dissipative heating of thin-wall structures containing piezoactive layers. In: Hetnarski, R.B. (ed.) Encyclopedia of Thermal Stresses, vol. D, pp. 971–985. Springer, Dordrecht-Heidelberg-New York-London (2014)
20. Zhuk, Y.A., Guz, I.A., Sands, C.M.: Analysis of the vibrationally induced dissipative heating of thin-wall structures containing piezoactive layers. Int. J. Non Linear Mech. **47**, 105–116 (2012)

Chapter 21
Topology Optimization of Adhesively Bonded Double Lap Joint

S. Kurennov, K. Barakhov, and I. Taranenko

Abstract In this paper, a one-dimensional mathematical model of stress state is proposed for a symmetrical double-shear adhesive joint with doublers having varying thicknesses along the joint length. This model is an extension of the classical Holland–Reissner model. Due to geometrical symmetry, load-carrying plates do not bear bending, but doublers are subjected to bending due to the eccentricity of the applied load. The suggested mathematical model is used to solve the problem of topological optimization of doubler shape and to find the length of the adhesive film with soft and rigid adhesives used in the same joint. The doubler shape is described by reducing doubler thickness along the joint length using the Fourier series with cosines. The optimization problem requires finding doubler length and Fourier coefficients, and both joint length and doubler cross-sectional area can be selected as the objective function. Constraints are applied on maximum stress in the adhesive film, stress in doublers, and minimum and maximum doubler thickness. The direct problem of determining the joint stress state at given geometrical parameters is solved using the method of finite differences. The optimization problem is solved using a genetic algorithm. To improve the convergence of this method, an island model of genetic algorithm is used. The distinctive feature of the suggested algorithm model is that mutations occur more often and with greater dispersion in one of the "islands" compared to the other two "islands." Such a solution ensures both the quickness of evolution selection and the stability of the obtained results. Two model problems are solved.

S. Kurennov · K. Barakhov (✉)
Department of Higher Mathematics and System Analysis, National Aerospace University "Kharkiv Aviation Institute", 17 Chkalov str., UKR-61070 Kharkiv, Ukraine
e-mail: kpbarakhov@gmail.com

I. Taranenko
Department of Composite Structures and Aviation Materials, National Aerospace University "Kharkiv Aviation Institute", 17 Chkalov str., UKR-61070 Kharkiv, Ukraine
e-mail: igor.taranenko@khai.edu

H. Altenbach et al. (eds.), *Selected Problems of Solid Mechanics and Solving Methods*, Advanced Structured Materials 204, https://doi.org/10.1007/978-3-031-54063-9_21

21.1 Introduction

Adhesive joints are one of the most crucial parts of up-to-date structures made of composites. The wide application of adhesive joints in composite structures is stipulated by their high manufacturability, air tightness, low weight, and high aerodynamic efficiency [15]. Adhesive joints do not disturb composite structure; therefore, they can allow the realization of high structural strength and mechanical properties. However, one of the well-known disadvantages of lapped joints is stress concentration near the joint edge [5, 33]. To reduce stress concentration near the joint edge, one can increase adhesive thickness at this zone [3], reduce joining plates thickness near the joint edge [9] (the order of composite layers stacking sequence is important [24]), use two or more different types of adhesives [29], introduce transversal elements to the joint [14], and use other methods [18, 32]. It is known that the application of symmetrical double-side joints allows the exclusion of structural bending and reduces peeling-off stress in the adhesive film [1, 22].

To describe the stress state of lapped joints, generally, mathematical models of sandwich rods or beams with thin, compliant filler [5, 33] are used. If elastic and geometrical parameters of layers are constant along the joint length, then the joint stress–strain state can be described analytically. However, even in the simplest case, if the thicknesses of the joining layers change linearly, then such a problem does not have an analytical solution. This is why, to estimate the stress state of a joint with a variable thickness along the joint length, one can use such numerical methods as the finite-difference method [14], the Ritz method, and the finite element method [17].

The topological optimization problem is more complicated than the classical parametric optimization problem. The reason for this is as follows. The objective function is the distribution of the material throughout the volume of the structure and not a set of a small number of unknown parameters. As a rule, topological optimization of lapped joints consists of finding the optimal length of the joint and the dependence of the joined plates' thickness on the joint's length. One possible way to solve this problem is to discretize objective function. In this case, the problem is to find the thickness of structural elements in a system of points [21]. Using discrete functions instead of continuous ones makes it possible to optimize joints with stepwise changes in thickness [12, 25, 37] (such joints are often used when joining layered composites).

If the objective function is considered to be continuous, then it can be described by a well-known system of points with splines [8], Bezier functions [2, 38], or trigonometric series [19]. Also, the following optimization problems are considered: optimization parameters of scarf joint [16] or thickness of adhesive film near joint edge [6, 31], dimension and shape of the squeezed recess of adhesive near the edge of the joint [4], and optimization of the composite structure [10, 36].

Finding a relatively large quantity of unknown parameters utilizing gradient methods meets some difficulties. Therefore, other optimization methods are used for considered problem-solving, such as genetic algorithm or method of sliding asymptotes

[11]. These approaches assume finding the optimal parameters of a problem through solving the consequences of direct problems.

As a rule, the direct problem of assessing a structure's stress–strain state for given parameters can be solved in a two-dimensional formulation using the finite element method [6, 8, 16]. A common disadvantage of using the finite element method when solving direct problems of determining the stress state of joints is the relatively long time required to find a solution. Therefore, to solve the optimization problem, it is more expedient to create a one-dimensional model of the stress state of the joint, which has a satisfactory accuracy for engineering analysis. In addition, the one-dimensional problem formulation allows the use of a relatively simple and fast method of finite differences to determine the stress state of the joint. This method successfully solves one-dimensional problems of joint mechanics [14, 21] and two-dimensional ones [30]. This method effectively solves one-dimensional [14] and two-dimensional problems of joint mechanics [21, 30].

21.2 Mathematical Model of Joint Stress State

Consider a structure consisting of two plates joined by two symmetrical doublers, as shown in Fig. 21.1a. This structure is widely used in engineering as it does not withstand bending during tension-compression joints. Due to its geometric symmetry, only a quarter of this structure can be analyzed. The main load-carrying plates, i.e., the central layer, have zero transversal displacements. If we analyze this structure within the framework of the theory of rods, only the adhesive joint zone, as shown in Fig. 21.1b, can be considered. The structure is subjected to longitudinal forces of $2F$.

Adhesive film thickness is assumed to be constant and the same in each section. Doubler thickness varies along the joint length. The length of the adhesive joint is equal to L.

A representative differential element of the joining zone and the force acting on it are shown in Fig. 21.2. The difference from classical mathematical models of connections is the doubler's variable thickness along the connection's length. Due to these longitudinal forces $N_1(x)$, eccentricity occurs in the doubler within

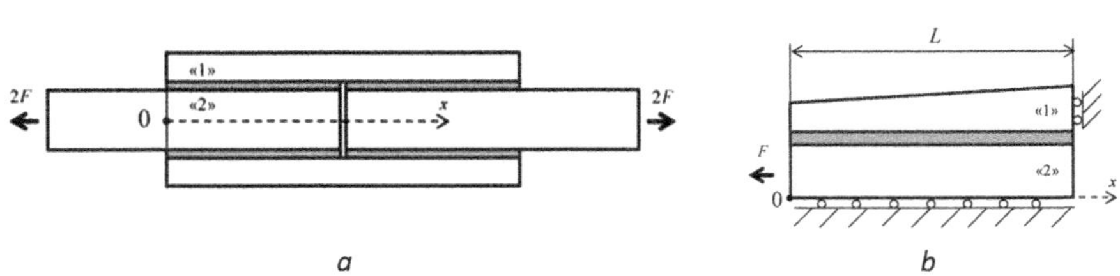

Fig. 21.1 Scheme of the adhesive joint

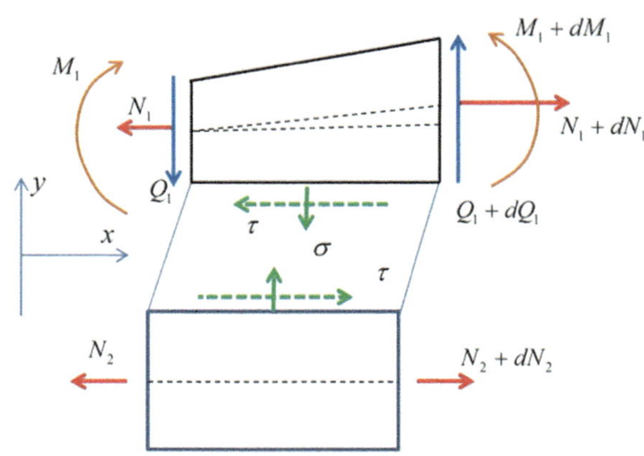

Fig. 21.2 Differential element of joint

the differential element, affecting the doubler's bending moment. The equilibrium equations for load-bearing layers (plates) are given below.

$$\begin{aligned} &\frac{dN_1}{dx} = -\tau, \quad \frac{dN_2}{dx} = \tau, \quad \frac{dQ_1}{dx} = \sigma, \\ &\frac{dM_1}{dx} - s_1(x)\,\tau - N_1\frac{ds_1}{dx} + Q_1 = 0, \end{aligned} \tag{21.1}$$

here N_1 and N_2 are longitudinal forces in the layers; Q_1 is the transversal force in the doubler; M_1 is the bending moment in doubler distributed along the width; τ and σ are shear and normal stresses in the adhesive film; s_1 is the distance from neutral axis to the adhesive film, in case of symmetrical (including uniform) doubler structure $s_1(x) = 0.5\delta_1(x)$, where $\delta_1(x)$ is doubler thickness.

The following equations can describe displacements of load-carrying layers:

$$N_1 = B_1\frac{dU_1}{dx}, \quad N_2 = B_2\frac{dU_2}{dx}, \quad D_1\frac{d^2W_1}{dx^2} = M_1, \tag{21.2}$$

where U_1, U_2 are the longitudinal displacements of load-carrying layers; W_1 are the doubler transversal translations; $B_1(x)$ and B_2 are the membrane rigidities of layers (if layers are uniform through the thickness, then $B_1(x) = \delta_1(x)E_1$, $B_2 = \delta_2 E_2$, where E_1, E_2 are the elasticity moduli of correspondent layer material); $D_1(x)$ is the doubler bending rigidity, $D_1(x) = \delta_1^3(x)E_1/12$.

Stress in adhesive film is assumed to be uniformly distributed through thickness and proportional to the difference in translations of load-carrying layers' inner surfaces:

$$\sigma = K \cdot w_1, \quad \tau = P\left(U_1 - U_2 + s_1(x)\frac{dW_1}{dx}\right), \tag{21.3}$$

where K and P are adhesive film rigidity at tension-compression and shear, respectively; these values can be calculated as $K = E_0/\delta_0$, $P = G_0/\delta_0$, where δ_0 is the adhesive film thickness, E_0 and G_0 are the elastic and shear moduli of adhesive substance.

The system of Eqs. (21.1)–(21.3) can be reduced to the system of three differential equations with respect to longitudinal translations of both load-carrying layers U_1, U_2 and doubler transversal translation W_1:

$$\frac{B_1}{P}\frac{d^2U_1}{dx^2} + \frac{1}{P}\frac{dB_1}{dx}\frac{dU_1}{dx} - U_1 + U_2 - s_1\frac{dW_1}{dx} = 0, \tag{21.4}$$

$$U_1 + \frac{B_2}{P}\frac{d^2U_2}{dx^2} - U_2 + s_1\frac{dW_1}{dx} = 0, \tag{21.5}$$

$$\begin{aligned} &\frac{D_1}{P}\frac{d^4w_1}{dx^4} + \frac{2}{P}\frac{dD_1}{dx}\frac{d^3w_1}{dx^3} + \left(\frac{1}{P}\frac{d^2D_1}{dx^2} - s_1^2\right)\frac{d^2w_1}{dx^2} \\ &- 2s_1\frac{ds_1}{dx}\frac{dw_1}{dx} + \frac{K}{P}w_1 - \frac{B_1}{P}\frac{ds_1}{dx}\frac{d^2U_1}{dx^2} + \frac{ds_1}{dx}U_2 \\ &+ \left(-\frac{1}{P}\frac{ds_1}{dx}\frac{dB_1}{dx} - s_1 - \frac{B_1}{P}\frac{d^2s_1}{dx^2}\right)\frac{dU_1}{dx} - \frac{ds_1}{dx}U_1 + s_1\frac{dU_2}{dx} = 0. \end{aligned} \tag{21.6}$$

Boundary conditions are selected as shown on Fig. 21.1b

$$\begin{array}{lll} U_1(L) = 0, & Q_1(0) = 0, & M_1(0) = 0, \\ N_2(0) = F, & N_2(L) = 0, & N_1(0) = 0, \\ Q_1(L) = 0, & \left.\dfrac{dw_1}{dx}\right|_{x=L} = 0. & \end{array} \tag{21.7}$$

Boundary conditions (21.1) can be written in terms of layers translations like it shown in (21.4)–(21.6).

21.3 Formulation of a Problem of Doubler Shape Optimization in Adhesive Joint

The problem of optimization can be formulated as follows: it is necessary to find joint length L and function $\delta_1(x)$, according to which doubler thickness changes along the joint length; moreover, these two parameters ensure extremum value of selected optimality criterion. As such a criterion, we can choose, for example, the mass of the doubler. Doubler mass is proportional to its cross-sectional area with precision up to an arbitrary multiplier. Therefore, the optimality criterion can be written as follows:

$$M = \int_0^L \delta_1 (x)\, dx \rightarrow \min . \tag{21.8}$$

On the other hand, structural cost can also be selected as an optimality criterion. Since quite valuable input to joint cost is defined by the adhesive substance cost and the cost of the manufacturing process of joining, joint length can be considered as an optimality criterion:

$$L \rightarrow \min . \tag{21.9}$$

Unknown variables L and function $\delta_1 (x)$ have to be found to guarantee structural strength and fulfill geometrical and technological restrictions.

The failure of joints can occur in such modes as doubler tearing, cohesive adhesive film failure, article delamination and peeling-off, composite article layers peeling-off, adhesive film peeling-off from joining articles, etc. However, if the adhesive substance is appropriately selected, then the most probable failure mode is the cohesive failure of the adhesive film without its delamination from joining the articles' surfaces. It is known that the adhesive film strength criterion and its degree of adhesion to article surfaces depend on such multiple factors as adhesive composition, joining articles' chemical composition and quality of joining surfaces preparation, temperature and duration of adhesive film curing, etc. The strength criterion of the created adhesive film can be determined using physical experiments. In this case, the adhesive film has to possess the required level of adhesion to joining surfaces to ensure that cohesive failure mode is probable, but not adhesive one. Paper [23] shows that the criterion of maximum normal stress (criterion of principal stress) can be selected as the adhesive film strength criterion.

$$\sigma_1 (x) = \frac{|\sigma (x)|}{2} + \frac{\sqrt{\sigma^2 (x) + 4\tau^2 (x)}}{2} \leqslant \sigma_{\max}, \tag{21.10}$$

where $x \in [0;\ L]$; $\sigma_1 (x)$ is the first principal stress in the adhesive film; $\sigma_{\max}$ is the adhesive strength, which is determined experimentally, for example, by tearing test.

Besides the abovementioned restrictions, function $\delta_1 (x)$ can also have upper and lower limits. In the first case, doubler thickness cannot be less than the given value:

$$\delta_1 (x) \geqslant \delta_{\min}, \tag{21.11}$$

where $\delta_{\min}$ is definite minimal technologically realizable double thickness, which can be equal to, for example, the thickness of a single monolayer of laminated composite.

Moreover, in some cases, restriction can be applied to maximum doubler thickness. Such restriction can be given by considerations of aerodynamic efficiency, possibility of quality control, distinctions of manufacturing technology, etc.

$$\delta_1 (x) \leqslant \delta_{\max}. \tag{21.12}$$

Besides restrictions (21.10)–(21.12), specific ones can be introduced on maximum stress inside the doubler because doubler tearing can also be the possible failure mode of the joint. Introducing similar auxiliary restrictions does not significantly influence the algorithm of solving the problem using a genetic one. It can be reflected only in composing the objective function (of fitness function by terminology of genetic algorithms).

21.4 Solution of a Direct Problems of Joint Stress State Estimation Using the Method of Finite Differences

Solving the direct problem on the determination of the stress–strain state of a joint assumes that joint length L and function $\delta_1(x)$ are known, and we are seeking translations of layers and stress in joint elements at given longitudinal force F.

The method of finite differences assumes discretization of a structure, in the current case—discretization of translation functions inside the segment $[0; L]$. Let us divide this section into equal intervals with enumeration from 0 to definite integer number N. The increment is equal to $h = L/N$. Besides inner and boundary nodes, we can introduce reference points with numbers -1 and $N + 1$, respectively (Fig. 21.3). Points with number i have coordinates $x_i = h \cdot i$. Layers translations at these points can be designated by the following rule:

$$U_1(x_i) = u_i^{(1)}, \quad U_2(x_i) = u_i^{(2)}, \quad W_1(x_i) = w_i^{(1)}.$$

Equations (21.4)–(21.6) have to be written in discrete form; coefficients in these equations also depend on longitudinal coordinate because doubler thickness changes along the joint length. The following notations can be introduced for convenience:

$$\begin{aligned} s_i^{(1)} &= s_1(x_i) = \delta_1(x_i)/2, \quad B_i^{(1)} = B_1(x_i) = E_1\delta_1(x_i), \\ B_i^{(2)} &= B_2 = E_2\delta_2, \quad D_i^{(1)} = D_1(x_i) = E_1\delta_1^3(x_i)/12. \end{aligned}$$

But Eqs. (21.4)–(21.6) contain these parameters and their derivatives. Moreover, function $\delta_1(x)$, from which the abovementioned parameters depend, is defined on

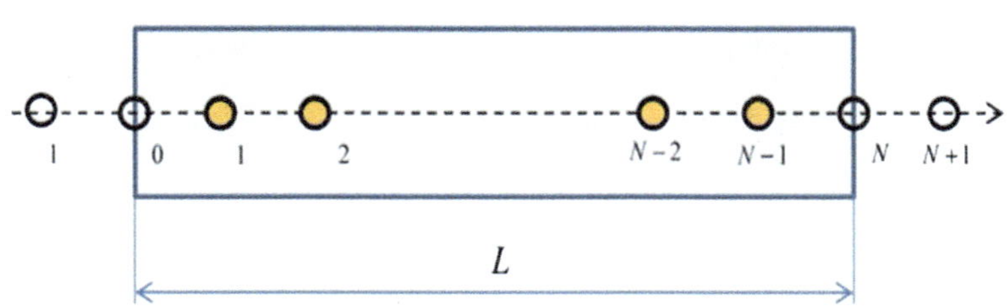

Fig. 21.3 Scheme of region division on reference (node) points

the interval $x \in [0; L]$ only. Thus, in calculating derivatives at inner points (with numbers from 1 to $N-1$) of the region, we can use classical symmetrical difference rules. However, we have to use single-side finite-difference rules near boundary points with numbers 0 and N. It concerns the calculation of equation coefficients only because the introduction of outer nodes allows the use of symmetrical finite-difference rules for all nodes with numbers from 0 to N included.

Schemes of implemented templates are shown below

$$
\begin{aligned}
z_i' &= \frac{z_{i+1} - z_{i-1}}{2h}, \quad z_i'' = \frac{z_{i+1} - 2z_i + z_{i-1}}{h^2} \quad (i = 1, 2, \ldots, N-1),\\
z_0' &= \frac{-z_2 + 4z_1 - 3z_0}{2h}, \quad z_N' = \frac{3z_N - 4z_{N-1} + z_{N-2}}{2h},\\
z_0'' &= z_1'' = \frac{z_2 - 2z_1 + z_0}{h^2}, \quad z_N'' = z_{N-1}'' = \frac{z_N - 2z_{N-1} + z_{N-2}}{h^2}.
\end{aligned}
$$

The following formulas are used for the determination of coefficients presented in Eqs. (21.4)–(21.6):

$$
\begin{aligned}
&\left.\frac{dB_1}{dx}\right|_{x=x_i} = DB_i^{(1)}, \quad \left.\frac{ds_1}{dx}\right|_{x=x_i} = DS_i^{(1)}, \quad \left.\frac{dD_1}{dx}\right|_{x=x_i} = DD_i^{(1)},\\
&\left.\frac{d^2D_1}{dx^2}\right|_{x=x_i} = D2D_i^{(1)}, \quad \left.\frac{d^2s_1}{dx^2}\right|_{x=x_i} = D2S_i^{(1)}.
\end{aligned}
$$

Equation (21.4) in the form of difference has the following form:

$$
\begin{aligned}
&\alpha_i^{(-1)} u_{i-1}^{(1)} + \alpha_i^{(0)} u_i^{(1)} + \alpha_i^{(+1)} u_{i+1}^{(1)} + h u_i^{(2)} + \beta_i^{(-1)} w_{i-1}^{(1)} + \beta_i^{(+1)} w_{i+1}^{(1)} = 0,\\
&\alpha_i^{(\pm 1)} = \frac{2B_i^{(1)} \pm DB_i^{(1)} h}{2Ph}, \quad \alpha_i^{(0)} = -h - \frac{2B_i^{(1)}}{Ph}, \quad \beta_i^{(\pm 1)} = \mp \frac{s_i^{(1)}}{2}.
\end{aligned}
\tag{21.13}
$$

By analogous way, Eq. (21.5) can be written as

$$
\begin{aligned}
&h u_i^{(1)} + \gamma_i^{(-1)} u_{i-1}^{(2)} + \gamma_i^{(0)} u_i^{(2)} + \gamma_i^{(+1)} u_{i+1}^{(2)} - \beta_i^{(+1)} w_{i+1}^{(1)} - \beta_i^{(-1)} w_{i-1}^{(1)} = 0\\
&\gamma_i^{(\pm 1)} = \frac{B_i^{(2)}}{Ph}, \quad \gamma_i^{(0)} = -h - \frac{2B_i^{(2)}}{Ph}, \quad \beta_i^{(\mp 1)} = \pm \frac{s_i^{(1)}}{2}.
\end{aligned}
\tag{21.14}
$$

Equation (21.6) in the form of difference can be derived as follows:

$$
\begin{aligned}
&\delta_i^{(-1)} u_{i-1}^{(1)} + \delta_i^{(0)} u_i^{(1)} + \delta_i^{(+1)} u_{i+1}^{(1)} + \varepsilon_i^{(-1)} u_{i-1}^{(2)} + \varepsilon_i^{(0)} u_i^{(2)} + \varepsilon_i^{(+1)} u_{i+1}^{(2)}\\
&\quad + \phi_i^{(-2)} w_{i-2}^{(1)} + \phi_i^{(-1)} w_{i-1}^{(1)} + \phi_i^{(0)} w_i^{(1)} + \phi_i^{(+1)} w_{i+1}^{(1)} + \phi_i^{(+2)} w_{i+2}^{(1)} = 0,
\end{aligned}
\tag{21.15}
$$

where in its turn

$$\delta_i^{(\mp 1)} = \pm\frac{s_i^{(1)}}{2} \pm \frac{1}{2P}\left(DS_i^{(1)} \cdot DB_i^{(1)} + D2S_i^{(1)} \cdot B_i^{(1)} \mp \frac{2 \cdot DS_i^{(1)} \cdot B_i^{(1)}}{h}\right),$$

$$\delta_i^{(0)} = \frac{2DS_i^{(1)}B_i^{(1)}}{Ph} - DS_i^{(1)}h, \varepsilon_i^{(\pm 1)} = \pm\frac{s_i^{(1)}}{2}, \quad \varepsilon_i^{(0)} = DS_i^{(1)}h,$$

$$\phi_i^{(\mp 1)} = \pm\frac{2DD_i^{(1)}}{Ph^2} + \frac{D2D_i^{(1)}}{Ph} \pm s_i^{(1)}DS_i^{(1)} - \frac{\left(s_i^{(1)}\right)^2}{h} - \frac{4D_i^{(1)}}{Ph^3},$$

$$\phi_i^{(0)} = \frac{2\left(s_i^{(1)}\right)^2}{h} - \frac{2 \cdot D2D_i^{(1)}}{Ph} + \frac{6D_i^{(1)}}{Ph^3} + \frac{K}{P}h, \quad \phi_i^{(\mp 2)} = \frac{D_i^{(1)}}{Ph^3} \pm \frac{DD_i^{(1)}}{Ph^2}.$$

Equations (21.13)–(21.15) are written inside the region for points with numbers $0, 1, \ldots, N$. However, we have unknown translations at the outer points, too. To close the system of linear equations, we have to write boundary conditions in the difference form. Finally, we can get the system of linear equations with respect to translations of division points. This system contains $3N + 11$ equations with respect to the same quantity of variables, i.e., the system is closed.

21.5 Genetic Algorithm for Doubler Shape Optimization

A genetic algorithm is used to solve a joint's topological optimization problem. The flexibility of genetic algorithms allows us to apply them to solve optimization problems that cannot be solved using classical gradient methods.

To realize a genetic algorithm, we have to describe a doubler shape with a definite set of parameters (genes) and find their optimal values, which is the optimization problem's objective. One of these parameters is the length of doubler L. An extension to the trigonometric series is used to describe the doubler shape function. The same conclusion can be made from considerations about smoothness and continuity of seeking function, i.e., it should not have any breaks, angular points, and racings. To be sure that function extension does not depend on joint length, we can apply extension in trigonometric series on the section $\xi \in [0; 1]$, and scale it on structure with length L, where length is independent optimization parameter. Such a form of presentation allows the definition of any precision of region division on finite differences, increasing the number of nodes N. The quantity of division points in the method of finite differences does not depend on M, i.e., the quantity of Fourier series members (this method is used for doubler shape description).

To determine thickness value at points of division at exact joint length L, both intervals $\xi \in [0; 1]$ and $x \in [0; L]$ have to be divided by $N + 1$ points ξ_i and x_i respectively, their enumeration varies from 0 to N. Doubler thickness at division points can be calculated as follows:

$$\delta_i^{(1)} = \frac{a_0}{2} + \sum_{n=1}^{M} a_n \cos \pi n \xi_n, \tag{21.16}$$

where a_i are the Fourier coefficients; M is the quantity of member in a series. Knowing thickness $\delta_i^{(1)}$ at points x_i, one can calculate all other necessary parameters of a doubler $s_i^{(1)}$, $B_i^{(1)}$, $D_i^{(1)}$ and also their derivatives. Description of a doubler geometrical shape by means of Fourier series (21.16) allows to estimate doubler mass (21.8)

$$M = \int_0^L \delta_1 (x) \, dx = \frac{a_0 L}{2} \to \min. \tag{21.17}$$

Thus, each individual in the genetic algorithm has to contain a set of numbers L and $a_0, a_1, a_2, \ldots, a_M$. To implement a genetic algorithm, we have to create a fitness function that allows us to range differently by quality sets of sought parameters L, $a_0, a_1, a_2, \ldots, a_M$. The objective function, by which individuals are arranged, has to depend on doubles mass (21.17) and possess terms or multipliers responsible for assigning penalties in violation of strength conditions or geometrical restrictions. If structural mass (21.8) is selected as the optimality criterion in the form of (21.17), restrictions are applied to maximum first principal stress in the adhesive film (21.10) and minimal doubler thickness (21.11), then fitness function can be written in the following way:

$$\begin{aligned} &\Phi = \frac{a_0 L}{2} + f_1 + f_2, \\ &f_1 = \begin{cases} Z_1 (S_1 - 1)^2, & S_1 > 1 \\ 0, & S_1 \leqslant 1 \end{cases}, \quad f_2 = \begin{cases} Z_2 (S_2 - 1)^2, & S_2 > 1 \\ 0, & S_2 \leqslant 1 \end{cases}, \\ &S_1 = \frac{\max\limits_i \left(\left| \sigma_i^{(1)} \right| \right)}{\sigma_{\max}}, \quad S_2 = \frac{\delta_{\min}}{\min\limits_i \left(\delta_i^{(1)} \right)} \end{aligned} \tag{21.18}$$

where Z_1 and Z_2 are certain large numbers, which define the level of penalty in case the solution exceeds its allowable margins; $\sigma_i^{(1)}$ are the first principal stress in adhesive film at points x_i, which can be calculated using (21.10); $\max(|\sigma_i{}^{(1)}|)$ are the maximum principal stress; $\min(\delta_i{}^{(1)})$ are the correspondent minimal valued of doubler thickness.

The suggested form of the objective function (21.18) ensures its smoothness near the boundary defined by restrictions. That is, in the case of negligible restriction violation, a penalty is also poor, but with violation degree, the growth penalty grows quadratically. Thus, if the solution (i.e., the set of optimization parameters L and the set $a_0, a_1, a_2, \ldots, a_M$) is obtained, then the value of fitness function is equal to a doubler cross-sectional area $a_0 L/2$. However, if at least at a single node, stress in adhesive film exceeds its allowable value or (and) if doubler thickness at least at a single node is less than the allowable one, then penalty terms are added to the given cross-sectional area; the greater the value of the term the more the violation of the correspondent restriction.

The algorithm for genetic optimization in a separated population has the following sequence:

1. **Composing of original population $\mathbf{h}^{(j)}$**, where $j = 1, \ldots, N_g$ (N_g is quantity of individuals in population). Each vector $\mathbf{h}^{(j)}$ (individual) contains components $L^{(j)}$ and $a_0^{(j)}, a_1^{(j)}, \ldots, a_M^{(j)}$. Using the set of selected parameters, we can determine correspondent values of Φ_j according to (21.18). For this, we have to solve the direct problem of finding translations in load-carrying layers, and then stress in the adhesive film can be calculated.
2. **Selection**. Range vectors $\mathbf{h}^{(j)}$ presented in population according to correspondent values of fitness function Φ_j. Select from the population $2k$ (where $2k < N_g$) elements $\mathbf{h}^{(j)}$. Here, the probability of being involved in a sample can depend on either the number in the range list or on values of Φ_j. The best individuals $\mathbf{h}^{(j)}$ have to be involved in a sample from the population that has fewer values of the fitness function.
3. **Division of parents on pairs**. Divide $2k$ selected individuals on pairs and got k pairs of "parents". Division on pairs can be realized by different strategies, by similarity, or vice versa, by differences of vectors $\mathbf{h}^{(j)}$. One possible measure of similarity-difference between two individuals, $\mathbf{h}^{(j)}$ and $\mathbf{h}^{(i)}$, could be the scalar product of the vectors divided by the product of the norms of both vectors:

$$\cos\theta_{i,j} = \frac{\left(\mathbf{h}^{(j)}, \mathbf{h}^{(i)}\right)}{\left|\mathbf{h}^{(j)}\right| \cdot \left|\mathbf{h}^{(i)}\right|}.$$

   One way to determine the similarity and difference between individuals $\mathbf{h}^{(j)}$ and $\mathbf{h}^{(i)}$ is by calculating the root of mean square deviation of functions (21.18) on the interval $\xi \in [0, 1]$.

$$\Delta_{i,j}^2 = \int_0^1 \left(\frac{a_0^{(j)}}{2} + \sum_{n=1}^{M} a_n^{(j)} \cos\pi n\xi\right)\left(\frac{a_0^{(i)}}{2} + \sum_{n=1}^{M} a_n^{(i)} \cos\pi n\xi\right) d\xi$$

   However, such a criterion does not consider differences in the length of individuals. For all individuals, the matrix of values of the selected criterion is determined. Then, k pairs from this set are selected. The authors of the paper used the principle of outbreeding.
4. **Crossing**. Arbitrary select parameters $L^{(\cdot)}$ and $a_0^{(\cdot)}, a_1^{(\cdot)}, \ldots, a_M^{(\cdot)}$ from both parental individuals for each pair for new individual. Each vector component is selected randomly from two analogous components of parents. As a result of this operation, one can get the population of k new individuals, i.e., "descendants".
5. **Mutations**. In the version of the algorithm implemented by authors, mutations occur in definite "descendants" only and only with quite small parts of components of vectors $\mathbf{h}^{(j)}$. Mutation is included in changing values of vector components on

definite negligible deviation. The value of random deviation is subjected to Gauss distribution with zero mean value.

6. **Returning of new individuals to population**. After embedding changes to gene code, new individuals are returned to the main population, and the quantity of individuals is increased from N_g to $N_g + k$ individuals.
7. **Extinction**. After new individuals return to population k, individuals are removed from the population. In the simplest case, one can remove k individuals with the worst fitness function values. It is also possible to select individuals randomly so that the worst fitness function values will increase the probability of an individual being removed from the population.
   The authors have adopted a hybrid approach in their optimization algorithm. During each breeding-extinction cycle, the algorithm either eliminates the worst k individuals in the population or creates a pool of individuals to be eliminated. The probability of an individual being added to this pool is determined by rank. In other words, the worst individuals are more likely to be included in the pool. This combination of deterministic and random methods has produced the best results in terms of the convergence of the optimization algorithm.
8. **Checking-up of stop criterion**. If the stop criterion (for example, given several breeding cycles K) is not reached, it is necessary to return to item 2.
   The paper utilizes an improved island model of the genetic algorithm [7, 28] to enhance the convergence rate of the genetic algorithm. The suggested variant of the island model of the genetic algorithm considers three islands. Mutations occur more frequently and with greater dispersion on one of the islands compared to the other two. After a certain number of cycles of generations, two islands are randomly selected, and individuals (migrants) from these islands reciprocally change. This approach ensures that new genetic information is introduced into the current population. The number of migrants should be significantly less than the total number of individuals on an island. One can select the fulfillment of a certain number of migrations as a criterion for stopping the change.

The Scheme of the island model of the evolution algorithm is shown in Fig. 21.4. It is worth noting that the migration model of individuals between separate populations in real nature is simulated by the model described below. This model demonstrates high variability on a single island and stability on two other ones. Regular migration of the best individuals between islands ensures a high operational rate of the evolution algorithm and stability of the results achieved. After the algorithm operation is complete, it is more efficient to select a definite quantity of the best individuals from the population to obtain optimal parameters. For instance, taking 20% of the total individuals and finding mean values of the seeking parameters is possible. This method minimizes the influence of "noise", which refers to arbitrary deviations of parameters from their optimal values.

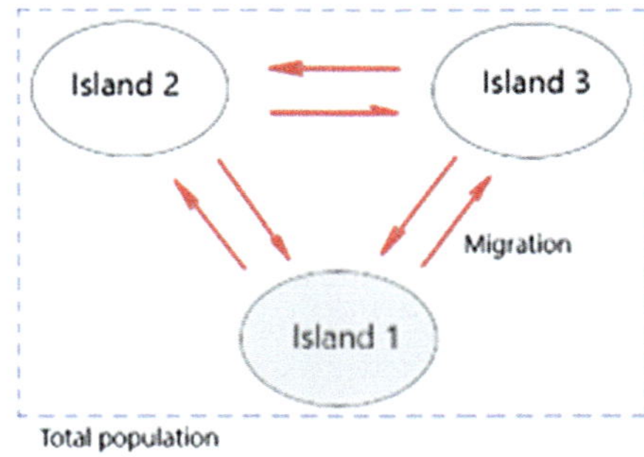

Fig. 21.4 Scheme of island model of genetic algorithm

21.6 Numerical Modeling

Let us consider the results of implementing the suggested algorithm of topological optimization of adhesive joints with exact parameters. Let our joint has following parameters: $E_1 = 100$ GPa, $E_2 = 70$ GPa, $\delta_2 = 3$ mm, $\delta_0 = 0.1$ mm, $E_0 = 2.274$ GPa, $G_0 = 0.54$ GPa, $\sigma_{\max} = 30$ MPa, $\delta_{\min} = 0.1$ mm. To investigate the influence of loading value on structural parameters, consider two load cases: $F_1 = 150$ kN/m and $F_2 = 300$ N/m.

The original population on each island is created by assigning a random joint length using a Gauss distribution with a mean value of $m_L = 20$ mm and a standard deviation of $\sigma_L = 4$ mm. The Fourier coefficients are calculated based on the linear relationship between double thickness and joint length, starting from a random value of $\delta_1(0)$ with an average value of $m_\delta = 1$ mm and a standard deviation of $\sigma_\delta = 0.1$ mm at $x = 0$, and $\delta_1(L) = 3$ mm. The number of Fourier series members is set at $M = 30$, and the joint stress state is determined by dividing the region into $N_{fem} = 100$ points. The population size on each island is $N_g = 120$, from which we select $2k = 40$ individuals for cross-breeding in each iteration. On two islands, the probability of joint length mutation during cross-breeding is assumed to be 0.2. Joint length changes to a random value with a Gauss distribution with a zero mean value and a root mean square deviation of 0.2 mm. The Fourier coefficients also mutate with a probability of 0.2. During mutations, they change to a random value with a Gauss distribution with a standard deviation of $\sigma_a = 2 \cdot 10^{-8}$ (if the corresponding Fourier coefficient is zero) and a coefficient of variation of $c_v = 0.02$. On one of the islands, these parameters are increased twice to ensure a wider difference between subpopulations on the island. The quantity of breeding and reproduction cycles on the interval between migrations is selected to be $K = 200$, i.e., for each $K = 200$ cycle of selection and reproduction, we arbitrarily select two islands, which rechange with $m = 10$ best individuals (migration).

21.6.1 The First Analysis Case

Analyzing optimal joint shape and loading with force $F_1 = 150$ kN/m, we got optimal joint length $L_1 = 12.09$ mm. The diagram of doubler thickness changing along the joint length is shown in Fig. 21.5. The straight line on the diagram corresponds to a half-thickness of the main load-carrying plate; this line is shown for diagram scale understanding.

Diagrams showing stress in the adhesive film are presented in Fig. 21.6. The diagrams are dimensional, and the current stress is compared to the maximum allowable stress of $\sigma_{\max} = 30$ MPa.

As it can be seen, at the right edge of a joint ratio of the first principal stress (21.10) to maximum allowable normal stress is equal to one. We draw attention

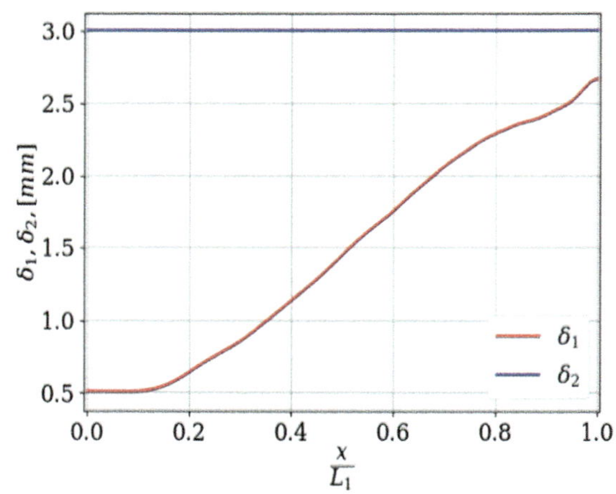

Fig. 21.5 Doubler thickness optimization at $F_1 = 150$ kN/m

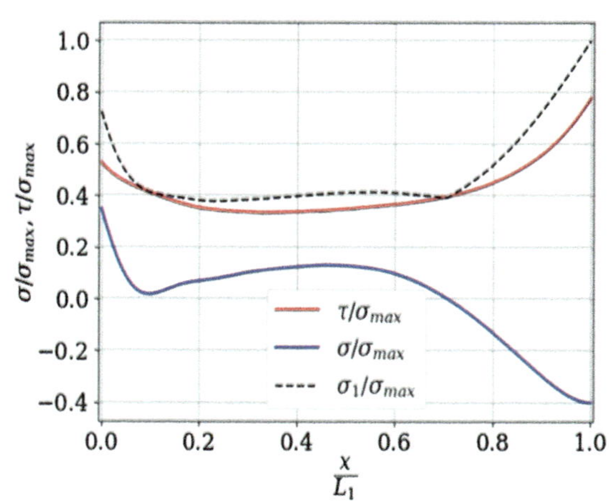

Fig. 21.6 Stress in adhesive film

to the following interesting distinction of doubler optimal shape: the presence of a section with constant thickness $\delta_1 = \delta_{\min}$ at the left side of a joint.

21.6.2 The Second Analysis Case

For the case of optimization problem-solving at twice $F_1 = 300$ kN/m load applied, we got joint length equal to $L_2 = 92.48$ mm, which is practically eight times more than in the first load case. The diagram of doubler thickness variation along the joint length is shown in Fig. 21.7. The straight line on the diagram corresponds to the half-thickness of the main load-carrying plate. At both edges of the joint, the first principal stress is equal to maximum allowable values $\sigma_{\max}$. We must draw attention to the fact that in the previous load case, stress reached its ultimate value at the right side of the joint only. Diagrams of stress inside the adhesive film are shown in Fig. 21.8.

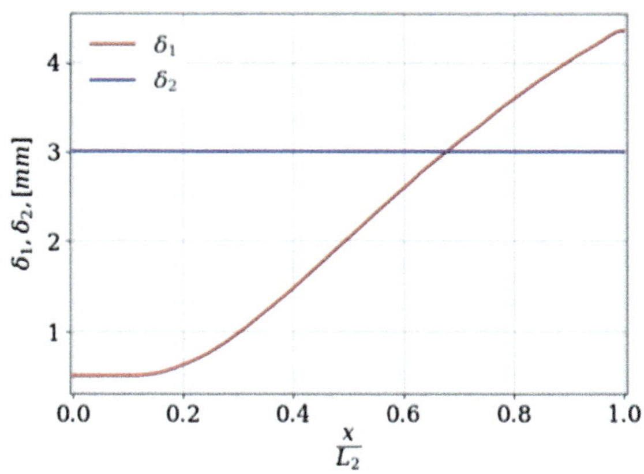

Fig. 21.7 Doubler optimal shape

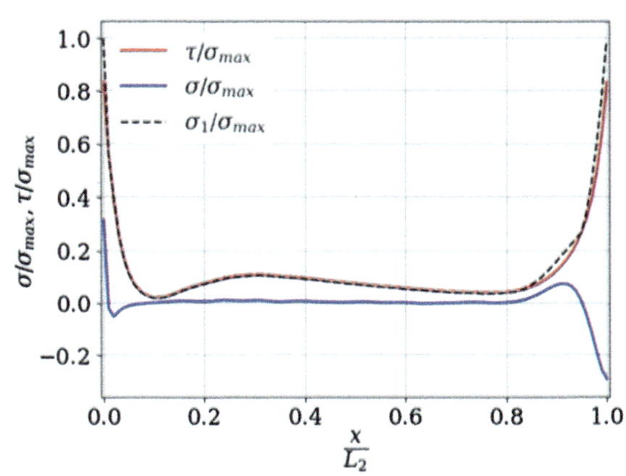

Fig. 21.8 Stress in adhesive film

21.6.3 Mathematical Model Verification. Comparison of Analytical Results with Analysis by Finite Element Methods

It is important to verify the results presented in the paper, as there may be errors in both the model and the software used for direct and optimization problem-solving. The software was developed using the Python coding language. The authors analyzed the stress state of the optimal joints found in the COMSOL Multiphysics 5.4 system. To better explain the modeling results, let us consider an example problem with loading of $F_1 = 300$ kN/m. The geometry of the two-dimensional model created for this example is shown in Fig. 21.9. The stress state of an adhesive film is influenced by structural features such as chamfers and influxes, particularly near its edges. This study modeled the adhesive film using finite elements with a minimum dimension of 0.25 δ_0. Chamfers were added to the faces of load-carrying plates, and an influx was added to the edge of the adhesive film. The figure below (Fig. 21.10) shows a fragment of the structure near the left joint edge, where the double has the minimum thickness. Figure 21.11 shows diagrams of normal and shear stress in adhesive film calculated by the analytical model suggested in the paper (designated with index "AM") and through the finite element model (designated with index "FEM").

The stress state can be estimated using two different methods, and the results are practically identical. However, there are some negligible deviations near joint edges, which is a common issue in all analytical models. These deviations occur because the hypotheses used to describe the deformation of both load-carrying layers and adhesive film are violated. Nonetheless, calculations show that the first principal stress calculated by the suggested one-dimensional model and used as a joint strength criterion is practically the same as the ones found by finite element modeling. Furthermore, the maximum values of these stresses are equal to the assigned ultimate values. Good convergence of stress–strain state estimation by two independent meth-

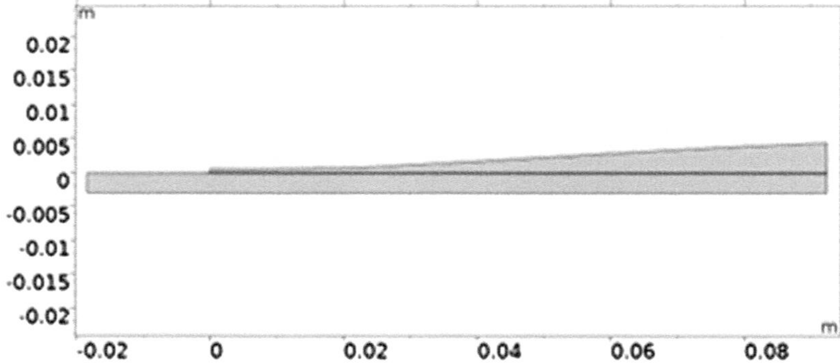

Fig. 21.9 Geometrical shape of optimal joint

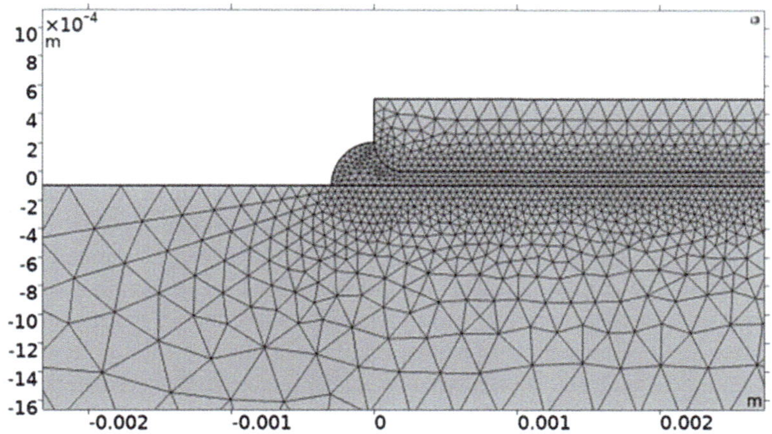

Fig. 21.10 Fragment of finite element model with grid shown

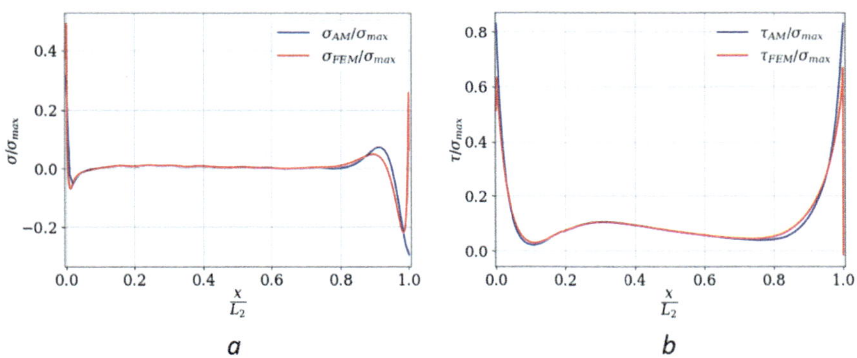

Fig. 21.11 Normal (**a**) and shear (**b**) stress in adhesive film in optimal adhesive joint, "AM" are the results obtained using the analytical model, "FEM" are the results calculated by the method of finite elements modeling

ods indicates the authenticity of the results found in the paper. Therefore, it allows us to make the following statements:

1. Authenticity of suggested mathematical models of joint stress state and numerical method of direct problem-solving;
2. Effective operation of suggested algorithm of solving the problem of joint topological optimization; this algorithm is based on the modified island model of genetic algorithm.

21.7 Conclusions

A paper suggests a mathematical model for a lapped double-side adhesive joint with variable doubler thickness along the joint length. It also proposes an improved genetic algorithm for joint length and doubler cross-section shape optimization. This approach is based on the classical one-dimensional model of structural stress state and genetic algorithm of optimization. The suggested genetic algorithm relates to island models of genetic algorithms, which realize the idea of parallel evolution processes with the migration of the best individuals. This approach has shown high efficiency and quickness. Implementing classical one-dimensional models of joint stress state allowed for a precise estimation of the structural stress state with a high rate of numerical analysis. This feature is crucially important for solving optimization problems using genetic algorithms. The suggested algorithm is highly flexible and can be generalized on other optimality criteria, strength criteria, and restrictions. After solving a series of problems and analyzing the results obtained, the following conclusions were formulated:

1. The dependence of doubler length and shape on the load transferred has a non-linear character.
2. The presence of a restriction on the minimal allowable doubler thickness in the optimization problem leads to a found optimal doubler shape that contains a horizontal section with minimal allowable thickness near the non-loaded doubler edge. This structural solution was not suggested previously.
3. It is impossible to reach uniform stress distribution in a joint under given conditions. The main obstacle is the constant thickness of the main load-carrying plate along the joint length. Therefore, the load-carrying ability of such a joint is limited. In the case of load F exceeding a specific value, which depends on adhesive strength, elasticity moduli of joint components, etc., the design problem has no solution at all.

The suggested approach can be developed and generalized in the following directions:

1. Application of finite-difference templates with elevated precision at direct problem-solving that allows conducting calculations with high precision at a lesser quantity of node points, increasing the calculation rate.
2. The number of restrictions in the problem can be increased. Besides restrictions on doubler thickness and adhesive film strength, one can add restrictions on sagging value, doubler strength, etc.
3. Topological optimization of joints in which quite compliant adhesive is used at joint edges and a more rigid one is used at the central joint zone [16, 25].
4. It is possible to apply the proposed genetic optimization algorithm to solve problems of topological optimization of connections in a two-dimensional formulation [20, 30].

The proposed optimization algorithm can be useful in the design of various real structures under static [13, 26] and dynamic loads [34, 35].

References

1. Amidi, S., Wang, J.: An analytical model for interfacial stresses in double-lap bonded joints. J. Adhes. **8464**, 1–25 (2018). https://doi.org/10.1080/00218464.2018.1464917
2. Ayaz Ümütlü, H.C., Kiral, Z.: Airfoil shape optimization using Bézier curve and genetic algorithm. Aviation **26**(1), 32–40 (2022). https://doi.org/10.3846/aviation.2022.16471
3. Barakhov, K.P., Taranenko, I.M.: Influence of joint edge shape on stress distribution in adhesive film. In: Nechyporuk, M., Pavlikov, V., Kritskiy, D. (eds.) Integrated Computer Technologies in Mechanical Engineering – 2021. ICTM 2021. Lecture Notes in Networks and Systems, vol. 367, pp. 123–132 (2022). https://doi.org/10.1007/978-3-030-94259-5_12
4. Belingardi, G., Goglio, L., Tarditi, A.: On the optimization of single lap metal/plastics adhesive joints. Key Eng. Mater. **221–222**, 161–172 (2001). https://doi.org/10.4028/www.scientific.net/kem.221-222.161
5. da Silva, L.F.M., das Neves, P.J.C., Adams, R.D., Spelt, J.K.: Analytical models of adhesively bonded joints: part I: literature survey. Int. J. Adhes. Adhes. **29**, 319–330 (2009). https://doi.org/10.1016/j.ijadhadh.2008.06.005
6. Ejaz, H., Mubashar, A., Ashcroft, I.A., Uddin, E., Khan, M.: Topology optimisation of adhesive joints using non-parametric methods. Int. J. Adhes. Adhes. **81**, 1–10 (2018). https://doi.org/10.1016/j.ijadhadh.2017.11.003
7. Gozali, A.A., Fujimura, S.: DM-LIMGA: dual migration localized island model genetic algorithm—a better diversity preserver Island model. Evol. Intell. **12**, 527–539 (2019). https://doi.org/10.1007/s12065-019-00253-2
8. Groth, H.L., Nordlund, P.: Shape optimization of bonded joints. Int. J. Adhes. Adhes. **11**(4), 204–212 (1991). https://doi.org/10.1016/0143-7496(91)90002-y
9. Haghani, R., Al-Emrani, M., Kliger, R.: Effect of laminate tapering on strain distribution in adhesive joints: experimental investigation. J. Reinf. Plast. Compos. **29**(7), 972–985 (2009). https://doi.org/10.1177/0731684408102698
10. Hassan Vand, M., Abbaszadeh, H., Shishesaz, M.: Optimization of adhesive single-lap joints under bending moment. J. Adhes. **98**(11), 1687–1712 (2021). https://doi.org/10.1080/00218464.2021.1932485
11. Hu, Z., Sun, S., Vambol, O., Tan, K.: Topology optimization of laminated composite structures under harmonic force excitations. J. Compos. Mater. **56**(3), 409–420 (2022). https://doi.org/10.1177/00219983211052605
12. Ichikawa, Kohei, Shin, Yuichiro, Sawa, Toshiyuki: A three-dimensional finite-element stress analysis and strength evaluation of stepped-lap adhesive joints subjected to static tensile loadings. Int. J. Adhes. Adhes. **28**(8), 464–470 (2008). https://doi.org/10.1016/j.ijadhadh.2008.04.011
13. Kantor, B.Y., Smetankina, N.V., Shupikov, A.N.: Analysis of non-stationary temperature fields in laminated strips and plates. Int. J. Solids Struct. **38**(48/49), 8673–8684 (2001). https://doi.org/10.1016/S0020-7683(01)00099-3
14. Karpov, Y.S.: Jointing of high-loaded composite structural components: part 2: modeling of stress-strain state. Strength Mater. **38**(5), 481–491 (2006). https://doi.org/10.1007/s11223-006-0067-9
15. Katsiropoulos, C.V., Chamos, A.N., Tserpes, K.I., Pantelakis, S.G.: Fracture toughness and shear behavior of composite bonded joints based on a novel aerospace adhesive. Compos. B Eng. **43**(2), 240–248 (2012). https://doi.org/10.1016/j.compositesb.2011.07.010
16. Kaye, R., Heller, M.: Through-thickness shape optimisation of typical double lap-joints including effects of differential thermal contraction during curing. Int. J. Adhes. Adhes. **25**(3), 227–238 (2005). https://doi.org/10.1016/j.ijadhadh.2004.07.006

17. Kumar, S., Pandey, P.C.: Behaviour of bi-adhesive joints. J. Adhes. Sci. Technol. **24**(7), 1251–1281 (2010). https://doi.org/10.1163/016942409x12561252291982
18. Kupski, J., Teixeira de Freitas, S.: Design of adhesively bonded lap joints with laminated CFRP adherends: review, challenges and new opportunities for aerospace structures. Compos. Struct. **268**, 113923 (2021). https://doi.org/10.1016/j.compstruct.2021.113923
19. Kurennov, S., Barakhov, K., Vambol, O.: Topological optimization of a symmetrical adhesive joint. Island model of genetic algorithm. Radioelectron. Comput. Syst. **3**, 67–83 (2021). https://doi.org/10.32620/reks.2022.3.05
20. Kurennov, S., Barakhov, K., Dvoretska, D., Poliakov, O.: Stress state of two glued coaxial tubes under nonuniform axial load. In: Nechyporuk, M., Pavlikov, V., Kritskiy, D. (eds.) Integrated Computer Technologies in Mechanical Engineering – 2020. ICTM 2020. Lecture Notes in Networks and Systems, vol. 188, pp. 389–402. Springer, Cham (2021). https://doi.org/10.1007/978-3-030-66717-7_33
21. Kurennov, S., Barakhov, K., Taranenko, I., Stepanenko, V.: A genetic algorithm of optimal design of beam at restricted sagging. Radioelectron. Comput. Syst. **2**, 83–91 (2022). https://doi.org/10.32620/reks.2022.1.06
22. Kurennov, S., Smetankina, N.: Stress-strain state of a double lap joint of circular form: axisymmetric model. In: Nechyporuk, M., Pavlikov, V., Kritskiy, D. (eds.) Integrated Computer Technologies in Mechanical Engineering – 2021. ICTM 2021. Lecture Notes in Networks and Systems, vol. 367, pp. 36–46 (2022). https://doi.org/10.1007/978-3-030-94259-5_4
23. Kurennov, S.S.: Refined mathematical model of the stress state of adhesive lap joint: experimental determination of the adhesive layer strength criterion. Strength Mater. **52**, 779–789 (2020). https://doi.org/10.1007/s11223-020-00231-5
24. Lee, H., Seon, S., Park, S., Walallawita, R., Lee, K.: Effect of the geometric shapes of repair patches on bonding strength. J. Adhes. **97**(3), 207–224 (2019). https://doi.org/10.1080/00218464.2019.1649660
25. Veisytabar, M., Reza, A., Shekari, Y.: Stress analysis of adhesively-bonded single stepped-lap joints based on three-parameter fractional viscoelastic foundation model. Proc. Inst. Mech. Eng., Part L: J. Mater.: Des. Appl. **236**(5), 933–949 (2022). https://doi.org/10.1177/14644207211062497
26. Misura, S., Smetankina, N., Misiura, I.: Optimal design of the cyclically symmetrical structure under static load. In: Nechyporuk, M., Pavlikov, V., Kritskiy, D. (eds.) ICTM 2020. Lecture Notes in Networks and Systems, vol. 188, pp. 256–266. Springer, Cham (2021). https://doi.org/10.1007/978-3-030-66717-7_21
27. Öz, Ö., Özer, H.: On the von Mises elastic stress evaluations in the bi-adhesive single-lap joint: a numerical and analytical study. J. Adhes. Sci. Technol. **28**(21), 2133–2153 (2014). https://doi.org/10.1080/01694243.2014.948110
28. Palomo-Romero, J.M., Salas-Morera, L., García-Hernández, L.: An Island model genetic algorithm for unequal area facility layout problems. Expert Syst. Appl. **68**, 151–162 (2017). https://doi.org/10.1016/j.eswa.2016.10.004
29. Ramezani, F., Ayatollahi, M.R., Akhavan-Safar, A., da Silva, L.F.M.: A comprehensive experimental study on bi-adhesive single lap joints using DIC technique. Int. J. Adhes. Adhes. 102674, (2020). https://doi.org/10.1016/j.ijadhadh.2020.102674
30. Rapp, P.: Mechanics of adhesive joints as a plane problem of the theory of elasticity: part II: displacement formulation for orthotropic adherends. Arch. Civ. Mech. Eng. **15**(2), 603–619 (2015). https://doi.org/10.1016/j.acme.2014.06.004
31. Rispler, A.R., Tong, L., Steven, G.P., Wisnom, M.R.: Shape optimisation of adhesive fillets. Int. J. Adhes. Adhes. **20**(3), 221–231 (2000). https://doi.org/10.1016/s0143-7496(99)00047-0
32. Shang, X., Marques, E.A.S., Machado, J.J.M., Carbas, R.J.C., Jiang, D., da Silva, L.F.M.: Review on techniques to improve the strength of adhesive joints with composite adherends. Compos. Part B: Eng. 107363, (2019). https://doi.org/10.1016/j.compositesb.2019.107363
33. Shishesaz, M., Hosseini, M.: Effects of joint geometry and material on stress distribution, strength and failure of bonded composite joints: an overview. J. Adhes. **96**(12), 1053–61121 (2018). https://doi.org/10.1080/00218464.2018.1554483

34. Smetankina, N.V., Shupikov, A.N., Sotrikhin, S.Y., Yareschenko, V.G.: Dynamic response of an elliptic plate to impact loading: theory and experiment. Int. J. Impact Eng. Impact Eng. **34**(2), 264–276 (2007). https://doi.org/10.1016/j.ijimpeng.2005.07.016
35. Smetankina, N.V., Shupikov, A.N., Sotrikhin, S.Y., Yareschenko, V.G.: A noncanonically shape laminated plate subjected to impact loading: theory and experiment. J. Appl. Mech., Trans. ASME **75**(5), 051004-1–051004-9 (2008). https://doi.org/10.1115/1.2936925
36. Symonov, V. S., Karpov, I. S., and Juračka, J.: Optimization of a panelled smooth composite shell with a closed cross-sectional contour by using a genetic algorithm. Mech. Compos. Mater. **49**(5), 563–570 (2013). https://doi.org/10.1007/s11029-013-9372-0
37. Wang, S., Xie, Z., Li, X.: A modified analytical model for stress analysis of adhesively bonded stepped-lap joints under tensile load. Eur. J. Mech. A. Solids **77**(103794), 103794 (2019). https://doi.org/10.1016/j.euromechsol.2019
38. Zou, P., Bricker, J., Uijttewaal, W.: Optimization of submerged floating tunnel cross section based on parametric Bézier curves and hybrid backpropagation - genetic algorithm. Mar. Struct. **74**, 102807 (2020). https://doi.org/10.1016/j.marstruc.2020.102807

Chapter 22
Research of Vibration Behavior of Porous FGM Panels by the Ritz Method

Lidiya Kurpa and Tetyana Shmatko

Abstract In this paper, free vibration analysis of the porous functionally graded shallow shells with a complex form of the plan and supported to elasticity foundation is conducted. A mathematical model of the problem is fulfilled by using the first shear deformation theory (FSDT). It is assumed that material properties vary continuously through the thickness of the shell according to a power law (P-FGM). Two types of models describing porosity distribution (even and uneven) are considered. The variational Ritz method combined with the R-functions theory is applied to analyze the free vibrations of the porous shallow shells with the complex plan form. A comparison of the obtained numerical results for the natural frequencies with known results has been done, and a good agreement is obtained. The investigation of the shallow shells with a complex geometric form of the plan is fulfilled to demonstrate the possibilities of the proposed approach. The influence of different geometric and mechanical parameters on the natural frequencies is studied.

22.1 Introduction

We devote this paper to an outstanding Ukrainian scientist, Academician of the National Academy of Science of Ukraine Yaroslav Grigorenko, who made a huge contribution to world science in the field of the plates and shells theory. He was a tireless scientist, a benevolent and tactful educator, noble, and bright person. He left a rich scientific legacy. Communication with him inspired us to solve various problems, gave us confidence, and taught us kindness. With many thanks for the constant support, we dedicate this work to a great scientist.

L. Kurpa (✉)
Department of Applied Mathematics, National Technical University "Kharkiv Polytechnic Institute", Kharkiv, Ukraine
e-mail: Lidiya.Kurpa@khpi.edu.ua

T. Shmatko
Department of Higher Mathematics, National Technical University "Kharkiv Polytechnic Institute", Kharkiv, Ukraine
e-mail: Tetyana.Shmatko@khpi.edu.ua

H. Altenbach et al. (eds.), *Selected Problems of Solid Mechanics and Solving Methods*, Advanced Structured Materials 204, https://doi.org/10.1007/978-3-031-54063-9_22

Today, one of the topical problems of plates and shells theory is the study of the static and dynamic behavior of these objects made of functionally gradient materials (FGM). As a rule, these materials are fabricated from a mixture of metal and ceramics. The specific characteristics of FGM can be obtained by changing the volume fraction index. It allows applying the FGMs in many fields of engineering and industry. Plates and shells are widely used as main components in modern constructions. So, the study of static, dynamic, and stability behaviors of these structure elements gains importance among researchers. A huge number of publications is devoted to this problem [1–5]. One of the important effects on the static and dynamic response of the plates and shells is the presence of porosities within the FGMs, which can change their mechanical properties. In addition, porous materials have higher stiffness and lower density, which have been widely applied in various engineering fields such as aerospace, aviation, and military. Therefore, it is significant for the designers to know about the vibration characteristics of FGM porous structures. In recent years, many works have appeared devoted to modeling and analyzing porous plates and shells. Wu et al. [6] present some reviews of publications about this problem. Free vibration analysis of the porous rectangular plates is carried out in [7–20]. Most researchers use the first-order shear deformation theory (FSDT). However, in [9–12, 17], the higher-order shear deformation theory is applied to study the vibration of the FG porous plates. Researchers study the vibration of the plates accounting for the effect of magneto-electro and thermomechanical fields on dynamic and static response [10–15]. The influence of the elastic foundation on the dynamic characteristics of the porous plates is reported [10, 11, 16–20]. Sandwich FG porous plates are considered in [12, 21].

It should be noted that vibrations of porous functionally graded shells are less studied than vibrations of plates. Vibrations of the cylindrical shells are investigated in [22, 23, 30, 31, 37]. The vibration behavior of the double curvature shells with porosity is presented in [24–27, 35, 36]. Nonlinear vibrations of the FGM plates and shells are discussed in [28–37].

The application of the R-functions theory to solve linear and geometrically nonlinear bending and vibration problems of the plate and shallow shells was discussed [38–42].

This paper uses the R-functions theory combined with the variational Ritz method to solve the vibration problems for porous FGM shallow shells resting on a two-parameter elastic foundation. These objects under consideration can have an arbitrary plan form and different boundary conditions.

22.2 FGM Shallow Shells with Porosities

A thin porous shallow shell resting on an elasticity foundation is considered. The shell has a constant thickness h; it is made of FGM (mixture of metal and ceramics). The plan form of the shallow shell and boundary conditions can be different. It is

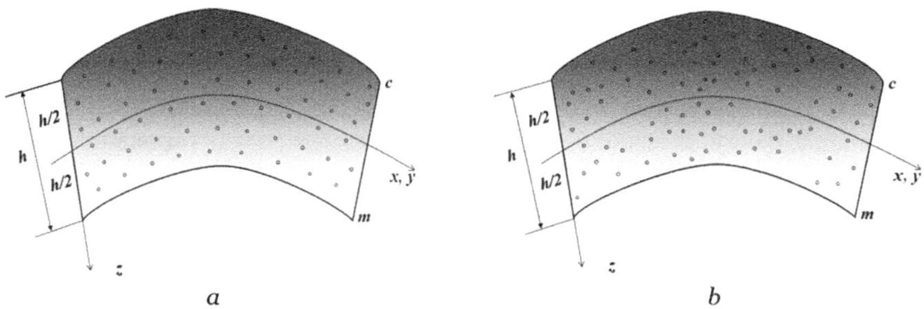

Fig. 22.1 Even distribution of porosity (Poro-I) (**a**), uneven distribution of porosity (Poro-I) (**b**)

assumed that porosities appear during the sintering process of metal and ceramics. Suppose that the volume fraction of metal V_m changes by power-law distribution:

$$V_m = \left(\frac{z}{h} + \frac{1}{2}\right)^p \tag{22.1}$$

The volume fraction of ceramics and metal satisfies the following requirement:

$$V_m + V_c = 1.$$

In (22.1), index p denotes the volume fraction exponent (gradient index); m, and c correspond to the designation of properties of metal and ceramics relatively. We study two types of porosities distribution in material: even (Fig. 22.1a) and uneven (Fig. 22.1b). The effective material properties of FGM $P_{\text{ef}}(z)(E_{\text{ef}}(z), \rho_{\text{ef}}(z))$ with even distributed porosity (Poro-I) are defined by the following relations [12]:

$$P_{\text{ef}}(z) = (P_m - P_c)V_m + P_c - \tfrac{1}{2}\alpha(P_m + P_c). \tag{22.2}$$

For uneven distribution (Poro-II) of porosities, the effective material properties vary across the thickness by a linear function:

$$P_{\text{ef}}(z) = (P_m - P_c)\,V_m + P_c - \tfrac{1}{2}\alpha\left(1 - 2\frac{|z|}{h}\right)(P_m + P_c). \tag{22.3}$$

Here index α characterizes a volume fraction of porosities. Values P_m and P_c denote corresponding properties (elastic modulus and density ρ) of the FGM constituents, and P is a general designation of the corresponding effective properties. Values P_m and P_c depend on the temperature of the environment [4, 14, 15]. We use Pasternak's model [19, 20] for the elastic foundation:

$$p_0 = k_W w - k_P \nabla^2 w, \tag{22.4}$$

where p_0 is the force per unit area, $\nabla^2 w = \partial^2 w/\partial x^2 + \partial^2 w/\partial y^2$; k_w and k_P are the Winkler foundation stiffness and shear stiffness of Pasternak foundation, respectively.

22.3 Mathematical Formulation of the Problem

The mathematical statement of the problem is carried out within the first-order shear deformation theory of shallow shells (FSDT). The variational formulation of the given problem is based on finding the extreme of the following functional:

$$I = U(u, v, w, \psi_x, \psi_y) + V_e\left(u, v, w, \psi_x, \psi_y\right) - \lambda^2 T\left(u, v, w, \psi_x, \psi_y\right), \quad (22.5)$$

where λ is a vibration frequency. Strain energy U, potential V_e and kinetic energy T in this case are defined by expressions:

$$U = \int_\Omega \big(N_{11}\varepsilon_{11} + N_{22}\varepsilon_{22} + N_{12}\varepsilon_{12} + M_{11}\chi_{11} \quad (22.6)$$

$$+ M_{22}\chi_{22} + M_{12}\chi_{12} + Q_1\gamma_{13} + Q_2\gamma_{23}\big)d\Omega,$$

$$V_e = \frac{1}{2}\int_\Omega \left(k_w w^2 + k_p (\nabla w)^2\right) d\Omega, \quad (22.7)$$

$$T = \frac{1}{2}\iint_\Omega \left(I_0\left(u^2 + v^2 + w^2\right) + 2I_1\left(u\psi_x + v\psi_y\right) + I_2\left(\psi_x^2 + \psi_y^2\right)\right) dxdy, \quad (22.8)$$

where $\varepsilon = \{\varepsilon_{11}, \varepsilon_{22}, \varepsilon_{12}\}^{\mathrm{T}}$ and $\chi = \{\chi_{11}, \chi_{22}, \chi_{12}\}^{\mathrm{T}}$ are strain components, $N = (N_{11}, N_{22}, N_{12})^{\mathrm{T}}$ is the force resultant vector in-plane; $M = (M_{11}, M_{22}, M_{12})^{\mathrm{T}}$ is the bending and twisting moment resultant vector. These values are computed by integration along the Oz-axes. According to the FSDT, constitutive relations read

$$[N] = [A]\{\varepsilon\} + [B]\{\chi\}, \quad [M] = [B]\{\varepsilon\} + [D]\{\chi\}. \quad (22.9)$$

Strain components $\varepsilon = \{\varepsilon_{11}, \varepsilon_{22}, \varepsilon_{12}\}^{\mathrm{T}}$ and $\chi = \{\chi_{11}, \chi_{22}, \chi_{12}\}^{\mathrm{T}}$ at an arbitrary point of the shallow shell are as follows:

$$\varepsilon_{11} = u_{,x} + w/R_x \quad \varepsilon_{22} = v_{,y} + w/R_y \quad \varepsilon_{12} = u_{,y} + v_{,x}, \quad (22.10)$$

$$\begin{aligned} &\varepsilon_{13} = w_{,x} + \psi_x, \quad \varepsilon_{23} = w_{,y} + \psi_y, \\ &\chi_{11} = \psi_{x,x}, \quad \chi_{22} = \psi_{y,y} \quad \chi_{12} = \psi_{x,y} + \psi_{y,x}, \end{aligned} \quad (22.11)$$

where u, v, and w are the displacement components of the middle surface in the Ox, Oy, and Oz directions; ψ_x and ψ_y are the independent rotation components of the transverse normal to the middle surface about the Oy and Ox axes, respectively.

The elements A_{ij}, B_{ij}, and D_{ij} $(i, j = 1, 2, 6)$ of matrices A, B, and D in (22.9) are calculated as

$$\left(A_{ij}, B_{ij} D_{ij}\right) = \int_{-h/2}^{h/2} Q_{ij}\left(1, z, z^2\right) dz, \quad (i, j = 1, 2, 6). \quad (22.12)$$

The values Q_{ij} $(i, j = 1, 2, 6)$ depend of the variable z and are defined by the following expressions:

$$Q_{11} = Q_{22} = \frac{E_{ef}(z)}{1-\nu^2}, \quad Q_{12} = \frac{\nu E_{ef}(z)}{1-\nu^2}, \quad Q_{66} = \frac{E_{ef}(z)}{2(1+\nu)}. \tag{22.13}$$

In the similar way, the values I_0, I_1 and I_2 in (22.8) are defined by the following relations:

$$(I_0, I_1, I_2) = \int_{-\frac{h}{2}}^{\frac{h}{2}} \rho_{ef}\left(1, z, z^2\right) dz, \tag{22.14}$$

Values E_{ef}, $\rho_{ef}(z)$ are defined by formulas (22.2) and (22.3). Transverse shear force resultants Q_x and Q_y have the following form:

$$Q_x = K_s^2 A_{66}\varepsilon_{13}, \quad Q_y = K_s^2 A_{66}\varepsilon_{23}, \tag{22.15}$$

where K_s^2 is a shear correction factor. If $\nu_m = \nu_c$, then coefficients A_{ij}, B_{ij}, and D_{ij} can be calculated in a direct way.

22.4 Solution Procedure

To solve the given problem, we propose to use the Ritz method. It means minimizing functional (22.5) is necessary. Unknown functions $u(x, y)$, $v(x, y)$, $w(x, y)$, $\psi_x(x, y)$, and $\psi_y(x, y)$ are presented as follows:

$$\begin{aligned}
u(x, y) &= \sum_{k=1}^{N_1} c_k u_k(x, y), \quad v(x, y) = \sum_{k=N_1+1}^{N_2} c_k v_k(x, y), \\
w(x, y) &= \sum_{k=N_2+1}^{N_3} c_k, w_k(x, y), \\
\psi_x(x, y) &= \sum_{k=N_3+1}^{N_4} c_k \psi_{xk}(x, y), \quad \psi_y(x, y) = \sum_{k=N_4+1}^{N_5} c_k \psi_{yk}(x, y),
\end{aligned} \tag{22.16}$$

where c_k $(k = 1, \ldots, N_5)$ are indetermined coefficients. These are found from the stationary condition of functional J (22.5). Sequences $\{u_k\}$, $\{v_k\}$, $\{w_k\}$, $\{\psi_{xk}\}$, and $\left\{\psi_{yk}\right\}$ $(k = 1, \ldots N_5)$ are set of admissible functions. The admissible functions should be linearly independent, differentiable to the required degree, and generate a complete system. They have to satisfy at least the geometrical boundary conditions. Construction of these functions is carried out by the R-functions theory [38] in case of the complicated shape of the shallow shell or complex boundary condition. A description of the composing procedure of the admissible function in detail is presented in [38, 39]. Below, we present these admissible functions for specific boundary conditions and geometry of the shallow shell shape.

22.5 Validation of the Proposed Approach and Created Software

The validation of the proposed approach and created software is confirmed by comparison of the obtained results with known ones.

Test 1. The free vibrations of simply supported FGM (Si_3N_4 /SUS304) square plate and double curved shells with porosity (Poro-I) are considered. The mechanical characteristics of the mixture are the following:

$$\begin{aligned} &Si_3N_4: && E = 322.27 \text{ GPa}, \quad \nu = 0.3, \quad \rho = 2370 \text{ kg/m}^3, \\ &SUS304: && E = 207.78 \text{ GPa}, \quad \nu = 0.3, \quad \rho = 8166 \text{ kg/m}^3. \end{aligned}$$

The comparison of the obtained results for different values of the porosity coefficient α and the gradient index p with results from [35] is shown in Table 22.1.

Test 2. The vibrations of the FG rectangular plate are studied. The FGM is assumed to be fabricated of aluminum and alumina (Al/Al_2O_3). The side of the square base is equal to 2a. Mechanical properties of the constituent materials:

$$\begin{aligned} E_m = 70\,\text{GPa}, \quad E_c = 380\,\text{GPa}, \quad \rho_m = 2707\,\text{kg/m}^3, \\ \rho_c = 3800\,\text{kg/m}^3, \quad \nu_m = 0.3, \quad \nu_c = 0.3. \end{aligned} \tag{22.17}$$

Comparison of the obtained results for the non-dimensional fundamental frequency $\Lambda = (\Omega\ (2a)^2/h)\sqrt{\rho_M/E_M}$ of the FG simply supported plate with known results are shown in Table 22.2 for two values of the gradient index $p = 1$ and $p = 5$. The thickness varies, parameters of the elastic foundations are designated: $\bar{K}_w = k_w(2a)^4/D_M$, $\bar{K}_P = k_p(2a)^2/D_M$, $D_M = E_M h^3/(12(1-\nu^2))$.

The results agreed with known ones, enabling the investigation of shells with complex plan forms.

Table 22.1 Comparison of non-dimensional fundamental frequency $\Lambda = \Omega(2a)^2 h\sqrt{\rho_c/E_c}$ for simply supported porous Si_3N_4/SUS304 FG plates and doubly curved shells $(a/b = 1, h/(2a) = 0.1)$ with results of [35]

	$2a/R_x$	$2a/R_y$	p	Method	Poro-I			Poro-II	
					$\alpha = 0$	$\alpha = 0.1$	$\alpha = 0.2$	$\alpha = 0.1$	$\alpha = 0.2$
Plate	0	0	0	[35]	0.0250	0.0241	0.0231	0.0249	0.0249
				RFM	0.0249	0.0241	0.0231	0.0249	0.0249
			1	[35]	0.0348	0.0348	0.0347	0.0352	0.0356
				RFM	0.0348	0.0347	0.0347	0.0352	0.0356
			5	[35]	0.0460	0.0477	0.0501	0.0473	0.0488
				RFM	0.0460	0.0477	0.0502	0.0472	0.0487
Sph	0.5	0.5	0	[35]	0.0329	0.0318	0.0304	0.0327	0.0324
				RFM	0.0326	0.0314	0.0301	0.0323	0.0321
			1	[35]	0.0461	0.0460	0.0460	0.0464	0.0467
				RFM	0.0459	0.0459	0.0457	0.0462	0.0465
			5	[35]	0.0613	0.0637	0.0670	0.0627	0.0645
				RFM	0.0610	0.0634	0.0699	0.0625	0.0643
Cyl	0	0.5	0	[35]	0.0272	0.0262	0.0251	0.0271	0.027
				RFM	0.0269	0.0260	0.0249	0.0268	0.0267
			1	[35]	0.0380	0.0379	0.0378	0.0383	0.0387
				RFM	0.0377	0.0377	0.0376	0.0388	0.0385
			5	[35]	0.0502	0.0522	0.0548	0.0516	0.0531
				RFM	0.0499	0.0518	0.0545	0.0512	0.0528
H/Par	0.5	−0.5	0	[35]	0.0250	0.0241	0.0231	0.0249	0.0249
				RFM	0.0247	0.0234	0.0225	0.0243	0.0243
			1	[35]	0.0348	0.0348	0.0347	0.0352	0.0356
				RFM	0.0339	0.0338	0.0337	0.0343	0.0347
			5	[35]	0.0460	0.0477	0.0501	0.0473	0.0488
				RFM	0.0447	0.0464	0.0478	0.0460	0.0475

22.6 Numerical Results for the Porous Shallow Shells with a Complex Plan Form

Let us investigate the shallow shells of the complex form made of different FGMs.

Geometric parameters are the following (Fig. 22.2):

$$\begin{gathered} h/2a = 0.1, \quad b/a = 1, \quad a_1/a = 0.6, \quad b_1/a = 0.3, \\ b_2/2a = 0.6, \quad b_3/2a = 0.1, \quad r/2a = 0.125, \\ k_1 = R_x/2a = 0.5, \quad k_2 = R_y/2a = 0.5. \end{gathered} \tag{22.18}$$

The following different boundary conditions are considered:

Table 22.2 Comparison of the present results for non-dimensionless frequency parameter of the simply supported square plate resting on elastic foundation with corresponding known results

$2a/h$	Method	$\bar{K}_w = 0$, $\bar{K}_p = 0$		$\bar{K}_w = 100$, $\bar{K}_P = 0$		$\bar{K}_w = 100$, $\bar{K}_P = 100$	
		$p = 1$	$p = 5$	$p = 1$	$p = 5$	$p = 1$	$p = 5$
5	RFM	8.135	6.884	8.567	7.413	14.689	14.635
	[18]	8.037	6.671	8.475	7.256	14.630	14.586
	[19]	8.012	6.668	8.452	7.253	14.479	14.584
	[20]	8.151	6.804	8.567	7.362	14.586	14.426
10	RFM	8.725	7.463	9.144	8.010	15.217	15.172
	[18]	8.069	7.403	9.111	7.952	15.189	15.150
	[19]	8.682	7.403	9.103	7.952	15.208	15.150
	[20]	8.818	7.553	9.228	8.087	15.150	15.167
20	RFM	8.901	7.657	9.317	8.196	15.380	15.352
	[18]	8.888	7.639	9.304	8.179	15.363	15.341
	[19]	8.886	7.639	9.302	8.179	15.429	15.341
	[20]	9.020	7.793	9.429	8.321	15.341	15.404

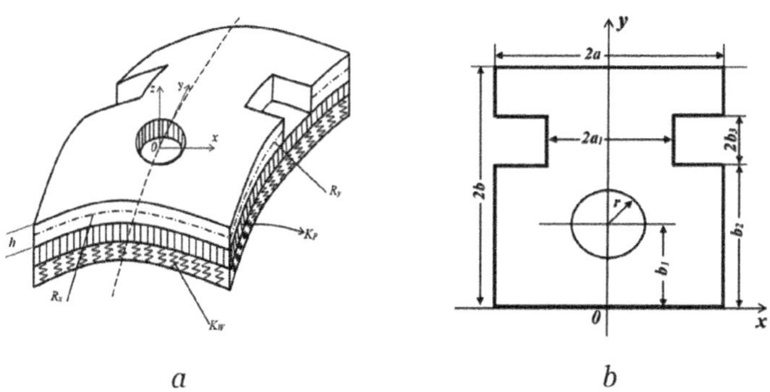

Fig. 22.2 Shallow shell with a complex shape and its plan form

1. C-C means that the shell is clamped on the whole boundary, including the cut-out;
2. C-F means that the cut-out is free and the external boundary is clamped.

To get the admissible functions by RFM, let us construct the corresponding structure of solutions [38, 39]. The solution structure for shells with complete clamped borders (C-C) can be taken in the form

$$w = \omega\Phi_1, \quad u = \omega\Phi_2, \quad v = \omega\Phi_3, \quad \psi_x = \omega\Phi_4, \quad \psi_y = \omega\Phi_5. \tag{22.19}$$

For boundary condition, C-F solution structure is the following:

Table 22.3 Non-dimensional natural frequency of the clamped SiN_4/SUS304 porous plate, cylindrical, spherical, parabolic shells

		Poro-I				Poro-II	
$2a/R_x$	$2a/R_y$	p	$\alpha = 0.1$	$\alpha = 0.1$	$\alpha = 0.2$	$\alpha = 0.1$	$\alpha = 0.2$
0	0	0	0.1207	0.1166	0.1116	0.1196	0.1185
		1	0.1689	0.1687	0.1685	0.1698	0.1708
		5	0.2246	0.2334	0.2455	0.2292	0.2359
0.5	0	0	0.1222	0.1180	0.1130	0.1211	0.1199
		1	0.1713	0.1711	0.1709	0.1722	0.1732
		5	0.2278	0.2367	0.2491	0.2332	0.2393
0.5	0.5	0	0.1246	0.1204	0.1153	0.1235	0.1222
		1	0.1749	0.1748	0.1746	0.1758	0.1768
		5	0.2326	*0.2418*	0.2546	0.2381	0.2443
0.5	–0.5	0	0.1231	0.1189	0.1139	0.1220	0.1207
		1	0.1724	0.1722	0.1719	0.1733	0.1743
		5	0.2293	*0.2383*	0.2507	0.2347	0.2408

$$\begin{aligned} w &= \omega_{\text{extern}}\Phi_1, \quad u = \omega_{\text{extern}}\Phi_2, \quad v = \omega_{\text{extern}}\Phi_3, \\ \psi_x &= \omega_{\text{extern2}}\Phi_4, \quad \psi_y = \omega_{\text{extern1}}\Phi_5, \end{aligned} \tag{22.20}$$

where Φ_i $(i = 1, \ldots, 5)$ are undetermined components of the solution structure. They are expanded in a power series. In (22.19), $\omega = 0$ is an equation of the whole border. In relations (22.20), the functions $\omega_{\text{extern}} = 0$ is equation of the outside border. These equations are constructed by the R-functions theory and have the following form:

$$\omega = (f_1 \wedge_0 f_2) \wedge_0 (f_3 \vee_0 f_4) \wedge_0 f_5, \quad \omega_{\text{extern}} = (f_1 \wedge_0 f_2) \wedge_0 (f_3 \vee_0 f_4). \tag{22.21}$$

Functions f_i $(i = 1, \ldots, 5)$ are defined as follows:

$$\begin{aligned} f_1 &= \frac{a^2 - x^2}{2a} \geqslant 0, \quad f_2 = \frac{b^2 - (y-b)^2}{2b} \geqslant 0, \quad f_3 = \frac{(y-y_0)^2 - b_3^2}{2b_3} \geqslant 0, \\ f_4 &= \frac{a_1^2 - x^2}{2a_1} \geqslant 0, \quad f_5 = \frac{x^2 + (y-b_1)^2 - r^2}{2r} \geqslant 0, \end{aligned} \tag{22.22}$$

where $y_0 = b_2 + b_3$.

Effect of volume fraction exponent p on non-dimensional fundamental frequency $\Lambda = \lambda(2a)^2\sqrt{\rho_c/E_c}\,h$ for porous Si_3N_4/SUS304 FG plates and shallow shells is shown in Table 22.3 and in Fig. 22.3a–d.

Figure 22.3 shows the effect of volume fraction index p on natural frequency parameters of clamped FG porous square plates and doubly curved shallow shells. Note that the ceramic portion increases with p. It leads to the increase in the value frequency parameter Λ. The effect of the curvatures on natural frequency is insignif-

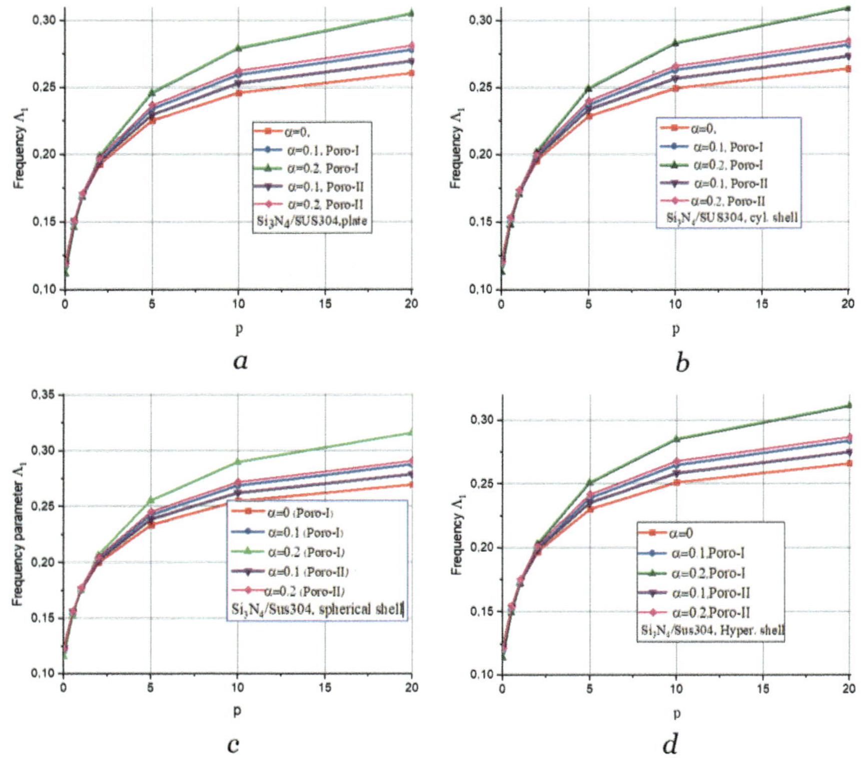

Fig. 22.3 Effect of porosity coefficient α and volume fraction exponent p on natural frequencies of the C-C plate, cylindrical, spherical, and hyperbolic paraboloid shells

icant for clamped boundary conditions. When the porosity coefficient α increases, it results in an increase in frequency parameters for all structures that are being considered. The increase in stiffness of the porous plates and shells can explain it. From Fig. 22.3, it follows that in the case of Poro-I, an increase in α affects natural frequencies more essentially than in the case of Poro-II.

Influence of the elastic foundation coefficients and volume fraction index p on natural frequency of the clamped spherical shell is illustrated in Fig. 22.4 for FGM Al/Al_2O_3 and in Fig. 22.5 for FGM $Si_3N_4/SUS304$. The case of Poro-2 with fixed porosity coefficient $\alpha = 0.2$ is analyzed. It has been noticed that the Winkler elasticity coefficient's impact on the frequency value is negligible, as the graphs for $K_w = 0$ and $K_w = 50$ are nearly identical. In contrast, the Pasternak K_p coefficient of elasticity significantly affects the frequency. Note that if $K_p = 100$ and K_w varies from 0 to 100, then curves for both FGM types also coincide.

Figure 22.6 shows the effect of porosity factor on natural frequency $\Lambda = \lambda(2a)^2\sqrt{\rho_c/E_c}h$ of the C-C spherical shells with fixed gradient index $p = 2$ for different types of the FGM and two types of porosity. For different boundary conditions, Fig. 22.7 demonstrates the comparison of natural frequency for porous

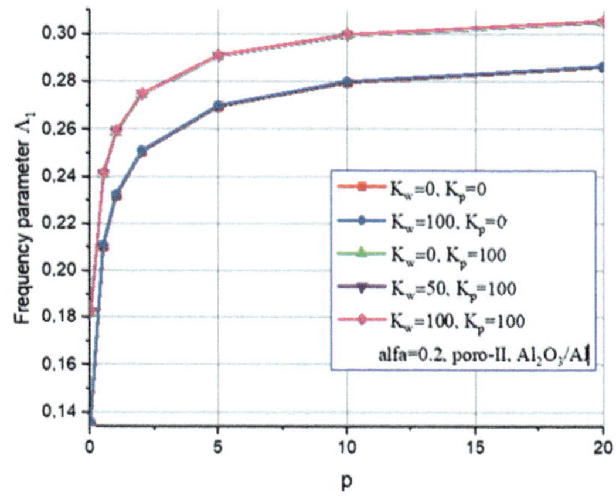

Fig. 22.4 Natural frequencies of C-C, spherical shells, made of Al/Al_2O_3

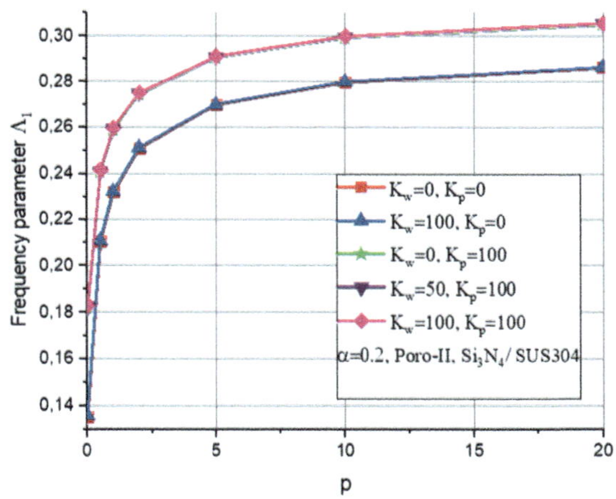

Fig. 22.5 Natural frequencies of C-C, spherical shells, made of Si_3N_4/SUS304

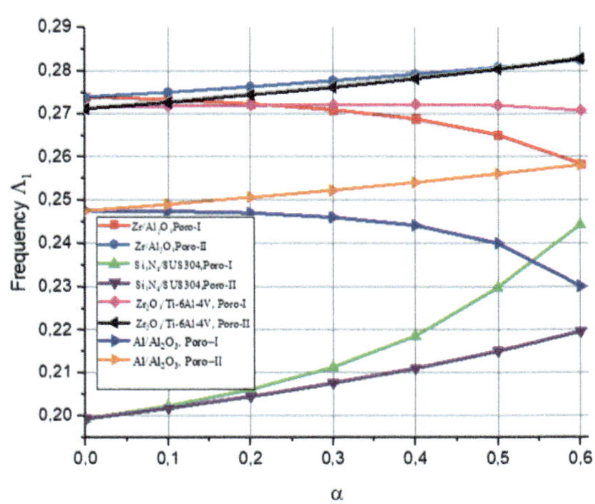

Fig. 22.6 Effect of porosity factor on the natural frequency of the C-C spherical shells for different types of the FGM

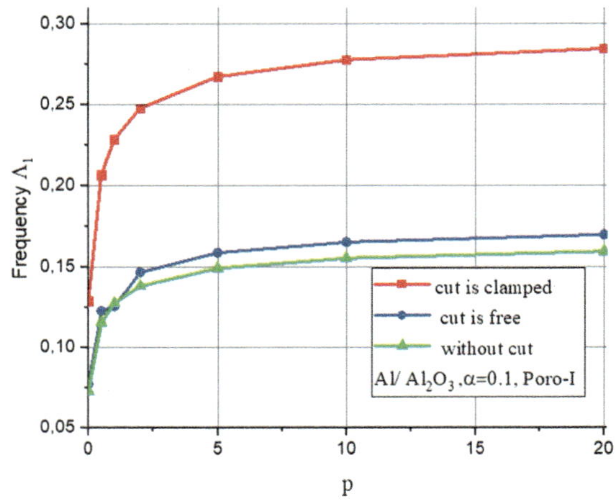

Fig. 22.7 Comparison of natural frequency for spherical shells with different boundary conditions

(Poro-I, $\alpha = 0.1$) spherical shells fabricated of Al/Al_2O_3. The natural frequency is highest in the clamped shell with a clamped cut-out. Clamped shell without cut has the smallest value of the frequency. Increasing gradient index p leads to the increase of the composite's ceramic portion and the decrease of the metal portion. This variation of material constituents increases the value of the parameter of the natural frequency Λ.

22.7 Conclusion

Investigation of the free linear vibrations of functionally graded shallow shells of a complex shape with porosity and elastic foundation was carried out by the R-functions method combined with the Ritz method. Considered test problems confirm the validation of the proposed approach and developed software. New results are obtained for porous FGM shallow shells of the complex shape. Effects of material types, power-law exponents, and porosity are studied for shells with different curvatures.

References

1. Thai, H.-T., Kim, S.-E.: A review of theories for the modeling and analysis of functionally graded plates and shells. J. Compos. Struct. **128**, 70–86 (2015)
2. Swaminathan, K., Naveenkumar, D.T., Zenkour, A.M., Carrera, E.: Stress, vibration and buckling analyses of FGV plates—a state-of-art review. J. Compos. Struct. **120**, 10–31 (2015)
3. Kumar, Y.: The Rayleigh-Ritz method for linear dynamic, static and buckling behavior of beams, shells and plates: a literature review. J. Vib. Control. **24**(7), 1205–1227 (2018)
4. Shen, H.S.: Functionally Graded Materials of Plates and Shells. CRC Press, Florida (2009)

5. Grigorenko, A.Y., Efimova, T.L., Korotkikh, Y.A.: Free axisymmetric vibrations of cylindrical shells made of functionally graded materials. Int. Appl. Mech. **51**, 654–663 (2015)
6. Wu, H., Yang, J., Kitipornchai, S.: Mechanical analysis of functionally graded porous structures: a review. Int. J. Struct. Stability Dyn. **20**, 2041015 (2020)
7. Rezaei, A.S.: Said, AR: Exact solution for free vibration of thick rectangular plates made of porous materials. J. Compos. Struct. **134**, 1051–1060 (2015)
8. Zhao, J., Wang, Q., Deng, X., Choe, K., Zhong, R., Shuai, C.: Free vibrations of functionally graded porous rectangular plate with uniform elastic boundary conditions. Compos. Part B Eng. **168**, 106–120 (2019)
9. Merdaci, S., Adda, H.M., Hakima, B., Dimitri, R., Tornabene, F.: Higher-order free vibration analysis of porous functionally graded plates. J. Compos. Sci. **5**(11), 305 (2021)
10. Duc, N.D., Quang, V.D., Nguyen, P.D., Chien, T.M.: Nonlinear dynamic response of functional graded porous plates on elastic foundation subjected to thermal and mechanical loads. J. Appl. Comput. Mech. **4**, 245–259 (2018)
11. Barati, M.R., Sadr, M.H., Zenkour, A.M.: Buckling analysis of higher-order graded smart piezoelectric plates with porosities resting on elastic foundation. Int. J. Mech. Sci. **117**, 309–320 (2016)
12. Zenkour, A.M.: A quasi-3D refined theory for functionally-graded single-layered and sandwich plates with porosities. Compos. Struct. **201**, 38–48 (2018)
13. Kiran, M.C., Kattimani, S.C., Vinyas, M.: Porosity influence on structural behaviour of skew functionally graded magneto-electro-elastic plate. Compos. Struct. **191**, 36–77 (2018)
14. Wang, Y.Q.: Zu, JW: Vibration behaviors of functionally graded rectangular plates with porosities and moving in thermal environment. Aerosp. Sci. Technol. **69**, 550–562 (2017)
15. Quang, V.D., Khoa, N.D., Duc, N.D.: The effect of structural characteristics and external conditions on the dynamic behavior of shear deformable FGM porous plates in thermal environment. J. Mech. Sci. Technol. **35**, 3323–3329 (2021)
16. Kumar, V., Singh, S.J., Saran, V., Harsha, S.: Vibration characteristics of porous FGM plate with variable thickness resting on Pasternak's foundation. Eur. J. Mech. A/Solids **85** (2021)
17. Kumar, R., Lal, A., Singh, B.N., Singh, J.: Meshfree approach on buckling and free vibration analysis of porous FGM plate with proposed IHHSDT resting on the foundation. Curve Layer Struct. **6**, 192–211 (2019)
18. Akavci, S.S.: An efficient shear deformation theory for free vibration of the functionally gradient thick rectangular plates on elastic foundations. Compos. Struct. **108**, 667–676 (2014)
19. Thai, H.T., Choi, D.H.: A refined shear deformation theory for free vibration of functionally graded plates on elastic foundation. Compos. Part B **43**, 2335–2347 (2012)
20. Mantari, J.L., Granados, E.V., Hinostroza, M.A., Soares, C.G.: Modeling advanced composite plates resting on elastic foundation by using a quasi-3D hybrid type HSDT. Compos. Struct. **118**, 455–471 (2014)
21. Vinh, P.V., Huy, L.Q.: Finite element analysis of functionally graded sandwich plates with porosity via a new hyperbolic shear deformation theory. Defence Technol. **18**, 490–508 (2021)
22. Wang, Y., Wu, D.: Free vibration of functionally graded porous cylindrical shell using a sinusoidal shear deformation theory. Aerosp. Sci. Technol. **66**, 83–91 (2017)
23. Li, H., Pang, F., Chen, H., Du, Y.: Vibration analysis of functionally graded porous cylindrical shell with arbitrary boundary restraints by using a semi-analytical method. Compos. Part B Eng. **164**, 249–264 (2019)
24. Li, H., Pang, F., Ren, Y., Miao, X., Ye, K.: Free vibration characteristics of functionally graded porous spherical shell with general boundary conditions by using first-order shear deformation theory. Thin-Walled Struct. **144**(5), 106331 (2019). https://doi.org/10.1016/j.tws.2019.10633125
25. Trinh, M.C., Kim, S.E.: A three variable refined shear deformation theory for porous functionally graded doubly curved shell analysis. Aerosp. Sci. Technol. **94**, 105356 (2019)
26. Jouneghani, F.Z., Dimitri, R., Bacciocchi, M., Tornabene, F.: Free vibration analysis of functionally graded porous doubly-curved shells based on the first-order shear deformation theory. Appl. Sci. **7**, 1252 (2017)

27. Zhao, J., Xie, F., Wang, A., Shuai, C., Tang, J., Wang, Q.: Vibration behavior of the functionally graded porous (FGP) doubly-curved panels and shells of revolution by using a semi-analytical method. Compos. Part B Eng. **157**, 219–238 (2019)
28. Alijani, F., Amabili, M., Karagiizis, K., Bakhtiari-Nejad, F.: Nonlinear vibration of functionally graded doubly curved shallow shells. J. Sound Vibr. **330**, 1432–1454 (2011)
29. Strozzi, M., Pellicano, F.: Nonlinear vibrations of functionally graded cylindrical shells. Thin-Walled Struct. **67**, 63–77 (2013)
30. Keleshteri, M.M., Jelovica, J.: Nonlinear vibration behavior of functionally graded porous cylindrical panels. Compos. Struct. **239**, 112028 (2020)
31. Liu, Y., Qin, Z., Chu, F.: Nonlinear forced vibrations of FGM sandwich cylindrical shells with porosities on an elastic substrate. Nonlin. Dyn. **104**, 1007–1021 (2021)
32. Ramteke, P.M., Panda, S.K., Patel, B.: Nonlinear eigenfrequency characteristics of multi-directional functionally graded porous panels. Compos. Struct. **279**, 114707 (2022)
33. Binh, C.T., Long, N.V., Tu, T.P., Minh, P.Q.: Nonlinear vibration of functionally graded porous variable thickness toroidal shell segments surrounded by elastic medium including the thermal effect. Compos. Struct. **255**, 112891 (2021)
34. Amir, M., Kim, S.-W., Talha, M.: Comparative study of different porosity models for the nonlinear free vibration analysis of the functionally graded cylindrical panels. Mech. Based Des. Struct. Mach. **50** (2022)
35. Trinh, M.C., Nguyen, D.-D., Kim, S.E.: Effect of porosity and thermomechanical loading on free vibration and nonlinear dynamic response of functionally graded sandwich shells with doubly curvature. J. Aerosp. Sci. Technol. **87**, 119–132 (2019)
36. Amir, M., Talha, M.: Nonlinear vibration characteristics of shear deformable functionally graded curved panels with porosity including temperature effects. Int. J. Press. Vessel. Pip. **172**, 28–41 (2019)
37. Gao, K., Gao, W., Wu, B., Wu, D., Song, C.: Nonlinear primary resonance of functionally graded porous cylindrical shells using the method multiple scales. Thin-Walled Struct. **125**, 282–293 (2018)
38. Rvachev, V.L.: The R-functions Theory and its Applications. Nauk, Dumka, Kyiv (1982). [in Russian]
39. Kurpa, L.V.: The R-functions method for solving linear bending and vibration problems of the shallow shells. NTU "KhPI", Kharkov (2009). [in Russian]
40. Kurpa, L.V., Shmatko, T.V., Awrejcewicz, J.: Linear and nonlinear free vibration analysis of laminated functionally graded shallow shells with complex plan form and different boundary conditions. Int. J. Nonlinear Mech. **107**, 161–169 (2018)
41. Kurpa, L.V., Shmatko, T.V.: Investigation of free vibrations and stability of functionally graded three-layer plates by using the R-functions theory and variational methods. J. Math. Sci. **249**, 496–520 (2020)
42. Awrejcewicz, J., Kurpa, L., Shmatko, T.: Application of the R-functions in free vibration analysis of FGM plates and shallow shells with temperature dependent properties. ZAMM Z. für Angew. Math. **101**, e202000080 (2021)

Chapter 23
Longitudinal Shear of Bimaterials with Interphase Thin Physically Nonlinear Layered Functional-Gradient Inhomogeneities

Roman Kushnir, Heorhiy Sulym, Yosyf Piskozub, and Roman Kaczynski

Abstract An analytical–numerical method for solving the class of problems of determining mechanical fields in composite structures with interfacial ribbon-like deformable multilayer physically nonlinear inhomogeneities under combined force and dislocation loading is proposed. The mathematical model of thin inclusion of material with arbitrary mechanical properties is constructed. The possibility of the existence of an imperfect contact along the part of the interface between the inclusion and the matrix, as well as between the layers of the inclusion, where there is surface energy or sliding with dry friction, is provided. Based on the method of jump functions, the stress–strain field in the vicinity of the inclusion at its interaction with concentrated forces and helical dislocations is calculated. The values of the generalized stress intensity coefficients for the asymptotic of the stress–strain fields in the vicinity of the ends of thin inhomogeneities are calculated, which can be used to calculate the stress concentration and local strength of the structure. Several new effects that can be used in the design of the structure of layers and modes of operation of such composites are revealed.

R. Kushnir · H. Sulym
Pidstryhach Institute for Applied Problems of Mechanics and Mathematics, National Academy of Science of Ukraine, Lviv, Ukraine
e-mail: dyrector@iapmm.lviv.ua

Y. Piskozub (✉)
Cracow University of Technology, Cracow, Poland
e-mail: yosyf.piskozub@pk.edu.pl

Ukrainian Academy of Printing, L'viv, Ukraine

R. Kaczynski
Bialystok University of Technology, Bialystok, Poland
e-mail: r.kaczynski@pb.edu.pl

H. Altenbach et al. (eds.), *Selected Problems of Solid Mechanics and Solving Methods*, Advanced Structured Materials 204, https://doi.org/10.1007/978-3-031-54063-9_23

23.1 Introduction

Microscopic, layered structures in fields such as microelectronics, biotechnology, energy, weaponry, etc. are gaining special attention in modern engineering and technology. Among the most important scientific projects, experts identify a significant increase in computer performance, restoration of human organs using reproduced tissues (obtained from 3D printers), and obtaining new structured materials created directly from given molecules and atoms. Quite often these inclusions are used as elements to reinforce structural parts of machines and structures or as fillers of composite materials. The theory and practice of the design and use of progressive composite materials with flat reinforcement provide indisputable evidence that their tensile strength in the transversal direction, in the case of unidirectional ribbon reinforcement, is between 50 and 75% of the strength in the longitudinal direction; the use of fibers typically yields only between 2 and 15% [5, 8, 41]. Reviews and monographs [5, 16, 24, 28, 30–32, 41] note the advantages of flat reinforcement, which improves manufacturability and the mechanical properties of the composite, increases the reinforcement factor and the resistance to leakage failure, and reduces the statistical variation of the designed properties. It is a testament to the big prospective application of composites with ribbon-like reinforcement. The use of external ribbon-like reinforcement in steel-reinforced concrete makes it possible to save between 15 and 45 percent of metal in comparison with reinforced concrete and pure metal structures. In addition, thin ribbon-like elements are a common phenomenon in micro- and nanostructures [5, 26, 27, 31, 35, 41, 46].

In the process of the exploitation of composites, the phenomenon of cracking and delamination is frequent, i.e., reinforcing heterogeneities can be in both ideal and non-ideal contact with the basic material, including at the interface of the media. Consideration of friction in the study of contact phenomena is one of the most pressing problems for mechanical engineering and materials science in the analysis of phenomena and processes occurring in the moving elements of machines, at various technological operations [11, 14, 17, 20, 25, 42]. Thus, friction may be accompanied by electrical, thermal, vibrational, and chemical processes that dampen internal dynamic processes, significantly affecting the intensity of the wear and tear of materials and, consequently, the reliability and durability of the structural elements made of them [10, 39].

Thin lamellar inhomogeneities are also a characteristic phenomenon at the interphase boundaries of crystalline grains arising during crystallization [24, 26, 28, 31, 45, 49]. In this regard, there is a need to provide mathematical modeling of nanostructure mechanics, which is still a pressing problem of materials science theory. At this stage of the development of mechanics, it is already possible to concentrate on the construction of the complex universal equations suitable for investigations of multiscale, including layered, structures, and the development of methods for their solution.

In such structures, each layer or their combination has its functional purpose [3–5, 9, 27, 33]. Thanks to multilayers [9], it is also possible to increase the service life of structures, and their use can significantly reduce material intensity and cost and increase the endurance of products. At the same time in a structure with thin layers, there is a concentration of stresses near places of change in the physical and mechanical characteristics of materials. Inhomogeneous structures with optimally varying physical and mechanical properties along with the thickness, known as functionally graded materials (FGMs) [14, 17, 21, 23, 25, 48], allow one to reduce such stress concentrations in the vicinity of the contact between the matrix and the interlayer by avoiding abrupt transitions in the properties of the components. A detailed review of the manufacturing techniques can be found in [2, 19, 42]. FGMs are often used in the coatings of structural elements to protect them from the harmful effects of temperature [3, 4, 6, 8, 13, 15, 22, 45, 47], etc. The complexity of the geometry of structural elements and consideration of imperfections in the contact of their components stimulate the process of improving mathematical models of FGMs to ensure their qualitative design both in terms of mechanical strength [1, 10–12, 18, 20, 27, 29, 30, 32, 33, 35–38, 40, 41, 49] and in terms of consideration of thermal, magnetic, and piezoelectric loading factors [29, 30]. The use of the FGMs seems to be one of the most effective materials in the realization of sustainable development in industries.

An important aspect of strength research, including tensile strength, for such structures, is to improve their strength criteria and to determine such key parameters as stress intensity factors (SIF) in the points of singularity. Moreover, since we consider thin inhomogeneities not only in the form of classical cracks but also thin cavities filled with an arbitrary elastic or nonlinearly elastic material, it makes sense to claim that the theory of thin inclusions is an essential generalization of the crack theory and the so-called generalized SIFs, which characterize the distribution of stress and displacement fields, is analogous parameters of fracture mechanics for the theory of thin inhomogeneities [34, 41].

The need to consider the aforementioned geometric nonlinearity (a priori unknown contact spots) significantly complicates the process of problem-solving and requires the use of various approximate methods, even for bodies of simple geometry [28, 49]. However, these methods do not always allow for the correct consideration of thin-walled heterogeneity, nor do they guarantee the accuracy of a solution for load optimization in future applications. During the operation of materials that have the structure that is being considered in this work, especially when loaded by concentrated force and dislocation factors, undesirable critical states of adhesion loss between the constituent parts may arise. This can be avoided by locating the loading points in the so-called "safety zone" when the shear stresses on the contact surfaces at each of their points do not exceed the critical value.

This work aims to develop a numerical–analytical method to study structures with ribbon-like nonlinear deformable elements, with possible frictional contact between the constituent elements; moreover, it aims to study the mechanical effects of loading by force and dislocation factors on its strength.

23.2 Formulation of the Problem

Consider an unbounded isotropic bulk consisting of two half-spaces with elastic constants E_k, ν_k, and G_k $(k = 1, 2)$, pressed to the interface by normal stresses $\sigma_{yy}^{\infty} < 0$, and under the action of uniformly distributed at infinity stresses σ_{xxk}^{∞} $(k = 1, 2)$. The external longitudinal shear load is determined by the stresses σ_{yz}^{∞} and σ_{xzk}^{∞} uniformly distributed at infinity, concentrated intensity forces Q_k, and screw dislocations with Burger's vector component b_k at points $\varsigma_{*k} \in S_k$ $(k = 1, 2)$, oriented along the axis in such a way that their action causes a quasi-static antiplane SSS in the body. To ensure the straightness of the material interface at infinity, the stresses must satisfy the conditions

$$\sigma_{xz2}^{\infty} G_1 = \sigma_{xz1}^{\infty} G_2, \quad \nu_2 \sigma_{yy}^{\infty} - (1 - \nu_2)\, \sigma_{xx2}^{\infty} / G_2 = \nu_1 \sigma_{yy}^{\infty} - (1 - \nu_1)\, \sigma_{xx1}^{\infty} / G_1 .$$

We will study the stress–strain state (SSS) of the body section with a plane xOy perpendicular to the direction Oz of its longitudinal displacement. The plane sections of half-spaces perpendicular to this axis form two half-planes S_k $(k = 1, 2)$ and the abscissa axis corresponds to the interface $L \sim x$ between them (Fig. 23.1). At the interface of half-spaces (plane xOz), there is a tunnel section $L' = [-a; a]$ in the direction of the shear axis Oz, in which a certain object of general thickness $2h$ $(h \ll a)$ is inserted.

In general, the formulated problem contains three component modules: an external problem (the SSS in the matrix), an internal problem (the SSS in the inclusion), and the contact conditions which relate them.

To solve the **external problem**, it is convenient to apply the method referenced in [34–38, 40, 41], which uses the well-known jump function method (JFM). According to its paradigm, the influence of thin inclusion in the matrix is modeled by jumps of the components of the stresses and displacement vectors on the line L' :

$$\begin{aligned} &\left[\sigma_{yz}\right]_h \cong \sigma_{yz}^{-} - \sigma_{yz}^{+} = f_3(x), \\ &\left[\frac{\partial w}{\partial x}\right]_h \cong \frac{\partial w^{-}}{\partial x} - \frac{\partial w^{+}}{\partial x} = \left[\frac{\sigma_{xz}}{G}\right]_h \equiv \frac{\sigma_{xz}^{-}}{G_1} - \frac{\sigma_{xz}^{+}}{G_2} = f_6(x), \quad x \in L', \end{aligned} \tag{23.1}$$

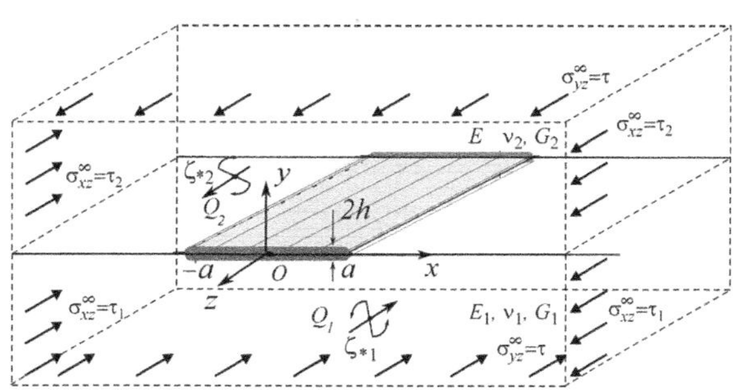

Fig. 23.1 Geometry and load pattern of the problem

$$f_3(x) = f_6(x) = 0, \text{ if } x \notin L'. \tag{23.2}$$

Hereinafter marked: $[\bullet]_h = \bullet(x, -h) - \bullet(x, +h)$, $\langle\bullet\rangle_h = \bullet(x, -h) + \bullet(x, +h)$; $[\bullet] = \bullet(x, -0) - \bullet(x, +0)$, $\langle\bullet\rangle = \bullet(x, -0) + \bullet(x, +0)$, superscripts "+" and "−" correspond to the limit values of functions at the upper and lower edges of the line L.

Applying the results of Sect. 20.2 in [41], we can obtain the dependencies according to which components of the stress tensor and the derivatives of the displacement vector inside an unbounded plane $S = S_1 \cup S_2$ obtain the form:

$$\begin{aligned}\sigma_{yzk}(\varsigma) + i\sigma_{xzk}(\varsigma) = {} & \sigma^0_{yzk}(\varsigma) + i\sigma^0_{xzk}(\varsigma) \\ & - C\frac{1}{\pi}\int\limits_{L'}\frac{f_6(\xi)d\xi}{\xi - \varsigma} + ip_k\frac{1}{\pi}\int\limits_{L'}\frac{f_3(\xi)d\xi}{\xi - \varsigma}, \\ p = \frac{1}{G_1 + G_2}, \quad & p_k = G_k p, \quad C = G_1 G_2 p, \\ & (\varsigma = x + iy, \ \varsigma \in S_k, \ r = 3, 6, \ k = 1, 2),\end{aligned} \tag{23.3}$$

and their boundary values on the upper and lower banks of the line L are the following:

$$\begin{aligned}\sigma^{\pm}_{yzk}(x, \pm h) &= \mp p_k f_3(x) - C\frac{1}{\pi}\int\limits_{L'}\frac{f_6(\xi)d\xi}{\xi - x} + \sigma^{0\pm}_{yz}(x, \pm h), \\ \sigma^{\pm}_{xzk}(x, \pm h) &= \mp p_k f_6(x) + C\frac{1}{\pi}\int\limits_{L'}\frac{f_3(\xi)d\xi}{\xi - x} + \sigma^{0\pm}_{xz}(x, \pm h).\end{aligned} \tag{23.4}$$

The values, further marked with the index "0" on top, correspond to the bulk's SSS model without inhomogeneities (inclusions, cracks, etc.) under the corresponding external load (homogeneous solution). Hereinafter, the notations [35, 40] are used:

$$\begin{aligned}& \sigma^0_{yzk}(\varsigma) + i\sigma^0_{xzk}(\varsigma) \\ & \quad = \tau + i\left\{\tau_k + D_k(\varsigma) + \frac{G_k - G_j}{G_1 + G_2}\overline{D_k}(\varsigma) + \frac{2G_k}{G_1 + G_2}D_j(\varsigma)\right\}, \\ & D_k(\varsigma) = -\frac{Q_k + iG_k b_k}{2\pi(\varsigma - \varsigma_{*k})} \quad (\varsigma \in S_k, \ k = 1, 2, \ j = 3 - k).\end{aligned} \tag{23.5}$$

Additional balance conditions must be imposed on the solution of the external problem:

$$\begin{aligned}\int\limits_{-a}^{a} f_3(\xi)d\xi &= 2h\left(\sigma^{\text{in}}_{xz}(a) - \sigma^{\text{in}}_{xz}(-a)\right), \\ \int\limits_{-a}^{a} f_6(\xi)d\xi &= [w](a) - [w](-a).\end{aligned} \tag{23.6}$$

The **contact** between the bimaterial medium components along a line $L\backslash L'$ and between the medium components and inclusion at L' is supposed to be different:

1. Mechanically perfect:

$$\begin{array}{lll} w(x,+0)=w(x,-0), & \sigma_{yz2}(x,+0)=\sigma_{yz1}(x,-0), & x\in L\backslash L' \\ w(x,-h)=w^{\text{in}}(x,-h), & \sigma_{yz}^{\text{in}}(x,-h)=\sigma_{yz1}(x,-h), & x\in L'\backslash L''^{-}, \\ w(x,h)=w^{\text{in}}(x,h), & \sigma_{yz}^{\text{in}}(x,h)=\sigma_{yz2}(x,h), & x\in L'\backslash L''^{+}. \end{array} \quad (23.7)$$

The superscript "in" denotes the values describing the inclusion material's SSS.

2. Contact with additional boundary tension:

$$\begin{array}{l} w^{\text{in}}(x,\pm h)=w(x,\pm h), \\ \sigma_{yz}^{\text{in}}(x,\pm h)=\sigma_{yzk}(x,\pm h)-T_k, \quad k=\{2,1\}. \end{array} \quad (23.8)$$

3. *Stick–slip contact with friction* between the inclusion and the matrix within $\{x, y-h\}$, $\{x, y+h\}$ in some a priori unknown areas $x\in L_f^{\pm}\subset L'$, $L_f^{\pm}=[b^{\pm};c^{\pm}]$ ($|b^{\pm}|\leqslant|a|$, $|c^{\pm}|\leqslant|a|$) on the upper and lower banks. At the contact areas $L_f{}^{\pm}$, we assume stick–slip contact conditions [9, 16, 18, 36], wherein mutual slippage of contacting body surfaces can start, causing heat release, energy dissipation, wear [36, 39, 47], etc., and that at all points of $L_f{}^{\pm}$ tangential stresses (friction forces) are equal to

$$\begin{array}{l} \sigma_{yz}^{\text{in}}(x,h)=\sigma_{yz2}(x,h)=-\operatorname{sign}\left(w^{\text{in}}(x,\pm)-w(x,h)\right)\tau_{yz}^{\max}(x), \\ \sigma_{yz}^{\text{in}}(x,-h)=\sigma_{yz1}(x,-h)=-\operatorname{sign}\left(w^{\text{in}}(x,-h)-w(x,-h)\right)\tau_{yz}^{\max}(x), \end{array} \quad (23.9)$$

where $\tau_{yz}^{\max}(x)=-\alpha\sigma_{yy}(x)$ ($\sigma_{yy}<0$); α is the sliding friction coefficient. Outside the area $L_f^{\pm}$, in the absence of a slip on the inclusion surface, the tangential stresses may not exceed the allowable maximum

$$\left|\sigma_{yz}(x,\pm h)\right|\leq\tau_{yz}^{\max}(x) \quad \left(\sigma_{yy}<0\right), \quad (23.10)$$

and there is no mutual displacement of contact surfaces (displacement jump). The sign (direction of action) of the tangential stresses is chosen depending on the sign of the displacement difference $w^{\text{in}}(x,\pm h)-w(x,\pm h)$ at the point in question at L''.

In the case of normal pressure in the first approximation, we obtain

$$\tau_{yz}^{\max}(x)=-\alpha\sigma_{yy}^{\infty}. \quad (23.11)$$

In the case $\tau_{yz}^{\max}=0$ there is smooth contact.

It is not difficult to obtain a more exact expression by constructing the solution of the corresponding plane problem. The application of friction law in classical form (23.11) gives an opportunity, of course, to simplify boundary conditions for the main problem, but the choice of complex friction models [36, 39, 40], considering wear, will not fundamentally complicate the solution process.

To solve the **internal problem**, it is necessary to build an appropriate mathematical model that most adequately reflects the physical and mechanical properties of the thin inclusion material. The construction of such a mathematical model of the thin inclusion layer should eventually reveal the relation between the stress–strain parameters inside the inclusion and on its external surface as the influence functions $\sigma_{yz}^{\text{in}}(x, \pm h)$ and $w^{\text{in}}(x, \pm h)$, which will be used in the further solution of the problem [34, 35, 41].

Let us consider a general mathematical model of thin inclusion. We suppose that inclusion material mechanical properties can be different (orthotropy) in different directions. They are characterized by a constitutive nonlinear equation of a general form:

$$\frac{\partial w^{\text{in}}}{\partial s} = \varpi_s\left(\sigma_{xz}^{\text{in}}, \sigma_{yz}^{\text{in}}\right), \quad s = \{x, y\} \tag{23.12}$$

where the monotonic function $\varpi_s(\sigma_{xz}^{\text{in}}, \sigma_{yz}^{\text{in}})$ is chosen either based on general theoretical arguments or as a function approximating the experimental results.

The mathematical model of thin inclusion is presented in the form of the so-called interaction conditions [41] equivalent to the conditions of imperfect contact between the matrix surfaces adjacent to the inclusion. The applied method of modeling the influence of thin object is based on the scheme of integration of the equations used to describe the physics and mechanical state of the material of inclusion over its volume with subsequent taking into account the smallness of the thickness of inclusion and the formulated constitutive relations as follows:

The balance equations at the inner points of the inclusion $\partial\sigma_{xz}^{\text{in}}/\partial x + \partial\sigma_{yz}^{\text{in}}/\partial y = 0$ are integrated over x within the limits $[-a, x]$ and average the result over the thickness $y \in [-h, h]$. Because of the smallness of the thickness of the inclusion

$$\frac{\partial w^{\text{in}}}{\partial y}(x, h) + \frac{\partial w^{\text{in}}}{\partial y}(x, -h) \simeq \frac{w^{\text{in}}(x, h) - w^{\text{in}}(x, -h)}{h} = -\frac{\left[w^{\text{in}}\right]_h}{h},$$

we get the following equations of the mathematical model of thin physically nonlinear inclusions:

$$\begin{gathered}
-\frac{\left[w^{\text{in}}\right]_h}{h} = \left\langle \varpi_s\left(\sigma_{xz}^{\text{in}}, \sigma_{yz}^{\text{in}}\right)\right\rangle, \\
\left\langle \varpi_x^{-1}\left(\frac{\sigma_{xz}^{\text{in}}}{G_x^{\text{in}}}, \frac{\sigma_{yz}^{\text{in}}}{G_y^{\text{in}}}\right)\right\rangle_h - 2\sigma_{xz}^{\text{in}}(-a) - \frac{1}{h}\int_{-a}^{x}\left[\sigma_{yz}^{\text{in}}\right]_h(\xi)d\xi = 0.
\end{gathered} \tag{23.13}$$

The function $\varpi_s(\sigma_{xz}^{\text{in}}, \sigma_{yz}^{\text{in}})$ can be simplified if we neglect the mutual influence of stresses: $\varpi_s(\sigma_{xz}^{\text{in}}, \sigma_{yz}^{\text{in}}) \sim \varpi_s(\sigma_{sz}^{\text{in}})$ (model of nonlinear Winkler-type springs). Then relations (23.12) can be rewritten in a simpler form as follows:

$$\sigma_{sz}^{\text{in}} = G_s^{\text{in}}(\sigma_{sz}^{\text{in}})\frac{\partial w^{\text{in}}}{\partial s}, \quad \{s = x, y\} \tag{23.14}$$

with given shear moduli $G_s^{\text{in}}(\sigma_{sz}^{\text{in}})$ varying in a certain way. Thus, we can use, e.g., Hooke's classical law of linear elasticity; the models of plastic deformation proposed by Bach–Schule, Sokolovskiy, Ramberg–Osgood, and Il'yushin; the model of deformation specified by a function approximating the empirical data obtained for a specific material; and the model of linear elastoplastic deformation with hardening [35, 36, 40]

$$\begin{aligned}
\frac{\partial w^{\text{in}}}{\partial s} &= \frac{\sigma_{sz}^{\text{in}}}{G_{0s}}, & \left|\sigma_{sz}^{\text{in}}\right| &< \tau_{\text{yield}}, \quad s = \{x, y\} \\
\frac{\partial w^{\text{in}}}{\partial s} &= \left(\sigma_{sz}^{\text{in}} - \tau_{\text{yield}}\right)\frac{G_{0s} - G_{1s}}{G_{0s}G_{1s}}, & \left|\sigma_{sz}^{\text{in}}\right| &\geqslant \tau_{\text{yield}}.
\end{aligned} \tag{23.15}$$

We also note that considering the G_s^{in} in the relation (23.14) to be dependent on the coordinates: $G_s^{\text{in}} = G_s^{\text{in}}(x)$ we can, using the mentioned model, simulate the inclusion made with FGM.

To illustrate, let us consider one of the particular variants of the functional gradient of the inclusion material in the transverse direction—the piecewise linear one:

$$G_x(x) = G_y(x) = \begin{cases} (G_{01} - G_0)\,x + G_{01}, & x \in [-a, 0] \\ (G_{02} - G_{01})\,x + G_{01}, & x \in [0, a], \end{cases} \tag{23.16}$$

where G_0 G_{01}, and G_{02} are some given constants.

For each indicated model of deformation of the form (23.14), given relation (23.13), we can get the following mathematical model of a physically orthotropic nonlinear thin inclusion:

$$\begin{aligned}
&G_x^{\text{in}}(\bullet)\left\langle\frac{\partial w^{\text{in}}}{\partial x}\right\rangle_h(x) - 2\sigma_{xz}^{\text{in}}(-a) - \frac{1}{h}\int_{-a}^{x}\left[\sigma_{yz}^{\text{in}}\right]_h(\xi)d\xi = 0, \\
&G_y^{\text{in}}(\bullet)\left[w^{\text{in}}\right]_h(x) + h\left\langle\sigma_{yz}^{\text{in}}\right\rangle_h(x) = 0, \quad (\bullet) = (\sigma_{xz}^{\text{in}},\ \ \sigma_{yz}^{\text{in}})(x).
\end{aligned} \tag{23.17}$$

The functional gradient of material properties along the thickness (y-coordinate) due to the thinness of inclusion is more difficult to model directly. Therefore, it is advisable to consider a multilayer thin inclusion, each of the layers of which has different physical and mechanical properties. By combining these properties in the desired sequence, it is possible to model any given functional gradient of the inclusion material along its thickness. Let us analyze the methodology for

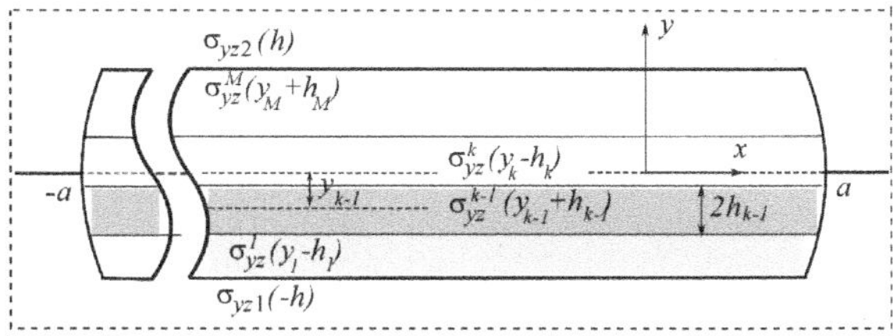

Fig. 23.2 Multilayered inclusion

constructing a mathematical model for the case of a multilayer package of thin inclusion layers. Consider a package of M different thin plane-parallel layers $\{x \in L';\ y \in [y_K - h_K;\ y_K + h_K],\ K = 1, \ldots, M\}$ of thickness $2h_K$ $(2h = 2\sum_{K=1}^{M} h_K)$, $y_1 - h_1 = -h$, $y_M + h_M = h$ with orthotropic mechanical properties $G_y^{\mathrm{in}K}$ and $G_x^{\mathrm{in}K}$ in the direction of two axes (Fig. 23.2).

Using the approach (23.12)–(23.17) to building a model of each of the layers of the multilayer inclusion, the following model can be written:

$$\frac{G_x^{\mathrm{in}K}}{2}\left\langle\frac{\partial w^{\mathrm{in}K}}{\partial x}\right\rangle_{y_K,h_K} - \sigma_{xz}^{\mathrm{in}K}(-a) - \frac{1}{2h_K}\int_{-a}^{x}\left[\sigma_{yz}^{\mathrm{in}K}\right]_{y_K,h_K}(\xi)d\xi = 0, \quad (23.18)$$

$$-\frac{\left[w^{\mathrm{in}K}\right]_{y_K,h_K}}{h_K} = \frac{\left\langle\sigma_{yz}^{\mathrm{in}K}\right\rangle_{y_K,h_K}}{G_y^{\mathrm{in}K}} \quad (K = 1, \ldots, M), \quad (23.19)$$

where $G_x^{\mathrm{in}K}$ and $G_y^{\mathrm{in}K}$ are the shear moduli of each of the layers given by the parameters y_K and h_K $(K = K = 1, \ldots, M)$ and which together with relations (23.19) fully describe the thin M-layered inclusion model written in the values of the stress–strain behavior of the inclusion package materials. Instead of the displacement jump, in many cases, it is convenient to use the formula for the jump of the strain components:

$$\left[w^{\mathrm{in}K}\right]_{y_K,h_K}(x) = \left[w^{\mathrm{in}K}\right]_{y_K,h_K}(-a) + \int_{-a}^{x}\left[\frac{\partial w^{\mathrm{in}K}}{\partial x}\right]_{y_K,h_K}(\xi)d\xi.$$

At each inclusion layer, the balance conditions must be satisfied:

$$\int_{-a}^{a} f_{3K}(\xi)d\xi = 2h_K\left(\sigma_{xzK}^{\mathrm{in}}(a) - \sigma_{xzK}^{\mathrm{in}}(-a)\right), \quad (23.20)$$

$$\int_{-a}^{a} f_{6K}(\xi)d\xi = \left[w^{\mathrm{in}K}\right]_{y_K,h_K}(a) - \left[w^{\mathrm{in}K}\right]_{y_K,h_K}(-a). \quad (23.21)$$

The partial cases of the model (23.13) and (23.17) of the following form are satisfied and coincide with those known in the literature.

- All layers are the same:

$$G_y^{\mathrm{in}K} = G_y^{\mathrm{in}}, \quad G_x^{\mathrm{in}K} = G_x^{\mathrm{in}}, \quad (K = 1, \dots, M);$$

- No inclusion or crack:

$$h_K \to 0, \quad G_y^{\mathrm{in}K} = G_y^{\mathrm{in}} \to 0, \quad G_x^{\mathrm{in}K} = G_x^{\mathrm{in}} \to 0 \quad (K = 1, \dots, M);$$

- Perfectly rigid homogeneous inclusion:

$$G_y^{\mathrm{in}K} = G_y^{\mathrm{in}} \to \infty, \quad G_x^{\mathrm{in}K} = G_x^{\mathrm{in}} \to \infty.$$

23.3 Materials and Methods

The above theory makes it possible to solve a large class of problems of studying the stress–strain state of structures with thin physically nonlinear, layered, or FGM inclusions. Considering possible nonlinear contact conditions of constituent elements of structures, such as frictional slipping or the presence of surface stresses, significantly expands this class of problems but also complicates the methods of their solution.

The classical approach, implemented in the JFM, is to substitute the boundary expressions for stresses and displacements of the structure constituent elements at their contact area into the specified contact conditions. This allows us to obtain a compact, but not always an easy-to-solve system of singular integral equations.

In general, the formulated problem contains three component modules: an internal problem (the stress–strain state in the inclusion), an external problem (the stress–strain state in the matrix), and the contact conditions, which relate them. We propose a different approach to solving such a problem, which can be called a structurally modified JFM. The idea is to combine all equations into a global system without substituting the boundary conditions (23.7)–(23.10) into the model equation (23.13), (23.17), (23.19), (23.20) and limit values of matrix components of the stress–strain relations (23.4). Only one of the model equations (23.13), (23.17), (23.19), (23.20), and one of the boundary conditions (23.7)–(23.10) can be fulfilled simultaneously.

Furthermore, it is convenient to solve this system of equations by any numerical–analytical method, for example, by the collocation method. Submitting the system of equations (23.1)–(23.12) in discrete form in the set of collocation points (x_n, ($n = 1, \dots, N$), we obtain the system of $6N$ linear algebraic equations (SLAE) for the determination of $6N$ unknowns $\sigma_{yz}^{\mathrm{in}}(x_n, \pm h)$, $\partial w^{\mathrm{in}}/\partial x(x_n, \pm h)$, $f_r(x_n)$ ($r = 3, 6, n = 1, \dots, N$). Of course, the number of unknowns for solving the problem increases, which is not crucial with modern computational capabilities. However, the construction of SLAEs is considerably simplified, as its modularity allows for making independent changes in separate modules with significantly less effort than constructing and solving new classes of problems with considerably more complicated parameters. To illustrate this approach, we consider the solutions to several special problems of determining the stress–strain state and strength parameters of structures with thin physically nonlinear, or FGM inclusions under possible frictional slip and surface stresses and shear loading.

An important aspect of strength research, including tensile strength, for such structures, is to improve their strength criteria. In fracture mechanics, it is common to use the stress intensity factor (SIF) K_3 to describe the asymptotic of the SSS in the vicinity of the crack tip [6–8, 22, 41, 43]. For the case of a thin elastic inclusion, this is not sufficient [34]. The introduction of a system of polar coordinates $(r, \theta)$with the origin near the right or the left tip of the inclusion $z = \pm r \exp(i\theta) \pm a$ makes it possible to obtain two-term asymptotic expressions for the distribution of the stresses and displacements in the vicinity of the tips ($|z_1| \ll 2a$) [34]). Consider the generalized stress intensity factors (GSIF) introduced by the expression:

$$K_{31} + i K_{32} = \lim_{r \to 0\,(\theta=0,\pi)} \sqrt{2\pi r}\left(\sigma_{yz} + i\sigma_{xz}\right). \tag{23.22}$$

Using these coefficients, it is easy to get asymptotic expansions for stresses in the vicinity of the end of the inclusion with the required accuracy. Thus, stresses in the nearest vicinity of the inclusion tips $x = \pm a$ may be obtained by the formula

$$\begin{aligned}\left\|\sigma_{yzk}, \sigma_{xzk}\right\|^{\pm}(r, \theta) = \frac{1}{2\sqrt{2\pi r}} \sum_{m=1}^{2} L_m^{\pm}\left(\theta, G_k, G_x^{\text{in}}(\pm a), G_y^{\text{in}}(\pm a)\right) K_{3m}^{\pm} \\ + R(\pm a) + O\left(r^{3/2}\right) \quad (k = 1, 2),\end{aligned} \tag{23.23}$$

where matrix parameters $L_m^{\pm}(\bullet)$ and $R(\pm a)$ are discussed in [34].

23.4 Numerical Results and Discussion

23.4.1 Longitudinal Shear of Bimaterial with Thin Interfacial Nonlinear Elastic or Transversely FGM Inclusion (Perfect Contact)

Suppose that the inclusion material (Fig. 23.3) obeys the laws of deformation diagrams (23.14) according to the relations (23.15) or (23.16) [35, 38, 40]. Assuming the inclusion contact with the matrix is mechanically perfect (23.7) and applying the mathematical model (23.17) we can obtain a system of singular integral equations (SSIE) with variable coefficients

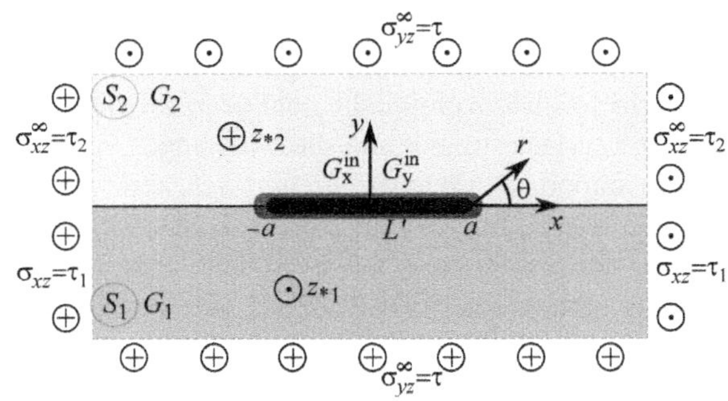

Fig. 23.3 Geometry and loading scheme of the problem

$$
\begin{aligned}
(p_2 - p_1) f_6(x) + 2pg_3(x) - \frac{1}{hG_x^{\text{in}}(\bullet)} \int_{-a}^{x} f_3(\xi)d\xi &= F_3(x) \\
(p_2 - p_1) f_3(x) + 2Cg_6(x) - \frac{G_y^{\text{in}}(\bullet)}{h} \int_{-a}^{x} f_6(\xi)d\xi &= F_6(x),
\end{aligned}
\tag{23.24}
$$

$$
\begin{aligned}
F_3(x) &= \frac{2}{G_x^{\text{in}}(\bullet)} \sigma_{xz}^{\text{in}}(-a) - \left(\sigma_{xz2}^{0}(x)/G_2 + \sigma_{xz1}^{0}(x)/G_1\right), \\
F_6(x) &= \left\langle \sigma_{yz}^{0} \right\rangle (x) - G_y^{\text{in}}(\bullet) \left\langle \frac{\sigma_{yzk}^{0}(x)}{G_k} \right\rangle - \frac{G_y^{\text{in}}(\bullet)}{h} \left[w^0\right](-a),
\end{aligned}
$$

with the additional conditions on power balance and unambiguity of displacements while going around the thin defect

$$
\begin{aligned}
\int_{-a}^{a} f_3(\xi)d\xi &= 2h\left(\sigma_{xz}^{\text{in}}(a) - \sigma_{xz}^{\text{in}}(-a)\right) \\
\int_{-a}^{a} f_6(\xi)d\xi &= [w](a) - [w](-a).
\end{aligned}
\tag{23.25}
$$

SSIE (23.24)–(23.25) is reduced to a system of linear algebraic equations to the unknown coefficients of the decomposition of the pre-dismeasured jump functions $f_r(x)$ $(r = 1, 2)$ in a series of orthogonal polynomials of Jacobi or Chebyshev and then solved by collocation method. Dependence $G_x^{\text{in}}(\sigma_{xz}^{\text{in}})$ and $G_y^{\text{in}}(\sigma_{yz}^{\text{in}})$ on current SSS generates serious complications in solving the problem because of its changeableness along L and step-by-step loading. This complication is solved by applying a recursively iterative approach when the stresses and strains obtained at each collocation point are refined to correspond to the deformation diagram (23.14) at each loading step. The computational procedure is stable and convergent. To achieve a relative solution accuracy of 1% in the worst case of rigid inclusion, six iteration steps were sufficient.

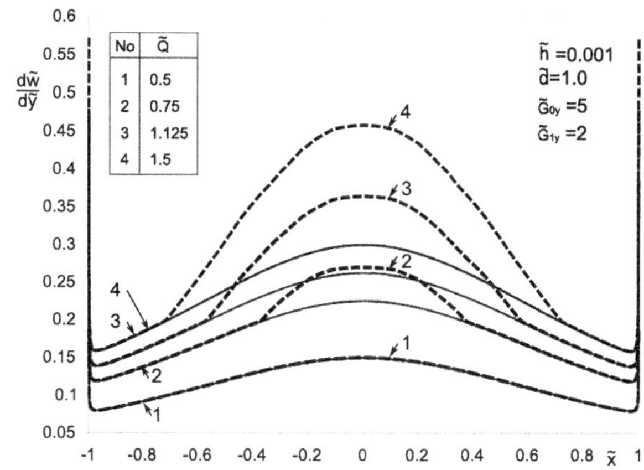

Fig. 23.4 Distribution of elastic-yield deformations in the inclusion (bar line) in comparison with perfectly elastic case (solid line) while loading

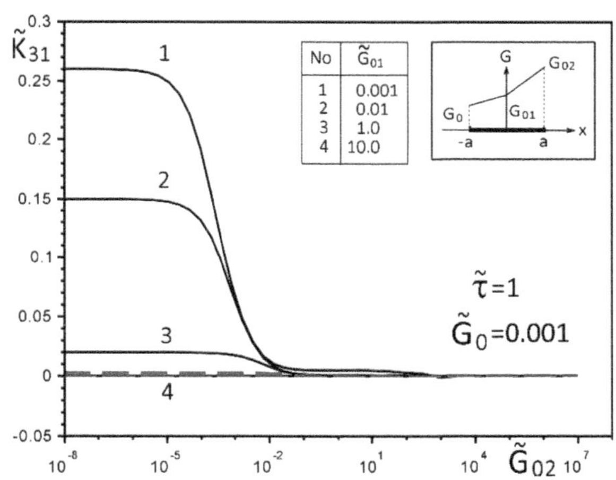

Fig. 23.5 Influence of the parameters $\tilde{G}_0$, $\tilde{G}_{01}$, $\tilde{G}_{02}$ on the GSIF $\tilde{K}_{31}^{+}$ under the load by uniformly distributed on infinity stress $\sigma_{yz}^{\infty} = \tau$

The numerical analysis of the solution to the posed problem is performed for a special case of equality of the elastic characteristics of half-spaces $G_1 = G_2 = G = G_{av}$ for the stepwise changes in the load by point forces $\tilde{Q} = Q/aG_{av}$ ($\tilde{Q}_2 = -\tilde{Q}_1 = \tilde{Q}$, $\tilde{z}_2 = -\tilde{z}_1 = i\tilde{d}$) in the global loading–unloading cycles caused by the changes in Q: $0, 1.5G_{av}, -1.5G_{av}, 1.5G_{av}, \ldots$. The comparison of the results of applications of the constitutive dependencies for the elastoplastic (dashed lines) deformation with hardening (23.15) and for the perfectly elastic deformation (continuous lines) is displayed in Fig. 23.4. It is well visible that, upon the attainment of a certain critical level of $\tilde{Q}$, we observe the onset of yield deformation localized along $L'' = [-b; b]$.

Figures 23.5 and 23.6 illustrate the dependence of the stress–strain behavior of the matrix in the inclusion vicinity on the variation of the parameters G_0, G_{01}, and G_{02}, the values of which were chosen to reveal a qualitative picture of the FGM effect on the stress–strain parameters. It can be immediately concluded from Fig. 23.5 that under the load $\tilde{\tau}$ the dimensionless K_{31}^{+} are expected to decrease with increasing shear

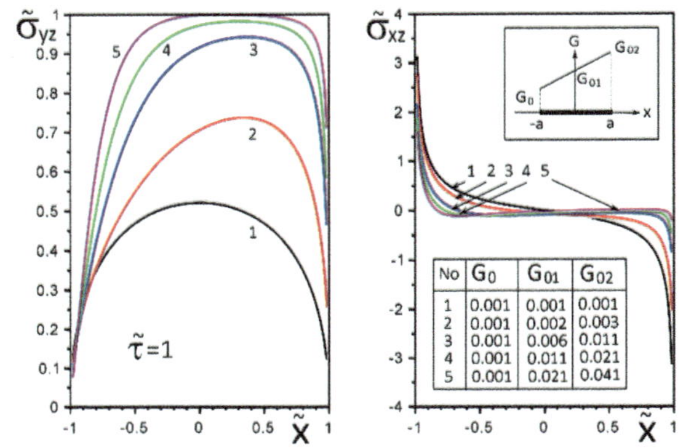

Fig. 23.6 Stress distribution along with the upper interface (inclusion–matrix half-space S_2) with a linear distribution of material stiffness

moduli of any part of the inclusion, while K_{32}^{+} appear and increase with increasing load $\tilde{\tau}_k$.

The effect of changes in the moduli $G_x(x)$ and $G_y(x)$ on the stresses $\tilde{\sigma}_{yz}$ and $\tilde{\sigma}_{xz}$ on the inclusion surface is more obvious if we choose a linear growth law for them along the inclusion axis (Fig. 23.6). The magnitude of the surface stresses increases significantly in the stiffer part of the inclusion. The larger the stiffness gradient, the more significant the increase.

The calculations of the stress–strain field components for simple cases of the functional dependence of the shear moduli of inclusion material have demonstrated the expected qualitative picture of their effect on the variation of the FGM parameters. In particular, (1) the stress magnitude increases significantly in the vicinity of the inclusion regions with increased stiffness; (2) the combination of the inclusion materials from parts with piecewise linear mechanical characteristics may lead to partial unloading of the inclusion and matrix in their softer part in the vicinity of the breaking point of the gradient dependence of the inclusion material parameters; (3) the contrast of the stress field changes of the inclusion and matrix is proportional to the increase in the gradient dependence.

23.4.2 Longitudinal Shear of Bimaterial with Thin Interfacial Nonlinear Elastic or Transversely FGM Microinclusion (Effect of Surface Tension)

To investigate the influence of the presence of surface stresses on the matrix–inclusion interface, the previous problem statement should be complicated by the contact conditions (23.8). The nature of the surface forces can be different at the macro-, micro-, or nanolevel and therefore can be modeled by different functional dependencies. In the work [36], the value of surface forces is taken as a dependence $\tilde{T}_2 = \tilde{T}_1 = k_T(\tilde{G}_{\text{in}})^{\alpha}$ which, due to its two-parameter nature, can approximate the value of surface forces of different natures. As a result of using model equations (23.17) with contact conditions (23.8), we obtain a similar to (23.24) and (23.25) SSIE, which differs only on the right-hand sides, in which there is an addition depending on the magnitude of the surface forces:

$$
\begin{aligned}
F_3(x) &= \frac{2}{G_x^{\text{in}}(\bullet)}\sigma_{xz}^{\text{in}}(-a) - \left(\frac{\sigma_{xz2}^0(x)}{G_2} + \frac{\sigma_{xz1}^0(x)}{G_1}\right) - \frac{x+a}{hG_x^{\text{in}}(\bullet)}(T_1 - T_2),\\
F_6(x) &= \left\langle \sigma_{yz}^0 \right\rangle(x) - G_y^{\text{in}}(\bullet)\left\langle \frac{\sigma_{yzk}^0(x)}{G_k} \right\rangle - \frac{G_y^{\text{in}}(\bullet)}{h}\left[w^0\right](-a) - T_1 - T_2,
\end{aligned}
\tag{23.26}
$$

and one of the additional conditions of balance

$$
\int_{-a}^{a} f_3(\xi)d\xi = 2h\left(\sigma_{xz}^{\text{in}}(a) - \sigma_{xz}^{\text{in}}(-a)\right) + 2a\left(T_1 - T_2\right). \tag{23.27}
$$

To assess the impact of the inclusion on SSS, it is convenient to analyze some characteristics, in particular, energy and stress intensity factors. The expression for the inclusion deformation energy is

$$
W^d = \int_{L'} \left(\sigma_{yz2}(x,h)w(x,h) - \sigma_{yz1}(x,-h)w(x,-h)\right)dx, \tag{23.28}
$$

and it can be somewhat simplified in the case of soft inclusion.

$$
\begin{aligned}
W^d &= W_0^d + W_L^d,\\
W^d &= -\frac{1}{2}\int_{L'} \left\langle \sigma_{yzk}\right\rangle_h (x)[w]_h(x)dx,\\
W_L^d &= -\frac{1}{2}(T_2 + T_1)\int_{L'} [w]_h(x)dx,\ W_0^d = -\frac{1}{2}\int_{L'} \left\langle \sigma_{yz}^{\text{in}}\right\rangle_h (x)[w]_h(x)dx,
\end{aligned}
\tag{23.29}
$$

where W_0^d is the deformation energy of the direct inclusion, W^d is the deformation energy in the matrix, and W_L^d is deformation energy at the inclusion–matrix boundary. It is also possible to use the above method to determine the forces acting on a dislocation with Burger's vector b in a point $z_{*k} \in S_k$ by the Peach–Koehler formulas [36, 41].

$$
F_x(z_{*k}) = b\hat{\sigma}_{yz}(z_{*k}),\quad F_y(z_{*k}) = -b\hat{\sigma}_{xz}(z_{*k}). \tag{23.30}
$$

For an illustration of the research technique, we will make a detailed analysis of the problem solution for a particular case of elastic characteristics equality of half-spaces ($G_1 = G_2 = G,\ T_1 = T_2 = T$) and the presence of b intensity dislocation at a point $\varsigma_2 = x_2 + iy_2$.

Fig. 23.7 Effect of the coefficient k_T on the energy of deformation of the matrix $\tilde{W}^d$ (solid line) and the inclusion $\tilde{W}^d_{\text{in}}$ (dashed-dotted line) when the point of application of the dislocation is shifted along the axis of the soft inclusion

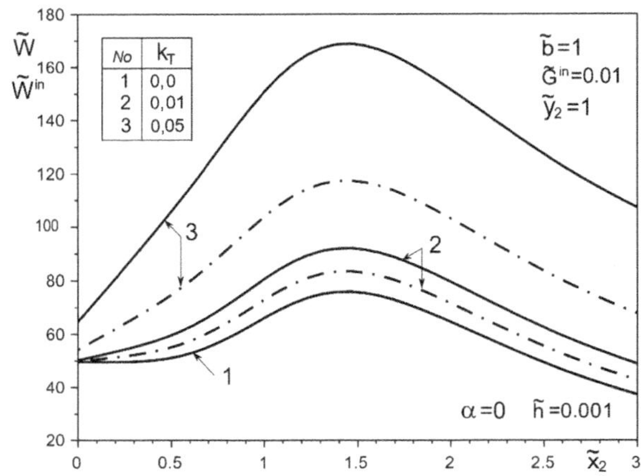

Fig. 23.8 Effect of the coefficient k_T on the SIFs $\tilde{K}_{31}$ (solid line) and $\tilde{K}_{32}$ (dashed-dotted line) when the dislocation is distanced from the inclusion axis

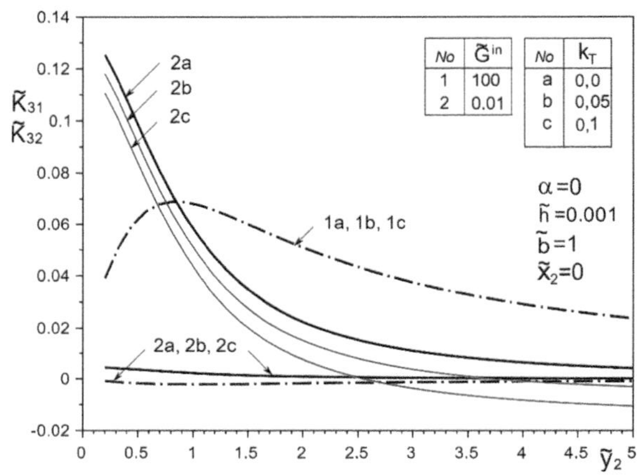

Thereinafter, the calculations were carried out for dimensionless values

$$\tilde{b} = b/\pi a, \quad \tilde{x}_2 x_2/a, \quad \tilde{y}_2 = y_2/a, \quad \tilde{T}_2 = \tilde{T}_1 = T/G, \quad \tilde{G}_{\text{in}} = G^{\text{in}}/G,$$
$$\tilde{W}^d = W^d/a^2 G, \quad \tilde{W}^d_{\text{in}} = W/a^2 G, \quad \tilde{K}_{31} = K_{31}/G\sqrt{\pi a},$$
$$\tilde{K}_{32} = K_{32}/G\sqrt{\pi a}, \quad F_x = F_x/a\pi G, \quad F_y = F_y/a\pi G.$$

Figures 23.7 and 23.8 show some illustrative cases of calculations of deformation energy change depending on the parameters of surface stresses. Figure 23.9 shows illustrative case of calculations of the forces acting on a dislocation with the Burgers vector b in a point $z_{*k} \in S_k$ by the Peach–Koehler formulas.

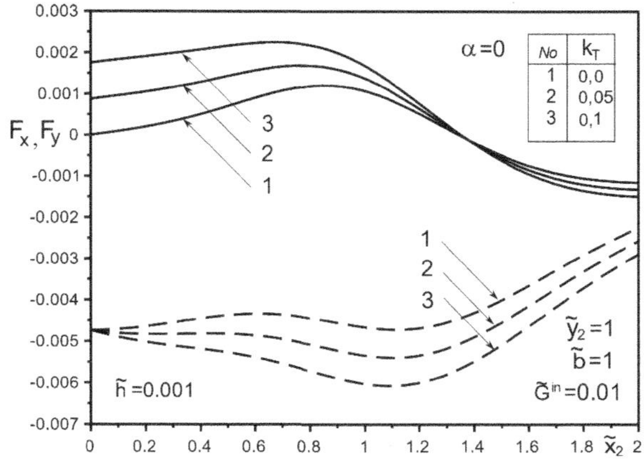

Fig. 23.9 Effect of the coefficient k_T on forces $\tilde{F}_x$ (solid line), $\tilde{F}_y$ (dashed line), acting on the dislocation when the point of application of the dislocation is shifted along the axis of the soft inclusion

23.4.3 Longitudinal Shear of Bimaterial with Thin Interfacial Nonlinear Elastic or Transversely FGM Microinclusion (Effect of Frictional Slipping on the Strength of Ribbon-Reinforced Composite)

To investigate the influence of the possible frictional slipping contact between the constituent elements on the matrix–inclusion interface, the problem formulated in paragraph 2 should be complicated by the contact conditions (23.9), (23.10) [37]. Consider an unbounded isotropic bulk consisting of two half-spaces with elastic constants E_k, ν_k, G_k $(k = 1, 2)$, pressed to the interface by normal stresses $\sigma_{yy}^{\infty} < 0$, and under the action of uniformly distributed at infinity stresses σ_{xxk}^{∞} $(k = 1, 2)$. Along the segment $L' = [-a; a]$ is a thin inclusion of thickness $2h$ $(h \ll a)$ (Fig. 23.10), of which the upper and lower banks may come in contact with the matrix non-ideally on the intervals $L''^{\pm} = [b^{\pm}, c^{\pm}]$, $\left(\left|b^{\pm}\right| \leqslant |a|, \left|c^{\pm}\right| \leqslant |a|\right)$, respectively. At the contact areas $L''^{\pm}$, we assume stick–slip contact conditions [37, 39, 47], wherein mutual slippage of contacting body surfaces can start, causing heat release, energy dissipation, wear [37], etc., and that all points of $L''^{\pm}$, tangential stresses (friction forces) are equal to

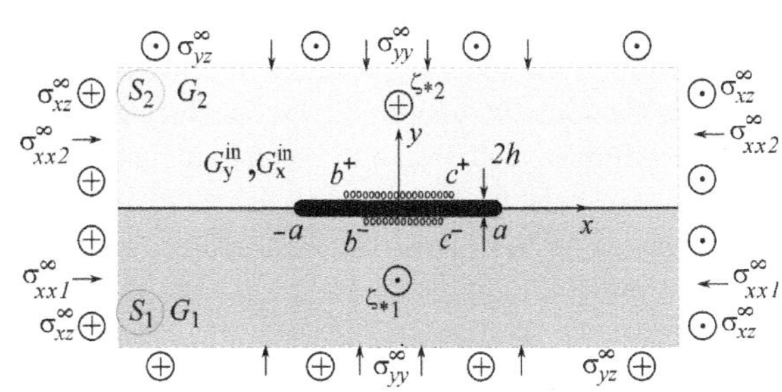

Fig. 23.10 Stick–slip contact. Geometry and loading scheme of the problem

$$\sigma_{yz}^{\text{in}}(x, \pm h) = \text{sign}_{yz2}(x, \pm h) = -\,\text{sign}\left(w^{\text{in}}(x, \pm h) - w(x, \pm h)\right) \tau_{yz}^{\max}(x),$$

where $\tau_{yz}^{\max}(x) = -\alpha\sigma_{yy}(x)$ $(\sigma_{yy} < 0)$; α is the sliding friction coefficient. Outside the area $L''^{\pm}$, in the absence of slip on the inclusion surface, the tangential stresses may not exceed the allowable maximum $|\sigma_{yz}(x, \pm h)| \leqslant \tau_{yz}^{\max}(x)$ $(\sigma_{yy} < 0)$ and there is no mutual displacement of contact surfaces (displacement jump). The sign (direction of action) of the tangential stresses is chosen depending on the sign of the displacement difference $w^{\text{in}}(x, \pm h) - w(x, \pm h)$ at the point in question at L''.

The accuracy of the solution in the case of non-ideal contact is very sensitive to the correct determination of the position and size of the slip zones. Within the framework of the proposed structural–modular JFM, we apply an iterative approach for this at each point of the interval $L''^{\pm}$: (1) gradually with growth, we apply a small load to check the condition $|\sigma_{yz}(x, \pm h)| = \tau_{yz}^{\max}(x)$ of the beginning of the slippage process; (2) as soon as this condition is satisfied at certain points x_n, we assign values $\tau_{yz}^{\max}$ to values $\sigma_{yz}^{\text{in}}(x_n, \pm h)$, $\sigma_{yzk}(x_n, \pm h)$ in all these points and resolve the SLAE; (3) we check whether constraint (23.10) is satisfied everywhere; if not, we assign values $\tau_{yz}^{\max}$ again to values $\sigma_{yz}^{\text{in}}(x_n, \pm h)$, $\sigma_{yzk}(x_n, \pm h)$ in those points x_n where (23.10) is not satisfied. The process is repeated until the condition (23.10) is satisfied at all points x_n $(n = 1, \ldots, N)$. It is proven that such an iterative algorithm is convergent under monotonically increasing non-contrast loading. The obtained results were validated by comparing them with known partial solutions for an interfacial crack and an interfacial thin rigid inclusion, as well as a thin elastic interfacial inclusion at its ideal contact with the matrix [18, 34, 35, 41].

Figures 23.11 and 23.12 show the effect of slip on the stress $\tilde{\sigma}_{yz} = \sigma_{yz}/G_{\text{av}}$ distribution along with the inclusion–matrix interface, as well as the growth of the slip zone size and its intensity $\tilde{w}_{sl} = w_{sl}/a$ depending on the problem parameters. Decreasing the distance of the application points from the inclusion axis, as well as increasing the load intensity, is expected to increase the slip area and its magnitude (Fig. 23.12). However, the displacement of the coordinates of the force application points along the inclusion axis to its apex (Fig. 23.11), as well as the distance from this axis, reduces the slip intensity. The loading by an applied screw dislocation with Burger's vector $\tilde{b}_2 = b_2 G_2/G_{\text{av}}$ in the point $x_{*2} = 0$, $y_{*2} = a$ leads to the appearance of two slip zones antisymmetric for the vertical axis of inclusion (Fig. 23.12).

It is revealed that with the friction coefficient, type, and place of load application unchanged, the slippage will start at a lower load and the slippage zone will be larger in the case of a harder than the matrix inclusion. The calculations confirmed the linear attenuation of the sizes of the safety and slip zones from a decrease in the intensity or distance from the inclusion of the concentrated factors. This is especially noticeable for the softer inclusion than the matrix inclusion.

The distribution of stresses in the inclusion vicinity, calculated to optimize the load regime of the investigated structure, makes it possible to conclude that the greatest influence on its stress–strain state is in the inclusion vicinity; accordingly, the appearance of slippage between the components has the case of applying concentrated forces in the point above the inclusion center.

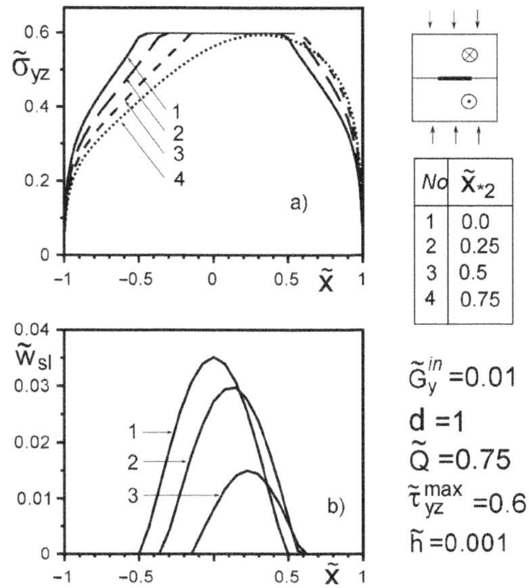

Fig. 23.11 Stress distribution along with the inclusion–matrix boundary (**a**) and the value of the slip zone (**b**) for a softer inclusion than the matrix depending on the change in the coordinates of the force application points along the inclusion axis

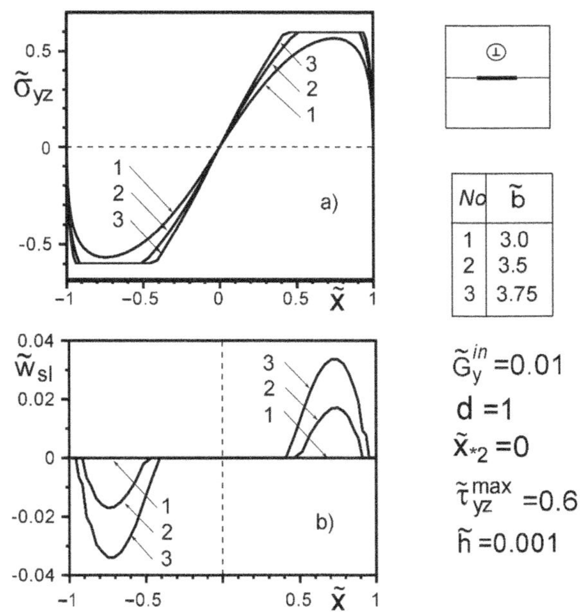

Fig. 23.12 Stress distribution along with the inclusion–matrix boundary (**a**) and the value of the slip zone (**b**) for a softer than matrix inclusion under screw dislocation loading

The obtained results and the proposed method can be used and developed to study the value of the energy dissipated on the inclusion, heat release due to friction, development of wear processes, and the interaction of band inhomogeneities in composite structures with subsequent determination of their effective mechanical characteristics, optimization of operating load modes, and the like.

23.4.4 Deformation and Strength Parameters of a Composite Structure with a Thin Multilayer Ribbon-Like Inclusion

The construction of a mathematical model of such a layered thin inclusion layer (internal problem) should eventually reveal the relation between the stress-strain parameters inside the inclusion and on its external surface as the influence functions $\sigma_{yz}^{\text{in}}(x, \pm h)$, $w^{\text{in}}(x, \pm h)$, which will be used in the further solution of the problem [12, 35, 38]. To connect the external and internal problems, one needs to use contact conditions between the components of the package. There are several variants of contact conditions between the layers and between the package and the matrix: perfect contact between all constituents of the package; nonperfect contact with additional tension between layers; contact with friction between the (K)th and $(K-1)$th layers at the boundary $\{x, y_K \pm h_K\}$ in some area $x \in L_f \subset L'$; perfect or nonperfect contact between the boundary components of the package and the matrix; contact with friction between the inclusion and the matrix within $\{x, y_1 - h_1\}, \{x, y_M + h_M\}$ in some area $x \in L_f \subset L'$.

Using relations

$$\left[\sigma_{yz}^{\text{in}K}\right]_{y_K, h_K} \cong \sigma_{yz}^{\text{in}K}(x, y_K - h_K) - \sigma_{yz}^{\text{in}K}(x, y_K + h_K) = f_{3K}(x), \quad \left(K = \overline{1, M}\right)$$

$$\left[\frac{\partial w^{\text{in}K}}{\partial x}\right]_{y_K, h_K} \cong \frac{\partial w^{\text{in}K}}{\partial x}(x, y_K - h_K) - \frac{\partial w^{\text{in}K}}{\partial x}(x, y_K + h_K) = f_{6K}(x), \quad x \in L';$$

$f_3(x) = f_6(x) = 0$, $f_{3K}(x) = f_{6K}(x) = 0$, if $x \notin L'$
and, for example, contact conditions

$$\begin{cases} w^{in(K-1)}(x, y_{K-1} + h_{K-1}) = w^{\text{in}K}(x, y_K - h_K) & \left(K = \overline{2, M}\right), \\ \sigma_{yz}^{\text{in}K}(x, y_K - h_K) = \sigma_{yz}^{in(K-1)}(x, y_{K-1} + h_{K-1}) - T_K. \end{cases}$$

we can obtain the expressions for the stresses and strains in the layers through the stress and strain limits for the inclusion package for the presence of interlayer tension:

$$\begin{aligned}
\sigma_{yz}^{\text{in}K}(x, y_K + h_K) &= \sigma_{yz}^{in1}(x, -h) - \sum_{j=1}^{K} f_{3,j} - \sum_{j=2}^{K} T_j = \\
&= \sigma_{yz}^{inM}(x, h) + \sum_{j=K+1}^{M} f_{3,j} + \sum_{j=K+1}^{M} T_j, \\
\frac{\partial w^{\text{in}K}}{\partial x}(x, y_K + h_K) &= \frac{\partial w^{in1}}{\partial x}(x, -h) - \sum_{j=1}^{K} f_{6,j} = \\
&= \frac{\partial w^{inM}}{\partial x}(x, h) + \sum_{j=K+1}^{M} f_{6,j} \qquad (x \in L') .
\end{aligned} \tag{23.31}$$

$$\begin{aligned}
\sigma_{yz}^{in1}(x, -h) - \sigma_{yz}^{inM}(x, h) &= \sum_{j=1}^{M} f_{3,j} + \sum_{j=2}^{M} T_j, \\
\frac{\partial w^{in1}}{\partial x}(x, -h) - \frac{\partial w^{inM}}{\partial x}(x, h) &= \sum_{j=1}^{M} f_{6,j} \quad (x \in L') .
\end{aligned}$$

The resulting limit stresses and strains for the inclusion package (23.31), the boundary values of the stresses and strains of the matrix (23.4), the boundary conditions (23.7)–(23.10), and mathematical model equations (23.13), (23.17) form a complete SSIE for the solution of the problem. Note that the dissimodularity of the inclusion layers in no way affects the peculiarity of the solution of the SSIE of the problem, which allows us to obtain a large variety of effects from manipulating the properties of the layers.

To illustrate the method, let us investigate the longitudinal shear of a structure in the form of a body with a thin two-layer inclusion with layers of thickness $2h_K$ $(K = 1, 2)$, $2h = 2h_1 + 2h_2$ and orthotropic mechanical properties $G_y^{\text{in}K}$, $G_x^{\text{in}K}$, respectively, under the condition of nonperfect mechanical contact with surface tension on the contact surfaces of the structural components under different kinds of loading (Fig. 23.3) when $\varsigma_{k*} = x_{k*} + iy_{k*}$ $(k = 1, 2)$.

The investigation of the influence of the inclusion layers' different modularity, external force factors at non-ideal contact with the surface tension of the structural components on the unmeasured stress–strain field parameters on the inclusion surfaces, and the dimensionless stress intensity factor$\tilde{K}_{31}$are illustrated in Figs. 23.13 and 23.14.

Figure 23.13 shows the results of the study of the effect of the level of dissimodularity on the GSIF $\tilde{K}_{31}$ under different external loadings and in the absence of surface forces.

Figure 23.13 confirms the known effect of the GSIF $\tilde{K}_{31}$ maximum under the loading by concentrated forces placed at a distance approximately $d \approx a$ from the inclusion axis, irrespective of the stiffness of the materials of the layers. However, it is more pronounced in the stiffness range of the materials of the layers softer than the matrix material. Moreover, if the material of one of the layers is equivalent to that of the matrix, we obtain the known results for a homogeneous elastic inclusion at the interface of the matrix materials [35, 40, 41].

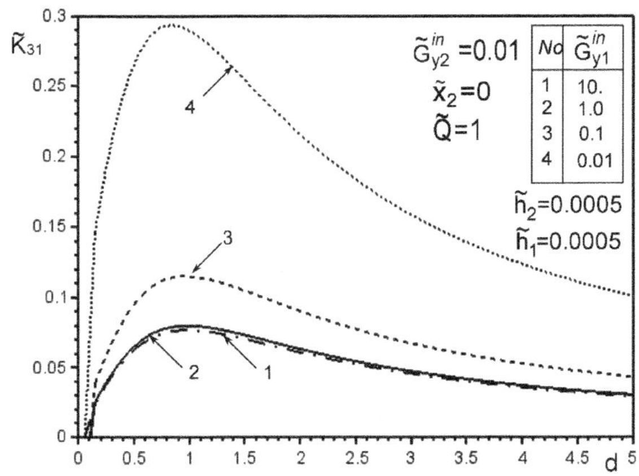

Fig. 23.13 The effect of changing the distance d of the point of application of concentrated forces from the inclusion and the level of dissimodularity to the GSIF $\tilde{K}_{31}$ in the absence of surface tension and layer 1 softer from the matrix (4—result for the case of the same layer materials, verified by comparison with [35, 40, 41])

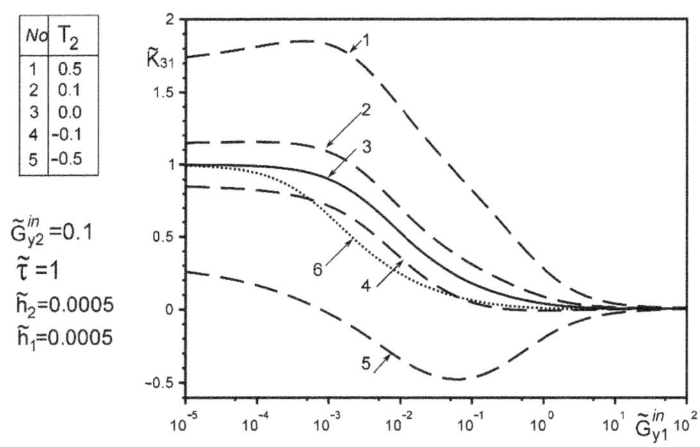

Fig. 23.14 The effect of the interlayer surface tension on the GSIF $\tilde{K}_{31}$ when layer 2 is softer than the matrix, and layer 1 is of the arbitrary material under the loading by stress uniformly distributed at infinity; (6—result for the case of the same layer materials verified by comparison with [42, 45])

Figure 23.14 shows the results of the study of the influence of the level of dissimodularity of the inclusion layers and the presence of surface forces under different external loads on the GSIF $\tilde{K}_{31}$. The dissimodularity of the materials of the layers significantly distorts this effect, which is especially noticeable when one of the layers is significantly softer than the matrix material. It is revealed that there are certain combinations of external load parameters, surface forces, and material properties of the layers, at which there are local SIF extremes. This effect can be useful in designing the modes of operation of structures with such a structure.

The solution method and results obtained for the two-layer inclusion have been verified by the coincidence of the numerical results with those known in literature [35, 40, 41] for a homogeneous thin elastic inclusion—the curves 4 in Fig. 23.13; 6 in Fig. 23.14.

23.5 Conclusions

We constructed a mathematical model of thin multilayer inclusions (layers) in the case of nonlinearity and anisotropy of their mechanical properties in the general form, taking into account the effect of frictional slipping and surface energy on their interfaces. On its basis, we derive a system of equations for solving the problems of antiplane deformation of a bimaterial with thin multilayer interfacial nonlinearly elastic or FGM inclusions under arbitrary force and dislocation loading in the case where the inclusion–matrix contact may be perfect or with surface tension or sliding (smooth or frictional). Such a model and methodology can be used to simulate a thin inclusion from a functionally graded material and to solve the corresponding problems of defining the stress–strain field of the corresponding micro- or nanostructures by efficient analytical–numerical methods (the jump function method and its modifications) without the need to involve purely numerical approaches (in particular FEM).

All these conclusions may be useful for the design of mechanical functionally graded parameters of inclusion and the modes of optimal operation of such structures under the given stresses. The proposed method has proved to be effective for solving a whole class of strain problems for bodies with thin deformable inclusions of finite length and may be used for the SSS calculation for various FGM inclusions.

References

1. Benveniste, Y., Baum, G.: An interface model of a graded three-dimensional anisotropic curved interphase. Proc. R. Soc. A. **463**, 419–434 (2007)
2. Bishop, A., Lin, C.Y., Navaratnam, M., Rawlings, R.D., McShane, H.B.: A functionally gradient material produced by a powder metallurgical process. J. Mater. Sci. Lett. **12**(1516–1518), 21–28 (1993)
3. Chen, X., Liu, Q.: Thermal stress analysis of multi-layer thin films and coatings by an advanced boundary element method. Comput. Model. Eng. Sci. **2** 337–349 (2001)
4. Chen, Y.Z.: Study of multiply-layered cylinders made of functionally graded materials using the transfer matrix method. J. Mech. Mater. Struct. **6**, 641–657 (2011)
5. Davim, J.P., Charitidis, C.A. (eds.).: Nanocomposites. Materials, Manufacturing and Engineering, p. 211. Walter de Gruyter GmbH, Berlin, Germany (2013)
6. Ding, S.H., Li, X.: Thermal stress intensity factors for an interface crack in a functionally graded layered structure. Arch. Appl. Mech. **81**, 943–955 (2011)
7. El-Borgi, S., Erdogan, F., Ben Hatira, F.: Stress intensity factors for an interface crack between a functionally graded coating and a homogeneous substrate. Int. J. Fract. **123**, 139–162 (2003)

8. Erdogan, F., Wu, B.H.: Crack problem in FGM layers under thermal stresses. J. Therm. Stresses **1955**, 237–265 (1996)
9. Goryacheva, I.G.: Contact Mechanics in Tribology; Series: Solid Mechanics and Its Applications, vol. 61, p. 346. Springer, Berlin (1998)
10. Gurtin, M.E., Murdoch, A.I.: A continuum theory of elastic material surfaces. Arch. Ration. Mech. Anal. **57**, 291–323 (1975)
11. Hashin, Z.: Thin interphase/imperfect interface in elasticity with application to coated fiber composites. J. Mech. Phys. Solids **50**, 2509–2537 (2002)
12. Hutsaylyuk, V., Piskozub, Y., Piskozub, L., Sulym, H.: Deformation and strength parameters of a composite structure with a thin multilayer ribbon-like inclusion. Materials **15**, 1435 (2022). https://doi.org/10.3390/ma15041435
13. Hsueh, C.H.: Thermal stresses in elastic multilayer systems. Thin Solid Film **418**, 182–188 (2002)
14. Ichikawa, K. (ed.).: Functionally Graded Materials in the 21ST Century, A Workshop on Trends and Forecasts, p. 264. Kluver Academic Publishers, Dordrecht, The Netherlands (2000)
15. Jin, Z.H.: An asymptotic solution of temperature field in a strip of a functionally graded material. Int. Commun. Heat Mass Transf. **29**, 887–895 (2002)
16. Kalker, J.J.: A survey of the mechanics of contact between solid bodies. Z. Angew. Math. Mech. **57**, T3–T17 (1977)
17. Kashtalyan, M., Menshykova, M.: Three-dimensional analysis of a functionally graded coating/substrate system of finite thickness. Phil. Trans. R. Soc. A Math. Phys. Eng. Sci. **366**, 1821–1826 (2008)
18. Kharun, I.V., Loboda, V.V.: A set of interface cracks with contact zones in combined tension-shear field. Acta Mech. **166**, 43–56 (2003)
19. Kieback, B., Neubrand, A., Riedel, H.: Processing techniques for functionally graded materials. Mater. Sci. Eng. A **362**, 81–105 (2003)
20. Kim, C.I., Schiavone, P., Ru, C.-Q.: The effects of surface elasticity on an elastic solid with Mode-III crack: Complete solution. Trans. ASME J. Appl. Mech. **77**, 021011 (2010)
21. Koizumi, M.: The concept of FGM. Ceram. Trans. Funct. Graded Mater. **34**, 3–10. NII Article ID (NAID), 10014680673 (1993)
22. Lee, Y.-D., Erdogan, F.: Interface cracking of FGM coatings under steady-state heat flow. Eng. Fract. Mech. **59**, 361–380 (1998)
23. Li, Y., Feng, Z., Hao, L., Huang, L., Xin, C., Wang, Y., Bilotti, E., Essa, K., Zhang, H., Li, Z., et al.: A review on functionally graded materials and structures via additive manufacturing: from multi-scale design to versatile functional properties. Adv. Mater. Technol. **5**, 1900981 (2020). https://doi.org/10.1002/admt.201900981
24. Mencik, J. Mechanics of Components with Treated or Coated Solids, p. 360. Kluwer Academic Publishing, Dordrecht, The Netherlands (1996)
25. Miyamoto, Y., Kaysser, W.A., Rabin, B.H., Kawasaki, A., Ford, R.G.: Functionally graded materials: Design, processing and applications. Mater. Technol. Ser. **5**, 247–313 (1999)
26. Mura, T.: Micromechanics of Defects in Solids, 2nd ed., p. 601. Martinus Nijhoff, Dordrecht, The Netherlands (1987)
27. Naik, R.A.: Simplified micromechanical equations for thermal residual stress analysis of coated fiber composites. J. Compos. Technol. Res. **14**, 182–186 (1992)
28. Nemat-Nasser, S., Hori, M.: Micromechanics: Overall Properties of Heterogeneous Materials, North-Holland Series in Applied Mathematics and Mechanics, vol. 37, p. 687. Elsevier, Amsterdam, The Netherlands (1993)
29. Pasternak, I.M., Pasternak, R.M., Sulym, H.T.: 2D boundary element analysis of defective thermoelectroelastic bimaterial with thermally imperfect but mechanically and electrically perfect interface. Eng. Anal. Bound. Elem. **61**, 94–206 (2015)
30. Pasternak, I.M., Sulym, H.T., Ilchuk, N.I.: Interaction of physicomechanical fields in bodies with thin structural inhomogeneities: A survey. J. Math. Sci. **253**, 63–83 (2021). https://doi.org/10.1007/s10958-021-05213-9

31. Paulo Davim, J., Constantinos A. (ed.).: Nanocomposites. Materials, Manufacturing and Engineering, p. 211. Walter de Gruyter GmbH, Berlin (2013)
32. Peng, X.-L., Lee, X.-F.: Thermoelastic analysis of functionally graded annulus with arbitrary gradient. Appl. Math. Mech. **30**, 1211–1220 (2009)
33. Perkowski, D.M., Kulchytsky-Zhyhailo, R., Kolodziejczyk, W.: On axisymmetric heat conduction problem for multilayer graded coated half-space. J. Theor. Appl. Mech. **56**, 147–156 (2018)
34. Piskozub, I.Z., Sulym, H.T.: Asymptotics of stresses in the vicinity of a thin elastic interphase inclusion. Mater. Sci. **32**, 421–432 (1996). https://doi.org/10.1007/BF02538967
35. Piskozub, I.Z., Sulym, H.T.: Nonlinear deformation of a thin interface inclusion. Mater. Sci. **53**, 600–608 (2018). https://doi.org/10.1007/s11003-018-0114-2
36. Piskozub, Y.: Effect of surface tension on the antiplane deformation of bimaterial with a thin interface microinclusion. Math. Model. Comput. **8**(1), 69–77 (2021). https://doi.org/10.23939/mmc2021.01.069
37. Piskozub, Y., Sulym, H.: Effect of frictional slipping on the strength of ribbon-reinforced composite. Materials **14**, 4928 (2021). https://doi.org/10.3390/ma14174928
38. Piskozub, Y., Piskozub L., Sulym, H.: Effect of the transverse functional gradient of the thin interfacial inclusion material on the stress distribution of the bimaterial under longitudinal shear. Materials **15**, 8591 (2022). https://doi.org/10.3390/ma15238591
39. Pyriev, S.Y., Yevtushenko, A.A., Sulym, G.T.: Thermomechanical wear during quasistationary heat generation by friction. J. Frict. Wear **33**, 315–321 (2012)
40. Sulym, H., Piskozub, Y., Polanski, J.: Antiplane deformation of a bimaterial with thin interfacial nonlinear elastic inclusion. Acta Mech. Autom. **12**, 190–195 (2018). https://doi.org/10.2478/ama-2018-0029
41. Sulym, H.T.: Bases of Mathematical Theory of Thermo-elastic Equilibrium of Solids Containing Thin Inclusions; Research and Publishing Center of NTSh, p. 716. L'viv, Ukraine (2007) (In Ukrainian). Available online: https://ua1lib.org/book/665574/5c937e
42. Uchida, Y.: Properties of functionally graded materials, Manufactured by progressive lamination method for applications. Aichi Inst. Technol. Res. Rep. **39–B**, 39–51 (2004)
43. Wang, B.L., Mai, Y.W., Noda, N.: Fracture mechanics analysis model for functionally graded materials with arbitrarily distributed properties. Int. J. Fract. **116**, 161–177 (2002)
44. Wang, J., Karihaloo, B.L.E., Duan, H.L.: Nano-mechanics or how to extend continuum mechanics to nano-scale. Bull. Pol. Acad. Sci. Tech. Sci. **55**, 133–140 (2007)
45. Wang, X., Sudak, L.J.: Three-Dimensional analysis of multi-layered functionally graded anisotropic cylindrical panel under thermomechanical loading. Mech. Mater. **40**, 235–254 (2008)
46. Wang, Y., Huang, Z.M.: Analytical micromechanics models for elastoplastic behavior of long fibrous composites: A critical review and comparative study. Materials **11**, 1919 (2018)
47. Yevtushenko, A.A., Rozniakowska, M., Kuciej, M.: Transient temperature processes in composite strip and homogeneous foundation. Int. Commun. Heat Mass Transf. **34**, 1108–1118 (2007)
48. Zhang, N., Khan, T., Guo, H., Shi, S., Zhong, W., Zhang, W.: Functionally graded materials/an overview of stability, buckling, and free vibration analysis. Adv. Mater. Sci. Eng. 1354150 (2019). https://doi.org/10.1155/2019/1354150
49. Zhou, K., Hoh, H.J., Wang, X., Keer, L.M., Pang, J.H., Song, B., Wang, Q.J.: A review of recent works on inclusions. Mech. Mater. **60**, 144–158 (2013)

Chapter 24
Dynamics of Three-Layer Spherical Shells with Heterogeneous Filler Under Nonstationary Loads

Peter Lugovyy

Abstract The dynamics of a three-layer spherical shell with a discrete-symmetric lightweight, rib-reinforced aggregate under nonstationary loads were studied. In analyzing the elements of the elastic structure, the model of the theory of shells and rods by Tymoshenko under independent static and kinematic hypotheses for each layer was used. The equation of motion of a three-layer spherical shell with a discrete-symmetric light, rib-reinforced filler is obtained according to the Hamilton–Ostrogradsky variational principle. Numerical results illustrating the nature of vibrations of a three-layer elastic structure are obtained by the finite-element method. The effect of the geometrical and physical–mechanical parameters of the symmetric layers of the shell on its dynamic behavior under axisymmetric internal impulse loading was investigated.

24.1 Introduction

The widespread use of layered structures in creating modern supersonic (hypersonic) aircraft and reusable space transport systems has led to the appearance of theories for studying their mechanical behavior under various loads. In [8], a broken line theory for solving problems for elastic layered plates is proposed. In [1], a model for the package of layers was built within the framework of the classical theory of shells, and in [2], the well-known Reissner–Mindlin theory of layered anisotropic plates and shells is widespread. Also worthy of attention is the approach proposed in [14, 22]. In these books, for the derivation of the equations of layered shells of arbitrary thickness, the assumption is made of the presence in each layer of local angles of rotation caused by transverse shear deformations (broken line hypothesis) when the conditions of continuity of displacements and tangential stresses on the contact surface of the layers are satisfied. At the same time, the movement and rotation angles of each of the layers are expressed through the values of one of them, and the continuity of the tangential stresses is ensured by the assumption that the

P. Lugovyy (✉)
S.P. Timoshenko Institute of Mechanics, National Academy of Science of Ukraine, Kyiv, Ukraine
e-mail: plugovyy@ukr.net

H. Altenbach et al. (eds.), *Selected Problems of Solid Mechanics and Solving Methods*, Advanced Structured Materials 204, https://doi.org/10.1007/978-3-031-54063-9_24

shear angles of the adjacent layers are inversely proportional to the shear modulus. The total number of independent kinematic parameters is five, as in the case when the hypothesis of a straight line is introduced for the package of layers as a whole.

A significant number of works [4–6, 9, 12–14] are dedicated to the study of the dynamics of three-layer spherical shells. Nevertheless, the creation of advanced technologies, special purpose objects, and other engineering applications often leads to the need to develop structural three-layer shell elements with an aggregate of a complex geometric structure. At the same time, these elements are subjected to dynamic loads of various types, including nonstationary loads. The dynamic behavior of such shells has not been sufficiently studied.

The current work considers three-layer spherical shells with a discrete-symmetric light reinforced with rib filler. Reinforcing ribs are located along the lines of the main curvatures and connect the bearing layers. The distances between the reinforcing ribs are much larger than the dimensions of their cross-sections. In this case, it is advisable to apply the theories of layered shells using independent hypotheses for each layer [8]. This increases the general order of the system of equations but allows for a more detailed study of the dynamic behavior of the three-layer structure under forced dynamic loads. The beginning of studies of the dynamics of three-layer spherical shells with a discrete heterogeneous aggregate under nonstationary loads was initiated in works [10, 11, 15, 20], in which such problems were solved by the finite-difference method. The presence of light aggregate located between the bearing layers and the reinforcing ribs significantly complicates the task. Therefore, the finite-element method [19] was used to study the dynamics of three-layer spherical shells with a discrete-symmetric light, reinforcing rib aggregate.

The solution to the problem is based on the theory of shells and rods, based on the shear model of Timoshenko [21], with the use of independent static and kinematic hypotheses for each layer [18]. The variational principle of the Hamilton–Ostrogradsky stationarity is used to derive the equations of oscillations of a three-layer non-homogeneous structure. Numerical modeling of the dynamics of three-layer spherical shells with a discrete-symmetric lightweight, rib-reinforced aggregate was carried out using the finite-element method. Numerical results of solving specific problems are given, and new mechanical effects are revealed.

24.2 Formulation of the Problem. Basic Equations

A three-layer spherical shell with light aggregate reinforced with discrete symmetrical ribs is an elastic structure consisting of internal (index 1), external (index 2) bearing layers, light aggregate (index t), and a set of discrete ribs (index j), rigidly connected with the specified bearing layers. The shell has a constant overall h and is assigned to coordinates αz. The coordinate line $R\alpha$ on the shell's middle surface at $z = 0$ coincides with the generating line; the coordinate line z is a straight line orthogonal to the median surfaces. We will consider the value of z to be positive if the point is on the side of the convexity of the middle surface. Discrete ribs and light aggregate rigidly connect the shells. The type of deformed state of the inner and outer bearing layers can be determined through the components of the generalized displacement vector $\bar{U}_1 = (u_s^1, u_3^1, \varphi_1^1)^T$ and $\bar{U}_2 = (u_s^2, u_3^2, \varphi_1^2)^T$ [21]. The displacement fields for a light ribbed aggregate are determined by the generalized displacement vector $\bar{U}_t = (u_s^t, u_3^t, \varphi_1^t)^T$ according to the model proposed in [3]. The coefficients of the first quadratic form and the curvature of the coordinate surface are written as follows: $A_1 = R_i, A_2 = R_i, k_1 = 1/R_i, k_1 = 1/R_i$.

The deformed state of the reinforcing rib directed along the circumferential coordinate will be determined by the generalized displacement vector $\bar{U}_j = (u_1^j, u_3^j, \varphi_1^j)^T$. Based on the assumptions of a rigid connection of reinforcing ribs with spherical bearing layers, the contact conditions of the centers of gravity of the ribs with the bearing layers are written [15, 20]:

$$u_1^j = u_1^{jk}(s_j) \mp \frac{H_j}{2}\varphi_1^{jk}(s_j), \quad u_3^j = u_3^{jk}(s_j), \quad \varphi_1^j = \varphi_1^{jk}(s_j) \quad (k = 1, 2), \tag{24.1}$$

where $s_j = R\alpha_j$ is the coordinate of the line of the sets of points of the projections of the centers of gravity of the cross-sections of the jth rib onto the corresponding median surface of the bearing layer; $h_j^i = 0.5h_i + 1/2H_i$ and h_i $(i = 1, 2)$ are thicknesses of spherical bearing layers; $1/2H_j$ is the distance from the axis of the jth edge to the surface of the smooth shells; $h_t = Hj$ thickness of light aggregate.

Based on the theory of shear deformation in shells [21], the displacements u_1^i and u_3^i in the bearing layers in the direction α (longitudinal), z (thickness), and t (time) with small linear displacements are expressed through the following dependencies:

$$\begin{aligned} u_1^i(s, z, t) &= u_0^i(s, t) + z_i\varphi_1^i(s, t), \\ u_3^i(s, z, t) &= u_{03}^i(s, t) \quad (i = 1, 2), \end{aligned} \tag{24.2}$$

where φ_1^i is the angle of rotation of the normal to the middle thickness h with a smooth middle surface in the orthogonal coordinate system of the surface of the bearing layers; $s = R\alpha$.

At the same time, the deformation ratios for the load-bearing layers and the jth edge are taken in the following form:

$$\begin{aligned} \varepsilon_{11}^i &= \frac{\partial u_0^i}{\partial s_i} + \frac{u_{03}^i}{R_i}, \quad \varepsilon_{22}^i = \frac{u_0^i}{R_i}\cot\alpha + \frac{u_{03}^i}{R_i}, \quad \varepsilon_{13}^i = \varphi_1^i \\ \kappa_{11}^i &= \frac{\partial \varphi_1^i}{\partial s_i}, \quad \kappa_{22}^i = \frac{\varphi_1^i}{R_i}\cot\alpha, \quad \varepsilon_{22j} = \frac{u_{3j}}{R_j}. \end{aligned} \tag{24.3}$$

Expressions for movements of light filler are written according to the model [3]:

$$\begin{aligned} u_1^t(s, z, t) &= \left(1 + \frac{z_t}{R_t}\right) u_0^t(s, t) + z_t u_1^t(s, t), \\ u_3^t(s, z, t) &= u_{03}^t(s, t). \end{aligned} \tag{24.4}$$

Kinematic dependences for the aggregate are taken based on small deformations:

$$\varepsilon_{11}^{t} = \frac{1}{(1 + z_t/R_t)}\left(\frac{\partial u_0^t}{\partial s} + \frac{u_{03}^i}{R_t}\right), \quad \varepsilon_{22}^{t} = \frac{u_{03}^t}{R_t + z_t};$$
$$2\varepsilon_{13}^{t} = \frac{1}{(1 + z_t/R_t)}\left(\frac{\partial u_{03}^t}{\partial s} - \frac{u_0^i}{R_t}\right) + u_1^t. \tag{24.5}$$

The compatibility conditions, which provide for an ideal connection between the aggregate and the bearing layers without separation and slippage, are presented in the following form [7]:

$$u_1^t(z = z_t^1) = u_0^i + 1/2(-1)^k h_i \varphi_1^i,$$
$$u_{03}^t = u_{03}^i, \tag{24.6}$$

where

$$k = \begin{cases} 0, & i = 1 \\ 1, & i = 2 \end{cases}, \quad z_t^i = \begin{cases} -1/2h_t, & i = 1 \\ 1/2h_t, & i = 2 \end{cases}$$

$$u_0^t = \frac{u_0^1 + u_0^2}{2} - \frac{1}{4}\left(h_2\varphi_1^2 - h_1\varphi_1^1\right),$$
$$u_1^t = \frac{u_0^1 - u_0^2}{h_t} - \frac{1}{2h_t}\left(h_2\varphi_1^2 + h_1\varphi_1^1\right), \tag{24.7}$$
$$u_{03}^t = \frac{1}{2}\left(u_{03}^1 + u_{03}^2\right).$$

The equations of motion for bearing layers and light aggregate are derived using the Hamilton–Ostrogradsky variational principle of stationarity, according to which

$$\delta \int_{t_1}^{t_2} (K - \Pi + A)dt = 0, \tag{24.8}$$

where Π is the total potential energy of the elastic system, K is the total kinetic energy of the elastic system, A is the work of external forces, and $t_{1,2}$ are fixed moments of time. When deriving the equations of oscillations of three-layer shells with a light aggregate, the components of the movements of the load-bearing layers, reinforcing ribs, and aggregate made of light material are subject to independent variation.

Expressions for variations of the total potential and kinetic energy of the indicated components are written in the form:

$$\delta\Pi = \delta\sum_{i=1}^{2}\Pi^i + \delta\sum_{j=1}^{J}\Pi^j + \delta\sum_{S_t}\Pi^t$$
$$\delta K = \delta\sum_{i=1}^{2}K^i + \delta\sum_{j=1}^{J}K^j + \delta\sum_{S_t}K^t, \tag{24.9}$$

where

$$\delta \Pi^{i} = \int_{S_i} \left[\int_{-1/2h}^{1/2h} \left(T_{11}^{i} \delta \varepsilon_{11}^{i} + T_{22}^{i} \delta \varepsilon_{22}^{i} + T_{13}^{i} \delta \varepsilon_{13}^{i} + M_{11}^{i} \delta \kappa_{11}^{i} + M_{22}^{i} \delta \kappa_{22}^{i} \right) dz_i \right] dS_i, \quad (24.10)$$

$$\delta \Pi^{t} = \int_{S_t} \left[\int_{-1/2h_t}^{1/2h_t} \left(T_{11}^{t} \delta \varepsilon_{11}^{t} + T_{22}^{t} \delta \varepsilon_{22}^{t} + T_{13}^{t} \delta \varepsilon_{13}^{t} + M_{11}^{t} \delta \kappa_{11}^{t} + M_{22}^{t} \delta \kappa_{22}^{t} \right) dz_t \right] dS_t, \quad (24.11)$$

$$\delta \Pi^{j} = \int_{L_J} \left(T_{11j} \delta \varepsilon_{11j} + T_{22j} \delta \varepsilon_{22j} + T_{13j} \delta \varepsilon_{13j} + M_{11j} \delta \kappa_{11j} \right) dL_J, \quad (24.12)$$

$$\delta K^{i} = \int_{S_i} \left\{ \int_{-1/2h_i}^{1/2h_i} \left[\rho_i h_i \left(\frac{\partial^2 u_1^i}{\partial t^2} \delta u_1^i + \frac{\partial^2 u_{03}^i}{\partial t^2} \delta u_3^i \right) + \rho_i \frac{h_i^3}{12} \left(\frac{\partial^2 \varphi_1^i}{\partial t^2} \delta \varphi_1^i \right) \right] dz_i \right\} dS_i, \quad (24.13)$$

$$\delta K^{t} = \int_{S_t} \left[\int_{-1/2h_t}^{1/2h_t} \rho_t h_t \left(\frac{\partial^2 u_0^t}{\partial t^2} \delta u_0^t + \frac{\partial^2 u_1^t}{\partial t^2} \delta u_1^t + \frac{h_t^2}{12} \frac{\partial^2 u_{03}^t}{\partial t^2} \delta u_{03}^t \right) dz_t \right] dS_t, \quad (24.14)$$

$$\delta K^{j} = \int_{L_j} \left[\rho_j F^j \left(\frac{\partial^2 u_1^j}{\partial t^2} \delta u_1^j + \frac{\partial^2 u_3^j}{\partial t^2} \delta u_3^j \right) + \rho_j \left(I_{kr}^j \frac{\partial^2 \varphi_1^j}{\partial t^2} \delta \varphi_1^j \right) \right] dL_j. \quad (24.15)$$

In (24.10)–(24.15), the values F^j and I_{kr}^j correspond to the geometric characteristics of the transverse reinforcing ribs, ρ_j is the specific weight of the material of the reinforcing rib, ρ_i and ρ_t are the specific weights of the materials of the bearing layers and light aggregate, respectively.

The three-layer spherical structure is assumed to be loaded with an internal axisymmetric distributed nonstationary normal load $P_1(s, t)$, where s and t are spatial and temporal coordinates.

It should be noted that when calculating the potential and kinetic energy for light aggregate in the expressions δ^t and δ^t, integration is carried out by the volume, the value of which is increased by the value of the volume of the reinforcing ribs. However, this fact practically does not affect the general error of the theory of shells since the volume of reinforcing ribs in the volume of light aggregate for three-layer shells of rotation is less than 5%.

After the standard transformations in the variational Eq. (24.8), taking into account relations (24.10)–(24.15), we obtain a system of hyperbolic equations of motion of the ninth order for a three-layer spherical shell with a light aggregate reinforced with discrete ribs under axisymmetric impulse loading and the corresponding boundary and initial conditions:

$$\begin{aligned}
&\frac{1}{\sin\alpha}\frac{\partial}{\partial s_i}(\sin\alpha T_{11}^i) - \frac{1}{R_i}(\cot\alpha T_{22}^i - T_{13}^i) - \frac{1}{R_t}T_{13}^i = \rho_i h_i \frac{\partial^2 u_0^i}{\partial t^2}\\
&\frac{1}{\sin\alpha}\frac{\partial}{\partial s_i}(\sin\alpha T_{13}^i)\\
&\quad -\frac{1}{R_i}(T_{11}^i + T_{22}^i) - \frac{1}{R_t h_t}(M_{11}^t + M_{22}^t) + P_3 = \rho_i h_i \frac{\partial^2 u_{03}^i}{\partial t^2}\\
&\quad \frac{1}{\sin\alpha}\frac{\partial}{\partial s_i}(\sin\alpha M_{11}^i) - \frac{ctg\alpha}{R_i}M_{22}^i - T_{13}^i = \frac{\rho_i h_i^3}{12}\frac{\partial^2 \varphi_1^i}{\partial t^2}\\
&\frac{1}{\sin\alpha}\frac{\partial}{\partial s_t}(\sin\alpha T_{11}^t) + \frac{1}{R_t}(T_{11}^i - T_{13}^i) + \frac{8}{h_t^2}M_{13}^t = \rho_t h_t \frac{\partial^2 u_0^i}{\partial t^2}\\
&\quad \frac{1}{\sin\alpha}\frac{\partial}{\partial s_t}(\sin\alpha M_{11}^t) - T_{13}^i + \frac{1}{R_{t1}}M_{13}^t = \rho_t h_t \frac{\partial^2 u_1^i}{\partial t^2}\\
&\quad \frac{1}{\sin\alpha}\frac{\partial}{\partial s_t}(\sin\alpha T_{13}^t) - \frac{1}{R_t}(T_{11}^t + T_{22}^t) = \rho_t h_t \frac{\partial^2 u_{03}^i}{\partial t^2}\\
&\qquad \left[T_{11}^{i\pm}\right]_j = \rho_j F_j \frac{\partial^2 u_{1j}}{\partial t^2}, \quad \left[T_{13}^{i\pm}\right]_j = \rho_j F_j \frac{\partial^2 u_{3j}}{\partial t^2}\\
&\qquad \left[M_{11}^{i\pm}\right]_j = \rho_j I_{krj} \frac{\partial^2 \varphi_{1j}}{\partial t^2}
\end{aligned} \tag{24.16}$$

$(i = 1, 2)$. These equations of oscillations of three-layer elastic structures are described by two systems of hyperbolic equations of the ninth order, which are formed by taking into account the discontinuous coefficients of the bearing layers-reinforcing elements. On the lines of breaks in the oscillation equations (24.16), $[T_{11}^{i\pm}]_j$, $[T_{13}^{i\pm}]_j$, and $[M_{11}^{i\pm}]_j$ correspond to the forces–moments acting on the jth discrete element from the bearing layers.

The relationship between the magnitudes of forces and moments and the corresponding magnitudes of deformations for bearing layers and reinforcing ribs have the form

$$
\begin{aligned}
T_{11}^{i} &= \frac{E_i h_i}{1-v_i^2}(\varepsilon_{11}^{i} + v_i \varepsilon_{22}^{i}), \quad T_{22}^{i} = \frac{E_i h_i}{1-v_i^2}(\varepsilon_{22}^{i} + v_i \varepsilon_{11}^{i}) \\
T_{13}^{i} &= k^2 G_{13}^{i} \varepsilon_{13}^{i}, \quad T_{22j} = E_j F_j \varepsilon_{22j} \\
M_{11}^{i} &= \frac{E_i h_i^3}{12(1-v_i^2)}(\kappa_{11}^{i} + v_i \kappa_{22}^{i}), \quad M_{22}^{i} = \frac{E_i h_i^3}{12(1-v_i^2)}(\kappa_{22}^{i} + v_i^2 \kappa_{11}^{i})
\end{aligned}
\tag{24.17}
$$

$(i = 1, 2)$; E_i, G_{13}^{i}, and v_i are physical and mechanical parameters of the material of the bearing layers; k^2 is the integral transverse shear coefficient of the theory of plates and shells; E_j and F_j are the modulus of elasticity of the material and the cross-sectional area of the jth rib, respectively.

Forces and moments for light aggregate can be determined as follows:

$$
\begin{aligned}
T_{11}^{t} &= \int_{-1/2h_t}^{1/2h_t} \left(1 + \frac{z_t}{R_t}\right) \sigma_{11}^{t} dz_t, \quad T_{13}^{t} = \int_{-1/2h_t}^{1/2h_t} \sigma_{13}^{t} dz_t \\
M_{11}^{t} &= \int_{-1/2h_t}^{1/2h_t} z_t \left(1 + \frac{z_t}{R_t}\right) \sigma_{11}^{t} dz_t.
\end{aligned}
\tag{24.18}
$$

The equations of oscillations (24.16) are supplemented by the corresponding boundary and initial conditions.

24.3 Numerical Results

The problem of dynamic deformation of a three-layer spherical dome is considered. The problem was solved on the spatial interval $[-\pi/2, \pi/2]$. Due to symmetry concerning the top of the sphere at $\alpha = 0$, the interval $[0, \pi/2]$ was considered. At the top of the sphere, there is a feature the need to reveal which made it possible to write down the boundary conditions at $\alpha = 0$ in the following form:

$$
u_1^{i} = \varphi_1^{i} = 0, \quad 2\frac{\partial T_{13}^{i}}{\partial s} - P_3(s_0, t)\delta_{21} = \rho h \frac{\partial^2 u_3^{i}}{\partial t^2}; \tag{24.19}
$$

at $\alpha = \pi/2$, we have rigid clamping

$$
u_1^{i} = u_3^{i} = \varphi_1^{i} = 0 \quad (i = 1, 2). \tag{24.20}
$$

In (24.19), δ_{2i} is the Kronecker symbol.

The external nonstationary load $P_3(s, t)$ was set in the form

$$
P_3 = A \sin \frac{\pi t}{T} \left[\eta(t) - \eta(t - T)\right], \tag{24.21}
$$

where A is the load amplitude; T is the duration of the load pulse; $T = R/c$, where c is the speed of sound in the metal of the bearing layers; $\eta(t)$ is the Heaviside function. The calculations depended on it: $A = 10^6$ Pa, $T = R_1/c = 5 \cdot 10^{-5}$ s. The initial conditions are zero for the bearing layers at $t = 0$:

$$u_1^i = u_3^i = \varphi_1^i = 0, \quad \frac{\partial u_1^i}{\partial t} = \frac{\partial u_3^i}{\partial t} = \frac{\partial \varphi_1^i}{\partial t} = 0 \quad (i = 1, 2). \tag{24.22}$$

The corresponding initial boundary value problem (24.16), (24.19)–(24.22) is solved using the finite-element method. A finite-element model of the shell was created, reflecting the relationship between the potential energy of deformations in the body and the potential of applied forces:

$$\Pi = E - W, \tag{24.23}$$

where E is the potential energy of deformations, and W is the potential of applied forces. After dividing the solid region into separate elements, dependence (24.23) takes the following form:

$$\Pi = \sum_{e=1}^{E} \left(E^{(e)} - W^{(e)} \right) = \sum_{e=1}^{E} \pi^{(e)}. \tag{24.24}$$

The global stiffness matrix and global column vector in the matrix equation read

$$[K]\{U\} = \{F\} \tag{24.25}$$

correspond to the ratios:

$$[K] = \sum_{e=1}^{E} \left[k^{(e)} \right], \quad \{F\} = -\sum_{e=1}^{E} \left\{ f^{(e)} \right\}.$$

Calculations were carried out with the same geometric and physical–mechanical parameters as for finite-difference calculations of three-layer spherical shells with discrete-symmetric aggregate [11]: $E_1^1 = E_1^2 = E_j = 7 \cdot 10^{10}$ Pa, $\nu_j = \nu_1^1 = \nu_1^2 = 0.3$, $\rho_1^1 = \rho_1^2 = \rho_j = 2.7 \cdot 10^3$, $h_1 = h_2 = 0.01$ m, $h_j = 2h_2$, $R_1 = 0.3$ m, $F_J = 2 \cdot 10^{-4}$ m^2 and $h/R = 0.13$. The physical and mechanical parameters for light aggregate are as follows: $E_{1,2}/E_t = 50$, $E_{1,2}/E_t = 500$, $\rho_t = 25$ kg/m^3, and $\upsilon_t = 0.27$. Discrete reinforcing ribs (parallels) were located at the following points:

$$\alpha_k = [41 + 40(k-1)]\, \Delta\alpha_1 \quad (k = 1, 2, 3).$$

This approach corresponds to the form of differential Eq. (24.16) and the fulfillment of the law of conservation of total mechanical energy of an asymmetric three-layer structure at the finite-element level. The development of finite-element models was carried out using a three-dimensional volumetric finite element of the solid type, which, in terms of the degree of elongation, narrowing and curvature, and

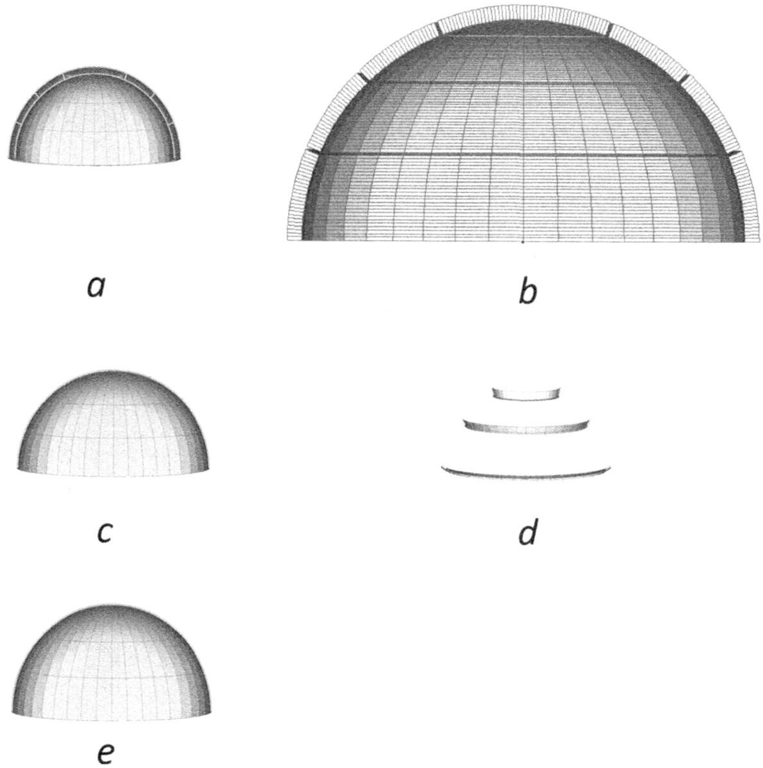

Fig. 24.1 The FE model

other indicators, met the requirements for ensuring the quality of the finite-element grid [19].

The created finite-element model (Fig. 24.1) included 32240 finite elements and 49680 nodes. In Fig. 24.1 and in the cross-section XZ, a model of a hemisphere with three edges is presented—foam plastic. In Fig. 24.1b, the cross-section XZ presents a foam model. In Fig. 24.1c, the model of the internal bearing layer is presented in the XZ-section. Figure 24.1d presents a model of the three ribs in the cross-section XZ. In Fig. 24.1e, the outer bearing layer is depicted.

The obtained numerical results allow the analysis of the stress–strain state of the three-layer elastic structure of the spherical type at any time (calculations were performed at $0 \leqslant t \leqslant 40T$). Figure 24.2 shows the dependence of the values of normal deflections $u_3{}^1(1)$ and $u_3{}^2(2)$ in the middle surfaces of the bearing layers on the angular coordinate α. Here, and in subsequent graphs, the curve with index 1 corresponds to the value $u_3{}^1$ of the internal bearing layer of the spherical shell, and with index 2—the value $u_3{}^2$ of the external bearing layer of the spherical shell at the time $t = 1.92T$ (the time of reaching the maximum value of quantities (1) and (2)). There is no light aggregate. The points of connection of curves 1 and 2 indicate the loca-

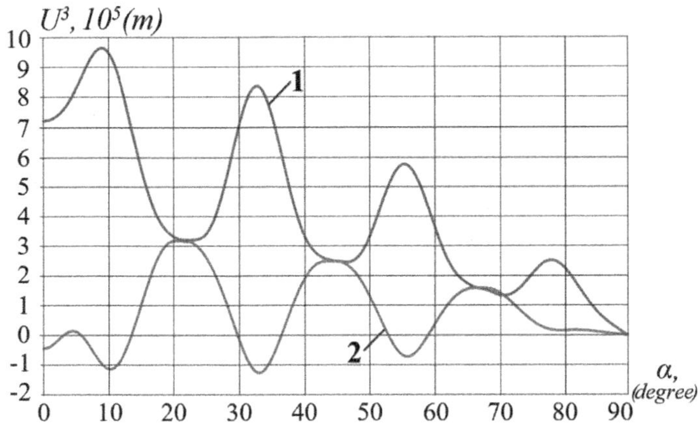

Fig. 24.2 Relationship between normal deflections in the middle surfaces of the bearing layers and the angular coordinate

Fig. 24.3 Relationship between normal stresses in the middle surfaces of the bearing layers and the angular coordinate

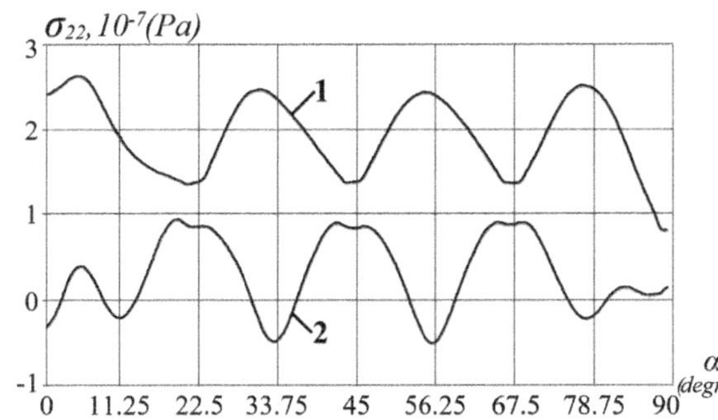

tion of discrete edges. From the presented graphic material, it is possible to visually determine the influence of the sphericity of the structure on the antisymmetry of the distribution of values $u_3{}^1(1)$ and $u_3{}^2(2)$ along the spatial coordinate. The first natural frequency of the design is 1441 Hz.

In Fig. 24.3, in the given scale, the dependences of the values of normal stresses $\sigma_{22}{}^1(1)$ and $\sigma_{22}{}^2(2)$ in the middle surfaces of the bearing layers on the angular coordinate α are shown. The curve with index 1 corresponds to the value $\sigma_{22}{}^1$ of the internal bearing layer of the spherical shell, and with index 2—the value $\sigma_{22}{}^2$ of the external bearing layer of the spherical shell at the moment of time $t = 1.92T$ (time of reaching the maximum value of values (1) and (2)). There is no light aggregate. The convergence points of curves 1 and 2 indicate the location of discrete edges. Based on the presented graphic material, it is possible to visually determine the influence of the sphericity of the structure on the antisymmetry of the distribution of values $\sigma_{22}^1(1)$ and $\sigma_{22}^2(2)$. The first natural frequency of the design is 1441 Hz.

In Fig. 24.4, in the reduced scale, the dependences of the values of normal deflections $u_3^1(1)$ and $u_3^2(2)$ in the middle surfaces of the bearing layers on the angular

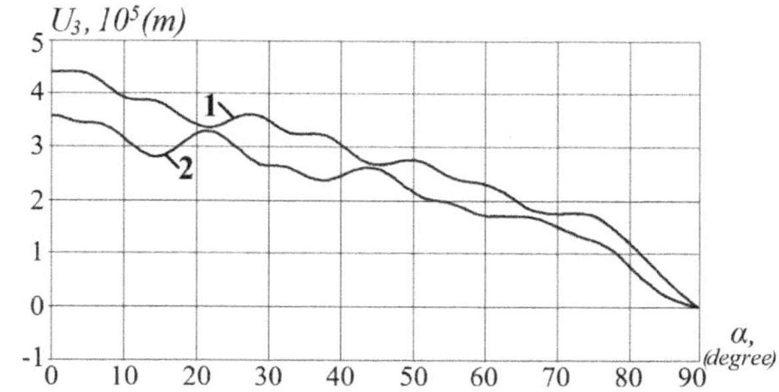

Fig. 24.4 Relationship between normal deflections in the middle surfaces of the bearing layers and the angular coordinate

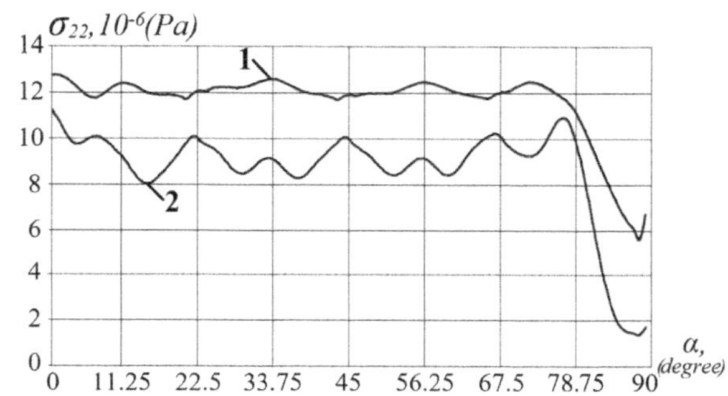

Fig. 24.5 Relationship between normal stresses in the middle surfaces of the bearing layers and the angular coordinate

coordinate α are given. The curve with index 1 corresponds to the value $u_3{}^1$ of the internal bearing layer of the spherical shell and with index 2—the value $u_3{}^2$ of the external bearing layer of the spherical shell at the moment of time $t = 1.53T$ (time of reaching the maximum value of values $u_3^1(1)$ and $u_3^2(2)$). The ratio of the modulus of elasticity of the bearing layers to the modulus of elasticity of the light aggregate $E_{1,2}/E_t = 50$. The points of connection of curves 1 and 2 indicate the location of discrete ribs. Based on the presented graphic material, it is possible to visually determine the influence of the sphericity of the structure on the antisymmetry of the distribution of values $u_3^1(1)$ and $u_3^2(2)$ along the angular coordinate α. The first natural frequency of the design is 1516 Hz.

In Fig. 24.5, in the given scale, the dependences of the values of normal stresses $\sigma_{22}{}^1(1)$ and $\sigma_{22}{}^2(2)$ in the middle surfaces of the bearing layers on the angular coordinate α are shown. The ratio of the bearing layers' modulus of elasticity to the light aggregate $E_{1,2}/E_t = 50$. The curve with index 1 corresponds to the value $\sigma_{22}{}^1$ of the internal bearing layer of the spherical shell, and with index 2—the value $\sigma_{22}{}^2$ of the external bearing layer of the spherical shell at the moment of time $t = 1.53T$ (the time of reaching the maximum value of quantities $\sigma_{22}^1(1)$ and $\sigma_{22}^2(2)$). Based on the presented graphic material, it is possible to visually determine the influence of the sphericity of the structure on the antisymmetry of the distribution of the quantity u3 along the spatial coordinate α. The first natural frequency of the design is 1516 Hz.

In Fig. 24.6, in the given scale, the dependences of the values of normal deflections $u_3{}^1(1)$ and $u_3{}^2(2)$ in the middle surfaces of the bearing layers on the angular

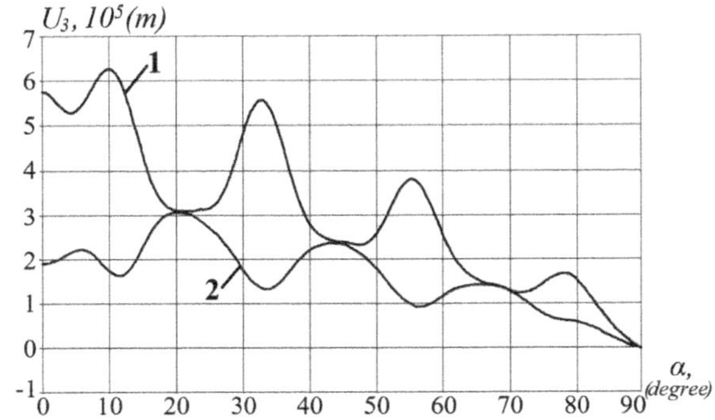

Fig. 24.6 Relationship between normal deflections in the middle surfaces of the bearing layers and the angular coordinate

coordinate α are given. The curve with index 1 corresponds to the value $u_3{}^1$ of the internal bearing layer of the spherical shell, and with index 2—the value u_3^2 of the external bearing layer of the spherical shell at the moment of time $t = 1.6T$ (time of reaching the maximum value of values $u_3{}^1(1)$ and $u_3{}^2(2)$). The ratio of the modulus of elasticity of the bearing layers to the modulus of elasticity of the light aggregate $E_{1,2}/Et = 500$. The points of connection of curves 1 and 2 indicate the location of discrete ribs. Based on the presented graphic material, it is possible to visually determine the influence of the sphericity of the structure on the antisymmetry of the distribution of quantities $u_3{}^1(1)$ and $u_3{}^2(2)$ along the spatial coordinate α. The first natural frequency of the structure is 1492 Hz.

In Fig. 24.7, in the given scale, the dependences of the values of normal stresses $\sigma_{22}{}^1(1)$ and $\sigma_{22}{}^2(2)$ on the middle surfaces of the bearing layers on the angular coordinate are shown. The curve with index 1 corresponds to the value $\sigma_{22}{}^1$ of the internal bearing layer of the spherical shell, and with index 2—the value $\sigma_{22}{}^2$ of the external bearing layer of the spherical shell at the moment of time $t = 1.6T$ (time of reaching the maximum value of values $\sigma_{22}^1(1)$ and $\sigma_{22}^2(2)$). The ratio of the modulus of elasticity of the bearing layers to the modulus of elasticity of the light aggregate is $E_{1,2}/E_t = 500$. Based on the presented graphic material, it is possible to visually determine the influence of the sphericity of the structure on the antisymmetry of the distribution of quantities $u_3^1(1)$ and $u_3^2(2)$ along the spatial coordinate. The first natural frequency of the structure is 1492 Hz.

Figures 24.8 and 24.9 show the dependences of the normal deflections in the middle surfaces of the inner $u_3{}^1(1)$ and outer $u_3{}^2(2)$ bearing layers, and their normal stresses $\sigma_{22}{}^1(1)$ and $\sigma_{22}{}^2(2)$ on time, respectively; a symmetric three-layer spherical shell with a discrete-symmetric aggregate at its top under the action of an impulse load. Light filler is absent. Index 1 indicates the values of the inner bearing layer, and index 2 represents the outer bearing layer. It should be noted that the graphs in Figs. 24.8 and 24.9 agree well with the corresponding graphs given in [11].

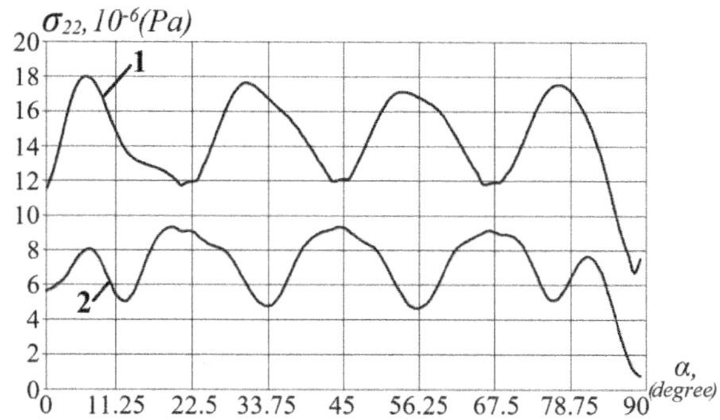

Fig. 24.7 Relationship between normal stresses in the middle surfaces of the bearing layers and the angular coordinate

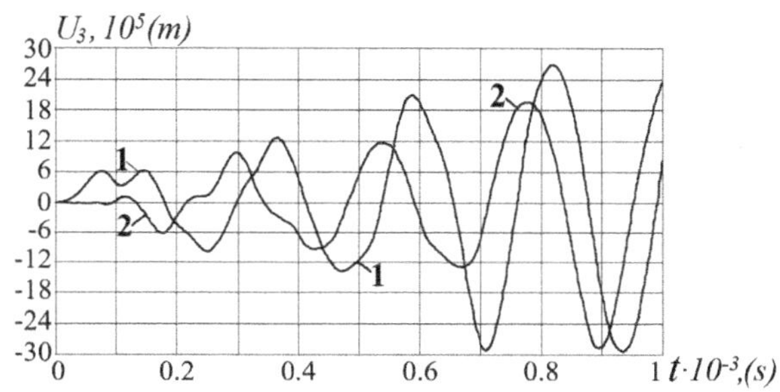

Fig. 24.8 Relationship between the normal deflections in the middle surfaces of the inner and outer bearing layers and time

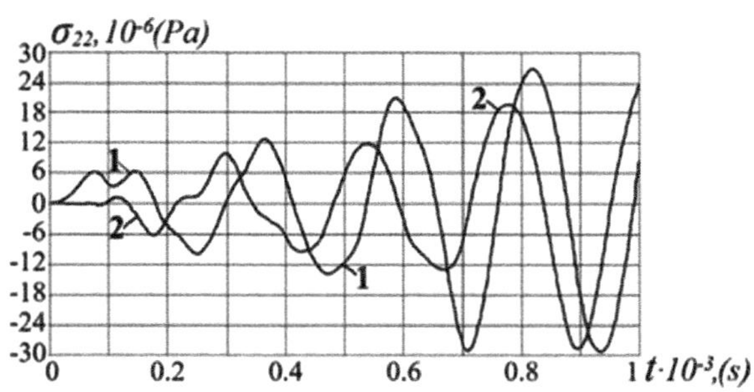

Fig. 24.9 Relationship between the normal stresses in the middle surfaces of the inner and outer bearing layers and time

24.4 Conclusions

In this article, independent static and kinematic hypotheses of Tymoshenko were applied for each layer of three-layer spherical shells. This increased the order of the system of solving equations of oscillations of three-layer structures with a discrete-symmetric lightweight, rib-reinforced aggregate. It made it possible to study the dynamic processes in such shells in more detail. The finite-element calculations show that the geometric and physical–mechanical parameters of the shell layers significantly affect the quantitative and qualitative characteristics of vibrations in three-layer spherical shells. Given that the original problem is multi-parametric (at different moments of time t, the kinematic and force parameters take different values along the coordinate s), in the future, we will consider the dependence of the initial values at the moments of time when they reach their maximum values.

In a symmetrical three-layer spherical shell without foam plastic, the maximum deflections of the inner bearing layer ${u_3}^1$ are equal to $9.6 \cdot 10^{-5}$ m, and in a three-layer shell with foam plastic $E_{1,2}/E_t = 50$, they decrease by 2.18 times and the first natural frequency increases compared to the three-layer spherical shell without foam plastic by 5.2%. In a three-layer spherical shell with foam $E_{1,2}/E_t = 500$, the maximum deflections ${u_3}^1$ decrease by 1.52 times, and the first natural frequency increases in comparison with a three-layer spherical shell without foam by 3.5%.

In a symmetrical three-layer spherical shell without foam plastic, the maximum deflections of the outer bearing layer ${u_3}^2$ are equal to $3.2 \cdot 10^{-5}$ m, and in a three-layer shell with foam plastic $E_{1,2}/E_t = 50$, they increase by 12.5%. In a three-layer spherical shell with foam $E_{1,2}/E_t = 500$, the maximum deflections ${u_3}^2$ remain almost unchanged, compared to a three-layer shell without foam.

References

1. Ambartsumyan, S.A.: Some basic equations of the theory of a thin layered shell. Dokl AN Arm SSR. **5**, 203–210 (1948). [in Russian]
2. Ambartsumyan, S.A.: Theory of Anisotropic Shells. Fizmatgiz, Moscow (1961). [in Russian]
3. Frostig, Y., Thomsen, O.T.: Higher-order free vibration of sandwich panels with a flexible core. Int. J. Solids Struct. **41**, 1697–1724 (2004)
4. Hause, T., Librescu, L.: Dynamic response of doubly-curved anisotropic sandwich panels impacted by blast loadings. Int. J. Solids Struct. **44**(20), 6678–6700 (2007)
5. Hohe, J., Becker, W.: Effective stress-strain relations for two-dimensional cellular sandwich cores: homogenization, material models, and properties. Appl. Mech. Rev. **55**(1), 61–87 (2002)
6. Hohe, J., Librescu, L.: Nonlinear theory for double-curved anisotropic sandwich shells with cross-compressed cores. Int. J. Solid Struct. **40**(5), 1059–1088 (2003)
7. Kheirikhah, M.M., Khalili, S.M.R., Malekzadeh, Fard K.: Biaxial buckling analysis of soft-core composite sandwich plates using improved high-order theory. Eur. J. Mech. A Solids. **31**, 54–66 (2012)
8. Lekhnytskyi, S.G.: Anisotropic Plates. OGIZ, Moscow (1947). [in Russian]
9. Lugovoi, P.Z., Meish, V.F.: Dynamics of inhomogeneous shell systems under nonstationary loading (suevey). Int. Appl Mech. **53**(5), 481–537 (2017)
10. Lugovoi, P.Z., Meish, V.F., Meish, Y.A., Orlenko, S.P.: Dynamic design of compound shell structures of revolution under nonstationary loads. Int. Appl. Mech. **56**(1), 22–32 (2020)
11. Lugovoi, P.Z., Meish, V.F., Orlenko, S.P.: Numerical simulation of the dynamics of spherical sandwich shells reinforced with discrete ribs under a shockwave. Int. Appl. Mech. **56**(5), 590–598 (2020)
12. Lychev, S.A., Sayfutdinov, Y.N.: Dynamics of a three-layer non-sloping spherical shell. Vestnik ChSPU im I. Ya Yakovleva. Mech Limit State. **2**, 54–90 (2007). [in Russian]
13. Lychev, S.A., Sidorov, Y.A.: Nonstationary oscillations of three-layer spherical shells with a multiple spectrum. Izv Univer Constr. **4**, 31–39 (2001). [in Russian]
14. Malekzadeh Fard, K., Livani, M., Veisi, A., Gholami, M.: Improved high-order bending analysis of double curved sandwich panels subjected to multiple loading conditions. Latin Amer. J. Solids Struct. **11**, 1591–1614 (2014)
15. Meish, V.F., Shtantsel, S.E.: Dynamic problems in the theory of sandwich shells of revolution with a discrete core under nonstationary loads. Int. Appl. Mech. **38**(12), 1501–1507 (2002)

16. Müeller, W., Grigorenko, Ya.M., Grigorenko, A.Ya., Vlaikov, G.G.: Mechanics of anisotropic heterogeneous shells: fundamental relations for different models. In: Recent Developments in Anisotropic Heterogeneous Shell Theory (2016)
17. Orlenko, S.P.: Numerical modeling of the dynamics of a three-layer spherical shell with a discrete heterogeneous aggregate. Dopov. Nac. akad. nauk Ukr. **3**, 19–27 (2020)
18. Piskunov, V.G., Rasskazov, A.O.: Development of the theory of layered plates and shells. In: Guz, A.N. (ed.) Adv. Mech., pp. 141–175. ASK, Kyiv (2007) [in Russian]
19. Rychkov, S.P.: Structural Modeling in Femap with NX Nastran. DMK Press, Moscow (2013). [in Russian]
20. Shul'ga, N.A., Meish, V.F.: Forced vibration of three-layered spherical and ellipsoidal shells under axisymmetric loads. Mech. Compos. Mater. **39**(5), 439–446 (2003)
21. Timoshenko, S.P., Voinovsky-Kriger, S.: Plates and Shells. McGraw-Hill, New York (1959)
22. YaM, Grygorenko: Isotropic and Anisotropic Layered Shells of Rotation of Variable Stiffness. Nauk Dumka, Kyiv (1973). [in Russian]

Chapter 25
New Structural Approach for Determination of Effective Thermoelastic Modules of Discrete Composite Layers

Mykhaylo Marchuk and Mykola Khomyak

Abstract The rule of mixture (ROM) for the thermoelastic properties of layered composite materials is important if we apply the effective single-layer model (ELSM), and their refinement is an actual problem. In contrast to the classical linear plate theory (CLPT), real through-the-thickness strain distributions for thick and moderately thick plates may significantly deviate from the linear ones, which is demonstrated for the typical stacks of layers within the framework of 3-D linear elasticity theory using the finite-element method (FEM). If the linear strain distribution is interpreted as a statistical regression model, then the concept of ESLM may be proposed for laminated plates in a wider range of the length-to-thickness ratio. The idea of the offered new ROM within the structural approach, taking into account material properties, reinforcement orientation, thickness, and discrete layers position in a stack, is based on the separate classification of strain and stress contributions of stiffer or more flexible layers for each in-plane (normal and shear) component. Following a statistical approach, hypotheses regarding to through-the-thickness strain and stress distributions are formulated. The averaging technique is shown in detail for all in-plane strains and stresses. Similar to the CLPT, a matrix constitutive relation between forces-moments and averaged strains-curvatures are obtained. The stiffness matrices of the laminated composite are calculated using both arithmetic (direct) and geometric (inverse) averaging rules. In the future, verification, improvement, and implementation of the approach are planned.

25.1 Introduction

Laminated composite materials (CM) are widely used in modern industry and technology, particularly in aircraft and rocket engineering [4, 5, 11, 15]. Unlike traditional structural materials, composites can rationally combine their advantages (first of all, the ratio of weight to stiffness and strength, technological features of manufac-

M. Marchuk (✉) · M. Khomyak
Pidstryhach Institute for Applied Problems of Mechanics and Mathematics, National Academy of Science of Ukraine, Lviv, Ukraine
e-mail: marchuk@iapmm.lviv.ua

H. Altenbach et al. (eds.), *Selected Problems of Solid Mechanics and Solving Methods*, Advanced Structured Materials 204, https://doi.org/10.1007/978-3-031-54063-9_25

turing, small temperature expansion, etc.) with problem-oriented requirements, both at the stages of design and operation, for various load conditions and temperatures. One of the effective ways to implement the advanced functional features of composites is the formation of a layered structure, where individual layers or their bundles are responsible for the desired range of properties for the structure as a whole. As an example, fiber-reinforced polymer composites are often used to provide a response to certain types of loads within separate layers [19]. Various design methods and stacking of layers create new properties for the structure, which are often significantly different from the properties of constituent layers.

Thermoelastic properties are important to determine the stress-strain state of structural elements manufactured from laminated composites and predict their strength and failure behavior. We will limit ourselves to the consideration of plane structures or plates. The corresponding mathematical theory—mechanics of laminated composite plates and shells—is well developed and has in its arsenal both classical theory (CLPT), based on the Kirchhoff-Love hypothesis (the straight line perpendicular to the middle surface remains undeformed and unrotated; the plane stress state is assumed) for a thin layer or a package as a whole, as well as refined theories, which are more adequate to describe the properties of composites, for example, the first-order shear deformation theory (FSDT) [1, 6–8, 13, 17].

A significant increase in the power of modern computers makes it possible to apply the three-dimensional linear elasticity theory (3-D LET) or combine it with two-dimensional theories, which continue to develop, increasing their complexity but offering a higher-order approximation and accuracy comparable to 3-D LET. If the layers have distinct interlaminar surfaces (a physical surface simulates the interface between adjacent discrete layers), then the set of models and methods to study the composite plates is expanded due to a different approach to individual layers. Such models can more adequately take into account structural defects (such as delaminations at the interface, etc.) inherent in real composites [12].

At the same time, the practice of engineering calculations does not reject simplified approaches, such as using the monolayer with effective characteristics (effective single-layer model, ESLM). This makes it possible to use known analytical solutions and existing software without significant modifications to solve problems with layered plates (laminates) as an object of research. Unlike monolayer plates, the material for the ESLM cannot be directly classified as isotropic or even orthotropic, although we assume that the individual layer is. Various approaches to averaging and accentuation on the fiber or matrix properties are also possible.

The properties of laminated composites are averaged based on the properties of individual unidirectionally reinforced layers (or plies) utilizing the structural data about the material, thickness, fiber direction of each ply, and its place in the ply order. That is, we may talk about the meso-level of consideration of composite properties, but at the micromechanics level, the effective thermoelastic properties of a layer are determined based on the matrix and fiber data, their relative content, methods of spatial packing, and other factors. Here, we assume that each layer is orthotropic. The local (natural) coordinate system is aligned with the direction of reinforcement (direction 1, in the plate plane), and the transition to the global coordinate system (common for all layers) can be carried out by using the known formulas of rotation for strain and stress tensors. Let us consider that the layer properties are already defined, and the rule of the mixture (ROM) for the meso-level is interesting only.

Note that Voigt's approach [23] or arithmetic averaging (the direct ROM) is widely used at the micromechanics level for effective elastic modulus E_1 in the direction of reinforcement 1, as well as the inverse ROM— A. Reuss' approach [18] or geometric averaging—for E_2 in the transverse direction 2, respectively:

$$E_1 = \omega^{(1)} E^{(1)} + (1 - \omega^{(1)}) E^{(2)},$$
$$\frac{1}{E_2} = \frac{\omega^{(1)}}{E^{(1)}} + \frac{(1 - \omega^{(1)})}{E^{(2)}},$$

where $E^{(1)}$ and $E^{(2)}$ are Young's moduli of the matrix and fibers, and $\omega^{(1)}$, $(1 - \omega^{(1)})$ are their relative volume contributions. Similarly, the inverse rule is more adequate for the shear modulus G_{12}. More details about the averaging of the engineering constants of Hooke's law can be found, for example, in the monographs [1, 11, 22] or in some recent papers [2, 3].

The direct ROM may be interpreted as a parallel combination of springs (the hypothesis of constant strains), and the inverse ROM—as a serial combination (the hypothesis of constant stresses). Applying both rules gives the upper and lower estimates, respectively. Modern structural approaches usually combine these hypotheses in order to clarify the effective values of thermoelastic parameters [9, 10]. More complex formulas can also be used, which take into account experimental data, exact solutions of the 3-D LET, etc. It is known that direct averaging gives overestimated values of the CM stiffness and, therefore, underestimates the deflections of a laminated plate with such effective characteristics.

On the contrary, at the meso-level, it is customary to use integration by thickness (properties of discrete layers are considered constant), and the direct ROM gives results well agreed with the experiments in some cases [11, 21]. The direct approach is inherent to the classical and many refined theories of layered structures (plates and shells) [17]. However, laminated composites contain layers with different stiffness in the directions of the global coordinate system. And probably they respond in a different way to both in-plane and transverse loads, causing the interaction of in-plane and moment stresses and strains. By analogy with micromechanics approaches, even within the framework of the classical theory of laminated plates (CLPT), which is chosen for simplicity of illustration of the main ideas of the proposed new approach, it is possible to use mixed ROM within the ESLM. The aim of this study is to propose a fairly simple and practical structural approach for determining the effective thermoelastic moduli of laminated composites that takes into account the differences in the deformation of relatively stiffer or compliant layers. Also, the thickness of each layer and its position in the entire stack are important. This approach is addressed both for unidirectional reinforced material oriented in different directions for different layers, as well as hybrid composites composed of different materials.

25.2 CLPT and Statistical Interpretation of Through-the-Thickness Strain Distributions

Let us consider a laminated plate of thickness h containing N layers, which is subjected to mechanical and temperature loads. We introduce a Cartesian coordinate system (x, y, z) common to all layers, where axes x and y are parallel to the plate edges and axis z is perpendicular to the layers' plane (x, y). The coordinate $z = 0$ defines the middle surface and varies from $-h/2$ to $h/2$ on the lower and upper faces, respectively (see Fig. 25.1). According to the ratio $s = L/h$, where L is the characteristic linear size along x or y, the plates are classified as thin ($s > 20$), medium thickness ($s \sim 10$), and thick ($s < 10$). This distinction is quite conditional and does not limit the application of specific hypotheses related to the thin-walled plates, with approaches to the laminate in all. For example, a sufficiently thick plate can be composed of many thin layers, etc.

Thin layers (or plates as a whole) have linear distributions of the stress-strain state characteristics across the thickness. Existing refined theories for layered plates in a wide range of thicknesses (from thin to thick) offer different options to describe displacements, strains, and stresses (both piecewise linear and nonlinear distributions by z). The generally accepted assumptions are (1) the continuity of displacements and strains across the surfaces between the layers (i.e., on the interface) and (2) a jump-like change of stresses at the interface. Thermoelastic properties for each layer are piecewise constant [11, 17], Fig. 25.1. We will explicitly or implicitly adhere to these assumptions, as they are sufficient for the further presentation of the ideas of a new mixed rule of thickness-averaged properties of laminate composites, i.e., within the framework of the ESLM.

In the framework of the CLPT, the conditions of the plane stress state are assumed to be fulfilled $\sigma_{xz} = \sigma_{yz} = \sigma_{zz} = 0$. If ith layer is sufficiently thin, then we can accept the hypothesis of linear through-the-thickness distributions:

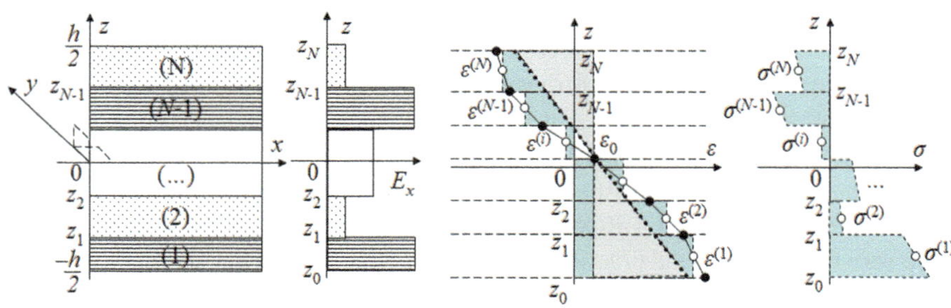

Fig. 25.1 Laminated plate: layer numbering and (typical) through-the-thickness distributions of elastic moduli E, strains ε, and stresses σ (illustration). Circles show the average values for the layer

$$\begin{aligned} \{\mathbf{e}^{(i)}(z)\} &= \{\boldsymbol{\varepsilon}_0^{(i)}\} + z\{\boldsymbol{\kappa}_0^{(i)}\}, \\ \{\boldsymbol{\sigma}^{(i)}(z)\} &= \{\boldsymbol{\sigma}_0^{(i)}\} + z\{\boldsymbol{\sigma}_1^{(i)}\} \quad (i = 1, \ldots, N), \end{aligned} \tag{25.1}$$

where

$$\{\mathbf{e}^{(i)}\} = \{e_{xx}^{(i)}, e_{yy}^{(i)}, \gamma_{xy}^{(i)}\}^{\mathrm{T}}$$

are total (elastic and temperature) strains,

$$\{\boldsymbol{\varepsilon}_0^{(i)}\} = \{\varepsilon_{xx}^{(i)}, \varepsilon_{yy}^{(i)}, \gamma_{xy}^{(i)}\}^{\mathrm{T}}, \quad \{\boldsymbol{\kappa}_0^{(i)}\} = \{\kappa_{xx}^{(i)}, \kappa_{yy}^{(i)}, \kappa_{xy}^{(i)}\}^{\mathrm{T}}$$

are mean strains and curvatures of the ith layer,

$$\{\boldsymbol{\sigma}^{(i)}\} = \{\sigma_{xx}^{(i)}, \sigma_{yy}^{(i)}, \sigma_{xy}^{(i)}\}^{\mathrm{T}}$$

are stresses (in-plane components), and

$$\begin{aligned} \{\boldsymbol{\sigma}_0^{(i)}\} &= \{(\sigma_{xx}^0)^{(i)}, (\sigma_{yy}^0)^{(i)}, (\sigma_{xy}^0)^{(i)}\}^{\mathrm{T}}, \\ \{\boldsymbol{\sigma}_1^{(i)}\} &= \{(\sigma_{xx}^1)^{(i)}, (\sigma_{yy}^1)^{(i)}, (\sigma_{xy}^1)^{(i)}\}^{\mathrm{T}} \end{aligned}$$

are mean stresses and stress gradients (or membrane and bending components), respectively. Of course, the truth of the hypothesis also depends on the load, boundary conditions, characteristics of the layers and their order in the package, etc.

Based on the representations of displacements and (in-plane) strains (first derivatives of displacements noted by the subscripts after the comma)

$$\begin{aligned} U_x(x, y, z) &= u_x(x, y) + z\theta_x(x, y), \quad x \to y, \\ W(x, y, z) &= w(x, y), \\ e_{xx}(x, y, z) &= u_{x,x}(x, y) + z\theta_{x,x}(x, y), \quad x \to y, \\ \gamma_{xy}(x, y, z) &= u_{x,y}(x, y) + u_{y,x}(x, y) + z\theta_{x,y}(x, y) + \theta_{y,x}(x, y), \end{aligned}$$

it is possible to conclude about the continuity of in-plane components along the thickness of the layered plate (in z-direction). On the contrary, differences in stiffness matrices of adjacent layers result in finite jumps of stress components at the interface between layers.

Next, let each layer be elastic (orthotropic). The axes of the local (natural) coordinate system $(1, 2, 3)$ of the layer are aligned to the axes of orthotropy, axis 3 of the local coordinate system and z of the global one are coincident, and the angle between axis 1 and x is $\alpha \in (-90°, 90°]$. Let such layers be unidirectionally or cross-reinforced composites, the elastic properties of which are determined using micromechanics formulas taking into account the properties of the matrix and fibers [1, 13, 17]. With respect to the axes of orthotropy of any layer and taking into account temperature expansions, Hooke's law becomes

$$\{\boldsymbol{\sigma}^{(i)}(z)\} = [\mathbf{Q}^{(i)}]\left(\{\mathbf{e}^{(i)}(z)\} - \{\boldsymbol{\alpha}_0^{(i)}\}\left(T_0^{(i)} + \Delta T_1^{(i)} z\right)\right) \tag{25.2}$$

$(i = 1, \ldots, N)$, where

$$\{\mathbf{e}^{(i)}(z)\} = \{e_{11}^{(i)}, e_{22}^{(i)}, \gamma_{12}^{(i)}\}^{\mathrm{T}}, \ \{\boldsymbol{\sigma}^{(i)}(z)\} = \{\sigma_{11}^{(i)}, \sigma_{22}^{(i)}, \sigma_{12}^{(i)}\}^{\mathrm{T}},$$

$$\{\boldsymbol{\alpha}_0^{(i)}\} = \{\alpha_1^{(i)}, \alpha_2^{(i)}, 0\}^{\mathrm{T}}, \qquad [\mathbf{Q}^{(i)}] = \begin{bmatrix} Q_{11}^{(i)} & Q_{12}^{(i)} & 0 \\ & Q_{22}^{(i)} & 0 \\ \text{Symm} & & Q_{66}^{(i)} \end{bmatrix}.$$

Here, ${\alpha_1}^{(i)}$ and ${\alpha_2}^{(i)}$ are coefficients of thermal expansion (CTE) in directions 1 and 2, respectively, ${T_0}^{(i)} = (T^{+(i)} + T^{-(i)})/2$ is the mean temperature over the layer, and $\Delta T^{(i)} = (T^{+(i)} - T^{-(i)})/h$ is the temperature gradient in the direction from its lower to the upper interface surface. Note that we assume the continuity of the temperature field along the thickness and on the interfaces of the ith layer. As a temperature value, we measure its difference from the reference level that does not cause stresses.

For averaging across the thickness, the global coordinate system will be used. Let us transform stresses and strains according to the known rotation rules from the local to the global coordinate system:

$$\{\boldsymbol{\sigma}^{(i)}\}_{\mathrm{gl}} = [T_\sigma^{(i)}]\{\boldsymbol{\sigma}^{(i)}\}_{\mathrm{loc}}, \quad \{\mathbf{e}^{(i)}\}_{\mathrm{gl}} = [T_\varepsilon]\{\mathbf{e}^{(i)}\}_{\mathrm{loc}}, \tag{25.3}$$

where

$$\{\boldsymbol{\sigma}^{(i)}\}_{\mathrm{gl}} = \{\sigma_{xx}^{(i)}, \sigma_{yy}^{(i)}, \sigma_{xy}^{(i)}\}^{\mathrm{T}}, \quad \{\mathbf{e}^{(i)}\}_{\mathrm{gl}} = \{e_{xx}^{(i)}, e_{yy}^{(i)}, \gamma_{xy}^{(i)}\}^{\mathrm{T}},$$

$$[T_\sigma] = \begin{bmatrix} c^2 & s^2 & 2cs \\ s^2 & c^2 & -2cs \\ -cs & cs & c^2 - s^2 \end{bmatrix}, \quad [T_\varepsilon] = \begin{bmatrix} c^2 & s^2 & cs \\ s^2 & c^2 & -cs \\ -2cs & 2cs & c^2 - s^2 \end{bmatrix},$$

$$[T_\sigma]^{-1} = [T_\varepsilon]^{\mathrm{T}}, \quad [T_\varepsilon]^{-1} = [T_\sigma]^{\mathrm{T}}, \quad c = \cos\alpha^{(i)}, \quad s = \sin\alpha^{(i)}.$$

If multiply the matrix constitutive Eq. (25.2) on the left by $[T_\sigma]$, take into account the transformation (25.3) and the fact that

$$[\mathbf{T}_\varepsilon]^{-1}[\mathbf{T}_\varepsilon] = [\mathbf{T}_\sigma]^{\mathrm{T}}[\mathbf{T}_\varepsilon]$$

is a $[3 \times 3]$-unity matrix, we obtain Hooke's law in the global coordinate system:

$$\{\boldsymbol{\sigma}^{(i)}\}_{\mathrm{gl}} = [\bar{\mathbf{Q}}^{(i)}]\{\mathbf{e}^{(i)}\}_{\mathrm{gl}} - \{\boldsymbol{\beta}^{(i)}\}\left(T_0^{(i)} + \Delta T_1^{(i)} z\right), \tag{25.4}$$

where

$$[\bar{\mathbf{Q}}^{(i)}] = [T_\sigma][\mathbf{Q}^{(i)}][T_\sigma]^{\mathrm{T}}$$

is the symmetric stiffness matrix,

$$\{\boldsymbol{\beta}^{(i)}\} = \left[\beta_{xx}^{(i)}, \beta_{yy}^{(i)}, \beta_{xy}^{(i)}\right]^{\mathrm{T}} = \left[\bar{\mathbf{Q}}^{(i)}\right]\{\boldsymbol{\alpha}^{(i)}\}$$

is thermal stiffness vector of the layer (all in the global coordinate system), and

$$\{\boldsymbol{\alpha}^{(i)}\} = \left\{\alpha_{xx}^{(i)}, \alpha_{yy}^{(i)}, \alpha_{xy}^{(i)}\right\}^{\mathrm{T}} = [\mathbf{T}_{\varepsilon}]\{\boldsymbol{\alpha}_0^{(i)}\}.$$

Hypothesis. In-plane strains $(25.1)_1$ have a linear through-the-thickness distribution for the entire package of layers:

$$\{\mathbf{e}(z)\} = \{\boldsymbol{\varepsilon}_0\} + z\{\boldsymbol{\kappa}\}, \tag{25.5}$$

where $\{\mathbf{e}(z)\} = \{e_{xx}(z), e_{yy}(z), \gamma_{xy}(z)\}^{\mathrm{T}}$ are total strains for the plane stress state, $\{\boldsymbol{\varepsilon}_0\} = \{\bar{\varepsilon}_x, \bar{\varepsilon}_y, \bar{\gamma}_{xy}\}^{\mathrm{T}}$ are mean strains of laminated plates brought to the middle surface, and $\{\boldsymbol{\kappa}\} = \{\bar{\kappa}_x, \bar{\kappa}_y, \bar{\kappa}_{xy}\}^{\mathrm{T}}$ are curvatures.

Equation (25.5) may be interpreted statistically as a linear regression model for the "true" distribution (25.1). According to the least squares method (LSM), regression (25.5) can be found from the minimum conditions of the residual function:

$$F(\{\boldsymbol{\varepsilon}_0\}, \{\boldsymbol{\kappa}\}) = \int_{-h/2}^{h/2} [\{\boldsymbol{\varepsilon}_0\} + z\{\boldsymbol{\kappa}\} - \{\mathbf{e}(z)\}]^2 dz \to \min. \tag{25.6}$$

From here we obtain a system of two equations

$$\begin{bmatrix} \partial F/\partial\{\boldsymbol{\varepsilon}_0\} \\ \partial F/\partial\{\boldsymbol{\kappa}\} \end{bmatrix} = \begin{bmatrix} h & 0 \\ 0 & d \end{bmatrix} \begin{bmatrix} \{\boldsymbol{\varepsilon}_0\} \\ \{\boldsymbol{\kappa}\} \end{bmatrix} - \begin{bmatrix} \sum_{i=1}^{N} \{\boldsymbol{\varepsilon}_0^{(i)}\} h_i + \sum_{i=1}^{N} \{\boldsymbol{\kappa}^{(i)}\} b_i \\ \sum_{i=1}^{N} \{\boldsymbol{\varepsilon}_0^{(i)}\} b_i + \sum_{i=1}^{N} \{\boldsymbol{\kappa}^{(i)}\} d_i \end{bmatrix} = 0, \tag{25.7}$$

where

$$\begin{gathered} h_i = z_i - z_{i-1}, \quad b_i = \frac{1}{2}\left(z_i^2 - z_{i-1}^2\right) = h_i \bar{z}_i, \quad \bar{z}_i = \tfrac{1}{2}\left(z_{i-1} + z_i\right), \\ d_i = \frac{1}{3}\left(z_i^3 - z_{i-1}^3\right), \quad h = \sum_{i=1}^{N} h_i, \quad 0 = \sum_{i=1}^{N} b_i, \quad d = \frac{1}{12}h^3 = \sum_{i=1}^{N} d_i. \end{gathered} \tag{25.8}$$

We will use the known result from mathematical statistics about two sets of numbers $\{x_1, x_2, \ldots, x_n\}$ and $\{y_1, y_2, \ldots, y_n\}$, which are a sample from the general set of independent quantities x and y. Then their covariance is zero:

$$\operatorname{cov}(x, y) \equiv \langle (x - \langle x \rangle)(y - \langle y \rangle) \rangle = \langle xy \rangle - \langle x \rangle \langle y \rangle = 0,$$

where $\langle \cdot \rangle$ is a mean (mathematical expectation) operator. That is, the mean of products is equal to the product of means (in the statistical sense),

$$\langle xy \rangle = \langle x \rangle \langle y \rangle.$$

If, for example, we consider pairs of values $\{\boldsymbol{\varepsilon}_0^{(i)}\}$ and b_i or $\{\boldsymbol{\kappa}^{(i)}\}$ and b_i as independent random variables, then

$$\sum_{i=1}^{N}\{\boldsymbol{\varepsilon}_0^{(i)}\}b_i = \sum_{i=1}^{N}\langle\{\boldsymbol{\varepsilon}_0^{(i)}\}\rangle b_i = \{\boldsymbol{\varepsilon}_0\}\sum_{i=1}^{N} b_i = 0,$$
$$\sum_{i=1}^{N}\{\boldsymbol{\kappa}^{(i)}\}b_i = \sum_{i=1}^{N}\langle\{\boldsymbol{\kappa}^{(i)}\}\rangle b_i = \langle\{\boldsymbol{\kappa}^{(i)}\}\rangle\sum_{i=1}^{N} b_i = 0,$$

and from (25.7), we get

$$\{\boldsymbol{\varepsilon}_0\} = \frac{1}{h}\sum_{i=1}^{N}\{\boldsymbol{\varepsilon}_0^{(i)}\}h_i, \quad \{\boldsymbol{\kappa}\} = \frac{1}{d}\sum_{i=1}^{N}\{\boldsymbol{\kappa}^{(i)}\}d_i. \tag{25.9}$$

Thereby, within the framework of the statistical approach, we may interpret a wider class of problems (and, accordingly, nonlinear through-the-thickness distributions of strains), slightly complicating the CLPT formalism, which may then be considered as a partial (deterministic) case. Values $\{\boldsymbol{\varepsilon}_0\}$ and $\{\boldsymbol{\kappa}\}$ may be interpreted as weighted means for a specific distribution of strains in layers (25.1). On the one hand, if we assume that all components of layers $\{\boldsymbol{\varepsilon}_0{}^{(i)}\}$ and $\{\boldsymbol{\kappa}^{(i)}\}$ are constant, then we obtain the case of the CLPT, and on the other hand, Eq. (25.9) may be postulated as some rule of statistical averaging that satisfies the LSM condition (25.6).

The advantages of the statistical approach include the fact that strain continuity is not mandatory here. Potentially, this opens up the possibility of applying the statistical approach and the averaging schemes proposed below to layered structures with non-ideal contact between layers. For example, if there are finite jumps of strains at the interfaces, then we may also construct the ESLM.

Although the stresses within CLPT have a piecewise continuous distribution over the thickness (jumps on interface surfaces are assumed to be finite), similar to (25.9) averaging schemes may be proposed. The average (linearized) stresses across the thickness of the plate

$$\{\boldsymbol{\sigma}\} = \{\sigma_{xx}(z), \sigma_{yy}(z), \sigma_{xy}(z)\}^{\mathrm{T}} = \frac{1}{h}\{\mathbf{N}\} + \frac{12z}{h^3}\{\mathbf{M}\} \tag{25.10}$$

exist in the LSM sense:

$$F(\{\mathbf{N}\}, \{\mathbf{M}\}) = \int_{-h/2}^{h/2}[\{\mathbf{N}\} + z\{\mathbf{M}\} - \{\boldsymbol{\sigma}(z)\}]^2 dz \to \min,$$

where integrated factors (forces and moments) are introduced,

$$\begin{bmatrix} \{\mathbf{N}\} \\ \{\mathbf{M}\} \end{bmatrix} = \int_{-h/2}^{h/2} \{\sigma(z)\} \begin{bmatrix} 1 \\ z \end{bmatrix} dz, \\ \{\mathbf{N}\} = \{N_x, N_y, N_{xy}\}^{\mathrm{T}}, \quad \{\mathbf{M}\} = \{M_x, M_y, M_{xy}\}^{\mathrm{T}}. \tag{25.11}$$

Therefore, taking into account the linear distribution $(25.1)_2$, and on the basis of statistical interpretation and the property of covariance, it is possible to propose an averaging rule for stresses

$$\{\mathbf{N}\} = \sum_{i=1}^{N} \{\sigma_0^{(i)}\} h_i, \quad \{\mathbf{M}\} = \sum_{i=1}^{N} \{\sigma_1^{(i)}\} d_i \tag{25.12}$$

for which is performed:

$$\sum_{i=1}^{N} \sigma_0^{(i)} b_i = 0, \quad \sum_{i=1}^{N} \sigma_1^{(i)} b_i = 0.$$

It should be noted that when choosing a statistical approach, in the narrow sense, we cannot interpret individual layers, for example, to restore stresses by layers, having the averaged strain distributions (25.5), substituting it into Eq. (25.4). However, this is a reasonable compromise when using the ESLM.

Hypothesis. The temperature distribution across the thickness of the entire package of layers is linear:

$$T(z) = T_0 + \Delta T_1 z. \tag{25.13}$$

We may also interpret the distribution (25.13) in the sense of the LSM:

$$F(T_0, \Delta T_1) = \int_{-h/2}^{h/2} [T_0 + \Delta T_1 z - T(z)]^2 dz \to \min,$$

where

$$T_0 = \frac{1}{h} \int_{-h/2}^{h/2} T(z) dz = \frac{1}{h} \sum_{i=1}^{N} T_0^{(i)} h_i, \ \Delta T_1 = \frac{1}{d} \int_{-h/2}^{h/2} T(z) z dz = \frac{1}{d} \sum_{i=1}^{N} \Delta T_1^{(i)} d_i,$$
$$\sum_{i=1}^{N} T_0^{(i)} b_i = 0, \ \sum_{i=1}^{N} \Delta T_1^{(i)} b_i = 0.$$

Then the temperature strains $\{\mathbf{e}_0^{\mathrm{T}}(z)\} = \{e_{xx}^{\mathrm{T}}(z), e_{yy}^{\mathrm{T}}(z), \gamma_{xy}^{\mathrm{T}}(z)\}^{\mathrm{T}}$ are approximated by a linear distribution

$$\{\mathbf{e}_0^{\mathrm{T}}(z)\} = \{\boldsymbol{\varepsilon}_0^{\mathrm{T}}\} + z\{\boldsymbol{\kappa}^{\mathrm{T}}\}, \tag{25.14}$$

where

$$\begin{bmatrix} h & 0 \\ 0 & d \end{bmatrix} \left[\begin{Bmatrix} \boldsymbol{\varepsilon}_0^{\mathrm{T}} \\ \boldsymbol{\kappa}^{\mathrm{T}} \end{Bmatrix} \right] = \sum_{i=1}^{N} \int_{z_{i-1}}^{z_i} \left[\{\boldsymbol{\alpha}^{(i)}\} \left(T_0^{(i)} + \Delta T_1^{(i)} z \right) \right] \begin{bmatrix} 1 \\ z \end{bmatrix} dz.$$

According to the statistical approach, we replace the average of the products of CTE $\{\boldsymbol{\alpha}^{(i)}\}$ and temperature $T^{(i)}(z)$ in layers with the product of their averages. Then we obtain for (25.14)

$$\{\boldsymbol{\varepsilon}_0^{\mathrm{T}}\} = T_0\{\boldsymbol{\alpha}\}, \quad \{\boldsymbol{\kappa}^{\mathrm{T}}\} = \Delta T_1\{\boldsymbol{\alpha}\},$$

where

$$\{\boldsymbol{\alpha}\} = \frac{1}{h} \int_{-h/2}^{h/2} \begin{bmatrix} \alpha_{xx}(z) \\ \alpha_{yy}(z) \\ \alpha_{xy}(z) \end{bmatrix} dz = \frac{1}{h} \sum_{i=1}^{N} \int_{z_{i-1}}^{z_i} \{\boldsymbol{\alpha}^{(i)}\} dz = \frac{1}{h} \sum_{i=1}^{N} \{\boldsymbol{\alpha}^{(i)}\} h_i,$$

and

$$\sum_{i=1}^{N} \{\boldsymbol{\alpha}^{(i)}\} b_i = 0.$$

Next, let us integrate Eq. (25.4) for the entire package of layers:

$$\begin{bmatrix} \{\mathbf{N}\} \\ \{\mathbf{M}\} \end{bmatrix} = \begin{bmatrix} [\mathbf{A}] & [\mathbf{B}] \\ [\mathbf{B}] & [\mathbf{D}] \end{bmatrix} \begin{bmatrix} \{\boldsymbol{\varepsilon}_0\} - \{\boldsymbol{\varepsilon}_0^{\mathrm{T}}\} \\ \{\boldsymbol{\kappa}\} - \{\boldsymbol{\kappa}^{\mathrm{T}}\} \end{bmatrix}, \tag{25.15}$$

where $[\mathbf{A}]$, $[\mathbf{B}]$, and $[\mathbf{D}]$ are the symmetric stiffness matrices,

$$\{[\mathbf{A}], [\mathbf{B}], [\mathbf{D}]\} = \sum_{i=1}^{N} [\bar{\mathbf{Q}}^{(i)}]\{h_i, b_i, d_i\}. \tag{25.16}$$

25.3 Some Numerical Experiments

Let us assume that strains and stresses of layered CM are linearly dependent on the thickness coordinate z according to (25.1). In practical applications, attention is focused on the study of stresses that affect the assessment of strength and failure of structures, and information about strains is of general features, that is, they are considered linear (and continuous) across the thickness of layers. However, for different layered structures (depending on the directions of reinforcement, thicknesses, and thermoelastic properties of the material), these stress and strain distributions can have both quantitative and qualitative features, which should be taken into account in the ESLM.

The statistical approach involves a certain sample from the general population of possible results for laminated plates (in particular, it can be numerical experiments).

Therefore, we present the results of some specific problems, from which we will try to draw both local and general conclusions (precisely, hypotheses) regarding the though-the-thickness distributions of strains or stress.

Many problems are known in the literature, where examples of such distributions are given, for example [14, 16]. They have been solved analytically or numerically, within the framework of various 2-D theories of laminated plates and approximations of the 3-D LET, in particular by FEM. A typical test problem is the bending of a simple supported rectangular (square) plate by the constant transverse load p. To solve it, our in-house code of FEM has been implemented based on governing equations of the 3-D LET and by using Serendip-type quadratic polynomials on the brick element with 20 nodes [24]. The finite-element model represents a 12×12-elements mesh and 1 element across the thickness of a layer. The symmetry of the problem is not taken into account since the distributions in the center of the plate will be analyzed, and it is desirable to remove the influence of any boundary conditions.

A square plate with length $L = 0.2$ m is considered, and the thicknesses of the layers are assumed to be the same and are calculated for the specified thin-walled parameter $s = L/h \in \{100, 10, 5\}$, as an example of the thin, medium, and thick plates, respectively. The constant load is chosen depending on the thickness of the plate, $p = 0.05 \cdot (100/s)^2$ MPa, in order to obtain a commensurate level within the limits of infinitesimally small strains in 3-D LET. Each layer is made of material with the following characteristics: $E_1 = 172.5$ GPa, $E_1/E_2 = 25$, $G_{12} = G_{13} = E_2/2$, $G_{23} = E_2/5$, and $\nu_{12} = \nu_{13} = \nu_{23} = 0.25$.

In partial cases, the numeric solutions have been controlled and practically coincided with results in the plate center known in the literature [14, 16, 17, 20]. For example, the errors both in deflections and stresses for the laminate (0/90/0) are within 1–2% (a reduced integration scheme with $2 \times 2 \times 2$ Gauss points is used).

Figures 25.2, 25.3, 25.4, and 25.5 show some of the obtained through-the-thickness distributions for strains and stresses in the plate center. Strains are given in percent, stresses are normalized according to the formula $\sigma = \sigma_{\max}/(s^2 p)$, where $\sigma_{\max} = |\sigma(x_0, y_0, \pm h/2)|$ is the maximum stress in the center of the plate. The figures are arranged as follows: the upper row—strains and stresses for a thin plate ($s = 100$, $p = 0.05$ MPa), the middle row—similarly for plates of a middle thickness ($s = 10$, $p = 5$ MPa), and the lower row—for thick plates ($s = 5$, $p = 20$ MPa). The two left columns present normal strains e_{xx}, e_{yy} and stresses σ_{xx}, σ_{yy}, and the two right columns represent the in-plane shear strain e_{xy} and stress σ_{xy}, respectively. The normalized thickness z/h varies from -0.5 (bottom surface) to 0.5 (the top surface where the load p is applied) and is plotted along the vertical axis in the graphs. The drawings are sufficiently simplified (non-detailed) and serve mainly for qualitative comparative analysis. Note that stress jumps on the interfaces are shown by horizontal lines, which should not be interpreted as continuity.

Several layered structures are considered:

1. $0°/90°/0°$—three-layer stack sequence with cross-ply reinforcement. It is also a test for software verification according to known results. The structure is symmet-

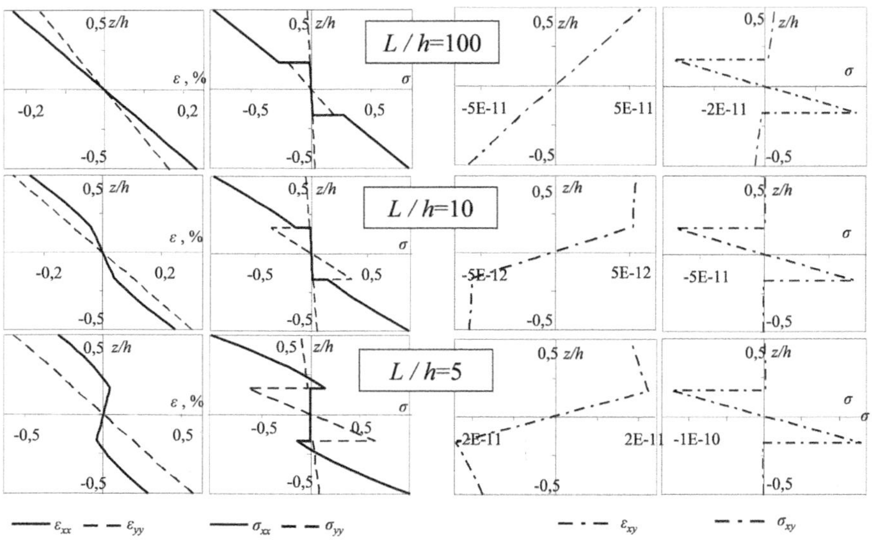

Fig. 25.2 Through-the-thickness distribution of in-plane strains and stresses in the center of a square plate, which is simply supported on all edges, and under the action of the constant transverse load. The layer stacking sequence in the global coordinate system (from bottom to top) is 0°/90°/0°

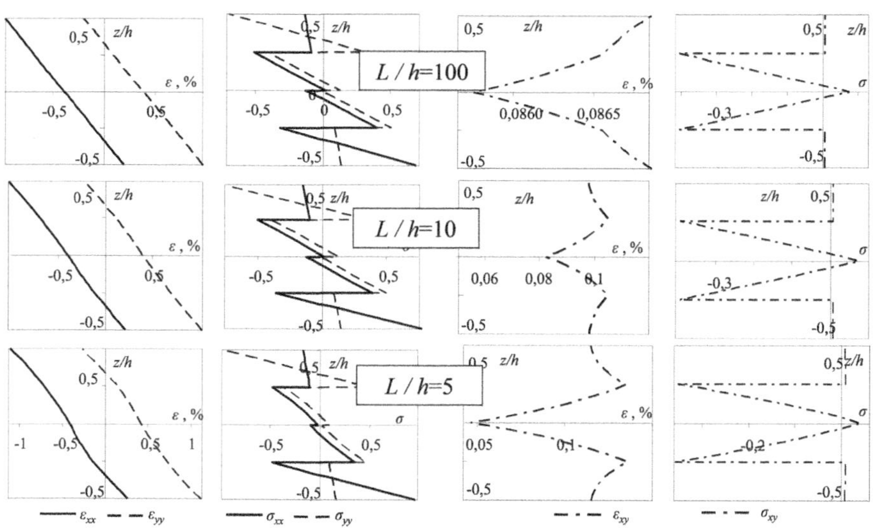

Fig. 25.3 Through-the-thickness distribution of in-plane strains and stresses in the center of a square plate, which is simply supported on all edges, and under the action of the constant transverse load. The layer stacking sequence is 0°/45°/−45°/90°

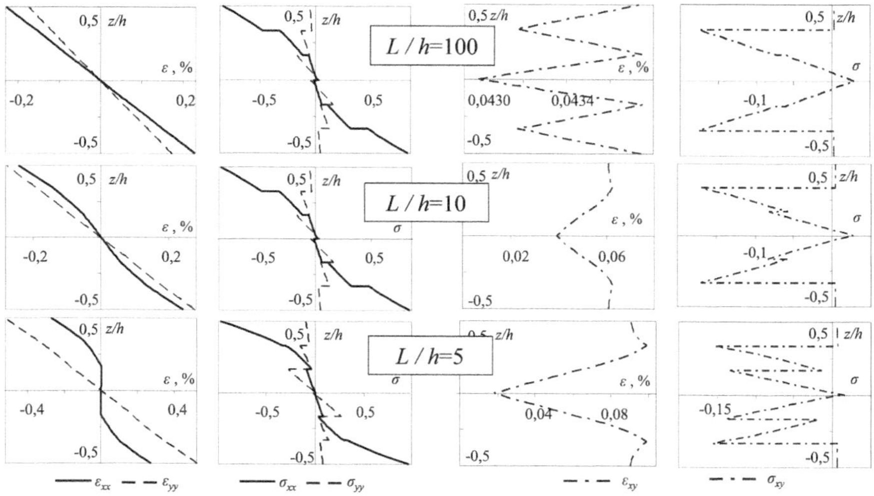

Fig. 25.4 Through-the-thickness distribution of in-plane strains and stresses in the center of a square plate, which is simply supported on all edges, and under the action of the constant transverse load. The layer stacking sequence is 0°/30°/60°/−60°/−30°/0°

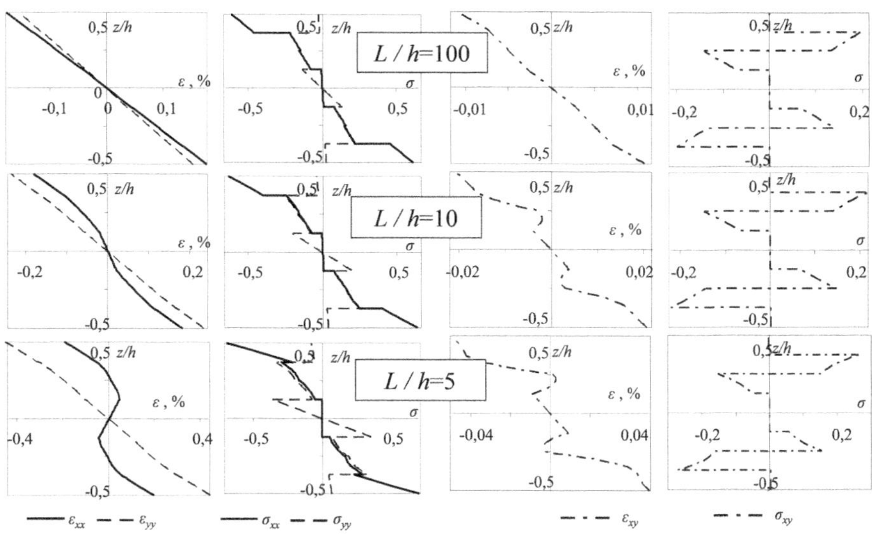

Fig. 25.5 Through-the-thickness distribution of in-plane strains and stresses in the center of a square plate, which is simply supported on all edges, and under the action of the constant transverse load. The layer stacking sequence is $(0°/45°-45°/90°)_S$

rical, with predominant reinforcement along the x-axis (two outer layers versus an inner layer oriented perpendicularly).

The load is taken by the upper unidirectional reinforced layer, oriented along the x-axis (hereinafter, we will say simply: the 0°-oriented layer, and so on);

2. $0°/45°/-45°/90°/0°$—asymmetric four-layer stacking, including cross- and angle-ply reinforcement. Here we focus on the inner layers, oriented at angles of $\pm 45°$;
3. $0°/30°/60°/-60°/-30°/0°$—six-layer sequence with angle-ply reinforcement at angles of $\pm 30°$ and $\pm 60°$. We will be interested in whether the gradient depends on the reinforcement angles;
4. $0°/45°/-45°/90°/90°/-45°/45°/0° = (0°/45°/-45°/90°)_S$—symmetrical eight-layer sequence.

Let us study the case of multi-layering.

The general qualitative characteristics of the stress-strain state show that the bending-related components with a significant through-thickness gradient are dominant. If graphs pass through the origin of coordinates, then membrane components (averaged strains or stresses) are equal to zero. Distributions of strains with increasing thickness become nonlinear (at least for some components), and distributions of stresses are affected by both the orientation and the order of alternating layers. In particular, one of the determining factors is the orientation of the upper layer, which takes the load. However, the distribution of all components within individual layers is linear or almost linear. Moreover, some of the stresses in the layers are close to zero (or have a much lower level and gradient than in other layers), although the corresponding strains are significant.

For a given load and boundary conditions, the normal components (e_{xx}, e_{yy} and σ_{xx}, σ_{yy}) prevail, and shear strains e_{xy} and stresses σ_{xy} have an insignificant effect, at least in the central part of the plate. Although near the corner points their influence becomes more significant, they qualitatively demonstrate the same effects (nonlinearity for non-thin plates, much lower values of stresses and gradients in some layers relative to others).

Analytical results are known for stacking variant 1 ($0°/90°/0°$) [14, 16]. Let us point out only the following. For thin plates, all strains e_{xx}, e_{yy}, and e_{xy} have linear through-the-thickness distributions, as predicted by the CLPT. Also, we show linear distributions of strains e_{yy} for all three cases of the thin, medium, and thick plates. For a thick plate, strains e_{xx} form a different order of alternating tension and compression zones compared to a plate of medium thickness. Accordingly, the stresses change the sign in the outer layers (closer to the interface with the 90°-oriented inner layer). Let us note that layers oriented with the same angles have (approximately) the same stress gradient. A similar conclusion is true for strain gradients.

Variant 2 of asymmetric stacking of layers ($0°/45°/-45°/90°$) has non-zero strains opposite in sign $e_{xx} \approx -e_{yy}$ on the middle surface ($z = 0$). Between 45°- and $-45°$-oriented layers, although there is a jump in stresses σ_{xx} and σ_{yy}, but their gradient is approximately (visually, according to the graph) the same. The jump for the shear stress σ_{xy} is not observed, and gradients have opposite signs but the same absolute value. Shear strain e_{xy}, in contrast to stacking $0°/90°/0°$, have the same order as e_{xx} or e_{yy}.

Variant 3 with angle-ply stacking of layers $(0°/30°/60°/-60°/-30°/0°)$ shows that layers with reinforcement angles of $\pm 30°$ "gravitate" to 0°-oriented layers. Let us explain it graphically as follows. If we replace the neighboring graph segments with a single one averaged in some sense, then the overall through-the-thickness distribution will not change significantly. We may observe that situation for the strain e_{xx} between layers with 0° and 30° orientations from the bottom or between layers with $-30°$ and 0° orientations from the top of the stacking. But even for stresses σ_{xx}, their jumps on the interface surfaces can be easily smoothed out over averaging. In the adjacent internal layers with orientations of 60°, strains e_{xx} and stresses σ_{xx} are distributed linearly (but with a significantly different gradient than for 0° and $\pm 30°$-oriented layers). For a thick plate, they are practically zero, and also stress jumps are absent.

For shear strains and stresses, on the contrary, $\pm 30°$- and $\pm 60°$-oriented layers have some common features of behavior. We will single out 0°-oriented layers, where the stresses are practically zero. Thin plates and plates of medium thickness are characterized by linear stress distribution, separately for pairs of layers (30°, 60°) and $(-60°, -30°)$.

A similar note is actually on the strain distribution e_{xy}, but already for non-thin plates with $s = 10$ and $s = 5$. Strains for a thin plate ($s = 100$) and stresses for a thick plate ($s = 5$) are characterized by a rapid, jump-like change from layer to layer. Totally, the graphs presented in the two left and the two right columns of Fig. 25.3 (in-plane normal strains and stresses—on the left vs. shear components—on the right) demonstrate significantly different behavior regarding the distribution in layers.

Variant 4 of symmetrical stacking of layers $(0°/45°/-45°/90°)_S$ reveals some features of multilayer plates: this is an amplification of nonlinear effects already for medium ($s = 10$) and especially for thick plates ($s = 5$). The presence of $\pm 45°$-oriented layers introduces some duality regarding the possibility of their grouping with layers with orientations 0° or 90°. We may average (replace with one segment) the strains of the bottom three $(0°/45°-45°)$ and the top three $(-45°/45°/0°)$ layers or the stresses for the six inner layers $(45°/-45°/90°/90°/-45°/45°)$, except for the 0°-oriented layers. If we compare with the graphs for shear stresses (in the right column of Fig. 25.4), both reinforcement directions 0° and 90° are characterized by practically zero values of the stress σ_{xy}. Moreover, layers with an orientation of 45° or $-45°$ have the opposite signs of the shear stress gradients σ_{xy}. The inversion of the gradient is also characteristic for shear strains, but it does not appear for thin plates ($s = 100$), and thus a classical linear distribution is available.

Summing up the numerical experiments with four variants of stacking layers, there are established the following.

1. The linear through-the-thickness distribution of strains is performed not only for individual layers or thin layered plates as a whole ($s > 20$), as is customary in the CLPT, but also in other cases, sometimes even for thick plates.
2. Averaging distributions, it is possible to group layers with close reinforcement angles. These are groups with orientations 0° or 90° for normal components, and

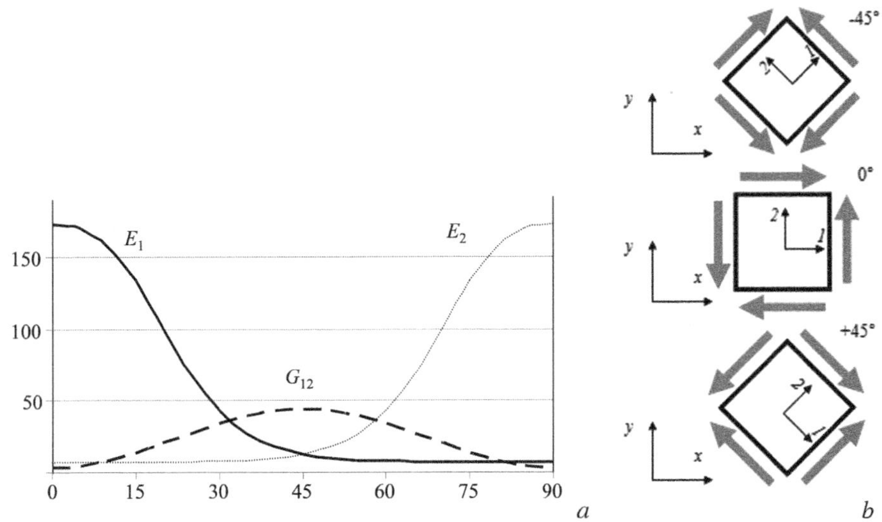

Fig. 25.6 **a**—dependence of Young's moduli E_1, E_2, and shear modulus G_{12} on the reinforcement angle (on the example of the test material); **b**—a scheme to determine the signs of the projection of shear strains (stresses) during a plane rotation of ±45° in the global coordinate system (x, y)

for shear ones—groups with orientations ±45°, as well as a united group with 0° or 90°-oriented layers.

3. Stresses in some layers have much smaller values, regardless of their localization in the layer's package. For example, the stress σ_{yy} is smaller for 0°-oriented layers (almost obviously!). Also, relatively small shear stress σ_{xy} is characteristic of layer orientations with angles of 0° and 90°.

Above we considered only some of the possible directions of reinforcement angles. Important information is contained in the continuous dependence of the modulus of elasticity on the reinforcement angle (Fig. 25.6a). The maximum value of Young's modulus E_1 will be when the reinforcing fibers are laid along the axis x of the global coordinate system, and for angles, say, greater than 45°, Young's modulus E_1 is much smaller. For Young's modulus E_2, the situation is the opposite: if angles of reinforcement are close to 90° (i.e., fibers oriented along y) then it is the largest, and if angles are less than 45° then it is much smaller. The shear modulus has two maxima for the orientation of the reinforcement of the layers −45° and 45°, which must be distinguished, because the shear strains and shear stresses will have the opposite sign in the projection on the axes of the global coordinate system (Fig. 25.6b). On other hand, the minimum of shear modulus G_{12} corresponds to angles 0° or 90°.

These facts will be utilized in the classification scheme of layers, which should be the basis of a new scheme for stiffness averaging across the thickness of laminated composites. A separate classification of layers is required in relation to the nature of through-the-thickness distributions of normal strains e_{xx}, e_{yy} and stresses σ_{xx}, σ_{yy}, as well as in relation to shear components e_{xy} and σ_{xy}.

25.4 Hypotheses of the Mixed Approach

Two families (sets) of layers that give a dominant contribution to the stiffnesses of the layers package relative to the axes x,y of the global coordinate system are distinguished. Families S_1 or S_2 contain layers in which prevails the stiffness in directions x or y, respectively, where $S_1 \cup S_2 = S$ is the entire package of layers, $S_1 \cap S_2 = \oslash$. Accordingly, the layers of the family S_1 are more compliant in the direction y, and the layers of the family S_2 are more compliant in the direction x. In practice, such classification (separation) may be performed, for example, by comparing Young's moduli (or stiffnesses) of unidirectionally reinforced composites after rotation to the global coordinate system: if the reinforcement angle is less than 45°, then we assign the layer to the family S_1, otherwise—to the family S_2. Choosing a threshold value of 45° allows you to assign each layer to one of the specified families unambiguously, except for the reinforcement angles themselves ±45°. Therefore, the classification procedure allows variations of the approach (both for composites made of the same material but with different reinforcement angles, as well as hybrid composites made of different materials). Or the question: to which family should be assigned isotropic layers?

More detailed classification algorithms may be used to better approximate the effective properties compared to experimental data. Still, here we will limit ourselves only to the above general recommendations. We assume that each layer of the plate belongs to one of the families S_1 or S_2, and these families do not intersect, and their union includes all the layers for the given stacking sequence. A partial case will be a single-layer unidirectional composite, then $S_1 = \oslash$ or $S_2 = \oslash$.

The classification of layers to describe the features of the through-the-thickness distribution of shear strains and stresses is independent of S_1 and S_2. Let us select a family of layers whose orientation is close to 0° or 90°, which form a combined family $S_\oplus$. But layers reinforced at angles of 45° and −45° need separated families (families S_+ and S_-, respectively). Some of these three families may be missing ($S_\oplus = \oslash$, or $S_+ = \oslash$, or $S_- = \oslash$), but their union should also be equivalent to the entire layer package ($S_\oplus \cup S_+ \cup S_- = S$).

Note that the classification of layers into "hard" and "flexible" may be carried out independently for each direction of reinforcement. Then four sets should be considered: S_{11}, S_{12} and S_{21}, S_{22} for classification in the x and y directions, respectively. Sets S_{11} and S_{22} determine dominant layers according to direction x or y, $S_{11} \cup S_{12} = S$, $S_{21} \cup S_{22} = S$. This classification is more complex but has an advantage when there is no dominant direction of reinforcement. For example, in the case of transversally isotropic layers, the in-plane properties are the same. On the contrary, classification only by families S_1 and S_2 is a partial case and mutually exclusive: $S_1 = S_{11} = S_{21}$, $S_2 = S_{12} = S_{22}$. Here it is considered only the variant with two mutually exclusive sets S_1 and S_2: if a layer is included in the set S_1, then it does not belong to the set S_2, and vice versa. The variant with S_1 and S_2 demonstrates the main ideas of the proposed approach sufficiently fully. Also, this variant is characteristic in the case of layers with unidirectional reinforcement.

In the remaining part of the work superscripts in angle brackets are used to indicate averaged characteristics (strains, stresses, stiffnesses) or geometric parameters related to one of the layer families: $\langle 1\rangle$ or $\langle 2\rangle$ means belonging to S_1 or S_2, $\langle +\rangle$, $\langle -\rangle$ or $\langle \oplus \rangle$—to families S_+, S_-, or $S_\oplus$, respectively. Then geometric characteristics of layer families similar to (25.8) can be written as follows:

$$[\mathbf{H}] = \sum_{i\in S} \int [\mathbf{Z}]dz = \begin{bmatrix} h & 0 \\ 0 & d \end{bmatrix}, \quad [\mathbf{H}^{\langle k\rangle}] = \sum_{i\in S_k} \int [\mathbf{Z}]dz = \begin{bmatrix} h^{\langle k\rangle} & b^{\langle k\rangle} \\ b^{\langle k\rangle} & b^{\langle k\rangle} \end{bmatrix}, \tag{25.17}$$

where

$$[\mathbf{Z}] = \begin{bmatrix} 1 & z \\ z & z^2 \end{bmatrix}, \quad k = 1, 2 \text{ or } \{+, -, \oplus\}.$$

The condition of completeness and consistency of the division of layers into families also is that

$$\left[\mathbf{H}^{\langle 1\rangle}\right] + \left[\mathbf{H}^{\langle 2\rangle}\right] = [\mathbf{H}], \quad \left[\mathbf{H}^{\langle\oplus\rangle}\right] + \left[\mathbf{H}^{\langle +\rangle}\right] + \left[\mathbf{H}^{\langle -\rangle}\right] = [\mathbf{H}]. \tag{25.18}$$

Hypotheses for layers of the family S_1. Assume that the averaging of strains $e_{xx}(z)$ across the thickness for the family S_1 determines the average deformations of the entire package of layers. That is, layers from S_1 "form" the character of strains $e_{xx}(z)$, but layers from S_2 in this direction are deformed subordinately.

For layers from S_1 we accept the hypothesis about a common statistically averaged across the thickness linear distribution of strains ${e_{xx}}^{(i)}$ $(i \in S_1)$:

$$e_{xx}^{(i)} = \bar{\varepsilon}_x + z\bar{\kappa}_x. \tag{25.19}$$

These layers are characterized by a low level of stresses $\sigma_{yy}^{(i)}$, $i \in S_1$, and we average them as follows:

$$\sigma_{yy}^{(i)} = \bar{\sigma}_y^0 + z\bar{\sigma}_y^1. \tag{25.20}$$

Hypotheses for layers of the family S_2. They are similar to hypotheses for layers S_1 with interchange of directions x and y. Strains $e_{yy}^{(i)}$ and stresses $\sigma_{xx}^{(i)}$ can be written as

$$e_{yy}^{(i)} = \bar{\varepsilon}_y + z\bar{\kappa}_y, i \in S_2, \tag{25.21}$$

$$\sigma_{xx}^{(i)} = \bar{\sigma}_x^0 + z\bar{\sigma}_x^1, i \in S_2. \tag{25.22}$$

Assume the parameters $\bar{\varepsilon}_x$, $\bar{\kappa}_x$, $\bar{\varepsilon}_y$, $\bar{\kappa}_y$ and $\bar{\sigma}_x^0$, $\bar{\sigma}_x^1$, $\bar{\sigma}_y^0$, $\bar{\sigma}_y^1$ are constant. Further, the last group should be excluded during the construction of refined Hooke's law relations for all layers as a whole, provided that hypotheses (25.19)–(25.22) are fulfilled.

Hypotheses for shear strains and stresses. In general, three families of layers are provided: $S_\perp$ (0° and 90°-oriented layers, combined), S_+ (layers orientation close to 45°), and S_- (layers orientation close to −45°).

Layers of the family $S_\perp$ are characterized by a relatively lower (or practically zero) level of stresses σ_{xy} compared to the rest layers. By analogy to normal components the stress ${\sigma_{xy}}^{(i)}$ ($i \in S_\perp$) can be written as

$$\sigma_{xy}^{(i)} = \bar{\sigma}_{xy}^{0} + z\bar{\sigma}_{xy}^{1}. \tag{25.23}$$

Note that the CLPT uses only averaging of strains, i.e., the direct ROM.

For the families of layers S_+ and S_-, we accept the hypotheses of the distribution of shear strains

$$\begin{aligned} \gamma_{xy}^{(i)}(z) &= \bar{\gamma}_{xy}^{(i)} + z\bar{\kappa}_{xy}^{\langle+\rangle},\ i \in S_+, \\ \gamma_{xy}^{(i)}(z) &= \bar{\gamma}_{xy}^{(i)} + z\bar{\kappa}_{xy}^{\langle-\rangle},\ i \in S_-. \end{aligned} \tag{25.24}$$

Next, an averaged linear through-the-thickness distribution

$$\gamma_{xy}(z) = \bar{\gamma}_{xy} + z\bar{\kappa}_{xy} \tag{25.25}$$

is defined by all layers from S_+ and S_-, $S_\pm = S_+ \cup S_-$, if there are

$$\begin{aligned} (h^{\langle+\rangle} + h^{\langle-\rangle})\bar{\gamma}_{xy} &= \textstyle\sum_{i\in S_\pm} \bar{\gamma}_{xy}^{(i)} h_i, \\ (d^{\langle+\rangle} + d^{\langle-\rangle})\bar{\kappa}_{xy} &= d^{\langle+\rangle}\bar{\kappa}_{xy}^{\langle+\rangle} + d^{\langle-\rangle}\bar{\kappa}_{xy}^{\langle-\rangle}. \end{aligned}$$

We assumed common curvatures $\bar{\kappa}_{xy}^{\langle+\rangle}$ and $\bar{\kappa}_{xy}^{\langle-\rangle}$ for both families S_+ and S_-, which take into account one slope of the line on the graphs (see the third column in Figs. 25.2, 25.3, 25.4, and 25.5). The averaged ${\bar{\gamma}_{xy}}^{(i)}$ for each layer remains different but lies between parallel segments on the graphs. Sometimes the sections of these graphs may be approximated by one common segment, as for neighboring layers with reinforcement angles close to 45° or −45° (see Fig. 25.4 for pairs (30°/60°) and (60°/30°), but not for the pair (45°/−45°), Figs. 25.2 or 25.5). Overall the data available from numerical experiments or other theoretical premises, at least, are not sufficient for more "strong" hypotheses or deterministic conclusions.

Hypotheses (25.24) are quite general and can simplify to the strain averaging rule in the CLPT in some cases, for example, if the family $S_\perp$ is absent. For some sequences of layers (stacking (0°/30°/60°/−60°/−30°/0°) is performed ${\bar{\kappa}_{xy}}^{\langle+\rangle} = -{\bar{\kappa}_{xy}}^{\langle-\rangle}$ (see Fig. 25.4), but for stacking $(0°/45°/-45°/90°)_S$ it is not (see Fig. 25.5).

Therefore, in this work, let us combine S_+ and S_- into the family $S_\pm$, for which the direct rule of strain averaging is declared. Assume that these shear strains are average for the entire package of layers also:

$$\langle\gamma_{xy}^{(i)}(z)\rangle = \bar{\gamma}_{xy} + z\bar{\kappa}_{xy}, \quad i \in S_\pm, \tag{25.26}$$

where

$$\begin{aligned}\bar{\gamma}_{xy} &= \frac{1}{h^{\langle\pm\rangle}} \sum_{i\in S_\pm} \bar{\gamma}_{xy}^{(i)} h_i = \frac{1}{h} \sum_{i=1}^{N} \bar{\gamma}_{xy}^{(i)} h_i, \\ \bar{\kappa}_{xy} &= \frac{1}{d^{\langle\pm\rangle}} \sum_{i\in S_\pm} \bar{\kappa}_{xy}^{(i)} d_i = \frac{1}{d} \sum_{i=1}^{N} \bar{\kappa}_{xy}^{(i)} d_i.\end{aligned} \tag{25.27}$$

Note that, unlike the classical theory, where the hypotheses of linear distribution of strains determine a direct averaging rule (ROM), we propose both the averaging of strains—hypotheses (25.19), (25.21), and (25.26), as well as the averaging of stresses—hypotheses (25.20), (25.22), and (25.23). Therefore, we call the proposed approach a *combined* (or *mixed*) rule to average the stiffness of laminated composites.

Temperature strains and stresses could also use an independent classification of layers by the CTE $\alpha_1^{(i)}$ and $\alpha_2^{(i)}$ (analogous to Young's moduli E_1 and E_2). For the sake of simplicity, note that such classification is qualitatively consistent with the separation into families S_1 and S_2, as well as into $S_\oplus$ and $S_\pm$.

25.5 Strains Averaging

Let us integrate strains $\{\mathbf{e}(z)\}$ across the thickness, taking into account the linear distribution $(25.1)_1$ and belonging to the proposed families of dominant or subordinate layers. The linear distribution of strains for the ESLM is determined by the formula (25.5).

For the strains $e_{xx}(z)$, LSM-averaging, according to hypothesis (25.19), gives

$$\begin{aligned}\int e_{xx}(z)\begin{bmatrix}1\\ z\end{bmatrix}dz &= \int (\bar{\varepsilon}_x + z\bar{\kappa}_x)\begin{bmatrix}1\\ z\end{bmatrix}dz = [\mathbf{H}]\begin{bmatrix}\bar{\varepsilon}_x\\ \bar{\kappa}_x\end{bmatrix} \\ &= \sum_{i\in S_1}\int (\bar{\varepsilon}_x + z\bar{\kappa}_x)\begin{bmatrix}1\\ z\end{bmatrix}dz + \sum_{i\in S_2}\int e_{xx}^{(i)}(z)\begin{bmatrix}1\\ z\end{bmatrix}dz \\ &= \left[\mathbf{H}^{\langle 1\rangle}\right]\begin{bmatrix}\bar{\varepsilon}_x\\ \bar{\kappa}_x\end{bmatrix} + \sum_{i\in S_2}\int e_{xx}^{(i)}(z)\begin{bmatrix}1\\ z\end{bmatrix}dz,\end{aligned}$$

from which, taking into account entered notations (25.17) and (25.18), it follows:

$$\sum_{i\in S_2}\int e_{xx}^{(i)}(z)\begin{bmatrix}1\\ z\end{bmatrix}dz = \left[\mathbf{H}^{\langle 2\rangle}\right]\begin{bmatrix}\bar{\varepsilon}_x\\ \bar{\kappa}_x\end{bmatrix}. \tag{25.28}$$

There are two choices to evaluate strains on the left side of Eq. (25.28):

1. Use Hooke's law, solved for strains, i.e., it is necessary to find the inverse matrix of stiffness—matrix of compliance. However, then the situation is the opposite: it is necessary to evaluate stress due to the parameters included in accepted hypotheses;
2. Determine desired strains locally (and approximately) from individual Eq. (25.4).

In accepted hypotheses, we operate with strains and stresses at the same time, so these two options have no special advantages.

The first equation of Hooke's law (25.4) can be written as

$$\begin{aligned} e_{xx}^{(i)}(z) &= (\bar{\sigma}_x^0 + z\bar{\sigma}_x^1)/\bar{Q}_{11}^{(i)} \\ &\quad -\bar{\mu}_{12}^{(i)}(\bar{\varepsilon}_y + z\bar{\kappa}_y) - \bar{\mu}_{16}^{(i)}\gamma_{xy}^{(i)} + \bar{\alpha}_{11}^{(i)}(T_0^{(i)} + \Delta T_1^{(i)} z), \end{aligned} \tag{25.29}$$

where the following notations are introduced:

$$\bar{\mu}_{1k}^{(i)} = \bar{Q}_{1k}^{(i)}/\bar{Q}_{11}^{(i)}, \quad k = 2 \text{ or } 6, \quad \bar{\alpha}_{xx}^{(i)} = \beta_{xx}^{(i)}/\bar{Q}_{11}^{(i)}, \quad i \in S_2.$$

Here $\bar{\mu}_{12}^{(i)}$ and $\bar{\mu}_{16}^{(i)}$ are the "local" interaction coefficients, analogous to Poisson's ratios in the case of the orthotropic material, and $\bar{\alpha}_{xx}{}^{(i)}$ is the "local" or "incomplete" CTE. (Named "local" since it is determined "locally" from the first equation, without inversion of the stiffness matrix.)

To evaluate the contribution of shear strains $\gamma_{xy}^{(i)}$, it needs to consider the subfamilies $S_2^{\oplus} = S_2 \cap S_{\oplus}$ and $S_2^{\pm} = S_2 \cap S_{\pm}$ separately. Using the third equation of Hooke's law (25.4) can be express $\gamma_{xy}{}^{(i)}$, taking into account the hypothesis (25.23):

$$\begin{aligned} \gamma_{xy}^{(i)} &= (\bar{\sigma}_{xy}^0 + z\bar{\sigma}_{xy}^1)/\bar{Q}_{66}^{(i)} \\ &\quad \underline{-\bar{\mu}_{61}^{(i)} e_{xx}^{(i)} - \bar{\mu}_{62}^{(i)}(\bar{\varepsilon}_y + z\bar{\kappa}_y) + \bar{\alpha}_{xy}^{(i)}(T_0^{(i)} + \Delta T_1^{(i)} z)}, \end{aligned} \tag{25.30}$$

where

$$\bar{\mu}_{6k}^{(i)} = \bar{Q}_{6k}^{(i)}/\bar{Q}_{66}^{(i)}, \quad k = 1 \text{ or } 2, \quad \bar{\alpha}_{xy}^{(i)} = \beta_{xy}^{(i)}/\bar{Q}_{66}^{(i)}, \quad i \in S_2^{\oplus}.$$

Now let us integrate (25.29) over the layers of the family S_2. Considering the factor $\bar{\mu}_{16}^{(i)}$ before $\gamma_{xy}^{(i)}$ in (25.29), the underlined expressions in (25.30) are ignored in the first approximation. Substituting the result into (25.28), we obtain

$$\begin{aligned} [\mathbf{H}^{\langle 2\rangle}]\begin{bmatrix}\bar{\varepsilon}_x\\ \bar{\kappa}_x\end{bmatrix} &= \frac{1}{\bar{Q}_{11}^{\langle 2\rangle}}[\tilde{\mathbf{H}}^{\langle 2\rangle}]\begin{bmatrix}\bar{\sigma}_x^0\\ \bar{\sigma}_x^1\end{bmatrix} - \bar{\mu}_{12}^{\langle 2\rangle}[\tilde{\tilde{\mathbf{H}}}^{\langle 2\rangle}]\begin{bmatrix}\bar{\varepsilon}_y\\ \bar{\kappa}_y\end{bmatrix} \\ &\quad - \bar{\mu}_{16}^{\langle 2\rangle}[\tilde{\mathbf{H}}_{\pm}^{\langle 2\rangle}]\begin{bmatrix}\bar{\gamma}_{xy}\\ \bar{\kappa}_{xy}\end{bmatrix} - \frac{\bar{\mu}_{16}^{\langle 2\rangle}}{\bar{Q}_{66}^{\langle 2\rangle}}[\tilde{\mathbf{H}}_{\oplus}^{\langle 2\rangle}]\begin{bmatrix}\bar{\sigma}_{xy}^0\\ \bar{\sigma}_{xy}^1\end{bmatrix} + \bar{\alpha}_{11}^{\langle 2\rangle}[\mathbf{H}_T^{\langle 2\rangle}]\begin{bmatrix}T_0\\ \Delta T_1\end{bmatrix}, \end{aligned} \tag{25.31}$$

where

$$
\begin{aligned}
\frac{1}{\bar{Q}_{11}^{\langle 2\rangle}}[\tilde{\mathbf{H}}^{\langle 2\rangle}] &= \sum_{i\in S_2}\int \frac{1}{\bar{Q}_{11}^{(i)}}[\mathbf{Z}]dz,\\
\bar{\mu}_{12}^{\langle 2\rangle}[\tilde{\tilde{\mathbf{H}}}^{\langle 2\rangle}] &= \sum_{i\in S_2}\int \bar{\mu}_{12}^{(i)}[\mathbf{Z}]dz,\\
\bar{\mu}_{16}^{\langle 2\rangle}[\tilde{\mathbf{H}}_{\pm}^{\langle 2\rangle}] &= \sum_{i\in S_2^{\pm}}\int \bar{\mu}_{16}^{(i)}[\mathbf{Z}]dz,\\
\frac{\bar{\mu}_{16}^{\langle 2\rangle}}{\bar{Q}_{66}^{\langle 2\rangle}}[\tilde{\mathbf{H}}_{\oplus}^{\langle 2\rangle}] &= \sum_{i\in S_2^{\oplus}}\int \frac{\bar{\mu}_{16}^{(i)}}{\bar{Q}_{66}^{(i)}}[\mathbf{Z}]dz,\\
\bar{\alpha}_{11}^{\langle 2\rangle}[\mathbf{H}_{T}^{\langle 2\rangle}] &= \sum_{i\in S_2}\int \bar{\alpha}_{11}^{(i)}[\mathbf{Z}]dz.
\end{aligned}
\tag{25.32}
$$

Similarly, let us average strains $e_{yy}(z)$. In the result, we obtain

$$
\begin{aligned}
[\mathbf{H}^{\langle 1\rangle}]\begin{bmatrix}\bar{\varepsilon}_y\\ \bar{\kappa}_y\end{bmatrix} &= \frac{1}{\bar{Q}_{22}^{\langle 1\rangle}}[\tilde{\mathbf{H}}^{\langle 1\rangle}]\begin{bmatrix}\bar{\sigma}_y^0\\ \bar{\sigma}_y^1\end{bmatrix} - \bar{\mu}_{21}^{\langle 1\rangle}[\tilde{\tilde{\mathbf{H}}}^{\langle 1\rangle}]\begin{bmatrix}\bar{\varepsilon}_x\\ \bar{\kappa}_x\end{bmatrix}\\
&- \bar{\mu}_{26}^{\langle 1\rangle}[\tilde{\mathbf{H}}_{\pm}^{\langle 1\rangle}]\begin{bmatrix}\bar{\gamma}_{xy}\\ \bar{\kappa}_{xy}\end{bmatrix} - \frac{\bar{\mu}_{26}^{\langle 1\rangle}}{\bar{Q}_{66}^{\langle 1\rangle}}[\tilde{\mathbf{H}}_{\oplus}^{\langle 1\rangle}]\begin{bmatrix}\bar{\sigma}_{xy}^0\\ \bar{\sigma}_{xy}^1\end{bmatrix} + \bar{\alpha}_{22}^{\langle 1\rangle}[\mathbf{H}_{T}^{\langle 1\rangle}]\begin{bmatrix}T_0\\ \Delta T_1\end{bmatrix},
\end{aligned}
\tag{25.33}
$$

where

$$
\begin{aligned}
\frac{1}{\bar{Q}_{22}^{\langle 1\rangle}}[\tilde{\mathbf{H}}^{\langle 1\rangle}] &= \sum_{i\in S_1}\int \frac{1}{\bar{Q}_{22}^{(i)}}[\mathbf{Z}]dz,\\
\bar{\mu}_{21}^{\langle 1\rangle}[\tilde{\tilde{\mathbf{H}}}^{\langle 1\rangle}] &= \sum_{i\in S_1}\int \bar{\mu}_{21}^{(i)}[\mathbf{Z}]dz,\\
\bar{\mu}_{26}^{\langle 1\rangle}[\tilde{\mathbf{H}}_{\pm}^{\langle 1\rangle}] &= \sum_{i\in S_1^{\pm}}\int \bar{\mu}_{26}^{(i)}[\mathbf{Z}]dz,\\
\frac{\bar{\mu}_{26}^{\langle 1\rangle}}{\bar{Q}_{66}^{\langle 1\rangle}}[\tilde{\mathbf{H}}_{\oplus}^{\langle 1\rangle}] &= \sum_{i\in S_1^{\oplus}}\int \frac{\bar{\mu}_{26}^{(i)}}{\bar{Q}_{66}^{(i)}}[\mathbf{Z}]dz,\\
\bar{\alpha}_{22}^{\langle 1\rangle}[\mathbf{H}_{T}^{\langle 1\rangle}] &= \sum_{i\in S_1}\int \bar{\alpha}_{22}^{(i)}[\mathbf{Z}]dz.
\end{aligned}
\tag{25.34}
$$

A similar but slightly different procedure is proposed for averaging shear strains. According to (25.26) and assuming that $S_{\oplus}$ and $S_{\pm}$ are not empty, strains $\gamma_{xy}(z)$ yield

$$
[\mathbf{H}]\begin{bmatrix}\bar{\gamma}_{xy}\\ \bar{\kappa}_{xy}\end{bmatrix} = \int e_{xy}(z)\begin{bmatrix}1\\ z\end{bmatrix}dz = [\mathbf{H}^{\langle\pm\rangle}]\begin{bmatrix}\bar{\gamma}_{xy}\\ \bar{\kappa}_{xy}\end{bmatrix} + \sum_{i\in S_{\oplus}}\int \gamma_{xy}^{(i)}(z)\begin{bmatrix}1\\ z\end{bmatrix}dz.
\tag{25.35}
$$

To evaluate strains $\gamma_{xy}^{(i)}$, the third equation of Hooke's law (25.4) can be written as

$$\gamma_{xy}^{(i)} = \frac{\bar{\sigma}_{xy}^{0} + z\bar{\sigma}_{xy}^{1}}{\bar{Q}_{66}^{(i)}} - \bar{\mu}_{61}^{(i)} e_{xx}^{(i)} - \bar{\mu}_{62}^{(i)} e_{yy}^{(i)} + \bar{\alpha}_{xy}^{(i)} (T_0^{(i)} + \Delta T_1^{(i)} z), \tag{25.36}$$

where

$$\bar{\mu}_{6k}^{(i)} = \bar{Q}_{6k}^{(i)} / \bar{Q}_{66}^{(i)}, \quad k = 1 \text{ or } 2, \quad \bar{\alpha}_{xy}^{(i)} = \beta_{xy}^{(i)} / \bar{Q}_{66}^{(i)}, \quad i \in S_{\oplus}.$$

To evaluate terms $\bar{\mu}_{61}^{(i)} e_{xx}^{(i)}$ and $\bar{\mu}_{62}^{(i)} e_{yy}^{(i)}$, consider two subfamilies: $S_1^{\oplus} = S_{\oplus} \cap S_1$ and $S_2^{\oplus} = S_{\oplus} \cap S_2$. According to hypotheses (25.19) and (25.21) normal strains yield

$$e_{xx}^{(i)} = \bar{\varepsilon}_x + z\bar{\kappa}_x, \; i \in S_1^{\oplus},$$
$$e_{yy}^{(i)} = \bar{\varepsilon}_y + z\bar{\kappa}_y, \; i \in S_2^{\oplus}.$$

For two more contributions, namely, $\bar{\mu}_{61}^{(i)} e_{xx}^{(i)}(z)$ $(i \in S_2^{\oplus})$ and $\bar{\mu}_{62}^{(i)} e_{yy}^{(i)}(z)$ $(i \in S_1^{\oplus})$, we use the first two equations of Hooke's law (25.4):

$$e_{xx}^{(i)}(z) = \frac{\bar{\sigma}_{x}^{0} + z\bar{\sigma}_{x}^{1}}{\bar{Q}_{11}^{(i)}} \underline{- \bar{\mu}_{12}^{(i)} (\bar{\varepsilon}_y^0 + z\bar{\kappa}_y) + \bar{\mu}_{16}^{(i)} \gamma_{xy}^{(i)} + \bar{\alpha}_{xx}^{(i)} (T_0^{(i)} + \Delta T_1^{(i)} z)}, \tag{25.37}$$

$$e_{yy}^{(i)}(z) = \frac{\bar{\sigma}_{y}^{0} + z\bar{\sigma}_{y}^{1}}{\bar{Q}_{22}^{(i)}} \underline{- \bar{\mu}_{21}^{(i)} e_{yy}^{(i)}(z) + \bar{\mu}_{26}^{(i)} \gamma_{xy}^{(i)} + \bar{\alpha}_{yy}^{(i)} (T_0^{(i)} + \Delta T_1^{(i)} z)}, \tag{25.38}$$

where

$$\bar{\mu}_{1k}^{(i)} = \bar{Q}_{1k}^{(i)} / \bar{Q}_{11}^{(i)}, \quad k = 2 \text{ or } 6, \quad \bar{\alpha}_{xx}^{(i)} = \beta_{xx}^{(i)} / \bar{Q}_{11}^{(i)},$$
$$\bar{\mu}_{2k}^{(i)} = \bar{Q}_{2k}^{(i)} / \bar{Q}_{22}^{(i)}, \quad k = 1 \text{ or } 6, \quad \bar{\alpha}_{yy}^{(i)} = \beta_{yy}^{(i)} / \bar{Q}_{22}^{(i)},$$

respectively.

In the first approximation, the underlined terms in (25.37) and (25.38) can be ignored. Integrate over the thickness of the layers $S_{\oplus}$:

$$\begin{aligned}
\left[\mathbf{H}^{\langle\oplus\rangle}\right] \begin{bmatrix} \bar{\gamma}_{xy} \\ \bar{\kappa}_{xy} \end{bmatrix} &= \frac{1}{\bar{Q}_{66}^{\langle\oplus\rangle}} \left[\tilde{\mathbf{H}}^{\langle\oplus\rangle}\right] \begin{bmatrix} \bar{\sigma}_{xy}^0 \\ \bar{\sigma}_{xy}^1 \end{bmatrix} - \bar{\mu}_{61}^{\langle\oplus\rangle} \left[\tilde{\mathbf{H}}_1^{\langle\oplus\rangle}\right] \begin{bmatrix} \bar{\varepsilon}_x \\ \bar{\kappa}_x \end{bmatrix} \\
&\quad - \bar{\mu}_{62}^{\langle\oplus\rangle} \left[\tilde{\mathbf{H}}_2^{\langle\oplus\rangle}\right] \begin{bmatrix} \bar{\varepsilon}_y \\ \bar{\kappa}_y \end{bmatrix} - \frac{\bar{\mu}_{61}^{\langle\oplus\rangle}}{\bar{Q}_{11}^{\langle\oplus\rangle}} \left[\tilde{\tilde{\mathbf{H}}}_2^{\langle\oplus\rangle}\right] \begin{bmatrix} \bar{\sigma}_x^0 \\ \bar{\sigma}_x^1 \end{bmatrix} \\
&\quad - \frac{\bar{\mu}_{62}^{\langle\oplus\rangle}}{\bar{Q}_{22}^{\langle\oplus\rangle}} \left[\tilde{\tilde{\mathbf{H}}}_1^{\langle\oplus\rangle}\right] \begin{bmatrix} \bar{\sigma}_y^0 \\ \bar{\sigma}_y^1 \end{bmatrix} + \bar{\alpha}_{xy}^{\langle\oplus\rangle} \left[\mathbf{H}_T^{\langle\oplus\rangle}\right] \begin{bmatrix} T_0 \\ \Delta T_1 \end{bmatrix},
\end{aligned} \tag{25.39}$$

where

$$
\begin{aligned}
\frac{1}{\bar{Q}_{66}^{\langle\oplus\rangle}}\left[\tilde{\mathbf{H}}^{\langle\oplus\rangle}\right] &= \sum_{i\in S_{\oplus}} \int \frac{1}{\bar{Q}_{66}^{(i)}}[\mathbf{Z}]dz,\\
\bar{\mu}_{61}^{\langle\oplus\rangle}\left[\tilde{\mathbf{H}}_1^{\langle\oplus\rangle}\right] &= \sum_{i\in S_1^{\oplus}} \int \mu_{61}^{(i)}[\mathbf{Z}]dz,\\
\frac{\bar{\mu}_{61}^{\langle\oplus\rangle}}{\bar{Q}_{11}^{\langle\oplus\rangle}}\left[\tilde{\tilde{\mathbf{H}}}_2^{\langle\oplus\rangle}\right] &= \sum_{i\in S_2^{\oplus}} \int \frac{\mu_{61}^{(i)}}{\bar{Q}_{11}^{(i)}}[\mathbf{Z}]dz,\\
\bar{\mu}_{62}^{\langle\oplus\rangle}\left[\tilde{\mathbf{H}}_2^{\langle\oplus\rangle}\right] &= \sum_{i\in S_2^{\oplus}} \int \mu_{62}^{(i)}[\mathbf{Z}]dz,\\
\frac{\bar{\mu}_{62}^{\langle\oplus\rangle}}{\bar{Q}_{22}^{\langle\oplus\rangle}}\left[\tilde{\tilde{\mathbf{H}}}_1^{\langle\oplus\rangle}\right] &= \sum_{i\in S_1^{\oplus}} \int \frac{\mu_{62}^{(i)}}{\bar{Q}_{22}^{(i)}}[\mathbf{Z}]dz,\\
\bar{\alpha}_{xy}^{\langle\oplus\rangle}\left[\mathbf{H}_T^{\langle\oplus\rangle}\right] &= \sum_{i\in S_{\oplus}} \bar{\alpha}_{xy}^{(i)}[\mathbf{Z}]dz.
\end{aligned}
\tag{25.40}
$$

Equations (25.31), (25.33), and (25.39) make it possible to express stress parameters $\bar{\sigma}_x^0, \bar{\sigma}_x^1, \bar{\sigma}_y^0, \bar{\sigma}_y^1$, and $\bar{\sigma}_{xy}^0, \bar{\sigma}_{xy}^1$ through $\bar{\varepsilon}_x, \bar{\kappa}_x, \bar{\varepsilon}_y, \bar{\kappa}_y$, and $\bar{\gamma}_{xy}, \bar{\kappa}_{xy}$ and further exclude them from consideration during calculating the effective characteristics within the framework of the ESLM. After rearranging stress and strain parameters, so that there are some on the left, and others on the right, the matrix equation is obtained:

$$
[\mathbf{M}_{\varepsilon}]\{\boldsymbol{\varepsilon}_{PAR}\} = [\mathbf{M}_{\sigma}]\{\boldsymbol{\sigma}_{PAR}\} + [\mathbf{M}_T]\{\mathbf{T}_{PAR}\}, \tag{25.41}
$$

where

$$
\begin{gathered}
[\mathbf{M}_{\varepsilon}] = \begin{bmatrix} \left[\mathbf{H}^{\langle 2\rangle}\right] & \bar{\mu}_{12}^{\langle 2\rangle}\left[\tilde{\tilde{\mathbf{H}}}^{\langle 2\rangle}\right] & \bar{\mu}_{16}^{\langle 2\rangle}\left[\tilde{\mathbf{H}}_{\pm}^{\langle 2\rangle}\right] \\ \bar{\mu}_{21}^{\langle 1\rangle}\left[\tilde{\tilde{\mathbf{H}}}^{\langle 1\rangle}\right] & \left[\mathbf{H}^{\langle 1\rangle}\right] & \bar{\mu}_{26}^{\langle 1\rangle}\left[\tilde{\mathbf{H}}_{\pm}^{\langle 1\rangle}\right] \\ \bar{\mu}_{61}^{\langle\oplus\rangle}\left[\tilde{\mathbf{H}}_1^{\langle\oplus\rangle}\right] & \bar{\mu}_{62}^{\langle\oplus\rangle}\left[\tilde{\mathbf{H}}_2^{\langle\oplus\rangle}\right] & \left[\mathbf{H}^{\langle\oplus\rangle}\right] \end{bmatrix},\\
[\mathbf{M}_{\sigma}] = \begin{bmatrix} \frac{1}{\bar{Q}_{11}^{\langle 2\rangle}}\left[\tilde{\mathbf{H}}^{\langle 2\rangle}\right] & \mathbf{0}_{2\times 2} & -\frac{\bar{\mu}_{16}^{\langle 2\rangle}}{\bar{Q}_{66}^{\langle 2\rangle}}\left[\tilde{\mathbf{H}}_{\oplus}^{\langle 2\rangle}\right] \\ \mathbf{0}_{2\times 2} & \frac{1}{\bar{Q}_{22}^{\langle 1\rangle}}\left[\tilde{\mathbf{H}}^{\langle 1\rangle}\right] & -\frac{\bar{\mu}_{26}^{\langle 1\rangle}}{\bar{Q}_{66}^{\langle 1\rangle}}\left[\tilde{\mathbf{H}}_{\oplus}^{\langle 1\rangle}\right] \\ -\frac{\bar{\mu}_{61}^{\langle\oplus\rangle}}{\bar{Q}_{11}^{\langle\oplus\rangle}}\left[\tilde{\tilde{\mathbf{H}}}_2^{\langle\oplus\rangle}\right] & -\frac{\bar{\mu}_{62}^{\langle\oplus\rangle}}{\bar{Q}_{22}^{\langle\oplus\rangle}}\left[\tilde{\tilde{\mathbf{H}}}_1^{\langle\oplus\rangle}\right] & \frac{1}{\bar{Q}_{66}^{\langle\oplus\rangle}}\left[\tilde{\mathbf{H}}^{\langle\oplus\rangle}\right] \end{bmatrix},\\
[\mathbf{M}_T] = \begin{bmatrix} \bar{\alpha}_{11}^{\langle 2\rangle}\left[\mathbf{H}_T^{\langle 2\rangle}\right] \\ \bar{\alpha}_{22}^{\langle 1\rangle}\left[\mathbf{H}_T^{\langle 1\rangle}\right] \\ \bar{\alpha}_{xy}^{\langle\oplus\rangle}\left[\mathbf{H}_T^{\langle\oplus\rangle}\right] \end{bmatrix}, \quad \{\mathbf{T}_{PAR}\} = \begin{bmatrix} T_0 \\ \Delta T_1 \end{bmatrix},\\
\{\boldsymbol{\sigma}_{PAR}\} = \left[\bar{\sigma}_x^0, \bar{\sigma}_x^1, \bar{\sigma}_y^0, \bar{\sigma}_y^1, \bar{\sigma}_{xy}^0, \bar{\sigma}_{xy}^1\right]^{\mathrm{T}},\\
\{\boldsymbol{\varepsilon}_{PAR}\} = \left[\bar{\varepsilon}_x, \bar{\kappa}_x, \bar{\varepsilon}_y, \bar{\kappa}_y, \bar{\gamma}_{xy}, \bar{\kappa}_{xy}\right]^{\mathrm{T}}.
\end{gathered}
\tag{25.42}
$$

For some configurations of composite plates (i.e., sequence of layers and materials), matrices in Eq. (25.41) can be simplified. In particular, the matrix $[\mathbf{M}_\sigma]$ in the case of orthotropic layers in the global coordinate system has a block-diagonal form. Peculiarities of matrices (25.42) for partial cases (in particular, if some of the layer families are empty) still require their investigation. Here, we limit ourselves to a general remark about the solvability of the system (25.41), that is, assuming that the inverse matrix $[\mathbf{M}_\sigma]^{-1}$ exists.

25.6 Stresses Averaging

To find forces and moments (25.11) or (25.12) within the framework of the introduced hypotheses (25.19)–(25.26), as for strains above, a component-wise approach is utilized.

Let us average the stresses $\sigma_{xx}(z)$. Taking into account the linear distribution (25.22), we obtain

$$\begin{aligned}\begin{bmatrix} N_x \\ M_x \end{bmatrix} &= \int_{-h/2}^{h/2} \sigma_{xx}(z) \begin{bmatrix} 1 \\ z \end{bmatrix} dz \\ &= \sum_{i\in S_1} \int_{z_{i-1}}^{z_i} \sigma_{xx}(z) \begin{bmatrix} 1 \\ z \end{bmatrix} dz + \sum_{i\in S_2} \int_{z_{i-1}}^{z_i} (\bar{\sigma}_x^0 + z\bar{\sigma}_x^1) \begin{bmatrix} 1 \\ z \end{bmatrix} dz.\end{aligned} \tag{25.43}$$

The estimation of the first term is based on the first equation of Hooke's law (25.4), taking into account the hypothesis (25.19):

$$\begin{aligned}\sigma_{xx}^{(i)}(z) &= \bar{Q}_{11}^{(i)}(\bar{\varepsilon}_x^0 + z\bar{\kappa}_x) \\ &\quad + \bar{Q}_{12}^{(i)} e_{yy}^{(i)}(z) + \bar{Q}_{16}^{(i)} \gamma_{xy}^{(i)} - \beta_{xx}^{(i)}(T_0^{(i)} + \Delta T_1^{(i)} z), \quad i \in S_1.\end{aligned} \tag{25.44}$$

In turn, from the second equation of Hooke's law (25.4), it is possible to express

$$\begin{aligned}e_{yy}^{(i)}(z) &= (\bar{\sigma}_y^0 + z\bar{\sigma}_y^1)/\bar{Q}_{22}^{(i)} \\ &\quad \underline{-\bar{\mu}_{21}^{(i)}(\bar{\varepsilon}_x + z\bar{\kappa}_x) - \bar{\mu}_{26}^{(i)}\gamma_{xy}^{(i)} + \bar{\alpha}_{yy}^{(i)}(T_0^{(i)} + \Delta T_1^{(i)} z)},\end{aligned}$$

where

$$\bar{\mu}_{21}^{(i)} = \bar{Q}_{21}^{(i)}/\bar{Q}_{22}^{(i)}, \quad k = 1 \text{ or } 6, \quad \bar{\alpha}_{yy}^{(i)} = \beta_{yy}^{(i)}/\bar{Q}_{22}^{(i)}, \quad i \in S_1.$$

Products of the underlined terms by $\bar{Q}_{12}^{(i)}$ have the character of the products of Poisson's ratios. Therefore, they can be ignored in the first approximation.

Let us perform the evaluation of $\gamma_{xy}{}^{(i)}$ on the subfamilies $S_1^{\oplus} = S_1 \cap S_{\oplus}$, and $S_1^{\pm} = S_1 \cap S_{\pm}$, taking into account hypotheses (25.23) and (25.26), respectively. Using the third equation of Hooke's law (25.4), we obtain (see notation to Eq. (25.36))

$$
\begin{aligned}
\gamma_{xy}^{(i)} = & \left(\bar{\sigma}_{xy}^{0} + z\bar{\sigma}_{xy}^{1}\right)/\bar{Q}_{66}^{(i)} \\
& \underline{-\bar{\mu}_{61}^{(i)}(\bar{\varepsilon}_x + z\bar{\kappa}_x) - \bar{\mu}_{62}^{(i)} e_{yy}^{(i)} + \bar{\alpha}_{xy}^{(i)}\left(T_0^{(i)} + \Delta T_1^{(i)} z\right)}, \quad i \in S_1^{\oplus},
\end{aligned} \tag{25.45}
$$

$$
\gamma_{xy}^{(i)} = \bar{\gamma}_{xy} + z\bar{\kappa}_{xy}, \quad i \in S_1^{\pm}. \tag{25.46}
$$

The underlined terms are also not taken into account in the first approximation. They contain the products of Poisson's ratios.

So, Eq. (25.43) can be written as follows:

$$
\begin{aligned}
\begin{bmatrix} N_x \\ M_x \end{bmatrix} = & \bar{Q}_{11}^{\langle 1\rangle}[\hat{\mathbf{H}}^{\langle 1\rangle}]\begin{bmatrix} \bar{\varepsilon}_x \\ \bar{\kappa}_x \end{bmatrix} + [\mathbf{H}^{\langle 2\rangle}]\begin{bmatrix} \bar{\sigma}_x^0 \\ \bar{\sigma}_x^1 \end{bmatrix} + \bar{\mu}_{12}^{\langle 1\rangle}[\hat{\hat{\mathbf{H}}}^{\langle 1\rangle}]\begin{bmatrix} \bar{\sigma}_y^0 \\ \bar{\sigma}_y^1 \end{bmatrix} \\
& + \bar{\mu}_{16}^{\langle 1\rangle}[\hat{\mathbf{H}}_{\oplus}^{\langle 1\rangle}]\begin{bmatrix} \bar{\sigma}_{xy}^0 \\ \bar{\sigma}_{xy}^1 \end{bmatrix} + \bar{Q}_{16}^{\langle 1\rangle}[\hat{\mathbf{H}}_{\pm}^{\langle 1\rangle}]\begin{bmatrix} \bar{\gamma}_{xy} \\ \bar{\kappa}_{xy} \end{bmatrix} - \bar{\beta}_{xx}^{\langle 1\rangle}[\hat{\mathbf{H}}_{T}^{\langle 1\rangle}]\begin{bmatrix} T_0 \\ \Delta T_1 \end{bmatrix},
\end{aligned} \tag{25.47}
$$

where

$$
\begin{aligned}
\bar{Q}_{11}^{\langle 1\rangle}[\hat{\mathbf{H}}^{\langle 1\rangle}] &= \sum_{i\in S_1} \int \bar{Q}_{11}^{(i)}[\mathbf{Z}]dz, \\
\bar{\mu}_{12}^{\langle 1\rangle}[\hat{\hat{\mathbf{H}}}^{\langle 1\rangle}] &= \sum_{i\in S_1} \int \frac{\bar{Q}_{12}^{(i)}}{\bar{Q}_{22}^{(i)}}[\mathbf{Z}]dz, \\
\bar{\mu}_{16}^{\langle 1\rangle}[\hat{\mathbf{H}}_{\oplus}^{\langle 1\rangle}] &= \sum_{i\in S_1^{\oplus}} \int \frac{\bar{Q}_{16}^{(i)}}{\bar{Q}_{66}^{(i)}}[\mathbf{Z}]dz, \\
\bar{Q}_{16}^{\langle 1\rangle}[\hat{\mathbf{H}}_{\pm}^{\langle 1\rangle}] &= \sum_{i\in S_1^{\pm}} \int \bar{Q}_{16}^{(i)}[\mathbf{Z}]dz, \\
\bar{\beta}_{xx}^{\langle 1\rangle}[\hat{\mathbf{H}}_{T}^{\langle 1\rangle}] &= \sum_{i\in S_1} \int \beta_{xx}^{(i)}[\mathbf{Z}]dz.
\end{aligned} \tag{25.48}
$$

By analogy, the equation for the direction y is

$$
\begin{aligned}
\begin{bmatrix} N_y \\ M_y \end{bmatrix} = & \bar{Q}_{22}^{\langle 2\rangle}[\hat{\mathbf{H}}^{\langle 2\rangle}]\begin{bmatrix} \bar{\varepsilon}_y \\ \bar{\kappa}_y \end{bmatrix} + [\mathbf{H}^{\langle 1\rangle}]\begin{bmatrix} \bar{\sigma}_y^0 \\ \bar{\sigma}_y^1 \end{bmatrix} + \bar{\mu}_{21}^{\langle 1\rangle}[\hat{\hat{\mathbf{H}}}^{\langle 2\rangle}]\begin{bmatrix} \bar{\sigma}_x^0 \\ \bar{\sigma}_x^1 \end{bmatrix} \\
& + \bar{\mu}_{26}^{\langle 2\rangle}[\hat{\mathbf{H}}_{\oplus}^{\langle 2\rangle}]\begin{bmatrix} \bar{\sigma}_{xy}^0 \\ \bar{\sigma}_{xy}^1 \end{bmatrix} + \bar{Q}_{26}^{\langle 1\rangle}[\hat{\mathbf{H}}_{\pm}^{\langle 2\rangle}]\begin{bmatrix} \bar{\gamma}_{xy} \\ \bar{\kappa}_{xy} \end{bmatrix} - \bar{\beta}_{yy}^{\langle 2\rangle}[\hat{\mathbf{H}}_{T}^{\langle 2\rangle}]\begin{bmatrix} T_0 \\ \Delta T_1 \end{bmatrix},
\end{aligned} \tag{25.49}
$$

where

$$
\begin{aligned}
\bar{Q}_{22}^{\langle 2\rangle}\left[\hat{\mathbf{H}}^{\langle 2\rangle}\right] &= \sum_{i\in S_2}\int \bar{Q}_{22}^{(i)}[\mathbf{Z}]dz,\\
\bar{\mu}_{21}^{\langle 1\rangle}\left[\hat{\mathbf{H}}^{\langle 2\rangle}\right] &= \sum_{i\in S_2}\int \frac{\bar{Q}_{21}^{(i)}}{\bar{Q}_{11}^{(i)}}[\mathbf{Z}]dz,\\
\bar{\mu}_{26}^{\langle 2\rangle}\left[\hat{\mathbf{H}}_{\oplus}^{\langle 2\rangle}\right] &= \sum_{i\in S_2^{\oplus}}\int \frac{\bar{Q}_{26}^{(i)}}{\bar{Q}_{66}^{(i)}}[\mathbf{Z}]dz,\\
\bar{Q}_{26}^{\langle 1\rangle}\left[\hat{\mathbf{H}}_{\pm}^{\langle 2\rangle}\right] &= \sum_{i\in S_2^{\pm}}\int \bar{Q}_{26}^{(i)}[\mathbf{Z}]dz,\\
\bar{\beta}_{yy}^{\langle 2\rangle}\left[\hat{\mathbf{H}}_{T}^{\langle 2\rangle}\right] &= \sum_{i\in S_2}\int \beta_{yy}^{(i)}[\mathbf{Z}]dz.
\end{aligned}
\tag{25.50}
$$

Finally, integration for shear stresses $\sigma_{xy}(z)$ taking into account the third line of Hooke's law (25.4) gives

$$
\begin{aligned}
\begin{bmatrix} N_{xy} \\ M_{xy} \end{bmatrix} &= \int_{-h/2}^{h/2} \sigma_{xy}(z)\begin{bmatrix} 1 \\ z \end{bmatrix} dz\\
&= \sum_{i\in S_{\oplus}}\int \left(\bar{\sigma}_{xy}^{0} + z\bar{\sigma}_{xy}^{1}\right)\begin{bmatrix} 1 \\ z \end{bmatrix} dz + \sum_{i\in S_{\pm}}\int \sigma_{xy}^{(i)}(z)\begin{bmatrix} 1 \\ z \end{bmatrix} dz,
\end{aligned}
\tag{25.51}
$$

where the evaluation of terms is as follows:

$$
\begin{aligned}
\sigma_{xy}^{(i)} = \bar{Q}_{61}^{(i)}e_{xx}^{(i)} + \bar{Q}_{62}^{(i)}e_{yy}^{(i)}&\\
&+ \bar{Q}_{66}^{(i)}(\bar{\gamma}_{xy} + z\bar{\kappa}_{xy}) - \beta_{xy}^{(i)}(T_0 + \Delta T_1 z), \quad i \in S_{\pm}.
\end{aligned}
\tag{25.52}
$$

To evaluate the first two terms of (25.52), we also use Hooke's law (see also Eqs. (25.37) and (25.38))

- for $\bar{Q}_{61}^{(i)}e_{xx}^{(i)}$ $(i \in S_2^{\pm} = S_{\pm} \cap S_2)$:

$$
\begin{aligned}
e_{xx}^{(i)} &= \left(\bar{\sigma}_x^{0} + z\bar{\sigma}_x^{1}\right)/\bar{Q}_{11}^{(i)}\\
&\underline{-\frac{\bar{Q}_{12}^{(i)}}{\bar{Q}_{11}^{(i)}}(\bar{\varepsilon}_y + z\bar{\kappa}_y) - \frac{\bar{Q}_{16}^{(i)}}{\bar{Q}_{11}^{(i)}}(\bar{\gamma}_{xy} + z\bar{\kappa}_{xy}) + \frac{\beta_{xx}^{(i)}}{\bar{Q}_{11}^{(i)}}\left(T_0^{(i)} + \Delta T_1^{(i)} z\right);}
\end{aligned}
\tag{25.53}
$$

- for $\bar{Q}_{62}^{(i)}e_{yy}^{(i)}$ $(i \in S_1 = S_{\pm} \cap S_1)$:

$$
\begin{aligned}
e_{yy}^{(i)} &= \left(\bar{\sigma}_y^{0} + z\bar{\sigma}_y^{1}\right)/\bar{Q}_{22}^{(i)}\\
&\underline{-\frac{\bar{Q}_{21}^{(i)}}{\bar{Q}_{22}^{(i)}}(\bar{\varepsilon}_x + z\bar{\kappa}_x) - \frac{\bar{Q}_{26}^{(i)}}{\bar{Q}_{22}^{(i)}}(\bar{\gamma}_{xy} + z\bar{\kappa}_{xy}) + \frac{\beta_{yy}^{(i)}}{\bar{Q}_{22}^{(i)}}\left(T_0^{(i)} + \Delta T_1^{(i)} z\right).}
\end{aligned}
\tag{25.54}
$$

As before, in the first-order approximation, the underlined terms are ignored (they contain the products of Poisson's ratios). For two more variants of subfamilies $S_1{}^{\pm}$ and $S_2{}^{\pm}$, accepting hypotheses (25.19) and (25.21) for strains $e_{xx}{}^{(i)}$ and $e_{yy}{}^{(i)}$, respectively, and substituting (25.52)–(25.54) into (25.51) give

$$\begin{aligned}\begin{bmatrix} N_{xy} \\ M_{xy} \end{bmatrix} &= [\mathbf{H}^{\langle \oplus \rangle}] \begin{bmatrix} \bar{\sigma}_{xy}^{0} \\ \bar{\sigma}_{xy}^{1} \end{bmatrix} + \bar{Q}_{66}^{\langle \pm \rangle} [\hat{\mathbf{H}}^{\langle \pm \rangle}] \begin{bmatrix} \bar{\gamma}_{xy} \\ \bar{\kappa}_{xy} \end{bmatrix} + \bar{Q}_{61}^{\langle \pm \rangle} [\hat{\mathbf{H}}_1^{\langle \pm \rangle}] \begin{bmatrix} \bar{\varepsilon}_{x} \\ \bar{\kappa}_{x} \end{bmatrix} \\ &+ \bar{Q}_{62}^{\langle \pm \rangle} [\hat{\mathbf{H}}_2^{\langle \pm \rangle}] \begin{bmatrix} \bar{\varepsilon}_{y} \\ \bar{\kappa}_{y} \end{bmatrix} + \bar{\mu}_{61}^{\langle \pm \rangle} [\hat{\hat{\mathbf{H}}}_2^{\langle \pm \rangle}] \begin{bmatrix} \bar{\sigma}_{x}^{0} \\ \bar{\sigma}_{x}^{1} \end{bmatrix} \\ &+ \bar{\mu}_{62}^{\langle \pm \rangle} [\hat{\hat{\mathbf{H}}}_1^{\langle \pm \rangle}] \begin{bmatrix} \bar{\sigma}_{y}^{0} \\ \bar{\sigma}_{y}^{1} \end{bmatrix} - \bar{\beta}_{xy}^{\langle \pm \rangle} [\hat{\mathbf{H}}_T^{\langle \pm \rangle}] \begin{bmatrix} T_0 \\ \Delta T_1 \end{bmatrix},\end{aligned} \tag{25.55}$$

where

$$\begin{aligned} \bar{Q}_{66}^{\langle \pm \rangle} [\hat{\mathbf{H}}^{\langle \pm \rangle}] &= \sum_{i \in S_{\pm}} \int \bar{Q}_{66}^{(i)} [\mathbf{Z}] dz, \\ \bar{\beta}_{xy}^{\langle \pm \rangle} [\hat{\mathbf{H}}_T^{\langle \pm \rangle}] &= \sum_{i \in S_{\pm}} \int \beta_{xy}^{(i)} [\mathbf{Z}] dz, \\ \bar{Q}_{6k}^{\langle \pm \rangle} [\hat{\mathbf{H}}_k^{\langle \pm \rangle}] &= \sum_{i \in S_1^{\pm}} \int \bar{Q}_{6k}^{(i)} [\mathbf{Z}] dz, \\ \bar{\mu}_{6k}^{\langle \pm \rangle} [\hat{\hat{\mathbf{H}}}_{3-k}^{\langle \pm \rangle}] &= \sum_{i \in S_{3-k}^{\pm}} \int \frac{\bar{Q}_{6k}^{(i)}}{\bar{Q}_{66}^{(i)}} [\mathbf{Z}] dz, \quad k = 1, 2. \end{aligned} \tag{25.56}$$

Equations (25.47), (25.49), and (25.55) may be combined into a matrix equation:

$$[\mathbf{F_{N,M}}] = [\mathbf{K}_{\sigma}] \{\boldsymbol{\sigma}_{PAR}\} + [\mathbf{K}_{\varepsilon}] \{\boldsymbol{\varepsilon}_{PAR}\} - [\mathbf{K}_T] \{\mathbf{T}_{PAR}\}, \tag{25.57}$$

where, in addition to (25.42), the notation is introduced

$$[\mathbf{K}_{\varepsilon}] = \begin{bmatrix} \bar{Q}_{11}^{\langle 1 \rangle} [\hat{\mathbf{H}}^{\langle 1 \rangle}] & \mathbf{0}_{2\times 2} & \bar{Q}_{16}^{\langle 1 \rangle} [\hat{\mathbf{H}}_{\pm}^{\langle 1 \rangle}] \\ \mathbf{0}_{2\times 2} & \bar{Q}_{22}^{\langle 2 \rangle} [\hat{\mathbf{H}}^{\langle 2 \rangle}] & \bar{Q}_{26}^{\langle 2 \rangle} [\hat{\mathbf{H}}_{\pm}^{\langle 2 \rangle}] \\ \bar{Q}_{61}^{\langle \pm \rangle} [\hat{\mathbf{H}}_1^{\langle \pm \rangle}] & \bar{Q}_{62}^{\langle \pm \rangle} [\hat{\mathbf{H}}_2^{\langle \pm \rangle}] & \bar{Q}_{66}^{\langle \pm \rangle} [\hat{\mathbf{H}}^{\langle \pm \rangle}] \end{bmatrix},$$

$$[\mathbf{K}_{\sigma}] = \begin{bmatrix} [\mathbf{H}^{\langle 2 \rangle}] & \bar{\mu}_{12}^{\langle 1 \rangle} [\hat{\hat{\mathbf{H}}}^{\langle 1 \rangle}] & \bar{\mu}_{16}^{\langle 1 \rangle} [\hat{\mathbf{H}}_{\oplus}^{\langle 1 \rangle}] \\ \bar{\mu}_{21}^{\langle 2 \rangle} [\hat{\hat{\mathbf{H}}}^{\langle 2 \rangle}] & [\mathbf{H}^{\langle 1 \rangle}] & \bar{\mu}_{26}^{\langle 2 \rangle} [\hat{\mathbf{H}}_{\oplus}^{\langle 2 \rangle}] \\ \bar{\mu}_{61}^{\langle \pm \rangle} [\hat{\hat{\mathbf{H}}}_2^{\langle \pm \rangle}] & \bar{\mu}_{62}^{\langle \pm \rangle} [\hat{\hat{\mathbf{H}}}_1^{\langle \pm \rangle}] & [\mathbf{H}^{\langle \oplus \rangle}] \end{bmatrix},$$

$$[\mathbf{K}_T] = \begin{bmatrix} \bar{\beta}_{xx}^{\langle 1 \rangle} [\hat{\mathbf{H}}_T^{\langle 1 \rangle}] \\ \bar{\beta}_{yy}^{\langle 2 \rangle} [\hat{\mathbf{H}}_T^{\langle 2 \rangle}] \\ \bar{\beta}_{xy}^{\langle \pm \rangle} [\hat{\mathbf{H}}_T^{\langle \pm \rangle}] \end{bmatrix},$$

$$[\mathbf{F_{N,M}}] = [N_x, M_x, N_y, M_y, N_{xy}, M_{xy}]^{\mathrm{T}}.$$

If solving (25.42) by

$$\{\sigma_{PAR}\} = [\mathbf{M}_\sigma]^{-1}[\mathbf{M}_\varepsilon]\{\boldsymbol{\varepsilon}_{PAR}\} - [\mathbf{M}_\sigma]^{-1}[\mathbf{M}_T]\{\mathbf{T}_{PAR}\}, \tag{25.58}$$

and substituting into (25.57), then follows

$$\begin{aligned}[\mathbf{F}_{\mathbf{N},\mathbf{M}}] = &\left\{[\mathbf{K}_\sigma][\mathbf{M}_\sigma]^{-1}[\mathbf{M}_\varepsilon] + [\mathbf{K}_\varepsilon]\right\}\{\boldsymbol{\varepsilon}_{PAR}\} \\ &- \left\{[\mathbf{K}_\sigma][\mathbf{M}_\sigma]^{-1}[\mathbf{M}_T] + [\mathbf{K}_T]\right\}\{\mathbf{T}_{PAR}\}.\end{aligned} \tag{25.59}$$

Here, the vector of parameters $\{\boldsymbol{\varepsilon}_{PAR}\}$ corresponds to a statistically averaged linear distribution of strains across the thickness of layers, and the left side is the integral stress characteristics (as predicted by the constitutive equation of the monolayer (25.15)). Therefore, after rearranging the rows and columns, Eq. (25.59) can be written in the form:

$$\begin{bmatrix}\{\mathbf{N}\}\\ \{\mathbf{M}\}\end{bmatrix} = \begin{bmatrix}[\tilde{\mathbf{A}}] & [\tilde{\mathbf{B}}_1]\\ [\tilde{\mathbf{B}}_2] & [\tilde{\mathbf{D}}]\end{bmatrix}\begin{bmatrix}\{\boldsymbol{\varepsilon}_0\}\\ \{\boldsymbol{\kappa}\}\end{bmatrix} - [\boldsymbol{\beta}_{6\times 2}]\begin{bmatrix}T_0\\ \Delta T_1\end{bmatrix}. \tag{25.60}$$

Note that the matrices $[\tilde{\mathbf{A}}]$, $[\tilde{\mathbf{B}}_1]$, $[\tilde{\mathbf{B}}_2]$, and $[\tilde{\mathbf{D}}]$ in (25.60), obtained according to the proposed mixed averaging scheme, are not symmetric, in contrast to the matrices of the CLPT. Among the reasons, there are the ignoring of some terms in the first approximation, different classifications of layers with respect to normal and shear strains, and using Hooke's law in the form (25.4) for "local" evaluation of the strain components. Therefore, for compatibility with the CLPT need to apply a symmetrisation for these matrices. The simplest method consists in averaging the off-diagonal elements. On other hand, it is also possible to give preference to some coefficients and, for example, to determine over-diagonal elements only (matrices are considered symmetric).

25.7 Conclusions and Future Work

Due to its simplicity and available software, the ESLM remains an important approach to solving the practical problems of the stress-strain state studying of composite plates. At the level of individual unidirectional reinforced layers, the structural approach to averaging makes it possible to obtain the effective thermoelastic properties of the entire package of layers quite simply. In contrast to the CLPT, this work proposes a new approach to calculating the stiffness matrix for the package of layers as a whole, which includes both the direct rule of mixture (based on the hypothesis of averaged strains) and the inverse one (based on the hypothesis of averaged stresses). The assumption of a linear distribution of strains across the thickness of the plate is proposed to be interpreted in a statistical sense as a regression model. This makes it possible to expand the class of problems, in particular, to propose the ESLM for layered plates in a wide range of the length-to-thickness ratio. Some numerical experiments are performed for typical layered structures using finite-element analysis within the framework of the 3-D LET. Also, the dependences of elastic moduli by the angle of in-plane rotation for the orthotropic layers of the plate are analyzed. The obtained results provide grounds to propose the criteria for the classification of layers, according to which it is recommended to choose a direct ROM (stiffnesses) or an inverse ROM (compliances). Stiffnesses in each direction of the global coordinate system in the plane of the plate are considered separately, and also, layers are classified independently with respect to shear strains.

The main idea of the proposed approach may be formulated in a simplified way as follows. Let us associate directions with high values of elasticity modulus (stiffness is greater) with arithmetic averaging (the direct ROM for strains). And on the contrary, directions with low elasticity moduli (compliance is greater) are associated with geometric averaging (the inverse ROM for stresses). The situation is similar to the rules of the mixture at the micromechanics level regarding the averaging of longitudinal and transverse Young's moduli of unidirectionally reinforced composite (the "fiber in matrix" system). It is expected that the overall effective stiffness of the composite as a planar system of layers will be somewhat reduced, and thereby, the understated plate deflections (it is typical for the CLPT), for example, will be in better agreement with the 3-D LET results.

The technique of averaging strains and stresses has been performed in sufficient detail, and the constitutive equations of state have been obtained in a matrix form. Similar to the CLPT model, forces and moments are related to linearly averaged strains. In the future, we plan to develop research on the mixed scheme of averaging thermoelastic properties in several directions:

1. Quantitative verification of the proposed mathematical model is necessary for its agreement with the CLPT and 3-D LET (or a full-scale experiment), for which it is necessary to create software, perform calculations of the effective properties of the monolayer, and analyze the stress-strain state of laminated plates for typical problems;
2. Finalize the mixed averaging scheme to take into account transverse shear strains to match the level of the so-called first-order shear deformation theories, which are often used for composite plates;
3. Integrate this approach into the software, especially FEM.

References

1. Altenbach, H., Altenbach, J., Kissing, W.: Mechanics of Composite Structural Elements, 2nd edn. Springer Nature Singapore Pte Ltd (2018)
2. Aßmus, M., Altenbach, H.: Elastic properties of polycrystalline silicon: experimental findings, effective estimates, and their relations. Contin. Mech. Thermodyn. **35**(2), 595–624 (2023)
3. Aßmus, M., von Zabiensky, M., Weber, M., Altenbach, H.: Size effects in the elastic properties of polycrystalline silicon. Appl. Res. e202300014 (2023)
4. Barbero, E.J.: Introduction to composite materials design, 3rd edn. Taylor & Francis, CRC Press, Boca Raton, FL (2017)

5. Clyne, T.W., Hull, D.: An Introduction to Composite Materials, 3rd edn. Cambridge University Press, Cambridge (2019)
6. Grigorenko, YaM., Grigorenko, A.Ya., Vlaikov, G.G.: Problems of Mechanics for Anisotropic Inhomogeneous Shells on the Basis of Different Models. Akademperiodika, Kyiv (2009)
7. Grigorenko, A.Ya, Müller, W.H., Grigorenko, YaM., Vlaikov, G.G.: Recent developments in anisotropic heterogeneous shell theory. General Theory and Applications of Classical Theory, vol. I. Springer, Berlin (2016)
8. Grigorenko, A.Ya, Müller, W.H., Grigorenko, Ya.M., Vlaikov, G.G.: Recent developments in anisotropic heterogeneous shell theory. Applications of Refined and Three-dimensional Theory, vol. II A. Springer, Berlin (2016)
9. Hashin, Z., Shtrikman, S.: A variational approach to the theory of the elastic behaviour of multiphase materials. J. Mech. Phys. Solids **11**(2), 127–140 (1963)
10. Hill, R.: A self-consistent mechanics of composite materials. J. Mech. Phys. Solids **13**(4), 213–222 (1965)
11. Jones, R.: Mechanics of Composite Materials, 2nd edn. Taylor & Francis, Philadelphia, PA (1999)
12. Li, D.: Analysis of Composite Laminates: Theories and Their Applications, 1st edn. Elsevier (2022)
13. Kollár, L.P., Springer, G.S.: Mechanics of Composite Structures. Cambridge University Press, Cambridge (2003)
14. Pagano, N.J.: Exact Solutions for Rectangular Bidirectional Composites and Sandwich Plates. J. Compos. Mater. **4**, 21–34 (1970)
15. Rana, S., Fangueiro, R.: Advanced Composite Materials for Aerospace Engineering: Processing, Properties and Applications. Woodhead Publishing (2016)
16. Reddy, J.N., Barberro, E.J., Teply, J.I.: A plate bending element based on a generalized laminate plate theory. Int. J. Numer. Methods Eng. **28**, 2275–2292 (1989)
17. Reddy, J.N.: Mechanics of Laminated Composite Plates and Shells: Theory and Analysis, 2nd edn. CRC Press LLC, Boca Raton, FL (2004)
18. Reuss, A.: Berechnung der Flie ßgrenze von Mischkristallen auf Grund der Plastizitätsbedingung für Einkristalle. Z. Angew. Math. Mech. **9**(1), 49–58 (1929)
19. Seydibeyoğlu, M.Ö., Mohanty, A.K., Misra, M.: Fiber Technology for Fiber-Reinforced Composites. Woodhead Publishing (2017)
20. Shah, P.H., Batra, R.C.: Through-the-thickness stress distributions near edges of composite laminates using stress recovery scheme and third order shear and normal deformable theory. Compos. Struct. **131**, 397–413 (2015)
21. Tsai, S.W.: Structural behavior of composite materials. NASA CR-71 (1964)
22. Tsai, S.W.: Introduction to Composite Materials. Technomic Publishing, Westport, CT (1980)
23. Voigt, W.: Lehrbuch der Kristallphysik (mit Ausschluss der Kristalloptik). Springer, Wiesbaden (1910)
24. Zienkiewicz, O.C., Taylor, R.L., Zhu, J.Z.: The Finite Element Method: Its Basis and Fundamentals, 7th edn. Butterworth-Heinemann (2013)

Chapter 26
Modeling of the Thermally Stressed State of Functionally Graded Shallow Shells in a Three-Dimensional Formulation

Alexandr Marchuk

Abstract The problems of the thermal stress state of layered functionally graded shallow shells based on a 3D problem statement are considered. Two approaches for solving the problems are presented. The first approach is based on the Reissner variational principle, and a system of integro-differential equilibrium equations and the corresponding boundary conditions are obtained without introducing an approximation. The second approach performs the differential equations of 3D thermoelastic equilibrium and a polynomial approximation of sought-for functions across the structure thickness. Its salient feature is an assignment of the functions to the outer surfaces of layers, which allows splitting the layers into sublayers with a corresponding increase in the accuracy of calculation results. For the case of hinge-supported shallow shells with a thermal load distributed according to a trigonometric law, the integro-differential equilibrium equations obtained for the first approach and the differential equations of the second approach allow an analytical implementation. In this statement, the proposed models are implemented analytically for the particular case of hinged support. The same result obtained based on the two methods may indicate reliability.

26.1 Introduction

A relatively large number of works are devoted to the problem of calculating rectangular plates for static and thermal loading in an accurate three-dimensional formulation. Their review is given, in particular, in the [2, 5]. The calculation of shallow shells in this formulation is not sufficiently considered. In [4], shallow shells' thermally stressed state is studied using a high-precision numerical approach. The exact solution of the three-dimensional problem of the thermal stress state of orthotropic shallow shells with rigid and sliding contacts of layers was obtained in [6]. The thermoelastic state of functionally graded shallow shells in the exact three-dimensional formulation was not considered.

A. Marchuk (✉)
National Transport University, Kyiv, Ukraine
e-mail: mksmntu386@gmail.com

H. Altenbach et al. (eds.), *Selected Problems of Solid Mechanics and Solving Methods*, Advanced Structured Materials 204, https://doi.org/10.1007/978-3-031-54063-9_26

26.2 Problem Statement

The approach for investigating the thermal stress state of layered functionally graded plate problems was proposed in [5]. This paper aims to extend this method to the calculation of the thermoelastic state of layered functionally graded shallow shells. The developed method will be implemented analytically for the partial case of boundary conditions (Navier). Different conditions on the bottom surface (free, rigidly fixed, and resting on an elastic base) will be considered. This problem has not previously been considered in the exact three-dimensional formulation.

The layered structure in the Cartesian coordinate system is considered. The outer surfaces of the layers have zero torsional curvature, $k_{12} = 0$, the change of principal curvatures is neglected, $k_{11}, k_{22} = \text{const}$. The construction is very shallow, i.e., the coefficients of the first quadratic form are taken as $A_1, A_2 \approx 1$. It is assumed that if the radii of curvature are much larger than the thickness of the structure, and its outer surfaces have the same curvature, it can be written as $1 + k_{11}z, 1 + k_{22}z \approx 1$. The introduced constraints allow us to identify the curvilinear system of orthogonal coordinates with the planar system of orthogonal coordinates. The x—(lower index 1) and y-axis (lower index 2) are directed within the shell plane, z-axis (lower index 3)—downward over the thickness; a comma in subscripts means differentiation operation. The following relations define the deformations of the shell layers of a functionally graded material:

$$
\begin{aligned}
e_{ii}^{(k)} &= \frac{1}{E^{(k)}(z)}\left(\sigma_{ii}^{(k)} + \nu^{(k)}\left[\sigma_{jj}^{(k)} + \sigma_{33}^{(k)}\right]\right),\\
e_{33}^{(k)} &= \frac{1}{E^{(k)}(z)}\left(\sigma_{33}^{(k)} + \nu^{(k)}\sigma_{ii}^{(k)}\right),\\
2e_{i3}^{(k)} &= \frac{\sigma_{i3}^{(k)}}{G^{(k)}(z)}, \quad 2e_{12}^{(k)} = \frac{\sigma_{12}^{(k)}}{G^{(k)}(z)} \quad (i \neq j = 1, 2).
\end{aligned}
\tag{26.1}
$$

26.3 Exact Analytical Three-Dimensional Solution of the Thermal Stress State of Functionally Graded Shallow Shells (A)

This approach is based on the separation of variables. As in [3, 9, 10], we present the displacements and stresses, as well as the temperature distribution, in the following form:

$$
\begin{aligned}
U_i^{(k)}(x, y, z) &= V_i^{(k)}(x, y) f_i^{(k)}(z),\\
\sigma_{i3}^{(k)}(x, y, z) &= \tau_{i3}^{(k)}(x, y) f_{i+3}^{(k)}(z),\\
T^{(k)}(x, y, z) &= t^{(k)}(x, y) f_7^{(k)}(z) \quad (i = 1, 2, 3).
\end{aligned}
$$

The following relations define the strain tensor:

$$\begin{aligned} e_{ii}^{(k)} &= V_{i,i}^{(k)} f_i^{(k)} + k_{ii} V_3^{(k)} f_3^{(k)}, \\ e_{33}^{(k)} &= V_3^{(k)} f_{3,3}^{(k)}, \quad 2e_{12}^{(k)} = U_{1l,2}^{(k)} f_{1l}^{(k)} + U_{2l,1}^{(k)} f_{2l}^{(k)}, \\ 2e_{i3}^{(k)} &= V_i^{(k)} f_{i,3}^{(k)} + V_{3,i}^{(k)} f_3^{(k)} \quad (i = 1, 2). \end{aligned}$$

The remaining components of the stress tensor are derived from the expressions:

$$\begin{aligned} \sigma_{ii}^{(k)} &= B_{ii}^{(k)}\left[V_{i,i}^{(k)} f_i^{(k)} + k_{ii} V_3^{(k)} f_3^{(k)}\right] \\ &\quad + B_{ij}^{(k)}\left[V_{j,j}^{(k)} f_j^{(k)} + k_{jj} V_3^{(k)} f_3^{(k)}\right] + B_{i3}^{(k)} \tau_{33}^{(k)} f_6^{(k)} \\ &\quad - \left[\alpha_i^{(k)} B_{ii}^{(k)} + \alpha_j^{(k)} B_{ij}^{(k)}\right] t^{(k)}(x, y) f_7^{(k)}(z) \quad (i \neq j = 1, 2), \\ \sigma_{12}^{(k)} &= G_{12}^{(k)}\left[V_{1,2}^{(k)} f_1^{(k)} + V_{2,1}^{(k)} f_2^{(k)}\right]. \end{aligned}$$

The ratios for determining the coefficients $B_{ii}^{(k)}$, $B_{ij}^{(k)}$, $B_{i3}^{(k)}$, and $B_{33}^{(k)}$ can be found in [8].

The resolving system of equations and the corresponding boundary conditions are found using the Reissner variational principle. For the pivoting bearing (the Navier limit conditions), the distribution of the functions to be sought in the plane has a trigonometric form.

Let us present the sought-for temperature as

$$T^{(k)}(x, y, z) = \sin \frac{\pi m x}{a} \sin \frac{\pi n y}{b} f_7^{(k)}(z).$$

Then, the equation of equilibrium of the layered shell and the known stationary thermal conductivity equation, which, under the assumptions of flatness, coincides with the known equation for the plate

$$\lambda_1^{(k)} T_{,11}^{(k)} + \lambda_2^{(k)} T_{,22}^{(k)} - Q_{,3}^{(k)} = 0, \quad Q_{,3}^{(k)} = -\lambda_3^{(k)} T_{,3}^{(k)},$$

are as follows:

$$f_{i,3}^{(k)} = -f_3^{(k)}(\bar{a}_i) + f_{(i+3)}^{(k)}\left(\frac{1}{\bar{G}_{i3}^{(k)} e^{z\gamma^{(k)}}}\right),$$

$$\begin{aligned} f_{3,3}^{(k)} &= f_i^{(k)} B_{i3}^{(k)}(\bar{a}_i) - f_3^{(k)}\left(k_{ii} B_{i3}^{(k)}\right) \\ &\quad + f_6^{(k)}\left(\frac{\bar{B}_{33}^{(k)}}{e^{z\gamma^{(k)}}}\right) + f_7^{(k)}\left(B_{i3}^{(k)} \frac{\bar{\alpha}_i^{(k)}}{e^{z\gamma^{(k)}}} + \frac{\bar{\alpha}_3^{(k)}}{e^{z\gamma^{(k)}}}\right), \end{aligned}$$

$$
\begin{aligned}
f_{(i+3),3}^{(k)} &= f_i^{(k)} \left[\bar{B}_{ii}^{(k)} (\bar{a}_i)^2 + \bar{G}_{12}^{(k)} (\bar{a}_j)^2 \right] e^{z\gamma^{(k)}} \\
&\quad + f_j^{(k)} \left(\bar{B}_{ij}^{(k)} + \bar{G}_{12}^{(k)} \right) e^{z\gamma^{(k)}} (\bar{a}_i) (\bar{a}_j) \\
&\quad - f_3^{(k)} \left(k_{ii} B_{ii}^{(k)} (\bar{a}_i) + k_{jj} B_{ij}^{(k)} (\bar{a}_i) \right) e^{z\gamma^{(k)}} \\
&\quad - f_6^{(k)} B_{i3}^{(k)} (\bar{a}_i) + f_7^{(k)} \left(\bar{B}_{ii}^{(k)} \bar{\alpha}_i^{(k)} + \bar{B}_{ij}^{(k)} \bar{\alpha}_j^{(k)} \right) (\bar{a}_i), \\
f_{6,3}^{(k)} &= - f_i^{(k)} \left(k_{ii} \bar{B}_{ii}^{(k)} (\bar{a}_i) + k_{jj} \bar{B}_{ij}^{(k)} (\bar{a}_i) \right) e^{z\gamma^{(k)}} \\
&\quad + f_3^{(k)} \left(k_{ii}^2 \bar{B}_{ii}^{(k)} + k_{ii} k_{jj} \bar{B}_{ij}^{(k)} \right) e^{z\gamma^{(k)}} \\
&\quad + f_{(i+3)}^{(k)} (\bar{a}_i) - f_6^{(k)} \left(k_{ii} B_{i3}^{(k)} \right), \\
f_{7,3}^{(k)} &= - \frac{e^{z\gamma^{(k)}}}{\bar{\lambda}^{(k)}} f_8^{(k)}, \\
f_{8,3}^{(k)} &= - f_7^{(k)} \frac{\bar{\lambda}^{(k)}}{e^{z\gamma^{(k)}}} (\bar{a}_i)^2 \quad (i \neq j = 1, 2).
\end{aligned}
\tag{26.2}
$$

Here,

$$
E^{(k)}(z) = \bar{E}^{(k)} e^{z\gamma^{(k)}}, \quad a^{(k)}(z) = \frac{\bar{a}^{(k)}}{e^{z\gamma^{(k)}}}, \quad \lambda^{(k)}(z) = \frac{\bar{\lambda}^{(k)}}{e^{z\gamma^{(k)}}},
$$
$$
\bar{a}_1 = \frac{\pi m}{a}, \quad \bar{a}_2 = \frac{\pi n}{b}.
$$

When the solution of system (26.2) is sought in the form

$$
\begin{aligned}
f_i^{(k)} &= \mu_i^{(k)} e^{(\beta^{(k)} - \gamma^{(k)}) z}, & f_{i+3}^{(k)} &= \mu_{i+3}^{(k)} e^{\beta^{(k)} z} \quad (i = 1, 2, 3), \\
f_7^{(k)} &= \mu_7^{(k)} e^{\beta^{(k)} z}, & f_8^{(k)} &= \mu_8^{(k)} e^{(\beta^{(k)} - \gamma^{(k)}) z},
\end{aligned}
$$

we obtain a system of homogeneous algebraic equations:

$$
\begin{aligned}
& - \mu_i^{(k)} (\beta^{(k)} - \gamma^{(k)}) - \mu_3^{(k)} (\bar{a}_i) + \mu_{i+3}^{(k)} \left(\frac{1}{\bar{G}_{i3}^{(k)}} \right) = 0, \\
& \mu_i^{(k)} B_{i3}^{(k)} (\bar{a}_i) - \mu_3^{(k)} \left(\beta^{(k)} - \gamma^{(k)} + \left(k_{ii} B_{i3}^{(k)} \right) \right) \\
& \qquad + \mu_6^{(k)} \bar{B}_{33}^{(k)} + \mu_7^{(k)} \left(B_{i3}^{(k)} \bar{\alpha}_i^{(k)} + \bar{\alpha}_3^{(k)} \right) = 0, \\
& \mu_i^{(k)} \left(\bar{B}_{ii}^{(k)} (\bar{a}_i)^2 + \bar{G}_{12}^{(k)} (\bar{a}_j)^2 \right) + \mu_j^{(k)} \left(\bar{B}_{ij}^{(k)} + \bar{G}_{12}^{(k)} \right) (\bar{a}_i) (\bar{a}_j) \\
& \qquad - \mu_3^{(k)} \left(k_{ii} \bar{B}_{ii}^{(k)} (\bar{a}_i) + k_{jj} \bar{B}_{ij}^{(k)} (\bar{a}_j) \right) - \mu_{(i+3)}^{(k)} \beta^{(k)} \\
& \qquad - \mu_6^{(k)} B_{i3}^{(k)} (\bar{a}_i) + \mu_7^{(k)} \left(\bar{B}_{ii}^{(k)} \bar{\alpha}_i^{(k)} + \bar{B}_{ij}^{(k)} \bar{\alpha}_j^{(k)} \right) (\bar{a}_i) = 0, \\
& - \mu_i^{(k)} \left(k_{ii} \bar{B}_{ii}^{(k)} (\bar{a}_i) + k_{jj} \bar{B}_{ij}^{(k)} (\bar{a}_i) \right) + \mu_3^{(k)} \left(k_{ii}^2 \bar{B}_{ii}^{(k)} + k_{ii} k_{jj} \bar{B}_{ij}^{(k)} \right)
\end{aligned}
$$

$$+\mu_{i+3}^{(k)}(\bar{a}_i) - \mu_6^{(k)}\left(\beta^{(k)} + \left(k_{ii}B_{i3}^{(k)}\right)\right) = 0,$$
$$-\mu_7^{(k)}\beta^{(k)} - \frac{1}{\bar{\lambda}^{(k)}}\mu_8^{(k)} = 0,$$
$$-\mu_7^{(k)}\bar{\lambda}^{(k)}(\bar{a}_i)^2 - \mu_8^{(k)}(\beta^{(k)} - \gamma^{(k)}) = 0.$$

Roots of the system of characteristic equations $\beta^{(k)}$ can be real and complex. Now, the solution of the system differential equations (26.2) can be presented.

$$\begin{aligned}
f_i^{(k)} &= \mu_{ij}^{(k)}\bar{C}_j^{(k)}e^{\left(\beta_j^{(k)}-\gamma\right)z},\\
f_{i+3}^{(k)} &= \mu_{(i+3)j}^{(k)}\bar{C}_j^{(k)}e^{\beta_j^{(k)}z} \quad (i = 1, 2, 3;\ j = 1, \ldots, 6),\\
f_7^{(k)} &= \bar{C}_7^{(k)}\mu_{77}^{(k)}e^{\beta_7^{(k)}z} + \bar{C}_8^{(k)}\mu_{78}^{(k)}e^{-\beta_7^{(k)}z},\\
f_8^{(k)} &= C_7^{(k)}\mu_{87}^{(k)}e^{\left(\beta_7^{(k)}-\gamma^{(k)}\right)z} + \bar{C}_8^{(k)}\mu_{88}^{(k)}e^{\left(\beta_8^{(k)}-\gamma^{(k)}\right)z}.
\end{aligned}$$

26.4 Approximate Three-dimensional Solution of the Thermal Stress State of Functionally Graded Shallow Shells (P)

In this approach, approximation over the shallow shells thickness is performed by polynomials using the well-known iterative approach [1, 7, 8].

The power function is assumed to be able to approximate the physical and mechanical characteristics through the thickness of the structure:

$$E^{(k)}(z) = \bar{E}^{(k)}f_e^{(k)}, \quad \alpha^{(k)}(z) = \bar{\alpha}^{(k)}f_\alpha^{(k)}, \quad \lambda^{(k)}(z) = \bar{\lambda}^{(k)}f_\lambda^{(k)}.$$

Here, $f_e^{(k)}$, $f_\alpha^{(k)}$, and $f_\lambda^{(k)}$ are given polynomials of the fourth degree.

Let us use the known approximation of sought-for displacements across the shallow shell thickness:

$$\begin{aligned}
U_i^{(k)}(x, y, z) &= U_{il}^{(k)}(x, y)f_{il}^{(k)}(z),\\
U_3^{(k)}(x, y, z) &= \bar{W}_p^{(k)}(x, y)\beta_p^{(k)}(z) \quad (i = 1, 2;\ l = 1, 2, 3;\ p = 1, 2, 3)
\end{aligned} \tag{26.3}$$

Here, ${U_{i1}}^{(k)}(x, y)$ and ${U_{i2}}^{(k)}(x, y)$ are the tangential displacements on surfaces of a kth layer of the shell; ${U_{i3}}^{(k)}(x, y)$ are the functions of shear in direction of the x- and y-axes; ${\bar{W}_1}^{(k)}(x, y)$ and ${\bar{W}_2}^{(k)}(x, y)$ are the normal displacements on surfaces of a kth layer of the plate; ${\bar{W}_3}^{(k)}(x, y)$, are the functions of thickness reduction; ${f_{i1}}^{(k)}(z)$, ${f_{i2}}^{(k)}(z)$, ${\beta_1}^{(k)}(z)$, and ${\beta_2}^{(k)}(z)$ are given polynomials of the first degree; ${\beta_3}^{(k)}(z)$ are polynomial of the second degree; ${f_{i3}}^{(k)}(z)$ are polynomials of the third degree.

The strain tensor is determined from (26.3)

$$e_{ii}^{(k)} = U_{il,1}^{(k)} f_{il}^{(k)} + k_{ii} \bar{W}_p^{(k)} \beta_p^{(k)},\ 2e_{12}^{(k)} = U_{1l,2}^{(k)} f_{1l}^{(k)} + U_{2l,1}^{(k)} f_{2l}^{(k)},$$
$$e_{33}^{(k)} = \bar{W}_p^{(k)} \beta_{p,3}^{(k)}, \qquad 2e_{i3}^{(k)} = U_{il}^{(k)} f_{il,3}^{(k)} + \bar{W}_{p,i}^{(k)} \beta_p^{(k)}. \tag{26.4}$$

The components of the stress tensor with the expressions for deformations (26.4) are derived via Hooke's law

$$\begin{aligned}
\sigma_{ii}^{(k)} &= \bar{C}_{ii}^{(k)}(z)(U_{il,i}^{(k)} f_{1l}^{(k)} + k_{ii} \bar{W}_p^{(k)} \beta_p^{(k)}) \\
&\quad + \bar{C}_{ij}^{(k)}(z) \left(U_{jl,2}^{(k)} f_{jl}^{(k)} + k_{jj} \bar{W}_p^{(k)} \beta_p^{(k)} \right) + \bar{C}_{i3}^{(k)}(z) \bar{W}_p^{(k)} \beta_{p,3}^{(k)} \\
&\quad - \left(\bar{C}_{ii}^{(k)} \alpha_i^{(k)} + \bar{C}_{ij}^{(k)} \alpha_j^{(k)} + \bar{C}_{i3}^{(k)} \alpha_3^{(k)} \right) t^{(k)} f_7^{(k)}, \\
\sigma_{33}^{(k)} &= \bar{C}_{3i}^{(k)}(z) \left(U_{il,1}^{(k)} f_{il}^{(k)} + k_{ii} \bar{W}_p^{(k)} \beta_p^{(k)} \right) + \bar{C}_{33}^{(k)}(z) \bar{W}_p^{(k)} \beta_{p,3}^{(k)} \\
&\quad - \left(\bar{C}_{3i}^{(k)} \alpha_i^{(k)} + \bar{C}_{33}^{(k)} \alpha_3^{(k)} \right) t^{(k)} f_7^{(k)}, \\
\sigma_{12}^{(k)} &= G^{(k)}(z) \left(U_{1l,2}^{(k)} f_{1l}^{(k)} + U_{2l,1}^{(k)} f_{2l}^{(k)} \right), \\
\sigma_{i3}^{(k)} &= G^{(k)}(z) \left(U_{il}^{(k)} f_{il,3}^{(k)} + \bar{W}_{p,i}^{(k)} \beta_p^{(k)} \right) \quad (i \neq j = 1, 2),
\end{aligned} \tag{26.5}$$

where

$$f_7^{(k)} = a_0^{(k)} + a_1^{(k)} z + a_2^{(k)} z^2$$

and the coefficients $a_0^{(k)}$, $a_1^{(k)}$, and $a_2^{(k)}$ can be found by solving the equation of stationary heat conductivity.

Using the expressions of strains (26.4) and stresses (26.5) as well as the Lagrange variational equation, we obtain the equilibrium equations:

$$\begin{aligned}
&- \left(B_{ii\bar{l}l}^{(k)} U_{il,ii}^{(k)} + B6_{ii\bar{l}l}^{(k)} U_{il,22}^{(k)} - TU_{i\bar{l}l}^{(k)} U_{il}^{(k)} \right) - \left(B_{ij\bar{l}l}^{(k)} + B6_{ij\bar{l}l}^{(k)} \right) U_{jl,ij}^{(k)} \\
&\quad - \left(SD_{i\bar{l}p}^{(k)} - CUW_{i\bar{l}p}^{(k)} + BR_{ii\bar{l}p}^{(k)} k_{ii} + BR_{ij\bar{l}p}^{(k)} k_{jj} \right) \bar{W}_{p,i}^{(k)} \\
&\quad\quad + BBT_{i\bar{l}}^{(k)} t_{,i}^{(k)} - q_{i\bar{l}}^{(k)} = 0, \\
&\left(SD_{i\bar{p}l}^{\prime(k)} - CUW_{i\bar{p}l}^{\prime(k)} + BR_{ii\bar{p}l}^{\prime(k)} k_{ii} + BR_{ij\bar{p}l}^{\prime(k)} k_{jj} \right) U_{il,i}^{(k)} \\
&- \left(CC_{i\bar{p}p}^{(k)} \bar{W}_{p,ii}^{(k)} - ZZ_{\bar{p}p}^{(k)} \bar{W}_p^{(k)} \right) + \left(R_{ii\bar{p}p}^{(k)} k_{ii} + RR_{ii\bar{p}p}^{(k)} k_{ii} \right) \bar{W}_p^{(k)} \\
&+ \left(ZR_{ii\bar{p}p}^{(k)} k_{ii}^2 + ZR_{ij\bar{p}p}^{(k)} k_{ii} k_{jj} \right) \bar{W}_p^{(k)} + \left(ZR_{ii\bar{p}p}^{(k)} k_{ii}^2 + ZR_{ij\bar{p}p}^{(k)} k_{ii} k_{jj} \right) \bar{W}_p^{(k)} \\
&\quad\quad + ZZT_{\bar{p}}^{(k)} t^{(k)} - q_{3\bar{p}}^{(k)} = 0
\end{aligned}$$

$(i \neq j = 1, 2)$.

In the case of hinged support, the stationary thermal conductivity equation is transformed to the following form:

$$-\lambda_1^{(k)}\left(\frac{\pi n x}{a}\right)^2-\lambda_2^{(k)}\left(\frac{\pi n y}{b}\right)^2+\lambda_{3,3}^{(k)}f_{7,3}^{(k)}+\lambda_3^{(k)}f_{7,33}^{(k)}=0.$$

To tackle this equation effectively, the grid method proves to be the most suitable. This method ensures the exact satisfaction of boundary conditions on layer surfaces for both temperature and heat flux, minimizing complications. The second derivative of the temperature distribution function across the thickness of the shallow shell is expressed using the following difference equation:

$$f_{7,33}^{(k)}=\frac{f_{7(i-1)}^{(k)}-2f_{7(i)}^{(k)}+f_{7(i+1)}^{(k)}}{h^{(k)}}.$$

The temperatures on layer surfaces and at their center found in such a way allow one to determine coefficients of the polynomial:

$$f_7^{(k)}=a_0^{(k)}+a_1^{(k)}z+a_1^{(k)}z^2.$$

26.5 Calculation Results

We will consider the thermoelastic equilibrium of a two-layer composite shallow shell of their functionally graded material, which is hinge-supported along its contour. The physicomechanical characteristics of the layers are as follows:

$$\bar{\alpha}=1/\text{deg.},\quad \bar{E}=1\text{MPa},$$

$$\alpha^{(1)}=23.8\cdot10^{-6}\cdot\bar{\alpha}/e^{\gamma(1)},\quad \lambda^{(1)}=1/e^{\gamma(1)z}\text{W/m}\cdot\text{deg.},\quad E^{(1)}(z)=\bar{E}e^{\gamma^{(1)}z},$$
$$\alpha^{(2)}=23.8\cdot10^{-6}\cdot\bar{\alpha}/e^{\gamma(2)z},\quad \lambda^{(2)}=1/e^{\gamma(2)z}\text{W/m}\cdot\text{deg.},\quad E^{(2)}(z)=\bar{E}e^{\gamma^{(2)}z}$$

$$\gamma^{(1)}=\{-5;-1\},\quad \gamma^{(2)}=-\gamma^{(1)},\quad \nu^{(1)}=\nu^{(2)}=0.3,$$
$$h=1\text{ m},\quad h^{(1)}=h^{(2)}=h/2.$$

The thickness of the layers is the same. The shell curvature:

$$k_{11}=\frac{8/5h}{(a+b)a},\quad k_{22}=\frac{8/5h}{(a+b)b}.$$

The shallow shell is square: $a=b=L$. The contact of layers is rigid. Consider the shallow shell with different conditions on the bottom surface: the free bottom surface, the rigidly fixed bottom surface, and support on an elastic foundation.

The elastic foundation is modeled with a finite thickness layer with the following physicomechanical characteristics:

$$\alpha^{(3)} = 23.8 \cdot 10^{-6} \cdot \bar{\alpha}/e^{\gamma(3)z}/100, \quad E^{(3)}(z) = \bar{E}e^{\gamma^{(3))}z}/100,$$
$$\lambda^{(3)} = 1/e^{\gamma(3)z}/100 \text{ W/m} \cdot \text{deg.}, \quad \gamma^{(3)} = \{5; 1\}, \quad \nu^{(3)} = 0.3, \quad h^{(3)} = 10h$$

(elastic foundation is the third layer). The third layer is split into 20 sublayers. The bottom surface of the elastic foundation is rigidly fixed. The shallow shell is loaded on its top and bottom surfaces by a sinusoidal temperature:

$$T_{q31} = 50T_0, \quad T_{q33} = 0 \cdot T_0 \quad (T_0 = 1 \text{ deg.})$$

In the case of an elastic foundation

$$T_{q31} = 50T_0, \quad T_{q34} = 0T_0.$$

A similar problem for plates was previously considered in [5].

Table 26.1 (obtained for $\gamma^{(2)} = -\gamma^{(1)} = 5$) and Table 26.2 ($\gamma^{(2)} = -\gamma^{(1)} = 1$) show the calculation results of a shallow shell with a free bottom surface. Table 26.3 ($\gamma^{(2)} = -\gamma^{(1)} = 5$) and Table 26.4 ($\gamma^{(2)} = -\gamma^{(1)} = 1$) show the calculation results of a shallow shell with a rigidly fixed bottom surface. Table 26.5 ($\gamma^{(3)} = \gamma^{(2)} = -\gamma^{(1)} = 5$) Furthermore, Table 26.6 ($\gamma^{(3)} = \gamma^{(2)} = -\gamma^{(1)} = 1$) shows the calculation results of a shallow shell on an elastic foundation. At the layer borders, the solutions

$$U_3 = \frac{U_3(a/2, b/2, z)}{h\bar{\alpha}T_0 \cdot 10^{-4}}, \quad \bar{\sigma}_{11} = \frac{\sigma_{11}(a/2, b/2, z)}{\bar{\alpha}T_0\bar{E} \cdot 10^{-4}}, \quad \text{and } T$$

were obtained.

Table 26.1 Comparison of calculation results obtained by two proposed approaches for a two-layer shallow shell with a free bottom surface at $\gamma^{(2)} = -\gamma^{(1)} = 5$ (N is layer number)

N	$-\bar{U}_3$			$\bar{\sigma}_{11}$			T		
	P1	P4	E	P1	P4	E	P1	P4	E
$L/h = 10$									
1	7.229	7.242	7.253	2.075	2.131	2.148	50.00	50.00	50.00
	5.553	5.432	5.442	−6.89	−6.85	−6.85	23.46	23.39	23.39
2	5.553	5.432	5.442	−6.89	−6.85	−6.85	23.46	23.39	23.39
	4.275	4.126	4.137	9.526	9.577	9.559	0	0	0
$L/h = 5$									
1	2.789	2.781	2.777	0.9594	1.017	1.030	50.00	50.00	50.00
	1.360	1.229	1.225	−5.94	−5.80	−5.79	19.78	19.56	19.54
2	1.360	1.229	1.225	−5.94	−5.80	−5.79	19.78	19.56 0	19.54 0
	0.3317	0.1779	0.1743	8.392	8.603	8.592	0	0	0

Table 26.2 Comparison of calculation results obtained by two proposed approaches for a two-layer shallow shell with a free bottom surface at $\gamma^{(2)} = -\gamma^{(1)} = 1$

N	$-\bar{U}_3$			$\bar{\sigma}_{11}$			T		
	P_1	P_4	E	P_1	P_4	E	P_1	P_4	E
$L/h = 10$									
1	44.26	44.25	44.26	−5.61	−5.60	−5.60	50.00	50.00	50.00
	39.83	39.82	39.83	−4.31	−4.28	−4.28	24.31	24.28	24.28
2	39.83	39.82	39.83	−4.31	−4.28	−4.28	24.31	24.28	24.28
	38.17	38.16	38.16	2.001	2.017	2.018	0	0	0
$L/h = 5$									
1	13.66	13.70	13.71	−5.81	−5.76	−5.75	50.00	50.00	50.00
	9.463	9.488	9.494	−4.04	−3.95	−3.94	22.44	22.34	22.33
2	9.463	9.488	9.494	−4.04	−3.95	−3.94	22.44	22.34	22.33
	7.979	7.998	8.004	1.875	1.920	1.921	0	0	0

Table 26.3 Comparison of calculation results obtained by two proposed approaches for a two-layer shallow shell with a rigidly fixed bottom surface at $\gamma^{(2)} = -\gamma^{(1)} = 5$

N	$-\bar{U}_3$			$\bar{\sigma}_{11}$			T		
	P_1	P_4	E	P_1	P_4	E	P_1	P_4	E
$L/h = 10$									
1	3.556	3.678	3.674	−10.8	−9.28	−9.16	50.00	50.00	50.00
	1.565	1.598	1.597	−7.70	−7.61	−7.61	23.46	23.39	23.39
2	1.565	1.598	1.597	−7.70	−7.61	−7.61	23.46	23.39	23.39
	0	0	0	0.2612	0.2319	.2297	0	0	0
$L/h = 5$									
1	2.829	2.926	2.923	−3.06	−1.72	−1.64	50.00	50.00	50.00
	1.252	1.274	1.272	−6.29	−6.15	−6.14	19.78	19.56	19.54
2	1.252	1.274	1.272	−6.29	−6.15	−6.14	19.78	19.56	19.54
	0	0	0	0.3910	.2921	0.2863	0	0	0

Calculations were carried out by the exact model (E), by the approximate model (P) with consideration of each layer within one sublayer (P_1), and by splitting each layer into four sublayers (P_4). We used the exponential distribution of physicomechanical characteristics across the layer thickness when calculating according to the model (P) and presented it by a fourth-degree polynomial.

In the case of calculating a shallow shell with a free bottom surface at $\gamma^{(2)} = -\gamma^{(1)} = 5$ and $\gamma^{(2)} = -\gamma^{(1)} = 1$, the proposed model with a polynomial approximation of the sought-for functions across the structure thickness provides sufficient accuracy with considering each layer within one sublayer (P_1). Further splitting of the layers into sublayers (P_4) leads to the refinement of the calculation results.

Table 26.4 Comparison of calculation results obtained by two proposed approaches for a two-layer shallow shell with a rigidly fixed bottom surface at $\gamma^{(2)} = -\gamma^{(1)} = 1$

N	$-\bar{U}_3$			$\bar{\sigma}_{11}$			T		
	P_1	P_4	E	P_1	P_4	E	P_1	P_4	E
$L/h = 10$									
1	8.163	8.147	8.145	−14.4	−14.4	−14.4	50.00	50.00	50.00
	2.439	2.446	2.447	−7.32	−7.31	−7.31	24.31	24.28	24.28
2	2.439	2.446	2.447	−7.32	−7.31	−7.31	24.31	24.28	24.28
	0	0	0	0.3173	0.3146	0.3144	0	0	0
$L/h = 5$									
1	7.113	7.097	7.096	−10.5	−10.5	−10.5	50.00	50.00	50.00
	2.168	2.189	2.190	−5.49	−5.44	−5.44	22.44	22.34	22.33
2	2.168	2.189	2.190	−5.49	−5.44	−5.44	22.44	22.44	22.33
	0	0	0	0.6803	0.6677	0.6670	0	0	0

Table 26.5 Comparison of calculation results obtained by two proposed approaches for a two-layer shallow shell on an elastic foundation at $\gamma^{(2)} = -\gamma^{(1)} = 5$

N	$-\bar{U}_3$			$\bar{\sigma}_{11}$			T		
	P_1	P_4	E	P_1	P_4	E	P_1	P_4	E
$L/h = 10$									
1	3.185	3.378	3.384	9.912	10.04	10.06	50.00	50.00	50.00
	0.4093	0.4484	0.4538	−12.8	−12.7	−12.7	43.57	43.40	43.39
2	0.4093	0.4484	0.4538	−12.8	−12.7	−12.7	43.57	43.40	43.39
	−2.29	−2.39	−2.38	12.15	12.18	12.18	42.85	42.76	42.75
$L/h = 5$									
1	2.323	2.333	2.325	4.919	5.020	5.023	50.00	50.00	50.00
	0.2668	0.1486	0.1399	−9.41	−9.17	−9.16	31.43	31.02	30.99
2	0.2668	0.1486	0.1399	−9.41	−9.17	−9.16	31.43	31.02	30.99
	−1.57	−1.78	−1.79	10.65	10.94	10.94	29.46	29.31	29.30

In the case of calculating a shallow shell with a rigidly fixed bottom surface, the proposed model with a polynomial approximation of the sought-for functions across the structure thickness provides acceptable accuracy when considering each layer within one sublayer (P_1) at $\gamma^{(2)} = -\gamma^{(1)} = 1$. If $\gamma^{(2)} = -\gamma^{(1)} = 5$, the calculation when considering each layer within one sublayer (P_1) contains a rather significant error. Further splitting of the layers into sublayers (P_4) leads to the refinement of the calculation results.

In the case of calculating a shallow shell on an elastic foundation at $\gamma^{(2)} = -\gamma^{(1)} = 5$ and at $\gamma^{(2)} = -\gamma^{(1)} = 1$, the proposed model with a polynomial approximation of the sought-for functions across the structure thickness provides sufficient accuracy with considering each layer within one sublayer (P_1). Splitting the layers

Table 26.6 Comparison of calculation results obtained by two proposed approaches for a two-layer shallow shell on an elastic foundation at $\gamma^{(2)} = -\gamma^{(1)} = 1$

N	$-\bar{U}_3$			$-\bar{\sigma}_{11}$			T		
	P_1	P_4	E	P_1	P_4	E	P_1	P_4	E
$L/h = 10$									
1	9.587	9.598	9.643	3.751	3.730	3.713	50.00	50.00	50.00
	3.660	3.667	3.715	8.165	8.117	8.107	46.35	46.29	46.28
2	3.660	3.667	3.715	8.165	8.117	8.107	46.35	46.29	46.28
	−1.99	−1.99	−1.94	2.951	2.925	2.915	45.32	45.30	45.30
$L/h = 5$									
1	6.712	6.717	6.721	4.896	4.823	4.812	50.00	50.00	50.00
	1.444	1.441	1.446	6.706	6.544	6.530	38.12	37.91	37.90
2	1.444	1.441	1.446	6.706	6.544	6.530	38.12	37.91	37.90
	−2.85	−2.87	−2.86	1.154	1.055	1.042	34.94	34.85	34.85

into sublayers (P_4) leads to the refinement of the calculation results. The stress-strain state of an elastic foundation is not considered.

26.6 Conclusions

The investigation into the thermal stress state of layered functionally graded shallow shells involved the application of two distinct approaches. The first approach, which avoids approximation errors in the distribution of sought-for functions across the shell thickness, involves the analytical solution of corresponding systems of differential equations. In the second approach, sought-for functions are approximated using polynomials, with a unique feature being the assignment of parameters to the outer surfaces of the shallow shell. This allows for potential sublayering, minimizing approximation errors. These two approaches complement each other, providing a comprehensive justification for calculation results across all considered problems.

In this study, the proposed models were analytically implemented for the specific case of hinged support with a sinusoidal thermal load distribution. Test studies on layered shallow shells were conducted under different conditions for the bottom surface: free, rigidly fixed, and supported on an elastic foundation. The applied model demonstrated high accuracy in calculation results without the need for sublayering, except in the case of a shallow shell with a rigidly fixed lower surface at $\gamma^{(2)} = -\gamma^{(1)} = 5$. To ensure sufficient accuracy, it became necessary to split each layer into four sublayers. Further sublayering yielded results coinciding with those given by the analytical model (omitted for brevity in the tables).

References

1. Ambartsumian, S.A.: On a general theory of anisotropic shells. J. Appl. Math. Mech. **22**(2), 305–319 (1958)
2. Brischetto, S.: Exact three-dimensional static analysis of single- and multi-layered plates and shells. Compos. Part B **119**, 230–252 (2017)
3. Galerkin, B.G.: Collection of Works, vol. 1, pp. 347–362. Izd AN SSSR, Moscow (1952). (in Russian)
4. Grigorenko, Y.M., Vasilenko, A.T., Pankratova, N.D.: Problems of the Theory of Elasticity of Inhomogeneous Solids. Naukova Dumka, Kyiv (1991). [in Russian]
5. Marchuk, A.V.: Analytical solution of the problem on the thermally stressed state of functionally graded plates based on the 3D elasticity theory. Compos.: Mech. Comput. Applicat. An. Int. J. **2**(4), 37–6 (2021)
6. Marchuk, A.V., Vasylevs'kyi, NYu., Lobashov, D.I.: Solution to the three-dimensional problem on the thermal stress state of hollow shells with rigid and sliding contacts of layers. J. Strength Mater. **53**(5), 744–750 (2021)
7. Piskunov, V.G., Verizhenko, V.E.: Linear and Nonlinear Problems of Calculation of Layered Structures. Budivel'nik, Kyiv (1986). [in Russian]
8. Rasskazov, A.O., Sokolovskaja, I.I., Shulga, N.A.: Theory and Calculation of Layered Orthotropic Plates and Shells. Vishcha Shkola, Kyiv (1986). [in Russian]
9. Timoschenko, S.P., Goodier, J.N.: Theory of Elasticity, 3rd edn. McCraw-Hill, NY (1970)
10. Vlasov, B.F.: On case of bending of a rectangular thick plate. Vest MGU **2**, 25–34 (1957)

Chapter 27
Solving of Stress State Problems of Anisotropic Thick Noncircular Cylindrical Shells with Different Nonhomogeneous Structures Based on Discrete Continual Approach

Liliya Rozhok, Svetlana Sperkach, and Larisa Vasil'eva

Abstract In this study, we introduce a novel approach utilizing discrete continual numerical methods to tackle the strain state problem of anisotropic noncircular cylinders with corrugated cross-sections and varying nonhomogeneous structures. Our investigation employs a comprehensive framework that integrates the separation of variables, discrete Fourier series, and a robust numerical discrete orthogonalization technique. Specifically, we focus on the intricate scenario of three-layer and continuously nonhomogeneous materials within thick, noncircular cylindrical shells. Our analysis delves into understanding the distribution of displacement and stress fields as they are influenced by the cross-sectional shape of the reference surface and the material's mechanical properties.

27.1 Introduction

Hollow noncircular cylinders serve as essential structural components in diverse fields of modern industry and construction, providing reliable high-strength solutions. Addressing stringent strength analysis requirements, incorporating accurate material properties, and accounting for three-dimensional effects in thick-walled structural elements necessitates a comprehensive study of hollow cylindrical structures in three dimensions.

L. Rozhok (✉)
National Transport University, Kyiv, Ukraine
e-mail: r.l.s@ua.fm

S. Sperkach
Technical Center of NAS of Ukraine, Kyiv, Ukraine
e-mail: svetlana@nasu.kiev.ua

L. Vasil'eva
V.O. Sukhomlynskyi Mykolaiv National University, Mykolaiv, Ukraine
e-mail: lara@vasiliev.mk.ua

H. Altenbach et al. (eds.), *Selected Problems of Solid Mechanics and Solving Methods*,
Advanced Structured Materials 204, https://doi.org/10.1007/978-3-031-54063-9_27

Analyzing stress and strain within thick-walled structures, guided by three-dimensional elasticity theory, presents challenges due to complex sets of partial differential equations and the need to satisfy boundary conditions on an elastic body's surface. These challenges are amplified when considering cylindrical components crafted from anisotropic and inhomogeneous materials, such as functionally graded materials (FGM) with varying elastic properties. Technological advances enable the production of structures with precisely tailored elastic moduli, and the physical and mechanical attributes of FGMs have been explored across different compositions [1, 2].

An area of significant practical interest and fundamental research lies in understanding the stress-strain state of structural members composed of FGMs, including noncircular cylinders. Thus, it becomes imperative to determine the strain state characteristics of these structural elements in three dimensions. While three-dimensional stress-strain analysis of elastic bodies made from FGMs poses considerable computational challenges, limited publications exist addressing this topic [3, 4, 8].

In addition to the widely-used universal methods such as finite-difference and finite-element techniques, discrete-continual approaches are gaining attention. These methods involve transforming original two-dimensional (or three-dimensional) problems into linear one-dimensional boundary-value problems. These problems can be numerically solved using the discrete orthogonalization method [5, 6]. Discrete-continual approaches serve as alternatives to universal numerical methods and are particularly suited for analyzing specific classes of shell objects. The solutions derived from these approaches may offer improved effectiveness and accuracy, making them valuable tools for testing and modifying various discrete techniques.

In this article, we present a variant of the discrete-continual approach. Our approach involves separating variables within the initial system of partial differential equations along the longitudinal coordinate. This reduction transforms the problem into a two-dimensional one, which is subsequently solved using a discrete Fourier series and a stable numerical discrete orthogonalization method. The study considers variants of nonhomogeneous cylinder materials: those with piecewise constant properties and continuously nonhomogeneous materials, such as functionally graded materials (FGMs). Our investigation delves into the strain state problem of anisotropic noncircular cylinders featuring corrugated cross sections and diverse nonhomogeneous structures.

27.2 Statement of the Problem

Based on the elasticity theory, we will study the strain state problem of the thick, noncircular cylindrical shells. Consider the curvilinear coordinate system (s, t, γ) where s, t, γ are the orthogonal curvilinear coordinates, s is the arclength along a generatrix, t is the polar angle in the cross-section, γ is the normal coordinate to the reference surface. In the coordinate system chosen, the squared length of a linear element of the cylinder is the form:

$$dS^2 = H_1^2 ds^2 + H_2^2 dt^2 + H_3^2 d\gamma^2, \tag{27.1}$$

where $H_1 = H_3 = 1$, $H_2 = 1 + [\gamma/R(t)]$ are Lamé parameters, $R(t) = R_t$ is the radius of curvature of the mid-surface directrix .

The multilayer cylindrical shells of the rigidly fixed layers interacting without sliding or tearing off are considered. The material is nonhomogeneous in thickness and homogeneous in generatrix and guide in each layer. In this case, the contact conditions of the layers are

$$\begin{aligned}
&\sigma_\gamma^i = \sigma_\gamma^{i+1}, \quad \tau_{s\gamma}^i = \tau_{s\gamma}^{i+1}, \quad \tau_{t\gamma}^i = \tau_{t\gamma}^{i+1}, \\
&u_\gamma^i = u_\gamma^{i+1}, \quad u_s^i = u_s^{i+1}, \quad u_t^i = u_t^{i+1},
\end{aligned} \tag{27.2}$$

where i is the layer number.

The basic relationships of the elasticity theory that describe the stress-state of a noncircular laminated hollow cylinder for each layer based on (27.1) are the following

- linear kinematic relations

$$\begin{aligned}
&e_s^i = \frac{\partial u_s^i}{\partial s}, \quad e_t^i = \frac{1}{H_2^i}\frac{\partial u_t^i}{\partial t} + \frac{1}{H_2^i}\frac{\partial H_2^i}{\partial \gamma} u_\gamma^i, \quad e_\gamma^i = \frac{\partial u_\gamma^i}{\partial \gamma}, \\
&e_{st}^i = \frac{1}{H_2^i}\frac{\partial u_s^i}{\partial t} + \frac{\partial u_t^i}{\partial s}, \quad e_{s\gamma}^i = \frac{\partial u_\gamma^i}{\partial s} + \frac{\partial u_s^i}{\partial \gamma}, \\
&e_{t\gamma}^i = H_2^i \frac{\partial}{\partial \gamma}\left(\frac{u_t^i}{H_2^i}\right) + \frac{1}{H_2^i}\frac{\partial u_\gamma^i}{\partial t};
\end{aligned} \tag{27.3}$$

- equilibrium equations

$$\begin{aligned}
&H_2^i \frac{\partial \sigma_s^i}{\partial s} + \frac{\partial \tau_{st}^i}{\partial t} + \frac{\partial}{\partial \gamma}\left(H_2^i \tau_{s\gamma}^i\right) = 0, \\
&\frac{\partial \sigma_t^i}{\partial t} + \frac{\partial}{\partial \gamma}\left(H_2^i \tau_{t\gamma}^i\right) + H_2^i \frac{\partial \tau_{st}^i}{\partial s} + \frac{\partial H_2^i}{\partial \gamma}\tau_{t\gamma}^i = 0, \\
&\frac{\partial}{\partial \gamma}\left(H_2^i \sigma_\gamma^i\right) + H_2^i \frac{\partial \tau_{s\gamma}^i}{\partial s} + \frac{\partial \tau_{t\gamma}^i}{\partial t} - \frac{\partial H_2^i}{\partial \gamma}\sigma_t^i = 0;
\end{aligned} \tag{27.4}$$

- Hooke's law for the more general orthotropic case

$$\begin{aligned}
&e_s^i = a_{11}^i \sigma_s^i + a_{12}^i \sigma_t^i + a_{13}^i \sigma_\gamma^i, \quad e_t^i = a_{12}^i \sigma_s^i + a_{22}^i \sigma_t^i +_{23}^i \sigma_\gamma^i, \\
&e_\gamma^i = a_{13}^i \sigma_s^i + a_{23}^i \sigma_t^i + a_{33}^i \sigma_\gamma^i, \quad e_{t\gamma}^i = a_{44}^i \tau_{t\gamma}^i, \\
&e_{s\gamma}^i = a_{55}^i \tau_{s\gamma}^i; \; e_{st}^i = a_{66}^i \tau_{st}^i,
\end{aligned} \tag{27.5}$$

where

$$
\begin{aligned}
& a_{11}^i=\frac{1}{E_s^i}, \quad a_{12}^i=-\frac{\nu_{st}^i}{E_t^i}=-\frac{\nu_{ts}^i}{E_s^i}, \quad a_{13}^i=-\frac{\nu_{s\gamma}^i}{E_\gamma^i}=-\frac{\nu_{\gamma s}^i}{E_s^i}, \\
& a_{22}^i=\frac{1}{E_t^i}, \quad a_{23}^i=-\frac{\nu_{\gamma t}^i}{E_t^i}=-\frac{\nu_{t\gamma}^i}{E_\gamma^i}, \quad a_{33}^i=\frac{1}{E_\gamma^i}, \\
& a_{44}^i=\frac{1}{G_{t\gamma}^i}, \quad a_{55}^i=\frac{1}{G_{s\gamma}^i}, \quad a_{66}^i=\frac{1}{G_{st}^i},
\end{aligned} \tag{27.6}
$$

and E_s^i, E_t^i, and E_γ^i are the elastic moduli in the directions axis of coordinates; $G_{s\gamma}^i$, $G_{t\gamma}^i$, and G_{st}^i are the shear modulus; $\nu_{t\gamma}^i$, $\nu_{s\gamma}^i$, ν_{st}^i, $\nu_{\gamma t}^i$, $\nu_{\gamma s}^i$, and ν_{ts}^i are Poisson's ratios.

Relations (27.3)–(27.5) are a closed system of partial differential equations describing the stress state of non-regular non-circular layered orthotropic cylindrical shells for such parameters that $\{0 \leqslant s \leqslant l, t_1 \leqslant t \leqslant t_2, \gamma_0 \leqslant \gamma \leqslant \gamma_P\}$. The boundary conditions on the outer and inner surfaces of the shell are:

$$
\begin{aligned}
& \sigma_\gamma^P=q_\gamma^+, \quad \tau_{s\gamma}^P=q_s^+, \quad \tau_{t\gamma}^P=q_t^+ \text{ on the outer surfaces } (\gamma=\gamma_P), \\
& \sigma_\gamma^0=q_\gamma^-, \quad \tau_{s\gamma}^0=q_s^-, \quad \tau_{t\gamma}^0=q_t^- \text{ on the inner surfaces } (\gamma=\gamma_0).
\end{aligned} \tag{27.7}
$$

After the transformation of Eqs. (27.3)–(27.5) we obtain a system of the resolving partial differential equations with the variable coefficients in terms of the displacements u_γ^i, u_s^i, u_t^i and of the stresses σ_γ^i, $\tau_{s\gamma}^i$, $\tau_{t\gamma}^i$ in the form

$$
\begin{aligned}
\frac{\partial \sigma_\gamma^i}{\partial \gamma}= & \left(c_2^i-1\right) \frac{1}{H_2^i} \frac{\partial H_2^i}{\partial \gamma} \sigma_\gamma^i-\frac{\partial \tau_{s\gamma}^i}{\partial s}-\frac{1}{H_2^i} \frac{\partial \tau_{t\gamma}^i}{\partial t}+b_{22}^i\left(\frac{1}{H_2^i} \frac{\partial H_2^i}{\partial \gamma}\right)^2 u_\gamma^i+ \\
& +b_{12}^i \frac{1}{H_2^i} \frac{\partial H_2^i}{\partial \gamma} \frac{\partial u_s^i}{\partial s}+b_{22}^i \frac{1}{H_2^{i2}} \frac{\partial H_2^i}{\partial \gamma} \frac{\partial u_t^i}{\partial t}, \\
\frac{\partial \tau_{s\gamma}^i}{\partial \gamma}= & -c_1^i \frac{\partial \sigma_\gamma^i}{\partial s}-\frac{1}{H_2^i} \frac{\partial H_2^i}{\partial \gamma} \tau_{s\gamma}^i-b_{12}^i \frac{1}{H_2^i} \frac{\partial H_2^i}{\partial \gamma} \frac{\partial u_\gamma^i}{\partial s}-b_{11}^i \frac{\partial^2 u_s^i}{\partial s^2}- \\
& -b_{66}^i \frac{1}{H_2^i} \frac{\partial}{\partial t}\left(\frac{1}{H_2^i} \frac{\partial u_s^i}{\partial t}\right)-\left(b_{12}^i+b_{66}^i\right) \frac{1}{H_2^i} \frac{\partial^2 u_t^i}{\partial s \partial t}, \\
\frac{\partial \tau_{t\gamma}^i}{\partial \gamma}= & -c_2^i \frac{1}{H_2^i} \frac{\partial \sigma_\gamma^i}{\partial t}-\frac{2}{H_2^i} \frac{\partial H_2^i}{\partial \gamma} \tau_{t\gamma}^i-b_{22}^i \frac{1}{H_2^i} \frac{\partial}{\partial t}\left(\frac{1}{H_2^i} \frac{\partial H_2^i}{\partial \gamma} u_\gamma^i\right)- \\
& -\left(b_{12}^i+b_{66}^i\right) \frac{1}{H_2^i} \frac{\partial^2 u_s^i}{\partial s \partial t}-b_{22}^i \frac{1}{H_2^i} \frac{\partial}{\partial t}\left(\frac{1}{H_2^i} \frac{\partial u_t^i}{\partial t}\right)-b_{66}^i \frac{\partial^2 u_t^i}{\partial s^2}, \\
\frac{\partial u_\gamma^i}{\partial \gamma}= & c_4^i \sigma_\gamma^i-c_2^i \frac{1}{H_2^i} \frac{\partial H_2^i}{\partial \gamma} u_\gamma^i-c_1^i \frac{\partial u_s^i}{\partial s}-c_2^i \frac{1}{H_2^i} \frac{\partial u_t^i}{\partial t}, \\
\frac{\partial u_s^i}{\partial \gamma}= & a_{55}^i \tau_{s\gamma}^i-\frac{\partial u_\gamma^i}{\partial s}; \quad \frac{\partial u_t^i}{\partial \gamma}=a_{44}^i \tau_{t\gamma}^i-\frac{1}{H_2^i} \frac{\partial u_\gamma^i}{\partial t}+\frac{1}{H_2^i} \frac{\partial H_2^i}{\partial \gamma} u_t^i,
\end{aligned} \tag{27.8}
$$

where

$$
\begin{aligned}
&b_{11}^i = a_{22}^i a_{66}^i / \Omega^i, \quad b_{12}^i = -a_{12}^i a_{66}^i / \Omega^i, \quad b_{22}^i = a_{11}^i a_{66}^i / \Omega^i, \\
&b_{66}^i = \left(a_{11}^i a_{22}^i - a_{12}^{i2}\right) / \Omega^i, \quad \Omega^i = \left(a_{11}^i a_{22}^i - a_{12}^{i2}\right) a_{66}^i, \\
&c_1^i = -\left(b_{11}^i a_{13}^i + b_{12}^i a_{23}^i\right), \quad c_2^i = -\left(b_{12}^i a_{13}^i + b_{22}^i a_{23}^i\right), \\
&c_4^i = a_{33}^i + c_1^i a_{13}^i + c_2^i a_{23}^i.
\end{aligned}
$$

In addition to the conditions on the outer and the inner surfaces, it is necessary to satisfy the boundary conditions at the ends of the shell l (l is the length of the shell).

The diaphragm takes place at the ends of the shell, which is absolutely rigid in its plane and flexible from it; that is, the following boundary conditions are

$$
\sigma_s^i = u_t^i = u_\gamma^i = 0 \quad \text{at} \quad s = 0, \quad s = l. \tag{27.9}
$$

The shells are closed along the guide so that the resolving functions will satisfy the conditions of the periodicity:

$$
\psi\left(s, t, \gamma\right) = \psi\left(s, t + T, \gamma\right), \tag{27.10}
$$

where $\psi = \{\sigma_\gamma^i, \tau_{s\gamma}^i, \tau_{t\gamma}^i, u_\gamma^i, u_s^i, u_t^i\}$, T is the perimeter of the cross-section of the reduction surface.

27.3 Discrete-Continual Analytical-Numerical Approach

Let us present the resolving functions and load components into the Fourier series along the generatrix of the shell (hereinafter, the i index is omitted for ease of presentation):

$$
X\left(s, t, \gamma\right) = \sum_{n=1}^{N} X_n\left(t, \gamma\right) \sin \lambda_n s, \quad Y\left(s, t, \gamma\right) = \sum_{n=0}^{N} Y_n\left(t, \gamma\right) \cos \lambda_n s, \tag{27.11}
$$

where

$$
\begin{aligned}
&X = \left\{\sigma_\gamma, \tau_{t\gamma}, u_\gamma, u_t, q_\gamma, q_t\right\}, \\
&Y = \left\{\tau_{s\gamma}, u_s, q_s\right\}, \quad \lambda_n = \pi n / l, \quad (0 \leqslant s \leqslant l).
\end{aligned}
$$

After substituting (27.11) in (27.8) and the boundary conditions (27.7), we get the following two-dimensional boundary value problem:

$$
\begin{aligned}
\frac{\partial \sigma_{\gamma,n}}{\partial \gamma} &= (c_2-1)\frac{1}{H_2}\frac{\partial H_2}{\partial \gamma}\sigma_{\gamma,n} + \lambda_n \tau_{s\gamma,n} - \frac{1}{H_2}\frac{\partial \tau_{t\gamma,n}}{\partial t} + \\
&\quad + b_{22}\left(\frac{1}{H_2}\frac{\partial H_2}{\partial \gamma}\right)^2 u_{\gamma,n} + b_{12}\lambda_n\frac{1}{H_2}\frac{\partial H_2}{\partial \gamma}u_{s,n} + b_{22}\frac{1}{H_2^2}\frac{\partial H_2}{\partial \gamma}\frac{\partial u_{t,n}}{\partial t},\\
\frac{\partial \tau_{s\gamma,n}}{\partial \gamma} &= -c_1\lambda_n\sigma_{\gamma,n} - \frac{1}{H_2}\frac{\partial H_2}{\partial \gamma}\tau_{s\gamma,n} - b_{12}\lambda_n\frac{1}{H_2}\frac{\partial H_2}{\partial \gamma}u_{\gamma,n} + b_{11}\lambda_n^2 u_{s,n} - \\
&\quad - b_{66}\frac{1}{H_2}\frac{\partial}{\partial t}\left(\frac{1}{H_2}\frac{\partial u_{s,n}}{\partial t}\right) - (b_{12}+b_{66})\lambda_n\frac{1}{H_2}\frac{\partial u_{t,n}}{\partial t},\\
\frac{\partial \tau_{t\gamma,n}}{\partial \gamma} &= -c_2\frac{1}{H_2}\frac{\partial \sigma_{\gamma,n}}{\partial t} - \frac{2}{H_2}\frac{\partial H_2}{\partial \gamma}\tau_{t\gamma,n} - b_{22}\frac{1}{H_2}\frac{\partial}{\partial t}\left(\frac{1}{H_2}\frac{\partial H_2}{\partial \gamma}u_{\gamma,n}\right) + \\
&\quad + (b_{12}+b_{66})\lambda_n\frac{1}{H_2}\frac{\partial u_{s,n}}{\partial t} - b_{22}\frac{1}{H_2}\frac{\partial}{\partial t}\left(\frac{1}{H_2}\frac{\partial u_{t,n}}{\partial t}\right) + b_{66}\lambda_n^2 u_{t,n},\\
\frac{\partial u_{\gamma,n}}{\partial \gamma} &= c_4\sigma_{\gamma,n} - c_2\frac{1}{H_2}\frac{\partial H_2}{\partial \gamma}u_{\gamma,n} + c_1\lambda_n u_{s,n} - c_2\frac{1}{H_2}\frac{\partial u_{t,n}}{\partial t},\\
\frac{\partial u_{s,n}}{\partial \gamma} &= a_{55}\tau_{s\gamma,n} - \lambda_n u_{\gamma,n},\\
\frac{\partial u_{t,n}}{\partial \gamma} &= a_{44}\tau_{t\gamma,n} - \frac{1}{H_2}\frac{\partial u_{\gamma,n}}{\partial t} + \frac{1}{H_2}\frac{\partial H_2}{\partial \gamma}u_{t,n}
\end{aligned}
\tag{27.12}
$$

$(n = 0, \ldots, N)$ with the boundary conditions

$$
\begin{aligned}
&\sigma_{\gamma,n} = q_{\gamma,n}^-, \quad \tau_{s\gamma,n} = q_{s,n}^-, \quad \tau_{t\gamma,n} = q_{t,n}^- \quad \text{on the inner surfaces } (\gamma = \gamma_0),\\
&\sigma_{\gamma,n} = q_{\gamma,n}^+, \quad \tau_{s\gamma,n} = q_{s,n}^+, \quad \tau_{t\gamma,n} = q_{t,n}^+ \quad \text{on the outer surfaces } (\gamma = \gamma_p).
\end{aligned}
\tag{27.13}
$$

The reducing of the resolving the boundary value two-dimensional problem to the one-dimensional problem, based on the method of separating variables along the guide shell which interfere with separation of variables along the cylinder directrix, with new additional functions, was carried out.

The obtained system of the differential equations, as well as (27.12), contains the terms that are products of the unknown functions and coefficients dependent on two coordinates (running along the directrix and thickness) and make it impossible to separate the directrix variables. At the second stage, these members in the resolving system of the equations are replaced by additional functions (hereafter the index n in the notation of the unknown functions and the loading components is omitted). These functions, include geometrical parameters are presented and have the form:

$$
\begin{aligned}
\varphi_1^j &= \frac{1}{H_2}\frac{\partial H_2}{\partial \gamma}\left\{\sigma_\gamma;\, \tau_{s\gamma};\, u_\gamma;\, u_s;\, \frac{1}{H_2}\frac{\partial H_2}{\partial \gamma}u_\gamma\right\}, \quad (j = 1, \ldots, 5),\\
\varphi_2^j &= \frac{1}{H_2}\frac{\partial H_2}{\partial \gamma}\left\{\tau_{t\gamma};\, u_t\right\} \quad (j = 1, 2),\\
\varphi_3^j &= \frac{1}{H_2}\left\{\frac{\partial \sigma_\gamma}{\partial t};\, \frac{\partial u_\gamma}{\partial t};\, \frac{\partial u_s}{\partial t}\right\} \quad (j = 1, 2, 3),\\
\varphi_4^j &= \frac{1}{H_2}\left\{\frac{\partial \tau_{t\gamma}}{\partial t};\, \frac{\partial u_t}{\partial t};\, \frac{1}{H_2}\frac{\partial H_2}{\partial \gamma}\frac{\partial u_t}{\partial t}\right\} \quad (j = 1, 2, 3),\\
\varphi_5 &= \frac{1}{H_2}\frac{\partial}{\partial t}\varphi_1^3, \quad \varphi_6 = \frac{1}{H_2}\frac{\partial}{\partial t}\varphi_3^3, \quad \varphi_7 = \frac{1}{H_2}\frac{\partial}{\partial t}\varphi_4^2.
\end{aligned}
\tag{27.14}
$$

After substituting additional functions (27.14) into the system (27.12) have at the following resolving system of the equations:

$$
\begin{aligned}
\frac{\partial \sigma_\gamma}{\partial \gamma} &= (c_2 - 1)\ \varphi_1^1 + \lambda_n \tau_{t\gamma} - \varphi_4^1 + b_{22}\varphi_1^5 + b_{12}\lambda_n\varphi_1^4 + b_{22}\varphi_4^3,\\
\frac{\partial \tau_{s\gamma}}{\partial \gamma} &= -c_1\lambda_n\sigma_\gamma - \varphi_1^2 - b_{12}\lambda_n\varphi_1^3 + b_{11}\lambda_n^2 u_s - b_{66}\varphi_6 - (b_{12} + b_{66})\,\lambda_n\varphi_4^2,\\
\frac{\partial \tau_{t\gamma}}{\partial \gamma} &= -c_2\varphi_3^1 - 2\varphi_2^1 - b_{22}\varphi_5 + (b_{12} + b_{66})\,\lambda_n\varphi_3^3 - b_{22}\varphi_7 - b_{66}\lambda_n^2 u_t,\\
\frac{\partial u_\gamma}{\partial \gamma} &= c_4\sigma_\gamma - c_2\varphi_4^2 + c_1\lambda_n u_s - c_2\varphi_1^3, \quad \frac{\partial u_s}{\partial \gamma} = a_{55}\tau_{s\gamma} - \lambda_n u_\gamma,\\
\frac{\partial u_t}{\partial \gamma} &= a_{44}\tau_{t\gamma} - \varphi_3^2 + \varphi_2^2
\end{aligned}
\tag{27.15}
$$

with the boundary conditions (27.14).

After that, represent of all the functions entering into the system (27.15) and boundary conditions (27.14) as the expansions into the Fourier series along the coordinate t:

$$
\tilde{X}(t, \gamma) = \sum_{k=0}^{K} \tilde{X}_k(\gamma)\cos\lambda_k t, \quad \tilde{Y}(t, \gamma) = \sum_{k=1}^{K} \tilde{Y}_k(\gamma)\sin\lambda_k t, \tag{27.16}
$$

$$
\begin{aligned}
\tilde{X} &= \left\{\sigma_\gamma, \tau_{s\gamma}, u_\gamma, u_s, \varphi_1^j, \varphi_4^j, \varphi_6, q_\gamma, q_s\right\},\\
\tilde{Y} &= \left\{\tau_{t\gamma}, u_t, \varphi_2^j, \varphi_3^j, \varphi_5, \varphi_7, q_t\right\} \quad (\lambda_k = 2k\pi/T).
\end{aligned}
$$

The reducing of the resolving the boundary value two-dimensional problem to the one-dimensional problem, based on the method of separating variables along the guide shell which interfere with separation of variables along the cylinder directrix, with new additional functions, was carried out.

Having substituted the series (27.16) into the equations of system (27.15) and boundary conditions (27.14) and having separated the variables, we obtain at the system of ordinary diferential equations with respect to the amplitude values of the functions, that enter into system (27.15):

$$
\begin{aligned}
\frac{d\sigma_{\gamma,k}}{d\gamma} &= (c_2-1)\ \overset{1}{\varphi}_{1,k} + \lambda_n \tau_{s\gamma,k} - \overset{1}{\varphi}_{4,k} + \\
&\qquad + b_{22}\overset{5}{\varphi}_{1,k} + b_{12}\lambda_n \overset{4}{\varphi}_{1,k} + b_{22}\overset{3}{\varphi}_{4,k},\\
\frac{d\tau_{s\gamma,k}}{d\gamma} &= -c_1\lambda_n\sigma_{\gamma,k} - \overset{2}{\varphi}_{1,k} - \\
&\qquad - b_{12}\lambda_n\overset{3}{\varphi}_{1,k} + b_{11}\lambda_n^2 u_{s,k} - b_{66}\varphi_{6,k} - (b_{12}+b_{66})\,\lambda_n\overset{2}{\varphi}_{4,k},\\
\frac{d\tau_{t\gamma,k}}{d\gamma} &= -c_2\overset{1}{\varphi}_{3,k} - 2\overset{1}{\varphi}_{2,k} - \\
&\qquad - b_{22}\varphi_{5,k} + (b_{12}+b_{66})\,\lambda_n\overset{3}{\varphi}_{3,k} - b_{22}\varphi_{7,k} + b_{66}\lambda_n^2 u_{t,k},\\
\frac{du_{\gamma,k}}{d\gamma} &= c_4\sigma_{\gamma,k} - c_2\overset{2}{\varphi}_{4,k} + c_1\lambda_n u_{s,k} - c_2\overset{3}{\varphi}_{1,k},\\
\frac{du_{s,k}}{d\gamma} &= a_{55}\tau_{s\gamma,k} - \lambda_n u_{\gamma,k},\\
\frac{du_{t,k}}{d\gamma} &= a_{44}\tau_{t\gamma,k} - \overset{2}{\varphi}_{3,k} + \overset{2}{\varphi}_{2,k}
\end{aligned}
\tag{27.17}
$$

$(k = 0, \ldots, K)$ with the following boundary conditions:

- on the inner surfaces

$$
(\gamma = \gamma_0):\quad \sigma_{\gamma,k} = q^{-}_{\gamma,k},\quad \tau_{s\gamma,k} = q^{-}_{s,k},\quad \tau_{t\gamma,k} = q^{-}_{t,k}; \tag{27.18}
$$

- on the outer surfaces

$$
(\gamma = \gamma_p):\quad \sigma_{\gamma,k} = q^{+}_{\gamma,k},\quad \tau_{s\gamma,k} = q^{+}_{s,k},\quad \tau_{t\gamma,k} = q^{+}_{t,k}.
$$

The one-dimensional boundary-value problem (27.17)–(27.18) can be solved with the stable numerical discrete-orthogonalization method for all the k harmonics of the Fourier series (27.16) simultaneously. At each step of the integration, the amplitudes of the additional functions are calculated from the current amplitudes of the unknown functions by approximating them by the discrete Fourier series. At the beginning of integration, the additional functions are determined from the boundary conditions and the initial values of the unknown functions.

27.4 Numerical Calculation Result

The choice of the main independent functions by which the contact conditions of adjacent layers and boundary conditions are formulated allows the automatic continuous solution of problems for a given number of non-uniform layers and the calculation of the values of all stress-strain state factors at specified points of thickness and, accordingly, along the guide and generatrix of the shell.

This algorithm allows you to take into account the monolayer and multilayer of the shells, different thickness of the layers, anisotropy and heterogeneity of the material, the arbitrary of the surface loads, the different configuration of the cross-section with a significant deviation from the circular shape.

The reliability and accuracy of the solutions obtained based on the above approach is determined by conducting test examples using various inductive techniques.

For example, the problem of the stress-strain state of the orthotropic thick cylindrical shells with a cross section in the form of corrugated ellipses under the influence of a uniform load $q_\gamma = q_0 \sin(\pi s/l)$ $(q_0 = \text{const})$ applied on the outer surface of the shell is considered. The convergence of the solution of this problem based on the discrete Fourier series was performed.

The reference surface of the shell in the polar coordinate system have the following form [7]:

$$\rho(\psi) = \frac{a}{(1 - e^2 \cos^2 \psi)^{1/2}} + \alpha \cos m\psi \quad (0 \leqslant \psi \leqslant 2\pi), \quad e = \frac{2\sqrt{\Delta}}{1 + \Delta},$$

the corrugation of the amplitudes

$$a = r_0(1 - \Delta)/f, \quad f = 1 + \frac{\Delta^2}{4} + \frac{\Delta^4}{64} + \frac{\Delta^6}{256} + \cdots \tag{27.19}$$

where α is the corrugation amplitude, m is the frequency of the corrugation, e is the eccentricity of the ellipse, Δ is the parameters of the ellipse geometry.

The geometrical parameters of the shells: the length is $l = 60$; the radius of the original circle is $r_0 = 40$; the parameters of the ellipse geometry is $\Delta = 0.20$; the thickness is $h = 8$; corrugation amplitude is $\alpha = 3$; the frequency of the corrugation is $m = 4$. The mechanical parameters of the orthotropic shell are $E_s = 3.68E_0$, $E_\psi = 2.68E_0$, $E_\gamma = 1.1E_0$, $\nu_{s\psi} = 0.105$, $\nu_{s\gamma} = 0.405$, $\nu_{\psi\gamma} = 0.431$, $G_{s\psi} = 0.5E_0$, $G_{s\gamma} = 0.45E_0$, and $G_{\psi\gamma} = 0.41E_0$.

Hereinafter, all the linear dimensions and the displacements are referred to unite the length, and stresses – to unit of the force. The table values of the additional functions are defined at the number points $R = 20, 40, 60, 80$; the number of the discrete Fourier series members is $M = 4, 6, 8, 10, 12$. Table 27.1 shows the value of the normal displacements of the reference surfaces and Table 27.2 shows the value of the stresses on the inner and the outer surfaces in the middle section of the length at some points of the guide shell. Due to the symmetry of the cross section for the guide, the interval values $0 \leqslant \psi \leqslant \pi/2$ is considered.

The high accuracy of the calculation results can be obtained by increasing of the points at which the table values of additional functions are calculated, as well as, by increasing the retained terms of discrete Fourier series. Table 27.1 and 27.2 show the values of the displacement and of the stresses at the different the values of the additional functions and of the number of the discrete Fourier series members.

The problem of the stress state of the hollow cylinders of the symmetrical structure with corrugated elliptical cross section under the action of internal pressure based

Table 27.1 Influence of accuracy of calculation of discrete Fourier series on convergence of displacement $u_\gamma E_0/q_0$

M	R	$\psi = 0$	$\psi = \pi/4$	$\psi = \pi/2$
4	20	32.517	35.633	38.749
	40	31.421	36.837	40.886
	60	31.445	36.841	40.896
6	20	36.392	135.399	63.446
	40	31.421	150.076	66.496
	60	31.445	150.023	68.448
8	20	−29.399	206.575	168.052
	40	−25.937	208.280	162.349
	60	−25.921	206.290	162.387
10	80	−25.920	206.290	162.388
	20	−76.347	274.645	168.097
	40	−56.619	271.893	162.350
	60	−56.731	271.531	169.186
	80	−56.371	271.931	169.188
12	20	−76.358	282.813	192.254
	40	−56.599	275.549	162.340
	60	−56.601	275.551	169.185
	80	−56.371	275.530	169.189
14	60	−56.369	275.529	169.188
	80	−56.371	275.530	169.189

on the proposed approach was solved. The middle layer is made of the orthotropic material with the mechanical parameters: $E_s = 3.68E_0$, $E_t = 2.68E_0$, $E_\gamma = 1.1E_0$, $\nu_{st} = 0.105$, $\nu_{s\gamma} = 0.405$, $\nu_{t\gamma} = 0.431$, $G_{st} = 0.5E_0$, $G_{s\gamma} = 0.45E_0$, and $G_{t\gamma} = 0.41E_0$. The outer layers are isotropic with the mechanical parameters: $E_s = E_t = E_\gamma = E_0$ and $\nu_{st} = \nu_{t\gamma} = \nu_{s\gamma} = 0.3$.

The shell geometric parameters are: the length of the cylinder is $l = 60$; the radius of the original circle is $r_0 = 40$; the parameters of the ellipse geometry are $\Delta = 0.1, 0.2$; the thickness of outer and inner of the cylinder layers are $h_1 = 1$, $h_3 = 1$, the thickness of the middle layer of the cylinder is $h_2 = 2$, the corrugation of the amplitudes are $\alpha = 1, 3$ and corrugation frequency is $m = 3$, respectively.

The results of the problem are presented for the values of displacements u_γ in Fig. 27.1, for the values of the stresses or σ_s stresses are presented in Fig. 27.2 and for the values of the stresses σ_ψ are presented—in Fig. 27.3 in the middle section of the shell length. The continuous lines correspond the values of the displacement and the stresses to amplitude $\alpha = 1$, the shaped line correspond the values of the displacement and the stresses to amplitude $\alpha = 3$ along a guide in the range of $0 \leqslant \psi \leq \pi$ (symmetry condition).

Table 27.2 Influence of accuracy of determination of discrete Fourier series on convergence of stresses σ_ψ / q_0

M	R	$\gamma = -h/2$			$\gamma = h/2$		
		$\psi = 0$	$\psi = \pi/4$	$\psi = \pi/2$	$\psi = 0$	$\psi = \pi/4$	$\psi = \pi/2$
4	20	−1.270	−2.701	−1.437	1.885	1.312	2.215
	40	−1.272	−2.762	−1.481	1.888	1.339	2.312
	60	−1.272	−2.762	−1.481	1.888	1.339	2.312
6	20	−1.672	−6.960	−2.221	1.476	4.881	2.665
	40	−1.607	−7.595	−2.274	1.151	5.392	2.862
	60	−1.607	−7.598	−2.275	1.149	5.393	2.864
8	20	−0.443	−9.979	−5.041	−2.382	7.204	7.859
	40	−0.538	−9.939	−4.910	−2.145	7.203	7.736
	60	−0.539	−9.939	−4.911	−2.144	7.204	7.737
	80	−0.539	−9.939	−4.911	−2.144	7.204	7.737
10	20	0.691	−12.423	−4.228	−4.742	10.058	7.640
	40	0.603	−12.320	−4.107	−4.372	9.946	7.411
	60	0.604	−12.322	−4.107	−4.373	9.947	7.411
	80	0.604	−12.322	−4.107	−4.373	9.947	7.411
12	20	1.123	−12.755	−5.085	−4.497	10.408	8.180
	40	0.255	−12.514	−3.790	−4.023	10.076	7.348
	60	0.248	−12.512	−3.792	−4.030	10.075	7.350
	80	0.248	−12.512	−3.792	−4.030	10.075	7.350
14	60	0.248	−12.488	−3.810	−4.005	10.033	7.389
	80	0.248	−12.488	−3.810	−4.005	10.033	7.389

It was assumed that the perimeter of the surface reduction cross section for an ellipse is equal to the length of a circumference of the radius r_0.

Figures 27.1, 27.2, and 27.3 show the results of the calculation of the displacements u_γ and of the stresses σ_s and σ_ψ respectively in the middle section of the shell length.

The continuous lines correspond the values of the displacement and the stresses to amplitude $\alpha = 1$, the shaped line correspond the values of the displacement and the stresses to amplitude $\alpha = 3$along a guide in the range of $0 \leqslant \psi \leq \pi$ (symmetry condition). Figure 27.1 shows for the values of the displacements u_γ. As can be seen, the maximum values of displacements are have with the increase of the corrugation depth at $\psi = \pi/3$. The increase of the displacement values is observed by 1.8 times for the amplitude $\alpha = 1$ and by 1.5 times for the amplitude $\alpha = 3$ with increase of the parameters of the ellipse geometry. The values of the displacements increase by 2.6 – times for shells with $\Delta = 0.1$ and by 2.3 times for $\Delta = 0.2$ with increasing the corrugation amplitude. Figure 27.2 shows the maximum values of the stresses σ_s along of the guide of the shell at the outer boundary of the middle layer ($\gamma = h/2$) in the corrugation depth ($\psi = \pi/3$). As it the increase of the value of the parameter Δ, can see the increase of the values of the stresses by 2.1 times for value of the

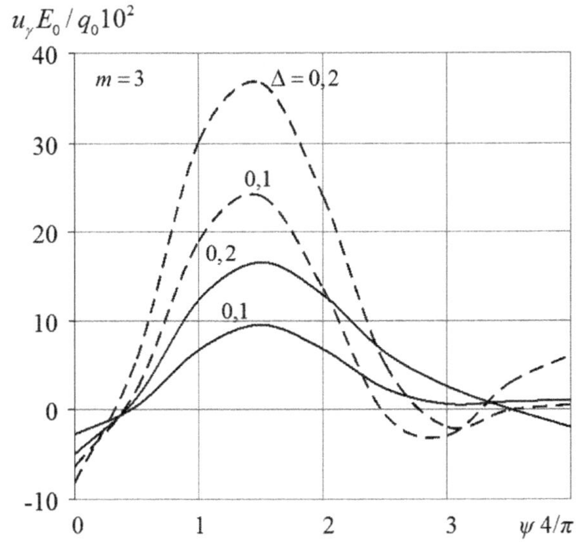

Fig. 27.1 Distribution of the displacement uγ depending on degree of the ellipticity and the corrugation amplitude

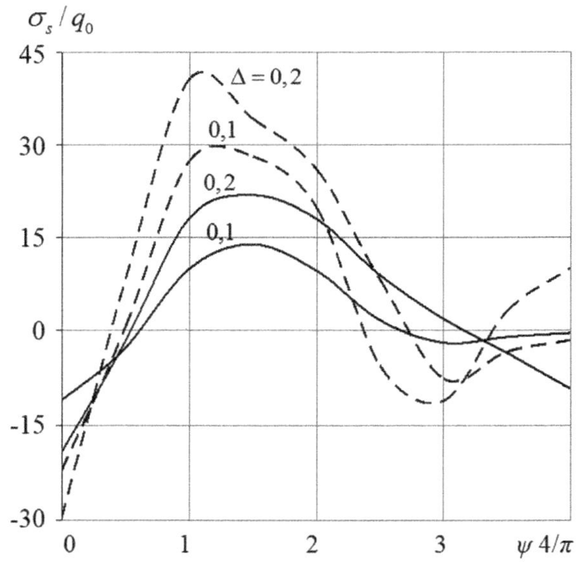

Fig. 27.2 Distribution of the stresses σs depending on the degree of the ellipticity and the corrugation amplitude

corrugation amplitude $\alpha = 1$ and by 1.8 times for the corrugation amplitude $\alpha = 3$. Figure 27.2 shows the maximum values of the stresses σ_ψ along of the guide of the shell at the outer boundary of the middle layer ($\gamma = h/2$). As it the increase of the value of the corrugation amplitude, can see the increase of the values of the stresses by 1.8 times for value of the parameters $\Delta = 0.1$ and the increase the values of the stresses by 1.6 times for the value of the parameters $\Delta = 0.2$. The values of the

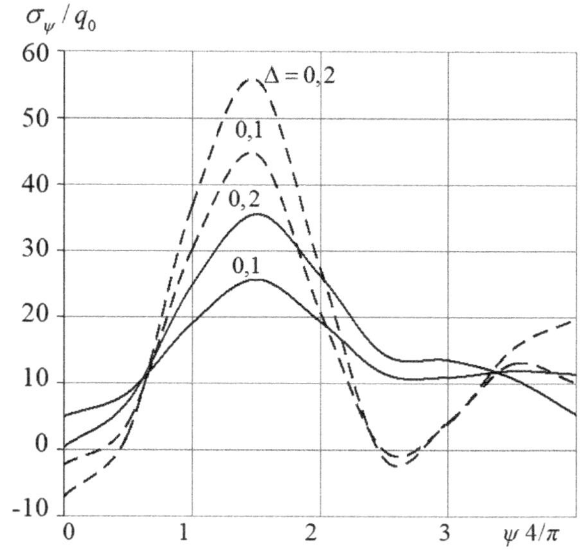

Fig. 27.3 Distribution of the stresses σ ψ depending on degree of ellipticity and corrugation amplitude

stresses increase by 1.6 times for amplitude $\alpha = 1$ and by 1.25 times for amplitude $\alpha = 3$ with the change of the parameter Δ.

The newly formed and actively developing theory of functional gradient materials (FGM) concentrates researchers' attention at artificially created heterogeneous composite materials with smoothly changing physical properties in a given direction without any transitional layers and interfaces. The mechanical parameters of such materials can be controlled by specifying the necessary distribution in one of the directions. FGMs are widely used in engineering, radio industry, instrumentation, medicine.

The problem of the stress-strain state of the elliptical cylindrical non-thin shells made of the continuously nonhomogeneous material (FGMs) is investigated. The composition of the material of the shell consist of the aluminum and of the carborundum. The mechanical properties of the composition materials are: the aluminium is $E = 69$ GPa and $\nu = 0.3$; the carborundum is $E = 427$ GPa and $\nu = 0.17$.

The relationship between the modulus of elasticity E and the Poisson coefficient ν of the continuously inhomogeneous material with the corresponding the elasticity parameters of the materials included in the composition is determined by the formulas:

$$E = (E_2 - E_1)V + E_1, \quad \nu = (\nu_2 - \nu_1)V + \nu_1,$$

where E_1, E_2, ν_1, and ν_2 are mechanical parameters of the aluminium and the carborundum respectively, V is the concentration of the carborundum material, depending on the thickness coordinate. The power law of change of elastic properties in calculations is following:

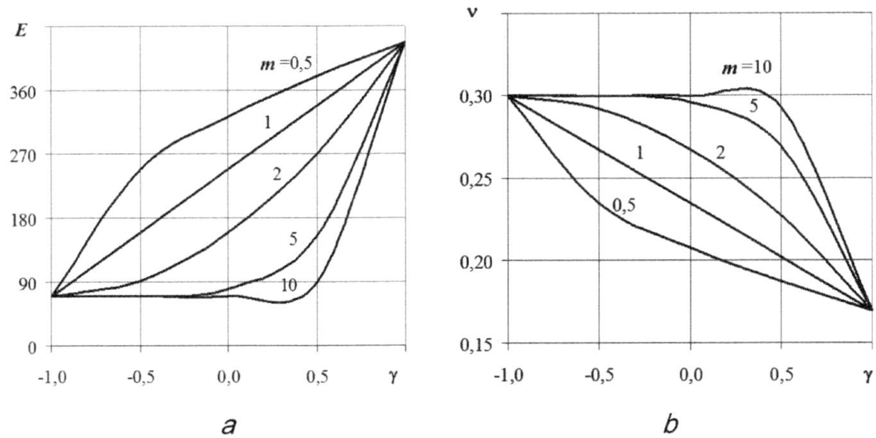

Fig. 27.4 Distribution of the modulus of the elasticity (*a*) and Poisson ratio (*b*) by the shell thickness depending on the parameter *m*

$$V = \left(\frac{\gamma}{h} + 0.5\right)^m \quad (-h/2 \leqslant \gamma \leqslant h/2.$$

The parameters have the values are $m = 0.5, 1, 2, 5, 10$.

Figure 27.4 shows the distribution of the modulus of elasticity and Poisson's ratio over the thickness of the shell for the different values of the parameter m.

The parametrically form of the cross-section of the reference surface of elliptic shells are $x = b\cos\psi$, $z = a\sin\psi$ $(0 \leqslant \psi \leqslant 2\pi)$, where a and b are ellipse half-axis, ψ is angular parameter in cross-section.

The elliptical shells with the perimeter of the cross section of the middle reference surface is equal to the length of the circle of radius R are considered. In this case have the relation:

$$\pi\,(a+b)\left(1+\frac{\Delta^2}{4}+\frac{\Delta^4}{64}+\frac{\Delta^6}{256}+\ldots\right)=2\pi R,\quad \Delta=\frac{b-a}{b+a}. \tag{27.20}$$

Solving the problem, the number of members of the series is limited by Δ^6. Then we have:

$$a=\frac{R}{f}(1-\Delta),\quad b=\frac{R}{f}(1+\Delta),\quad f=1+\frac{\Delta^2}{4}+\frac{\Delta^4}{64}+\frac{\Delta^6}{256},\quad \frac{a}{b}=\frac{1-\Delta}{1+\Delta}.$$

The geometrical parameters of the shell are: length is $L = 20l_0$ the thickness is $h = 2l_0$, the radius of the circle is $R = 10l_0$, the parameters of the ellipse geometry is $\Delta = 0, 0.1$.

Figures 27.5 and 27.6 shows the values of the normal stresses σ_ψ and σ_ψ respectively along the thickness depending on the of the distribution of elastic properties

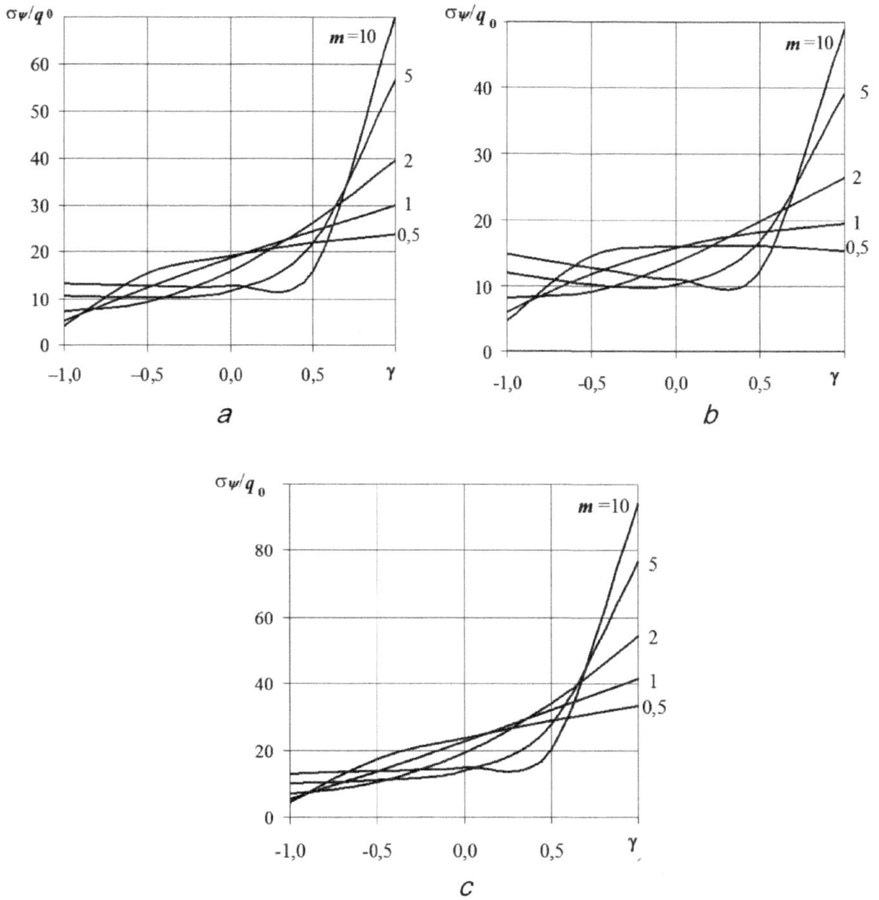

Fig. 27.5 Stress distribution by the thickness in the circular σ_ψ $\Delta = 0$ **a** and the elliptical $\Delta = 0, 1,$ in the section $\psi = 0$ **b** and $\psi = \pi/2$ **c** of the cylinders

of the material. Figures 27.5a and 27.6a show the distribution of the stresses for the circular shell ($\Delta = 0$) and the elliptical shell with the parameters of the ellipse geometry is $\Delta = 0.1$. Figures 27.5b and 27.6b shows the distribution of the stresses in the area of the greater rigidity and Figs. 27.5c and 27.6c show the distribution of the stresses in the area of the lower rigidity. The distributions of the elastic properties of the functional gradient material along the thickness in the cylindrical circular and the elliptical shells influences on the magnitude of the normal stresses.

Thus, for example, the elastic characteristics with the value of the thickness is $-h/2 \leqslant \gamma \leqslant 0$ for $m = 10$ are practically unchanged; if the value of the thickness is $0 \leqslant \gamma \leqslant h/2$ the elastic modulus interval increases rapidly, and the Poisson ratio decreases. At the same time, the maximum value of the stresses on the outer surface of the shell increases more than 5 times for the circular shell and 3 times for the elliptical

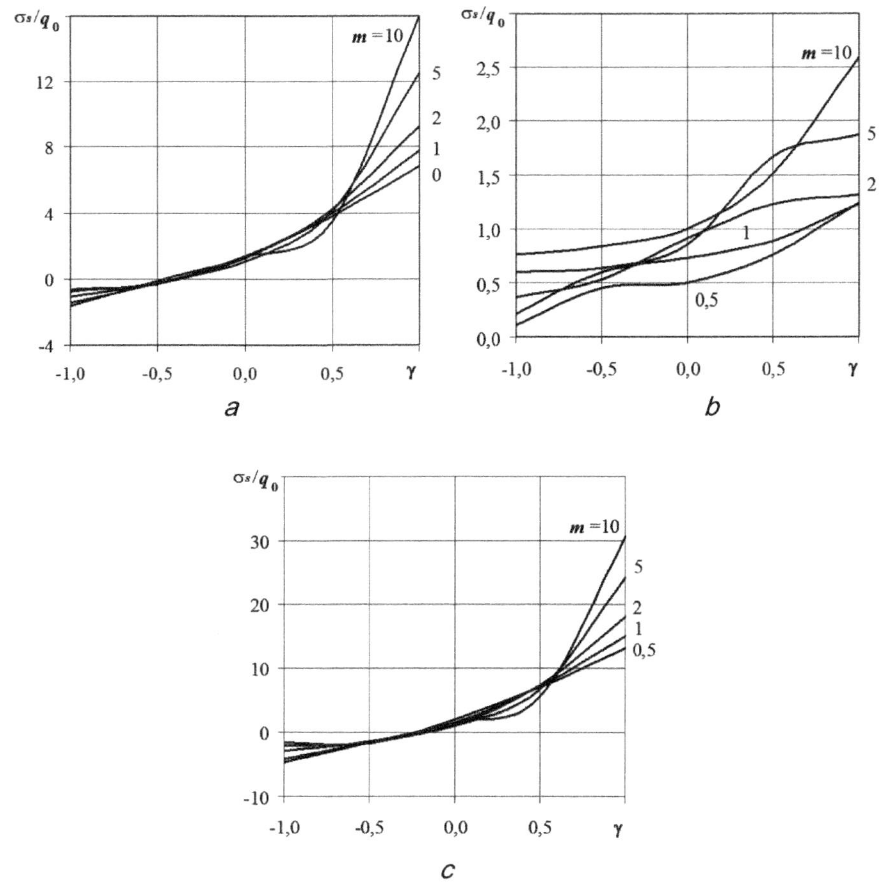

Fig. 27.6 Stress distribution σ_s by the thickness in the circular $\Delta = 0$ (a) and the elliptical $\Delta = 0, 1$, in the section $\psi = 0$ (b) and $\psi = \pi/2$ (c) of the cylinders

in section $\psi = 0$ and in section $\psi = \pi/2$—more than 7 times. The maximum value of the stresses σ_s can see on the outer surface and increases in comparison with the inner surface almost 10 times for a circular shell and in section $\psi = \pi/2$—almost 2 times for an elliptical shell compared with a circular shell.

27.5 Conclusions

In this study, our focus was on addressing the intricate stress and strain challenges encountered within cylindrical bodies characterized by complex geometries, all within the framework of elasticity theory. To tackle these challenges, we introduced a discrete-continual approach that proved to be highly effective. This approach involved the transformation of the initial two-dimensional (or three-dimensional) problems into more manageable one-dimensional boundary-value problems, which we subsequently addressed through numerical solutions facilitated by the discrete-orthogonalization method. Our efforts centered on solving the stress state problems inherent in three-layered noncircular cylindrical shells. These shells exhibited a symmetrical structure with a pivotal orthotropic middle layer. Our analysis extended to encompass corrugated ellipse cross sections, featuring deviations in shape from circularity governed by two distinct parameters. Furthermore, we ventured into solving the complex stress-strain issues within elliptical cylindrical shells, distinguished by their non-thin nature and composition of continuously nonhomogeneous materials–specifically, a combination of aluminum and carborundum.

Throughout our study, we meticulously examined the distribution patterns of displacement and stress fields. These distributions, we found, are intricately influenced by the specific shape of the reference surface's cross-section as well as the underlying mechanical properties of the material under scrutiny. Notably, our exploration was underpinned by a steadfast commitment to ensuring the trustworthiness of our results through rigorous numerical computations and validation.

The insights garnered from our investigation hold significant practical implications. By effectively addressing the intricate complexities of stress and strain within cylindrical forms characterized by noncircular geometries and varying nonhomogeneous structures, our findings offer a robust foundation for optimizing the selection of both geometrical and mechanical parameters in structural elements. These results stand to make meaningful contributions to the advancement of engineering design and performance, particularly in scenarios where such intricate configurations are encountered.

References

1. Birman, V., Byrd, L.W.: Modeling and analysis of functionally graded materials and structures. Appl. Mech. Rev. **60**, 195–215 (2007)
2. Koizumi, M.: The concept of FGM. Ceram. Trans. Funct. Gradient Mater. **34**, 3–10 (1993)
3. Kashtalyan, M.: Three-dimensional elasticity solution for bending of functionally graded rectangular plates. Eur. J. Mech. - A/Solids **23**(5), 853–864 (2004)
4. Nguyen-Ngoc, H., Cuong-Le, T., Nguyen, K.D., Nguyen-Xuan, H., Abdel-Wahab, M.: Three-dimensional polyhedral finite element method for the analysis of multi directional functionally graded solid shells. Compos. Struct. **305**, 116538 (2023)
5. Grigorenko, Y.M., Rozhok, L.S.: Equilibrium of elastic hollow inhomogeneous cylinders of corrugated elliptic cross-section. J. Eng. Math. **54**, 145–157 (2006)
6. Grigorenko, Y.M., Grigorenko, A.Y., Rozhok, L.S.: Stress state of non-thin nearly circular cylindrical shells made of continuously inhomogeneous materials. Int. App. Mech. **58**(4), 381–388 (2022)
7. Korn, G.A., Korn, T.M.: Mathematical Handbook for Scientists and Engineers. McGraw-Hill, New York (1961)
8. Zafarmand, H., Kadkhodayan, M.: Three dimensional elasticity solution for static and dynamic analysis of multi-directional functionally graded thick sector plates with general boundary conditions. Compos. Part B: Eng. **69**, 592–602 (2015)

Chapter 28
Transition from the Classical Biot Problem on Cylindrical Wave to the Case of Transversely Isotropic Medium

Jeremiah Rushchitsky

Abstract The problem of the surface harmonic elastic wave propagating along the free surface of a cylindrical cavity in the direction of the cavity axis is considered. In the case of isotropic medium, this problem is classical Biot's problem of 1952. First, Biot's pioneer work is described: the analytical part of Biot's findings is shown in the main fragments. The features: using two potentials and representation of the solution by the Macdonald functions of different indexes. Then, the new direct generalization of Biot's problem for the case of the transversely isotropic medium is proposed. It includes the new statement and new mathematical procedures. The transition to the transverse isotropy leads to the necessity of using the more complex representations of displacements through two potentials. As a result, the newly stated problem is solved in its analytical part.

28.1 Introduction

The subject of this study is the cylindrical surface elastic wave and the mathematical specificities of its analysis when transiting from the model of isotropic medium to the model of transversally isotropic one. In the framework of both models, a difference in the statement of the problem on the wave motion is only in properties of symmetry of elastic medium. The statement uses the cylindrical coordinates $Or\vartheta z$ and assumes that the infinite elastic medium with the cylindrical circular cavity having the symmetry axis Oz and constant radius is loaded especially. This load excites the attenuating in the depth of medium surface harmonic wave, which propagates along the cavity surface in a direction Oz. As a result, the problem becomes mathematically the axisymmetric one.

This problem is the classical one, solved by Biot in 1952 [3] with the assumption that the medium is isotropic, and named Biot's problem.

J. Rushchitsky (✉)
S.P. Timoshenko Institute of Mechanics, National Academy of Science of Ukraine, Kyiv, Ukraine
e-mail: rushch@inmech.kiev.ua

H. Altenbach et al. (eds.), *Selected Problems of Solid Mechanics and Solving Methods*, Advanced Structured Materials 204, https://doi.org/10.1007/978-3-031-54063-9_28

28.2 On the Transition from the Isotropic Medium to the Transversely Isotropic One

The history of the theory of elasticity [8] gives interesting information about the transition from the classical case of isotropic medium to the mediums with the more complex properties of symmetry. For a long time, this problem was beside the main direction of the development of the theory. Nevertheless, the anisotropy of the medium was studied in different partial problems [2, 11, 12]. Two lines of the introduction of anisotropy into the linear theory can be outlined.

The first line is more known and linked with the general representation of constitutive equations in the form

$$\sigma_{ik} = C_{iknm}\varepsilon_{nm}, \qquad (28.1)$$

where $\sigma_{ik}$is the second-rank symmetric stress tensor, ε_{nm} is the linear second-rank symmetric strain tensor, and C_{iknm} is the four-rank tensor of elastic constants of different kinds of symmetry (36 independent constants). The division of elastic media into hyperelastic, hypoelastic, and simply elastic ones permits the narrowing of the number of independent constants for the hyperelastic media to 21. But anew, the theory of elasticity long time was interested only in three types of symmetry of properties—isotropic medium with two independent constants, transversely isotropic medium with five independent constants, and orthotropic medium with nine independent constants. The top-rated solid mechanics book by Lekhnitsky [17] also accented to these three types of symmetry. Even the rapid development of the theory of composite materials did not destroy the interest in the three noted types of medium symmetry and did not lead to an analysis of the variety of medium symmetry. This variety came to mechanics from the theory of crystalline media and the theory of waves in such media. It can be said that the most significant publication is here the book by Dieulesaint and Royer (1974) [5] on the wave theory.

The second line is less known and linked with the development of the linearized theory of elasticity by two world-leading scientists—Biot and Guz. Here, the way of introducing anisotropy into the basic equations is slightly different. The process of linearization is realized with the use of two different states of deformations. The reference (non-perturbed) state is described mainly by characteristics

$$u_m^o\left(x_1, x_2, x_3, t\right),\ \varepsilon_{nm}^o\left(x_1, x_2, x_3, t\right),\ \sigma_{iik}^o\left(x_1, x_2, x_3, t\right), \ldots, e\left(\varepsilon_{nm}^o\right), \ldots \qquad (28.2)$$

and the actual (perturbed) state is described mainly by characteristics

$$u_m = u_m^o + u_m', \quad \varepsilon_{nm} = \varepsilon_{nm}^o + \varepsilon_{nm}', \quad \sigma_{ik} = \sigma_{ik}^o + \sigma_{ik}', ..., e\left(\varepsilon_{nm}\right), \ldots \qquad (28.3)$$

where $u_m', \varepsilon_{nm}', \sigma_{ik}', \ldots$ are the perturbances.

Two sets of components of stress tensors are used. First, the Piola–Kirchhoff stress tensor is used in the linearized equations of motion in the classical form

$$\rho \ddot{u}_i = t_{ik,k} + F_i. \tag{28.4}$$

But it so happened that the Cauchy–Lagrange stress tensor is usually used in the linearized theory of elasticity, and then the formula of the link between the Piola–Kirchhoff and Lagrange–Cauchy stress tensors is needed

$$t_{ik} = \sigma_{ij}\left(\delta_{nj} + u^o_{n,j}\right) + \sigma^o_{ij} u_{n,j}.$$

Then, the motion equations can be written in a new form that can be considered as the new linearized form

$$\rho \ddot{u}_i = \left[\sigma_{ij}\left(\delta_{nj} + u^o_{n,j}\right) + \sigma^o_{ij} u_{n,j}\right]_{,k} + F_i. \tag{28.5}$$

Thus, the choice of symmetry of two states generates new symmetries of the elastic medium, which are wider than the known symmetries of the hyperelastic medium.

28.3 On the Transversely Isotropic Medium

Consider the Cartesian coordinates $Ox_1x_2x_3$, and choose the axis Ox_3 as the axis of symmetry, and Ox_1x_2 as the plane of isotropy. Note that this is the case of symmetry corresponding to the hexagonal crystalline system or simply the case of transversal isotropy. Then the matrix of elastic properties is characterized by 5 independent elastic constants C_{11}, C_{12}, C_{13}, C_{33}, C_{44} and 12 non-zero components [17, 23]

$$C_{IK} = \begin{pmatrix} C_{11} & C_{12} & C_{13} & 0 & 0 & 0 \\ C_{12} & C_{11} & C_{13} & 0 & 0 & 0 \\ C_{13} & C_{13} & C_{33} & 0 & 0 & 0 \\ 0 & 0 & 0 & C_{44} & 0 & 0 \\ 0 & 0 & 0 & 0 & C_{44} & 0 \\ 0 & 0 & 0 & 0 & 0 & \frac{1}{2}(C_{11} - C_{12}) \end{pmatrix}. \tag{28.6}$$

The constitutive relations components $\sigma \sim \varepsilon$ are as follows [17, 23]:

$$\begin{aligned} \sigma_{11} &= C_{11kl}\varepsilon_{kl} = C_{11}\varepsilon_{11} + C_{12}\varepsilon_{22} + C_{13}\varepsilon_{33}, \\ \sigma_{22} &= C_{22kl}\varepsilon_{kl} = C_{12}\varepsilon_{11} + C_{22}\varepsilon_{22} + C_{13}\varepsilon_{33}, \\ \sigma_{33} &= C_{33kl}\varepsilon_{kl} = C_{13}\varepsilon_{11} + C_{13}\varepsilon_{22} + C_{33}\varepsilon_{33}, \\ \sigma_{12} &= (C_{11} - C_{12})\,\varepsilon_{12}, \quad \sigma_{13} = 2C_{44}\varepsilon_{13}, \quad \sigma_{23} = 2C_{44}\varepsilon_{23}, \end{aligned} \tag{28.7}$$

or in components $\sigma \sim u$

$$\begin{aligned}
\sigma_{11} &= C_{11}u_{1,1} + C_{12}u_{2,2} + C_{13}u_{3,3},\\
\sigma_{22} &= C_{12}u_{1,1} + C_{11}u_{2,2} + C_{13}u_{3,3},\\
\sigma_{33} &= C_{13}u_{1,1} + C_{13}u_{2,2} + C_{11}u_{3,3},\\
\sigma_{12} &= \frac{1}{2}\left(C_{11} - C_{12}\right)\left(u_{1,2} + u_{2,1}\right),\\
\sigma_{13} &= C_{44}\left(u_{1,3} + u_{3,1}\right), \quad \sigma_{23} = \frac{1}{2}C_{44}\left(u_{2,3} + u_{3,2}\right).
\end{aligned} \tag{28.8}$$

Also, the five independent so-called elastic technical constants are often used

$$E_x = E_y, \quad E_z, \quad G_{xy}, \quad G_{xz} = G_{yz}, \quad \nu_{xy}, \quad \nu_{xz} = \nu_{yz}, \quad G_{xy} = \frac{E_x}{1 + 2\nu_{xy}}.$$

They can be evaluated through C_{NM} by the known formulas:

- The longitudinal Young modulus

$$E_z = C_{33} - \frac{2\left(C_{13}\right)^2}{C_{11} + C_{12}}$$

that corresponds to tension along the symmetry axis Oz.
- The transverse Young modulus

$$E_x = \left(C_{11} - C_{12}\right)\frac{\left(C_{11} + C_{12}\right)C_{33} - 2\left(C_{13}\right)^2}{C_{11}C_{33} + \left(C_{13}\right)^2}$$

that corresponds to tension in the isotropy plane Oxy.
- The shear modulus $G_{xy} = C_{66} = \frac{1}{2}\left(C_{11} - C_{12}\right)$ that corresponds to the shear along the isotropy plane Oxy.
- The shear modulus $G_{xz} = C_{44}$ that corresponds to the shear along the symmetry axis Oz.
- The Poisson ratio $\nu_{xz} = C_{13}/\left(C_{11} + C_{12}\right)$ that corresponds to the shear along the symmetry axis Oz under tension in the isotropy plane and characterizes the shortening in this plane.

Sometimes, corresponding Lamé's moduli are used:

$$\begin{aligned}
&\lambda_{xy} + 2\mu_{xy} = C_{11}, \quad \lambda_{xy} = C_{12}, \quad \mu_{xy} = \frac{1}{2}\left(C_{11} - C_{12}\right)\\
&\lambda_{xz} + 2\mu_{xz} = C_{33}, \quad \lambda_{xz} = C_{13}, \quad \mu_{xz} = C_{44}.
\end{aligned}$$

The Poisson ratio ν_{xz} is determined by the known formula of isotropic theory,

$$\nu_{xz} = \frac{\lambda_{xz}}{2\left(\lambda_{xy} + \mu_{xy}\right)}.$$

The Poisson ratio ν_{xy} that corresponds to the shear along the symmetry axis Oz under tension along the isotropy plane is also determined by the classical formula:

$$\nu_{xy} = \frac{\lambda_{xy}}{\lambda_{xy} + \mu_{xy}}.$$

The constants C_{11}, C_{12}, C_{13}, C_{33}, and C_{44} are represented through the technical constants E, E', ν, ν', and G' by the formulas:

$$\begin{aligned}
C_{11} &= \frac{1-(\nu')^2(E/E')}{1-\nu^2+(1+2\nu)(\nu')^2(E/E')}E,\\
C_{12} &= \frac{\upsilon-(\nu')^2(E/E')}{1-\nu^2+(1+2\nu)(\nu')^2(E/E')}E,\\
C_{13} &= \frac{\nu'(1-\nu)}{1-\nu^2+(1+2\nu)(\nu')^2(E/E')}E,\\
C_{33} &= \frac{1-\nu^2}{1-\nu^2+(1+2\nu)(\nu')^2(E/E')}E',\\
C_{44} &= G'.
\end{aligned}$$

The real transversally isotropic materials need some special attention to their specificities. First, they can be divided into natural and artificial ones. An example of a classical natural material is a rock in its different kinds. An example of modern materials is a family of fibers "Kevlar". Kevlar ®KM2 [4] is characterized by elastic constants $E_x = 1.34$ GPa, $E_z = 84.62$ GPa, $G_{xz} = 24.40$ GPa, $\nu_{xy} = 0.24$, $\nu_{xz} = 0.60$. An example of composite materials can be four fibrous composites of micro- and nanolevels, which are described in [13]. The corresponding elastic constants for some variants of these materials are as follows:

- 10% of carbon microfibers $E_x = 3.59$ GPa, $E_z = 25.22$ GPa, $G_{xz} = 1.17$ GPa, $\nu_{xy} = 0.39$, $\nu_{xz} = 0.58$,
- 10% of graphite microwhiskers $E_x = 3.69$ GPa, $E_z = 102.4$ GPa, $G_{xz} = 1.14$ GPa, $\nu_{xy} = 0.39$, $\nu_{xz} = 0.62$,
- 10% of zig-zag carbon nanotubes $E_x = 3.70$ GPa, $E_z = 67.21$ GPa, $G_{xz} = 1.14$ GPa, $\nu_{xy} = 0.39$, $\nu_{xz} = 0.62$,
- 10% of chiral carbon nanotubes $E_x = 3.67$ GPa, $E_z = 126.4$ GPa, $G_{xz} = 1.14$ GPa, $\nu_{xy} = 0.39$, $\nu_{xz} = 0.62$.

The shown above values are typical for transversely isotropic materials, and therefore, they should be briefly commented on below.

Comment 1. The Young modulus in the direction along the symmetry axis E_z exceeds essentially the Young modulus in the isotropy plane E_x (from 6 to 34 times in the examples above, but can in some cases exceed hundreds of times).

Comment 2. The Lamé moduli λ_x and λ_z repeat the relations between E_x and E_z.

Comment 3. The Poisson ratio ν_{xz} along the symmetry axis Oz exceeds the classical red line in 0.5 for values of this ratio.

Comment 4. The shear moduli G_{xy} and G_{xz} differ quite moderately.

28.4 First Steps of Analyzing Classical Biot's Problem

Let us start with the statement of the problem on surface wave along the cylindrical cavity and wave equations in potentials. As before, use the cylindrical coordinate system $Or\vartheta z$ and consider the elastic harmonic wave. Assume that this wave has the phase variable $\sigma = k\,(z - vt)$, unknown wavenumber $k = (\omega/v)$, unknown phase velocity v, and arbitrary (but given) frequency ω and amplitude A.

Let this wave propagate in the direction of the vertical coordinate z in an infinite medium with a cylindrical cavity of constant radius r_o and possibly attenuate in the direction of the radial coordinate r. In this linear statement (deformations are small), the problem is axisymmetric, and deformations are described by only two displacements $(u_r(r,z,t)$, $u_\varphi(r,z,t) = 0$, and $u_z(r,z,t)$ and two Lamé's equations of the form:

$$\frac{C_{11} - C_{12}}{2}\left(\Delta_{rz}u_r - \frac{1}{r^2}u_r\right) + \frac{C_{11} + C_{12}}{2}\left(u_{r,r} + \frac{1}{r}u_r + u_{z,z}\right)_{,r} = \rho u_{r,tt},$$

$$\tfrac{1}{2}\left(C_{11} - C_{12}\right)\Delta_{rz}u_z + \tfrac{1}{2}\left(C_{11} + C_{12}\right)\left(u_{r,r} + \frac{1}{r}u_r + u_{z,z}\right)_{,z} = \rho u_{z,tt}$$

or

$$\begin{aligned}(\lambda + 2\mu)\left(u_{r,rr} + \frac{1}{r}u_{r,r} - \frac{1}{r^2}u_r + u_{z,rz}\right) &= \rho u_{r,tt}, \\ +\mu\left(u_{r,zz} - u_{z,rz}\right) &\end{aligned} \tag{28.9}$$

$$\begin{aligned}(\lambda + 2\mu)\left(u_{r,rz} + \frac{1}{r}u_{r,z} + u_{z,zz}\right) &= \rho u_{z,tt}. \\ -\mu\left[\frac{1}{r}\left(u_{r,z} - u_{z,r}\right) + \left(u_{r,rz} - u_{z,rr}\right)\right] &\end{aligned} \tag{28.10}$$

The important element of Biot's approach to the analysis is that the potentials $\Phi\,(r,z,t)$ and $\Psi\,(r,z,t)$ are introduced:

$$\begin{aligned}u_r &= \Phi_{,r} - \Psi_{,z}, \\ u_z &= \Phi_{,z} + \Psi_{,r} + (1/r)\,\Psi.\end{aligned} \tag{28.11}$$

When (28.11) is substituted into Eqs. (28.9) and (28.10), two uncoupled and rather different linear wave equations are obtained

$$\Delta_{rz}\Phi - (1/v_L)^2\,\Phi_{,tt} = 0, \tag{28.12}$$

$$\Delta_{rz}\Psi - \left(1/r^2\right)\Psi - (1/v_T)^2\,\Psi_{,tt} = 0, \tag{28.13}$$

where the standard notations of the velocities of longitudinal and transverse waves in the isotropic elastic medium

$$v_L = \sqrt{(\lambda + 2\mu)/\rho},$$

$$v_T = \sqrt{\mu/\rho}$$

are used.

The solution of equations (28.12) and (28.13) is found in the form of harmonic waves in the direction of vertical coordinate:

$$\begin{aligned}\Phi(r,z,t) &= \Phi^*(r)\,e^{i(kz-\omega t)},\\ \Psi(r,z,t) &= \Psi^*(r)\,e^{i(kz-\omega t)},\\ \Phi(r,z,t) &= \Phi^*(r)\cos k(z - vt),\\ \Psi(r,z,t) &= \Psi^*(r)\sin k(z - vt).\end{aligned} \tag{28.14}$$

A substitution of representations (28.14) into the wave equations (28.12) and (28.13) gives the equations relative to the unknown amplitudes

$$\begin{aligned}\Phi^*(r),\ \Psi^*(r)\ \Phi^*_{,rr} + (1/r)\,\Phi^*_{,r} - \left(k^2 - k_L^2\right)\Phi^* &= 0,\\ \Phi^*_{,rr} + (1/r)\,\Phi^*_{,r} - k^2\left(1 - (v/v_L)^2\right)\Phi^* &= 0,\\ \Psi^*_{,rr} - (1/r)\,\Psi^*_{,r} - \left[k^2 - k_T^2 + \left(1/r^2\right)\right]\Psi^* &= 0,\end{aligned} \tag{28.15}$$

$$\Psi^*_{,rr} - (1/r)\,\Psi^*_{,r} - \left\{k^2\left[1 - (v/v_T)^2\right] + \left(1/r^2\right)\right\}\Psi^* = 0. \tag{28.16}$$

Both equations correspond to the Bessel equation for the modified Bessel functions of the second kind [10] (Macdonald functions)

$$K_\lambda(x)\ y'' + (1/x)\,y' - \left[1 + \left(\lambda^2/x^2\right)\right]y = 0.$$

Equations (28.15) and (28.16) have the solutions in the form of Macdonald functions if the conditions

$$\begin{aligned}k^2 - k_L^2 &> 0\\ k^2 - k_T^2 &> 0\\ k^2\left(1 - (v/v_L)^2\right) &> 0\\ k^2\left(1 - (v/v_T)^2\right) &> 0\end{aligned} \tag{28.17}$$

are fulfilled. According to these conditions, the wavenumber of the cylindrical wave must be real, and the wave velocity must be less than the velocities of classical longitudinal and transverse plane waves.

Consider now the wave equations (28.15), (28.16) separately.

The first equation can be written in the form

$$\begin{gathered}\Phi^*_{,rr} + (1/r)\,\Phi^*_{,r} - m_L^2\Phi^* = 0,\\ m_L = k\sqrt{1-(v/v_L)^2}.\end{gathered} \tag{28.18}$$

This solution of equation (28.18) is expressed through the Macdonald function of zeroth order and unknown argument $x = m_L r$, which includes the unknown velocity v

$$\Phi^*(r) = A_\Phi K_0(m_L r)\,. \tag{28.19}$$

The second equation can be written in the following form:

$$\begin{gathered}\Psi^*_{,rr} - (1/r)\,\Psi^*_{,r} - \left\{m_T^2 + \left(1/r^2\right)\right\}\Psi^* = 0,\\ m_T = k\sqrt{1-(v/v_T)^2}.\end{gathered} \tag{28.20}$$

The corresponding solution is expressed by the Macdonald function of the first order and unknown argument $x = m_T r$ and the unknown wave velocity:

$$\Psi^*(r) = A_\Psi K_1(m_T r)\,. \tag{28.21}$$

The amplitude coefficients A_Φ and A_Ψ are assumed to be constant and arbitrary.

Note that the Macdonald functions have the property of attenuation with increasing their arguments [19]. Therefore, the propagating along the vertical coordinate z waves (28.14) can be considered as the waves with amplitudes $\Phi^*(r)$ and $\Psi^*(r)$, which attenuate with increasing the radius r. This means that the amplitudes can decrease essentially with increasing the distance from the surface of the cylindrical cavity. In this sense, the waves (28.14) are the surface ones. This also forms the sense of conditions (28.17). The same conditions are used in the analysis of the classical Rayleigh surface wave, which propagates along the plane surface of an isotropic elastic medium. However, the Rayleigh wave attenuates as an exponential function when being moved from the free surface, whereas the cylindrical surface Biot's wave attenuates as the Macdonald functions. At that, the arguments in the exponential function and Macdonald functions are identical and depend on the wave velocity.

At least three *facts* follow from the analysis above.

1. The introduced potentials $\Phi(r, z, t)$ and $\Psi(r, z, t)$ are determined in the form of harmonic waves.
2. The displacements do not perform the pure harmonic wave like the potentials. They perform the motion in the form of a modulated wave consisting of the sum of two different amplitude and phase harmonic waves.

3. The potentials include the unknown wave amplitudes and velocities, which have to be determined additionally from the boundary conditions.

28.5 First Steps of Analyzing the Transition from Biot's Problem to the Case of Transversely Isotropic Medium with Axial Symmetry

The case of axial symmetry means that the medium has the symmetry axis Oz, and all characterizing this medium functions do not depend on the coordinate ϑ. The displacements are characterized by two only components $u_r\,(r,z,t)$ and $u_z\,(r,z,t)$. The motion equations in stresses have the form

$$\sigma_{rr,r} + \sigma_{rz,z} + (1/r)\left(\sigma_{rr} - \sigma_{\varphi\varphi}\right) = 0,$$
$$\sigma_{rz,r} + (1/r)\,\sigma_{\varphi z,\varphi} + \sigma_{zz,z} + (1/r)\,\sigma_{rz} = 0.$$

The first difference from Biot's problem is in the more complicated form of the constitutive equations:

$$\begin{aligned}
\sigma_{rr} &= C_{11}u_{r,r} + C_{12}\,(1/r)\,u_r + C_{13}u_{z,z},\\
\sigma_{zz} &= C_{13}u_{r,r} + C_{13}\,(1/r)\,u_r + C_{33}u_{z,z},\\
\sigma_{rz} &= \tfrac{1}{2}C_{44}\left(u_{z,r} + u_{r,z}\right),\\
\sigma_{\varphi z} &= \sigma_{r\varphi} = 0.
\end{aligned}$$

In this case, the motion equations in displacements are as follows:

$$\begin{matrix} C_{11}\left[u_{r,rr} + (1/r)\,u_{r,r} - \left(1/r^2\right)u_r\right] \\ +C_{44}u_{r,zz} + \left[C_{13} + C_{44}\right]u_{z,rz} \end{matrix} = \rho u_{r,tt}, \qquad (28.22)$$

$$\begin{matrix} C_{44}\left(u_{z,rr} + (1/r)\,u_{z,r}\right) + C_{33}u_{z,zz} \\ +\left[C_{13} + C_{44}\right]\left(u_{r,rz} + (1/r)\,u_{r,z}\right) \end{matrix} = \rho u_{z,tt}. \qquad (28.23)$$

The next step is to use the potential. Note that the potentials are introduced in the theory of elasticity mainly for static problems. Transition to dynamic problems is associated with complication that is sometimes impassable. Because the problem of waves is studied, we can predict the new complications with the introduction of the potentials. It seems to be worthy to describe briefly three classical methods of introducing the potentials in transversely isotropic elasticity.

Method 1. [17] It is proposed for the axisymmetric problems of equilibrium, characteristic for representations in the transversely isotropic elasticity, and based on introducing of one only potential $\varphi\,(r,z)$ as the function of stresses

$$\begin{aligned}\sigma_{rr} &= -\left\{\varphi_{,rr} + b\,(1/r)\,\varphi_{,r} + a\varphi_{,zz}\right\}_{,z}, \\ \sigma_{\theta\theta} &= -\left\{b\varphi_{,rr} + (1/r)\,\varphi_{,r} + a\varphi_{,zz}\right\}_{,z},\end{aligned} \tag{28.24}$$

$$\begin{aligned}\sigma_{zz} &= -\left\{c\varphi_{,rr} + c\,(1/r)\,\varphi_{,r} + d\varphi_{,zz}\right\}_{,z}, \\ \sigma_{rz} &= -\left\{\varphi_{,rr} + (1/r)\,\varphi_{,r} + a\varphi_{,zz}\right\}_{,r}.\end{aligned} \tag{28.25}$$

Thus, formulas (28.24) and (28.25) include four unknown parameters a, b, c, and d.

In the next step, formulas (28.24) and (28.25) have to be substituted into the equation (28.22) (in the static variant) and the equations which are obtained from the linear Cauchy relations and formulas for the strain tensor. This permits a determination of the unknown parameters through the elastic constants represented in the equilibrium equations. Further, Eq. (28.23) gives the biharmonic equation for finding the potential

$$\Delta_{s1}\Delta_{s2}\varphi = 0, \tag{28.26}$$

where

$$\Delta_{sN}\varphi = \varphi_{,rr} + (1/r)\,\varphi_{,r} + \left(1/\left(s_N\right)^2\right)\varphi_{,zz} \quad (N = 1, 2)$$

are some "complicated" copies of classical expressions

$$\Delta\varphi = \varphi_{,rr} + (1/r)\,\varphi_{,r} + \varphi_{,zz},$$

associated with the Laplace operator. Two constants s_N are determined from the algebraic equations

$$\begin{aligned}s^4 - [(a + c)/d]\,s^2 + (1/d) = 0, \\ s_{1,3} = \pm\sqrt{\frac{a + c + \sqrt{(a + c)^2 - 4d}}{2d}}, \\ s_{2,4} = \pm\sqrt{\frac{a + c - \sqrt{(a + c)^2 - 4d}}{2d}}.\end{aligned}$$

Thus, a transition from the isotropic case to the transversally isotropic one complicates solving the static problems. Here, the necessity of solving the classical biharmonic equation is changed to the necessity of solving some generalization of this equation in the form (28.26).

Method 2 [7, 17]. This method is also proposed for static problems. Here, two potentials are introduced, which are linked immediately with the displacements

$$u_r = \phi_{1,r} + \phi_{2,r}, \quad u_z = k_1\phi_{1,z} + k_2\phi_{2,z}. \tag{28.27}$$

A substitution of potentials (28.27) into Eqs. (28.22) and (28.23) (in the equilibrium variant)

$$C_{11}\left[u_{r,rr}+(1/r)\,u_{r,r}-\left(1/r^2\right)u_r\right]+C_{44}u_{r,zz}+[C_{13}+C_{44}]\,u_{z,rz}=0,$$
$$C_{44}\left(u_{z,rr}+(1/r)\,u_{z,r}\right)+C_{33}u_{z,zz}+[C_{13}+C_{44}]\left(u_{r,rz}+(1/r)\,u_{r,z}\right)=0$$

allows us to determine the unknown constants k_1, k_2. The idea is that both equations must be transformed into identical equations relative to potentials by comparing some coefficients

$$\frac{k_{1(2)}\,(C_{13}+C_{44})+C_{44}}{C_{11}}=\frac{kC_{33}}{k_{1(2)}C_{44}+(C_{13}+C_{44})}=V.$$

This expression gives the quadratic equation for $k_{1(2)}$ and V

$$V^2+\frac{C_{13}\,(2C_{44}+C_{33})-C_{11}C_{33}}{C_{11}C_{44}}V+\frac{C_{33}}{C_{11}}=0. \tag{28.28}$$

So, these potentials fulfill the linear equations

$$\Delta_{rzN}\varphi_N=\varphi_{N,rr}+(1/r)\,\varphi_{N,r}+\left((1/V_N)^2\right)\varphi_{N,zz}. \tag{28.29}$$

Note that the simple link $V_N=(1/s_N)$ exists between the constants V_N and s_N, which makes the ways 1 and 2 very similar.

The stresses are expressed through the new potentials in the following way:

$$\sigma_{rr}=\begin{array}{c}-(C_{11}-C_{12})\,(1/r)\left(\phi_{1,rr}+\phi_{2,rr}\right)\\-C_{44}\left((1+k_1)\,\phi_{1,zz}+(1+k_2)\,\phi_{2,zz}\right),\end{array}$$
$$\sigma_{\theta\theta}=\begin{array}{c}-(C_{11}-C_{12})\,(1/r)\left(\phi_{1,rr}+\phi_{2,rr}\right)\\-\left((C_{13}k_1-C_{12}V_1)\,\phi_{1,zz}+(C_{13}k_2-C_{12}V_2)\,\phi_{2,zz}\right),\end{array}$$
$$\sigma_{zz}=(C_{33}k_1-C_{13}V_1)\,\phi_{1,zz}+(C_{33}k_2-C_{13}V_2)\,\phi_{2,zz},$$
$$\sigma_{rz}=C_{44}\left((1+k_1)\,\phi_{1,rz}+(1+k_2)\,\phi_{2,rz}\right).$$

Method 3 [3, 7]. This method is proposed for equations of motion but only for the isotropic theory of elasticity. It can be used for the static problems of the transversely isotropic theory of elasticity. The initial equations here are the equations of motion (28.22) and (28.23) without inertial summands

$$C_{11}\left[u_{r,rr}+(1/r)\,u_{r,r}-\left(1/r^2\right)u_r\right]+C_{44}u_{r,zz}+[C_{13}+C_{44}]\,u_{z,rz}=0, \tag{28.30}$$

$$C_{44}\left(u_{z,rr}+(1/r)\,u_{z,r}\right)+C_{33}u_{z,zz}+[C_{13}+C_{44}]\left(u_{r,rz}+(1/r)\,u_{r,z}\right)=0. \tag{28.31}$$

The potentials are introduced like (28.27), but the representations are complicated by the necessity of introducing two new unknown parameters n, m

$$u_r = \Phi_{,r} - \Psi_{,z}, \quad u_z = n\Phi_{,z} + m\Psi_{,r} + m\,(1/r)\,\Psi. \tag{28.32}$$

These potentials can be obtained by substitution of representations (28.32) into equations of motion (28.30) and (28.31). The first equation gives two equations:

$$\Phi_{,rr} + (1/r)\,\Phi_{,r} + \frac{C_{44} + n\,(C_{13} + C_{44})}{C_{11}}\Phi_{,zz} = 0, \tag{28.33}$$

$$\Psi_{,rr} + (1/r)\,\Psi_{,r} - \left(1/r^2\right)\Psi + \frac{C_{44}}{C_{11} - m\,(C_{13} + C_{44})}\Psi_{,zz} = 0, \tag{28.34}$$

whereas the second one gives three equations

$$\Phi_{,rr} + (1/r)\,\Phi_{,r} + \frac{nC_{33}}{nC_{44} + (C_{13} + C_{44})}\Phi_{,zz} = 0, \tag{28.35}$$

$$\Psi_{,rrz} + (1/r)\,\Psi_{,rz} - (1/r)^2\,\Psi_{,z} + \frac{C_{33}m - (C_{13} + C_{44})}{C_{44}m}\Psi_{,zzz} = 0, \tag{28.36}$$

$$\Psi_{,rr} + (1/r)\,\Psi_{,r} - (1/r)^2\,\Psi + \frac{C_{33}m - (C_{13} + C_{44})}{C_{44}m}\Psi_{,zz} = 0. \tag{28.37}$$

The last two equations are identical. Equations (28.33) and (28.35), and Eqs. (28.34) and (28.37) have to be identical (that is, the coefficients in these equations have to be identical). As a result, two equations can be obtained for the determination of the unknown constants n and m

$$\frac{C_{44} + n\,(C_{13} + C_{44})}{C_{11}} = \frac{nC_{33}}{nC_{44} + (C_{13} + C_{44})} \rightarrow$$

$$\rightarrow n^2 - n\frac{C_{11}C_{33} - (C_{44})^2 - (C_{13} + C_{44})^2}{C_{44}\,(C_{13} + C_{44})} + 1 = 0, \tag{28.38}$$

$$n_{1,2} = \frac{C_{11}C_{33} - (C_{44})^2 - (C_{13} + C_{44})^2}{2C_{44}\,(C_{13} + C_{44})} \times \left(1 \pm \sqrt{1 - 4\left[\frac{C_{44}\,(C_{13} + C_{44})}{C_{11}C_{33} - (C_{44})^2 - (C_{13} + C_{44})^2}\right]^2}\right), \tag{28.39}$$

$$\frac{C_{44}}{C_{11} - m\left(C_{13} + C_{44}\right)} = \frac{C_{33}m - \left(C_{13} + C_{44}\right)}{C_{44}m}$$

$$\rightarrow\ m^2 + m\left[\frac{\left(C_{44}\right)^2 + C_{11}C_{33} + \left(C_{13} + C_{44}\right)^2}{C_{33}\left(C_{13} + C_{44}\right)}\right] + \frac{C_{11}}{C_{33}} = 0, \qquad (28.40)$$

$$m_{1,2} = -\frac{\left(C_{44}\right)^2 + C_{11}C_{33} + \left(C_{13} + C_{44}\right)^2}{2C_{33}\left(C_{13} + C_{44}\right)} \times \left\{1 \pm \sqrt{1 - 4\frac{C_{11}}{C_{33}}\left[\frac{C_{33}\left(C_{13} + C_{44}\right)}{\left(C_{44}\right)^2 + C_{11}C_{33} + \left(C_{13} + C_{44}\right)^2}\right]^2}\right\}. \qquad (28.41)$$

The unknown potentials $\Phi\left(r, z\right)$ and $\Psi\left(r, z\right)$ have to be determined from the simple equations (28.34) and (28.35) which are the classical Bessel equations of orders 0 and 1 and arguments depending on some rational combination of elastic constants.

Thus, three methods of introduction of potentials in the static problems of the transversely isotropic theory of elasticity are shown. The different attempts to transfer these ways into the dynamic problems meet some troubles—the presence of inertial summands generates new additional conditions for the unknown constants in representations of potentials. Introducing the new constants does not help—the number of conditions is still more than the number of all constants.

Return now to the equations of motion (28.22) and (28.23) and use the potentials from method 3 in the new form:

$$\begin{aligned} u_r\left(r, z, t\right) &= \Phi_{,r}\left(r, z, t\right) - \Psi_{,z}\left(r, z, t\right), \\ u_z\left(r, z, t\right) &= n\Phi_{,z}\left(r, z, t\right) + m\Psi_{,r}\left(r, z, t\right) + m\left(1/r\right)\Psi\left(r, z, t\right). \end{aligned} \qquad (28.42)$$

It is assumed here that these new potentials also depend on time.

Substitute the representation (28.42) into the equations of motion. As a result, five equations relative to the potentials can be obtained. The first equation gives two equations:

$$\Phi_{,rr} + \left(1/r\right)\Phi_{,r} + \frac{C_{44} + n\left(C_{13} + C_{44}\right)}{C_{11}}\Phi_{,zz} = \frac{\rho}{C_{11}}\Phi_{,tt}, \qquad (28.43)$$

$$\Psi_{,rr} + \left(1/r\right)\Psi_{,r} - \left(1/r^2\right)\Psi + \frac{C_{44}}{C_{11} - m\left(C_{13} + C_{44}\right)}\Psi_{,zz} = \frac{\rho}{C_{11} - m\left(C_{13} + C_{44}\right)}\Psi_{,tt}. \qquad (28.44)$$

The second equation gives three equations

$$\Phi_{,rr} + \frac{\Phi_{,r}}{r} + \frac{nC_{33}}{nC_{44} + (C_{13} + C_{44})}\Phi_{,zz} = \frac{n\rho}{nC_{44} + (C_{13} + C_{44})}\Phi_{,tt}, \tag{28.45}$$

$$\Psi_{,rrz} + \frac{\Psi_{,rz}}{r} - (1/r)^2\,\Psi_{,z} + \frac{C_{33}m - (C_{13} + C_{44})}{C_{44}m}\Psi_{,zzz} = \frac{\rho}{C_{44}}\Psi_{,ztt}, \tag{28.46}$$

$$\Psi_{,rr} + (1/r)\,\Psi_{,r} - (1/r)^2\,\Psi + \frac{C_{33}m - (C_{13} + C_{44})}{C_{44}m}\Psi_{,zz} = \frac{\rho}{C_{44}}\Psi_{,tt}. \tag{28.47}$$

The two last equations are identical. Also, the new potentials can be used for analysis in the case only when the equations for potential Φ are identical as well as the equations for potential Ψ are identical too.

Let us assume at the next step additionally that the problem in hand considers the solution in the form of the harmonic in time cylindrical wave with the unknown wavenumber k and known frequency ω

$$\Phi\,(r,z,t) = \widehat{\Phi}\,(r)\,e^{i(kz-\omega t)}, \quad \Psi\,(r,z,t) = \widehat{\Psi}\,(r)\,e^{i(kz-\omega t)}. \tag{28.48}$$

Note that because the waves are the surface ones, then the unknown functions $\widehat{\Phi}\,(r)\,,\widehat{\Psi}\,(r)$ must describe an attenuation of wave depth down. They have to be found from equations, which are obtained by substitution of representations (28.48) into Eqs. (28.44), (28.45)

$$\widehat{\Phi}_{,rr} + (1/r)\,\widehat{\Phi}_{,r} - \left[\frac{C_{44} + n\,(C_{13} + C_{44})}{C_{11}}k^2 - k^2_{L(11)}\right]\widehat{\Phi} = 0,$$
$$k_{L(11)} = \left(\omega/v_{L(11)}\right), \quad v_{L(11)} = \sqrt{C_{11}/\rho},$$
$$\widehat{\Phi}_{,rr} + (1/r)\,\widehat{\Phi}_{,r} - \frac{n}{nC_{44} + (C_{13} + C_{44})}\left(C_{33}k^2 - C_{11}k^2_{L(11)}\right)\widehat{\Phi} = 0,$$
$$\widehat{\Psi}_{,rr} + (1/r)\,\widehat{\Psi}_{,r} - \left(1/r^2\right)\widehat{\Psi} - \frac{C_{44}}{C_{11} - m\,(C_{13} + C_{44})}\left(k^2 - k^2_{T(44)}\right)\widehat{\Psi} = 0,$$
$$k_{T(44)} = \left(\omega/v_{L(44)}\right), \quad v_{L(44)} = \sqrt{C_{44}/\rho},$$
$$\widehat{\Psi}_{,rr} + (1/r)\,\widehat{\Psi}_{,r} - (1/r)^2\,\widehat{\Psi} - \left[\frac{C_{33}m - (C_{13} + C_{44})}{C_{44}m}k^2 - k^2_{T(44)}\right]\widehat{\Psi} = 0.$$

Finally, two equations can be obtained from a comparison of the last four equations that permits to determine the constants n and m from the following quadratic algebraic equations:

$$n^2 - 2N_1 n + N_2 = 0, \quad m^2 + 2M_1 m + M_2 = 0, \tag{28.49}$$

$$N_{\pm}(M_{\pm}) = N_1(M_1) \pm \sqrt{[N_1(M_1)]^2 - N_2(M_2)},$$

$$2N_1 = \frac{\left[C_{11}C_{33} - (C_{13}+C_{44})^2\right]k^2 - C_{11}\left[C_{11} - C_{44}\right]k^2_{L(11)} - (C_{44})^2}{C_{44}(C_{13}+C_{44})k^2},$$

$$N_2 = \frac{C_{44} - C_{11}k^2_{L(11)}}{C_{44}k^2} = 0,$$

$$2M_1 = \frac{\left[(C_{44})^2 - C_{11}C_{33} - (C_{13}+C_{44})^2\right]k^2 - \left[(C_{44})^2 - C_{11}C_{44}\right]k^2_{T(44)}}{(C_{13}+C_{44})\left(C_{33}k^2 - C_{44}k^2_{T(44)}\right)},$$

$$M_2 = \frac{C_{11}}{\left(C_{33}k^2 - C_{44}k^2_{T(44)}\right)}k^2.$$

Note that the restriction on the kind of solution (it has to be a wave) allows us to unite two different conditions into one—conditions for equaling the coefficients in summands with the second derivative by time t and vertical coordinate z. In this case, the number of unknown constants coincides with the number of conditions that are necessary for the determination of potentials. As a result, the shown wave attenuation new transformed potentials $\widehat{\Phi}(r)$, $\widehat{\Psi}(r)$ can be determined from the equations of the Bessel type

$$\widehat{\Phi}_{,rr} + (1/r)\widehat{\Phi}_{,r} - M^2_{L(11)}\widehat{\Phi} = 0,$$
$$\widehat{\Psi}_{,rr} + (1/r)\widehat{\Psi}_{,r} - \left[\left(1/r^2\right) + M^2_{T(44)}\right]\widehat{\Psi} = 0, \tag{28.50}$$

$$M_{L(11)} = \sqrt{\frac{C_{44} + n(C_{13}+C_{44})}{C_{11}}k^2 - k^2_{L(11)}},$$

$$M_{T(44)} = \sqrt{\frac{C_{44}}{C_{11} - m(C_{13}+C_{44})}\left(k^2 - k^2_{T(44)}\right)}.$$

Success in the determination of transformed potentials is accompanied by a complication of conditions that provide wave attenuation. They have the form

$$\frac{C_{44} + n(C_{13}+C_{44})}{C_{11}}k^2 - k^2_{L(11)} > 0, \quad \frac{C_{33}m - (C_{13}+C_{44})}{C_{44}m}k^2 - k^2_{T(44)} > 0. \tag{28.51}$$

Let us recall similar conditions for the case of an isotropic medium.

$$k^2 - k_L^2 > 0, \quad k^2 - k_T^2 > 0$$

are simpler and coincide with the corresponding conditions of the classical Rayleigh surface wave. The complexity of conditions (28.51) is increased by the complex form of dependence of constants n, m on the wavenumber k.

If the conditions (28.51) are fulfilled, then the solution of wave equations for the potentials can be written:

$$\widehat{\Phi}(r) = \widehat{A_\Phi} K_0 \left(M_{L(11)} r\right), \quad \widehat{\Psi}(r) = \widehat{A_\Psi} K_1 \left(M_{T(44)} r\right). \tag{28.52}$$

With allowance for formulas (28.52), the representations of potentials take the more definite view

$$\Phi(r, z, t) = \widehat{A_\Phi} K_0 \left(M_{L(11)} r\right) e^{i(kz-\omega t)}, \quad \Psi(r, z, t) = \widehat{A_\Psi} K_1 \left(M_{T(44)} r\right) e^{i(kz-\omega t)}. \tag{28.53}$$

The formulas (28.53) complete the first analytical part of solving the problem on the cylindrical surface wave.

28.6 Using the Boundary Conditions for Determination of the Unknown Wavenumbers

In this wave problem, the boundary conditions correspond to the absence of stresses on the surface$r = r_o$ and have the form

$$\sigma_{rr}(r = r_o, t) = 0, \quad \sigma_{rz}(r = r_o, t) = 0. \tag{28.54}$$

First, these stresses

$$\sigma_{rr} = 2\mu u_{r,r} + \lambda\left((u_r/r) + u_{r,r} + u_{z,z}\right), \quad \sigma_{rz} = \mu\left(u_{r,z} + u_{z,r}\right)$$

should be written through the potentials

$$\sigma_{rr} = \begin{array}{l} (\lambda + 2\mu)\left(\Phi_{,rr} - \Psi_{,rz}\right) + \lambda\{(1/r)\left(\Phi_{,r} - \Psi_{,z}\right) \\ \qquad + \Phi_{,zz} + \Psi_{,rz} + (1/r)\Psi_{,z}\}, \end{array} \tag{28.55}$$

$$\sigma_{rz} = \mu\left[\left(\Phi_{,rz} - \Psi_{,zz}\right) + \Phi_{,zr} + \Psi_{,rr} + (1/r)\Psi_{,r} - \left(1/r^2\right)\Psi\right]. \tag{28.56}$$

Second, the boundary conditions (28.54) are represented through the potentials

$$\left[2\mu\left(\Phi_{,rr} - \Psi_{,rz}\right) + \lambda\Delta\Phi\right]_{r=r_o} = 0, \quad \mu\left[2\left(\Phi_{,rz} - \Psi_{,zz}\right) + \Delta\Psi - \left(1/r^2\right)\Psi\right]_{r=r_o} = 0. \tag{28.57}$$

In the work [3], Biot used the expressions

$$\Delta\Phi - (1/v_L)^2\Phi_{,tt} = 0, \quad \Delta\Psi - \left(1/r^2\right)\Psi - (1/v_T)^2\Psi_{,tt} = 0$$

and rewrite equations (28.57) to the form

$$\left[\left(\Phi_{,rr}-\Psi_{,rz}\right)+\left(\frac{\lambda}{2\mu}\right)\left(\frac{1}{v_L}\right)^2\Phi_{,tt}\right]_{r=r_o}=0,\ \left[2\left(\Phi_{,rz}-\Psi_{,zz}\right)+\left(\frac{1}{v_T}\right)^2\Psi_{,tt}\right]_{r=r_o}=0.$$

The substitution of solutions (28.14) into the boundary conditions (28.57) gives two homogeneous algebraic equations relative to the unknown constant amplitude coefficients

$$\left[1-(v/v_L)^2-\frac{\lambda}{\mu}(v/v_L)^2\frac{K_0(m_Lr_o)}{K_0(m_Lr_o)+K_2(m_Lr_o)}\right]A_\Phi--\sqrt{1-(v/v_L)^2}A_\Psi=0,$$

$$2\sqrt{1-(v/v_L)^2}A_\Phi+\left(2-(v/v_T)^2\right)\frac{K_1(m_Tr_o)}{K_1(m_Lr_o)}A_\Psi=0.$$

These equations describe the cylindrical surface wave. Its analysis is similar to the analysis that has been carried out by Rayleigh for the classical wave propagating along the plane surface. Some novelty in the analysis of the system (28.57) is the consideration of the system of relative quantities $K_1(m_Lr_o)A_\Phi$ and $K_1(m_Sr_o)A_\Psi$

$$\left\{\left(1-(v/v_L)^2\right)\left[\frac{K_0(m_Lr_o)}{K_1(m_Lr_o)}+\frac{1}{m_Lr_o}\right]-\frac{\lambda}{2\mu}(v/v_L)^2\frac{K_0(m_Lr_o)}{K_1(m_Lr_o)}\right\}K_1(m_Lr_o)A_\Phi$$
$$+\sqrt{1-(v/v_S)^2}\left[\frac{K_0(m_Tr_o)}{K_1(m_Tr_o)}+\frac{1}{m_Tr_o}\right]K_1(m_Tr_o)A_\Psi=0, \tag{28.58}$$

$$2\sqrt{1-(v/v_L)^2}K_1(m_Lr_o)A_\Phi+\left(2-(v/v_T)^2\right)K_1(m_Tr_o)A_\Psi=0. \tag{28.59}$$

Two results follow from the solution of system (28.58) and (28.59). First, the solution can be found accurate within one amplitude factor. Second, an equation for the determination of the phase velocity of the cylindrical surface wave can be obtained in an explicit form.

In his work of 1952, Biot demonstrated some art in handling the Macdonald functions and written equation (28.58) through only functions of the zeroth and first orders. For that, the formulas $K_0'(x)=-K_1(x)$, $K_1'(x)=-K_0''(x)$, $K_0''(x)+(1/x)K_0'(x)=K_0(x)$, $K_0''(x)=(1/x)K_1(x)+K_0(x)$ were used [10, 19]. As a result, the equation for determination of the velocity of the cylindric cal surface wave has the form

$$\left(2-(v/v_T)^2\right)\left\{\left[2-(v/v_T)^2\right]\frac{K_0(m_Lr_o)}{K_1(m_Lr_o)}+\frac{1-(v/v_L)^2)}{m_Lr_o}\right\}$$
$$-\ 4\sqrt{1-(v/v_L)^2}\sqrt{1-(v/v_T)^2}\left[\frac{K_0(m_Tr_o)}{K_1(m_Tr_o)}+\frac{1}{m_Tr_o}\right]=0. \tag{28.60}$$

For comparison, write the corresponding equation for the Rayleigh wave [1, 18, 24, 25]

$$4\sqrt{1-(v/v_L)^2}\sqrt{1-(v/v_S)^2}-\left[2-(v/v_S)^2\right]^2=0. \tag{28.61}$$

So, the presence of Macdonald functions in Eq. (28.60) complicates an analysis essentially because these functions have an unknown velocity in their argument according to relations

$$m_L = k\sqrt{1-(v/v_L)^2}, \quad m_S = k\sqrt{1-(v/v_T)^2}.$$

Let us note that if the cavity radius is not small, then the Macdonald function is represented by the simple formula

$$K_0(r) = K_1(r) = e^{-r}\sqrt{\pi/2r},$$

and Eq. (28.60) is reduced to the Rayleigh equation (28.61).

Strictly speaking, the analytical part of the analysis is ended by obtaining the Eq. (28.60). Further analysis can be continued with the aim of the numerical methods.

Return now to the case of a transversely isotropic medium and consider the possibilities of continuation of the analytical study using the boundary conditions.

First, note that these conditions are identical for all kinds of symmetry of properties. That is, they have the form (28.54).

The expressions for stresses through the potentials reflect now the features of introducing the potentials. In this case, they have the form

$$\sigma_{rr} = \begin{array}{c}(\lambda+2\mu)\left(\Phi_{,rr}-\Psi_{,rz}\right)+\lambda\left[(1/r)\left(\Phi_{,r}-\Psi_{,z}\right)\right.\\ \left.+n\Phi_{,zz}+m\Psi_{,rz}+m(1/r)\Psi_{,z}\right],\end{array} \tag{28.62}$$

$$\sigma_{rz} = \mu\left[\left(\Phi_{,rz}-\Psi_{,zz}\right)+n\Phi_{,zr}+m\Psi_{,rr}+\frac{m\Psi_{,r}}{r}-m\left(1/r^2\right)\Psi\right]. \tag{28.63}$$

Substitute now the last representations into the boundary conditions and take into account the formulas on differentiation of Macdonald functions

$$\begin{aligned}\left[dK_0\left(M_{L(11)}rx\right)/dr\right] &= -M_{L(11)}K_1\left(M_{L(11)}rx\right),\\ \left[d^2K_0\left(M_{L(11)}r\right)/dr^2\right] &= M_{L(11)}(1/r)K_1\left(M_{L(11)}r\right)+\left(M_{L(11)}\right)^2K_0\left(M_{L(11)}r\right),\\ \left[dK_1\left(M_{T(44)}r\right)/dr\right] &= -(1/r)K_1\left(M_{T(44)}r\right)-M_{T(44)}K_0\left(M_{T(44)}r\right).\end{aligned}$$

Then, the boundary conditions are transformed into algebraic equations relative to the quantities

$$K_1\left(M_{L(11)}r_o\right)\widehat{A_\Phi},\ K_1\left(M_{T(44)}r_o\right)\widehat{A_\Psi}$$

$$\begin{aligned}
&\times\left[M_{L(11)}\frac{1}{r_o}+\frac{v_L^2}{v_T^2}\left(M_{L(11)}-\frac{v_L^2-v_T^2}{v_L^2}nk^2\right)^2\frac{K_0\left(M_{L(11)}r_o\right)}{K_1\left(M_{L(11)}r_o\right)}\right]\\
&\qquad\times\widehat{A_\Phi}K_1\left(M_{L(11)}r_o\right)-ik\frac{v_L^2-v_T^2}{v_T^2}=0,\\
&\times\left[\left(2(1-m)+\frac{v_T^2}{v_L^2-v_T^2}\right)\frac{1}{r_o}+\left((1-m)+\frac{v_T^2}{v_L^2-v_T^2}\right)\right.\\
&\qquad\left.\times M_{T(44)}\frac{K_0\left(M_{T(44)}r_o\right)}{K_1\left(M_{T(44)}r_o\right)}\right]\widehat{A_\Psi}K_1\left(M_{T(44)}r_o\right)\\
&\qquad(1+n)ik\frac{K_0\left(M_{L(11)}r_o\right)}{K_1\left(M_{L(11)}r0\right)}K_1\left(M_{L(11)}r_o\right)\widehat{A_\Phi}\\
&\qquad+\left[m\left(M_{T(44)}\right)^2+k^2\right]K_1\left(M_{T(44)}r\right)\widehat{A_\Psi}=0.
\end{aligned}\tag{28.64}$$

Then the determinant of the linear homogeneous system of equations (28.64) is equaled to zero, and the equation for the unknown wavenumbers k is obtained

$$\begin{aligned}
&(1+n)k^2\frac{v_L^2-v_T^2}{v_T^2}\frac{K_0\left(M_{L(11)}r_o\right)}{K_1\left(M_{L(11)}r0\right)}\left[\left(2(1-m)+\frac{v_T^2}{v_L^2-v_T^2}\right)(1/r_o)\right.\\
&\left.+\left((1-m)+\frac{v_T^2}{v_L^2-v_T^2}\right)M_{T(44)}\frac{K_0\left(M_{T(44)}r_o\right)}{K_1\left(M_{T(44)}r_o\right)}\right]-\left[m\left(M_{T(44)}\right)^2+k^2\right]\\
&\times\left[M_{L(11)}\frac{1}{r_o}+\frac{v_L^2}{v_T^2}\left(\left(M_{L(11)}\right)^2-\frac{v_L^2-v_T^2}{v_L^2}nk^2\right)\frac{K_0\left(M_{L(11)}r_o\right)}{K_1\left(M_{L(11)}r_o\right)}\right]=0.
\end{aligned}\tag{28.65}$$

Note that in the coefficients $M_{L(11)}$ and $M_{T(44)}$ of Macdonald functions arguments

$$\frac{K_0\left(M_{L(11)}r_o\right)}{K_1\left(M_{L(11)}r0\right)},\quad\frac{K_0\left(M_{T(44)}r_o\right)}{K_1\left(M_{T(44)}r_o\right)}$$

the sufficiently complex expression relative to the wavenumber is hidden. Therefore, the analytical part of the analysis is finished on these formulas. Further, numerical approaches have to be utilized.

Note also that the simple and convenient condition from analysis of the classical surface Rayleigh wave, when the wavenumber depends only on ratio $\left(v_L^2/v_T^2\right)$, does not exist in the analysis of cylindrical surface wave. Here, the new parameters $M_{L(11)}$ and $M_{T(44)}$ depend in the complicated form on all elastic constants. Of course, the Macdonald functions can be represented approximately in series through their arguments. But only the numerical methods can give the final result—the value of wavenumber or phase velocity.

28.7 Conclusions

The old classical Biot problem of 1952 on the surface harmonic wave propagating in the isotropic elastic medium along the free surface of the cylindrical cavity is revisited. The main goal is to generalize the case of transversally isotropic medium of propagation. The first conclusion is that a restriction of the general dynamic analysis to the analysis of harmonic waves permits the utilization of the representation of displacements through the potentials. The second conclusion consists of that the obtained for the static problems of the theory of transversally isotropic elasticity representation can be generalized on the case of dynamical problems, namely, on the case of harmonic surface waves. The main conclusion is that the stated problem of generalization of Biot's problem is solved theoretically.

References

1. Achenbach, J.D.: Wave Propagation in Elastic Solids. North-Holland, Amsterdam (1973)
2. Ambartsumian, S.A.: Theory of Anisotropic Plates. Nauka, Moscow (1967). [In Russian]
3. Biot, M.A.: Propagation of elastic waves in a cylindrical bore containing a fluid. J. Appl. Phys. **23**(9), 997–1005 (1952). https://doi.org/10.1063/1.1702365
4. Cheng, M., Chen, W., Weerasoorlya, T.: Mechanical properties of kevlar SKM2. J. Eng. Mater. Technol. **127**(2), 197–203 (2005). https://doi.org/10.1115/1.1857937
5. Dieulesaint, E., Royer, D.: Ondes elastiques dans les solides. Application au traitement du signal, Masson et Cie, Paris (1974). [In French]
6. Dixit, U.S., Hazarika, M., Davim, J.P.: A Brief History of Mechanical Engineering. Springer International Publishing, Cham (2017). https://doi.org/10.1007/978-3-319-42916-8
7. Elliot, H.A.: Three-dimensional stress distribution in hexagonal aelotropic crystals. Math. Proc. Camb. Phil. Soc. **44**(N4), 522–533 (1948). https://doi.org/10.1017/50305004100024531
8. Truesdell, C.: Essays in the History of Mechanics. Springer, Berlin (1968)
9. Fedorov, F.I.: Theory of Elastic Waves in Crystals. Academic, New York (1975)
10. Gradstein, I., Ryzhik, I.: Table of Integrals, Series, and Products. 7th revised edn., Jeffrey, A., Zwillinger, D., (edn.). Academic Inc. New York (2007)
11. Grigorenko, Y.M., Vasilenko, A.T., Bespalova, E.I., Pankratova, N.D., Polishchuk, T.I., lacinnik, IF.: Numerical Solution of Problems of Statics of Orthotropic Shells with Variable Parameters. Naukova Dumka, Kyiv (1975). [In Russian]
12. Grigorenko, Y.M., Vasilenko, A.T., Golub, G.P.: Statics of Anisotropic Shells with Finite Shear Stiffness. Naukova Dumka, Kyiv (1987). [In Russian]
13. Guz, A.N., Rushchitsky, J.J.: Short Introduction to Mechanics of Nanocomposites. Scientific & Academic Publishing, Rosemead, CA (2013)
14. Guz, A.N.: Elastic Waves in Bodies with Initial Stresses. In: 2 Parts. Part 1. General Questions. Waves in Infinite Bodies and Surface Waves. LAP LAMBERT Academic Publishing, Saarbrucken (2016). [In Russian]

15. Guz, A,N.: Elastic Waves in Bodies with Initial Stresses. In: 2 Parts. Part 2. Waves in Partially Bounded Bodies. LAP LAMBERT Academic Publishing, Saarbrucken (2016). [In Russian]
16. Kiselev, A.P.: Rayleigh wave with a transverse structure. Proc. Roy. Soc. Lond. A. **460**(2050), 3059–3064 (2004). https://doi.org/10.1098/rspa.2004.1353
17. Lekhnitsky, S.G.: Theory of Elasticity of Anisotropic Elastic Body. Golden Day Inc., San Francisco (1963). https://doi.org/10.1137/1009023
18. Nowacki, W.: Theoria Sprazystosci. PWN, Warszawa (1970). [In Polish]
19. Olver, F.W.J., Lozier, D.W., Bousvert, R.F., Clark, C.W. (eds.): NIST (National Institute of Standards and Technology). Handbook of Mathematical Functions. Cambridge University Press, Cambridge (2010)
20. Royer, D., Dieulesaint, E.: Elastic Waves in Solids, vol. I. II. Advanced Texts in Physics, Springer, Berlin (2000)
21. Rushchitsky, J.J.: Theory of Waves in Materials. Ventus Publishing ApS, Copenhagen (2011). https://doi.org/10.13140/RG.2.1.3162.8647
22. Rushchitsky, J.J.: Nonlinear Elastic Waves in Materials. Springer, Heidelberg (2014). https://doi.org/10.1007/978-3-319-00464-8
23. Rushchitsky, J.J.: Foundations of Mechanics of Materials. Ventus Publishing ApS, Copenhagen (2021)
24. Sedov, L.I.: A Course in Continuum Mechanics. vols. I–IV. Wolters Noordhoff Publishing, Amsterdam (1971)
25. Viktorov, I.A.: Rayleigh and Lamb Waves. Plenum Press, New York (1967)

Chapter 29
Stress State of the Solid Fuel of a Rocket Engine Supported by the Shell Under Pulse Loading

Ihor Senchenkov, Olha Chervinko, and Nina Yakovenko

Abstract The stress state of the solid fuel and the shell of the rocket cylindrical engine demonstration version under impulse normal loading on the internal or external surfaces is investigated in the axisymmetric statement. The fuel material is assumed to be an isotropic linearly viscoelastic material. The solution methodology is based on the approximate Schapery method. The problem is solved by FEM in combination with the Newmark method of time integration. The maximum tensile stresses at the fuel–shell interface and maximum tensile and compressive stresses in the shell in the transient wave process are obtained.

29.1 Introduction

One of the problems arising in the design of rocket engine with solid fuel (RESF) is the calculation of the stress–strain state of the charge and the body of the RESF under dynamic loading [1–3]. During fusion, the charge is subjected to sharply increasing internal pressure. If the rocket is mounted in a shaft, the charge is subjected to a high external pressure. The stresses in the waves generated in this case can significantly exceed, as well as differ in signs from the stresses under the corresponding static loads.

This paper presents the results of the numerical simulation of thermomechanical processes in RESF under mechanical loading in the framework of the axisymmetric formulation. A comparison of analytical and numerical solutions is given, which substantiates the validity of the approach developed.

I. Senchenkov · O. Chervinko (✉) · N. Yakovenko
S.P. Timoshenko Institute of Mechanics, National Academy of Science of Ukraine, Kyiv, Ukraine
e-mail: chop497@gmail.com

I. Senchenkov
e-mail: term@inmech.kiev.ua

H. Altenbach et al. (eds.), *Selected Problems of Solid Mechanics and Solving Methods*, Advanced Structured Materials 204, https://doi.org/10.1007/978-3-031-54063-9_29

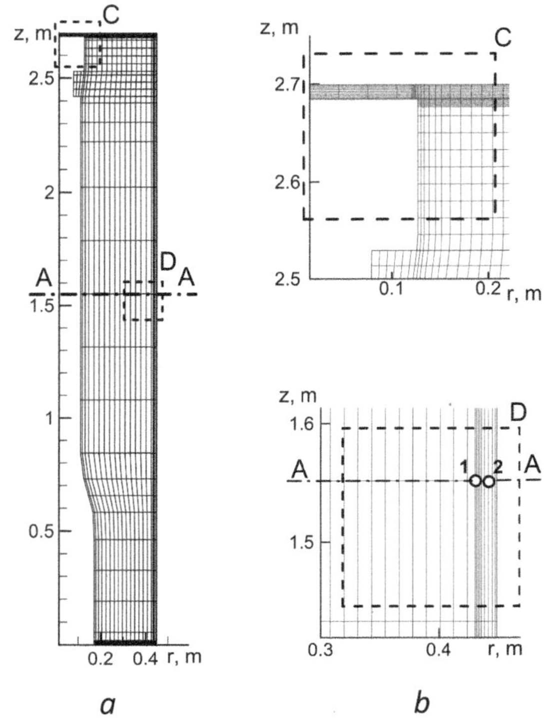

Fig. 29.1 Finite-element breakdown of the meridional cross-section of the demonstration version of the RESF, **a**—full cross-section; **b**—fragments C and D and position of points 1 and, in which time curves are given below, point 1—is in the fuel, point 2—is in the shell

29.2 Schematic Drawing of the Demonstration Model of the RESF

Figure 29.1 shows a finite-element breakdown of the meridional cross-section of the demonstration variant of the RESF with a circular internal hole with radial layers: fuel (0, 1085 m $< r <$ 0, 434 m), thermal insulation (0.434 m $< r <$ 0.437 m), shell (0.437 m $< r <$ 0.447 m), thermal insulation (0.437 m $< r <$ 0.45 m).

29.3 Formulation of the Axisymmetric Dynamic Problem of Thermomechanics

The formulation of the dynamic axisymmetric problem in a cylindrical coordinate system $Orz\varphi$ for an isotropic material includes the following relations.

- Cauchy equations

$$\begin{aligned}
\varepsilon_{zz} &= \frac{\partial u_z}{\partial z},\\
\varepsilon_{rr} &= \frac{\partial u_r}{\partial r},\\
\varepsilon_{\varphi\varphi} &= \frac{u_r}{r},\\
\varepsilon_{rz} &= \frac{1}{2}\left(\frac{\partial u_z}{\partial r} + \frac{\partial u_r}{\partial z}\right),
\end{aligned} \tag{29.1}$$

  where u_r and u_z are components of displacement vector; ε_{rr}, ε_{zz}, ε_{rz}, and $\varepsilon_{\varphi\varphi}$ are components of strain tensor;
- equations of motion:

$$\begin{aligned}
\frac{\partial \sigma_{rr}}{\partial r} + \frac{1}{r}\left(\sigma_{rr} - \sigma_{\varphi\varphi}\right) + \frac{\partial \sigma_{rz}}{\partial z} &= \rho \ddot{u}_r,\\
\frac{\partial \sigma_{rz}}{\partial r} + \frac{1}{r}\sigma_{rz} + \frac{\partial \sigma_{zz}}{\partial z} &= \rho \ddot{u}_z,
\end{aligned} \tag{29.2}$$

  where σ_{rr}, σ_{zz}, σ_{rz}, and $\sigma_{\varphi\varphi}$ are components of stress tensor; ρ is the material density;
- the heat conduction equation:

$$\frac{\partial}{\partial z}\left(\lambda \frac{\partial T}{dz}\right) + \frac{1}{r}\frac{\partial}{\partial r}\left(\lambda r \frac{\partial T}{dr}\right) = c\frac{\partial T}{dt}. \tag{29.3}$$

- The temperature boundary conditions on the surface of the body look like:

$$-\left(\lambda \frac{\partial T}{dz} n_z + \lambda \frac{\partial T}{dr} n_r\right) = \alpha_s (T - T_C). \tag{29.4}$$

  where T—temperature, $T_C$is the ambient temperature, λ is the coefficient of heat conduction, α_s is the coefficient of heat exchange.

Appropriated initial conditions to the Eqs. (29.2) and (29.3) have to be added.

The solid fuel is assumed to be a linearly viscoelastic isotropic thermally simple material. For such materials, the relationship between the deviator and spherical components of the stress and strain tensors is accepted in the form [4]:

$$\begin{aligned}
s_{ij}(x_i, t) &= 2\int_{-\infty}^{t} G(\xi - \xi')(\partial e_{ij}(x_i, \tau)/\partial \tau d\tau,\\
\sigma_{kk}(x_i, t) &= 3\int_{-\infty}^{t} K(\xi - \xi')\left[\partial \varepsilon_{kk}(x_i, \tau)/\partial \tau - \partial \varepsilon_{kk}^{T}(x_i, \tau)/\partial \tau\right] d\tau,
\end{aligned} \tag{29.5}$$

where the so-called reduced times ξ and ξ' are defined as follows:

$$\xi = \int_0^t \frac{d\eta}{a_T(T(x_i,\eta))}, \quad \xi' = \int_0^\tau \frac{d\eta}{a_T(T(x_i,\eta))}. \tag{29.6}$$

Here $a_T(T)$ is the function defining the horizontal shift of the isothermal relaxation function along the scale $\log t$ for different temperatures. According to Williams–Landel–Ferry, we have

$$\log a_T = \frac{-C_1(T - T_{\text{ref}})}{C_2 + (T - T_{\text{ref}})}, \tag{29.7}$$

where T_{ref} is reference temperature, C_1 and C_2 are material coefficients.

The materials of the outer insulation and shell layers (indices q and c) are assumed to be isotropic linearly elastic. For them, Hooke's law, written for the deviator and spherical components of the stress and strain tensors, reads

$$s_{ij} = 2G_\beta e_{ij}, \quad \sigma_{kk} = 3K_\beta\left(\varepsilon_{kk} - \varepsilon_{kk}^T\right),$$

where G_β and K_β are the shear and bulk moduli, respectively; ε_{kk}^T is the thermal strain, which is determined by the equation $\varepsilon_{ij}^T = \alpha_\beta(T - T_0)\delta_{ij}$, where α_β is the linear thermal expansion coefficient, $\beta = q, c$.

Thermal and mechanical boundary conditions are formulated as follows:

- On the surface between the layers k and $k+1$, conditions of ideal thermomechanical contact take place

$$\begin{aligned} &T^{(k)} = T^{(k+1)}, \qquad q_n^{(k)} = q_n^{(k+1)}, \\ &\sigma_{ij}^{(k)} n_j = \sigma_{ij}^{(k+1)} n_j, \; u_i^{(k)} = u_i^{(k+1)} \quad (i, j = r, z, \varphi). \end{aligned} \tag{29.8}$$

- On the internal or external surface of the cylinder s_i or s_e, the pressure and temperature are given as

$$\begin{aligned} \sigma_{ij} n_i n_j &= \begin{cases} l - p_0 t / t_\sigma, & 0 < t < t_\sigma \\ -p_0, & t \geqslant t_\sigma \end{cases} \\ T &= T_1, \end{aligned} \tag{29.9}$$

  where t_σ is the rise time of loading, T_1 is the surface temperature, and p_0 is the pressure. Note that tangential stresses are absent.
- The remaining surfaces s_l, excepting the lower end, are free from load and exchange heat with the environment according to Newton's law (29.4).
- At the lower end of the cylinder $z = 0$, the conditions of smooth support and similar conditions of heat exchange are given

$$\sigma_{rz} = 0, \quad u_z = 0, \quad \lambda \partial T / \partial z = -\alpha_z (T - T_C),$$

  where α_z is the heat transfer coefficient at the lower end.

- The formulation of the problem also includes the initial conditions

$$\dot{u}_r = \dot{u}_z = 0, \quad u_r = u_z = 0, \quad T = T_0 \ \text{at} \ t = 0,$$

where T_0 is the initial temperature, $T_C = T_0$.

To simplify the problem, some assumptions are made:

1. The volume deformation of fuel is assumed to be linearly elastic

$$K = \text{const}, \tag{29.10}$$

then the second equation (29.5) is written as in elasticity theory:

$$\sigma_{kk} = 3K(\varepsilon_{kk} - \varepsilon_{kk}^T), \tag{29.11}$$

where $\varepsilon_{ij}^T = \alpha_p(T - T_0)$, α_p is the coefficient of linear thermal expansion of the fuel.

2. A simplified viscoelastic model is used.
Following Schapery [5], we take the fuel relaxation function E_{rel} in the form of a Prony series:

$$E_{\text{rel}}(t) = E_\infty + \sum_{k=1}^{n} A_k \exp\left(-\frac{t}{2a_T \tau_k}\right), \tag{29.12}$$

where E_∞ is equilibrium modulus, τ_k are relaxation times, A_k are some constants. The Laplace transform for (29.12) gives the so-called operational modulus:

$$E(s) = s\bar{E}_{\text{rel}}(s) = E_\infty + \sum_{k=1}^{n} \frac{sA_k}{s + \dfrac{1}{2a_T \tau_k}}. \tag{29.13}$$

Using (29.10), we obtain the expressions for the Poisson's operational coefficients and the shear modulus

$$\nu(s) = \frac{1}{2}\left[1 - \frac{E(s)}{3K}\right], \quad G(s) = \frac{3KE(s)}{9K - E(s)}. \tag{29.14}$$

For any time moment, substituting the value $s = 1/2t$ into Eqs. (29.13) and (29.14), we obtain $G(t)$, $\nu(t)$, and $E(t)$, which can be used in the associated elastic solution. It can be considered as an approximate time-dependent solution to the viscoelasticity problem in the Schapery sense.

29.4 Material Characteristics and Loading Parameters

In expressions (29.13), the numerical values of τ_k and A_k are taken in [6] with $n = 16$. For the equilibrium E_∞, instantaneous $E(0)$ (according to (29.12)), and reference E_r (used in elastic calculations) values of the modulus E, as well as the bulk modulus K, we assume the values

$$E_\infty = 2\,\text{MPa}, \quad E(0) = 159.6\,\text{MPa}, \quad E_r = 5\,\text{MPa}, \quad K = 3530\,\text{MPa}. \qquad (29.15)$$

For the thermal values, we have [7]

$$\alpha = 10^{-4}\,1/^\circ\text{C}, \quad \lambda = 0.15\,\text{W/m}^\circ\text{C}, \quad c = 1.98 \cdot 10^6\,\text{J/m}^{3\circ}\text{C}. \qquad (29.16)$$

The parameters in the temperature shift coefficient a_T (29.7) are the following:

$$C_1 = 6, \quad C_2 = 157, \quad T_r = 193^\circ\text{K}(-80^\circ\text{C}). \qquad (29.17)$$

The insulation is assumed to be an elastic material with the following values of physical and mechanical characteristics

$$\begin{aligned} &E = 5.0\,\text{MPa}, \quad \nu = 0.4999, \quad \alpha = 3 \cdot 10^{-4}\,1/^\circ\text{C}, \\ &\lambda = 0.2\,\text{W/m}^\circ\text{C}, \quad c = 6.46 \cdot 10^6\,\text{J/m}^{3\circ}\text{C}. \end{aligned} \qquad (29.18)$$

The shell is assumed to be made of isotropic elastic material with the following physical and mechanical characteristics [8]:

$$\begin{aligned} &E = 38.29\,\text{GPa}, \quad \nu = 0.05, \quad \alpha = 2 \cdot 10^{-6}\,1/^\circ\text{C}, \\ &\lambda = 2.1\,\text{W/m}^\circ\text{C}, \quad c = 1.8 \cdot 10^6\,\text{J/m}^{3\circ}\text{C}. \end{aligned} \qquad (29.19)$$

The finite-element technique developed in [9] and adapted to the dynamic problems of inelastic bodies in work [10] is used to solve the problems of mechanics and heat conduction.

29.5 Test Problem About Stepwise Loading of a Cylinder with a Shell by External Pressure

As a test problem, we consider the problem of an elastic cylinder connected to a shell under stepwise pressure changes on the surface of the shell. For such a problem, an exact analytical solution has been obtained in [11], assuming the conditions of planar deformation and incompressibility of the cylinder material. The cross section of the cylinder is shown in Fig. 29.2a.

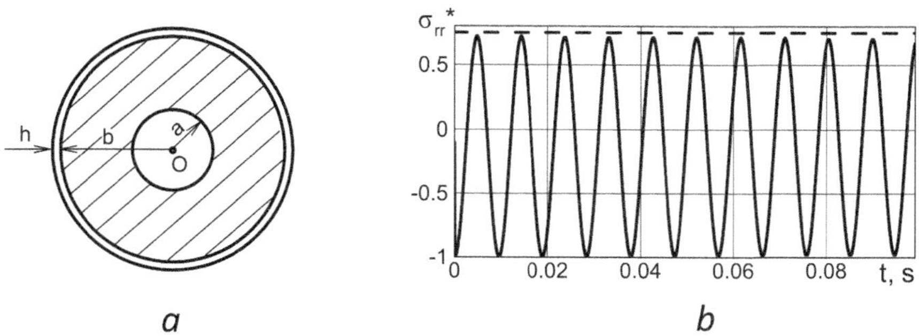

Fig. 29.2 Test example, **a** axial cross-section of the cylinder; **b** time dependences σ_{rr}^* in the fuel in contact with the shell

The cylinder loading is given by the following boundary conditions:

$$\begin{aligned} &\sigma_{rr} = -p_0 h(t), \; \sigma_{rz} = 0, \; r = b + h, \\ &\sigma_{rr} = 0, \qquad\qquad \sigma_{rz} = 0, \; r = a, \end{aligned} \tag{29.20}$$

where $h(t)$ is the Heaviside function. The plane deformation hypothesis accepted in [11] for the axisymmetric problem is modeled by the conditions at the ends of the cylinder $0 < z < l$

$$\sigma_{rz} = 0, \quad u_z = 0 \quad \text{at} \quad z = 0, l. \tag{29.21}$$

The calculations were performed at the following values of dimensionless parameters:

$$\rho_c/\rho_p = 0.2, \quad b/a = 3, \quad h/b = 0.01, \quad E_c/\left(1 - \nu_c^2\right)/G = 10^4,$$

where ρ_c and ρ_p are densities of the shell and cylinder; E_c and ν_c are Young's module and Poisson's ratio of the shell, respectively; G is fuel shear modulus, $K = 3.53 \cdot 10^3$ MPa.

During the finite-element calculation, the load was set by the smoothed pressure step on the external surface of the shell

$$\sigma_{rr} = \begin{cases} -\dfrac{p_0 t}{t_\sigma}, & 0 < t < t_\sigma \\ -p_0 & t \geq t_\sigma \end{cases} \quad r = b + h \quad \text{at} \quad t_\sigma = 10^{-4}\,\text{s}. \tag{29.22}$$

The calculation results are shown in Fig. 29.2b. Here and below σ_{ij}^* is normalized stresses, $\sigma_{ij}^* = \sigma_{ij}/p_0$. The solid line is the result of the calculation of the contact radial stresses according to the proposed methodology, and the dashed line corresponds to the amplitudes obtained on the analytical solution [11]. The difference in the dimensionless maximum tensile stresses does not exceed 1%.

29.6 Problem of Pulse Loading of RESF by External Pressure

We consider the problem of a rapid change in external pressure for the demonstration configuration shown in Fig. 29.1. The normal loading conditions correspond to (29.9) with the parameters $t_\sigma = 10^{-4}$ s and $p_0 = 8$ MPa. The convergence on the time step of integration by the Newmark method with 2% accuracy is achieved at $\Delta t = 10^{-6}$ s. The results of the calculation of the stresses at point 1 of the fuel–shell contact and at point 2 of the shell's median surface are shown in Fig. 29.3a, b, respectively. The dashed lines correspond to the static problem.

All components at point 1 σ_{rr}, σ_{zz}, and $\sigma_{\varphi\varphi}$ (Fig. 29.3a) are equal to each other, which indicates the hydrostatic stress state. As in the test example, tensile radial stresses occur on the contact surface at the external pressure, reaching the value of the applied pressure p_0. Such stresses can cause the breaking of the fuel from the shell. Preliminary compressive stresses should be used to compensate for them.

The stress state at point 2 of the shell (Fig. 29.3b) is characterized by significant compressive circumferential stresses $\sigma_{\varphi\varphi}$, much smaller axial stresses σ_{zz}, and radial stresses on the contact surface not exceeding $-p_0$. In the viscoelastic problem, as in the elastic test problem, the same level of tensile radial stresses on the contact surface

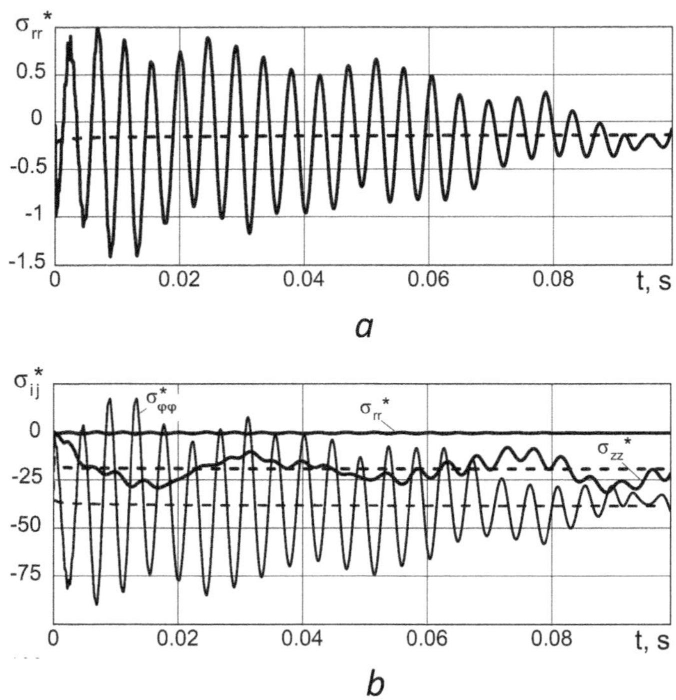

Fig. 29.3 Time dependences of stresses under external loading: **a**—at the point 1, **b**—at point 2

Fig. 29.4 Dependence of maximum stresses σ^*_{rr} at point 1 and stresses $\sigma^*_{\varphi\varphi}$ at point 2 on the external load pulse time

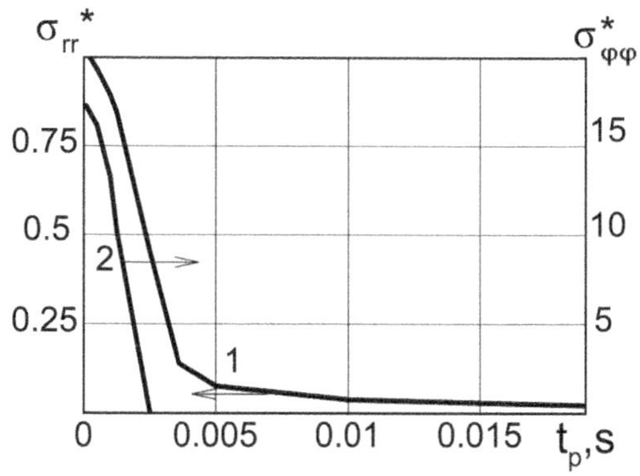

occurs at the initial moments but the oscillations are more complex and damped in time.

In the initial transition process of oscillations, tensile circumferential stresses in the shell are possible (Fig. 29.3b). Figure 29.4 shows the dependence of the maximum stress in point 1 and the stress in point 2 as a function of the pressure rise time t_p.

The higher the rate of pressure growth, the greater the possibility of breaking of the fuel from the body of the RESF, and $\max \sigma_{rr} \approx p_0$. The dependence of tensile stresses $\sigma_{\varphi\varphi}$ in the shell is similar, and the condition $\sigma_{\varphi\varphi} > 0$ is realized in the transient dynamic process at $t_p \leqslant 2 \cdot 10^{-3}$ s.

29.7 Problem of Pulse Loading of RESF by Internal Pressure

The problem of pressure loading on the internal surface for values of parameters $t_\sigma = 10^{-4}$ s, $p_0 = 8$ MPa is considered. The change in stresses at point 1 (Fig. 29.1b) is shown in Fig. 29.5a. The minimum value of the contact stress is almost equal to twice the pressure $\min \sigma_{rr} \approx -2p_0$, and for the tensile, we have $\max \sigma_{rr} \approx 0.37p_0$. As in the problem of the external impulse, the condition

$$\sigma_{rr} \approx \sigma_{zz} \approx \sigma_{\varphi\varphi}$$

is satisfied at point. In this case, the tensile stresses are reached during the first 3 cycles of oscillation. There is a damping of the waves in time $t > 0.1$ s due to energy dissipation in the viscoelastic material of the fuel. As a result, the stresses approach the static solution (dashed lines).

Note that in the problem [11] for a circular cylinder under internal loading, the stepwise pressure increase does not lead to tensile stresses on the contact surface. The presence of tensile stresses in the RESF is probably due to the influence of the uppercase cover, which leads to a violation of the plane deformation condition adopted in [11].

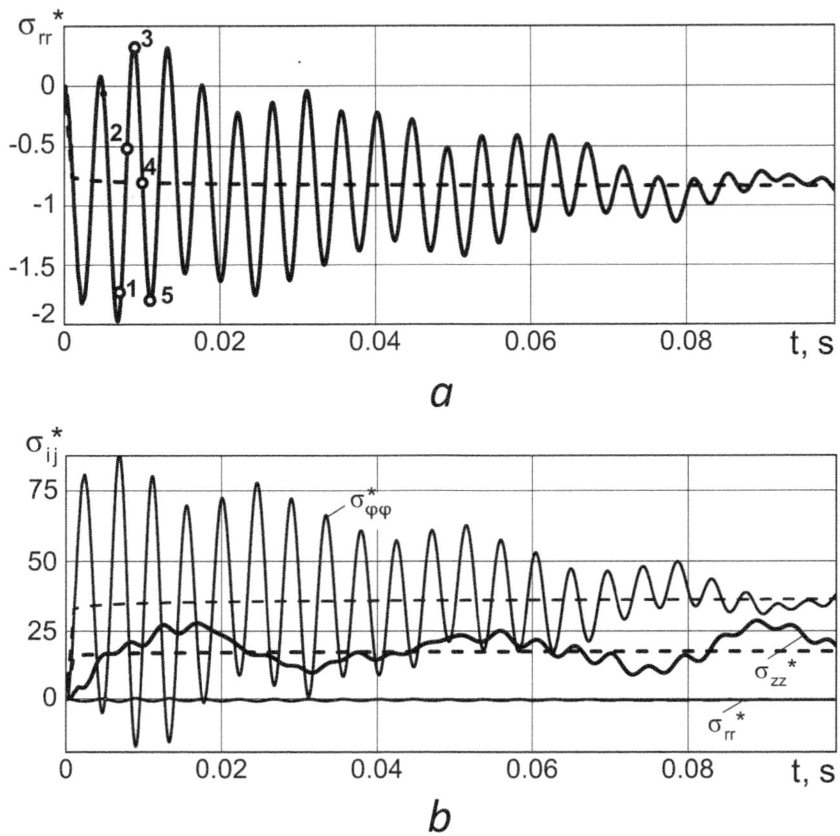

Fig. 29.5 Time dependence of the stress components at point 1 (fuel–shell boundary) and at the point 2 (shell)

Figure 29.5b shows a picture of the damping of vibrations at point 2 located in the shell. The circumferential stresses $\sigma_{\varphi\varphi}$ significantly dominate the components σ_{rr} and σ_{zz}, providing resistance of the cylinder to internal pressure. Due to the compressibility of the fuel material ($\nu \neq 0.5$, $K \neq \infty$), a transient process takes place during pulse loading, the essence of which consists of the propagation of the compression wave front from the loading surface. In the problem considered in [11], due to the hypothesis of incompressibility of the inner cylinder material, there is no transient process, and the oscillation mode is established immediately.

The propagation of the compression wave from the internal surface to the external one and back is shown in Fig. 29.6a (incident wave) and b (reflected wave). The numbers indicate points in time. The arrows show the direction of wave propagation. In the area occupied by the fuel $\sigma_{rr}(r) \approx \sigma_{\varphi\varphi}(r)$.

Figure 29.7 shows the stress distribution $\sigma_{rr}(r)$ in the middle axial section of the cylinder $z = 1.55$ m, the curve numbers correspond to the points in Fig. 29.5*a*:

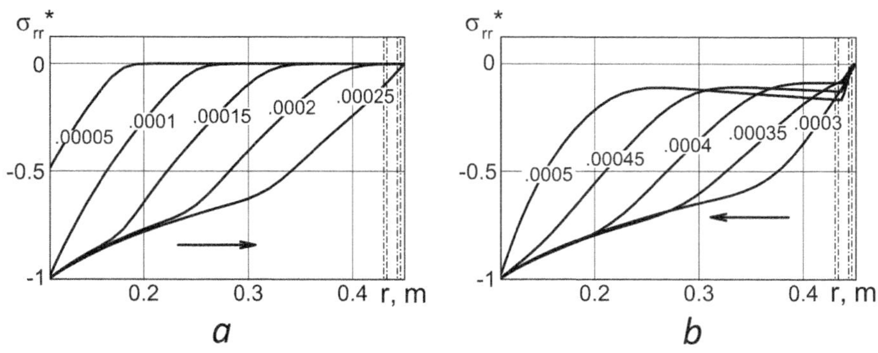

Fig. 29.6 Radial stress σ_{rr}^* of incident (**a**) and reflected (**b**) waves in the average cross-section $z = 1.55$ m at the time moments indicated by the numbers

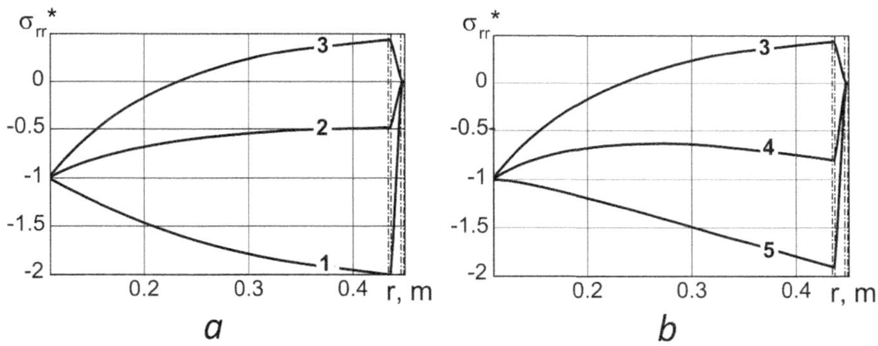

Fig. 29.7 Radial stress σ_{rr}^* distributions in the middle section at the time points shown in Fig. 29.5*a*. The dashed dotted lines correspond to the boundaries of the layers

$$1 - t = 0.07 \text{ s}, 2 - t = 0.08 \text{ s}, 3 - t = 0.09 \text{ s}, 4 - t = 0.10 \text{ s}, 5 - t = 0.11 \text{ s}.$$

According to the data in Fig. 29.5a, b, the most dangerous are stresses σ_{rr} at the fuel–shell contact and circumferential stresses $\sigma_{\varphi\varphi}$ in the shell. It is of interest to establish the dependence of the maximum values of the indicated tensile stresses on the thickness h of the shell. The calculation results are shown in Fig. 29.8.

Note that σ_{rr}^* and $\sigma_{\varphi\varphi}^*$ values noticeably decrease with increasing thickness of the body, which is composed of cylindrical shell and cover (dashed lines). If, however, only the thickness of the cylindrical part of the casing is varied (solid lines), the dependence $\sigma_{rr}^*(h)$ becomes weakly increasing, and the dependence $\sigma_{\varphi\varphi}^*(h)$ remains practically unchanged. Thus, the construction of the top cover can have a noticeable qualitative and quantitative effect on the values of σ_{rr}^*. The results in Fig. 29.8 allow selecting a thickness h depending on p_0, which provides the necessary safety margin under dynamic loading of the charge under pulsed internal pressure, as well as

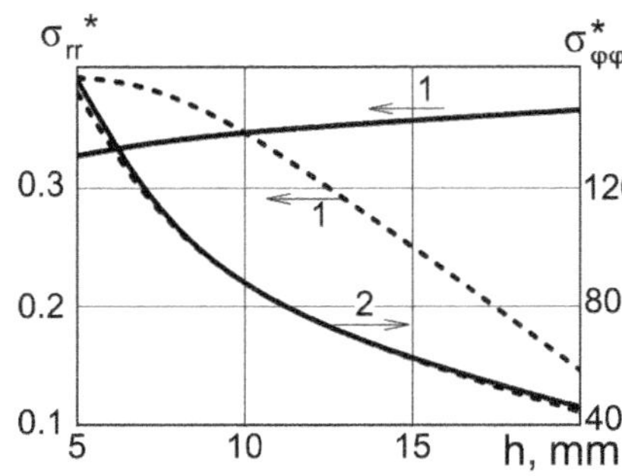

Fig. 29.8 Dependence of maximum stresses σ_{rr}^* and $\sigma_{\varphi\varphi}^*$ on the thickness of the shell. Solid lines denote the thickness of the cover is 10 mm, and the thickness of the side shell varies. Dashed lines denote the thickness of the whole shell (including the cover) varies. 1, 2 are numbers of points

ensuring the strength of the connection at the fuel–shell contact. In the viscoelastic problem, as in the test problem, the level of tensile radial stresses on the contact surface is maintained, but the oscillations are more complex and damped in time.

29.8 Evaluation of Thermal Effects

According to the data presented above, the characteristic damping time t_f of oscillations in the cylinder is $t_f \approx 0.1$ s. Let us estimate the depth of heating on the internal surface of the cylinder from the fuel during this time. For this purpose, we consider the heat conduction problem for a half-space $x > 0$.

$$\begin{aligned} &\frac{\partial T}{\partial t} = \kappa \frac{\partial^2 T}{\partial x^2}, \\ &T = T_1 \quad \text{at} \quad t > 0, \quad x = 0, \\ &T = T_0 \quad \text{at} \quad t = 0, \quad 0 < x < \infty, \end{aligned}$$

where $\kappa = k/c_V$ is the fuel thermal diffusivity.

The solution to the problem is as follows [12]:

$$T = T_1 \Phi^*(x/2\sqrt{\kappa t}) \tag{29.23}$$

with

$$\Phi^*(z) = \frac{2}{\pi} \int_z^\infty e^{-\xi^2} d\xi.$$

For estimation, consider the depth Δx, to which the material will warm up $0.01 T_1$ in a time $t = 0.1$ s. Taking into account (29.23), the required value is determined from equation

$$\frac{T_1 - T}{T_1} = 1 - \Phi^*(x/2\sqrt{\kappa t}) = 0.01 \quad \text{at} \quad t = 0.1\,\text{s}.$$

The calculation gives $\Delta x = 3 \cdot 10^{-4}$ m. For the cylinder in consideration, the thickness of the fuel wall $\Delta R = 0.325$ m , i.e., $\Delta x/\Delta R = 2.7 \cdot 10^{-5} = 2.7 \cdot 10^{-3}\%$. The above data show that in the dynamic stress damping time interval, the heating more $20°C$ occurs in an extremely thin layer of fuel, which is $3 \cdot 10^{-3}\%$ its thickness. Consequently, the effect of fuel heating on the dynamic stresses in the fuel can be neglected.

29.9 Conclusions

The calculation of dynamic stresses in the RESF at pulse pressure on the internal or external surfaces has been carried out:

- The possibility of tensile radial stresses on the fuel–shell contact surface at the initial stage of the process was established for the internal pressure pulse.
- The influence of the shell thickness on the maximum breaking contact stresses at the fuel–shell interface, as well as on the maximum circumferential tensile and compressive stresses in the shell, has been investigated.
- The presence of significant tensile radial stresses at the surface of the fuel–shell contact is shown during the external pressure pulse. Insignificant tensile circumferential stresses are possible in the shell during the transient oscillatory process.
- The thermal effects of heat conduction in parabolic statements are evaluated. In the dynamic stress damping time interval, the heating occurs in an extremely thin layer of fuel. Therefore, its effect on the dynamic stresses in the fuel can be neglected.

References

1. Moskvitin, V.V.: Resistance of Viscoelastic Materials. In: Relation to the Charges of Rocket Engines on Solid Fuel. Nauka, Moscow (1972) [in Russian]
2. Sinyukov, A.M., Volkov, P.I., Lvov, A.I., Mishkovich, A.M.: Solid Propellant Ballistic Missile. Military Publishing House, Moscow (1972). [in Russian]
3. Fakhrutdinov, IKh., Kotelnikov, A.V.: Structures and Design of Solid Propellant Rocket Engines. Mashinostroenie, Moscow (1987). [in Russian]
4. Christensen, R.M.: Theory of Viscoelasticity: An Introduction. Academic (1971)
5. Schapery, R.A.: Approximate methods of transform inversion of viscoelastic stress analysis. Proc. U.S. Nath. Congr. Appl. Mech. **2**, 1075–1085 (1962)
6. Renganahan, K., Nageswara Rao, B., Jana, M.K.: Slump estimation of cylindrical segment grains of a typical rocket motor under vertical storage conditions. Trends Tech. Sci. Res. **1**, 97–104 (2006)
7. Marimuthu, R., Nageswara Rao, B.: Development of efficient finite elements for structural integrity analysis of solid rocket motor propellant grains. Int. J. Press. Vessels Pip. **111–112**, 131–145 (2013)
8. Jayakumar, K., Yadav, D.: Nageswara Rao: A multi-layer cylindrical shell under electro-thermo-mechanical loads. Trends Tech. Sci. Res. **1**, 386–401 (2006)

9. Motovilovets, I.A., Kozlov, V.I.: Thermoelasticity. Mechanics of coupled fields in materials and structural elements. Vol. 1. Nauk. Dumka, Kyiv (1987) [in Russian]
10. Zhuk, Y.A., Senchenkov, I.K., Kozlov, V.I., Tabieva, G.A.: Axisymmetric dynamic problem of coupled thermoviscoplasticity. Int. Appl. Mech. **37**, 1311–1317 (2001)
11. Achenbach, J.D.: Dynamic Response of a Long Case-Bonded Viscoelastic cylinder. AIAA J. **3**, (1965)
12. Carslaw, H.S., Jaeger, J.C.: Conduction of Heat in Solids, 2nd edn. Oxford University Press (1986)

Chapter 30
Dynamics of Shell with Structural Features

Yurii Skosarenko

Abstract A technique is described for studying the stress–strain state of a conical shell reinforced with longitudinal ribs and containing concentrated masses under short-term axisymmetric pressure, the front of which moves from the smaller base of the shell to the larger one. The effect of ribs, masses, and the position of the wave front on deflections and bending moments in the shell is studied.

30.1 Introduction

As structural components, shells find applications across various technological fields, including aircraft construction, rocket science, and energy-related applications. Real-world shells invariably incorporate various technological and structural elements. These elements come in the form of one-dimensional components and attached bodies, exhibiting properties such as variable stiffness, multilayer structures, interaction with the external environment, and susceptibility to temperature variations. The problems surrounding the statics and dynamics of shells featuring diverse types of singularities have been the focus of extensive study by numerous scientists. For instance, an extensive presentation of research outcomes is given in books [1, 2], which delve into calculating stress and strain within shells while considering the discrete arrangement of one-dimensional elements. These publications provide a comprehensive list of references and analyze works published since the 1950s. Notably, papers like [6, 11, 12] delineate the methodologies developed by the authors and present the findings from numerous investigations, including examining shells with geometric and physical inhomogeneities. In the contemporary research landscape, researchers are growing interested in problems concerning the dynamics of plates, shells, and shell structures. These studies address technological and geometric features pertinent to various scientific and technical domains [3–5, 8–10, 13–15, 18].

Y. Skosarenko (✉)
S.P. Timoshenko Institute of Mechanics, National Academy of Science of Ukraine, Kyiv, Ukraine
e-mail: yuskos@ukr.net

H. Altenbach et al. (eds.), *Selected Problems of Solid Mechanics and Solving Methods*, Advanced Structured Materials 204, https://doi.org/10.1007/978-3-031-54063-9_30

30.2 Research Methodology

Below we describe a method for studying the dynamics of cylindrical and truncated conical multilayer shells with structurally orthotropic layers supported by longitudinal and annular ribs with attached bodies. The shell interacts with the elastic base on a part or its entire surface.

The stress–strain state of the shell and reinforcing ribs is determined within the framework of the theory that takes into account transverse shear deformations (based on the beam model of taking into account transverse shear deformations by S. P. Timoshenko). The equations of motion are obtained in the displacements of the points of the middle surface and the shear angles, which are the same for all layers of the shell and for the stiffeners. The relations connecting the displacements of the centers of the shell layers (ribs) with the displacements of the points of the middle surface of the shell were obtained taking into account the rigid connection of the indicated elements to each other [1, 2, 10].

Using the relationship between internal forces and deformations, as well as between deformations and displacements, expressions are obtained for the potential energy of the system, as the sum of the potential energies of elastic deformation of the shell itself, longitudinal and annular ribs, and a two-parameter elastic base, and kinetic energy, as the sum of the energies of the shell, ribs, elastic foundation, and added masses.

It is assumed that the displacements of the points of the elastic base, when exposed to it as a result of the deformation of the shell, decay as they move away from its surface. The elastic base is axially symmetric and interacts with the shell along its entire length or in some given section from ξ_{1u} to ξ_{2u}. The connection between the displacements of the points of the elastic foundation in the direction normal to the middle surface of the shell w_f is given by the relation:

$$w_f = r(r+z)^{-1}\left(C_{1uo}w + C_{2uo}\frac{\partial w}{\partial \xi}\right), \tag{30.1}$$

where C_{1uo} and C_{2uo} are the dimensionless stiffness coefficients of the elastic foundation.

As can be seen from (30.1), the elastic base prevents normal displacements of the shell, as well as rotations around the ring coordinate. After substituting Eq. 30.1 into the expressions for the energies of the elastic foundation and integrating over z, we obtain expressions for the potential and kinetic energies of the elastic foundation in the displacements of the points of the middle surface of the shell.

Note that for shells of a cylindrical shape, within the framework of this technique, models of shell structures with plates embedded in various ways are considered [16].

Thus, it is accepted that the total potential energy of a given mechanical system in a deflected state in the displacements of the points of the middle surface of the shell and the shear angles is represented by the sum of the potential energies of the shell, longitudinal and annular ribs, the elastic base, and the work of external forces, and the kinetic energy is the sum of the kinetic energies of the indicated elements and added masses.

Based on the Hamilton principle, using expressions for energies, taking into account the independence and arbitrariness of displacement variations, an inhomogeneous system of equations of motion of a shell with the structural elements listed above and natural boundary conditions at its ends are obtained. Due to the cumbersomeness, this system of equations is not given in the article.

To find the solution of the obtained equations of motion, the desired displacements of the points of the middle surface of the shell are approximated by trigonometric series along the annular coordinate and by series using beta-splines of the third degree along the longitudinal coordinate:

$$\begin{aligned} u &= \sum_{m=-1}^{M+1} \sum_{n=0}^{N} \left[u_{1mn}^{1}(t) \cos n\theta + u_{1mn}^{2}(t) \sin n\theta\right] B_3^m(\xi) \\ v &= \sum_{m=-1}^{M+1} \sum_{n=0}^{N} \left[u_{2mn}^{1}(t) \cos n\theta + u_{2mn}^{2}(t) \sin n\theta\right] B_3^m(\xi) \\ w &= \sum_{m=-1}^{M+1} \sum_{n=0}^{N} \left[u_{3mn}^{1}(t) \cos n\theta + u_{3mn}^{2}(t) \sin n\theta\right] B_3^m(\xi) \\ \psi_1 &= \sum_{m=-1}^{M+1} \sum_{n=0}^{N} \left[u_{4mn}^{1}(t) \cos n\theta + u_{4mn}^{2}(t) \sin n\theta\right] B_3^m(\xi) \\ \psi_2 &= \sum_{m=-1}^{M+1} \sum_{n=0}^{N} \left[u_{5mn}^{1}(t) \sin n\theta + u_{5mn}^{2}(t) \cos n\theta\right] B_3^m(\xi). \end{aligned} \tag{30.2}$$

Here $u_{1mn}^{1}(t), u_{1mn}^{2}(t), \ldots u_{5mn}^{2}(t)$ are the unknown coefficients of the series, which are functions of time; N and M are the number of terms retained in trigonometric series and the number of collocation points along the coordinate ξ; $B_3^m(\xi)$ is beta-spline of the third degree, which is built in the points ξ_m.

After substituting (30.2) into the equations of motion and integrating over the coordinate (the Bubnov–Galerkin method), we write down the system of equations for each point (the collocation method [7]), including those at the ends of the shell, satisfying the given boundary conditions. As a result, we obtain a system of generally interconnected inhomogeneous ordinary differential equations with respect to the required order $2 \times (M + 4) \times N$ functions $u_{1mn}^{1}(t), u_{1mn}^{2}(t), \ldots u_{5mn}^{2}(t)$.

Note that the solution in the form (30.2) must be used when the shell is supported by an irregular system of longitudinal ribs or has an irregular system of added masses. If the shell is supported by identical longitudinal ribs evenly spaced along the circumference (in the presence of masses, they should not violate the regularity of the mechanical system), the solution to the problem can be found by approximating the displacements independently by the first and second terms of expressions (30.2). For a shell regular in the ring coordinate, the system of ordinary differential equations, using approximation (30.2), splits into two independent systems of equations for the functions $u_{1mn}^{1}(t), u_{2mn}^{1}(t), \ldots u_{5mn}^{1}(t)$ and $u_{1mn}^{2}(t), u_{2mn}^{2}(t), \ldots u_{5mn}^{2}(t)$. We also note that for a particular case of boundary conditions, namely when the ends are hinged ($v = w = \psi_2 = 0$, as well as the longitudinal force $T_1 = 0$ and bending moment $M_1 = 0$ at $\xi = 0; l/r$), it is more rational, with computational point of view, to use the approximation of displacements in the form of double trigonometric series (30.3):

$$\begin{aligned}
u &= \sum_{m=0}^{M} \sum_{n=0}^{N} \left[u_{1mn}^{1}(t) \cos n\theta + u_{1mn}^{2}(t) \sin n\theta \right] \cos d_m \xi \\
v &= \sum_{m=1}^{M} \sum_{n=0}^{N} \left[u_{2mn}^{1}(t) \sin n\theta + u_{2mn}^{2}(t) \cos n\theta \right] \sin d_m \xi \\
w &= \sum_{m=1}^{M} \sum_{n=0}^{N} \left[u_{3mn}^{1}(t) \cos n\theta + u_{3mn}^{2}(t) \sin n\theta \right] \sin d_m \xi \\
\psi_1 &= \sum_{m=0}^{M} \sum_{n=0}^{N} \left[u_{4mn}^{1}(t) \cos n\theta + u_{4mn}^{2}(t) \sin n\theta \right] \cos d_m \xi \\
\psi_2 &= \sum_{m=1}^{M} \sum_{n=0}^{N} \left[u_{5mn}^{1}(t) \sin n\theta + u_{5mn}^{2}(t) \cos n\theta \right] \sin d_m \xi .
\end{aligned} \tag{30.3}$$

Next, we consider the problem of the stress state of a conical hinged shell with regular elements. It is also assumed that the structural elements of the shell are symmetrical with respect to the origin of the ring coordinate. In this case, approximations (30.2) and (30.3) do not take into account the second terms, and the structure of the system of inhomogeneous ordinary differential equations, after integrating the system of equations of motion over spatial coordinates with respect to the desired functions $u_{1mn}^{1}(t)$, $u_{1mn}^{2}(t)$,... $u_{5mn}^{2}(t)$, becomes much simpler.

The solution of the system of ordinary differential equations obtained as a result of transformations in spatial coordinates makes it possible to obtain results under the action of non-stationary dynamic loads on the shell. The solution is found by an approximate numerical method of finite differences, using an explicit finite-difference scheme in time for given initial conditions. In a particular case, for example, for a multilayer cylindrical shell with regular reinforcement, the layers of which are symmetrical with respect to its middle surface, an exact solution was obtained by expanding it into eigenmodes [17].

30.3 Research Results

The dynamics of a three-layer ribbed conical shell is studied under an axisymmetric short-term pressure wave, the front of which moves along the shell axis from a smaller base to a larger one. Shell dimensions: height $H = 0.25$ m, the radius of the larger base $r = 0.25$ m, and the radius of the smaller base $r_1 = 0.175$ m. The thicknesses of the outer and inner layers are the same $h^{(1)} = h^{(3)} = 0.4 \cdot 10^{-3}$ m, and the thickness of the middle layer is $h^{(2)} = 0.2 \cdot 10^{-3}$ m. The layer material is isotropic. Physical constants of outer layers: modulus of elasticity $E^{(1)} = E^{(3)} = 0.7 \cdot 10^{11}$ Pa, density $\rho^{(1)} = \rho^{(3)} = 2.7 \cdot 10^{3}$ kg/m^3, and shear modulus $G^{(1)} = G^{(3)} = 0.26 \cdot 10^{11}$ Pa. Middle layer: $E^{(2)} = 0.4 \cdot 10^{11}$ Pa, $G^{(2)} = 0.2 \cdot 10^{11}$ Pa, and $\rho^{(2)} = 1.7 \cdot 10^{3}$ kg/m^3. Poisson's ratio is the same for all layers $\nu^{(1)} = \nu^{(2)} = \nu^{(3)} = 0.3$. The shell is reinforced with eight identical, evenly spaced longitudinal ribs, the material characteristics of which are the same as those of the outer layers. The bending stiffness of the $I_{i_1\theta} = 0.827 \cdot 10^{-10}$ m^4 ribs is taken into account.

The calculations were carried out while retaining in the approximating series (30.3) seventeen terms along the annular coordinate ($n = 0, 8, \ldots 128$) and twenty-one terms along the longitudinal coordinate ($m = 1, 2, \ldots 21$), which makes it possible to obtain reliable qualitative results.

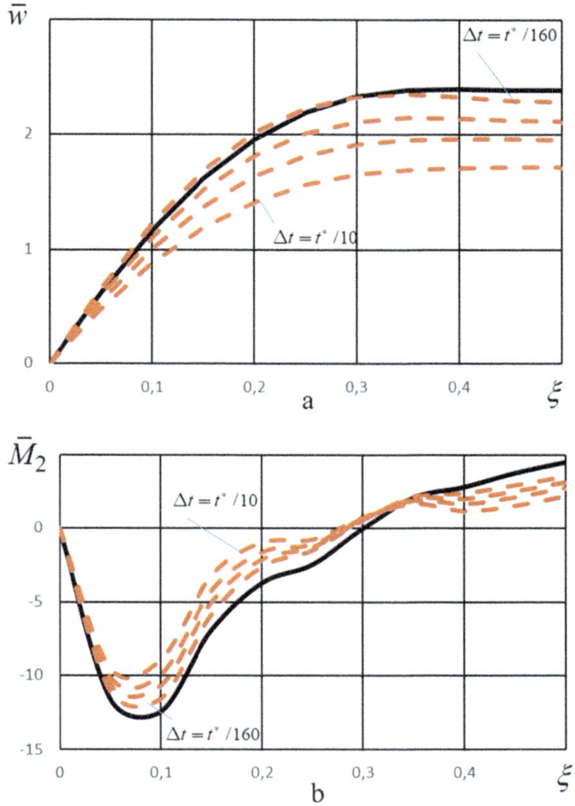

Fig. 30.1 Comparison of calculation results by different methods

At the first stage, on the example of a shell with the above parameters but of a cylindrical shape ($r_1 = r = 0.25$ m), comparative calculations of the characteristics of the stress–strain state under zero initial conditions were performed by expanding the solution in terms of natural vibrations and by the finite-difference method to determine a satisfactory time step. An axisymmetric pressure pulse acts along the entire shell length, changing according to a stepwise law. The duration of the impulse $t^* = 0.7355 \cdot 10^{-3}$ s.

The results of calculations at the current time $t = t^*$ are shown in Fig. 30.1. The dimensionless values of the deflection $\bar{w}$ (Fig. 30.1a) and the annular bending moment $\bar{M}_2$ acting on the platforms normal to the annular coordinate (Fig. 30.1b) are given in the longitudinal sections of the shell on the ribs, where the bending moment $\bar{M}_2$ takes on maximum values. The solid lines correspond to the calculation by the method of expansion in terms of the modes of natural oscillations; the dashed lines correspond to the finite difference method using an explicit difference scheme in time. As can be seen from the graphs, as the time step decreases, the dashed curves approach solid curves. The difference between the maximum values of the deflection at the line of connection of the shell with the rib at a time step $\Delta t = t^*/10$ is 18.3%, at

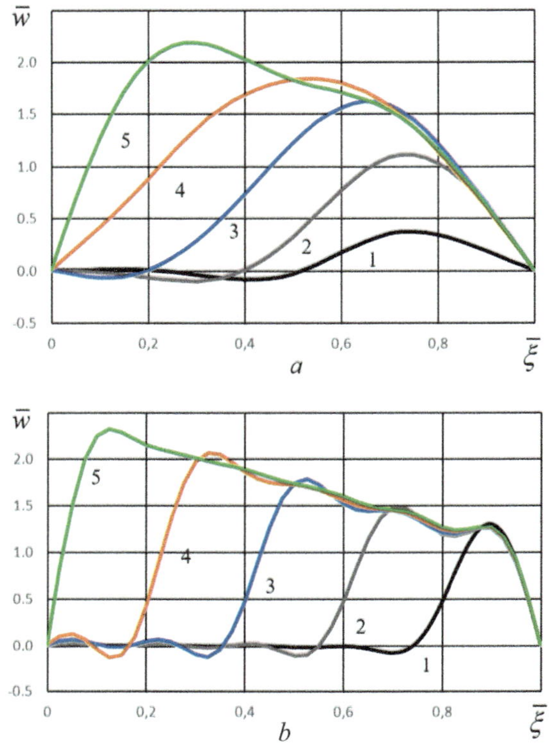

Fig. 30.2 Shell deflection with symmetrical layers

$\Delta t = t^*/20$—10.8%, at $\Delta t = t^*/40$—6.1%, at $\Delta t = t^*/160$—2.2%, between the maximum values of the annular bending moment $\bar{M}_2$, respectively, 18.4%, 13.6%, 9.9%, and 6.8%. The obtained estimate of the accuracy of the results allows us to choose the time step of the finite difference scheme.

Further calculations of the dynamic stress–strain state of the conical shell are performed with a time step $\Delta t = t^*/160$. In this case, the wave front moves from a smaller base to a larger one at a speed of 340 m/s. In subsequent calculations, the duration of the pulse is equal to the time t^* of the pressure wave front moving from the smaller base to some given section of the shell $\bar{\xi}^* = 1 - \bar{\xi}$, where $\bar{\xi} = \xi \cdot r/l$. The deformation forms along the line of contact of the ribs with the shell (Fig. 30.2*a*) and along the line located in the middle between the ribs (Fig. 30.2*b*) were obtained at the current time $t = 0.147 \cdot 10^{-3}$ s for five values of $\bar{\xi}^* = 0.2, 0.4, 0.6, 0.8, 1$, respectively, curves 1–5.

As can be seen from the graphs, the maximum deflection of the shell shifts from the smaller base to the larger one along with the movement of the wave front.

For a shell with asymmetric layers (the parameters of the outer layer are different from those given above, $E^{(3)}2.1 \cdot 10^{11}$ Pa, $G^{(3)} = 0.81 \cdot 10^{11}$ Pa, $\rho^{(3)} = 7.8 \cdot 10^3$ kg/m^3), Fig. 30.3 shows the bending moment M_2 at the line of contact of the ribs with the shell at different times at $t = 0.147 \cdot 10^{-3}$ s (Fig. 30.3a)

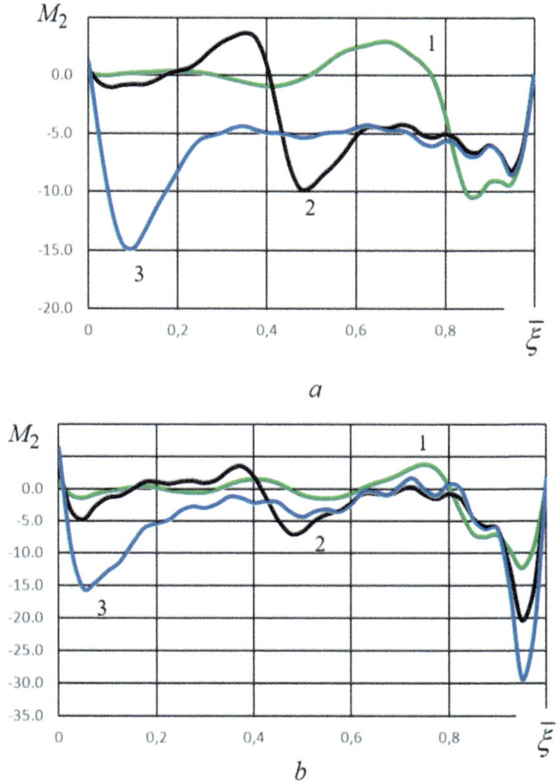

Fig. 30.3 Bending moment in a shell with asymmetrical layers

and $t = 0.735 \cdot 10^{-3}$ s (Fig. 30.3b). Curves 1–3 were obtained, respectively, at $\bar{\xi}^* = 0.2, 0.6, 1$. As can be seen from the graphs, an increase is observed M_2 at the locations of the wave front. However, its maximum value M_2 is reached in the zone of the edge effect near the smaller base (Fig. 30.3b, $\bar{\xi} \approx 0.95$).

With a symmetrical arrangement of layers, the dependences of the annular moment M_2 on $\bar{\xi}$ are similar to those shown in Fig. 30.3, but its maximum values are less than about 1.5 times.

A shell with symmetrical layers and attached masses located on each edge in the middle section of the shell is considered $\bar{\xi}_{i_m} = 0.5$. The value of each mass is $m_{i_m} = m_{\text{total}}/8$. The distance of the centers of mass from the middle surface of the shell is zero. Figure 30.4 shows the dependences of the deflection of the shell in sections along the ribs. It can be seen from the graphs that a "node" of the deformation form is formed at the location of the masses.

It should be noted that, as calculations have shown, the deflection of the shell in the middle between the ribs is close to the deflection of the shell without masses. Under a short-term external dynamic impact in some small neighborhood near the point of attachment of the masses, the normal displacements of the shell are very small compared to the maximum values.

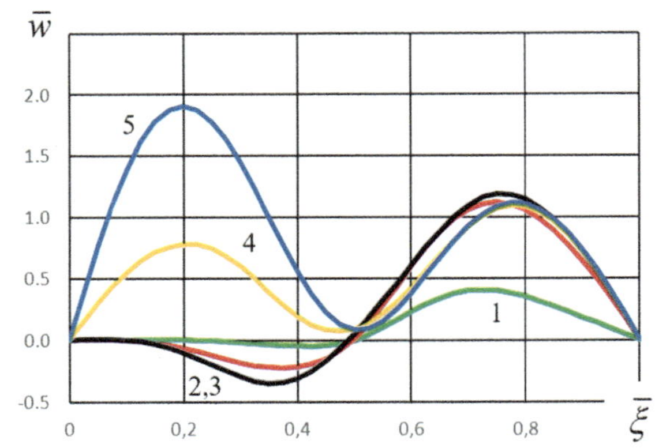

Fig. 30.4 Shell deflection with masses

30.4 Conclusions

Thus, a technique has been developed for determining the stress–strain state of ribbed layered conical shells with asymmetric layers interacting with masses and an elastic foundation under the action of dynamic loads. The equations of shell motion in the displacements of points of the middle surface were obtained in the framework of the linear shear theory of shells and ribs, based on the hypothesis of S. P. Timoshenko for beams. To solve the equations of motion in the displacements of the points of the middle surface of the shell, we used the approximation of the required displacements by trigonometric series or power series based on beta-splines, and an explicit finite-difference scheme in time.

On the example of a ribbed layered cylindrical shell, a comparison is made of the results obtained using an explicit finite-difference scheme in time and an exact solution of the problem by the method of its expansion in terms of eigenmodes. An estimate of the accuracy of calculations by a numerical method is obtained depending on the number of steps along the time coordinate.

The stress–strain state of a conical shell with symmetric and asymmetric layers, longitudinal ribs, and masses was studied under the action of a short-term axisymmetric external load distributed over parts of the shell, the front of which moves from the smaller base of the shell to the larger one. It has been found that the maximum deflection of the shell shifts from the smaller base to the larger one in accordance with the movement of the pressure wave.

For a shell with masses, the change in the deflection along the ribs, in the middle section of which the masses are located, differs significantly from the case of a shell without masses. With a short-term external dynamic action at the locations of the masses, the normal displacements of the shell are greatly reduced.

In shells with both symmetric and asymmetric layers, the annular bending moment on the ribs increases at the locations of the wave front. This moment reaches its maximum value for some values of the flow time in the zone of the edge effect at the smaller base.

References

1. Amiro I.Ya., Zarutsky V.A., Polyakov P.S.: Ribbed cylindrical shells. - Kyiv: Nauk. Dumka (1973)
2. Amiro I.Ya., Zarutsky V.A.: Theory of ribbed shells. Nauk. Dumka, Kyiv (1980). (Methods for calculating shells: In 5 volumes; Vol. 2)
3. Anik'ev I.I., Mikhailova M.I., Sushchenko E.A.: Dynamic loading of cylindrical and spherical bidies interacting with a shock wave. Int. Appl. Mech. **40**(12), 1405–1410 (2004)
4. Banerjee, J.R., Cheung, C.W., Morishima, R., Perera, M., Njuguna, J.: Free vibration of a three layered sandwich beam using the dynamic stiffness method and experiment. Int. J. of Solids Struct. **44**(22), 7543– 7563 (2004)
5. Davar, A., Azarafza, R., Fayez, M.S., Fallahi, S., Jam, J.E.: Dynamic response of a grid-stiffened composite cylindrical shell reinforced with carbon nanotubes to a radial impulse load. Mech. Compos. Mater. **57**, 181–204 (2021)
6. Grigorenko Ya.M., Vasilenko A.T.: Theory of shells with variable stiffness. Kyiv: Nauk. Dumka (1981). 544 p. (Methods for calculating shells: In 5 volumes; V.4)
7. Grigorenko Ya.M., Shevchenko Yu.N.: Numerical methods. K.: A.S.K. 448 p. (2002) (Mechanics of composites: in 12 volumes. Vol. 11)
8. Harvey, P.S., Jr., Virgin, L.N.: Effect of stiffener geometry on the response grid-stiffened panels. J. Eng. Mech. **144**(2), 06017021 (2017)
9. Lugovoi, P.Z., Meish, V.F., Meish, Yu.A., Orlenko, S.P.: On the calculation of the dynamics of composite shell structures of rotation under non-stationary loads. Int. Appl. Mech. **56**(1), 22 – 32 (2020)
10. Lugovoi, P.Z., Skosarenko, Yu.V., Orlenko, S.P., Shugailo, A.P.: Application of the spline-collocation method to solve problems of statics and dynamics for multilayer cylindrical shells with design and manufacturing features. Int. Appl. Mech. **55**(5), 524–533 (2019)
11. Guzya, A.N. (ed.) : Mechanics of composite materials and structural elements: Under the general, 3 volumes. Nauk. Dumka, Kyiv (1982–1983). Grigorenko, Ya.M.: 2 Mechanics of structural elements, 464 p. (1983)
12. Mechanics of composites: In 12 vols. V.8. Statics of structural elements / Ya. M. Grigorenko, A. T. Vasilenko, I. G. Emelyanov, N. N. Kryukov, Yu. N. Nemish, N.D. Pankratova, B. L. Pelekh, G. G. Vlaikov, A. V. Maksimuk, G. P. Urusova; Ed.: A. N. Guz; NAS of Ukraine. Institute of Mechanics. - K.: Nauk. Dumka, 1999
13. Photiadis, D.M., Houston, B.H., Williams, E.G., Bucaro, J.A.: Resonant response of complex shell structures. J. Acoust. Soc. Am. **108**, 1027–1035 (2000)
14. Qatu, M.S.: Vibration of Laminated Shells and Plates. Academic, New York (2004)
15. Qatu M.S., Asadi E., Wang W.: Review of recent literature on static analyses of composite shells: 2000–2100. Open J. Composit. Mater. **2**(3), 61–86 (2012)
16. Skosarenko Yu.V.: Natural vibrations of a cylindrical shell reinforced by rectangular plates. Int. Appl. Mech. **42**(3), 325–330 (2006)
17. Timoshenko S.P., Yang D.Kh., Weaver U.: Fluctuations in Engineering. Mashinostroenie, Moscow (1985)
18. Tran-Van-Nhieu, M.: Scattering from a ribbed finite cylindrical shell. J. Acoust. Soc. Am. **110**, 2858–2866 (2001)

Chapter 31
Stationary Temperature Fields in Radially Inhomogeneous Hollow Cylinders

Yuriy Tokovyy

Abstract A technique for solving two-dimensional steady-state heat-conduction problems for inhomogeneous hollow cylinders is discussed when the temperatures on both inner and outer surfaces exhibit arbitrary variation within the circumferential coordinate. Using the direct integration method, the original boundary value problems are reduced to an integral equation of the second kind. This allows for addressing arbitrary variation profiles of the thermophysical properties within the radial coordinate and deriving a solution in the form of explicit dependences on the applied thermal loadings. A modification of the solution technique allows for constructing a similar solution in the case of multilayer cylinders with perfectly contacting layers whose properties are arbitrary functions of the radial coordinate.

31.1 Introduction

The thermal impact is one of the most significant reasons for changing the internal stress pattern in structural members at operation, fabrication, or treatment [7]. This impact is crucial for composite and inhomogeneous structures where the thermal stresses arise even under uniform thermal fields due to the mismatch in the thermal expansion of the constituent materials within the composition [10, 22]. Thus, a new trend in material science and engineering is the use of continuously inhomogeneous materials (having nearly continuous distribution profiles of their material properties within the spatial coordinates at the macroscale), e.g., functionally graded materials (FGM), allowing for minimization of the residual stresses occurred otherwise on the interfaces of the dissimilar materials [3, 5, 8, 23, 25]. Considering the material properties' variability makes the relevant boundary value problems of heat transport and mechanics extremely difficult for accurate analysis [16–18, 26]. Furthermore, the vast majority of the methods for the analysis of continuously inhomogeneous or FGM solids are developed for the material profiles assumed to be monotonic

Y. Tokovyy (✉)
Pidstryhach Institute for Applied Problems of Mechanics and Mathematics, National Academy of Science of Ukraine, Lviv, Ukraine
e-mail: tokovyy@iapmm.lviv.ua

H. Altenbach et al. (eds.), *Selected Problems of Solid Mechanics and Solving Methods*, Advanced Structured Materials 204, https://doi.org/10.1007/978-3-031-54063-9_31

functions “from one surface to another” [22]. Such structures, however, are widely used as an intermediate layer between contrasting materials or within an assembly of layers forming a protective coating, etc. [6, 24]. In the latter case, the size effect (i.e., the terrific difference between the layer-base thicknesses) can be also a significant challenge for the analysis efficiency. Hence, the adequate analysis of their thermomechanical performance necessitates the development of efficient methods for multilayer structures under the assumption of all or some layers having arbitrary variation profiles with simultaneous accounting of possible mismatch of the layer thicknesses.

A dominant strategy in analyzing multilayer structures rests upon the concept of the by-the-layer treatment [1, 2, 9, 15]. Within the framework of this concept, a boundary value problem for a multilayer structure is reduced to a number of auxiliary boundary value problems formulated for each layer. The solutions for each layer are then composed within the context of the interface conditions and actual boundary conditions imposed on the surface. The main advantage of this solution strategy is its clarity and the fact that the auxiliary problems are usually of a conventional type and thus can be solved by classical methods. A strong argument against this strategy is the inefficiency in analyzing a significant number of inhomogeneous layers of dramatically different thicknesses.

The latter case is better addressed with an alternative strategy for solving the problems for multilayer structures. Within the framework of this strategy, the entire assembly is considered a single unity with piecewise variable properties implying discontinuities at the layer interfaces [19]. This fact makes the solution process less sensitive to the number and thickness of the layers. On the other hand, the method’s efficiency directly relies on the analysis method for an inhomogeneous solid with an arbitrary variation of material properties. Moreover, the latter is to be realized alongside the non-classical formulation of the boundary value problem in view of the discontinuity of the material moduli at the interfaces. This implies the formulation of the governing equations in terms of the generalized derivatives [11]. When addressing solids with thin multilayer coatings, this concept is realized, particularly, in the form of an elegant approach of generalized boundary conditions [14].

Making use of the direct integration method [23] allows for the construction of explicit solutions to the boundary value problems for continuously inhomogeneous solids with arbitrary dependence of material properties on one of the spatial coordinates. The method implies the reduction of the original problems to the integral equations of the second kind whose solutions are managed to be found in the form of explicit dependences on the boundary conditions. This feature was efficiently used for formulating and solving the inverse boundary value problems on the identification of unknown thermal loading on an inaccessible surface of an inhomogeneous solid [12]. This approach can be easily extended toward the analysis of multilayer structures within the single-solid concept.

Herein, an application of the direct integration method is presented for solving steady-state (stationary) heat-conduction problems for radially inhomogeneous hollow cylinders with thermal loading on the inner and outer cylindrical surfaces varying within the circumferential coordinate. Two different approaches are discussed with concern for the reduction of the original heat-conduction equation to a governing integral equation, whose solution is constructed through the resolvent kernel. An extension of this approach is presented for the analysis of cylinders consisting of a number of coaxial radially inhomogeneous layers under the condition of perfect thermal contact.

31.2 Continuously Inhomogeneous Cylinders: Reduction to an Integral Equation

Consider a two-dimensional heat-conduction problem for a cylinder with a plane cross-section $R_\text{i} \leqslant r \leqslant R_\text{o}$, $0 \leqslant \varphi \leqslant 2\pi$ in the cylindrical–polar coordinate system (r, φ), where R_i and R_o are the inner and outer radii of the cylinder, respectively. Introduce the dimensionless radial coordinate $\rho = r/R_\text{o}$ to vary within $k \leqslant \rho \leqslant 1$, where $k = R_\text{i}/R_\text{o}$. We consider temperature distribution within domain $\mathcal{D} = \{k \leqslant \rho \leqslant 1, 0 \leqslant \varphi \leqslant 2\pi\}$, the inner and outer circumferences of which are registered with the following stationary temperatures:

$$T(k, \varphi) = T_k(\varphi), \quad T(1, \varphi) = T_1(\varphi). \tag{31.1}$$

Due to the fact that the thermal field expects not to show significant variation along cylinder's axis in the absence of internal heat sources, the considered heat-conduction problem is governed with the following equation:

$$\operatorname{div}(\lambda(\rho)\,\operatorname{grad} T(\rho, \varphi)) = 0, \tag{31.2}$$

where $\lambda(\rho)$ is the heat-conduction coefficient exhibiting an arbitrarily continuous distribution profile within the radial coordinate and $\lambda(\rho) > 0$. For the considered polar coordinates, Eq. (31.2) has the following form:

$$\frac{1}{\rho}\frac{\partial}{\partial \rho}\left(\rho\lambda(\rho)\frac{\partial T(\rho, \varphi)}{\partial \rho}\right) + \frac{\lambda(\rho)}{\rho^2}\frac{\partial^2 T(\rho, \varphi)}{\partial \varphi^2} = 0. \tag{31.3}$$

To solve the heat-conduction problem (31.1), (31.3), we use the Fourier series for the quested-for and given functions in the form as follows:

$$\left\{\begin{matrix} T(\rho, \varphi) \\ T_k(\varphi) \\ T_1(\varphi) \end{matrix}\right\} = \sum_{n=0}^{\infty}\left(\left\{\begin{matrix} T_n^1(\rho) \\ T_{k,n}^1 \\ T_{1,n}^1 \end{matrix}\right\}\cos(n\varphi) + \left\{\begin{matrix} T_n^2(\rho) \\ T_{k,n}^2 \\ T_{1,n}^2 \end{matrix}\right\}\sin(n\varphi)\right), \tag{31.4}$$

where

$$\left\{\begin{array}{c} T_n^{\ell}(\rho) \\ T_{k,n}^{\ell} \\ T_{1,n}^{\ell} \end{array}\right\} = \frac{1}{\pi}\int_0^{2\pi}\left\{\begin{array}{c} T(\rho,\varphi) \\ T_k(\varphi) \\ T_1(\varphi) \end{array}\right\}\left(\delta_{\ell,1}\left(1-\frac{\delta_{n,0}}{2}\right)\cos(n\varphi)+\delta_{\ell,2}\sin(n\varphi)\right)d\varphi, \tag{31.5}$$

$\ell = 1, 2$, $n = 0, 1, 2, \ldots$, $\delta_{i,j}$ is the Kronecker delta, and $T_0^2(\rho)$, $T_{k,0}^2$, and $T_{1,0}^2$ are involved formally, while $T_0^1(\rho)$ corresponds to the one-dimensional problem with uniform tem-peratures $T_{k,0}^1$ and $T_{1,0}^1$ on the inner and outer circumferences of $\mathcal{D}$.

Using (31.4) toward (31.1) and (31.3) yields the following boundary-value problems:

$$\frac{d^2T_0^1(\rho)}{d\rho^2} + \frac{1}{\rho}\frac{dT_0^1(\rho)}{d\rho} = -\frac{d\ln\lambda(\rho)}{d\rho}\frac{dT_0^1(\rho)}{d\rho}, \quad T_0^1(k) = T_{k,0}^1, \quad T_0^1(1) = T_{1,0}^1 \tag{31.6}$$

for the one-dimensional case and

$$\frac{d^2T_n^{\ell}(\rho)}{d\rho^2} + \frac{1}{\rho}\frac{dT_n^{\ell}(\rho)}{d\rho} - \frac{n^2}{\rho^2}T_n^{\ell}(\rho) = -\frac{d\ln\lambda(\rho)}{d\rho}\frac{dT_n^{\ell}(\rho)}{d\rho},$$
$$T_n^{\ell}(k) = T_{k,n}^{\ell}, \quad T_n^{\ell}(1) = T_{1,n}^{\ell}, \quad n = 1, 2, \ldots, \quad \ell = 1, 2, \tag{31.7}$$

for coefficients of the angle-dependent parts of the temperature.

The left-hand sides of the heat-conduction equations in (31.6) and (31.7) are of the conventional form used of the homogeneous materials, while the variable heat-conduction coefficient is involved in the right-hand sides of these equations.

Integrating the heat-conduction equation in (31.6) with respect to its left-hand side within the context of the first boundary condition yields the following integral equation of the second kind:

$$T_0^1(\rho) = \left(1 + k\ln\frac{\rho}{k}\left.\frac{d\ln\lambda(\rho)}{d\rho}\right|_{\rho=k}\right)T_{k,0}^1 + k\ln\frac{\rho}{k}\left.\frac{dT_0^1(\rho)}{d\rho}\right|_{\rho=k}$$
$$+\int_k^{\rho} T_0^1(\eta)\mathcal{K}(\rho,\eta)d\eta, \tag{31.8}$$

where

$$\mathcal{K}(\rho,\eta) = \frac{d}{d\eta}\left(\eta\ln\frac{\rho}{\eta}\frac{d\ln\lambda(\eta)}{d\eta}\right). \tag{31.9}$$

A solution to Eq. (31.8) can be given through the resolvent-kernel [4]

$$\mathcal{L}_0(\rho,\eta) = \sum_{l=0}^{\infty}\mathcal{K}_0^{l+1}(\rho,\eta), \quad \mathcal{K}_0^1(\rho,\eta) = \mathcal{K}_0(\rho,\eta),$$
$$\mathcal{K}_0^{l+1}(\rho,\eta) = \int_{\eta}^{\rho}\mathcal{K}_0^1(\rho,t)\mathcal{K}_0^l(t,\eta)dt. \tag{31.10}$$

As a result of using the resolvent (31.10), we arrive at the following expression:

$$T_0^1(\rho) = \left(\psi_0(\rho) + k\,\frac{d\ln\lambda(\rho)}{d\rho}\bigg|_{\rho=k}\psi(\rho)\right)T_{k,0}^1 + k\,\frac{dT_0^1(\rho)}{d\rho}\bigg|_{\rho=k}\psi(\rho), \quad (31.11)$$

where $\psi_0(\rho) = 1 + \vartheta_0(\rho)$, $\psi(\rho) = \ln(\rho/k) + \vartheta(\rho)$, $\vartheta(\rho) = \vartheta_1(\rho) - \ln k\vartheta_0(\rho)$, and

$$\vartheta_0(\rho) = \int_k^\rho \mathcal{L}_0(\rho,\eta)d\eta, \quad \vartheta_1(\rho) = \int_k^\rho \ln\eta\mathcal{L}_0(\rho,\eta)d\eta.$$

Substituting Eq. (31.11) into the second boundary condition in (31.6) allows for eliminating the uncertainty $dT_0^1(\rho)/d\rho$ at $\rho = k$ to arrive at

$$T_0^1(\rho) = \frac{\psi(1)\psi_0(\rho) - \psi_0(1)\psi(\rho)}{\psi(1)}T_{k,0}^1 + \frac{\psi(\rho)}{\psi(1)}T_{1,0}^1. \quad (31.12)$$

A solution to the equation in (31.7) can be given in the following form:

$$T_n^\ell(\rho) = A_n^\ell\rho^n + B_n^\ell\rho^{-n} + \int_k^\rho T_n^\ell(\eta)\mathcal{K}_n(\rho,\eta)d\eta, \quad \ell = 1, 2, \quad (31.13)$$

where A_n^ℓ and B_n^ℓ are the unknown constants of integration, and

$$\mathcal{K}_n(\rho,\eta) = \frac{1}{2n}\frac{d}{d\eta}\left(\eta\left(\left(\frac{\rho}{\eta}\right)^n - \left(\frac{\eta}{\rho}\right)^n\right)\frac{d\ln\lambda(\eta)}{d\eta}\right), \quad n = 1, 2, \ldots \quad (31.14)$$

Making use of the resolvent kernel

$$\mathcal{L}_n(\rho,\eta) = \sum_{l=0}^\infty \mathcal{K}_n^{l+1}(\rho,\eta), \quad \mathcal{K}_n^1(\rho,\eta) = \mathcal{K}_n(\rho,\eta),$$

$$\mathcal{K}_n^{l+1}(\rho,\eta) = \int_\eta^\rho \mathcal{K}_n^1(\rho,t)\mathcal{K}_n^l(t,\eta)dt \quad (31.15)$$

allows for expressing $T_n^\ell(\rho)$ in the form as follows:

$$T_n^\ell(\rho) = A_n^\ell\psi_n^+(\rho) + B_n^\ell\psi_n^-(\rho), \quad (31.16)$$

where $\psi_n^\pm(\rho) = \rho^{\pm n} + \int_k^\rho \eta^{\pm n}\mathcal{L}_n(\rho,\eta)d\eta$.

Putting (31.16) into the boundary conditions in (31.7) yields

$$T_n^{\ell}(\rho) = \psi_{k,n}(\rho)T_{k,n}^{\ell} + \psi_{1,n}(\rho)T_{1,n}^{\ell}, \tag{31.17}$$

where

$$\begin{Bmatrix} \psi_{k,n}(\rho) \\ \psi_{1,n}(\rho) \end{Bmatrix} = \frac{1}{\psi_n}\left(\begin{Bmatrix} \psi_n^-(1) \\ -\psi_n^-(k) \end{Bmatrix}\psi_n^+(\rho) - \begin{Bmatrix} \psi_n^+(1) \\ -\psi_n^+(k) \end{Bmatrix}\psi_n^-(\rho)\right),$$

$$\psi_n = \psi_n^+(k)\psi_n^-(1) - \psi_n^+(1)\psi_n^-(k).$$

Evidently, the evaluation of the temperature implies the determination of constituents of the Fourier series (31.4) through the formulae (31.12) and (31.17). It is worth noting that the key point lies in the determination of resolvent kernels (31.10) and (31.15), generated using subsequent formulae through the kernels (31.9) and (31.14) of the integral equations (31.8) and (31.13), respectively. The resolvent kernels are independent of the thermal loadings. Instead, they depend on the heat conduction coefficient as a function of the radial coordinate.

For many practical cases, the resolvent kernels can be derived in a closed analytical form [13]. If the latter form is somehow unmanageable, the resolvents can be computed through the following approximate formulae:

$$\mathcal{L}_0(\rho,\eta) \approx \mathcal{L}_0^{N_0}(\rho,\eta) = \sum_{l=0}^{N_0} \mathcal{K}_0^{l+1}(\rho,\eta),$$

$$\mathcal{L}_n(\rho,\eta) \approx \mathcal{L}_n^{N_n}(\rho,\eta) = \sum_{l=0}^{N_n} \mathcal{K}_n^{l+1}(\rho,\eta) \tag{31.18}$$

substituting the original infinite series with the finite sums, where N_0 and N_n are constant parameters allowing for the required accuracy in satisfaction of the integral equations (31.8) and (31.13) by expressions (31.12) and (31.17). Actual numerical values for these parameters can be established by the series of numerical experiments implying the comparison of solutions (31.12) and (31.17) computed for N_0 and N_n with the ones computed for $N_0 - 1$ and $N_n - 1$ in (31.18); if the deviation between the solutions falls within the required accuracy for all $\rho \in [k, 1]$, then the values of N_0 and N_n are to be accepted as the sufficient ones.

The evaluation of sufficient values for N_0 and N_n can also be performed by minimizing the residual term arising in Eqs. (31.8) and (31.13) when using the approximate formulae. For example, the well-known formula

$$\mathcal{K}_1^0(\rho,\eta) + \int_{\eta}^{\rho} \mathcal{K}_1^0(\rho,t)\mathcal{L}_0(t,\eta)dt = \mathcal{L}_0(\rho,\eta) \tag{31.19}$$

follows from (31.10). This formula, also known as the resolvent equation [4], guaranties the fact that integral equation (31.8) is identically satisfied by the expression (31.12) with the resolvent (31.10). When using formula (31.18) instead of (31.10), formula (31.19) takes the following form:

$$\int\limits_{\eta}^{\rho} \mathcal{K}_1^0(\rho, t)\mathcal{L}_0^{N_0}(t, \eta)dt = \mathcal{L}_0^{N_0}(\rho, \eta) - \mathcal{K}_1^0(\rho, \eta) + \mathcal{K}_{N_0+2}^0(\rho, \eta).$$

Then, obviously, the sufficient accuracy in the satisfaction of integral equation (31.8) can be achieved by minimizing the kernel $|\mathcal{K}_{N_0+2}^0(\rho, \eta)| \leqslant \iota_{N_0}$ (where $\iota_{N_0}$is a given accuracy) under which the residual term in Eq. (31.10) has the following form:

$$\Psi_n = \left| \int\limits_{\eta}^{\rho} \left(T_{k,0}^1 + \frac{T_{1,0}^1 - T_{k,0}^1 \psi_0(1)}{\psi_l(1)} \ln \frac{\eta}{k} \right) \mathcal{K}_{N+2}^0(\rho, \eta) d\eta \right| \leqslant \iota_{N_0} \mu_{N_0},$$

where

$$\mu_{N_0} = \int_k^{\rho} \left| T_{k,0}^1 + \frac{T_{1,0}^1 - T_{k,0}^1 \psi_0(1)}{\psi_l(1)} \ln \frac{\eta}{k} \right| d\eta.$$

In such manner, expressions (31.12) and (31.17) for the constituents of temperature (31.4) present a solution to the heat-conduction problem constructed in the form of explicit dependences on the boundary temperatures (31.1) for the case of an arbitrary continuous dependence of the heat-conduction coefficient on the radial coordinate. It is worth noting that the in the case of $\lambda = \text{const}$, the kernels (31.9) and (31.14) and, consequently, the resolvents (31.10) and (31.15) equal to zero.

31.3 An Alternative Approach

When trying to extend the foregoing technique for solving a similar problem for a multilayer cylinder, it appears to be challenging to engage equations in (31.6) and (31.7) directly due to the fact that the heat conduction coefficient, involved in the right-hand side, may be a discontinuous function of the radial coordinate. For that reason, we consider a modified approach.

Let us represent the boundary value problem (31.6) in the following form:

$$\frac{d}{d\rho}\left(\rho\lambda(\rho)\frac{dT_0^1(\rho)}{d\rho}\right) = 0, \quad T_0^1(k) = T_{k,0}^1, \quad T_0^1(1) = T_{1,0}^1. \tag{31.20}$$

By the consequent integration of equation in (31.20) by the radial coordinate within the context of involved boundary conditions, we derive the following formula:

$$T_0^1(\rho) = \frac{\Lambda(\rho)}{\Lambda(1)} T_{1,0}^1 + \frac{\Lambda(1) - \Lambda(\rho)}{\Lambda(1)} T_{k,0}^1, \tag{31.21}$$

where

$$\Lambda(\rho) = \int_k^\rho \frac{d\xi}{\xi \lambda(\xi)}. \tag{31.22}$$

It is worth noting that in contrast to (31.12), an alternative solution (31.21) is given in an explicit form without application of the resolvent kernel.

The boundary-value problem (31.7) can also be given in the following form:

$$\frac{d}{d\rho}\left(\rho\lambda(\rho)\frac{dT_n^\ell(\rho)}{d\rho}\right) - \frac{\lambda(\rho)}{\rho} n^2 T_n^\ell(\rho) = 0, \quad T_n^\ell(k) = T_{k,n}^\ell, \quad T_n^\ell(1) = T_{1,n}^\ell. \tag{31.23}$$

Integrating equation in (31.23) with account for the first boundary condition yields

$$\frac{dT_n^\ell(\rho)}{d\rho} = \frac{k\lambda(k)}{\rho\lambda(\rho)} \left.\frac{dT_n^\ell(\rho)}{d\rho}\right|_{\rho=k} + \frac{n^2}{\rho\lambda(\rho)} \int_k^\rho \frac{\lambda(\eta)}{\eta} T_n^\ell(\eta) d\eta. \tag{31.24}$$

The subsequent integration of (31.24) with account for the second condition in (31.23) allows for deriving the following integral equation:

$$T_n^\ell(\rho) = T_{k,n}^\ell + k\lambda(k) \left.\frac{dT_n^\ell(\rho)}{d\rho}\right|_{\rho=k} \Lambda(\rho) + \int_k^\rho T_n^\ell(\eta) \tilde{\mathcal{K}}_n(\rho, \eta) d\eta. \tag{31.25}$$

Here,

$$\tilde{\mathcal{K}}_n(\rho, \eta) = \frac{n^2}{\eta} \left(\Lambda(\rho) - \Lambda(\eta)\right) \lambda(\eta), \tag{31.26}$$

and function $\Lambda(\rho)$ is expressed in the form (31.22).

For solving the integral equation (31.25), we employ the resolvent kernel

$$\tilde{\mathcal{L}}_n(\rho, \eta) = \sum_{l=0}^{\infty} \tilde{\mathcal{K}}_n^{l+1}(\rho, \eta), \tag{31.27}$$

where the recurring kernels are derived in the form (31.15) where symbols $\mathcal{K}$ are substituted with $\tilde{\mathcal{K}}$. As a result, we arrive at the following expression:

$$T_n^\ell(\rho) = T_{k,n}^\ell \varphi_n(\rho) + k\lambda(k) \left.\frac{dT_n^\ell(\rho)}{d\rho}\right|_{\rho=k} \psi_n(\rho), \tag{31.28}$$

where

$$\begin{Bmatrix} \varphi_n(\rho) \\ \psi_n(\rho) \end{Bmatrix} = \begin{Bmatrix} 1 \\ \Lambda(\rho) \end{Bmatrix} + \int_k^\rho \begin{Bmatrix} 1 \\ \Lambda(\eta) \end{Bmatrix} \tilde{\mathcal{L}}_n(\rho, \eta) d\eta,$$

$$T_n^\ell(\rho) = \frac{\psi_n(\rho)}{\psi_n(1)} T_{1,n}^\ell + \frac{\psi_n(1)\varphi_n(\rho) - \varphi_n(1)\psi_n(\rho)}{\psi_n(1)} T_{k,n}^\ell. \tag{31.29}$$

Making use of the presented approach, which is oriented toward the integration of the heat conduction equation in terms of the heat flux, allows for deriving the solution in the form (31.4) with the constituents given by formulae (31.21) and (31.29). In contrast to the approach considered in the previous section, a closed-form solution is obtained in the case of the one-dimensional temperature field. It is worth noting that in the case of $\lambda = \text{const}$, the kernel of integral equation (31.26) takes the form $\tilde{\mathcal{K}}_n(\rho, \eta) = \eta^{-1} n^2 \ln(\eta/\rho) \neq 0$. Hence, the resolvent (31.27) differs, in general, from zero even for the case of a homogenous isotropic material. Finally, formula

$$\tilde{\mathcal{L}}_n(\rho, \eta) \approx \tilde{\mathcal{L}}_n^N(\rho, \eta) = \sum_{l=0}^{N} \tilde{\mathcal{K}}_n^{l+1}(\rho, \eta) \tag{31.30}$$

can be used for the practical computation instead of (31.27).

31.4 Multilayer Hollow Cylinders

Let the considered cylinder consists of M coaxial layers separated with interfaces $\rho = \rho_j, j = 1, 2, \ldots, M-1, k = \rho_0 < \rho_1 < \rho_2 < \cdots < \rho_{M-1} < \rho_M = 1$. To simulate the radial variation of the heat-conduction coefficient, we introduce the formula

$$\lambda(\rho) = \sum_{j=1}^{M} \lambda_{[j]}(\rho) S_j(\rho), \tag{31.31}$$

where $S_j(\rho) = \mathrm{H}(\rho - \rho_{j-1}) - \mathrm{H}(\rho - \rho_j)$, $\mathrm{H}(\rho)$ is the Heaviside function, $\lambda_{[j]}(\rho)$ is the coefficient of the jth layer (ρ_{j-1}, ρ_j), $j = 1, 2, \ldots, M$,

$$\lambda_{[j]}(\rho_j) = \lim_{\rho \to \rho_j - 0} \lambda_{[j]}(\rho), \quad \lambda_{[j]}(\rho_{j-1}) = \lim_{\rho \to \rho_{j-1} + 0} \lambda_{[j]}(\rho).$$

The temperature, naturally, can be presented in a form similar to (31.31). The inner and outer circumferences of the cylinder are subject to boundary conditions (31.1) and the layers are perfectly connected implying

$$T_{[j]}(\rho_j, \varphi) = T_{[j+1]}(\rho_j, \varphi),$$
$$\lambda_{[j]}(\rho_j)\frac{\partial T_{[j]}(\rho_j, \varphi)}{\partial \rho} = \lambda_{[j+1]}(\rho_j)\frac{\partial T_{[j+1]}(\rho_j, \varphi)}{\partial \rho}, \quad j = 1, 2, \ldots, M-1. \quad (31.32)$$

Within the context of the conditions (31.32), Eq. (31.3) and, consequently (31.20) and (31.23) are formulated in terms of classical derivatives by the radial coordinate. Thus, in the considered case of the multilayer cylinder under conditions (31.32), a solution to the problem (31.20) can be given in the form (31.21), where function (31.22) reads

$$\Lambda(\rho) = \sum_{j=1}^{M} \Lambda_{[j]}(\rho) S_j(\rho), \quad \Lambda_{[j]}(\rho) = \int_{\rho_{j-1}}^{\rho} \frac{d\eta}{\eta\lambda_{[j]}(\eta)} + \sum_{i=1}^{j-1} \int_{\rho_{i-1}}^{\rho_i} \frac{d\eta}{\eta\lambda_{[i]}(\eta)}. \quad (31.33)$$

A solution to problem (31.23), in this case, has the form (31.29), where

$$\varphi_n(\rho) = \sum_{I=1}^{M} \varphi_{[I],n}(\rho) S_I(\rho), \quad \psi_n(\rho) = \sum_{I=1}^{M} \psi_{[I],n}(\rho) S_I(\rho),$$
$$\begin{Bmatrix} \varphi_{[I],n}(\rho) \\ \psi_{[I],n}(\rho) \end{Bmatrix} = \begin{Bmatrix} 1 \\ \Lambda_{[I]}(\rho) \end{Bmatrix} + \int_{\rho_{I-1}}^{\rho} \begin{Bmatrix} 1 \\ \Lambda_{[I]}(\eta) \end{Bmatrix} \widehat{\mathfrak{L}}_{[I,I],n}(\rho, \eta) d\eta$$
$$+ \sum_{i=1}^{I-1} \int_{\rho_i - 1}^{\rho_i} \begin{Bmatrix} 1 \\ \Lambda_{[1]}(\eta) \end{Bmatrix} \widehat{\mathfrak{L}}_{[I,i],n}(\rho, \eta) d\eta. \quad (31.34)$$

Herein, the resolvent kernel takes the form

$$\widehat{\mathfrak{L}}_n(\rho, \eta) = \sum_{I=1}^{M} \sum_{J=1}^{I} \widehat{\mathfrak{L}}_{[I,J],n}(\rho, \eta) S_I(\rho) S_J(\eta), \quad (31.35)$$

where

$$\widehat{\mathcal{L}}_{[I,J],n}(\rho,\eta) = \sum_{l=0}^{\infty} \widehat{\mathcal{K}}^{l+1}_{[I,J],n}(\rho,\eta),$$

$$\widehat{\mathcal{K}}^{l+1}_{[I,J],n}(\rho,\eta) = \int_{\rho_{I-1}}^{\rho} \widehat{\mathcal{K}}^{1}_{[I,I],n}(\rho,\xi)\widehat{\mathcal{K}}^{l}_{[I,J],n}(\xi,\eta)d\xi$$
$$- \int_{\rho_{J-1}}^{\eta} \widehat{\mathcal{K}}^{1}_{[I,J],n}(\rho,\xi)\widehat{\mathcal{K}}^{l}_{[J,J],n}(\xi,\eta)d\xi$$
$$+ \sum_{i=J}^{I-1} \int_{\rho_{i-1}}^{\rho_i} \widehat{\mathcal{K}}^{1}_{[I,i],n}(\rho,\xi)\widehat{\mathcal{K}}^{l}_{[i,J],n}(\xi,\eta)d\xi,$$

$$\widehat{\mathcal{K}}^{1}_{[I,J],n}(\rho,\eta) = \frac{n^2}{\eta}\lambda_{[J]}(\eta)\left(\int_{\rho_{I-1}}^{\rho} \frac{d\xi}{\xi\lambda_{[I]}(\xi)} - \int_{\rho_{J-1}}^{\eta} \frac{d\xi}{\xi\lambda_{[J]}(\xi)} + \sum_{i=J}^{I} \int_{\rho_{i-1}}^{\rho_i} \frac{d\xi}{\xi\lambda_{[i]}(\xi)}\right).$$

Like in the cases presented in the foregoing sections, practical computation of the resolvent can be realized by using the approximate formula

$$\widehat{\mathcal{L}}_{[I,J],n}(\rho,\eta) \approx \widehat{\mathcal{L}}^{N}_{[I,J],n}(\rho,\eta) = \sum_{l=0}^{N} \widehat{\mathcal{K}}^{l+1}_{[I,J],n}(\rho,\eta). \tag{31.36}$$

In such manner, the solution scheme is extended toward solving the heat-conduction problem for a multilayer cylinder in the form (31.4) with coefficients (31.22) and (31.29) within the context of formulae (31.33)–(31.35).

31.5 Numerical Examples and Discussion

Consider a case study where the boundary conditions (31.1) are realized in the following form [20]:

$$T_k(\varphi) = T_0, \quad T_1(\varphi) = T_0\left(d_0 + d_2\cos(2\varphi)\right), \tag{31.37}$$

where T_0 is a constant in the dimension of temperature, and d_0 and d_2 are dimensionless constants allowing control of distribution (31.37). At $d_2 = 0$, the temperatures on the boundaries are uniform and, thus, the temperature resulted from the boundary value problem (31.1) and (31.3) is one-dimensional. Making use of formulae (31.5) yields $T^1_{k,0} = T_0$, $T^1_{1,0} = d_0T_0$, $T^1_2 = d_2T_0$, and $T^2_2 = T^\ell_n = 0$ ($n \neq 2$, $\ell = 1, 2$).

Assuming the thermal conductivity to be constant under the considered loading (31.37) implies the following distribution of the temperature

$$T(\rho,\varphi) = T^1_0(\rho) + T^1_2(\rho)\cos(2\varphi), \tag{31.38}$$

$T_0^1(\rho) = T_0 d_0$, $T_2^1(\rho) = T_0 d_2(\rho^4 - k^4)/(1-k^4)\rho^2$, which is irrespective of the material properties. The distribution of temperature with the profile (31.38) was discussed by Tokovyy and Ma [20, 21].

Allowing the thermal conductivity to vary within the radial coordinate, for example, in the form of the power-law distribution

$$\lambda(\rho) = \lambda_0 \rho^{\varkappa} \quad (\varkappa \in \mathbb{R}), \quad \lambda_0 = \text{const } [°\text{K/m}^2] \tag{31.39}$$

yields the solution (31.38), where

$$\begin{aligned} \frac{T_0^1(\rho)}{T_0} &= \frac{d_0 - k^{\varkappa}}{1-k^{\varkappa}} + \frac{1-d_0}{1-k^{\varkappa}}\left(\frac{k}{\rho}\right)^{\varkappa}, \\ \frac{T_2^1(\rho)}{T_0} &= \frac{k^{\frac{\varkappa}{2}}\rho^{\frac{\varkappa}{2}}}{k^{\frac{\gamma_2}{2}} - k^{-\frac{\gamma_2}{2}}}\left(\left(\frac{k}{\rho}\right)^{\frac{\gamma_2}{2}} - \left(\frac{\rho}{k}\right)^{\frac{\gamma_2}{2}}\right) d_2, \end{aligned} \tag{31.40}$$

and $\gamma_2 = \sqrt{\varkappa^2 + 16}$. Figure 31.1 presents the distribution of the dimensionless temperature (31.38), (31.40) under $d_0 = 1$, $d_2 = 2$. Evidently, being qualitatively close, the higher conductivity near the outer periphery causes deeper temperature propagation along the radial coordinate.

It can be easily shown that solution (31.21) coincides with the exact one given by the first formula in (31.40), while the resolvent solution (31.12), for the considered constant parameters, falls within 99.6% of accuracy under $N_0 = 1$ in (31.13). Figure 31.2 demonstrates the high convergence rate of the resolvent solution (31.29)

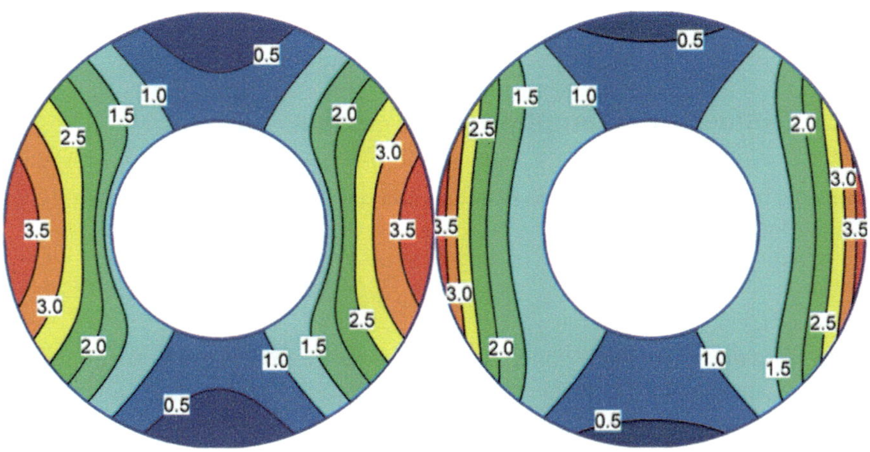

Fig. 31.1 Full-field distribution of the dimensionless temperature $T(\rho, \varphi)/T_0$ computed through the formulae (31.38) and (31.40) for a radially in-homogeneous cylinder with $k = 0.5$ and the heat conduction coefficient (31.39) under $\varkappa = 3$ (left-hand side) and $\varkappa = -3$ (right-hand side)

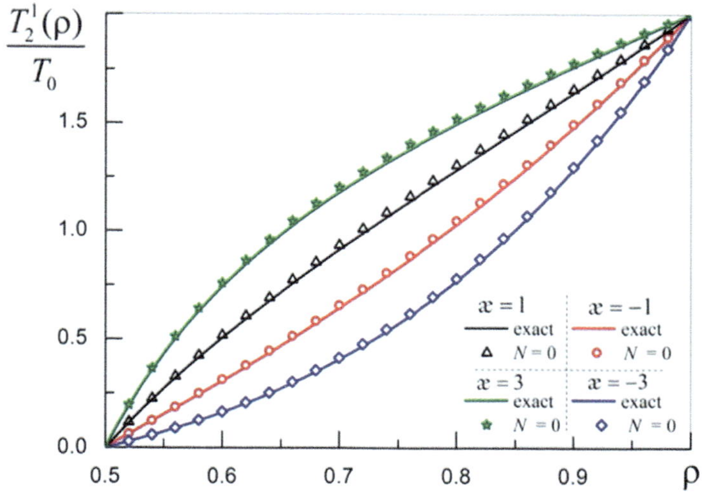

Fig. 31.2 Comparison of the dimensionless coefficient $T_2^1(\rho)/T_0$ computed by the exact formula in (31.40) and the solution (31.29) when using only the first term ($N = 0$) in formula (31.30) under $\varkappa = \pm 1; \pm 3$ in (31.39) and $d_0 = 1$, $d_2 = 2$ in (31.37)

with the approximate resolvent kernel (31.30) to the exact one given by the second formula in (31.40). Evidently, keeping one term only ($N = 0$) in the approximate formula for the resolvent (31.30), i.e., just using the original kernel (31.26) of the integral equation (31.25) allows for achieving 97% of accuracy.

Consider the temperature field for in a two-layer cylinder $k = \rho_0 < \rho_1 < \rho_M = 1$, which is thermally loaded according to (31.1) and (31.37) and having the heat-conduction coefficient assumed in the form as follows:

$$\lambda(\rho) = \lambda_0 \begin{cases} \rho_0^{\varkappa}, & \rho_0 < \rho < \rho_1, \\ \rho_M^{\varkappa}, & \rho_1 < \rho < \rho_M. \end{cases} \tag{31.41}$$

Note that the profile (31.41) is a kind of step-wise realization of the continuous distribution (31.39) realizing through the extreme values with the layers.

Figure 31.3 presents the full-field distribution of the temperature (31.38) with the coefficients computed through the formulae (31.22) and (31.29) under (31.33)–(31.35), and (31.41). The higher conductivity of the external layer compared to the inner one, the heating of the outer periphery affects the entire assembly. This holds for the different thicknesses of the layers; the second row demonstrates the effect of coatings.

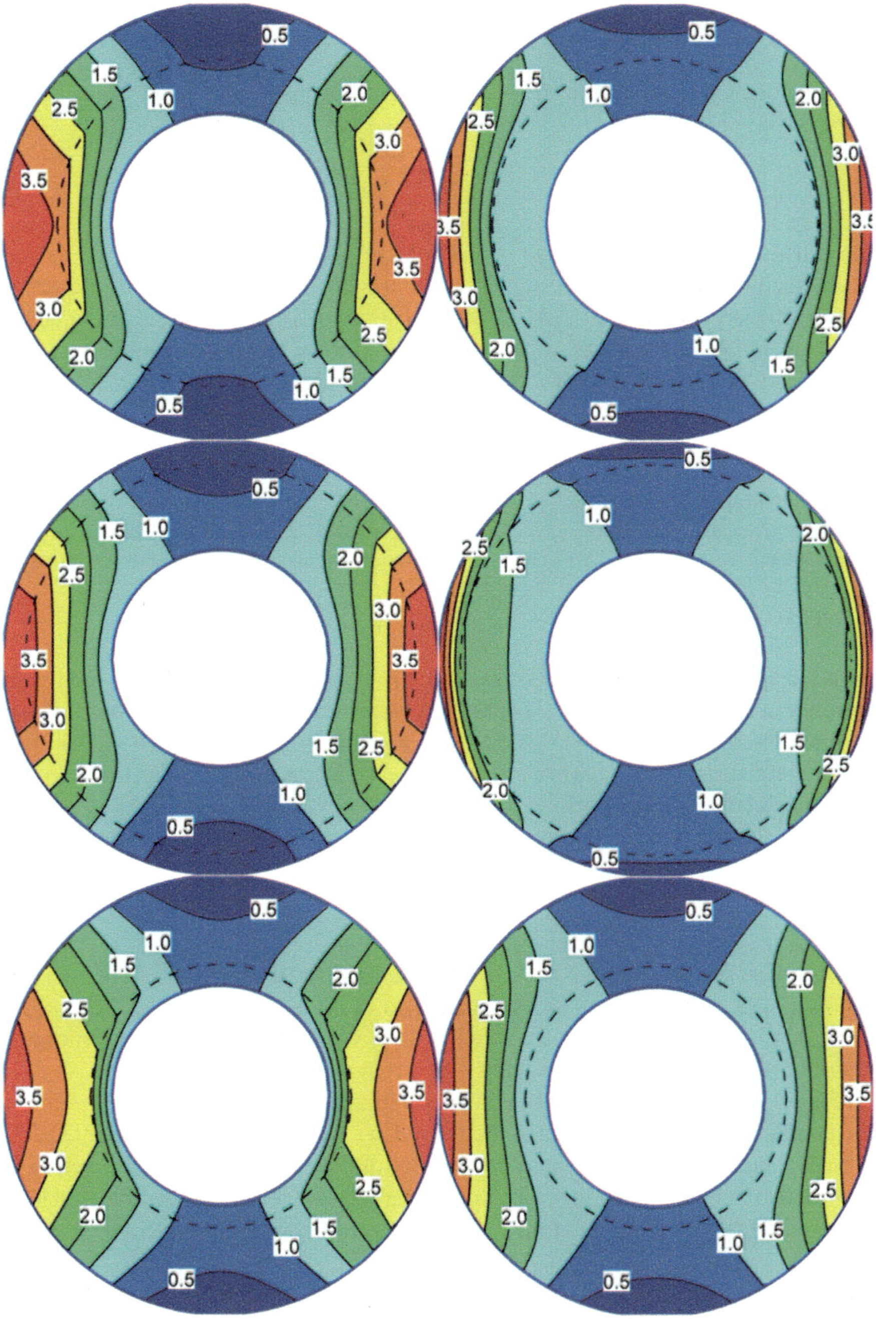

Fig. 31.3 Full-field distribution of the temperature (31.38) with the coefficients computed through the formulae (31.21) and (31.29) under (31.33)–(31.35), and (31.41) at $\varkappa = 3$ (left-hand sides) and $\varkappa = -3$ (right-hand sides) for $k = 0.5$and $\rho_1 = 0.75$ (first row), $\rho_1 = 0.90$ (second row), and $\rho_1 = 0.60$ (third row)

31.6 Conclusions

Using the direct integration method [22, 23] toward the heat-conduction equation allowed for constructing explicit analytical solutions to the two-dimensional problems for radially inhomogeneous hollow cylinders exhibiting arbitrary radial variation of the heat-conduction coefficient. Solutions are constructed by reducing the original equation to the Volterra integral equations of the second kind, which are then solved using the resolvent-kernel technique.

The method is extended toward analyzing thermal fields in multilayer cylinders with an arbitrary radial variation of the layer properties. The only difference in applying this technique is that the key functions depend on the radial coordinate stepwise. Under such an approach, the number of layers or thickness does not significantly affect the solution's efficiency.

Acknowledgements This research was partially supported by the Joint Ukrainian-Slovak Research and Development Project under the grant No. 0123U103235.

References

1. Abramidze, E., Grigorenko, Y.: Research of the stress state of flexible multilayered corrugated cylindrical shells of rotation according to a refined theory. Bull. TICMI **26**(1), 19–26 (2022)
2. Amiri Delouei, A., Emamian, A., Sajjad, I.H., Atashafrooz, M., Li, Y., Wang, L.P., Jing, D.: A comprehensive review on multi-dimensional heat conduction of multi-layer and composite structures: analytical solutions. J. Thermal Sci. **30**, 1875–1907 (2021)
3. Barbaros, I., Yang, Y., Safaei, B., Yang, Z., Qin, Z., Asmael, M.: State-of-the-art review of fabrication, application, and mechanical properties of functionally graded porous nanocomposite materials. Nanotechnol. Rev. **11**(1), 321–371 (2022)
4. Brunner, H.: Volterra Integral Equations. An Introduction to Theory and Applications. Cambridge University Press, Cambridge (2017)
5. Charan, M.S., Naik, A.K., Kota, N., Laha, T., Roy, S.: Review on developments of bulk functionally graded composite materials. Int. Mater. Rev. **67**(8), 797–863 (2022)
6. Chen, X.W., Wang, S., Yue, Z.Q.: Mechanics of multilayered or functionally graded material pressed onto an imperfect rigid base with a circular opening by localized surface loading. Int. J. Solids Struct. **236–237**, 111335 (2022)
7. Chen, Z., Akbarzadeh, A.: Advanced Thermal Stress Analysis of Smart Materials and Structures. Springer, Cham (2020)
8. Ghayesh, M.H., Farajpour, A.: A review on the mechanics of functionally graded nanoscale and microscale structures. Int. J. Engin. Sci. **137**, 8–36 (2019)
9. Grigorenko, Y., Vasilenko, A., Pankratova, N.: Problems of the Elasticity Theory of Inhomogeneous Solids. Naukova Dumna, Kyiv [in Ukrainian] (1991)
10. Hemath, M., Mavinkere Rangappa, S., Kushvaha, V., Dhakal, H.N., Siengchin, S.: A comprehensive review on mechanical, electromagnetic radiation shielding, and thermal conductivity of fibers/inorganic fillers reinforced hybrid polymer composites. Polymer Compos. **41**, 3940–3965 (2020)
11. Kushnir, R.M.: Application of the method of generalized coupling problems in the thermoelasticity of piecewise-homogeneous bodies under nonideal contact. J. Math. Sci. **97**(1), 3854–3861 (1999)

12. Kushnir, R.M., Yasinskyy, A.V., Tokovyy, Y.V.: Effect of material properties in the direct and inverse thermomechanical analyses of multilayer functionally graded solids. Adv. Engin. Mater. **24**(5), 2100875 (2022)
13. Sharma, D.C., Goyal, M.C.: Integral Equations. PHI Learning Privete Ltd, Delhi (2017)
14. Shevchuk, V.A.: Modeling and computation of heat transfer in a system "body-multylayer coating". Heat Trans. Res. **37**(5), 421–433 (2006)
15. Soldatos, K.P., Ye, J.Q.: Three-dimensional static, dynamic, thermoelastic and buckling analysis of homogeneous and laminated composite cylinders. Compos. Struct. **29**(2), 131–143 (1994)
16. Swaminathan, K., Sangeetha, D.M.: Thermal analysis of FGM plates - A critical review of various modeling techniques and solution methods. Compos. Struct. **160**, 43–60 (2017)
17. Tanigawa, Y.: Some basic thermoelastic problems for nonhomogeneous structural materials. Appl. Mech. Revs. **48**(6), 287–300 (1995)
18. Thai, H.T., Kim, S.E.: A review of theories for the modeling and analysis of functionally graded plates and shells. Compos. Struct. **128**, 70–86 (2015)
19. Tokova, L., Yasinskyy, A., Ma, C.C.: Effect of the layer inhomogeneity on the distribution of stresses and displacements in an elastic multilayer cylinder. Acta Mech. **228**, 2865–2877 (2017)
20. Tokovyy, Y.V., Ma, C.C.: Thermal stresses in anisotropic and radially inhomogeneous annular domains. J. Thermal Stress. **31**(9), 892–913 (2008)
21. Tokovyy, Y.V., Ma, C.C.: Analytical solutions to the planar non-axisymmetric elasticity and thermoelasticity problems for homogeneous and inhomogeneous annular domains. Int. J. Engin. Sci. **47**(3), 413–437 (2009)
22. Tokovyy, Y.V., Ma, C.C.: Elastic analysis of inhomogeneous solids: history and development in brief. J. Mech. **35**(5), 613–626 (2019)
23. Tokovyy, Y.V., Ma, C.C.: The Direct Integration Method for Elastic Analysis of Nonhomogeneous Solids. Cambridge Scholars Publishing, Newcastle (2021)
24. Toktaş, S.E., Dag, S.: Multi-layer model for moving contact problems of functionally graded coatings with general variations in physical properties. Proc. Inst. Mech. Eng., Part L: J. Mater.: Design Appl. **236**(10), 1967–1980 (2022)
25. Vitucci, V., Mishuris, G.: Analysis of residual stresses in thermoelastic multilayer cylinders. J. Europ. Ceramic Soc. **36**(9), 2411–2417 (2016)
26. Zhong, Z., Nie, G.: Analytical or Semi-analytical Solutions of Functionally Graded Material Structures. Springer, Singapore (2021)

Chapter 32
Influence of Delamination Defects on the Dynamic Stress-Strain State of Composite Elements of Launch Vehicles

Borys Zaitsev, Natalia Smetankina, Tetiana Protasova, Dmytro Klymenko, and Dmytro Akimov

Abstract The application of the methodology for modeling cracks in the scheme of the finite element method is considered for evaluating the effect of delamination in composite elements of launch vehicles on the dynamic stress-deformed state. Cracks are modeled by modifying the finite element mesh and applying formalized operations to the stiffness and mass matrices of the initially intact body. The problem of calculating oscillations is reduced to a system of differential equations of the finite element method in time for a body with curvilinear anisotropy of properties and stiffness and mass matrices reflecting the presence of cracks. The solution to the initial problem is performed according to Newmark's and Wilson's finite-difference schemes. A computational analysis of the dynamic stress-deformed state of the rocket fairing in the process of its separation by the pyrotechnic system is carried out. Delamination in irregular fiberglass-winding zones is simulated in the composite shell, which is part of the assembled structures of launch vehicle fairing. It is shown that external delamination from the shell edge is more dangerous than internal delamination and leads to a significant increase in dynamic stresses. The problem of lateral oscillations of a rod with a near-surface crack, taking into account the local buckling of delamination, is posed and solved. A model of delamination is constructed, the stiffness of which depends on the amount of deformation and the fulfillment of the condition for buckling, which leads to the nonlinearity of oscillations. The effects of nonlinear oscillations associated with the nonisochronism of free oscillations and super- and sub-harmonic resonances are investigated.

B. Zaitsev · N. Smetankina (✉) · T. Protasova
A. Pidgornyi Institute of Mechanical Engineering Problems, National Academy of Sciences of Ukraine, Kharkiv, Ukraine
e-mail: nsmetankina@ukr.net

D. Klymenko · D. Akimov
Yuzhnoye State Design Office, Dnipro, Ukraine
e-mail: akimovdv@kbu.net

H. Altenbach et al. (eds.), *Selected Problems of Solid Mechanics and Solving Methods*, Advanced Structured Materials 204, https://doi.org/10.1007/978-3-031-54063-9_32

32.1 Introduction

With the development of technologies for obtaining polymer composite materials (PCM), and the creation of modern equipment for producing large-sized PCM-based products, the predominant trend is to turn to composite structures in creating space-rocket hardware, such as fairings. The use of PCMs allows one to reduce weight characteristics and achieve high volume parameters for placing useful load [1]. Achievements and prospects of using PCMs in the creation of launch vehicle (LV) fairings are dealt with in [2, 3].

Since the processes of creating a PCM-based structure and its material are combined, important and possible, with account taken of various kinds of constraints, are the optimization measures to reduce the weight of the structure [4, 5]. Different aspects of the development of the PCM-based space-rocket hardware optimization concept are considered in [6, 7]. When designing composite structures, a wide range of PCM characteristics (such as strength, deformation resistance, and fracture toughness) is used [8, 9]. Modern methods and means for determining these characteristics are presented in [10, 11].

For composite structures, as well as for more traditional metal structures, the tasks of ensuring the strength, stability, and thermal protection of LV compartments are urgent. These tasks are due to the effects of aerodynamic loads, thermal heating in the dense layers of the atmosphere, as well as dynamic forces during LV-compartment separations [12]. A specific type of fracture of composite layered structures is the structural delamination whose main cause, as is commonly believed, is impact effects [13]. Delamination, during their development, can cause macroscopic destruction of entire structures. Delamination causes stress redistribution, which can also lead to the buckling of structures and the appearance of nonlinear oscillations [14, 15].

Different approaches are possible to assess the risk of delamination and its effect on the bearing capacity of a structure. In particular, it is important to evaluate the conditions that are favorable for the appearance of delaminations, which are most likely to occur in irregular, i.e. transition zones of PCM lay-up, where the layered composite lay-up structure is variable. Due to the change in the structure, the stresses between the layers are redistributed, which leads to the appearance of interlayer shear and separation stresses [16, 17]. In the presence of micro- or macroscopic delaminations revealed by non-destructive flaw detection methods, it becomes necessary either to predict their development to a critical level or to assess the effect of the delamination-induced stress redistribution on the limiting state.

32.2 Method for Modeling Structural Defects of Composite Structures of Space-Rocket Hardware

Modeling a delamination (or crack) in a body is performed according to the cut introduction technique in the scheme of the finite element method. The technique was proposed in [18] and tested in addressing a number of research and practical problems [19].

Much attention is paid to taking into account structural cracks and modeling the stress-strain states at crack tips to determine their local criterial characteristics. There is a significant number of publications in domestic and foreign editions, as well as a number of monographs [20, 21]. The main attention in them is paid to the quality of modeling the stress-strain states in the vicinities of crack tips, bringing the modeling closer to a specific asymptotic distribution. For this purpose, special finite elements are used, for example, isoparametric quadratic elements with displaced nodes, special singular elements, and macroelements with a crack.

Little attention is paid to issues of the crack accounting technology, which ensures displacement discontinuity along crack faces in an arbitrary location of a structure without significantly revising the technique or creating a special finite element. At the same time, when designing structures, it is of interest to assess the effect of cracks both on stress redistribution and changes in the spectrum of natural vibrations. The methods and computation programs used should allow simulating cracks in an arbitrary element, and the amount of necessary information about it should be easily formalized.

The developed method of accounting for a crack in the form of a cut is integrated into the computer complex that is built on the regular discretization of three-dimensional structures with a multilinear isoparametric finite element. This complex performs the necessary transformations for the generated stiffness matrices that are obtained as a result of regular discretization. These transformations are equivalent to the introduction of cracks without involving additional deformation parameters, i.e. the procedure for introducing cracks is reduced to changing the structure and elements of the mass and stiffness matrices.

When modeling finite or zero-thickness discontinuities, one assumes that they are oriented along the surface of topological faces. Their introduction is carried out according to the method based on the transformation of the initial finite element mesh by displacing nodes to the surface of the discontinuities.

The essence of this approach is illustrated by the example of a crack located in one topological plane, schematically shown in Fig. 32.1, where CS is the crack surface and CL is the crack line.

Along the crack plane, two rows of A-, B-, C-, and D-type finite elements are introduced on each side. Q-nodes are excluded from consideration, and the corresponding variables are not taken into account in the stiffness matrices. The equations corresponding to the Q-nodes have non-zero diagonal elements (the rest of the elements are equal to zero). Equal to zero are elements in other equations related to the variables of Q-nodes. S-nodes are shifted toward the crack, forming a double-node layer.

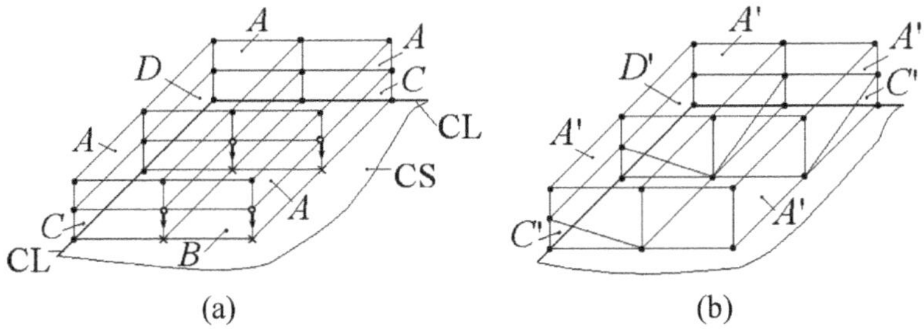

Fig. 32.1 A scheme of mesh transformation with the introduction of a crack: **a**—the original regular grid without a crack; **b**—the resulting irregular mesh with a crack; *A*, *C*, and *D*—transformable finite elements; *B*—disregarded finite elements; o—shifted *S*-nodes; ×—disregarded *Q*-nodes

As a result of the mesh transformation, elements *B* are actually no longer considered; elements *A* are transformed, changing the position of some of the nodes; elements *C* degenerate into triangular prisms; and in element *D* two quadrangular faces degenerate into triangular ones. The formation of the stiffness matrices of degenerated finite elements is formalized at the level of matrix operations.

The stiffness matrices of elements *A* are formed when a number of coordinates change, and those of elements *B* are formed when the *Q*-nodes are isolated by diagonalizing the matrices relative to the indicated nodes.

The stiffness matrices of elements *A* are formed when a number of coordinates change, and those of elements *B* are formed when the *Q*-nodes are isolated by diagonalizing the matrices relative to the indicated nodes.

The stiffness matrix of element *B* can be written using the dyadic form of tensor (matrix) representation [22] in the form

$$[K_B] = N \cdot i_j i_j,$$

where i_j is a vector in which only the *j* component is non-zero and equal to one; *j* is a group of *Q*-nodes and displacement components in these nodes; *N* is a large arbitrary number.

Shape functions of an element in the form of a triangular prism are obtained from the hexagon shape function by reducing the gap between the *S*- and *Q*-nodes and equating the corresponding variables at these nodes. The stiffness matrix of elements *C* can be written in accordance with the specified properties

$$[K_c'] = [A]^T \cdot [K_c] \cdot [A] + N \cdot i_j i_j,$$

where $[A] = E - i_j i_j + i_j i_k$ is the transformation matrix; $E = \sum_S i_S i_S$ is the identity matrix; *k* is the index indicating *S*-nodes and components of displacements in these nodes (the indices *j* and *k* are consistent with each other); $[K_c]$ is the stiffness matrix with offset nodes.

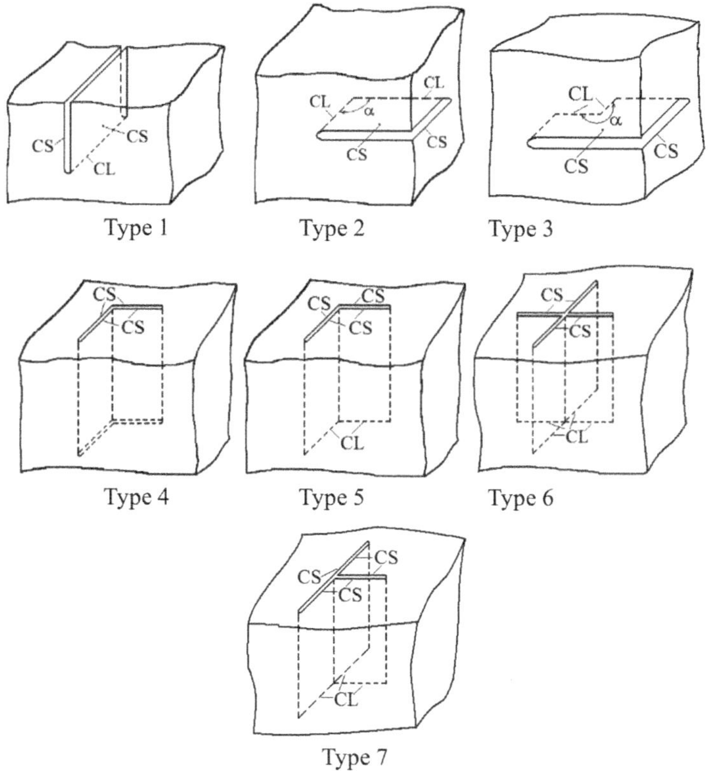

Fig. 32.2 Types of cracks taken into account: 1—through-thickness flat surface or internal crack; 2—non-through-thickness flat crack ($\alpha < 180°$), surface or internal one; 3—non-through-thickness flat crack ($\alpha > 180°$), surface or internal one; 4—L-shaped through-thickness crack; 5—L-shaped non-through-thickness crack, internal or surface one; 6—X-shaped crack, surface, internal, or through-thickness one ; 7—T-shaped crack, surface, internal, or through-thickness one

For the element D', the transformation formulas are similar to those for C', but the transformation matrix is different.

Figure 32.2 shows the most critical types and characteristics of the cracks taken into account.

According to the introduced types of cracks, schemes have been developed for transforming a regular finite element mesh into an irregular one, using transformed and degenerated finite elements. The general method for obtaining the stiffness matrix of the degenerated finite elements used for crack types 4 and 5, where S-nodes merge, is complemented by the procedure of the penalty function method.

When the nodes S_1 and S_2 merge, it is necessary to ensure equality of the variables $U_i^{S_k}$ of the nodes S_1 and S_2.

We introduce the function

$$F = N_1 \sum_{i=1}^{3} \left(U_i^{S_1} - U_i^{S_2} \right)^2,$$

where N_1 is an arbitrary sufficient number (penalty factor), which, when obtaining the stiffness matrix of a finite element, is added to the finite element strain energy, expressed by the quadratic form of the nodal parameters U_i

$$\Pi_e^* = \Pi_e + F.$$

Here, Π_e^* is a modified expression of the strain energy of the finite element e, the expression being a positive definite quadratic form.

Applying the usual procedure for obtaining the stiffness matrix of a finite element

$$\frac{\partial^*}{\partial u_i} = K u_i \quad ({}^* = u^T K^* u)$$

leads to obtaining a positive definite stiffness matrix K^*, in which N_1 is added to the diagonal elements of the variable nodes S_1 and S_2, and $-N_1$ is added to the corresponding coupling elements. Field tests of the penalty function procedure efficiency for various, but rather large values of N_1, gave positive results. The values of the variables of the nodes S_1 and S_2 showed good agreement, and no distortions were noticed in solving the complete system of the finite element method.

The method does not limit the number of cracks under consideration, and the formalization of the assignment of the initial information on the cracks ensures the efficiency of its introduction in case of variant studies.

32.3 A Method for Calculating the Dynamics of Space-Rocket Hardware Structures with Composite Structure Imperfections and Account Taken of the Curvilinear Anisotropy of Properties

The method for computing the dynamic stress-strain state and movement of an LV fairing is based on the application of the finite element method in three-dimensional statement where a volumetric poly-linear finite element with a topologically regular sampling system is used. When modeling the material of structural elements, both continuous heterogeneity or piecewise homogeneity and curvilinear anisotropy are allowed to be present, which makes it possible to compute both assembled and composite structures.

Applying the finite element method procedure, which is based on the use of the Lagrange–D'Alembert kineto-static variational principle, leads to a mathematical model represented by a system of ordinary differential equations

$$[M]\ddot{u} + [D]\dot{u} + [K]u = F \tag{32.1}$$

where u is the displacement vector for finite element mesh nodes; F is the vector of a given time-varying load; $[M]$, $[D]$, and $[K]$ are, respectively, the mass, damping, and stiffness matrices modified taking into account the presence of delaminations in the structure of composite elements.

Note that in dynamic problems with impulsive action, where the process is investigated over a relatively short period of time, the effect of damping is insignificant. In addition, damping values are usually unknown or determined with low accuracy. Therefore, the effect of damping is not taken into account here ($[D] = 0$).

Equation (32.1) reflect the displacements of the points of a body as rigid-body displacements, as well as mutual ones, that is, oscillations accompanied by deformations. The displacement vector u, and, accordingly, the equations themselves refer, usually, to the Cartesian coordinate system, which is the same for all finite elements, that is, the global coordinate system. One of the peculiarities of this problem is the presence in space-rocket hardware structures, in particular in fairings, of anisotropic elements, which can be conveniently simulated in a system of local coordinates x', y', z' in accordance with the principal axes of anisotropy (Fig. 32.3). In this case, the direction of the axes changes from element to element. That is, there is a case of curvilinear anisotropy, which introduces difficulties in determining and using the stiffness matrix. It is necessary to compute this matrix in a convenient local coordinate system, but use it in the general equation (32.1), which are written in the global Cartesian coordinate system.

In the local coordinate system x', y', z', there can be written the relationship

$$[K']u' = F' \tag{32.2}$$

where $[K']$ is the stiffness matrix; u' is the vector of nodal displacements; F' is the vector of nodal forces in local coordinates.

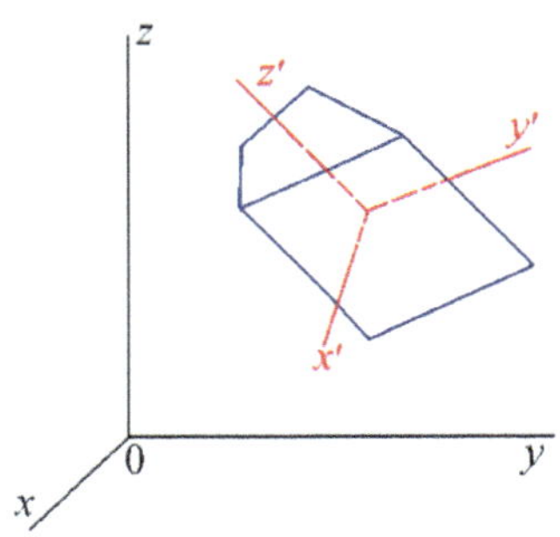

Fig. 32.3 A structural element in a local coordinate system associated with the principal axes of anisotropy

Nodal displacements and forces in the local and global coordinate systems are related as follows:

$$u' = [T]\,u, \;\; F' = [T]\,F, \tag{32.3}$$

where $[T]$ is the transformation matrix of vectors from the global to the local coordinate system (the cosine matrix).

We introduce transformations (32.3) into relation (32.2) and obtain

$$\left[K'\right][T]\,u = [T]\,F, \tag{32.4}$$

and then, on the left and right sides, we multiply relation (32.4) by the matrix T^{-1}, inverse to $[T]$, which leads to

$$[T]^{-1}\left[K'\right][T]\,u = [T]^{-1}\,[T]\,F. \tag{32.5}$$

Considering that $[T]^{-1}[T] = I$, where I is the identity matrix, and the transition matrix $[T]$ is orthogonal, and, therefore, $T^{-1} = T^{T}$, we obtain

$$[T]^{B}\left[K'\right][T]\,u = F. \tag{32.6}$$

Relation (32.6) is the relationship between the vectors of nodal displacements and forces in the global coordinate system, and the matrix expression on the left-hand side is the stiffness matrix in the global coordinate system

$$[K] = [T]^{\mathrm{B}}\left[K'\right][T]\,. \tag{32.7}$$

Relation (32.7) makes it possible to use stiffness matrices obtained in local axes, which refer to the principal axes of anisotropy, in order to form the main system of equations (32.1).

The solution of the equation (32.1) is carried out according to Newmark's and Wilson's implicit finite-difference schemes [23], which are second-order accuracy unconditionally stable. According to these schemes, the acceleration at the time step Δt is a linear function, and the Eq. (32.1) in the Newmark scheme are written for the time $t + \Delta t$, while in the Wilson scheme, for the time $t + \theta\Delta t$ ($\theta = 1.4$). In this case, there are no restrictions on the choice of the step Δt, which is mainly determined by the requirement for the accuracy and efficiency of computations.

In the problems of evaluating overloads under impulsive loading, where accelerations are taken as criterion parameters, or during collisions between structures, where a more accurate determination of velocities is important, the use of the Wilson-theta method is preferable.

32.4 Dynamic Stress-Strain State of a Monolithic-Composite Fairing Assembly During Its Separation from the LV

When the LV fairing is separated, a number of requirements are put forward due to the need to safely remove the fairing or its components from the LV trajectory. The fulfillment of these requirements is due to the implementation of the kinematics of the components being separated with account taken of the aerodynamic effect, as well as ensurance of their structural integrity during separation, which is determined by the fulfillment of strength conditions.

The fairing separation system considered in this work belongs to the class of pyromechanical systems. Its operation requires that the fairing has a greater speed in relation to that of the LV, which is achieved due to the impulse from the activation of the pyrotechnic separation system that is structurally connected with the fairing. When the fairing is accelerated, there develop significant dynamic loads causing unsteady oscillations and dynamic stresses, the assessment of which is necessary during design. Dynamic deformation processes during LV fairing separations are investigated in [24].

The LV fairing with elements of the pyrotechnic separation system is a composite structure whose design model is shown in Fig. 32.4.

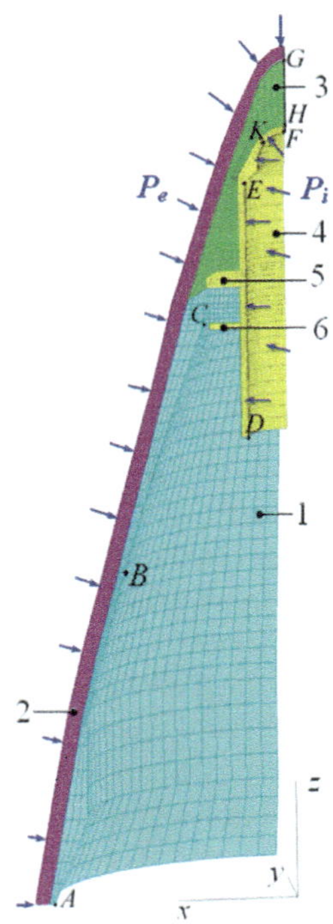

Fig. 32.4 Fairing design model: 1—shell; 2—heat-protection layer; 3—filler; 4—cylinder; 5—flange; 6—cover plate

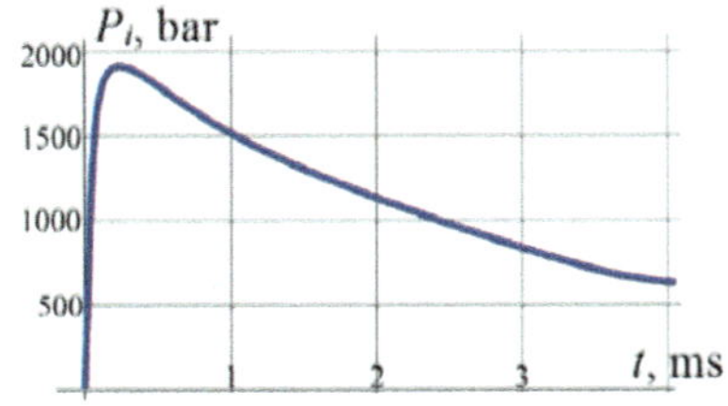

Fig. 32.5 Pressure P_i change

The model represents the metal elements (a cylinder, a flange, and a cover plate) related to the pyrotechnic separation system. The cover plate is not explicitly represented in the original design, but is introduced to connect the fiberglass shell with the cylinder instead of the bolted connection. The filler made of the DSV-2 material, partially carries the acting load, while performing the function of a supporting base. In the design model, the heat-protection layer is represented by almost only its mass characteristics (surface density 7.3 kg/m^2).

The fairing design is close to axisymmetric, if we do not take into account the presence of a nozzle whose influence is of a local nature. In this regard, in the model, one fourth of the whole structure is represented, and symmetry displacement conditions are introduced. The load on the fairing consists of the internal P_i and external P_e pressures caused, in the first case, by gases from powder sample combustion, and in the second one, by aerodynamic drag during motion. The pressures are known and set in tabular form. The curve of time-dependent pressure P_i changes is shown in Fig. 32.5. One of the peculiarities of the problem is the free movement of the fairing structure under the action of high internal pressure until the speed required for the actuation of the pyrotechnic fairing separation system is reached. Thus, there are no axial anchors in the fairing design model.

Computational studies were carried out for the main design variant of the shell made of anisotropic material (fiberglass), with and without taking into account the external aerodynamic pressure effect, and also, for comparison, for the variant of the shell made of isotropic material with averaged characteristics. The characteristics for the anisotropic material, corresponding to the mechanical properties of the shell fiberglass (lengthwise and crosswise), were taken as follows: $E_t = 18.3$ GPa; $E_\theta = E_n = 16.0$ GPa; $\nu_t = 0.15$; $\nu_\theta = \nu_n = 0.1$; $G_{t\theta} = G_{\theta n} = G_{tn}$=1.83 GPa. The values $E_t = E_\theta = E_n = 18.3$ GPa; $\nu_t = \nu_\theta = \nu_n = 0.3$, where the indices t, θ, n refer, respectively, to the meridional, circumferential, and normal directions in the shell, are taken as averaged characteristics.

Kinematic quantities (the axial and radial displacements, as well as axial velocities at point A on the shell edge) are shown in Fig. 32.6, where 1 refers to the isotropic shell; 2 refers to the anisotropic shell with no account taken of the external pressure P_e; 3 refers to the anisotropic shell with account taken of P_e. The law of axial displacement variation (close to parabolic) and axial velocity (close to linear) reflects a combination of movements caused by the action of an impulse of the internal pressure P_i. The total motion consists of the translational rigid-body motion of the entire structure under the action of a force that smoothly decreases in time after reaching the maximum value

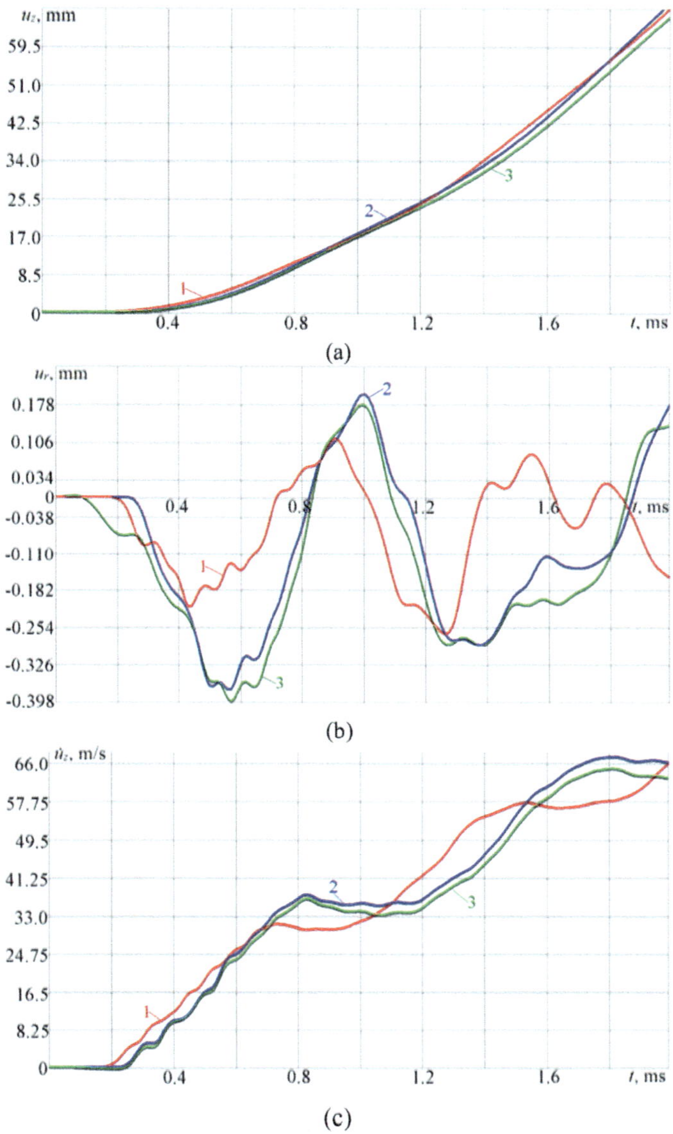

Fig. 32.6 Kinematic parameters of the shell at point *A*: **a**—axial displacement; **b**—radial movement; **c**—axial speed

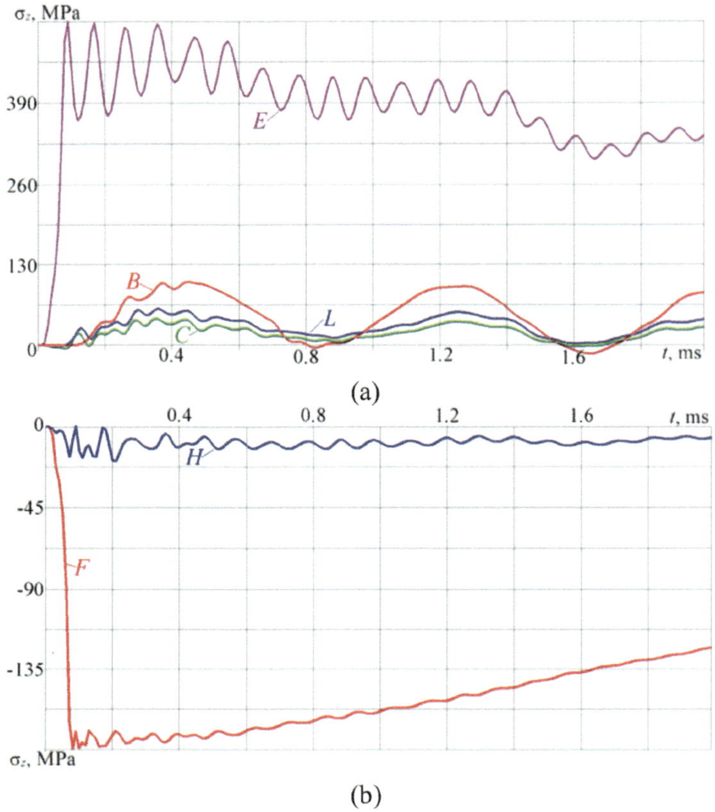

Fig. 32.7 Axial stresses at control points

(absolute motion), and oscillatory processes due to deformation (relative motion), where the dynamic effect manifests itself. A delay in the reaction of the stress-strain state at point A on the shell edge to the action of the internal pressure pulse P_i is observed, which is associated with the wave nature of the propagation of disturbances.

Stress computation data for the fairing made of the anisotropic elastic material are shown in Figs. 32.7, 32.8 and 32.9, which depict the results for the steel cylinder and fiberglass shell. The computed axial stress data at points F and H of the cylinder (Fig. 32.7) correspond with sufficient accuracy to the boundary values of the internal pressure P_i, which is a criterion for achieving the required accuracy of solving the entire problem. The stress values in the steel cylinder (points E, F, and K) are very moderate and far from the limiting values ($\sigma_y = 1300$ MPa, $\sigma_f = 1600$ MPa).

Comparison of the computed data for different shell variants shows that taking into account the external aerodynamic resistance has an insignificant effect on the values of the kinematic and power characteristics being determined. The anisotropy of properties is largely reflected in the radial displacements and stresses in the shell.

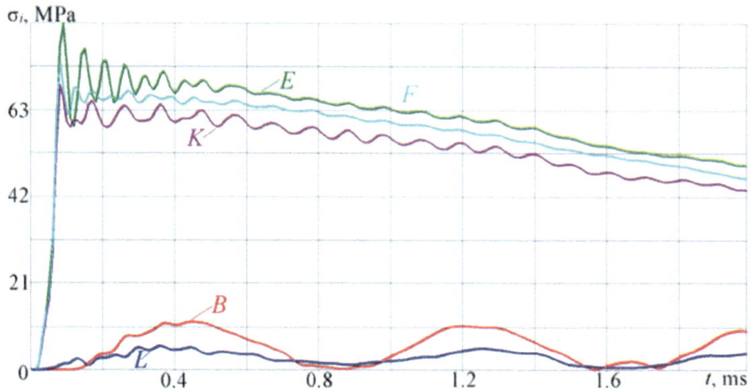

Fig. 32.8 Stresses intensity at control points

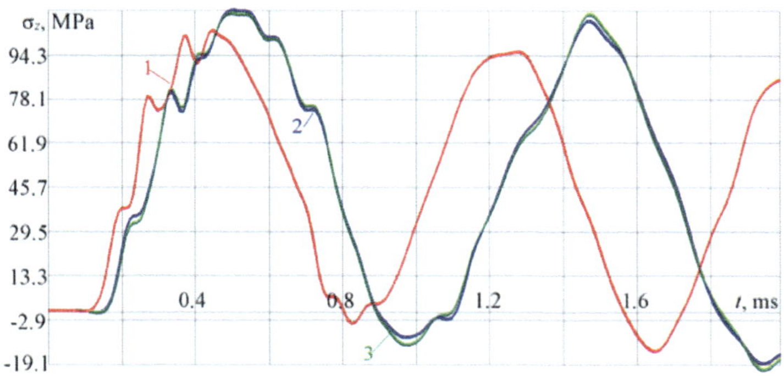

Fig. 32.9 Axial stresses at point *B* of the shell

The value of the maximum tensile stresses in the fiberglass shell (about 110 MPa) is much lower than that of the rupture stresses (480 and 220 MPa, respectively, lengthwise and crosswise), and the maximum values of spall stresses and interlayer shear stresses ($\tau_{tn} = 4.5$ MPa), which also does not exceed the ultimate shear stress (13 MPa).

32.5 Influence of Delamination Defects in the Fairing Composite Shell on the Dynamic Stress-Strain State of the Fairing During Its Separation from the LV

In studying laminated fiberglass shells, various layers are considered, in particular, those located in the upper part of the shell from the flange and in the lower part from the shell edge [24]. The location of delaminations, which are assumed to be axisymmetric, for simplicity, is depicted in Fig. 32.10, which also shows the control

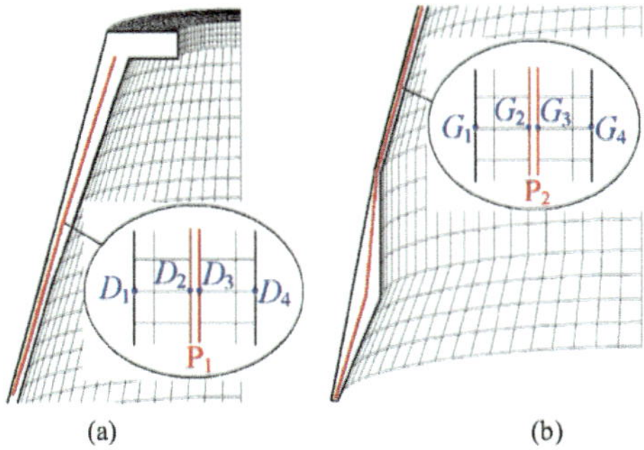

Fig. 32.10 Types of delamination in the shell: **a**—delamination P_1; **b**—delamination P_2

points on the outer shell surfaces and delamination surfaces. The dimensions of the delaminations are taken to be macroscopic, commensurate with the characteristic shell dimensions, in order to reveal the large effects of stress-state redistribution. The interaction of the resulting delamination surfaces among themselves was not taken into account.

Two types of delaminations are considered (P_1 and P_2, in Fig. 32.10), which initiate from irregular fiberglass-winding zones: the transition zone from the shell flange (P_1) and the zone from the shell edge (P_2), where delamination is most likely to occur. The main attention in the computational studies was paid to the redistribution of the nominal values of the stress-strain state due to delamination. The influence of the external pressure P_e on the aerodynamic resistance is small. In the study, only the impulse load from the propellant gas pressure P_i was taken into account.

The solution of equations (32.1) for the axial component of the displacement vector uz includes the rigid-body displacement whose value is usually determined by the methods of theoretical mechanics and characterizes the motion of the center of mass. In the free motion along the z-axis, these displacements are predominant and significantly exceed the relative axial displacements that characterize the axial oscillations caused by deformations. The relative displacements u_z are obtained by excluding the rigid-body displacement from the solution because the displacement was determined by separately computing the dynamics of the fairing with overestimated mechanical characteristics of the materials of its elements.

With the appearance of a delamination, the displacements at points on its surfaces (points D_2 and D_3, of the delamination P_1 and points G_2 and G_3 of the delamination P_2) are not the same, that is, they are undergoing rupture. Curves of changes in radial (u_r) and axial (u_z) displacements are shown in Figs. 32.11 and 32.12.

The relative displacements u_z, after excluding the rigid-body displacement from the solution, are computed starting from the center of mass of the fairing. It can be noted that the differences in the displacements along the P_1 delamination faces are significantly less than in the case of the P_2 ones. This is due to the constraint on the mutual displacements of the faces of the internal delamination P_1, in contrast to the delamination P_2, which extends beyond the boundary. It was pointed out that contact interaction is not taken into account, which is quite justified, since the

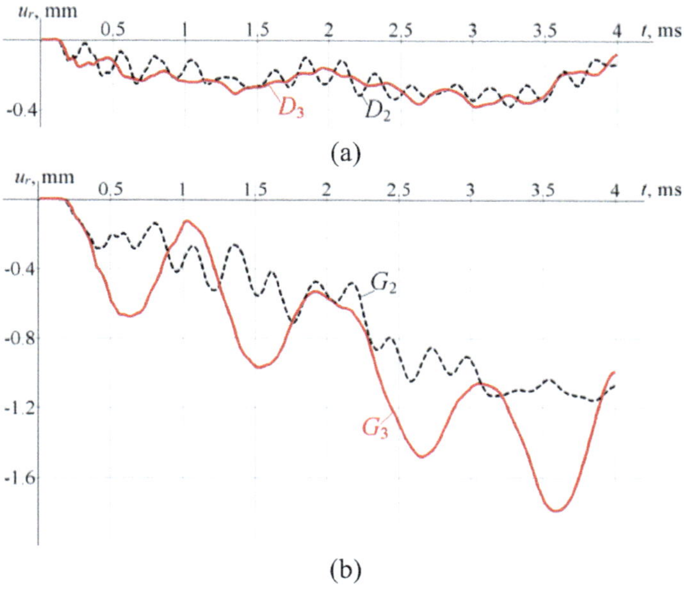

Fig. 32.11 Radial displacements (u_r) at points D_2, D_3 (at P_1) and at points G_2, G_3 (at P_2)

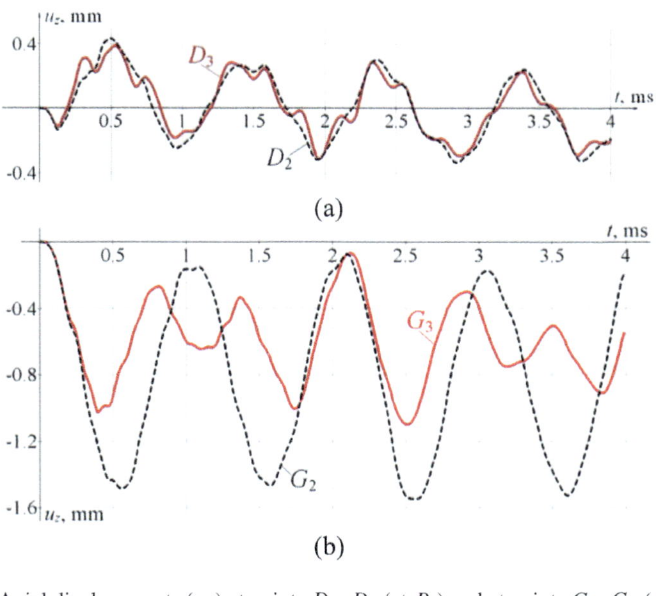

Fig. 32.12 Axial displacements (u_z) at points D_2, D_3 (at P_1) and at points G_2, G_3 (at P_2)

jumps in displacements along the normal to the delamination surface are small, and the gap between them does not disappear. The contact of the faces within the framework of the adopted methodology can be taken into account in accordance with [24]. The displacements for the delamination P_2 are maximum $(u_r)_{max} = 3.15$ mm; $(u_z)_{max} = 2.95$ mm, and occur at the lower shell edge on the outer face of the heat-protection layer. For a one-piece shell, the maximum displacements occur in the same place and are $(u_r)_{max} = 1.76$ mm and $(u_z)_{max} = 2.06$ mm.

The stresses σ_z in the fairing shell are maximum, and their redistribution due to delamination is of greatest interest. Their change in time for a one-piece shell and

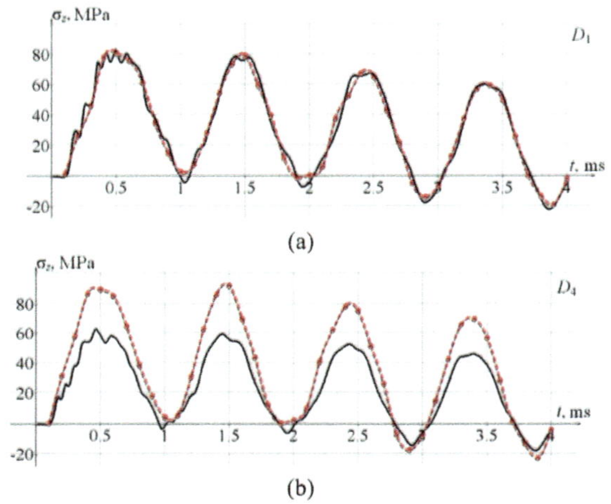

Fig. 32.13 Axial stresses at points D_1 (**a**) and D_4 (**b**): solid line—one-piece shell; dotted line—delamination P_1

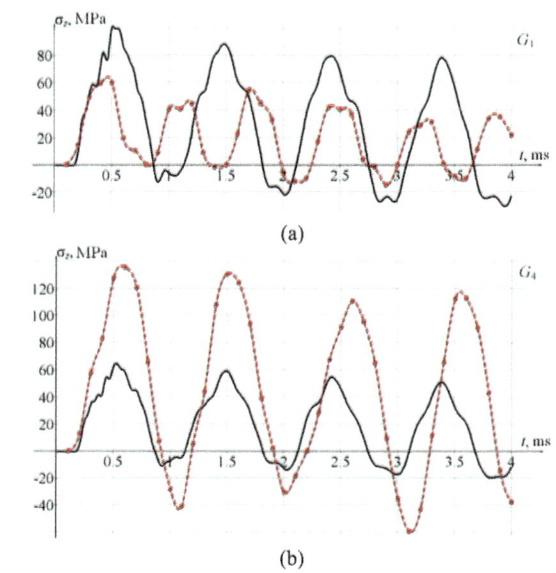

Fig. 32.14 Stresses σ_z at points G_1 (**a**) and G_4 (**b**): solid line—one-piece shell; dotted line—delamination P_2

a delaminated one at certain points on their surfaces (points D_1, D_4 and G_1, G_4) is shown in Figs. 32.13 and 32.14 It can be noted that the response delay at points G_1 and G_4 (P_2), which are more remote from the place of the impulse action, is greater than at points D_1 and D_4 (P_1), which is caused by the wave nature of the response at the beginning of the disturbance. For P_2, the stress redistribution over the delamination layers is much greater than that for P_1 because of the above reason. In addition, an important circumstance is the action of the inertial load from the side of the heat-protection layer directly on the outer layer of the delaminated shell, which predetermines a significant increase in the stress state. Based on the above results, it can be noted that the delaminations going beyond the edge of the fiberglass shell are more dangerous, because they significantly increase stresses due to their redistribution.

32.6 Oscillations of a Locally Buckled Structural Element with Near-Surface Delamination. Model of a Rod with a Longitudinal Crack and Calculation Method

The presence of longitudinal cracks in elements of thin-walled structures is quite likely, in particular for layered composites, where cracks occur due to delamination. When such elements are under alternating bending strain, there appear tensile and compressive stresses directed along the cracks. When the elements are under tensile stress, the cracks do not manifest themselves, and when they are under compressive stresses, there may appear phenomena related to the buckling of the structural equilibrium of material in the volume surrounding the crack. In this case, one of the main tasks of fracture mechanics is to assess the strength of bodies with longitudinally compressed cracks. In the most general setting, this problem was investigated in the papers reviewed in [25, 26]. According to the developed 3D linearized theory of stability of deformable bodies [27], once the compression stresses have reached their critical level, there occur, in the volume of material surrounding the crack, stress-strain state perturbations, or so-called "local forms of buckling". If the crack is close to the surface, near-surface buckling occurs. The buckling of material in the volume surrounding the crack precedes brittle fracture.

A particularly important task is to assess the strength of layered and fiber composites, which are destroyed through delamination during compression in the direction of reinforcement. In the context of applying the concept of local buckling, a significant number of publications are devoted to this problem, in particular [27, 28]. In contrast to the specified approach based on strict provisions, approximate beam models were developed [29, 30]. Despite the lack of estimates of the accuracy of the assumptions used, "beam" approaches are simpler, more pronounced, and can be successfully used to solve problems of applied mechanics.

The phenomenon of local buckling around a longitudinal crack does not always result in the destruction of material and a further advancement of the crack. For

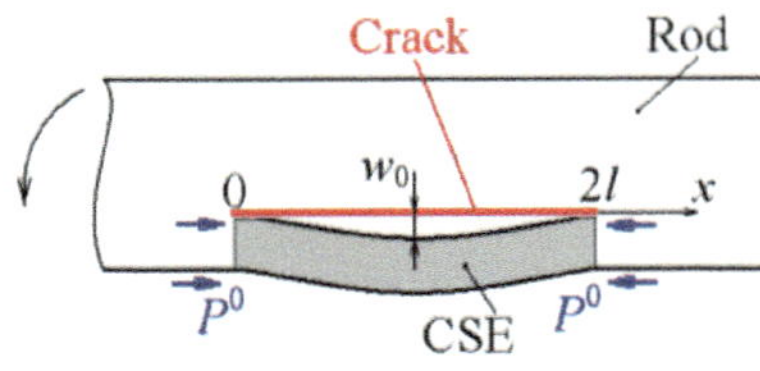

Fig. 32.15 A model of a rod with a near-surface crack

example, in the case of a near-surface crack in a rod, the local buckling of its part (the element between the crack and the rod surface, further referred to as the crack-to-surface element (CSE)) occurs once the local compression strain has reached its critical value. With further rod deformation, the CSE experiences supercritical deformation, i.e. it is both in a state of compression and bending. In this case, the bending stiffness of the rod changes depending on the magnitude of deformations, which determines the nonlinearity of oscillations.

The task of this study is to determine the influence of local buckling in a rod with a longitudinal near-surface crack on its characteristics during oscillations. In this case, a "beam" scheme is used to determine both the state of local buckling and supercritical deformation at small displacements [31].

A rod fragment with a longitudinal near-surface crack during bending is considered (Fig. 32.15), with a thin rod-like bucklable CSE of length $2l$ highlighted with a bold line. When the rod undergoes bending, the CSE ends experience the relative displacement u along the x-axis, the displacement determining the longitudinal force arising in the CSE. When the rod undergoes compression after the critical value of u_c (critical force) has been reached, the CSE buckles additionally undergoes bending. With a further bending of the rod (increase of u), there occurs the supercritical CSE deformation, which is considered at small deflections without taking into account the inertia properties during oscillations. Obviously, the rigidity of the CSE, as part of the rod, changes after buckling, i.e. it is a variable and determined by the dependence $P(u)$, where P is the force in the CSE with the relative displacement u. The $P(u)$ dependence is to be determined.

For definiteness, we will accept the boundary conditions for CSE displacements $(0, 2l)$ in the form of fixed-ends, although they have a more complex form, intermediate between that of fixed and free support. The displacement u is composed of the power component u_p and the geometric component u_g.

$$u = u_p + u_g. \tag{32.8}$$

The u_p and u_g components are defined by the formulas

$$u_p = \frac{2Pl}{EF}, \quad u_g = \frac{1}{2}\int_0^{2l} \left[w'(x)\right]^2 dx, \tag{32.9}$$

where F is the cross-sectional area of the CSE; w_x is the deflection of the CSE after buckling.

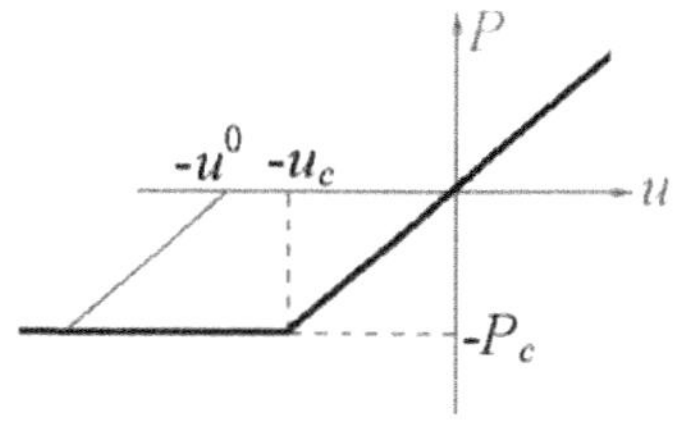

Fig. 32.16 $P(u)$ dependence for CSE

Given the boundary conditions, we take

$$w = \frac{1}{2} w_0 \left(1 - \cos \frac{\pi}{l} x\right),$$

where w_0 is the deflection of the CSE in the middle (Fig. 32.15).

In view of Eqs. (32.8) and (32.9), we get

$$u = \frac{2Pl}{EF} + \frac{\pi^2 w_0^2}{8l}. \tag{32.10}$$

The state of buckling $P = P_c$, where $P_c = \pi^2 EJ/l^2$ is the Euler force, is reached at $u_c = 2P_c l/(EF) = 2\pi^2 J/(Fl)$. In a subcritical state, at $u \leqslant u_c$ $(P < P_c)$, $w_0 = 0$ and the equilibrium form is rectilinear, at $u \leqslant u_c$ $(P < P_c)$, after buckling, $w_0 \neq 0$ and supercritical deformation occurs, controlled by the displacement u. After buckling at small deflections, the equilibrium state is indifferent, i.e. the supercritical deformation occurs at $P = P_c$. The $P(u)$ dependence corresponding to the above is presented in Fig. 32.16.

The variable CSE stiffness also causes the rod stiffness change, depending on the magnitude and sign of the deformation (displacement u). In the CSE deformation calculations, it is appropriate to perform the linearization of the nonlinear dependence $P(u)$ by the initial deformation method. The physical relation for the CSE is represented in the form $P = EF(\varepsilon - \varepsilon^0)$, where $\varepsilon^0 = u^0/(2l)$ is the additional deformation. The account of the additional deformation in the calculations is equivalent to the application of additional forces at CSE ends, $P^0 = u^0/\beta$, where $\beta = 2l/(EF)$ is the CSE compliance under static loading; $u^0 = u - u_c$.

In the case of several near-surface cracks, the buckling of the CSEs located between the cracks occurs sequentially from the external CSE, adjacent to the surface of the rod, to the internal ones. The calculation formulas and the algorithm of taking into account the CSE stiffness changes are similar to those given above.

The modeling of delamination in a structural element and the calculation of oscillations are performed according to the methodology presented above. A feature of its application in this case is the introduction of forces P_0 at CSE edges at each integration step, for which the displacement u can be represented in the form

$$u = u_t + \alpha P^0 \tag{32.11}$$

where u_t is the current step displacement in the Wilson scheme (excluding additional forces); α is the displacement of CSE edges in the rod due to the impulse of unit forces P_0 (dynamic compliance). Taking into account the expression for $P^0 = (u - u_c)/\beta$ by the method of additional deformations and excluding u from the Eq. (32.11), we obtain the expression for the edge forces

$$P^0 = \frac{u_t - u_c}{\beta - \alpha}. \tag{32.12}$$

Denominator in the formula (32.12) is non-zero, since the dynamic compliance is less than static. The additional forces P_0 at each step are uniquely determined by formula (32.12) and do not require iterations. A similar approach to take into account the nonlinearity of individual structural elements was used in [18].

The assessment of the influence of the buckling of near-surface delaminated elements on oscillations is considered for a number of problems for layered rods with defects in the form of various types of delaminations.

32.7 A Cantilever Rod with a Single Near-Surface Crack

A diagram of a rod with a through-thickness longitudinal crack is shown in Fig. 32.17.

The calculation data are as follows: $L = 80$ cm, $b = 1$ cm, $H = 4$ cm, $h_1 = 1$ cm, $l = 40$ cm, $E = 20.6$ MN $\cdot$ cm^2, and $m = 8.15$ kg. It is accepted that the rod is massless while performing the function of variable stiffness, and concentrated masses are placed at its end. In this case, the system can be reduced to a single-mass variable stiffness model and analyzed analytically.

Free oscillations of the presented nonlinear system were studied, and the dependence of the frequency of oscillations on their amplitude was determined. Note that due to the asymmetry of the dependence of the restoring force, the maximum deviations of the different signs A^+ and A^- differ. In this case, introduced are the following characteristics: the center of oscillations, $A = (A^- - A^+)/2$, and the half-amplitude of oscillations, $A = (A^- + A^+)/2$. The dependence of the restoring force, $F(w_m)$, where F and w_m are the force at the cantilever end and its displacement (mass m), is calculated for a static load, and the data are placed in Table 32.1.

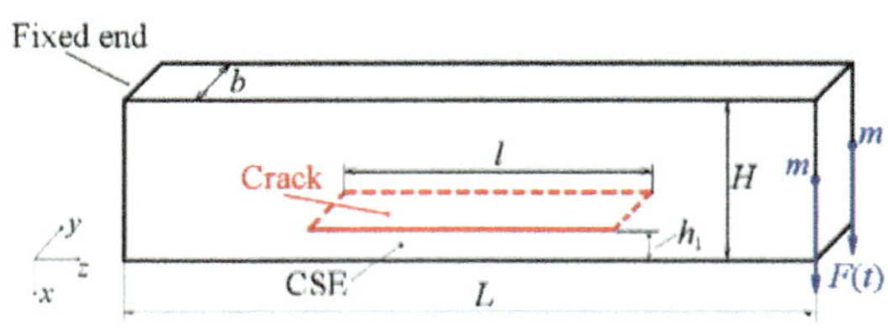

Fig. 32.17 A model of a rod with a near-surface crack

Table 32.1 Dependence of the restoring force on displacement

F, kN	2.452	0	−0.981	−1.020	−1.472	−1.962	−2.452	−2.943	−3.434
w_m, cm	3.801	0	−1.520	−1.582	−2.557	−3.618	−4.679	−5.740	−6.801

Table 32.2 Dependence of the frequency of free oscillations on amplitude. Proposed model

A, cm	1.582	1.977	2.270	3.047	3.865	4.673
Δ, cm	0.000	0.005	0.017	0.038	0. 106	0.161
p, Hz	10.00	9.914	9.833	9.699	9.590	9.537

The dependence $F(w_m)$ is bilinear and is represented as

$$F = \begin{cases} k_0 w_m, & w_m \geq w_c \\ -F_c + k_1(w_m + w_c), & w_m < -w_c, \end{cases} \tag{32.13}$$

where F_c and w_c are the values of the force at the cantilever end and its displacements at the moment when the CSE buckles ($F_c = 1.020$ kN, $w_c = 1.582$ cm), respectively; k_0 and k_1 are the coefficients characterizing the slope of lines in the diagram ($k_0 = 1.291$ kN/cm, $k_1 = 0.925$ kN/cm).

Free oscillations were caused by the initial deflection of the system during quasi-static loading by the forces $F(t)$ and subsequent drop of load in the presence of a small level of damping. The calculation results for the proposed model and methodology are presented in Table 32.2.

For comparison, we present data for the natural oscillation frequency of a solid rod, $p = 10.15$ Hz, and a rod with a crack, but without taking into account the CSE buckling, $p = 10.0$ Hz. The data in Table 32.2 determine the skeletal curve with soft nonlinearity, which occurs in this case.

To assess the accuracy of calculating nonlinear oscillations according to the presented method, the frequencies of free oscillations were calculated using the analytical dependence obtained by the direct linearization method [32]. In doing so, the bilinear dependence of the restoring force (32.13) obtained by calculation was used. The frequency of free oscillations is determined according to the formula

$$p^2 = \frac{5}{4mA^5} \int_{-A}^{A} F(u - \Delta)\, u^3 du, \tag{32.14}$$

where the function F is the restoring force characteristic (32.13).

The calculation data with the use of the formula (32.14) are presented in Table 32.3.

Comparison of the data in Tables 32.2 and 32.3, obtained respectively by calculation and analytically, are in good agreement. This indicates the acceptable accuracy of the calculations of nonlinear oscillations with the restoring force characteristic

Table 32.3 Dependence of the frequency of free oscillations on amplitude. Direct linearization method

A, cm	1.582	1.977	2.271	3.031	3.866	4.673
Δ, cm	0.000	0.006	0.015	0.053	0.105	0.160
p, Hz	10.010	9.952	9.886	9.741	9.633	9.562

determined by the proposed model. The change in natural oscillation frequencies for the selected range of amplitudes, limited by the magnitude of deflections and stresses to the yield strength, is up to 6%.

32.8 A Cantilever Rod with a System of Collinear Surface Cracks

The scheme of the rod is similar to the previous example, but additionally, through the thickness of the CSE shown in Fig. 32.17, one and two cracks of the same length l were introduced. Accordingly, the following design studies were considered: two cracks (two CSEs with thicknesses $h_2 = 0.5$ cm) and three cracks (three CSEs with thicknesses $h_3 = 1/3$ cm). Due to the complexity of the problem, a higher finite element discretization was used. When modeling the oscillations of the rod, a scheme of successive buckling from the external CSE to the internal one was used.

Studies of the nonisochronism of free oscillations were carried out, and amplitude-frequency characteristics were obtained with the identification of various resonances.

The dependencies of the frequency of free oscillations on amplitude for a different number of cracks in the form of skeletal curves (nonisochronism of oscillations) are shown in Fig. 32.18. According to the dependencies, with an increase in amplitude, the frequency of free oscillations for each of the considered options tends to a certain limit. With an increase in the number of cracks, and, correspondingly, a decrease in the h_i of the CSE, the values involved in the critical displacement u_c are matched with $F \sim h_i$, $J \sim h_i^3$, resulting in $u_c \sim h_i^2$. That is, with an increase in the number of cracks, the CSE resistance to compressive stresses rapidly decreases, which is almost equivalent to the absence of material resistance in the region occupied by the cracks. In the limit, the restoring force characteristic, referred to as the unit of mass, is bilinear with a kink at the origin of coordinates and the tilt angles α_1, α_2 of the lines in the diagram for positive and negative displacements, respectively. In this limiting case, the system is isochronous, and the frequency of free oscillations is determined by the formula

$$p_0 = \frac{p_1 \cdot p_2}{p_1 + p_2},$$

where $p_1^2 = \tan \alpha_1$, $p_2^2 = \tan \alpha_2$.

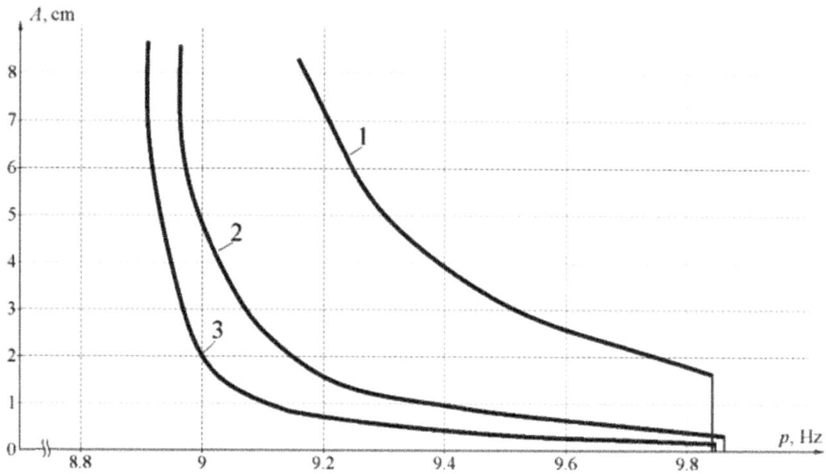

Fig. 32.18 Skeletal curves for the rod: 1—with one crack; 2—with two cracks; 3—with three cracks

The stiffness calculations of an inhomogeneous rod with underestimated and initial elastic moduli yielded the values $\tan\alpha_1 = 3.856\ \mathrm{s}^{-2}$, $\tan\alpha_2 = 2.557\ \mathrm{s}^{-2}$, which determines $p_1 = 9.883$ Hz, $p_2 = 8.048$ Hz and, accordingly, $p_0 = 8.872$ Hz. The value of p_0 is close to the limiting frequency value at large oscillation amplitudes for the rod with three cracks.

The amplitude-frequency characteristics for the rod with three cracks are constructed for harmonic action by the cantilever force $F = F_0 \sin(\omega t)$ with different fixed amplitude values $F_0 = 29.43$ N (3 kgf); 49.05 N (5 kgf); 98.1 N (10 kgf) for viscous damping with the logarithmic decrement of oscillations $\delta = 0.38$. Such a significant damping is adopted for a higher rate of oscillation stabilization. Figure 32.19 shows the amplitude-frequency characteristics in the region of the main resonance close to the natural frequency of a quasi-linear system. The parameter $\kappa = A/A_{st}$, where A is the half-amplitude of periodic forced oscillations at the excitation frequency ω and A_{st} is the half-amplitude of oscillations at the frequency $\omega = 1$ Hz (quasi-static loading). A distinctive feature of the amplitude-frequency characteristics of the rod with a system of three cracks is their dependence on the amplitude of the exciting effect. In particular, the maxima of the amplitude-frequency characteristics are biased, occurring at 9.0; 9.2; 9.4 Hz with a decrease in the amplitude of excitations, and accordingly the amplitude of oscillations. This result is due to the oscillatory properties of the considered nonlinear system, in particular its skeletal curve.

In a wider frequency range, the amplitude-frequency characteristic reveals features due to the presence in the periodic solution of various harmonics that are multiples of the period. Computational studies were carried out for the rod with three cracks at the excitation amplitude $F_0 = 0.981$ kN (100 kgf). In the region of lower excitation frequencies (with respect to the frequency of the main resonance), the

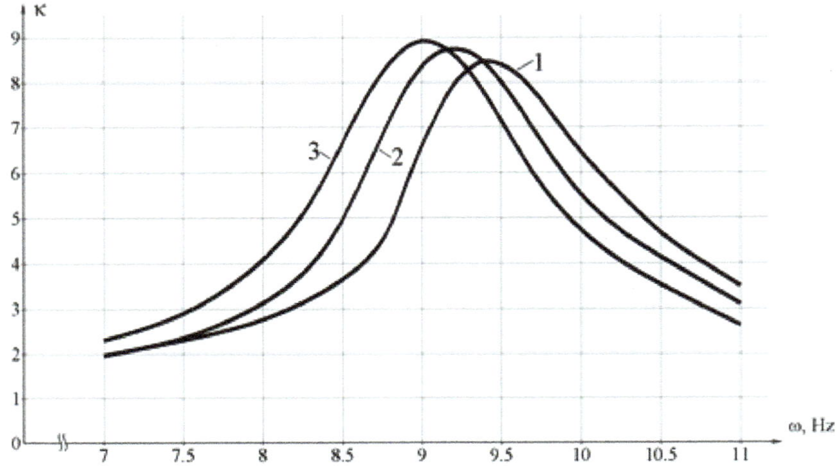

Fig. 32.19 Resonance curves at different amplitudes of excitation: 1—$F_0 = 29.43$ N; 2—$F_0 = 49.05$ N; 3—$F_0 = 98.1$ N

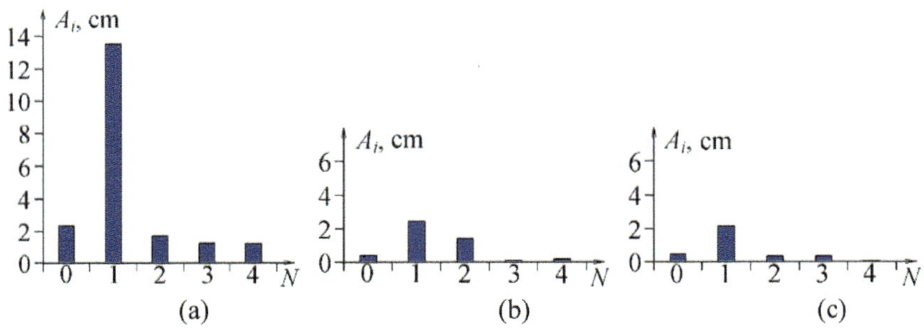

Fig. 32.20 Amplitudes of harmonics at resonant frequencies: **a**—$\omega = 8.9$ Hz (main resonance); **b**—$\omega = 4.5$ Hz (super-resonance 2/1); **c**—$\omega = 3.0$ Hz (super-resonance 3/1)

frequencies $\omega_{2/1} \approx 4.5$ Hz and $\omega_{3/1} \approx 3.0$ Hz were revealed, for which the second and third harmonics in the Fourier expansion, respectively, have a local extrema. The histograms of frequency analysis are presented in Fig. 32.20, where, also for comparison, a histogram is shown for the case of the main resonance ($\omega \approx 8.9$ Hz). The indicated frequencies correspond to the states of super-resonances 2/1 and 3/1, whose oscillatory mode features are illustrated in Fig. 32.21. Thus, with super-resonances, higher harmonics commensurate with the fundamental harmonic, especially with the 2/1 super-resonance, and the damping does not change the harmonic ratio over time.

Under high-frequency excitation, in particular at the doubled frequency of the main resonance, $\omega = 17.8$ Hz (oscillatory period $T_\omega = 5.618 \cdot 10^{-2}$ s), a periodic movement occurs with the frequency $\omega/2$ (oscillatory period $T_{\omega/2} = 1.124 \cdot 10^{-1}$ s). This can be considered as the appearance of the 1/2 sub-harmonic, which has a local extremum at the indicated frequency, i.e. the 1/2 sub-resonance takes place. The

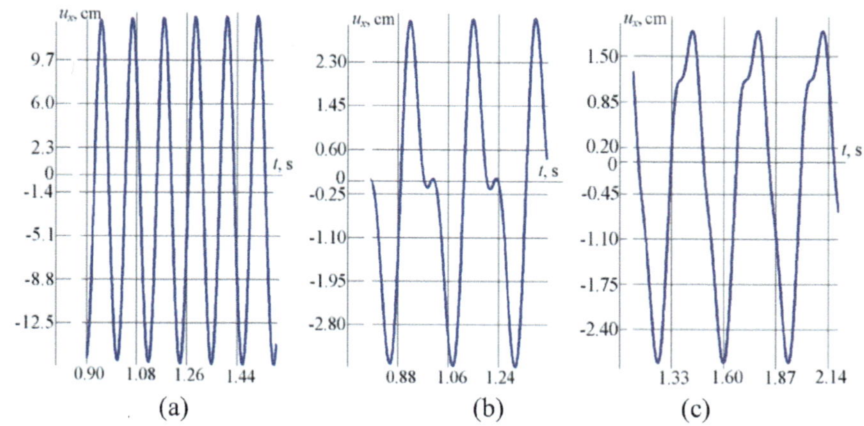

Fig. 32.21 Oscillatory modes at resonances: **a**—$\omega = 8.9$ Hz (main resonance); **b**—$\omega = 4.5$ Hz (super-resonance 2/1); **c**—$\omega = 3.0$ Hz (super-resonance 3/1)

Table 32.4 Change in the 1/2 sub-harmonic over time

Number of periods $T_{\omega/2}$	A_0, cm	$A_{1/2}$, cm	A_1, cm	A_2, cm	$A_{1/2}/A_1$
8	0.117	0.528	0.647	0.002	0.816
16	0.104	0.283	0.653	0.002	0.433
24	0.101	0.170	0.654	0.002	0.260
32	0.100	0.106	0.655	0.001	0.162

frequency analysis data are placed in Table 32.4, where the amplitude values of the harmonics are indicated for different instants of time (in periods of oscillation $T_{\omega/2}$).

It is obvious that over time, under the influence of damping, the ratio of harmonics $A_{1/2}$ and A_1 changes significantly, which is evident when oscillatory modes are considered. That is, under the influence of damping over time, the 1/2 sub-harmonic is suppressed.

32.9 Conclusion

The proposed mathematical support for the dynamic analysis of the effect of delamination in composite elements of space-rocket hardware has a sufficient range of capabilities and shows its effectiveness. In the example of an LV fairing during its separation, various types of delamination from irregular winding zones are investigated. Our calculations show that the most unfavorable delaminations are those going beyond the boundary of a composite element and leading to a significant increase in dynamic stresses due to their redistribution. In addition to the effect of delamination on the change in the stress-strain state, the phenomenon of local buckling for the case of near-surface delamination is considered.

A model has been built and a method has been developed to take into account the effect of local delamination-induced buckling on the stiffness of a structural element during bending oscillations. In the examples of a structure with one near-surface crack and a system of collinear cracks, its nonlinear oscillations are considered. The nonisochronism of free oscillations, the features of the main resonance depending on the excitation amplitude, the presence of super- and sub-resonances, and the effect of damping on sub-harmonic oscillations are shown numerically.

The proposed method can be used to study the oscillations of layered composite structures [33, 34], taking into account cracks and buckling [35, 36]. Further studies are needed on the delamination buckling interaction and analyses of the influence of delamination geometry and its position on strength.

References

1. Kondratiev, A., Gaidachuk, V., Nabokina, T., Tsaritsynskyi, A.: New possibilities of creating the efficient dimensionally stable composite honeycomb structures for space applications. In: Nechyporuk, M., Pavlikov, V., Kritskiy, D., (eds.) Integrated Computer Technologies in Mechanical Engineering. Advances in Intelligent Systems and Computing, vol. 1113. Springer, Cham (2020). https://doi.org/10.1007/978-3-030-37618-5_5
2. Kondratiev, A., Pištěk, V., Smovziuk, L., Shevtsova, M., Fomina, A., Kučera, P.: Stress-strain behaviour of reparable composite panel with step-variable thickness. Polymers **13**(21), 3830 (2021). https://doi.org/10.3390/polym13213830
3. Kondratiev, A.V., Gaidachuk, V.E.: Mathematical analysis of technological parameters for producing superfine prepregs by flattening carbon fibers. Mech. Compos. Mater. **57**(1), 91–100 (2021). https://doi.org/10.1007/s11029-021-09936-3
4. Gontarovskyi, P., Smetankina, N., Garmash, N., Melezhyk, I.: Numerical analysis of stress-strain state of fuel tanks of launch vehicles in 3D formulation. In: Nechyporuk, M., Pavlikov, V., Kritskiy, D., (eds.) Integrated Computer Technologies in Mechanical Engineering–2020. ICTM 2020. Lecture Notes in Networks and Systems, vol. 188, pp. 609–619. Springer, Cham (2021). https://doi.org/10.1007/978-3-030-66717-7_52
5. Vambol, O., Kondratiev, A., Purhina, S., Shevtsova, M.: Determining the parameters for a 3D-printing process using the fused deposition modeling in order to manufacture an article with the required structural parameters. Eastern-European J. Enterprise Technol. **2**(1), 70–80 (2021). https://doi.org/10.15587/1729-4061.2021.227075
6. Kondratyev, A.V., Kovalenko, V.A.: Optimization of design parameters of the main composite fairing of the launch vehicle under simultaneous force and thermal loading. Space Sci. Technol. **25**(4), 3–21 (2019). https://doi.org/10.15407/knit2019.04.003
7. Kulaga, Y.S., Olenin, I.G.: Development of head fairings made of composite materials. Collect. Sci. Tech. Develop. OKB-23 – Design Bureau "Salyut" **1**, 418–436 (2006)
8. Kondratiev, A., Gaidachuk, V.: Weight-based optimization of sandwich shelled composite structures with a honeycomb filler. Eastern-European J. Enterprise Technol. **1**(1), 24–33 (2019). https://doi.org/10.15587/1729-4061.2019.154928
9. Drozdov, A.V., Kharchenko, V.V., Dzyuba, V.S., Kravchuk, L.V., Potapov, A.M., Sirenko, V.N., Gusarova, I.A., Klimenko, D.V., Kharchenko, V.N., Samusenko, A.A.: Multifunctional data acquisition and control system rigging the beds and units for strength tests of models and structure elements from composite materials. Strength Mater. **48**(5), 632–638 (2016). https://doi.org/10.1007/s11223-016-9806-8

10. Drozdov, A.V., Kharchenko, V.V., Potapov, A.M., Klimenko, D.V., Kharchenko, V.N., Samusenko, A.A.: Computation software for strength and elastic characteristics of polymer composites. Strength Mater. **48**(6), 768–776 (2016). https://doi.org/10.1007/s11223-017-9823-2
11. Zaitsev, B.P., Protasova, T.V., Smetankina, N.V., Klymenko, D.V., Larionov, I.F., Akimov, D.V.: Oscillations of the payload fairing body of the cyclone-4M launch vehicle during separation. Strength Mater. **52**(6), 849–863 (2020). https://doi.org/10.1007/s11223-021-00239-5
12. Gontarovskyi, P., Garmash, N., Melezhyk, I.: Numerical modeling of dynamic processes of elastic-plastic deformation of axisymmetric structures. In: Advances in Mechanical and Power Engineering. CAMPE 2021. Lecture Notes in Mechanical Engineering. Springer, Cham, 334–342 (2023). https://doi.org/10.1007/978-3-031-18487-1_34.
13. Shupikov, A.N., Ugrimov, S.V., Kolodiazhny, A.V., Yareschenko, V.G.: High-order theory of multilayer plates. the impact problem. Int. J. Solids Struct. **35**(25), 3391–3403 (1998). https://doi.org/10.1016/S0020-7683(98)00020-1.
14. Ugrimov, S.V., Shupikov, A.N., Lytvynov, L.A., Yareshchenko, V.G.: Non-stationary response of sapphire rod on longitudinal impact. Theory and experiment. Int. J. Impact Eng. **104**, 55–63 (2017). https://doi.org/10.1016/j.ijimpeng.2017.02.005
15. Smetankina, N.V., Sotrikhin, S.Y., Shupikov, A.N.: Theoretical and experimental investigation of vibration of multilayer plates under the action of impulse and impact loads. Int. J. Solids Struct. **32**(8–9), 1247–1258 (1995). https://doi.org/10.1016/0020-7683(94)00132-G
16. Pagano, N.J.: Interlayer Effects in Composite Materials. Mir, Moscow (1993)
17. Sklepus, S.N.: Numerical-analytical method of studying creep and sustained strength characteristics of a multilayer shell. Strength Mater. **49**(2), 313–319 (2017). https://doi.org/10.1007/s11223-017-9871-7
18. Zaytsev, B., Asayenok, A., Protasova, T., Klimenko, D., Akimov, D., Sirenko, V.: Dynamic processes during the through-plastic-damper shock interaction of rocket fairing separation system components. J. Mech. Engin. - Problemy Mashynobuduvannia **21**(3), 19–30 (2018). https://doi.org/10.15407/pmach2018.03.019
19. Zaitsev, B., Protasova, T., Smetankina, N., Klymenko, D., Akimov, D.: Reduction of dynamic stresses and overloads using dampers in the rocket fairing separation system. In: Advances in Mechanical and Power Engineering. CAMPE 2021. Lecture Notes in Mechanical Engineering, pp. 323–333 . Springer, Cham (2023). https://doi.org/10.1007/978-3-031-18487-1_33
20. Shiratori, M., Miyoshi, T., Matsushita, H.: Computational Fracture Mechanics. Mir, Moscow (1986)
21. Atluri, S.N.: (Ed.) (1986) Computational Methods in the Mechanics of Fracture. Elsevier Science Ltd
22. Lurye, A.I.: Theory of Elasticity. Nauka, Moscow (1970)
23. Bathe, K.-J.: Finite Element Procedures. Prentice Hall Publishing (1996)
24. Zaytsev, B., Protasova, T., Klimenko, D., Akimov, D., Sirenko, V.: Dynamic analysis of a composite rocket dome during separating taking into account structure delaminating. Aerospace Tech. Technol. **8**(168), 19–26 (2020). https://doi.org/10.32620/aktt.2020.8.03
25. Guz, A.N.: Establishing the foundations of the mechanics of fracture of materials compressed along cracks (Review). Int. Appl. Mech. **50**, 1–57 (2014). https://doi.org/10.1007/s10778-014-0609-y
26. Bogdanov, V.L., Guz, A.N., Nazarenko, V.M.: Spatial problems of the fracture of materials loaded along cracks (Review). Int. Appl. Mech. **51**, 489–560 (2015). https://doi.org/10.1007/s10778-015-0710-x
27. Bogdanov, V.L., Guz, A.N., Nazarenko, V.M.: Nonclassical problems in the fracture mechanics of composites with interacting cracks. Int. Appl. Mech. **51**, 64–84 (2015). https://doi.org/10.1007/s10778-015-0673-y
28. Bolotin, V.V.: Stability Problems in Fracture Mechanics. Wiley, New York (1994)
29. Kachanov, L.M.: Delamination Buckling of Composite Materials. Kluwer Academic Publisher, Boston (1988)

30. Kienzler, R., Herrmann, G.: Mechanics in Material Space with Applications to Defect and Fracture Mechanics. Springer, Berlin (2000)
31. Zaitsev, B.P., Protasova, T.V.: Model and method to calculate nonlinear oscillations of a rod with a longitudinal near-surface crack, with account taken of local buckling. In: Hudramovich, V.S. (ed.) Mathematical and Computer Modeling of Engineering Systems, pp. 143–158. Baltija Publishing, Riga (2020)
32. Blekhman, I.I.: Vibrations in Engineering: A Handbook in 6 Vols, vol. 2. Oscillations of Nonlinear Mechanical Systems. Mashinostroeenie, Moscow (1979)
33. Miroshnikov, V., Younis, B., Savin, O., Sobol, V.: A linear elasticity theory to analyze the stress state of an infinite layer with a cylindrical cavity under periodic load. Computation **10**(9), 160 (2022). https://doi.org/10.3390/computation10090160
34. Shupikov, A.N., Smetankina, N.V., Svet, Y.V.: Nonstationary heat conduction in complex-shape laminated plates. J. Heat Trans., Trans. ASME **129**(3), 335–341 (2007). https://doi.org/10.1115/1.2427073
35. Voropay, A., Gnatenko, G., Yehorov, P., Povaliaiev, S., Naboka, O.: Identification of the pulse axisymmetric load acting on a composite cylindrical shell, inhomogeneous in length, made of different materials. Eastern-European J. Enterprise Technol. **5**(7–119), 21–34 (2022). https://doi.org/10.15587/1729-4061.2022.265356
36. Keshav, V., Patel, S.N.: Non-Linear dynamic pulse buckling of laminated composite curved panels. Struct. Eng. Mech. **73**(2), 181–190 (2021). https://doi.org/10.12989/SEM.2020.73.2.181

Printed by Printforce, the Netherlands